WORLD HEALTH ORGANIZATION

INTERNATIONAL AGENCY FOR RESEARCH ON CANCER

IARC MONOGRAPHS
ON THE
EVALUATION OF CARCINOGENIC RISKS TO HUMANS

Printing Processes and Printing Inks, Carbon Black and Some Nitro Compounds

VOLUME 65

This publication represents the views and expert opinions
of an IARC Working Group on the
Evaluation of Carcinogenic Risks to Humans,
which met in Lyon,

10–17 October 1995

1996

IARC MONOGRAPHS

In 1969, the International Agency for Research on Cancer (IARC) initiated a programme on the evaluation of the carcinogenic risk of chemicals to humans involving the production of critically evaluated monographs on individual chemicals. In 1980 and 1986, the programme was expanded to include evaluations of carcinogenic risks associated with exposures to complex mixtures and other agents.

The objective of the programme is to elaborate and publish in the form of monographs critical reviews of data on carcinogenicity for agents to which humans are known to be exposed and on specific exposure situations; to evaluate these data in terms of human risk with the help of international working groups of experts in chemical carcinogenesis and related fields; and to indicate where additional research efforts are needed.

This project is supported by PHS Grant No. 5-UO1 CA33193-14 awarded by the United States National Cancer Institute, Department of Health and Human Services. Additional support has been provided since 1986 by the European Commission.

©International Agency for Research on Cancer, 1996

IARC Library Cataloguing in Publication Data

IARC Working Group on the Evaluation of Carcinogenic Risks to Humans
   (1995 : Lyon, France)
Printing processes and printing inks, carbon black and some nitro compounds : views and expert opinions of an IARC Working Group on the Evaluation of Carcinogenic Risks to Humans which met in Lyon, 10–17 October 1995.

(IARC monographs on the evaluation of carcinogenic risks to humans ; 65)

1. Carcinogens – congresses   2. Chemical industry – congresses
3. Neoplasms – chemically induced   4. Nitro compounds – congresses
I. Series

ISBN 92 832 1265 7   (NLM Classification: W 1)

ISSN 0250-9555

Publications of the World Health Organization enjoy copyright protection in accordance with the provisions of Protocol 2 of the Universal Copyright Convention.

All rights reserved. Application for rights of reproduction or translation, in part or in toto, should be made to the International Agency for Research on Cancer.

Distributed by IARC*Press* (Fax: +33 72 73 83 02; E-mail: press@iarc.fr)
and by the World Health Organization Distribution and Sales, CH-1211 Geneva 27
(Fax: +41 22 791 4857)

PRINTED IN THE UNITED KINGDOM

CONTENTS

NOTE TO THE READER 1

LIST OF PARTICIPANTS 3

PREAMBLE
    Background 9
    Objective and Scope 9
    Selection of Topics for Monographs 10
    Data for Monographs 11
    The Working Group 11
    Working Procedures...11
    Exposure Data 12
    Studies of Cancer in Humans 14
    Studies of Cancer in Experimental Animals 17
    Other Data Relevant to an Evaluation of Carcinogenicity and Its Mechanisms 20
    Summary of Data Reported 21
    Evaluation 23
    References 27

THE MONOGRAPHS

Printing processes and printing inks 33
    1. Exposure data 33
        1.1 Historical overview...33
        1.2 Description of the industry 35
            1.2.1 Printing ink manufacture 35
            1.2.2 The printing processes 36
                (a) Lithography...36
                (b) Letterpress..38
                (c) Flexography..38
                (d) Gravure 39
                (e) Screen process...39
        1.3 Chemistry and uses of printing inks 40
            1.3.1 Pigments 40
            1.3.2 Vehicles and ink drying 42
            1.3.3 Printing inks and processes..42
                (a) Lithographic (offset) inks 42
                (b) Letterpress inks 44

			(c)	Flexographic inks	44
			(d)	Gravure inks	45
			(e)	Screen process inks	46
	1.4	Production			46
	1.5	Occupational exposures in the printing processes and to printing inks			47
		1.5.1	Exposures in printing ink manufacture		49
		1.5.2	Exposures in printing operations		54
	1.6	Regulations and guidelines			58
2.	Studies of cancer in humans				59
	2.1	Ecological studies			60
	2.2	Community-based studies			61
		2.2.1	Death certificate studies		61
		2.2.2	Record-linkage studies		63
		2.2.3	Case–control studies		66
			(a)	Cancer at multiple sites	66
			(b)	Urinary bladder cancer	72
			(c)	Lung cancer	77
			(d)	Lymphohaematopoietic neoplasms	86
			(e)	Oropharyngeal cancer	86
			(f)	Testicular cancer	87
			(g)	Other cancer sites	87
		2.2.4	Cohort studies		93
	2.3	Industry-based studies			93
	2.4	Childhood cancer in relation to parental exposure			106
3.	Studies of cancer in experimental animals				108
	Subcutaneous injection				108
4.	Other data relevant to an evaluation of carcinogenicity and its mechanisms				109
	4.1	Absorption, distribution, excretion and metabolism			109
	4.2	Toxic effects			109
		4.2.1	Humans		109
		4.2.2	Experimental systems		112
	4.3	Reproductive and developmental effects			113
		4.3.1	Humans		113
			(a)	Cohort study	113
			(b)	Case–control studies of congenital anomalies	114
			(c)	Case–control study of infertility	119
			(d)	Case–control study of mental retardation	119
		4.3.2	Experimental systems		120
	4.4	Genetic and related effects			120
		4.4.1	Humans		121
			(a)	Urinary mutagenicity	121
			(b)	Cytogenetic damage in lymphocytes	121
		4.4.2	Experimental systems		124

5. Summary of data reported and evaluation 125
   5.1 Exposure data 125
   5.2 Human carcinogenicity data 126
   5.3 Animal carcinogenicity data 130
   5.4 Other relevant data 130
   5.5 Evaluation 131
6. References 132

Carbon black 149

Some nitro compounds
   2-Chloronitrobenzene, 3-chloronitrobenzene and 4-chloronitrobenzene 263
   3,7-Dinitrofluoranthene and 3,9-dinitrofluoranthene 297
   2,4-Dinitrotoluene, 2,6-dinitrotoluene and 3,5-dinitrotoluene 309
   2-Nitroanisole 369
   Nitrobenzene 381
   2-Nitrotoluene, 3-nitrotoluene and 4-nitrotoluene 409
   Tetranitromethane 437
   2,4,6-Trinitrotoluene 449
   Musk ambrette and musk xylene 477

SUMMARY OF FINAL EVALUATIONS 497

APPENDIX 1. SUMMARY TABLES OF GENETIC AND RELATED EFFECTS 499

APPENDIX 2. ACTIVITY PROFILES FOR GENETIC AND RELATED EFFECTS 521

SUPPLEMENTARY CORRIGENDA TO VOLUMES 1–64 549

CUMULATIVE INDEX TO THE *MONOGRAPHS* SERIES 551

NOTE TO THE READER

The term 'carcinogenic risk' in the *IARC Monographs* series is taken to mean the probability that exposure to an agent will lead to cancer in humans.

Inclusion of an agent in the *Monographs* does not imply that it is a carcinogen, only that the published data have been examined. Equally, the fact that an agent has not yet been evaluated in a monograph does not mean that it is not carcinogenic.

The evaluations of carcinogenic risk are made by international working groups of independent scientists and are qualitative in nature. No recommendation is given for regulation or legislation.

Anyone who is aware of published data that may alter the evaluation of the carcinogenic risk of an agent to humans is encouraged to make this information available to the Unit of Carcinogen Identification and Evaluation, International Agency for Research on Cancer, 150 cours Albert Thomas, 69372 Lyon Cedex 08, France, in order that the agent may be considered for re-evaluation by a future Working Group.

Although every effort is made to prepare the monographs as accurately as possible, mistakes may occur. Readers are requested to communicate any errors to the Unit of Carcinogen Identification and Evaluation, so that corrections can be reported in future volumes.

IARC WORKING GROUP ON THE EVALUATION OF CARCINOGENIC RISKS TO HUMANS: PRINTING PROCESSES AND PRINTING INKS, CARBON BLACK AND SOME NITRO COMPOUNDS

Lyon, 10–17 October 1995

LIST OF PARTICIPANTS

Members

A. Aitio, Biomonitoring Laboratory, Institute of Occupational Health, Arinatie 3A, 00370 Helsinki, Finland

J.A. Bond, Chemical Industry Institute of Toxicology, PO Box 12137, 6 Davis Drive, Research Triangle Park, NC 27709, United States

A. Brøgger, Department of Genetics, Institute for Cancer Research, The Norwegian Radium Hospital, Montebello, 0310 Oslo, Norway (*Chairman*)

J.R. Bucher, Environmental Toxicology Program, National Institute of Environmental Health Sciences, PO Box 12233, Research Triangle Park, NC 27709, United States

K. Gardiner, Institute of Occupational Health, The University of Birmingham, PO Box 363, Edgbaston, Birmingham B15 2TT, United Kingdom

U. Heinrich, Department of Experimental and Environmental Hygiene, Fraunhofer-Institute for Toxicology and Aerosol Research, Nikolaï Fuchs Strasse, 30625 Hannover, Germany

J. Little, Institute of Medical Sciences, Department of Medicine and Therapeutics, University of Aberdeen Medical School, Polwarth Building, Foresterhill, Aberdeen AB9 2ZD, United Kingdom

A. Maekawa, Department of Pathology, Sasaki Institute, 2-2 Kanda-Surugadai, Chiyoda-ku, Tokyo 101, Japan

G. Oberdörster, Department of Environmental Medicine, School of Medicine and Dentistry, University of Rochester, 575 Elmwood Avenue, Box EHSC, Rochester, NY 14642, United States

S. Olin, Risk Science Institute, International Life Sciences Institute, 1126 Sixteenth Street NW, Washington, DC, 20036, United States

A.C. Pesatori, Institute of Occupational Health, Clinica del Lavoro 'Luigi Devoto', University of Milan, via S. Barnaba 8, 20122 Milan, Italy

J. Siemiatycki, Institut Armand-Frappier, Unit of Epidemiology and Biostatistics, Case Postale 100, Laval-des-Rapides, Québec H7V 1B7, Canada

S. Skerfving, Department of Occupational and Environmental Medicine, University Hospital, 221 85 Lund, Sweden

L.T. Stayner, Education and Information Division, National Institute for Occupational Safety and Health, 4676 Columbia Parkway, Cincinnati, OH 45226-1998, United States (*Vice-Chairman*)

M. Waters, National Institute for Occupational Safety and Health, 4676 Columbia Parkway, Cincinnati, OH 45226-1998, United States

C.H. Williams, 3 Bradbourne Court, Bradbourne Vale Road, Sevenoaks, Kent TN13 3QG, United Kingdom

F.E. Würgler, Institute of Toxicology, Swiss Federal Institute of Technology, Schorenstrasse 16, 8603 Schwerzenbach/Zürich, Switzerland

Representatives/Observers[1]

American Industrial Health Council

M.A. Martens, Europe/Africa, Monsanto Technical Center, Rue Laid-Burniat, 1348 Louvain-la-Neuve, Belgium

Commonwealth Department of Human Services and Health

D. Wagner, Chemicals Policy and Review Section, Environmental Health and Safety Unit (MDP 88), Commonwealth Department of Human Services and Health, GPO Box 9848, Canberra ACT 2601, Australia

European Carbon Black Centre

L.S. Levy, Institute of Occupational Health, The University of Birmingham, Edgbaston, Birmingham B15 2TT, United Kingdom

European Centre for Ecotoxicology and Toxicology of Chemicals

E. Löser, Bayer, Institute of Toxicology, Postfach 130105, 42096 Wuppertal, Germany

Secretariat

P. Boffetta, Unit of Environmental Cancer Epidemiology
D. Eaton, Unit of Environmental Carcinogenesis
M. Friesen, Unit of Environmental Carcinogenesis
A. Hallikainen, Unit of Carcinogen Identification and Evaluation
N. Jourenkova, Unit of Environmental Cancer Epidemiology
C. Malaveille, Unit of Endogenous Cancer Risk Factors
D. McGregor, Unit of Carcinogen Identification and Evaluation
A. Meneghel, Unit of Carcinogen Identification and Evaluation

---

[1] Unable to attend: S.M. Sieber, Division of Cancer Etiology, National Cancer Institute, Building 31, Room 11A03, Bethesda, MD 20892, United States

PARTICIPANTS

E. Merler, Unit of Environmental Cancer Epidemiology
D. Mietton, Unit of Carcinogen Identification and Evaluation
R. Montesano, Unit of Mechanisms of Carcinogenesis
W.G. Morgan, Edinburgh, United Kingdom
T. Partanen, Unit of Environmental Cancer Epidemiology
C. Partensky, Unit of Carcinogen Identification and Evaluation
E. Smith, IPCS, WHO, Geneva
J. Wilbourn, Unit of Carcinogen Identification and Evaluation

Secretarial assistance

M. Lézère
J. Mitchell
S. Reynaud
S. Ruiz

PREAMBLE

IARC MONOGRAPHS PROGRAMME ON THE EVALUATION OF CARCINOGENIC RISKS TO HUMANS[1]

PREAMBLE

1. BACKGROUND

In 1969, the International Agency for Research on Cancer (IARC) initiated a programme to evaluate the carcinogenic risk of chemicals to humans and to produce monographs on individual chemicals. The *Monographs* programme has since been expanded to include consideration of exposures to complex mixtures of chemicals (which occur, for example, in some occupations and as a result of human habits) and of exposures to other agents, such as radiation and viruses. With Supplement 6 (IARC, 1987a), the title of the series was modified from *IARC Monographs on the Evaluation of the Carcinogenic Risk of Chemicals to Humans* to *IARC Monographs on the Evaluation of Carcinogenic Risks to Humans*, in order to reflect the widened scope of the programme.

The criteria established in 1971 to evaluate carcinogenic risk to humans were adopted by the working groups whose deliberations resulted in the first 16 volumes of the *IARC Monographs series*. Those criteria were subsequently updated by further ad-hoc working groups (IARC, 1977, 1978, 1979, 1982, 1983, 1987b, 1988, 1991a; Vainio *et al.*, 1992).

2. OBJECTIVE AND SCOPE

The objective of the programme is to prepare, with the help of international working groups of experts, and to publish in the form of monographs, critical reviews and evaluations of evidence on the carcinogenicity of a wide range of human exposures. The *Monographs* may also indicate where additional research efforts are needed.

The *Monographs* represent the first step in carcinogenic risk assessment, which involves examination of all relevant information in order to assess the strength of the available evidence that certain exposures could alter the incidence of cancer in humans. The second step is quantitative risk estimation. Detailed, quantitative evaluations of epidemiological data may be made in the *Monographs*, but without extrapolation beyond

---

[1] This project is supported by PHS Grant No. 5-UO1 CA33193-14 awarded by the United States National Cancer Institute, Department of Health and Human Services. Since 1986, the programme has also been supported by the European Commission.

the range of the data available. Quantitative extrapolation from experimental data to the human situation is not undertaken.

The term 'carcinogen' is used in these monographs to denote an exposure that is capable of increasing the incidence of malignant neoplasms; the induction of benign neoplasms may in some circumstances (see p. 19) contribute to the judgement that the exposure is carcinogenic. The terms 'neoplasm' and 'tumour' are used interchangeably.

Some epidemiological and experimental studies indicate that different agents may act at different stages in the carcinogenic process, and several different mechanisms may be involved. The aim of the *Monographs* has been, from their inception, to evaluate evdence of carcinogenicity at any stage in the carcinogenesis process, independently of the underlying mechanisms. Information on mechanisms may, however, be used in making the overall evaluation (IARC, 1991a; Vainio *et al.*, 1992; see also pp. 25-27).

The *Monographs* may assist national and international authorities in making risk assessments and in formulating decisions concerning any necessary preventive measures. The evaluations of IARC working groups are scientific, qualitative judgements about the evidence for or against carcinogenicity provided by the available data. These evaluations represent only one part of the body of information on which regulatory measures may be based. Other components of regulatory decisions may vary from one situation to another and from country to country, responding to different socioeconomic and national priorities. **Therefore, no recommendation is given with regard to regulation or legislation, which are the responsibility of individual governments and/or other international organizations.**

The *IARC Monographs* are recognized as an authoritative source of information on the carcinogenicity of a wide range of human exposures. A users' survey, made in 1988, indicated that the *Monographs* are consulted by various agencies in 57 countries. Each volume is generally printed in 4000 copies for distribution to governments, regulatory bodies and interested scientists. The Monographs are also available from the International Agency for Research on Cancer in Lyon and via the Distribution and Sales Service of the World Health Organization.

3. SELECTION OF TOPICS FOR MONOGRAPHS

Topics are selected on the basis of two main criteria: (a) there is evidence of human exposure, and (b) there is some evidence or suspicion of carcinogenicity. The term 'agent' is used to include individual chemical compounds, groups of related chemical compounds, physical agents (such as radiation) and biological factors (such as viruses). Exposures to mixtures of agents may occur in occupational exposures and as a result of personal and cultural habits (like smoking and dietary practices). Chemical analogues and compounds with biological or physical characteristics similar to those of suspected carcinogens may also be considered, even in the absence of data on a possible carcinogenic effect in humans or experimental animals.

The scientific literature is surveyed for published data relevant to an assessment of carcinogenicity. The IARC information bulletins on agents being tested for carcinogenicity (IARC, 1973–1994) and directories of on-going research in cancer epidemiology (IARC, 1976–1994) often indicate those exposures that may be scheduled for future meetings. Ad-hoc working groups convened by IARC in 1984, 1989, 1991 and 1993 gave recommendations as to which agents should be evaluated in the IARC Monographs series (IARC, 1984, 1989, 1991b, 1993).

As significant new data on subjects on which monographs have already been prepared become available, re-evaluations are made at subsequent meetings, and revised monographs are published.

4. DATA FOR MONOGRAPHS

The *Monographs* do not necessarily cite all the literature concerning the subject of an evaluation. Only those data considered by the Working Group to be relevant to making the evaluation are included.

With regard to biological and epidemiological data, only reports that have been published or accepted for publication in the openly available scientific literature are reviewed by the working groups. In certain instances, government agency reports that have undergone peer review and are widely available are considered. Exceptions may be made on an ad-hoc basis to include unpublished reports that are in their final form and publicly available, if their inclusion is considered pertinent to making a final evaluation (see pp. 25–27). In the sections on chemical and physical properties, on analysis, on production and use and on occurrence, unpublished sources of information may be used.

5. THE WORKING GROUP

Reviews and evaluations are formulated by a working group of experts. The tasks of the group are: (i) to ascertain that all appropriate data have been collected; (ii) to select the data relevant for the evaluation on the basis of scientific merit; (iii) to prepare accurate summaries of the data to enable the reader to follow the reasoning of the Working Group; (iv) to evaluate the results of epidemiological and experimental studies on cancer; (v) to evaluate data relevant to the understanding of mechanism of action; and (vi) to make an overall evaluation of the carcinogenicity of the exposure to humans.

Working Group participants who contributed to the considerations and evaluations within a particular volume are listed, with their addresses, at the beginning of each publication. Each participant who is a member of a working group serves as an individual scientist and not as a representative of any organization, government or industry. In addition, nominees of national and international agencies and industrial associations may be invited as observers.

6. WORKING PROCEDURES

Approximately one year in advance of a meeting of a working group, the topics of the monographs are announced and participants are selected by IARC staff in consultation with other experts. Subsequently, relevant biological and epidemiological data are collected by IARC from recognized sources of information on carcinogenesis, including data storage and retrieval systems such as MEDLINE and TOXLINE, and EMIC and ETIC for data on genetic and related effects and reproductive and developmental effects, respectively.

For chemicals and some complex mixtures, the major collection of data and the preparation of first drafts of the sections on chemical and physical properties, on analysis, on production and use and on occurrence are carried out under a separate contract funded by the United States National Cancer Institute. Representatives from industrial associations may assist in the preparation of sections on production and use. Information on production and trade is obtained from governmental and trade publications and, in some cases, by direct contact with industries. Separate production data on some agents may not be available because their publication could disclose confidential information. Information on uses may be obtained from published sources but is often complemented by direct contact with manufacturers. Efforts are made to supplement this information with data from other national and international sources.

Six months before the meeting, the material obtained is sent to meeting participants, or is used by IARC staff, to prepare sections for the first drafts of monographs. The first drafts are compiled by IARC staff and sent, prior to the meeting, to all participants of the Working Group for review.

The Working Group meets in Lyon for seven to eight days to discuss and finalize the texts of the monographs and to formulate the evaluations. After the meeting, the master copy of each monograph is verified by consulting the original literature, edited and prepared for publication. The aim is to publish monographs within six months of the Working Group meeting.

The available studies are summarized by the Working Group, with particular regard to the qualitative aspects discussed below. In general, numerical findings are indicated as they appear in the original report; units are converted when necessary for easier comparison. The Working Group may conduct additional analyses of the published data and use them in their assessment of the evidence; the results of such supplementary analyses are given in square brackets. When an important aspect of a study, directly impinging on its interpretation, should be brought to the attention of the reader, a comment is given in square brackets.

7. EXPOSURE DATA

Sections that indicate the extent of past and present human exposure, the sources of exposure, the people most likely to be exposed and the factors that contribute to the exposure are included at the beginning of each monograph.

Most monographs on individual chemicals, groups of chemicals or complex mixtures include sections on chemical and physical data, on analysis, on production and use and on occurrence. In monographs on, for example, physical agents, occupational exposures and cultural habits, other sections may be included, such as: historical perspectives, description of an industry or habit, chemistry of the complex mixture or taxonomy.

Monographs on biological agents have sections on structure and biology, methods of detection, epidemiology of infection and clinical disease other than cancer.

For chemical exposures, the Chemical Abstracts Services Registry Number, the latest Chemical Abstracts Primary Name and the IUPAC Systematic Name are recorded; other synonyms are given, but the list is not necessarily comprehensive. For biological agents, taxonomy and structure are described, and the degree of variability is given, when applicable.

Information on chemical and physical properties and, in particular, data relevant to identification, occurrence and biological activity are included. For biological agents, mode of replication, life cycle, target cells, persistence and latency and host response are given. A description of technical products of chemicals includes trades names, relevant specifications and available information on composition and impurities. Some of the trade names given may be those of mixtures in which the agent being evaluated is only one of the ingredients.

The purpose of the section on analysis or detection is to give the reader an overview of current methods, with emphasis on those widely used for regulatory purposes. Methods for monitoring human exposure are also given, when available. No critical evaluation or recommendation of any of the methods is meant or implied. The IARC publishes a series of volumes, *Environmental Carcinogens: Methods of Analysis and Exposure Measurement* (IARC, 1978–93), that describe validated methods for analysing a wide variety of chemicals and mixtures. For biological agents, methods of detection and exposure assessment are described, including their sensitivity, specificity and reproducibility.

The dates of first synthesis and of first commercial production of a chemical or mixture are provided; for agents which do not occur naturally, this information may allow a reasonable estimate to be made of the date before which no human exposure to the agent could have occurred. The dates of first reported occurrence of an exposure are also provided. In addition, methods of synthesis used in past and present commercial production and different methods of production which may give rise to different impurities are described.

Data on production, international trade and uses are obtained for representative regions, which usually include Europe, Japan and the United States of America. It should not, however, be inferred that those areas or nations are necessarily the sole or major sources or users of the agent. Some identified uses may not be current or major applications, and the coverage is not necessarily comprehensive. In the case of drugs, mention of their therapeutic uses does not necessarily represent current practice nor does it imply judgement as to their therapeutic efficacy.

Information on the occurrence of an agent or mixture in the environment is obtained from data derived from the monitoring and surveillance of levels in occupational environments, air, water, soil, foods and animal and human tissues. When available, data on the generation, persistence and bioaccumulation of the agent are also included. In the case of mixtures, industries, occupations or processes, information is given about all agents present. For processes, industries and occupations, a historical description is also given, noting variations in chemical composition, physical properties and levels of occupational exposure with time and place. For biological agents, the epidemiology of infection is described.

Statements concerning regulations and guidelines (e.g. pesticide registrations, maximal levels permitted in foods, occupational exposure limits) are included for some countries as indications of potential exposures, but they may not reflect the most recent situation, since such limits are continuously reviewed and modified. The absence of information on regulatory status for a country should not be taken to imply that that country does not have regulations with regard to the exposure. For biological agents, legislation and control, including vaccines and therapy, are described.

8. STUDIES OF CANCER IN HUMANS

(a) Types of studies considered

Three types of epidemiological studies of cancer contribute to the assessment of carcinogenicity in humans — cohort studies, case–control studies and correlation (or ecological) studies. Rarely, results from randomized trials may be available. Case series and case reports of cancer in humans may also be reviewed.

Cohort and case–control studies relate individual exposures under study to the occurrence of cancer in individuals and provide an estimate of relative risk (ratio of incidence or mortality in those exposed to incidence or mortality in those not exposed) as the main measure of association.

In correlation studies, the units of investigation are usually whole populations (e.g. in particular geographical areas or at particular times), and cancer frequency is related to a summary measure of the exposure of the population to the agent, mixture or exposure circumstance under study. Because individual exposure is not documented, however, a causal relationship is less easy to infer from correlation studies than from cohort and case–control studies. Case reports generally arise from a suspicion, based on clinical experience, that the concurrence of two events — that is, a particular exposure and occurrence of a cancer — has happened rather more frequently than would be expected by chance. Case reports usually lack complete ascertainment of cases in any population, definition or enumeration of the population at risk and estimation of the expected number of cases in the absence of exposure. The uncertainties surrounding interpretation of case reports and correlation studies make them inadequate, except in rare instances, to form the sole basis for inferring a causal relationship. When taken together with case–control and cohort studies, however, relevant case reports or correlation studies may add materially to the judgement that a causal relationship is present.

Epidemiological studies of benign neoplasms, presumed preneoplastic lesions and other end-points thought to be relevant to cancer are also reviewed by working groups. They may, in some instances, strengthen inferences drawn from studies of cancer itself.

(b) Quality of studies considered

The Monographs are not intended to summarize all published studies. Those that are judged to be inadequate or irrelevant to the evaluation are generally omitted. They may be mentioned briefly, particularly when the information is considered to be a useful supplement to that in other reports or when they provide the only data available. Their inclusion does not imply acceptance of the adequacy of the study design or of the analysis and interpretation of the results, and limitations are clearly outlined in square brackets at the end of the study description.

It is necessary to take into account the possible roles of bias, confounding and chance in the interpretation of epidemiological studies. By 'bias' is meant the operation of factors in study design or execution that lead erroneously to a stronger or weaker association than in fact exists between disease and an agent, mixture or exposure circumstance. By 'confounding' is meant a situation in which the relationship with disease is made to appear stronger or to appear weaker than it truly is as a result of an association between the apparent causal factor and another factor that is associated with either an increase or decrease in the incidence of the disease. In evaluating the extent to which these factors have been minimized in an individual study, working groups consider a number of aspects of design and analysis as described in the report of the study. Most of these considerations apply equally to case–control, cohort and correlation studies. Lack of clarity of any of these aspects in the reporting of a study can decrease its credibility and the weight given to it in the final evaluation of the exposure.

Firstly, the study population, disease (or diseases) and exposure should have been well defined by the authors. Cases of disease in the study population should have been identified in a way that was independent of the exposure of interest, and exposure should have been assessed in a way that was not related to disease status.

Secondly, the authors should have taken account in the study design and analysis of other variables that can influence the risk of disease and may have been related to the exposure of interest. Potential confounding by such variables should have been dealt with either in the design of the study, such as by matching, or in the analysis, by statistical adjustment. In cohort studies, comparisons with local rates of disease may be more appropriate than those with national rates. Internal comparisons of disease frequency among individuals at different levels of exposure should also have been made in the study.

Thirdly, the authors should have reported the basic data on which the conclusions are founded, even if sophisticated statistical analyses were employed. At the very least, they should have given the numbers of exposed and unexposed cases and controls in a case–control study and the numbers of cases observed and expected in a cohort study. Further tabulations by time since exposure began and other temporal factors are also important. In a cohort study, data on all cancer sites and all causes of death should have been given, to reveal the possibility of reporting bias. In a case–control study, the effects of investigated factors other than the exposure of interest should have been reported.

Finally, the statistical methods used to obtain estimates of relative risk, absolute rates of cancer, confidence intervals and significance tests, and to adjust for confounding should have been clearly stated by the authors. The methods used should preferably have been the generally accepted techniques that have been refined since the mid-1970s. These methods have been reviewed for case–control studies (Breslow & Day, 1980) and for cohort studies (Breslow & Day, 1987).

(c) *Inferences about mechanism of action*

Detailed analyses of both relative and absolute risks in relation to temporal variables, such as age at first exposure, time since first exposure, duration of exposure, cumulative exposure and time since exposure ceased, are reviewed and summarized when available. The analysis of temporal relationships can be useful in formulating models of carcinogenesis. In particular, such analyses may suggest whether a carcinogen acts early or late in the process of carcinogenesis, although at best they allow only indirect inferences about the mechanism of action. Special attention is given to measurements of biological markers of carcinogen exposure or action, such as DNA or protein adducts, as well as markers of early steps in the carcinogenic process, such as proto-oncogene mutation, when these are incorporated into epidemiological studies focused on cancer incidence or mortality. Such measurements may allow inferences to be made about putative mechanisms of action (IARC, 1991a; Vainio et al., 1992).

(d) *Criteria for causality*

After the quality of individual epidemiological studies of cancer has been summarized and assessed, a judgement is made concerning the strength of evidence that the agent, mixture or exposure circumstance in question is carcinogenic for humans. In making their judgement, the Working Group considers several criteria for causality. A strong association (i.e. a large relative risk) is more likely to indicate causality than a weak association, although it is recognized that relative risks of small magnitude do not imply lack of causality and may be important if the disease is common. Associations that are replicated in several studies of the same design or using different epidemiological approaches or under different circumstances of exposure are more likely to represent a causal relationship than isolated observations from single studies. If there are inconsistent results among investigations, possible reasons are sought (such as differences in amount of exposure), and results of studies judged to be of high quality are given more weight than those of studies judged to be methodologically less sound. When suspicion of carcinogenicity arises largely from a single study, these data are not combined with those from later studies in any subsequent reassessment of the strength of the evidence.

If the risk of the disease in question increases with the amount of exposure, this is considered to be a strong indication of causality, although absence of a graded response is not necessarily evidence against a causal relationship. Demonstration of a decline in risk after cessation of or reduction in exposure in individuals or in whole populations also supports a causal interpretation of the findings.

Although a carcinogen may act upon more than one target, the specificity of an association (i.e. an increased occurrence of cancer at one anatomical site or of one morphological type) adds plausibility to a causal relationship, particularly when excess cancer occurrence is limited to one morphological type within the same organ.

Although rarely available, results from randomized trials showing different rates among exposed and unexposed individuals provide particularly strong evidence for causality.

When several epidemiological studies show little or no indication of an association between an exposure and cancer, the judgement may be made that, in the aggregate, they show evidence of lack of carcinogenicity. Such a judgement requires first of all that the studies giving rise to it meet, to a sufficient degree, the standards of design and analysis described above. Specifically, the possibility that bias, confounding or misclassification of exposure or outcome could explain the observed results should be considered and excluded with reasonable certainty. In addition, all studies that are judged to be methodologically sound should be consistent with a relative risk of unity for any observed level of exposure and, when considered together, should provide a pooled estimate of relative risk which is at or near unity and has a narrow confidence interval, due to sufficient population size. Moreover, no individual study nor the pooled results of all the studies should show any consistent tendency for relative risk of cancer to increase with increasing level of exposure. It is important to note that evidence of lack of carcinogenicity obtained in this way from several epidemiological studies can apply only to the type(s) of cancer studied and to dose levels and intervals between first exposure and observation of disease that are the same as or less than those observed in all the studies. Experience with human cancer indicates that, in some cases, the period from first exposure to the development of clinical cancer is seldom less than 20 years; latent periods substantially shorter than 30 years cannot provide evidence for lack of carcinogenicity.

9. STUDIES OF CANCER IN EXPERIMENTAL ANIMALS

All known human carcinogens that have been studied adequately in experimental animals have produced positive results in one or more animal species (Wilbourn *et al.*, 1986; Tomatis *et al.*, 1989). For several agents (aflatoxins, 4-aminobiphenyl, azathioprine, betel quid with tobacco, BCME and CMME (technical grade), chlorambucil, chlornaphazine, ciclosporin, coal-tar pitches, coal-tars, combined oral contraceptives, cyclophosphamide, diethylstilboestrol, melphalan, 8-methoxypsoralen plus UVA, mustard gas, myleran, 2-naphthylamine, nonsteroidal oestrogens, oestrogen replacement therapy/steroidal oestrogens, solar radiation, thiotepa and vinyl chloride), carcinogenicity in experimental animals was established or highly suspected before epidemiological studies confirmed the carcinogenicity in humans (Vainio *et al.*, 1995). Although this association cannot establish that all agents and mixtures that cause cancer in experimental animals also cause cancer in humans, nevertheless, **in the absence of adequate data on humans, it is biologically plausible and prudent to regard agents and mixtures for which there is sufficient evidence (see p. 24) of carcinogenicity in experimental animals as if they presented a carcinogenic risk to humans.** The possibility that a given agent may cause cancer through a species-specific mechanism which does not operate in humans (see p. 27) should also be taken into consideration.

The nature and extent of impurities or contaminants present in the chemical or mixture being evaluated are given when available. Animal strain, sex, numbers per group, age at start of treatment and survival are reported.

Other types of studies summarized include: experiments in which the agent or mixture was administered in conjunction with known carcinogens or factors that modify carcinogenic effects; studies in which the end-point was not cancer but a defined precancerous lesion; and experiments on the carcinogenicity of known metabolites and derivatives.

For experimental studies of mixtures, consideration is given to the possibility of changes in the physicochemical properties of the test substance during collection, storage, extraction, concentration and delivery. Chemical and toxicological interactions of the components of mixtures may result in nonlinear dose–response relationships.

An assessment is made as to the relevance to human exposure of samples tested in experimental animals, which may involve consideration of: (i) physical and chemical characteristics, (ii) constituent substances that indicate the presence of a class of substances, (iii) the results of tests for genetic and related effects, including genetic activity profiles, DNA adduct profiles, proto-oncogene mutation and expression and suppressor gene inactivation. The relevance of results obtained, for example, with animal viruses analogous to the virus being evaluated in the monograph must also be considered. They may provide biological and mechanistic information relevant to the understanding of the process of carcinogenesis in humans and may strengthen the plausibility of a conclusion that the biological agent that is being evaluated is carcinogenic in humans.

(a) *Qualitative aspects*

An assessment of carcinogenicity involves several considerations of qualitative importance, including (i) the experimental conditions under which the test was perormed, including route and schedule of exposure, species, strain, sex, age, duration of follow-up; (ii) the consistency of the results, for example, across species and target organ(s); (iii) the spectrum of neoplastic response, from preneoplastic lesions and benign tumours to malignant neoplasms; and (iv) the possible role of modifying factors.

As mentioned earlier (p. 11), the *Monographs* are not intended to summarize all published studies. Those studies in experimental animals that are inadequate (e.g. too short a duration, too few animals, poor survival; see below) or are judged irrelevant to the evaluation are generally omitted. Guidelines for conducting adequate long-term carcinogenicity experiments have been outlined (e.g. Montesano *et al.*, 1986).

Considerations of importance to the Working Group in the interpretation and evaluation of a particular study include: (i) how clearly the agent was defined and, in the case of mixtures, how adequately the sample characterization was reported; (ii) whether the dose was adequately monitored, particularly in inhalation experiments; (iii) whether the doses and duration of treatment were appropriate and whether the survival of treated animals was similar to that of controls; (iv) whether there were adequate numbers of animals per group; (v) whether animals of both sexes were used; (vi) whether animals were allocated randomly to groups; (vii) whether the duration of observation was adequate; and (viii) whether the data were adequately reported. If available, recent data on the incidence of specific tumours in historical controls, as well as in concurrent controls, should be taken into account in the evaluation of tumour response.

When benign tumours occur together with and originate from the same cell type in an organ or tissue as malignant tumours in a particular study and appear to represent a stage in the progression to malignancy, it may be valid to combine them in assessing tumour incidence (Huff *et al.*, 1989). The occurrence of lesions presumed to be preneoplastic may in certain instances aid in assessing the biological plausibility of any neoplastic response observed. If an agent or mixture induces only benign neoplasms that appear to be end-points that do not readily undergo transition to malignancy, it should nevertheless be suspected of being a carcinogen and requires further investigation.

(b) Quantitative aspects

The probability that tumours will occur may depend on the species, sex, strain and age of the animal, the dose of the carcinogen and the route and length of exposure. Evidence of an increased incidence of neoplasms with increased level of exposure strengthens the inference of a causal association between the exposure and the development of neoplasms.

The form of the dose–response relationship can vary widely, depending on the particular agent under study and the target organ. Both DNA damage and increased cell division are important aspects of carcinogenesis, and cell proliferation is a strong determinant of dose–response relationships for some carcinogens (Cohen & Ellwein, 1990). Since many chemicals require metabolic activation before being converted into their reactive intermediates, both metabolic and pharmacokinetic aspects are important in determining the dose–response pattern. Saturation of steps such as absorption, activation, inactivation and elimination may produce nonlinearity in the dose–response relationship, as could saturation of processes such as DNA repair (Hoel *et al.*, 1983; Gart *et al.*, 1986).

(c) Statistical analysis of long-term experiments in animals

Factors considered by the Working Group include the adequacy of the information given for each treatment group: (i) the number of animals studied and the number examined histologically, (ii) the number of animals with a given tumour type and (iii) length of survival. The statistical methods used should be clearly stated and should be the generally accepted techniques refined for this purpose (Peto *et al.*, 1980; Gart *et al.*, 1986). When there is no difference in survival between control and treatment groups, the Working Group usually compares the proportions of animals developing each tumour type in each of the groups. Otherwise, consideration is given as to whether or not appropriate adjustments have been made for differences in survival. These adjustments can include: comparisons of the proportions of tumour-bearing animals among the effective number of animals (alive at the time the first tumour is discovered), in the case where most differences in survival occur before tumours appear; life-table methods, when tumours are visible or when they may be considered 'fatal' because mortality rapidly follows tumour development; and the Mantel-Haenszel test or logistic regression, when occult tumours do not affect the animals' risk of dying but are 'incidental' findings at autopsy.

In practice, classifying tumours as fatal or incidental may be difficult. Several survival-adjusted methods have been developed that do not require this distinction (Gart et al., 1986), although they have not been fully evaluated.

10. OTHER DATA RELEVANT TO AN EVALUATION OF CARCINO-GENICITY AND ITS MECHANISMS

In coming to an overall evaluation of carcinogenicity in humans (see p. 25), the Working Group also considers related data. The nature of the information selected for the summary depends on the agent being considered.

For chemicals and complex mixtures of chemicals such as those in some occupational situations and involving cultural habits (e.g. tobacco smoking), the other data considered to be relevant are divided into those on absorption, distribution, metabolism and excretion; toxic effects; reproductive and developmental effects; and genetic and related effects.

Concise information is given on absorption, distribution (including placental transfer) and excretion in both humans and experimental animals. Kinetic factors that may affect the dose–response relationship, such as saturation of uptake, protein binding, metabolic activation, detoxification and DNA repair processes, are mentioned. Studies that indicate the metabolic fate of the agent in humans and in experimental animals are summarized briefly, and comparisons of data from humans and animals are made when possible. Comparative information on the relationship between exposure and the dose that reaches the target site may be of particular importance for extrapolation between species. Data are given on acute and chronic toxic effects (other than cancer), such as organ toxicity, increased cell proliferation, immunotoxicity and endocrine effects. The presence and toxicological significance of cellular receptors is described. Effects on reproduction, teratogenicity, fetotoxicity and embryotoxicity are also summarized briefly.

Tests of genetic and related effects are described in view of the relevance of gene mutation and chromosomal damage to carcinogenesis (Vainio et al., 1992). The adequacy of the reporting of sample characterization is considered and, where necessary, commented upon; with regard to complex mixtures, such comments are similar to those described for animal carcinogenicity tests on p. 17. The available data are interpreted critically by phylogenetic group according to the end-points detected, which may include DNA damage, gene mutation, sister chromatid exchange, micronucleus formation, chromosomal aberrations, aneuploidy and cell transformation. The concentrations employed are given, and mention is made of whether use of an exogenous metabolic system in vitro affected the test result. These data are given as listings of test systems, data and references; bar graphs (activity profiles) and corresponding summary tables with detailed information on the preparation of the profiles (Waters et al., 1987) are given in appendices.

Positive results in tests using prokaryotes, lower eukaryotes, plants, insects and cultured mammalian cells suggest that genetic and related effects could occur in mammals. Results from such tests may also give information about the types of genetic effect produced and about the involvement of metabolic activation. Some end-points described are clearly genetic in nature (e.g. gene mutations and chromosomal aberrations), while others are to a greater or lesser degree associated with genetic effects (e.g. unscheduled DNA synthesis). In-vitro tests for tumour-promoting activity and for cell transformation may be sensitive to changes that are not necessarily the result of genetic alterations but that may have specific relevance to the process of carcinogenesis. A critical appraisal of these tests has been published (Montesano et al., 1986).

Genetic or other activity manifest in experimental mammals and humans is regarded as being of greater relevance than that in other organisms. The demonstration that an agent or mixture can induce gene and chromosomal mutations in whole mammals indicates that it may have carcinogenic activity, although this activity may not be detectably expressed in any or all species. Relative potency in tests for mutagenicity and related effects is not a reliable indicator of carcinogenic potency. Negative results in tests for mutagenicity in selected tissues from animals treated *in vivo* provide less weight, partly because they do not exclude the possibility of an effect in tissues other than those examined. Moreover, negative results in short-term tests with genetic end-points cannot be considered to provide evidence to rule out carcinogenicity of agents or mixtures that act through other mechanisms (e.g. receptor-mediated effects, cellular toxicity with regenerative proliferation, peroxisome proliferation) (Vainio et al., 1992). Factors that may lead to misleading results in short-term tests have been discussed in detail elsewhere (Montesano et al., 1986).

When available, data relevant to mechanisms of carcinogenesis that do not involve structural changes at the level of the gene are also described.

The adequacy of epidemiological studies of reproductive outcome and genetic and related effects in humans is evaluated by the same criteria as are applied to epidemiological studies of cancer.

Structure–activity relationships that may be relevant to an evaluation of the carcinogenicity of an agent are also described.

For biological agents — viruses, bacteria and parasites — other data relevant to carcino-genicity include descriptions of the pathology of infection, molecular biology (integration and expression of viruses, and any genetic alterations seen in human tumours) and other observations, which might include cellular and tissue responses to infection, immune response and the presence of tumour markers.

11. SUMMARY OF DATA REPORTED

In this section, the relevant epidemiological and experimental data are summarized. Only reports, other than in abstract form, that meet the criteria outlined on p. 11 are considered for evaluating carcinogenicity. Inadequate studies are generally not summarized: such studies are usually identified by a square-bracketed comment in the preceding text.

(a) Exposures

Human exposure to chemicals and complex mixtures is summarized on the basis of elements such as production, use, occurrence in the environment and determinations in human tissues and body fluids. Quantitative data are given when available. Exposure to biological agents is described in terms of transmission, and prevalence of infection.

(b) Carcinogenicity in humans

Results of epidemiological studies that are considered to be pertinent to an assessment of human carcinogenicity are summarized. When relevant, case reports and correlation studies are also summarized.

(c) Carcinogenicity in experimental animals

Data relevant to an evaluation of carcinogenicity in animals are summarized. For each animal species and route of administration, it is stated whether an increased incidence of neoplasms or preneoplastic lesions was observed, and the tumour sites are indicated. If the agent or mixture produced tumours after prenatal exposure or in single-dose experiments, this is also indicated. Negative findings are also summarized. Dose–response and other quantitative data may be given when available.

(d) Other data relevant to an evaluation of carcinogenicity and its mechanisms

Data on biological effects in humans that are of particular relevance are summarized. These may include toxicological, kinetic and metabolic considerations and evidence of DNA binding, persistence of DNA lesions or genetic damage in exposed humans. Toxicological information, such as that on cytotoxicity and regeneration, receptor binding and hormonal and immunological effects, and data on kinetics and metabolism in experimental animals are given when considered relevant to the possible mechanism of the carcinogenic action of the agent. The results of tests for genetic and related effects are summarized for whole mammals, cultured mammalian cells and nonmammalian systems.

When available, comparisons of such data for humans and for animals, and particularly animals that have developed cancer, are described.

Structure–activity relationships are mentioned when relevant.

For the agent, mixture or exposure circumstance being evaluated, the available data on end-points or other phenomena relevant to mechanisms of carcinogenesis from studies in humans, experimental animals and tissue and cell test systems are summarized within one or more of the following descriptive dimensions:

(i) Evidence of genotoxicity (i.e. structural changes at the level of the gene): for example, structure–activity considerations, adduct formation, mutagenicity (effect on specific genes), chromosomal mutation/aneuploidy

(ii) Evidence of effects on the expression of relevant genes (i.e. functional changes at the intracellular level): for example, alterations to the structure or quantity of the product of a proto-oncogene or tumour suppressor gene, alterations to metabolic activation/inactivation/DNA repair

(iii) Evidence of relevant effects on cell behaviour (i.e. morphological or behavioural changes at the cellular or tissue level): for example, induction of mitogenesis, compensatory cell proliferation, preneoplasia and hyperplasia, survival of premalignant or malignant cells (immortalization, immunosuppression), effects on metastatic potential

(iv) Evidence from dose and time relationships of carcinogenic effects and interactions between agents: for example, early/late stage, as inferred from epidemiological studies; initiation/promotion/progression/malignant conversion, as defined in animal carcinogenicity experiments; toxicokinetics

These dimensions are not mutually exclusive, and an agent may fall within more than one of them. Thus, for example, the action of an agent on the expression of relevant genes could be summarized under both the first and second dimension, even if it were known with reasonable certainty that those effects resulted from genotoxicity.

12. EVALUATION

Evaluations of the strength of the evidence for carcinogenicity arising from human and experimental animal data are made, using standard terms.

It is recognized that the criteria for these evaluations, described below, cannot encompass all of the factors that may be relevant to an evaluation of carcinogenicity. In considering all of the relevant scientific data, the Working Group may assign the agent, mixture or exposure circumstance to a higher or lower category than a strict interpretation of these criteria would indicate.

(a) Degrees of evidence for carcinogenicity in humans and in experimental animals and supporting evidence

These categories refer only to the strength of the evidence that an exposure is carcinogenic and not to the extent of its carcinogenic activity (potency) nor to the mechanisms involved. A classification may change as new information becomes available.

An evaluation of degree of evidence, whether for a single agent or a mixture, is limited to the materials tested, as defined physically, chemically or biologically. When the agents evaluated are considered by the Working Group to be sufficiently closely related, they may be grouped together for the purpose of a single evaluation of degree of evidence.

(i) Carcinogenicity in humans

The applicability of an evaluation of the carcinogenicity of a mixture, process, occupation or industry on the basis of evidence from epidemiological studies depends on the variability over time and place of the mixtures, processes, occupations and industries. The Working Group seeks to identify the specific exposure, process or activity which is considered most likely to be responsible for any excess risk. The evaluation is focused as narrowly as the available data on exposure and other aspects permit.

The evidence relevant to carcinogenicity from studies in humans is classified into one of the following categories:

Sufficient evidence of carcinogenicity: The Working Group considers that a causal relationship has been established between exposure to the agent, mixture or exposure circumstance and human cancer. That is, a positive relationship has been observed between the exposure and cancer in studies in which chance, bias and confounding could be ruled out with reasonable confidence.

Limited evidence of carcinogenicity: A positive association has been observed between exposure to the agent, mixture or exposure circumstance and cancer for which a causal interpretation is considered by the Working Group to be credible, but chance, bias or confounding could not be ruled out with reasonable confidence.

Inadequate evidence of carcinogenicity: The available studies are of insufficient quality, consistency or statistical power to permit a conclusion regarding the presence or absence of a causal association, or no data on cancer in humans are available.

Evidence suggesting lack of carcinogenicity: There are several adequate studies covering the full range of levels of exposure that human beings are known to encounter, which are mutually consistent in not showing a positive association between exposure to the agent, mixture or exposure circumstance and any studied cancer at any observed level of exposure. A conclusion of 'evidence suggesting lack of carcinogenicity' is inevitably limited to the cancer sites, conditions and levels of exposure and length of observation covered by the available studies. In addition, the possibility of a very small risk at the levels of exposure studied can never be excluded.

In some instances, the above categories may be used to classify the degree of evidence related to carcinogenicity in specific organs or tissues.

(ii) Carcinogenicity in experimental animals

The evidence relevant to carcinogenicity in experimental animals is classified into one of the following categories:

Sufficient evidence of carcinogenicity: The Working Group considers that a causal relationship has been established between the agent or mixture and an increased incidence of malignant neoplasms or of an appropriate combination of benign and malignant neoplasms in (a) two or more species of animals or (b) in two or more independent studies in one species carried out at different times or in different laboratories or under different protocols.

Exceptionally, a single study in one species might be considered to provide sufficient evidence of carcinogenicity when malignant neoplasms occur to an unusual degree with regard to incidence, site, type of tumour or age at onset.

Limited evidence of carcinogenicity: The data suggest a carcinogenic effect but are limited for making a definitive evaluation because, e.g. (a) the evidence of carcinogenicity is restricted to a single experiment; or (b) there are unresolved questions regarding the adequacy of the design, conduct or interpretation of the study; or (c) the agent or mixture increases the incidence only of benign neoplasms or lesions of uncertain neoplastic potential, or of certain neoplasms which may occur spontaneously in high incidences in certain strains.

Inadequate evidence of carcinogenicity: The studies cannot be interpreted as showing either the presence or absence of a carcinogenic effect because of major qualitative or quantitative limitations, or no data on cancer in experimental animals are available.

Evidence suggesting lack of carcinogenicity: Adequate studies involving at least two species are available which show that, within the limits of the tests used, the agent or mixture is not carcinogenic. A conclusion of evidence suggesting lack of carcinogenicity is inevitably limited to the species, tumour sites and levels of exposure studied.

(b) *Other data relevant to the evaluation of carcinogenicity and its mechanisms*

Other evidence judged to be relevant to an evaluation of carcinogenicity and of sufficient importance to affect the overall evaluation is then described. This may include data on preneoplastic lesions, tumour pathology, genetic and related effects, structure–activity relationships, metabolism and pharmacokinetics, physicochemical parameters and analogous biological agents.

Data relevant to mechanisms of the carcinogenic action are also evaluated. The strength of the evidence that any carcinogenic effect observed is due to a particular mechanism is assessed, using terms such as weak, moderate or strong. Then, the Working Group assesses if that particular mechanism is likely to be operative in humans. The strongest indications that a particular mechanism operates in humans come from data on humans or biological specimens obtained from exposed humans. The data may be considered to be especially relevant if they show that the agent in question has caused changes in exposed humans that are on the causal pathway to carcinogenesis. Such data may, however, never become available, because it is at least conceivable that certain compounds may be kept from human use solely on the basis of evidence of their toxicity and/or carcinogenicity in experimental systems.

For complex exposures, including occupational and industrial exposures, the chemical composition and the potential contribution of carcinogens known to be present are considered by the Working Group in its overall evaluation of human carcinogenicity. The Working Group also determines the extent to which the materials tested in experimental systems are related to those to which humans are exposed.

(c) *Overall evaluation*

Finally, the body of evidence is considered as a whole, in order to reach an overall evaluation of the carcinogenicity to humans of an agent, mixture or circumstance of exposure.

An evaluation may be made for a group of chemical compounds that have been evaluated by the Working Group. In addition, when supporting data indicate that other, related compounds for which there is no direct evidence of capacity to induce cancer in humans or in animals may also be carcinogenic, a statement describing the rationale for this conclusion is added to the evaluation narrative; an additional evaluation may be made for this broader group of compounds if the strength of the evidence warrants it.

The agent, mixture or exposure circumstance is described according to the wording of one of the following categories, and the designated group is given. The categorization of an agent, mixture or exposure circumstance is a matter of scientific judgement, reflecting the strength of the evidence derived from studies in humans and in experimental animals and from other relevant data.

Group 1 — The agent (mixture) is carcinogenic to humans.
The exposure circumstance entails exposures that are carcinogenic to humans.

This category is used when there is *sufficient evidence* of carcinogenicity in humans. Exceptionally, an agent (mixture) may be placed in this category when evidence in humans is less than sufficient but there is *sufficient evidence* of carcinogenicity in experimental animals and strong evidence in exposed humans that the agent (mixture) acts through a relevant mechanism of carcinogenicity.

Group 2

This category includes agents, mixtures and exposure circumstances for which, at one extreme, the degree of evidence of carcinogenicity in humans is almost sufficient, as well as those for which, at the other extreme, there are no human data but for which there is evidence of carcinogenicity in experimental animals. Agents, mixtures and exposure circumstances are assigned to either group 2A (probably carcinogenic to humans) or group 2B (possibly carcinogenic to humans) on the basis of epidemiological and experimental evidence of carcinogenicity and other relevant data.

Group 2A — The agent (mixture) is probably carcinogenic to humans.
The exposure circumstance entails exposures that are probably carcinogenic to humans.

This category is used when there is *limited evidence* of carcinogenicity in humans and sufficient evidence of carcinogenicity in experimental animals. In some cases, an agent (mixture) may be classified in this category when there is inadequate evidence of carcinogenicity in humans and *sufficient evidence* of carcinogenicity in experimental animals and strong evidence that the carcinogenesis is mediated by a mechanism that also operates in humans. Exceptionally, an agent, mixture or exposure circumstance may be classified in this category solely on the basis of limited evidence of carcinogenicity in humans.

Group 2B — The agent (mixture) is possibly carcinogenic to humans.
The exposure circumstance entails exposures that are possibly carcinogenic to humans.

This category is used for agents, mixtures and exposure circumstances for which there is *limited evidence* of carcinogenicity in humans and less than *sufficient evidence* of carcinogenicity in experimental animals. It may also be used when there is *inadequate evidence* of carcinogenicity in humans but there is *sufficient evidence* of carcinogenicity in experimental animals. In some instances, an agent, mixture or exposure circumstance for which there is *inadequate evidence* of carcinogenicity in humans but *limited evidence* of carcinogenicity in experimental animals together with supporting evidence from other relevant data may be placed in this group.

Group 3 — The agent (mixture or exposure circumstance) is not classifiable as to its carcinogenicity to humans.

This category is used most commonly for agents, mixtures and exposure circumstances for which the evidence of carcinogenicity is inadequate in humans and inadequate or limited in experimental animals.

Exceptionally, agents (mixtures) for which the evidence of carcinogenicity is inadequate in humans but sufficient in experimental animals may be placed in this category when there is strong evidence that the mechanism of carcinogenicity in experimental animals does not operate in humans.

Agents, mixtures and exposure circumstances that do not fall into any other group are also placed in this category.

Group 4 — The agent (mixture) is probably not carcinogenic to humans.

This category is used for agents or mixtures for which there is *evidence suggesting lack of carcinogenicity* in humans and in experimental animals. In some instances, agents or mixtures for which there is *inadequate evidence* of carcinogenicity in humans but *evidence suggesting lack of carcinogenicity* in experimental animals, consistently and strongly supported by a broad range of other relevant data, may be classified in this group.

References

Breslow, N.E. & Day, N.E. (1980) *Statistical Methods in Cancer Research*, Vol. 1, *The Analysis of Case–Control Studies* (IARC Scientific Publications No. 32), Lyon, IARC

Breslow, N.E. & Day, N.E. (1987) *Statistical Methods in Cancer Research*, Vol. 2, *The Design and Analysis of Cohort Studies* (IARC Scientific Publications No. 82), Lyon, IARC

Cohen, S.M. & Ellwein, L.B. (1990) Cell proliferation in carcinogenesis. *Science*, **249**, 1007–1011

Gart, J.J., Krewski, D., Lee, P.N., Tarone, R.E. & Wahrendorf, J. (1986) *Statistical Methods in Cancer Research*, Vol. 3, *The Design and Analysis of Long-term Animal Experiments* (IARC Scientific Publications No. 79), Lyon, IARC

Hoel, D.G., Kaplan, N.L. & Anderson, M.W. (1983) Implication of nonlinear kinetics on risk estimation in carcinogenesis. *Science*, **219**, 1032–1037

Huff, J.E., Eustis, S.L. & Haseman, J.K. (1989) Occurrence and relevance of chemically induced benign neoplasms in long-term carcinogenicity studies. *Cancer Metastasis Rev.*, **8**, 1–21

IARC (1973–1994) *Information Bulletin on the Survey of Chemicals Being Tested for Carcinogenicity/Directory of Agents Being Tested for Carcinogenicity*, Numbers 1–16, Lyon

    Number 1 (1973)   52 pages
    Number 2 (1973)   77 pages
    Number 3 (1974)   67 pages

Number 4 (1974)   97 pages
Number 5 (1975)   8 pages
Number 6 (1976)   360 pages
Number 7 (1978)   460 pages
Number 8 (1979)   604 pages
Number 9 (1981)   294 pages
Number 10 (1983)   326 pages
Number 11 (1984)   370 pages
Number 12 (1986)   385 pages
Number 13 (1988)   404 pages
Number 14 (1990)   369 pages
Number 15 (1992)   317 pages
Number 16 (1994)   293 pages

IARC (1976–1994)

Directory of On-going Research in Cancer Epidemiology 1976. Edited by C.S. Muir & G. Wagner, Lyon

Directory of On-going Research in Cancer Epidemiology 1977 (IARC Scientific Publications No. 17). Edited by C.S. Muir & G. Wagner, Lyon

Directory of On-going Research in Cancer Epidemiology 1978 (IARC Scientific Publications No. 26). Edited by C.S. Muir & G. Wagner, Lyon

Directory of On-going Research in Cancer Epidemiology 1979 (IARC Scientific Publications No. 28). Edited by C.S. Muir & G. Wagner, Lyon

Directory of On-going Research in Cancer Epidemiology 1980 (IARC Scientific Publications No. 35). Edited by C.S. Muir & G. Wagner, Lyon

Directory of On-going Research in Cancer Epidemiology 1981 (IARC Scientific Publications No. 38). Edited by C.S. Muir & G. Wagner, Lyon

Directory of On-going Research in Cancer Epidemiology 1982 (IARC Scientific Publications No. 46). Edited by C.S. Muir & G. Wagner, Lyon

Directory of On-going Research in Cancer Epidemiology 1983 (IARC Scientific Publications No. 50). Edited by C.S. Muir & G. Wagner, Lyon

Directory of On-going Research in Cancer Epidemiology 1984 (IARC Scientific Publications No. 62). Edited by C.S. Muir & G. Wagner, Lyon

Directory of On-going Research in Cancer Epidemiology 1985 (IARC Scientific Publications No. 69). Edited by C.S. Muir & G. Wagner, Lyon

Directory of On-going Research in Cancer Epidemiology 1986 (IARC Scientific Publications No. 80). Edited by C.S. Muir & G. Wagner, Lyon

Directory of On-going Research in Cancer Epidemiology 1987 (IARC Scientific Publications No. 86). Edited by D.M. Parkin & J. Wahrendorf, Lyon

Directory of On-going Research in Cancer Epidemiology 1988 (IARC Scientific Publications No. 93). Edited by M. Coleman & J. Wahrendorf, Lyon

Directory of On-going Research in Cancer Epidemiology 1989/90 (IARC Scientific Publications No. 101). Edited by M. Coleman & J. Wahrendorf, Lyon

Directory of On-going Research in Cancer Epidemiology 1991 (IARC Scientific Publications No.110). Edited by M. Coleman & J. Wahrendorf, Lyon

Directory of On-going Research in Cancer Epidemiology 1992 (IARC Scientific Publications No. 117). Edited by M. Coleman, J. Wahrendorf & E. Démaret, Lyon

Directory of On-going Research in Cancer Epidemiology 1994 (IARC Scientific Publications No. 130). Edited by R. Sankaranarayanan, J. Wahrendorf & E. Démaret, Lyon

IARC (1977) *IARC Monographs Programme on the Evaluation of the Carcinogenic Risk of Chemicals to Humans*. Preamble (IARC intern. tech. Rep. No. 77/002), Lyon

IARC (1978) *Chemicals with Sufficient Evidence of Carcinogenicity in Experimental Animals — IARC Monographs Volumes 1–17* (IARC intern. tech. Rep. No. 78/003), Lyon

IARC (1978–1993) *Environmental Carcinogens. Methods of Analysis and Exposure Measurement*:

Vol. 1. *Analysis of Volatile Nitrosamines in Food* (IARC Scientific Publications No. 18). Edited by R. Preussmann, M. Castegnaro, E.A. Walker & A.E. Wasserman (1978)

Vol. 2. *Methods for the Measurement of Vinyl Chloride in Poly(vinyl chloride), Air, Water and Foodstuffs* (IARC Scientific Publications No. 22). Edited by D.C.M. Squirrell & W. Thain (1978)

Vol. 3. Analysis of Polycyclic Aromatic Hydrocarbons in Environmental Samples (IARC Scientific Publications No. 29). Edited by M. Castegnaro, P. Bogovski, H. Kunte & E.A. Walker (1979)

Vol. 4. *Some Aromatic Amines and Azo Dyes in the General and Industrial Environment* (IARC Scientific Publications No. 40). Edited by L. Fishbein, M. Castegnaro, I.K. O'Neill & H. Bartsch (1981)

Vol. 5. *Some Mycotoxins* (IARC Scientific Publications No. 44). Edited by L. Stoloff, M. Castegnaro, P. Scott, I.K. O'Neill & H. Bartsch (1983)

Vol. 6. N-*Nitroso Compounds* (IARC Scientific Publications No. 45). Edited by R. Preussmann, I.K. O'Neill, G. Eisenbrand, B. Spiegelhalder & H. Bartsch (1983)

Vol. 7. *Some Volatile Halogenated Hydrocarbons* (IARC Scientific Publications No. 68). Edited by L. Fishbein & I.K. O'Neill (1985)

Vol. 8. *Some Metals: As, Be, Cd, Cr, Ni, Pb, Se, Zn* (IARC Scientific Publications No. 71). Edited by I.K. O'Neill, P. Schuller & L. Fishbein (1986)

Vol. 9. *Passive Smoking* (IARC Scientific Publications No. 81). Edited by I.K. O'Neill, K.D. Brunnemann, B. Dodet & D. Hoffmann (1987)

Vol. 10. *Benzene and Alkylated Benzenes* (IARC Scientific Publications No. 85). Edited by L. Fishbein & I.K. O'Neill (1988)

Vol. 11. *Polychlorinated Dioxins and Dibenzofurans)* (IARC Scientific Publications No. 108). Edited by C. Rappe, H.R. Buser, B. Dodet & I.K. O'Neill (1991)

Vol. 12. *Indoor Air* (IARC Scientific Publications No. 109). Edited by B. Seifert, H. van de Wiel, B. Dodet & I.K. O'Neill (1993)

IARC (1979) *Criteria to Select Chemicals for* IARC Monographs (IARC intern. tech. Rep. No. 79/003), Lyon

IARC (1982) *IARC Monographs on the Evaluation of the Carcinogenic Risk of Chemicals to Humans*, Supplement 4, *Chemicals, Industrial Processes and Industries Associated with Cancer in Humans* (IARC Monographs, Volumes 1 to 29), Lyon

IARC (1983) *Approaches to Classifying Chemical Carcinogens According to Mechanism of Action* (IARC intern. tech. Rep. No. 83/001), Lyon

IARC (1984) *Chemicals and Exposures to Complex Mixtures Recommended for Evaluation in IARC Monographs and Chemicals and Complex Mixtures Recommended for Long-term Carcinogenicity Testing* (IARC intern. tech. Rep. No. 84/002), Lyon

IARC (1987a) *IARC Monographs on the Evaluation of Carcinogenic Risks to Humans*, Supplement 6, *Genetic and Related Effects: An Updating of Selected* IARC Monographs *from Volumes 1 to 42*, Lyon

IARC (1987b) *IARC Monographs on the Evaluation of Carcinogenic Risks to Humans*, Supplement 7, *Overall Evaluations of Carcinogenicity: An Updating of* IARC Monographs *Volumes 1 to 42*, Lyon

IARC (1988) *Report of an IARC Working Group to Review the Approaches and Processes Used to Evaluate the Carcinogenicity of Mixtures and Groups of Chemicals* (IARC intern. tech. Rep.No. 88/002), Lyon

IARC (1989) *Chemicals, Groups of Chemicals, Mixtures and Exposure Circumstances to be Evaluated in Future IARC Monographs, Report of an ad hoc Working Group* (IARC intern. tech. Rep. No. 89/004), Lyon

IARC (1991a) *A Consensus Report of an IARC Monographs Working Group on the Use of Mechanisms of Carcinogenesis in Risk Identification* (IARC intern. tech. Rep. No. 91/002), Lyon

IARC (1991b) *Report of an Ad-hoc* IARC Monographs *Advisory Group on Viruses and Other Biological Agents Such as Parasites* (IARC intern. tech. Rep. No. 91/001), Lyon

IARC (1993) *Chemicals, Groups of Chemicals, Complex Mixtures, Physical and Biological Agents and Exposure Circumstances to be Evaluated in Future* IARC Monographs, *Report of an ad-hoc Working Group* (IARC intern. Rep. No. 93/005), Lyon

Montesano, R., Bartsch, H., Vainio, H., Wilbourn, J. & Yamasaki, H., eds (1986) *Long-term and Short-term Assays for Carcinogenesis — A Critical Appraisal* (IARC Scientific Publications No. 83), Lyon, IARC

Peto, R., Pike, M.C., Day, N.E., Gray, R.G., Lee, P.N., Parish, S., Peto, J., Richards, S. & Wahrendorf, J. (1980) Guidelines for simple, sensitive significance tests for carcinogenic effects in long-term animal experiments. In: *IARC Monographs on the Evaluation of the Carcinogenic Risk of Chemicals to Humans*, Supplement 2, *Long-term and Short-term Screening Assays for Carcinogens: A Critical Appraisal*, Lyon, pp. 311–426

Tomatis, L., Aitio, A., Wilbourn, J. & Shuker, L. (1989) Human carcinogens so far identified. *Jpn. J. Cancer Res.*, **80**, 795–807

Vainio, H., Magee, P., McGregor, D. & McMichael, A., eds (1992) *Mechanisms of Carcinogenesis in Risk Identification* (IARC Scientific Publications No. 116), Lyon, IARC

Vainio, H., Wilbourn, J.D., Sasco, A.J., Partensky, C., Gaudin, N., Heseltine, E. & Eragne, I. (1995) Identification of human carcinogenic risk in *IARC Monographs*. *Bull. Cancer*, **82**, 339–348 (in French)

Waters, M.D., Stack, H.F., Brady, A.L., Lohman, P.H.M., Haroun, L. & Vainio, H. (1987) Appendix 1. Activity profiles for genetic and related tests. In: *IARC Monographs on the Evaluation of Carcinogenic Risks to Humans*, Suppl. 6, *Genetic and Related Effects: An Updating of Selected IARC Monographs from Volumes 1 to 42*, Lyon, IARC, pp. 687–696

Wilbourn, J., Haroun, L., Heseltine, E., Kaldor, J., Partensky, C. & Vainio, H. (1986) Response of experimental animals to human carcinogens: an analysis based upon the IARC Monographs Programme. *Carcinogenesis*, **7**, 1853–1863

THE MONOGRAPHS

PRINTING PROCESSES AND PRINTING INKS

1. Exposure Data

1.1 Historical overview

The origins of printing can be traced back several centuries. Pictorial prints were produced from cut wood blocks in Japan during the tenth century and probably earlier in China. The first movable type, moulded in clay, can be traced to China in the eleventh century, and wooden type appeared in China in the fourteenth century. In Europe, book production from wood blocks was seen early in the fifteenth century, and Gutenberg introduced cast metal type in the middle of the fifteenth century. These inventions were the basis of the original printing method, namely letterpress printing.

As the first printing was a development from writing and drawing/painting, it was natural that the first printing inks would be based on writing inks and paints. They were composed of lampblack or coloured minerals dispersed in water-soluble gum. However, Gutenberg soon found that the aqueous gum solution-based inks did not wet metal type surfaces satisfactorily. The composition of the inks developed by Gutenberg is not known with certainty but was probably derived from the artists' paints of the time. These were based upon vegetable oils, such as linseed or nut oil, which were heated to increase their viscosity and fortified with natural rosin; to accelerate drying, metal salts were added. The first clear records of compositions of printing inks date from the seventeenth century and are of this nature (Wood, 1994).

Until the middle of the eighteenth century, printers made their own inks. When the specialist industry of ink manufacture began to develop, the ink supplied was little more than a concentrated pigment dispersion. Any skilled printer considered that he was a craftsman and would modify the ink that he purchased with his own 'secret' additives to give the printing properties that he wanted.

During the eighteenth century, there were many publications of printing ink formulations. They all followed the same basic composition but included the use of other vegetable oils and natural resins, gave more details about the pigments used and focused on the details of the manufacturing methods. Throughout this period, a significant hazard to the ink makers (in terms of fires, vapours and spills) was in the heating of the various oils. Some of the processing even required the hot oils to be ignited and then extinguished with a metal cover.

The lithographic (or litho) process was introduced in 1796 in Germany by Alois Senefelder. This process relied upon a particular type of hydrophilic limestone upon which images were drawn with greasy inks. These images were then receptive to oil-based inks, while the remainder of the surface was not. The first lithographic inks were composed of beeswax, tallow soap and lampblack, again produced by heating and burning.

Gradually, the basic composition of letterpress and litho inks began to converge, with rosin-fortified linseed oil being the basis of most coloured inks and rosin oil or mineral oils being the basis of blacks. The difference between the inks for the two processes was minor, but important; the litho inks contained additives and had a substantially higher viscosity.

The invention of phenol–formaldehyde resins and the introduction of oil-soluble formulations in the 1920s initiated the era of synthetic resin media. Then in 1936, petroleum distillates were introduced to create the two-phase quick-setting mechanism that is the basis of the majority of conventional letterpress and litho inks used today (Wood, 1994).

Although printing was carried out by letterpress for many centuries, this process has declined rapidly in the last two decades and is now limited to a few specialist applications and those sectors where older equipment has yet to be replaced.

Rotary letterpress printing from rubber printing plates (stereos) originated around 1890 and took the name 'aniline printing' from the aniline-derived dyes that were dissolved in water or alcohol to make the ink. The crude process has been refined since then, particularly over the last 30 years, and has developed into a discrete process in its own right under the name of flexographic (or flexo) printing. The basic dyes still have limited use but modern inks are based upon synthetic pigments in a wide range of synthetic media dissolved in volatile solvents, such as industrial methylated spirits (denatured ethanol).

The intaglio process, in which the image is engraved as a recess in a metal plate, was probably first used for printing purposes in the fifteenth century. A very viscous ink, which was likely to be similar to the letterpress inks of the time, was wiped over the surface so that it filled the recesses but was removed from the surface. A much refined form of this basic process is still in limited use for high-security printing such as the printing of bank notes.

An evolution of the intaglio process occurred in 1852 with the introduction of a method of etching the image into the plate rather than relying on the highly skilled art of engraving. This has led to the rotogravure or gravure process of today. A coating of dichromated gelatine was exposed to sunlight and then etched with a solution of ferric chloride. A variety of techniques were used over the years to control the depth of the etching and thus the strength of the print produced. Modern production is by mechanical or laser engraving so that the light-sensitive coatings and etchants have largely fallen out of use.

In the early days, the gravure inks were similar to those of intaglio printing. By the end of the eighteenth century, metal 'doctor blades' had been introduced to replace the wiping of the surplus ink with a cloth, but the inks remained the same. It was not until the end of the nineteenth century that the 'liquid' inks were introduced. The earliest of these were water-based and probably similar to the aniline inks of the time. These were later abandoned in favour of inks containing hydrocarbon solvents and natural or synthetic resins.

Screen printing is a small segment of the printing industry, for which the history is less well recorded. As a development of stencilling, the process has been in use for many centuries, primarily for the decoration of textiles. In the 1920s, it started to attract attention as a convenient way of producing short runs of posters and for printing on difficult surfaces such as glass. The process has developed as a means of depositing heavy films of ink upon a wide variety of substrates, often of difficult shape. In the early days, no suitable inks were available for screen printing, and use was made of ordinary decorative paints. Modern screen printing inks are based on a wide range of synthetic resins and polymers in a range of solvents with suitable volatility.

At the end of the 1960s, a totally new technology, ultraviolet (UV) (see IARC, 1992) curing, was introduced into printing ink formulation (Carlick, 1971; Hargreaves, 1995). These 100%-solids systems polymerize by mechanisms of free radicals or acid catalysis initiated by irradiation with suitable wavelengths of UV radiation. The technology was initially developed for litho and letterpress printing but soon spread to screen ink formulations and is now being introduced to flexo and gravure printing.

Although the smallest companies and industries in lesser developed regions may still use older technologies, the printing and printing ink industries today are generally very different from those of 50 years ago. Rapidly changing technologies, automation and computer control, and safer materials and processes characterize the industries today in the developed countries.

For reviews on historical overview, see Leach and Pierce (1993), Wood (1994) and Taggi and Walker (1996).

1.2 Description of the industry

The manufacture of printing inks and the major printing processes have been described in several reviews (National Association of Printing Ink Manufacturers, 1988; Williams, 1992; Leach & Pierce, 1993; Taggi & Walker, 1996).

1.2.1 *Printing ink manufacture*

The development of printing inks followed somewhat different paths, depending upon the printing process. Letterpress and litho printing use high-viscosity inks that are formulated using low-volatility materials and are generically known as 'paste' or 'oil' inks. The flexo and gravure processes use low-viscosity inks formulated with volatile solvents and are known collectively as 'liquid' inks. Screen printing inks generally are of intermediate viscosity. Ink manufacturers tend to specialize in one of these three areas, with only the largest companies covering a broader field.

The companies in the industry range from large international groups (of which Japanese-based Dainippon Ink & Chemicals and French-based Coates Lorilleux (The Freedonia Group, 1995) are the biggest) to small companies with less than 10 employees. For example, there are thought to be over 200 ink manufacturers each in the United

States of America (National Association of Printing Ink Manufacturers, 1988) and in Europe.

Generally, the large groups have associations with chemical companies that provide resin or pigment manufacturing resources. It is usually these groups that develop the new technology that drives the industry forward; they manufacture their products from the raw materials of pigments, resins, oils and solvents, and supply the largest printing companies or groups. The smaller companies focus upon products of established technology supplied to small- to middle-sized users. Commonly, manufacture of their inks is from partly processed intermediates, such as varnishes and pigment dispersions, which are bought from specialist manufacturers. Their business is generally based upon providing good service in areas where the major ink manufacturing groups find it uneconomical to do so.

1.2.2 *The printing processes*

The printing industry itself is equally varied with a limited number of very large groups and many small printers. Whatever the size, there is a tendency to concentrate on one, or at the most two, of the different printing technologies described below. Each of the processes has its own characteristics, summarized in Table 1.

(a) Lithography

Lithography is the most widespread of the printing processes and is probably used by more individual companies than all the other processes combined.

The litho process depends upon the ability to make the non-image areas of the planar printing plate hydrophilic, and thus readily wetted by water, and to make the image areas oleophilic, and thus readily wetted by oil-based inks. The plate is then wetted with a water-based fountain (damping) solution and inked with the oil-based ink. The development of the modern litho process has been made possible by the introduction of 'offsetting', in which the ink film is transferred from the printing plate onto a rubber blanket and then offset onto the substrate. This offsetting process prolongs plate life, particularly when printing on metal, and also improves the quality of print on substrates of poor smoothness.

The printing plate most commonly consists of a thin flat sheet of aluminium that has been treated to make the surface hydrophilic. Onto this is placed a light-sensitive polymer coating that is exposed to UV radiation through a film (which is positive or negative depending on the coating). The non-image areas are then developed away to expose the hydrophilic surface. Modern processing is usually in self-contained equipment that restricts any exposure of the operator to the developing solutions involved. Aqueous developers are replacing solvent-based systems. The metal plate is replaced by paper or polyester materials for some short-run small-format printing.

The paper or other substrate being printed can be fed into the press as individual sheets and delivered as a pile of sheets (sheet-fed litho) or as a continuous reel (web offset) that is converted to the finished publication after printing. Web offset printing can be further subdivided into cold-set web offset (with inks drying by penetration) and heat-set web offset (with inks drying by evaporation in gas-heated ovens). The sheet-fed process is generally used for printing a wide variety of commercial/advertising print and cartons whereas web offset is used for newspapers and magazines.

Table 1. Main characteristics of the printing processes

Printing process		Ink film thickness (µm)	Substrate types	Typical applications
Offset litho	Sheet-fed	< 2	Wide range of paper and board, plastic sheet and metal	All general print, business forms, technical documentation, packaging, promotional, magazines, credit cards
	Web-fed			
	Heat-set	< 2	Wide range of coated and uncoated paper	Magazines and similar format products
	Cold-set	< 2	Newsprint	Newspapers
Letterpress	Rotary	3–4	Newsprint, self-adhesive materials	Newspapers, labels
	Offset	< 2	Plastic containers, metals	Dairy product and drinks containers
Flexography	Narrow web	0.75–2	Paper and plastic film	Labels, flexible packaging
	Wide web	0.75–2	Newsprint	Newspapers
		0.75–2	Wide range of plastic film, paper, corrugated cardboard	Boxes and many other types of packaging, sacks
Gravure	Large web	< 6	Coated or uncoated paper	Magazines and similar products, mail order catalogues, wood grain patterns
	Smaller web	< 6	Coated or uncoated paper, plastic films, board	Packaging (especially flexible), cigarette cartons, postage stamps
	Sheet	< 6	Paper	Fine art reproductions
Screen		< 30	Card, fabric, wide range of plastic, shaped containers	Point-of-sale displays, plastic containers, labels, T-shirts
Intaglio		< 30	Paper	Bank notes, security documents

From Leach & Pierce (1993), except letterpress data which have been compiled by the Working Group

The inks used are high-viscosity products based upon low-volatility resins and oils. The printing press is cleaned after use with solvent washes using volatile organic solvents, usually hydrocarbons, although there is now a move towards vegetable oil-based cleansers and even water-based cleansers (Searle, 1993).

A major segment of the lithographic printing industry is the cold-set web offset printing of newspapers using lower-viscosity inks that run on high-speed presses and are prone to misting, but less so than with rotary letterpress.

(b) Letterpress

Letterpress is a 'relief' printing process — that is, the image area is raised above the non-image area. The raised image area is inked before being brought into direct or indirect contact with the substrate to be printed. The more specialized applications that have developed over the last two decades have made greater use of offset letterpress, where the image is transferred via a rubber blanket.

Traditionally, the raised image was a lead alloy type or etched metal plates, but these have largely been replaced by photopolymer plates that are exposed to either negative or positive film and developed in dedicated equipment. Aqueous developers are replacing solvent-based systems.

Sheet-fed letterpress has declined to such a degree that it is of little more than historic interest today in industrialized countries. The same is true of the use of rotary letterpress in the printing of newspapers. The one sector in which letterpress is still of major importance is the printing of reel-fed self-adhesive labels, but even there its use is now in decline. The inks most widely used for rotary label printing are UV-cured, although there is still some use of heat-set inks. The inks used for letterpress printing are very similar, and sometimes identical, to those used for lithography.

(c) Flexography

Although the newest of the major print processes, flexography (or flexo) is now the most important process for the production of flexible packaging. A number of other applications include corrugated cases, tickets, forms, directories, paperback books, comics and newspapers.

The process is similar to letterpress in that the image area is raised above the non-image area. An inking roller is positioned in an ink duct and runs in contact with an engraved metering roller known as the 'anilox'. This controls the amount (film weight) of ink that is transferred to the surface of the printing plate. At one time, the two processes were easily differentiated because letterpress used metal type plates and flexo used rubber stereos. Now both may use photopolymer plates. Although the distinction between letterpress and flexo is becoming blurred in some areas, this is not the case in the most important sector, namely flexible packaging, where the use of letterpress is insignificant. The dry ink film weight deposited in flexo is similar to that of litho and thus generally lower than that of letterpress.

The majority of flexo printing is web-fed with the main exception being the printing of corrugated cartons that are fed in as sheets.

The inks used for flexo are of low viscosity and are based upon a wide range of synthetic resins dissolved in low-boiling solvents. The drying process requires the removal of these solvents, usually by evaporation. However, there is a rapidly increasing interest in UV-cured (Atkinson, 1995) and water-based inks (Williams, 1994a,b) in an effort to reduce the emissions of volatile organic compounds.

(d) Gravure

The printing image for gravure consists of recessed cells engraved in a metal cylinder. This image cylinder runs directly in the ink duct so that both the cells and the surface of the cylinder are flooded with ink. A reciprocating metal doctor blade held against the rotating cylinder scrapes the surplus ink from the non-image surface of the cylinder. The substrate, usually in web form, is held against the cylinder by an impression roller so that transfer of the ink takes place. The printed substrate then passes through a heated drying oven to remove the solvent by evaporation.

Until the mid 1980s, chemical etching of the copper cylinder through a gelatine image resist with ferric chloride solution was the usual process. Because of the difficulty in maintaining consistent quality and the potential hazard of ferric chloride, this has now been largely replaced by electromechanical or laser engraving.

The cells in the image vary in surface area according to the screen ruling being used. However, with a typical 60-lines/cm screening, the cells will be about 125 µm across and the depth will vary from 2 µm in the highlights to 40 µm or more in the shadows. The result is an ink film that varies in thickness according to the density of colour and can be up to three times as thick as that deposited by litho or flexo.

There are two sectors in the gravure printing industry: production of flexible packaging similar to that of flexo, and publication gravure that produces long-run magazines. With flexible packaging, gravure is generally preferred to flexo, since higher print quality is required and longer runs are used, and the high set-up costs can be recovered. In publication, gravure is again preferred to web offset for the long runs where it is more cost-effective.

The inks used are similar to those of flexo printing, although hydrocarbon solvents can be used as there are no rubber parts in the printing unit.

(e) Screen process

Unlike those used in other mainline printing processes, screen printing (sometimes called silk screen or screen process) inks do not have to transfer from an image surface to the substrate but actually pass through the image, which is the stencil on the screen.

The screen consists of a fine mesh, usually nylon or polyester, but sometimes stainless steel, stretched over a metal frame. The stencil is produced by coating the mesh with a photosensitive coating, exposing it to UV radiation through a positive film and developing away the unexposed image areas. The open areas of the screen around the stencil are blocked with screen filler, using volatile solvents such as xylene (see IARC, 1989a) or dichloromethane (see IARC, 1987a).

The inks used are intermediate in viscosity between those used for litho and those used for flexo or gravure. Evaporation, oxidation, chemical curing and UV curing methods are all used for drying in screen printing. As with the other printing processes, there is growing use of UV- and water-based inks (Pfirrmann, 1994).

Screen printing is the most versatile process and can cover much work that is impossible by other means — it can print on curved surfaces, even those that are not totally regular or symmetrical; it can deposit a heavy ink film that is very much more durable than those applied by other processes; and it can be used on a wide range of surfaces including fabrics, plastics, metals, glass and printed circuit laminates as well as on paper and board. Above all, origination costs and set-up times can be very low so that short-run work is economically viable. As a result, the importance of the process has grown steadily over recent years.

1.3 Chemistry and uses of printing inks

The chemistry and composition of printing inks have been reviewed (National Association of Printing Ink Manufacturers, 1988; Williams, 1992; Kübler, 1993; Leach & Pierce, 1993; Bassemir *et al.*, 1995).

Among other factors, the formulation of a printing ink must take into account the printing process, the type and speed of the press, the characteristics of the substrate (the base material which is coated or printed) and the end use and desired appearance of the final printed product. For this reason, most printing inks are custom-made for specific applications. It is estimated that nearly a million new ink formulae are prepared each year and that several millions are in use (National Association of Printing Ink Manufacturers, 1988).

Printing inks are broadly distinguished by the printing process in which they are used. Letterpress and lithographic inks are known as paste inks and are of higher viscosity than the flexographic and gravure inks, which are called liquid inks. Printing inks are mixtures of three basic types of ingredients: pigments, vehicles and additives. Pigments determine the colour of the ink, including its hue (shade) and strength, and also affect physical properties, such as flow characteristics (rheology), opacity (or transparency), fastness and bleed resistance. Vehicles serve as carriers for the pigment during the printing process and bind the pigment to the substrate. Additives may include any of a very large number of ingredients needed to impart specific characteristics to the ink, such as driers, waxes, plasticizers, antioxidants, lubricants, rheological agents and curing agents.

1.3.1 *Pigments*

Pigments used in printing inks include both inorganic and organic pigments. Opaque pigments reflect light from their surface and therefore can cover or hide the surface on which they are printed. Transparent pigments transmit the light and allow the background on which they are printed to be seen through the film. Some of the most commonly used pigments are described below (National Association of Printing Ink Manufacturers, 1988; Williams, 1992; Kübler, 1993; Bassemir *et al.*, 1995).

(a) *Inorganic pigments*

The only black pigment used to an appreciable extent is carbon black (Pigment Black 7), mainly furnace black and thermal black (see separate monograph, this volume). It is used in newsprinting and in publication, commercial and packaging printing, and is therefore used in large quantities. Different grades of carbon black, varying in particle size, surface condition and structure, are required depending on the application.

Titanium dioxide (Pigment White 6) (see IARC, 1989b) is by far the most commonly used opaque white pigment. Other opaque whites used in printing inks include zinc oxide, zinc sulfide and lithopones (mixtures of zinc sulfide and barium sulfate).

Transparent white pigments, in order of decreasing transparency, include alumina hydrate, magnesium carbonate, calcium carbonate (chalk), blanc fixe (precipitated barium sulfate), talc and clays. Transparent white pigments are also known as extenders, because they are used to reduce the colour strength of inks, as well as to modify ink dispersion and flow characteristics.

Inorganic colour pigments are generally complex mixtures of inorganic salts or minerals, manufactured to meet colour, rheological and other specifications for the particular application. In printing inks, some of these pigments include chrome yellow and chrome orange (mixtures of lead chromate (see IARC, 1990) and other lead compounds (see IARC, 1987b)), iron blue (a series of precipitated complex ferriferro-cyanide compounds) and several iron oxide (see IARC, 1987c) and iron silicate pigments (Pigment Red 101 and 102, Pigment Brown 6 and 7). Cadmium (see IARC, 1993a) yellow, orange and red (selenides) and cadmium-mercury red have been used in printing applications where their high stability to light and alkali are advantageous, although their use has reportedly been almost completely eliminated except where extreme chemical resistance properties are required (National Association of Printing Ink Manufacturers, 1988).

Metallic powders (bronzes) also find some use in producing 'silver' (aluminium powder) and 'gold' (copper-zinc alloy powder) inks for packaging and advertising.

(b) Organic pigments

The printing ink industry uses more organic pigments than any other industry. Approximately 100 organic pigments are suitable for use in printing inks. Most of these pigments are prepared from azo, anthraquinone and triarylmethane dyes, phthalocyanines and vat dyes. Organic pigments are characterized by their high colouring strength, pure shades and transparency. However, as with the inorganic pigments, they are seldom pure chemicals and are classified by colour rather than by chemical composition. Organic pigments known as 'lakes' may be insoluble organic salts or may be produced by depositing organic dyes on one of the transparent inorganic white pigments, usually alumina hydrate.

Yellows are mainly diarylide yellows, Hansa yellows and lake yellows. These yellows are frequently used in place of the chrome yellows to avoid the use of lead and chromate compounds. Diarylide yellows are the principal yellow pigments currently used in printing inks. Pigment Yellow 12, one of the most common diarylide yellows, is a *bis*-azo dye based on 3,3′-dichlorobenzidine (see IARC, 1987d). Hansa yellows are also azo pigments based on toluidine and β-naphthol (Lewis, 1993a).

The most common orange pigment in printing inks is Pigment Orange 34, a diarylide orange. Others include dianisidine orange and Persian orange lakes.

There are more red pigments used in inks than any other organic colour. Some of the more common are the para reds (lakes based on the azo dye from *para*-nitroaniline and β-naphthol), lithol reds (e.g. Pigment Red 57:1, a calcium salt of an azo dye, which is the most important magenta colourant in inks), rhodamine reds (e.g. Pigment Red 81, which is rhodamine 6G laked by precipitation with phosphotungstic and phosphomolybdic acids) and Red Lake C (Pigment Red 53:1 (see IARC, 1993b), which is the barium salt of the azo dye obtained from 2-amino-4-methyl-5-chlorobenzenesulfonic acid and β-naphthol).

Phthalocyanine blues, such as Pigment Blue 15, are the most common organic blue pigments in inks. Peacock blues (Lewis, 1993b) (salts of certain sulfonated triarylmethane dyes, sometimes laked on alumina hydrate) were used in the past. Other organic blue pigments include the victoria blues and the alkali blues.

Phthalocyanines also form green pigments (for example, Pigment Green 7, a chlorinated phthalocyanine). Triarylmethane dyes, the salts of which are used as pigments, include malachite green and methyl violet. Pigment Violet 19, a quinacridone structure, is an example of the anthraquinone-type pigments used in printing inks.

1.3.2 *Vehicles and ink drying*

In a printing ink, the fluid vehicle in which the pigments are dispersed serves not only as a carrier for the pigments but also effects the binding of the pigments to the printed surface ('drying'). The vehicle is also the primary determinant of the tack[1] and flow characteristics of the ink. The drying of printing inks is accomplished in most cases by one or more of the following physical/chemical processes: absorption, evaporation, precipitation, oxidation, quick-setting and radiation curing.

Table 2 gives an overview of the types of vehicles used for various drying systems and printing processes (National Association of Printing Ink Manufacturers, 1988; Williams, 1992; Kübler, 1993; Bassemir *et al.*, 1995). In practice, drying often involves a combination of these physical/chemical processes. For example, so-called quick-setting inks, currently used in most letterpress and offset litho processes where rapid drying is essential, contain a balanced combination of resin, drying oil and solvent; the solvent is rapidly absorbed by the paper, leaving a partially dry or set ink film of resin and oil that subsequently hardens by oxidation (National Association of Printing Ink Manufacturers, 1988).

1.3.3 *Printing inks and processes*

(a) *Lithographic (offset) inks*

The broad spectrum of applications for lithography requires a wide variety of inks to serve the needs of the offset printing industry (National Association of Printing Ink Manufacturers, 1988; Kübler, 1993; Bassemir *et al.*, 1995; Pardoen, 1995).

---

[1] Tack is a relative measurement of the cohesion of an ink film which is responsible for its resistance to splitting between two rapidly separating surfaces.

Table 2. Printing ink drying systems and vehicles

Drying system	Printing process	Type of vehicle	Examples
Absorption	Newspaper printing (letterpress or cold-set web offset)	Non-drying oil	High-boiling petroleum oils (mineral oil, b.p. 280–350 °C)
Solvent evaporation	Heat-set web offset	Solvent/resin	Lower-boiling petroleum distillates (b.p. within 240–300 °C) with resins and drying oils
	Gravure	Solvent/resin	Toluene (publication gravure), or ethyl acetate and alcohols (packaging gravure) with resins
	Flexography	Solvent/resin	Alcohol, ethyl acetate or water and resins
Oxidation	Letterpress, offset, intaglio	Drying oil/resin	Linseed oil varnish with resins and driers (cobalt or manganese soaps)
Precipitation	Letterpress (specialty)	Glycol/resin	Moisture-set inks (glycol solvent with water-insoluble resin binder)
Radiation curing (UV or electron beam)	All processes	Monomers	Acrylate or vinyl ether/epoxy resins for the oligomers, reactive resins and monomers (with photoinitiators for UV curing)
Quick-setting	Litho, letterpress	Drying oil/resins/ distillate	Linseed oil/resin with petroleum distillate (b.p. 260–280 °C)

Adapted from National Association of Printing Ink Manufacture (1988), Williams (1992), Kübler (1993) and Bassemir et al. (1995)
b.p., boiling point

Newspaper offset (cold-set) printing inks are typically very simple. Blacks are carbon black in a high-boiling mineral oil with asphaltic material (bitumen (see IARC, 1987e), gilsonite). Such inks could consist of up to 70% mineral oil (see IARC, 1987f) which, in the past, might have contained up to 15% aromatic hydrocarbons. During the last decade, these mineral oils have been replaced with grades that have about 5% aromatics, of which less than 0.1% are polycyclic aromatic hydrocarbons (PAHs) (see IARC, 1983). Coloured inks generally have a soya bean oil vehicle instead of mineral oil. For recycling of newsprint, the ink must be removable and therefore these inks do not contain binders that undergo significant oxidative cross-linking.

Heat-set web offset inks are designed to produce high-gloss printed images (magazines, books). They contain lower-boiling mineral oils that are removed (within 1 sec) as the printed roll (web) passes through a hot air oven. A typical formulation might be: organic pigments (15–25 wt %), hard resins (25–35 wt %), soft resins and drying oils (5–15 wt %), mineral oil (b.p. 240–260 °C; 25–40 wt %) and additives (5–10 wt %) (Williams, 1992; Kübler, 1993).

Sheet-fed offset inks are used in commercial litho presses for printing, for example, advertising brochures, business papers and packaging, on individual sheets rather than long rolls. Inks are based on phenolic or maleic acid-modified rosin ester and alkyd resins in vegetable drying oils (linseed, soya, tung) diluted with mineral oil. Inks dry by quick-setting, i.e. by absorption and oxidation. A typical formulation would be: organic pigments (12–20 wt %), hard resins (20–25 wt %), soft resins and drying oils (20–30 wt %), mineral oil (b.p. 250–300 °C; 20–30 wt %) and additives (5–10 wt %) (Kübler, 1993).

Radiation-cured offset inks, which are based on acrylate or vinyl/ether monomers, are becoming very important in both sheet-fed and web offset processes. The printed substrate is exposed to UV radiation or an electron beam at the end of the press, and the ink sets within a fraction of a second.

To exclude the oil-based ink from the hydrophilic areas of the printing plate, all offset litho printing processes require water-based fountain or dampening solutions. These solutions are typically slightly acidic aqueous solutions (pH 3.5–5.5) containing small amounts of buffers, alcohols, surfactants, hydrophilic polymers (gum arabic or cellulose derivatives), complexing agents (EDTA, ethylenediamine tetraacetic acid) and preservatives.

(b) Letterpress inks

While letterpress is being replaced by other printing processes, it is still used to a limited extent to produce newspapers, magazines, self-adhesive labels, packaging and other printed products (National Association of Printing Ink Manufacturers, 1988; Williams, 1992; Kübler, 1993; Bassemir *et al.*, 1995).

Letterpress news ink is similar to web offset inks used to print newspapers.

Moisture-set inks have been used for food packaging printing and contain maleic or fumaric acid-modified rosin products or modified phenolics as binders in glycol solvents. The printed surface is treated with steam or a fine mist of water, and the water-insoluble acidic binders precipitate, setting the ink.

Water-miscible inks maintain the stability of the ink through an organic base that evaporates or is neutralized to induce drying.

A variety of other ink types have been used in letterpress printing, including heat-set, quick-set, water-washable and high-gloss inks (National Association of Printing Ink Manufacturers, 1988).

(c) Flexographic inks

Flexographic inks are liquid inks, rather than pastes, and are designed to dry quickly primarily by evaporation. Both solvent- and water-based ink systems are used extensively in flexography.

Common solvents in the solvent-based inks include the lower alcohols (ethyl, *n*-propyl, isopropyl (see IARC, 1987g)), usually mixed with esters and sometimes small amounts of higher glycol ethers or aliphatic hydrocarbons to obtain optimum resin solubility, viscosity and drying speed. A wide variety of resins are used in solvent-based flexo inks, such as nitrocellulose, polyamides, cellulose esters, acrylics and various modified rosins. Although pigments are the most common colourants, there is some use of both basic and metal-complex dyes in flexographic inks. Because much of the flexographic printing is on non-absorbent flexible packaging (polyethylene (see IARC, 1987h), polypropylene (see IARC, 1987i), poly(vinyl chloride) (see IARC, 1987j)), ink additives may include plasticizers to promote formation of a flexible ink film and waxes to add rub resistance (National Association of Printing Ink Manufacturers, 1988; Kübler, 1993; Bassemir et al., 1995).

For environmental reasons, the solvents in ink are being increasingly replaced by water. Approximately 50% of all flexographic inks have water as their primary solvent (Bassemir et al., 1995). Resins in these water-based formulations are generally acidic acrylates or fumaric acid-modified rosin or shellac, neutralized with ammonia or volatile amines, which evaporate from the printed substrate and thereby set the ink film. A typical water-based flexographic ink for paper or paperboard contains: organic pigments (12–15%), resins (10–25%) and additives (5–7%); the remainder is water. For printing on plastics, the ink usually contains a small amount of alcohol (2–5%) and more additives (6–10%) (Kübler, 1993).

Water-based flexographic printing also finds some application in newspaper printing, but it still represents only a small segment of the market (6–7% in the United States in 1992) (Bassemir et al., 1995), and removal of the ink from newsprint before recycling the paper remains a problem.

UV-cured inks are also beginning to be used in flexographic printing. Their composition is similar to the UV-cured offset litho inks although they are less viscous (Kübler, 1993; Atkinson, 1995).

(d) Gravure inks

Gravure inks are similar to flexographic inks except that ketones and aromatic hydrocarbons can be used as solvents, providing a much greater latitude in the selection of binders (Bassemir et al., 1995). In the United States, gravure inks are divided into 10 categories according to the type of binder or solvent. For example, aliphatic hydrocarbons (hexane, VM&P (varnish makers' and painters') naphtha, mineral spirits) are used mainly in type A, B and D inks; aromatic hydrocarbons (toluene (see IARC, 1989c), xylene) are used in types B, D, M and T; and ketones (acetone, methyl ethyl ketone) and esters (ethyl, isopropyl, *n*-propyl and butyl acetates) are required for type C inks (Bassemir et al., 1995).

Gravure inks also are classified according to the printed product. 'Publication gravure' (magazines, catalogues) utilizes hard resins dissolved in toluene and/or aliphatic solvents. Resins include maleic acid-modified rosin and phenolic resins, calcium and zinc resinates, hydrocarbon resins and others. Polyethylene-based waxes are often added to improve abrasion resistance. A typical publication gravure ink might contain: pigments (8–15 wt %), resins (15–20 wt %), solvent (60–70 wt %) and additives (0.5–5 wt %) (Kübler, 1993).

'Packaging gravure' does not use hydrocarbon solvents but rather uses esters and alcohols (type C inks in the United States nomenclature). For various packaging substrates, resins may include cellulose nitrate, maleic resins, acrylate resins, polyurethane resins and polyamide resins, or mixed polymers of vinyl chloride/vinyl acetate/vinyl alcohol. Plasticizers (phthalates, citrates, adipates) also may be required, especially with cellulose nitrate. Basic dyes are used occasionally, in addition to pigments, in gravure inks. More recently, water-based gravure inks (type W), with formulations very similar to the water-based flexographic inks, are finding increasing use in packaging gravure (National Association of Printing Ink Manufacturers, 1988; Kübler, 1993; Bassemir et al., 1995).

A special application of gravure (as well as other processes) is in printing with transfer inks. Aqueous inks containing selected textile-disperse dyes are printed and dried on special papers. The printed image can then be transferred by sublimation to the textile materials (e.g. polyester fabric) by pressing at approximately 200 °C (Kübler, 1993).

Intaglio is another specialized process using a higher-viscosity ink in which a high-quality image is engraved on a steel plate. Inks often contain special high-durability pigments, fillers to increase viscosity, drying oil vehicle and a number of additives. This process is used in printing paper currency (bank-notes), postage stamps, stock certificates and similar products (Kübler, 1993; Bassemir et al., 1995).

(e) Screen process inks

Screen printing is a highly versatile process that can apply a thicker film of ink to the substrate than other printing processes. A very wide range of ink formulations is available, depending on the substrate and the requirements for the printed product. Drying may be by evaporation, oxidation, radiation-curing or other processes. Any of the resin types found in litho, flexo and gravure inks may be used, and the solvents can be of almost any type as long as they evaporate at a suitable rate, which is slower than that required for flexo and gravure. Solvents (propylene glycol ethers, aromatic and aliphatic hydrocarbons and cyclohexanone) typically have somewhat higher boiling-points than those used in gravure printing, and inks are more viscous (Kübler, 1993; Leach & Pierce, 1993; Bassemir et al., 1995).

Although newly introduced product ranges are usually lead-free, screen printing is the one process that still makes significant use of lead chromate pigments. The use of *N*-vinyl pyrrolidone (see IARC, 1987k) in UV-cured inks has declined.

1.4 Production

The manufacture of printing inks traditionally has been a batch process, and the large majority of all printing inks are still made in batches. Only a few high-volume, standardized inks (e.g. news inks) are made by continuous processes (National Association of Printing Ink Manufacturers, 1988; Williams, 1992).

Ink vehicles are usually produced in separate resin/varnish plants and may be received by the ink manufacturer as solid resins or fluid varnishes (Bassemir et al., 1995).

Pigments and other ink components are normally purchased by ink manufacturers from suppliers. Pigments are available as presscake, flushed colours (in which the water has been replaced by vehicle), dry colours (with virtually all of the water removed) and colour concentrates (liquids or pastes with 35–65% pigment). In paste inks, flushed colours are the principal form of colourant used in the United States. In Europe, dry colours are generally used because of the greater choice of vehicles available to the formulator (Bassemir *et al.*, 1995).

If very finely ground (predispersed) pigments or flushed pigment concentrates are used, paste inks can often be prepared simply by mixing pigment thoroughly with vehicles, solvents, oils and additives. Many sizes (5–1000 gallons [20–3790 L]) and types of mixers are used. Inks are sometimes filtered in a final step to remove residual particles (National Association of Printing Ink Manufacturers, 1988; Bassemir *et al.*, 1995).

Dry pigments or resin-coated pigments usually require a two-stage process — the pigments are first mixed with the other ink components and then thoroughly ground and dispersed using various types of ink mills. Milling may be done in three-roll mills, ball mills, sand mills, shot mills and others. Fluid inks (flexographic, gravure) must be milled in closed systems because of the volatility of the solvents, and, appropriately, controlled shot or sand mills, colloid mills and ball mills all are in use. Some large-volume inks, such as web offset process colours and coloured news inks, are manufactured by pump-filtration, in which a flushed colour is dispersed in the vehicle by high-speed high-shear mixers and then pumped through a series of filters (National Association of Printing Ink Manufacturers, 1988; Bassemir *et al.*, 1995). The major ink suppliers will carry out most or all of these processes themselves.

Inks can be packaged and shipped in metal cans, in metal or plastic pails or in metal or fibre drums. Large-volume fluid inks (news, flexo, gravure) may be delivered directly to the printer in tank trucks. Flexo, gravure and screen inks are shipped at higher viscosity, and solvent is added by the printer (National Association of Printing Ink Manufacturers, 1988; Bassemir *et al.*, 1995).

Worldwide printing ink production in 1994 was approximately 2.7 million tonnes, with the 15 largest producers accounting for 85% and the remaining 15% produced by several hundred smaller manufacturers (Table 3). European production levels in 1993 are shown by country in Table 4.

The importance of lithographic (offset) printing and the decline in market share for letterpress are evident from the data in Table 5. Table 6 shows the estimated worldwide distribution of printing inks by process.

1.5 Occupational exposures in printing processes and to printing inks

Occupational exposures in the printing industry and processes are discussed below according to their occurrence in printing ink manufacture and in printing operations such as letterpress, lithography, flexography, gravure and screen printing. Occupational exposure measurements from several of the studies described herein are presented in Table 7.

Table 3. 1994 World production of printing inks

Country/region	Thousands of tonnes	Percentage of total
United States	1020	38.0
Europe	730	27.0
Japan	500	18.5
Rest of the world	450	16.5
Total	2700	100.0

From European Confederation of Paint, Printing Ink and Artists' Colours Manufacturers' Association (1995)

Table 4. 1993 European production of printing inks

Country/region	Thousands of tonnes	Percentage of total
Germany	276	40.0
United Kingdom	114	16.5
Italy	72	10.5
France	70	10.5
Belgium	30	4.5
Netherlands	30	4.5
Others	93	13.5
Total	685	100.0

From European Confederation of Paint, Printing Ink and Artists' Colours Manufacturers' Association (1995)

Table 5. Percentage distribution of printing ink use by process in the United States in 1981–1994

Process	1981[a]	1985[b]	1989[b]	1994[b]
Lithography and offset	44	44.5	45	49
Flexography	14	19	21	22
Gravure	17	20	19	16
Letterpress	20	7.5	6	5
Screen printing and other	5	9	9	8

[a] From Bruno (1982)
[b] From The Freedonia Group (1995)

Table 6. Estimated worldwide distribution of printing ink use by process

End-use market	Printing process	Thousands of tonnes	Market (%)
Newspapers	Letterpress	30	1
	Web offset	350	13
Publications and magazines	Gravure	590	22
	Heat-set/offset	300	11
Packaging and commercial	Offset (Litho)	385	14
	Gravure	390	14
	Flexo (solvent)	150	5.5
	Flexo (water)	150	5.5
Others (posters, plastic containers)	Screen	85	3
Miscellaneous	All processes	270	11
Total		2700	100

From European Confederation of Paint, Printing Ink and Artists' Colours Manufacturers' Association (1995)

Some common job categories in printing operations include the following: typesetters (compositors), photoengravers and plate makers who set type or transfer images to plates; press room or machine room operators who operate and tend the printing machines and presses that produce the printed product; and publishing room workers, folding machine operators or collators who bundle or fold the printed product.

1.5.1 *Exposures in printing ink manufacture*

During manufacture of liquid and paste inks, powder pigments are transferred from drums or bags into rotary mixing vessels and mixed and blended with resins and solvents in a batch process. In the past, mixers were often operated open; however, closed systems have replaced many open mixers. Dust exposures to raw pigments may occur if bags or drums are emptied into the mixer by hand. Solvents are usually piped into the mixture and exposure to solvents during mixing depends on whether the mixer is closed or open.

Crude ink mixtures are then milled or dispersed for thorough mixing of pigments. Once crude ink has been prepared, potential for inhalation of pigment dust is lessened, although dermal contact with paste ink provides continued opportunity for exposures.

Liquid inks are further diluted with solvents by the printer for some applications where low-viscosity inks are required such as for flexography. Common solvent components of liquid inks include: toluene, ethanol, isopropanol, methyl ethyl ketone, ethyl acetate, xylene, isobutanol and acetone (Sakurai, 1982; Leach & Pierce, 1993). In addition, the use of methanol, dichloromethane, ethyl ether and 2-methoxyethanol has been reported (Sakurai, 1982). Benzene (see IARC, 1987l) was withdrawn from significant use in printing inks in Europe in 1950, although substantial contamination of

Table 7. Occupational exposures in printing processes and to printing inks

Industry	Process/operation	Sample type	No. of samples	Analyte	Air concentration				Year	Reference Country
					Mean		Range (or SD)			
					ppm	mg/m³	ppm	mg/m³		
Printing ink manufacture (Flexography)	NR	Personal	22	Ethanol	21	39.5	<1–89	<2–167	1983	Winchester (1985) New Zealand
			3	MIBK	23	94	1–59	4–236		
			24	n-Propylacetate	8	33	2–18	8.4–75		
			6	Isopropyl acetate	7	29.3	<1–14	<4–58.5		
			12	n-Propanol	4	9.8	<1–10	<2.5–25		
			14	Isopropanol	4	9.8	<1–11	<2.5–27		
			27	Toluene	4	15	<1–21	<4–79		
			27	Higher aromatics"	2		<1–6			
			7	MEK	1	3	<1–2	<3–6		
			3	Cyclohexanone	1	4	≤1	≤4		
			12	n-Hexane	1	3.5	<1–3	<3.5–10.5		
			18	2-Ethoxyethanol	1	3.7	<1–5	<3.7–18.5		
Printing ink manufacture	Liquid ink department	Personal	10	Toluene		>735		27–4595	1988	Lewis (1994) South Africa
			3	Xylene		>314		<LD–>783		
			8	Ethyl acetate		183		<LD–464		
			9	Ethanol		>164		<LD–>366		
			4	Isopropanol		>11		<LD–>22		
			1	n-Hexane		4		NA		
	Paste ink department	Personal	9	Toluene		27		<LD–42		
			3	Xylene		24		<LD–58		
			1	Ethanol		16		NA		
Printing industry	Various	NR	94	EGMEA		4.3 (GM)		3.9–4.7	1983–86	Veulemans et al. (1987) Belgium
				EGEE		9.8 (GM)		0.7–182.0		
				EGEEA		16.4 (GM)		0.3–186.8		
				EGBE		4.1 (GM)		1.5–17.7		
				EGBEA		12.7 (GM)		4.6–26.5		
Printing facility	Various	Personal	90	Isopropanol		208		8–647	Early 1980s	Brugnone et al. (1983) Italy

PRINTING PROCESSES AND PRINTING INKS   51

Table 7 (contd)

Industry	Process/operation	Sample type	No. of samples	Analyte	Air concentration				Year	Reference Country
					Mean		Range (or SD)			
					ppm	mg/m³	ppm	mg/m³		
Printing facility	Between press lines	Area	1	Aliphatic hydrocarbons	117		NA		1970	FIOH (1995) Finland
	Box lacquering and printing	Area	1	Acetone	15	35	NA	–	1970	
		Personal	18	Acetone	23	54	7–98	16–232	1970	
	Gravure	Area	5	Benzene	5	16	2–7	6–22	1962	
	Printing press	Personal	4	Butanol	28	84	ND–64	ND–192	1970	
	Silk screen printing	Personal	3	Hydrocarbons	41	–	18–82	–	1970	
	Gravure	Area	14	Xylene	19	82	2.5–70	11–304	1960–62	
	Lacquering	Personal	7	Xylene	3	13	ND–5	ND–22	1970	
	Printing press	Personal	3	Lead		0.113		0.05–0.17	1970	
	Printing press	Personal	3	Antimony		0.04		ND–0.09	1970	
	Silk screen	Personal	5	Toluene	16	60	7–28	26–105	1970	
	Printing press	Personal	21	Toluene	20	75	8–53	30–200	1970	
Printing facility	Rotogravure	Area	NR	Benzene	288	922	125–532	400–1702	1953	Forni et al. (1971) Italy
	Rotogravure	Area	NR	Toluene	NR	NR	0–240	0–905	1954–56	
	Rotogravure	Area	NR	Toluene	264	995	56–824	211–3106	1957–67	
Printing facility	Rotogravure	Area	NR	Toluene	100–200	377–754	NR–700	NR–2640	<1977	Funes-Cravioto et al. (1977) Sweden
Printing facility	Rotogravure	Area	NR	Toluene	NR	NR	200–300	754–1130	1970s	Bauchinger et al. (1982) Germany
Printing facility	Rotogravure	Personal	NR	Toluene	NR	NR	7–112	26–422	<1979	Mäki-Paakkanen et al. (1980) Finland
Printing facility	Rotogravure	Personal	30	Toluene		128		42–253	1980s	De Rosa et al. (1986) Italy

Table 7 (contd)

Industry	Process/operation	Sample type	No. of samples	Analyte	Air concentration				Year	Reference Country
					Mean		Range (or SD)			
					ppm	mg/m³	ppm	mg/m³		
Printing facility	Rotogravure	Personal	NR	Toluene		NR		30–600	<1993	Monster et al. (1993) Netherlands
Newspaper printing	Rotary letterpress	Area	NR	Dust		1.1		NR	1979	Lynge et al. (1995) Denmark
	Rotary letterpress	Area	NR	Dust		NR		1–4 (65–90% ink mist)	1986–87	
	Offset lithography	Area	NR	Dust		0.28		NR	1979	
	Offset lithography	Area	NR	Dust		NR		0.1–0.7 (25% ink mist)	1986–87	
Newspaper printing	Press operation	Personal	5	Oil mist		0.5		0.1–1.0	1976	Kronoveter & Gill (1977) USA
	Press operation	Personal	4	Stoddard solvent		6.5		5–9		
Newspaper printing	Press operation	Personal	2	Cyclohexanone	0.1	0.4	0.05–0.15	0.2–0.6	1981	Daniels (1981) USA
	Plate photocuring	Personal	2	Acetic acid	0.08	0.2	0.04–0.12	0.1–0.3		
Paper box printing	Colour offset printing, UV curing, press operation	Personal	2	DPGME	3.7	22.2	1.8–5.5	10.8–33.6	Early 1980s	Cullen et al. (1983) USA
Screen printing	Printing press	Personal	18	Isophorone	23	115	5.4 (SD)	27 (SD)	<1982	Samimi (1982) USA
				Cyclohexanone	28	112	5 (SD)	20 (SD)		
				Cellosolve acetate	18.5	100	4 (SD)	22 (SD)		
				Butyl acetate	11.5	55	0.8 (SD)	3.8 (SD)		
				Xylene	15	65	6.8 (SD)	29.5 (SD)		
				Diacetone alcohol	14	66	6.2 (SD)	29.5 (SD)		
				Petroleum distillate	85	–	14.5 (SD)	–		
Screen printing	Automatic dryer	Personal	19	Isophorone	9.5	48	3.3 (SD)	17 (SD)		
				Cyclohexanone	11	44	3 (SD)	12 (SD)		
				Cellosolve acetate	11	59	2.5 (SD)	13.5 (SD)		
				Butyl acetate	2.5	12	1.3 (SD)	6.2 (SD)		
				Xylene	4.5	19.5	1.5 (SD)	6.5 (SD)		
				Diacetone alcohol	3.5	16.6	1.5 (SD)	7.1 (SD)		
				Petroleum distillate	28.4	–	7.6 (SD)	–		

Table 7 (contd)

Industry	Process/operation	Sample type	No. of samples	Analyte	Air concentration				Year Reference Country
					Mean		Range (or SD)		
					ppm	mg/m³	ppm	mg/m³	
Screen printing	Manual drying	Personal	15	Isophorone	15	75	4.1 (SD)	21 (SD)	Samimi (1982) USA (contd)
				Cyclohexanone	18	72	3.8 (SD)	15 (SD)	
				Cellosolve acetate	18	97	3.5 (SD)	19 (SD)	
				Butyl acetate	4.5	21	1.5 (SD)	7 (SD)	
				Xylene	8.5	37	3.8 (SD)	16.5 (SD)	
				Diacetone alcohol	12	57	4.8 (SD)	23 (SD)	
				Petroleum distillate	49.5	–	12 (SD)	–	
Screen printing	Paint mixing	Personal	12	Isophorone	17.8	89	5.5 (SD)	28 (SD)	
				Cyclohexanone	8.5	34	4.1 (SD)	16.4 (SD)	
				Cellosolve acetate	10	54	2.1 (SD)	11.3 (SD)	
				Butyl acetate	3.4	16	1.1 (SD)	5 (SD)	
				Xylene	3.5	15	1.8 (SD)	7.8 (SD)	
				Diacetone alcohol	2.8	13	1.3 (SD)	6.2 (SD)	
				Petroleum distillate	56.5	–	18 (SD)	–	
Screen printing	Screen wash	Personal	14	Isophorone	8.3	42	5.6 (SD)	28 (SD)	
				Cyclohexanone	6	24	4.5 (SD)	18 (SD)	
				Cellosolve acetate	5	27	0.5 (SD)	2.7 (SD)	
				Butyl acetate	85	404	17 (SD)	81 (SD)	
				Xylene	35	152	8.9 (SD)	39 (SD)	
				Diacetone alcohol	6.8	32	1.5 (SD)	7 (SD)	
				Petroleum distillate	39	–	15.5 (SD)	–	

NR, not reported; MIBK, methyl isobutyl ketone; MEK, methyl ethyl ketone; LD, limit of detection; NA, not applicable; EGMEA, ethylene glycol monomethyl ether acetate; GM, geometric mean; EGEE, ethylene glycol monoethyl ether; EGEEA, ethylene glycol monoethyl ether acetate; EGBE, ethylene glycol monobutyl ether; EGBEA, ethylene glycol monobutyl ether acetate; UV, ultraviolet; DPGME, dipropylene glycol monomethyl ether; SD, standard deviation

[a] Higher aromatics, largely ethyl benzene and trimethylbenzenes

toluene with benzene remained until about 1960 (European Confederation of Paint, Printing Ink and Artists' Colours Manufacturers' Associations, 1995).

Few studies of exposures during ink manufacture have been published. In a 1983 study of 27 workers in five plants manufacturing colour flexographic ink, the highest personal solvent exposures occurred during the mixing of ingredients, the production of resin and quality control (Winchester, 1985). The most common solvents were ethanol (56% of total solvent volume), toluene (6.5%) and n-propyl acetate (5.4%). Full-shift average exposures to these and 14 other solvents ranged from 15% to 38% of their respective threshold limit values (TLV). The highest mean personal exposures were to ethanol (21 ppm [39.5 mg/m^3]) and methyl isobutyl ketone (23 ppm [94 mg/m^3]). Relatively lower exposures were found during dispersing, diluting and ink packing operations.

In a survey of a South African printing ink manufacturing plant in 1988, exposures to solvents were measured for 27 workers (Lewis, 1994). Personal partial-shift average exposures were higher in the liquid ink department than in the paste ink department and are shown in Table 7. Job titles in the liquid ink department include the mixer/weigher who operated the mixer for varnishes and inks, the pot washer who cleaned mixing drums with solvents and the weigher who weighed ingredients. Job titles in the paste ink department include the millhand who operated and cleaned the mill with solvents, the mixer/weigher who mixed and transported inks to the milling machine and the storeman who ran the pigment and varnish stockroom. This survey was conducted during a summer heat wave and the exposures probably represented their annual peak. Skin contact with solvents occurred in the liquid ink department during batch mixing, weighing and pouring of liquid inks into containers and during pot washing, where workers climbed into drums and scrubbed them with toluene-soaked rags. Gloves and paper filter masks were provided to the workers but were not used. At the time of this study, only general dilution ventilation was in use in the liquid ink department; however, local exhaust ventilation hoods over the mixing vessels to remove pigment dust existed in the paste ink department.

In the formulation of printing ink, estimated exposures to ink pigments were ranked for different job titles (Kay, 1976). These ranks were based on professional judgement rather than actual monitoring results. Exposures for weighers, mixers and laboratory staff were estimated to be the highest, followed by container washers, maintenance workers, millhands, porters, utility men and working foremen.

1.5.2 *Exposures in printing operations*

In newspaper production, exposures in the early twentieth century included lead dust and fumes, due to the use of lead stereotype plates. Lead alloy type contributed to exposures of manual typesetters (Hamilton, 1925). Mechanical typesetting somewhat reduced lead exposures in the printing industry, followed by further reduction in exposure potential with the introduction of the linotype. By the mid-1970s, computerized typesetting had eliminated these lead exposures arising from the handling of lead type (Michaels *et al.*, 1991; Kristensen & Anderson, 1992).

Other common exposures besides lead in older letterpress newspaper printing operations include noise, carbon tetrachloride (see IARC, 1987m), benzene, toluene, xylene and oil mist containing ink. Ink mist can be generated in the press room by high-speed rollers. The mean of 106 press room area samples taken in the 1970s was 1.1 mg/m^3 total aerosol, with 45% of the mass distribution of these particles as respirable particulates. Ink mist particulate may contain ink and oil mist and cellulose paper dust (Beaulieu & Anderson, 1978). The newspaper industry made a gradual process shift in the 1970s–80s and now primarily uses offset lithography rather than letterpress, resulting in lower lead and ink mist exposures (American Newspaper Publishers Association, 1988). Modern printing presses are fitted with closed ink-feed systems and point source extraction, and provide a higher level of automation and computerized control than previously.

Benzene was a common solvent in rotogravure processes from the 1930s up to the beginning of the 1960s in the United States but was eventually replaced by other solvents, primarily toluene (Greenburg et al., 1939; Lloyd et al., 1977; Svensson et al., 1990; European Confederation of Paint, Printing Ink and Artists' Colours Manufacturers' Associations, 1995). Benzene was used exclusively as an ink solvent and diluent in an Italian rotogravure plant up to 1953, when it was replaced by toluene following an epidemic of benzene poisoning (Forni et al., 1971). In a Swedish study of two rotoprinting factories [probably rotogravure], benzene was used as a solvent in the 1940s up to 1950, when toluene replaced it, and was a probable contaminant of toluene up to 1958 (Funes-Cravioto et al., 1977). In two Finnish rotogravure printing factories, benzene concentration in the toluene solvent has been controlled since 1962 and averaged 0.006% (always < 0.05%) (Mäki-Paakkanen, 1980). In a German rotogravure plant in the 1970s and early 1980s, the toluene printing ink solvent contained < 0.3% benzene (Bauchinger et al., 1982).

Polycyclic aromatic hydrocarbons (PAHs) such as benzo[a]pyrene adsorb onto the surface of carbon black particles and have been found in press room atmospheres (Casey et al., 1983). Although some yellow, red and orange pigments are manufactured from 3,3'-dichlorobenzidine and substituted derivatives of toluidine, laboratory analysis of 16 samples as part of a renal cancer study among paperboard printing workers did not detect either 3,3'-dichlorobenzidine or ortho-toluidine (see IARC, 1987n) in the bulk pigment (Sinks et al., 1992).

In newspaper plant surveys prior to 1976, Kay (1976) observed that ink was heavily dispersed into the air during printing and black ink oil mist readily settled on surfaces within the room. Higher-viscosity vegetable oil-based inks were introduced in the late 1980s to replace colour mineral oil inks, although black inks still contain mineral oil to disperse the darker pigments and carbon black (American Newspaper Publishers Association, 1988)

In a study of total dust concentrations in British press rooms between 1967 and 1981, ink mist droplet and cellulose paper fibre concentrations ranged from 0.62 to 2.16 mg/m^3 for 'particles < 60 μm' and from < 0.06 to 0.9 mg/m^3 for respirable particles of < 7 μm diameter (Casey et al., 1983). A study of a newspaper plant press room in the United

States detected air concentrations of oil mist ranging from 5 to 21 mg/m³ in the 1960s, with 15% of the mass distribution of particles in the respirable size range (Goldstein et al., 1970).

With the introduction of UV-cured acrylic resins and inks into colour offset printing in the 1970s, the use of high concentrations of glycol ethers increased. However, the trend has been towards the use of higher relative molecular mass glycol ethers, such as propylene and dipropylene glycol mono alkyl ethers (Cullen et al., 1983; Williams, 1992).

In a colour offset printing operation in a paper box printing company in the early 1980s, inks were both UV-cured and air-dried, the former accounting for 20–50% of the press operation time (Cullen et al., 1983). Solvent exposures occurred due to evaporation during both curing and mixing of solvents with inks. Exposure to inks and resins occurred by splashing and misting from press rollers, but the highest solvent exposures occurred during clean-up (25% of the work time), when rollers were wiped with press wash solutions. This latter operation provided the opportunity for dermal exposure as well as inhalation. Respiratory protection and gloves were seldom used. Solvents used in the printing process included substituted benzenes, dichloromethane, dichloroethane, 1-hexanol 2-terpinol, 1,2-dichloromethylene, 1,1,1-trichloroethane, methyl ethyl ketone and glycerol triacetate. UV-cured wash solutions contained glycol ethers diluted in *n*-propanol. Wash solutions for air-dried inks contained glycol ethers, mixed with aliphatic and aromatic hydrocarbons.

The use of gasoline (see IARC, 1989d) as a cleaning agent in typographic printing shops and by rotogravure pressmen, when cleaning rollers, has been reported in Finland (Partanen et al., 1991).

In a Taiwan colour printing factory examined in 1985, carbon tetrachloride was used to clean a machine pump associated with the printing machine, resulting in estimated levels of 300–500 ppm [1890–3150 mg/m³] carbon tetrachloride vapour (Deng et al., 1987).

An Italian study of rotogravure workers in the 1980s (when the printing ink was diluted with pure toluene) showed mean toluene personal exposures of 128 mg/m³, with no full-shift average exposure over the TLV of 375 mg/m³ (De Rosa et al., 1986). Personal toluene exposures correlated with post-shift urinary hippuric acid and *ortho*-cresol levels for six subjects over five days.

A Swedish study of the rotogravure industry found median toluene concentrations of 33 ppm [124 mg/m³] in two plants and 7 ppm [26 mg/m³] in a third, more modern plant between 1983 and 1986. Toluene concentrations ranged from 300 to 450 ppm [1130–1695 mg/m³] between 1920 and 1965 in six rotogravure factories, but dropped to average concentrations of less than 50 ppm [190 mg/m³] by 1985. Benzene concentrations (from contamined toluene) in one plant in 1960–62 ranged from 0 to 61 ppm [0–195 mg/m³] and averaged 3 ppm [9.6 mg/m³] (Svensson et al., 1990).

In a Danish study of 52 printers in six printing factories (letterpress and newspaper printing, offset printing and flexopress and rotogravures) in the late 1980s, personal time-weighted average (TWA) concentrations of toluene averaged 5.4 ppm [20 mg/m³] and were found in 88% of the samples (Baelum, 1990). An average of seven different solvents were found in the air samples (range, 3–16) with decane, ethanol, xylene and isopropanol following toluene in frequency of detection [actual concentrations not given].

In a Japanese study of solvent exposures in 169 workers in 52 printing facilities in the 1980s, 24% of area air concentrations of solvents were greater than their occupational exposure limits, although actual concentrations were not given (Ukai et al., 1986).

A Belgian study of ethylene glycol ethers and other solvent exposures in 24 printing plants during 1983–86 found detectable levels of several glycol ethers and other solvents (see Table 7; Veulemans et al., 1987). The most common solvents detected were toluene (in 26% of samples), xylenes (24%), ethanol (35%), n-butanol (31%), isopropanol (31%), iso- and tert-butanol (26%), ethyl acetate (66%), n-butyl acetate (32%) and methyl ethyl ketone (45%). Actual solvent concentrations were not reported except for the glycol ethers.

In a large Canadian study of seven commercial printing operations and two newspaper plants, exposure to total particulate and polycyclic aromatic hydrocarbons (PAHs) was determined (Purdham et al., 1993). Total particulate exposures for press room workers ranged from 0.1 to 4.7 mg/m^3 (mean, 0.63 mg/m^3) for lithographic and rotagravure commercial printers and from 0.14 to 0.31 mg/m^3 for newspaper plants. Personal PAH exposures ranged from none to 0.39 mg/m^3 (mean, 0.0165 mg/m^3), with no detectable PAH in 45% of samples. Naphthalene was the greatest constituent of PAH exposures. Companies classified as having poor ventilation had the greatest particulate and PAH exposures. The authors observed that second and third pressman (less senior) had the greatest exposure to oil and ink mist, whereas rollmen and catchers (at the end of the line) had particulate exposures comprised mostly of paper dust. The newspaper plants had lower particulate and PAH exposures than the other printing operations due to high-speed offset presses and relatively better ventilation.

In the same study, particulate exposures appeared to be inversely related to press speed, although the method of paper feeding may be part of the reason (sheet was dustier than roll or 'web'). The authors concluded that job category within the press rooms was not a factor in predicting total particulate exposure, while printing process type and impression area were. Gravure printing produced higher particulate exposures than lithography, and, the larger the image area, the more particulates were produced. Other important determinants of particulate exposure were effectiveness of ventilation, paper type and feed method.

In a Dutch study in the 1980s, substantial opportunity for dermal exposure to PAHs on the hands and through clothing was observed among press operators exposed to black offset ink from a newspaper printing industry during bulk material transfer (Jongeneelen et al., 1988). Urinary 1-hydroxypyrene was examined as a marker of exposure (see monograph on Carbon Black, this volume).

Acrylates and methacrylates are often found in printing processes, as a component of UV-cured inks (Krishnan et al., 1987; McCammon et al., 1987a,b). Various acrylates are used for the production of photoprepolymer relief printing plates, as are other contact sensitizers such as isocyanates and epoxy resins. It was well recognized by the 1970s that press operators had become sensitized to acrylates (Malten, 1982).

Samimi (1982) reported exposures to isophorone (3.5.5-trimethyl-Δ^2-cyclohexanone) and other solvents during screen printing in the United States. Isophorone was the most widely used ink thinner in the plant (comprising 75%), followed by cyclohexanone, petroleum distillates, butyl acetate, diacetone alcohol, cellosolve acetate (2-ethoxy ethylacetate) and xylene. Exposures for each of these solvents are listed in Table 7 for five different job classifications. Screen press operators had the highest exposures to organic vapours, followed by manual drying, paint mixing, screen washing and automatic drying. Workers involved in press operations, drying operations, ink formulations and screen washing all handled inks, solvents and freshly printed sheets, and were exposed more highly than workers in die cutting, finishing, packing and stencil making. The major exposure determinant was proximity of the solvent evaporating surfaces to the workers' breathing zones.

The National Occupational Exposure Survey conducted by the United States National Institute for Occupational Safety and Health (1995) between 1981 and 1983 indicated that 190 900 employees in the United States were potentially exposed to printing ink, with 29 300 employees potentially exposed specifically to lithographic inks and 8300 to screen process inks. These estimates were based on a survey of companies and did not involve measurements of actual exposures.

In industrial hygiene surveys conducted in the United States by the United States Occupational Safety and Health Administration (1995) between 1979 and 1994, the 12 most commonly monitored exposures in the printing industry were noise, toluene, isopropanol, xylenes, petroleum distillates, acetone, Stoddard solvent, methyl ethyl ketone, methylene chloride (dichloromethane), carbon monoxide, lead and benzene.

1.6 Regulations and guidelines

Exposures in the printing industry are regulated on a component-specific basis for some solvents and ink constitutents and also using non-specific measurements such as the measurement of aerosols and total hydrocarbons.

Oil mist, ink mist (a mist of ink dispersed in oil) and paper dust levels are all characterized in the occupational environment by measuring either total or respirable particulates. Although further analysis of the particulate filter sample for specific components such as PAHs and dyes is possible, it is generally not possible to differentiate between the individual components such as oil, ink solids and paper dust as fractions of the total particulate. Although analytical methods for solvent-soluble fractions of the particulate sample have been developed (such as the cyclohexane-soluble fraction), occupational exposure limits do not exist for these measurements.

Occupational exposure limits have been set for many individual substances found in printing ink manufacture and in printing processes. Examples include alcohols, esters, glycol ethers, ketones, aliphatic hydrocarbons, aromatic hydrocarbons, acrylates, carbon black and organic dyes. Standard compilations of occupational exposure limits may be consulted for individual chemicals (International Labour Office, 1991; American Conference of Governmental Industrial Hygienists, 1995).

2. Studies of Cancer in Humans

Many epidemiological studies contain some evidence concerning cancer risk in printing trades and printing industries. However, several problems compromised the value of these studies for the purpose of this monograph. The first major problem relates to so-called publication bias. Because of pressures relating to the publication of scientific results, it cannot be assumed that all of the evidence that exists concerning cancer risks for printing processes or printing inks would have been published and therefore available to the Working Group. Indeed, for many of the possible associations reviewed in this monograph, the published evidence is probably a biased sample. Namely, it is much more probable that so-called positive findings are published than so-called non-positive or negative findings. This is a particular problem for the evidence regarding occupation or industry titles. Many studies have collected data on the whole range of occupations or industries in a community (from death certificates or from questionnaires) and related these to occurrence of one or many types of cancer. The large mass of possible associations that can be analysed usually precludes the publication of all findings. Thus, there is often a tendency to publish results for a given occupation, printer for example, only if the association is statistically suggestive. In industrial cohort studies also, it is often the practice not to publish the results found for each site of cancer, but only for prominent sites (e.g. lung) and a selection of other sites based on statistical significance.

The magnitude of the problem varies both by site of cancer and by type of study design. Once there is a controversy about a topic, there is a greater tendency for results to be published irrespective of whether the evidence is positive or negative. Thus, because there has been some discussion in the literature of risks of urinary bladder cancer related to inks, but less of risks of lung cancer, the Working Group believed that, among community-based case–control studies of urinary bladder cancer that collected data on job histories, complete reporting of results on printers was more probable than among analogous studies of lung cancer. Therefore, the Working Group tended to consider the body of evidence from case–control studies of urinary bladder cancer as being more representative and valid than the body of evidence from case–control studies of lung cancer. On the other hand, when evaluating cohort studies of printers, the Working Group believed that it is more probable that investigators report their lung cancer results, irrespective of their being positive or negative, than that they report their urinary bladder cancer results. This is because the incidence of lung cancer is much higher and the results on urinary bladder cancer are often considered to be based on too few numbers of cases to be of interest, unless the result is statistically significant. The same problem holds for most rare sites of cancer in relation to cohort studies. For instance, studies on printers reported their results for renal cancer much more probably if the standardized mortality ratio (SMR) was positive than if it was negative.

Another major problem in the review of the epidemiological literature on printers and the printing industry is the poor specificity of the occupational information available for epidemiological analysis. As described in Section 1 of this monograph, there is a wide variety of printing processes and a wide variety of jobs in the printing industry. Each process and each job is associated with different types of exposures. The bulk of the epidemiological evidence for this monograph is based on broad and often poorly defined employment categories, such as printers or printing industry. Where community-based studies have been carried out using such broad categories, irrespective of whether the studies were based on interviews, record linkage or abstraction from death certificates, it is clear that the exposure variable encompassed a large variety of exposure circumstances. This detracts significantly from the ability of the study to detect any risk, should one be present in the printing industry, since it is improbable that the same risk would occur across the whole range of exposure circumstances. In addition to compromising the power to detect risks, however, this problem also subverts the opportunity to seek consistency of findings between studies, since it is quite possible that the group of printers in one study has a quite different distribution of processes and precise job titles than a group of printers in another study. Even among cohort studies, many of them were described as taking place among printers with no further specification of types of printing processes or precise job titles. This problem of lack of specificity also applied to those studies that attempted to attribute exposure to printing inks.

2.1 Ecological studies

Blot and Fraumeni (1978) correlated 3056 United States county urinary bladder cancer mortality rates for the period 1950–69 with demographic, socioeconomic, ethnic and industrial data at the county level. Using data derived from the 1963 Census of Manufacturers, counties were categorized into three groups (high, mid, low) according to the proportion (> 1%; 0.1–1%; < 0.1% of the total county population) of employees engaged in each of 18 selected industrial productions. Using multiple regression models, average age-adjusted rates were estimated for each county as functions of other variables in the models. For white men, there was a positive correlation between the proportion of workers in the printing industry in each county and the rate of urinary bladder cancer.

Greene *et al.* (1979a) evaluated 211 pathologically verified cases of mycosis fungoides that were reported to the Mycosis Fungoides Cooperative Study Group during 1973–76 and 1948 deaths from mycosis fungoides occurring between 1950 and 1975 in 3056 United States counties. Analysis by specific industries was presented for 1665 white male deaths. For each county, the population distribution by age, race and sex was obtained from the 1960 United States census. Using data derived from the 1963 Census of Manufacturers, counties were categorized as in Blot and Fraumeni (1978). For each group of counties, age-standardized rates were computed based on the 1960 United States population distribution; in addition, SMRs were calculated using the total population rates as reference. The SMRs were 0.9, 1.0 and 1.2 respectively for those counties with < 0.1%, 0.1–1% and > 1% of their population employed in printing industries. [The Working Group noted that numbers of observed deaths were not given.]

2.2 Community-based studies

2.2.1 *Death certificate studies*

Kennaway and Kennaway (1947) selected the death certificates for lung cancer (23 549 deaths) and laryngeal cancer (14 869 deaths) occurring in men aged 20 years or more in England and Wales between 1921 and 1938 and compared the observed distribution of occupations recorded on the certificates to those expected on the basis of the occupational groupings in the 1931 census. The SMRs for lung cancer were 1.2 ([95% confidence interval (CI), 1.0-1.4]; 177 observed deaths) among printers and 0.7 ([95% CI, 0.3–1.4]; 7 observed deaths) among lithographic and process engravers; the SMRs for laryngeal cancer were 1.1 ([95% CI, 0.9-1.4]; 104 observed deaths) and 1.4 ([95% CI, 0.6–2.7]; 8 observed deaths), respectively.

Menck and Henderson (1976) examined lung cancer rates in Los Angeles County, United States, by occupation and industry. The study included 2161 death certificates mentioning lung cancer in white men aged 20–64 for the period 1968–70 and 1777 incident cases of lung cancer in white men of the same age reported to the Los Angeles County Cancer Surveillance Program for 1972–73. Mortality and morbidity data were pooled. Occupations and industries of employment were abstracted from death certificates for deceased cases and from hospital admission records for living cases. No occupation was reported for 17.5% and no industry for 31% of the study population. The white male population at risk by age, occupation and industry was obtained from the 1970 census. Expected number of deaths and incident cases in each occupation were computed using the rates of lung cancer in the 1970 census population. Significantly increased SMRs were found among men classified as photoengravers (SMR, 3.2 [95% CI, 1.4–6.3] based on four deaths and four incident cases) and pressmen (SMR, 2.8 [95% CI, 1.7–4.3] based on 10 deaths and 10 incident cases). For the printing and newspaper industry, the SMR was 1.0 [95% CI, 0.7–1.3] (30 observed deaths and 16 incident cases). [The Working Group noted the high proportion of study subjects without information on occupation or industry.]

Petersen and Milham (1980) investigated the occupational and mortality patterns of white male residents in California, United States, for the period 1959–61, using the occupation reported on death certificates. The total number of deaths from all causes among pressmen and plate printers was 1144. No increased risk (proportionate mortality ratio [PMR]) was reported for any cancer site [specific figures not given].

Dubrow and Wegman (1984) examined cancer mortality patterns by occupation for white men in Massachusetts, United States, in the period 1971–1973. Using age-standardized mortality odds ratios (sMORs), 397 occupational categories derived from death certificate information were assessed for their association with 62 malignancies. Statistically significant increased sMORs were reported for cancer of the buccal cavity and pharynx in the printing industry (sMOR, 2.5; $p < 0.001$; 15 deaths), and for cancer of the trachea, bronchus and lung (sMOR, 2.2; $p < 0.05$; 12 deaths) and cancer of the prostate (sMOR, 3.8; $p < 0.01$; 6 deaths) in compositors and typesetters.

In an occupational mortality surveillance study, Dubrow (1986) examined cause-specific mortality patterns by occupation and industry among Rhode Island, United States, residents (≥ 16 years old) who died during the period 1968–78. He used age-standardized PMRs stratified by sex and race. Expected numbers of deaths within specific occupations and industries were calculated using the total Rhode Island mortality experience. Information on usual (longest) occupation and usual industry was derived from death certificates. Results for skin melanoma deaths were presented. Out of a total of 577 white male decedents whose usual industry was printing, six died from melanoma (PMR, 4.6 [95% CI, 2.1–10.2]). When the analysis was restricted to occupations in the printing industry, four melanoma deaths were observed (PMR, 5.7 [95% CI, 2.1–15.2]). Examination of death certificates revealed that three of the four decedents either were lithographers or had worked at companies where lithography was performed.

Mortality by occupation was studied in British Columbia, Canada (Gallagher et al., 1989). Cause of death and occupation were abstracted from death certificates from 1950 through to 1984. A total of 320 423 male and 216 213 female deaths were available for the analysis. Age-standardized PMRs were calculated. During the study period there were 1314 deaths in all printers. A larger number of deaths from colonic cancer was observed than that which was anticipated (PMR, 1.5; 95% CI, 1.1–2.1; 34 deaths). Mortality from lung cancer did not differ from expectation in printers (PMR, 1.0; 95% CI, 0.8–1.3; 71 deaths). The PMR for Hodgkin's disease was nonsignificantly elevated in all printers (PMR, 2.5; 95% CI, 0.8–5.8; 5 deaths). Four cases of pancreatic cancer occurred among printing press operators (PMR, 2.6; 95% CI, 0.7–6.7).

A proportionate mortality study using occupational data recorded in death certificates was conducted on 588 090 Washington State, United States, white male deaths between the years 1950 and 1989, and 88 071 white female deaths during 1974–89 (Milham, 1992). The PMRs were standardized for age and year of death. In men, there were 2775 deaths in printing pressmen, plate printers and typesetters. A significant excess from all cancers was observed for these occupations (PMR, 1.1 [95% CI, 1.0–1.1]; $p < 0.05$; 606 cases). Excesses for specific cancer sites were observed for malignancies of the oral mesopharynx (PMR, 4.4 [95% CI, 2.2–13.1]; $p < 0.01$; 6 cases), for rectal cancer (PMR, 1.7 [95% CI, 1.1–2.5]; $p < 0.01$; 27 cases) and for cancer of the bronchus, trachea and lung (PMR, 1.2 [95% CI, 1.1–1.4]; $p < 0.01$; 191 cases). No excess was found for urinary bladder cancer (PMR, 1.0 [95% CI, 0.7–1.5]; 21 cases), leukaemia (PMR, 1.2 [95% CI, 0.7–1.7]; 27 cases) or Hodgkin's disease (PMR, 0.7 [95% CI, 0.1–0.8]; 3 cases). Seven cases of malignant melanoma of the skin were observed (PMR, 1.3 [95% CI, 0.5–2.4]. In women, 153 deaths occurred in printers and typesetters; there was a significant excess of colonic cancer (PMR, 2.0 [95% CI, 0.9–4.0]; $p < 0.05$; 8 cases). No significant PMR was found either for cancer of the bronchus, trachea and lung (PMR, 0.9 [95% CI, 0.4–1.8]; 7 cases) or for malignancies of lymphatic and haematopoietic tissue (PMR, 1.5 [95% CI, 0.5–3.9]; 5 cases).

2.2.2 Record-linkage studies

A number of studies have been carried out linking job or industry titles derived from census data with subsequent cancer morbidity or mortality. Compared with studies based on information derived from death certificates alone, record-linkage studies provide higher-quality occupational data and follow-up data on identified cohorts. However, since the occupational data are collected at one point in time, this may not reflect accurately lifetime employment histories, particularly in occupationally mobile populations. In addition, the lack of longitudinal data precludes estimation of employment duration.

Malker and Gemne (1987) evaluated cancer risk among printing industry workers (24 652 men and 6450 women) using the Swedish Cancer-Environment Registry, which linked national cancer incidence for the period 1961–73 to 1960 census-derived data on occupation and industry. Standardized incidence ratios (SIRs) were based on birth cohort-specific incidence rates. Three different reference populations were used: the total Swedish population, all employed persons and blue-collar workers. Among male printing workers, the age- and region-adjusted SIR for lung cancer was 1.5 ([95% CI, 1.3–1.8]; 190 cases). The risk was similar (SIR, 1.6 [95% CI, 1.4–1.9]; 149 cases) in blue-collar workers employed in printing enterprises (newspaper, journal, book printing and other graphic enterprises). The excess risk for lung cancer, adjusted by region, was mainly evident in those born around 1900 (for birth cohort 1900–04: SIR, 1.9 [95% CI, 1.4–2.5]; 45 cases). Among female printing workers, an excess risk for cervical cancer was observed (age- and region-adjusted SIR, 1.3 [95% CI, 1.1–1.5]; 162 cases), but this disappeared when employed persons and blue-collar workers were used as reference. No other site was found to be at excess risk.

To describe the occurrence of cancer in occupational groups, Olsen and Jensen (1987) examined 90 651 primary tumours identified in the Danish Cancer Registry in 1970–79. Occupations were derived from computer-based national registries such as the Supplementary Pension Fund and the Central Population Registry. The risk for specific cancer sites by occupation or industry was estimated as the standardized proportionate incidence ratio (SPIR). Among men employed in the printing and publishing industries, statistically significantly increased risks were found for cancers of the renal pelvis and ureter (SPIR, 2.5; 95% CI, 1.4–4.7; 10 cases) and for urinary bladder cancer (SPIR, 1.4; 95% CI, 1.1–1.8; 55 cases). Slightly increased risks were also found for cancers of the lung and trachea (SPIR, 1.1; 95% CI, 0.9–1.3; 107 cases), pancreas (SPIR, 1.3; 95% CI, 0.9–2.1; 20 cases), testis (SPIR, 1.2; 95% CI, 0.8–1.8; 22 cases) and for non-melanoma skin cancer (SPIR, 1.2; 95% CI, 0.9–1.5; 58 cases). No increased risk was found for other cancer sites. In women employed in the printing, publishing and allied industries, a statistically significantly increased risk was found for cancer of the lung and trachea (SPIR, 1.4; 95% CI, 1.0–1.9; 44 cases).

Based on an extended follow-up of the aforementioned Swedish Cancer-Environment Registry, McLaughlin *et al.* (1988) investigated the risk for melanoma of the skin among male printing workers for the period 1961–79; a significantly elevated SIR was observed in the printing industry (SIR, 1.4 [95% CI, 1.1–1.7]; 91 cases). Men employed in newspaper printing (SIR, 1.9 [95% CI, 1.3–2.6]; 39 cases) and newspaper publishing

(SIR, 3.1 [95% CI, 1.2–6.4]; 7 cases) industries seemed to account for much of the association. Analysis by occupation within industry revealed a statistically significantly increased risk for typographers in the newspaper printing industry (SIR, 2.0 [95% CI, 1.2–3.1]; 19 cases), whereas the risk for this occupational group in any of the other printing industries was not increased. Risks were also significantly raised for other occupations within the printing industries: machine repairers (SIR, 14.5 [95% CI, 1.6–52.3]; 2 cases), journalists and editors (SIR, 2.4 [95% CI, 1.4–3.9]; 16 cases) and business executives (SIR, 9.1 [95% CI, 2.9–21.2]; 5 cases).

Aronson and Howe (1994) examined cancer-specific mortality patterns by occupation or industry in a cohort of 242 196 Canadian women identified through an employment survey of approximately 10% of the Canadian labour force between 1965 and 1971 in which employers were asked to provide information (name, surname, year of birth, current occupation and industry) for each of their workers. The mortality of the cohort in the period 1965–79 was determined by a computerized record-linkage to the Canadian National Mortality Data Base. SMRs by occupation or industry were computed using age- and calendar year-specific rates for the entire female cohort. The results were only published for significantly elevated SMRs. The only finding related to the printing and publishing industry concerned breast cancer in women under 64 years of age (SMR, 2.2; 95% CI, 1.1–3.9; 11 deaths).

A cohort of the residents of Turin, Italy, enrolled on the basis of the 1981 census record, was followed up until the end of 1989 (73 606 deaths recorded among 1 056 102 persons; 10 798 deaths among persons employed and aged 18–64 at the 1981 census) (Costa et al., 1995). SMRs were computed by applying age- and sex-specific death rates of the whole active population to the person-years accrued by the 'exposed' group under study. In men employed in printing and publishing, nonsignificant increases were shown for malignancies of the pleura (SMR, 6.0; 95% CI, 0.7–21.6; 2 cases), for colonic cancer (SMR, 2.1; 95% CI, 0.9–4.4; 7 deaths) and for malignancies of the haematopoietic system (SMR, 1.6; 95% CI, 0.6–3.3; 7 deaths). The SMR for lung cancer was 1.1 ([95% CI, 0.7–1.7]; 22 deaths) and that for urinary bladder cancer was 1.0 ([95% CI, 0.1–3.6]; 2 deaths). In women employed in printing and publishing, risks were raised for lung cancer (SMR, 2.6; 95% CI, 0.5–7.6; 3 deaths), colonic cancer (SMR, 2.7; 95% CI, 0.3–9.7; 2 deaths), ovarian cancer (SMR, 3.2; 95% CI, 0.6–9.3; 3 deaths) and haematopoietic system malignancies (SMR, 2.0; 95% CI, 0.2–7.2; 2 deaths). In male printers, an increased risk of liver cancer (SMR, 1.7; 95% CI, 0.3–5.0; 3 deaths), colonic cancer (SMR, 1.8; 95% CI, 0.4–5.3; 3 deaths) and lung cancer (SMR, 1.2 [95% CI, 0.6–2.1]; 12 deaths) was shown; there were also two deaths from multiple myeloma [SMR, 9.7; 95% CI, 1.1–33.1], but no death from urinary bladder cancer.

All Italian residents aged 18–74, identified through the 1981 census (which included 94 163 deaths, 15 734 of which were among employed individuals aged 18–64) were followed up from November 1981 to April 1982 (Costa et al., 1995). Cause-specific relative risks by occupation were estimated as age-adjusted Mantel-Haenszel odds ratios. An excess risk of renal cancer was found among male printing and publishing workers (odds ratio, 4.8; 95% CI, 1.7–13.4; 3 cases). The odds ratios in men were 1.1 ([95% CI,

0.8–1.4]; 11 cases) for lung cancer and for urinary bladder cancer 2.9 ([95% CI, 0.8–11.1]; 2 cases).

Lynge et al. (1995) examined the cancer risk of 15 534 men and 3593 women, 20–64 years old, living in Denmark and working in the printing and bookbinding industry according to the 1970 Census. Incident cancer cases were identified by linkage with the Danish Cancer Registry for the period 1970–87, 1970 being the year in which the change from rotary letterpress to offset lithography took place in newspaper production in Denmark. Measurements undertaken in Denmark in 1979 and 1986–87 detected dust concentrations of 1.1 mg/m^3 and 1–4 mg/m^3 (65–90% being ink mist), respectively, for rotary letterpress and 0.28 mg/m^3 and 0.1–0.7 mg/m^3 (25% being ink mist), respectively, for offset. Age-adjusted SIRs were calculated using the cancer incidence rates of all employed persons as reference. Male workers had significantly increased risks for all cancers (SIR, 1.2; 95% CI, 1.1–1.2; 1095 cancers) and for cancers of the liver (SIR, 1.9; 95% CI, 1.1–3.0; 19 cases), lung (SIR, 1.3; 95% CI, 1.1–1.4; 248 cases), renal pelvis (SIR, 1.9; 95% CI, 1.1–3.1; 15 cases) and urinary bladder (SIR, 1.3; 95% CI, 1.1–1.6; 109 cases). Excess risks of borderline statistical significance were found for cancers of the oral cavity (SIR, 1.9; 95% CI, 0.9–3.4; 11 cases) and colon (SIR, 1.2; 95% CI, 1.0–1.6; 80 cases) and non-melanoma skin cancer (SIR, 1.2; 95% CI, 1.0–1.4; 148 cases). The SIR for melanoma was 1.1 (95% CI, 0.8–1.6; 28 cases). Based on census-derived data on occupation and industry, analysis by occupational groups was performed. The risk for lung cancer was significantly increased in factory workers in the printing industry (SIR, 1.5; 95% CI, 1.1–1.9; 62 cases) and particularly among operators of rotary letterpress machines in newspaper and magazine production (SIR, 2.0; 95% CI, 1.3–3.0; 26 cases). An excess of borderline statistical significance was found among photo-engravers (SIR, 1.7; 95% CI, 0.9–2.8; 15 cases). The risk for urinary bladder cancer was elevated among typographers (SIR, 1.5; 95% CI, 1.1–2.0; 40 cases) and among factory workers in printing establishments (SIR, 1.6; 95% CI, 0.9–2.7; 14 cases). The risk for cancer of the renal pelvis was elevated in typographers both in printing establishments (SIR, 2.3; 95% CI, 0.8–5.0; 6 cases) and newspaper and magazine production (SIR, 2.2; 95% CI, 0.6–5.5; 4 cases); two cases among lithographers yielded a SIR of 3.4 (95% CI, 0.4–12.2), whereas three cases were observed in factory workers in the printing industry (SIR, 1.8; 95% CI, 0.4–5.3). Several occupational subgroups contributed to the excess risk for primary liver cancer, the risk being high among lithographers (SIR, 6.7; 95% CI, 2.2–15.8; 5 cases). Statistically significantly increased SIRs for cancer of the gall-bladder (SIR, 3.4; 95% CI, 1.1–7.9; 5 cases) and for male breast cancer (SIR, 7.9; 95% CI, 1.6–23.1; 3 cases) were found among typographers in newspaper and magazine production. Female workers showed significantly increased risks for all cancers (SIR, 1.2; 95% CI, 1.0–1.3; 284 cancers), lung cancer (SIR, 1.7; 95% CI, 1.2–2.5; 32 cases) and breast cancer (SIR, 1.4; 95% CI, 1.1–1.7; 88 cases). Cervical cancer was also elevated (SIR, 1.4; 95% CI, 0.9–2.0; 29 cases). No increased risk was found for melanoma (SIR, 1.1; 95% CI, 0.5–2.0; 9 cases). Smoking and drinking habits reported by members of the printing trade unions at a survey in 1972 were compared with habits reported by members of other trade unions. Based on the reported patterns of smoking and alcohol consumption in this population, the authors stated that the observed elevated risks were unlikely to be entirely attributable to confounding by these factors.

In Finland, the 1970 Population Census file was linked with the Finnish Cancer Registry file for the period 1971–85 (Pukkala, 1995). Crude occupation-specific SIRs and social class-adjusted SIRs were calculated. Expected numbers were based on the incidence for the total employed, active, Finnish population of the same sex during the same period. The total number of cases included in the analyses was 47 178 men and 46 853 women aged 35–64 years. A significant excess of colonic cancer in men was found for printing occupations (adjusted SIR, 2.2; 95% CI, 1.2–3.5; 16 cases). The excess was nonsignificant in women (adjusted SIR, 1.4; 95% CI, 0.7–2.5; 10 cases). No significant excess was observed in lung cancer incidence either in male (adjusted SIR, 0.8; 95% CI, 0.6–1.1; 50 cases) or in female printing occupations (adjusted SIR, 1.8; 95% CI, 0.9–3.2; 12 cases). The analysis by job title in men confirmed the results. A significant excess of breast cancer was observed in female printers (adjusted SIR, 1.4; 95% CI, 1.1–1.8; 74 cases), and there was also a statistically significant increase in ovarian cancer for the same group (adjusted SIR, 2.2; 95% CI, 1.5–3.1; 30 cases). No increase was shown for cancer of the urinary bladder in male printers (adjusted SIR, 1.1; 95% CI, 0.5–2.0; 9 cases), for cancer of the lung in male printers (adjusted SIR, 1.1; 95% CI, 0.7–1.7; 19 cases), for skin melanoma in printers of either sex (adjusted SIR, 1.1; 95% CI, 0.5–2.4; 7 male cases; adjusted SIR, 1.1; 95% CI, 0.3–2.5; 5 female cases) or for leukaemia in male printers (adjusted SIR, 0.4; 95% CI, 0.1–1.4; 2 cases). Female lithographers had a significant excess of basal-cell carcinoma of the skin (adjusted SIR, 4.4; 95% CI, 1.2–11.2; 4 cases).

A cancer registry-based study in Tianjin, China (Wang et al., 1995) identified 4806 male and 3595 female cases of incident lung cancers and 14 685 male and 13 010 female controls (all other cancers) ≥ 20 years old during 1981–87. The subjects' most-recent occupation and industry were abstracted from cancer notifications. For printers and related occupations, the odds ratio, when roughly adjusted for smoking, was 1.1 (95% CI, 0.6–1.8; 20 cases) in men and 1.5 (95% CI, 0.7–3.3; 9 cases) in women.

These studies are summarized in Table 8.

2.2.3 *Case–control studies*

Case–control studies are summarized in Tables 9–11.

(a) *Cancer at multiple sites*

Cancer cases (17 sites) and patients with non-neoplastic diseases admitted to the Roswell Park Memorial Institute in Buffalo, New York, United States, from 1956 to 1965 were compared with regard to occupations associated with either inhalation of combustion products or chemicals (Viadana et al., 1976). Information on lifetime occupational history, educational level and smoking habits was collected at the time of hospital admission. A total of 11 591 white male patients (cancer cases and non-cancer cases) were included in the analysis. Workers ever employed in specific occupations were compared with those in an unexposed clerical group. Eleven cases of cancer of the

Table 8. Record linkage studies among workers in the printing industry

Reference, country	Study subjects	Period of follow-up	Occupation/exposure	Cancer site/cause of death	No. obs.	RR	95% CI	Comments
Malker & Gemne (1987) Sweden	24 652 men and 6450 women registered at 1960 census as printing workers	1961–73	Printing workers (M)	Lung	190	1.5	[1.3–1.8]	Morbidity
			Blue-collar workers (M) in printing enterprises (newspaper, journal/book printing, others)		149	1.6	[1.4–1.9]	
			Birth cohort around 1990 (M)	Urinary bladder	45	1.9	1.4–2.5	
				Kidney	76	1.3	NG	$p > 0.01$
				Skin melanoma	48	1.1	NG	$p > 0.01$
					27	1.2	NG	$p > 0.01$
			Printing workers (F)	Lung	9	1.3	NG	$p > 0.01$
				Urinary bladder	5	0.8	NG	$p > 0.01$
				Kidney	7	1.1	NG	$p > 0.01$
				Skin melanoma	8	1.2	NG	$p > 0.01$
				Cervix/uteri	162	1.3	[1.1–1.5]	
Olsen & Jensen (1987) Denmark	90 651 primary tumours	1970–79	Printing and publishing industries (M)	Renal pelvis	10	2.5	1.4–4.7	Standardized proportionate incidence ratio
				Urinary bladder	55	1.4	1.1–1.8	
				Lung and trachea	107	1.1	0.9–1.3	
				Pancreas	20	1.3	0.9–2.1	
				Testis	22	1.2	0.8–1.8	
				Non-melanoma skin cancer	58	1.2	0.9–1.5	
			Printing, publishing and allied industries (F)	Lung and trachea	44	1.4	1.0–1.9	

Table 8 (contd)

Reference, country	Study subjects	Period of follow-up	Occupation/exposure	Cancer site/cause of death		No. obs.	RR	95% CI	Comments
McLaughlin et al. (1988) Sweden	Male printing workers at 1960 census; 91 melanomas	1961–79	Printing industry	Skin melanoma		91	1.4	[1.1–1.7]	Morbidity
			Newspaper printing industry			39	1.9	[1.3–2.6]	
			Newspaper publishing industry			7	3.1	[1.2–6.4]	
			Typographers in newspaper printing industry			19	2.0	[1.2–3.1]	
			Machine repairers in newspaper printing industry			2	14.5	[1.6–52.3]	
			Journalists/editors in newspaper printing industry			16	2.4	[1.4–3.9]	
			Business/executives in newspaper printing industry			5	9.1	[2.9–21.2]	
Aronson & Howe (1994) Canada	242 196 women identified through employment survey	1965–79	Printing and publishing industry	Breast		11	2.2	1.1–3.9	Mortality; other sites not significantly elevated
Costa et al. (1995) Italy	1981 population census of Turin, Italy, residents; 10 798 deaths among persons employed	1981–89	Printing and publishing industry	M	Pleura	2	6.0	0.7–22	Mortality
					Colon	7	2.1	0.9–4.4	
					Lung	22	1.1	[0.7–1.7]	
					Urinary bladder	2	1.0	[0.1–3.6]	
					Haematopoietic	7	1.6	0.6–3.3	
				F	Lung	3	2.6	0.5–7.6	
					Colon	2	2.7	0.3–9.7	
					Ovarian	3	3.2	0.6–9.3	
					Haematopoietic	2	2.0	0.2–7.2	
			Printers	M	Liver	3	1.7	0.3–5.0	
					Colon	3	1.8	0.4–5.3	
					Multiple myeloma	2	[9.7]	[1.1–33.1]	
					Lung	12	1.2	[0.6–2.1]	
					Urinary bladder	0	–	–	

Table 8 (contd)

Reference, country	Study subjects	Period of follow-up	Occupation/exposure	Cancer site/cause of death	No. obs.	RR	95% CI	Comments
Costa et al. (1995) Italy	1981 population census of Italian residents; 15 734 deaths among persons employed	1981–82	Printing and publishing industry	M Kidney Lung Urinary bladder	3 11 2	4.8 1.1 2.9	1.7–13.4 [0.8–1.4] [0.8–11.1]	Mortality
Lynge et al. (1995) Denmark	15 534 men and 3593 women working in printing and bookbinding industry based on 1970 Census	1970–87	*Men* Whole printing and book binding industry Factory workers in printing industry Factory workers in newspaper and magazine production Photoengravers Whole industry^a Lithographers Whole industry^a Typographers in printing establishment Factory workers in printing establishment Whole industry^a Typographers in printing establishments Typographers in newspaper and magazine production Whole industry^a	All cancers Lung Liver Urinary bladder Colon Renal pelvis Oral cavity	1095 248 62 26 15 19 5 109 40 14 80 15 6 4 11	1.2 1.3 1.5 2.0 1.7 1.9 6.7 1.3 1.5 1.6 1.2 1.9 2.3 2.2 1.9	1.1–1.2 1.1–1.4 1.1–1.9 1.3–3.0 0.9–2.8 1.1–3.0 2.2–15.8 1.1–1.6 1.1–2.0 0.9–2.7 1.0–1.6 1.1–3.1 0.8–5.0 0.6–5.5 0.9–3.4	(SIR)

Table 8 (contd)

Reference, country	Study subjects	Period of follow-up	Occupation/exposure	Cancer site/cause of death	No. obs.	RR	95% CI	Comments
Lynge et al. (1995) Denmark (contd)			Whole industry[a]	Skin melanoma	28	1.1	0.8–1.6	
				Non-melanoma skin cancer	148	1.2	1.0–1.4	
				Women				
				All cancers	284	1.2	1.0–1.3	
				Lung	32	1.7	1.2–2.5	
				Breast	88	1.4	1.1–1.7	
				Cervix uteri	29	1.4	0.9–2.0	
				Skin melanoma	9	1.1	0.5–2.0	
Pukkala (1995) Finland	1970 population census 47 178 men, 46 853 women	1971–85	Printing occupations	M Colon	16	2.2	1.2–3.5	Morbidity
				F Colon	10	1.4	0.7–2.5	
				M Lung	50	0.8	0.6–1.1	
				F Lung	12	1.8	0.9–3.2	
			Printers	F Breast	74	1.4	1.1–1.8	
				F Ovarian	30	2.2	1.5–3.1	
				F Skin melanoma	5	1.1	0.3–2.5	
				M Skin melanoma	7	1.1	0.5–2.4	
				M Urinary bladder	9	1.1	0.5–2.0	
				M Leukaemia	2	0.4	0.1–1.4	
				M Lung	19	1.1	0.7–1.7	
			Lithographers	F Skin basal-cell carcinoma	4	4.4	1.2–11.2	

RR, relative risk estimated by SMR (for mortality) or SIR (for morbidity); M, male; F, female
[a] Means whole printing and book binding industry

buccal cavity and pharynx among printers yielded a statistically significant age-adjusted odds ratio of 2.6 ($p < 0.05$), while the odds ratio adjusted for smoking was 2.1 ($p > 0.05$). A nonsignificant ($p > 0.05$) increased age-adjusted odds ratio of 1.5 was found for lung cancer (7 cases).

As part of the United States Third National Cancer Survey, 7518 incident cancer cases, representing a 57% response rate, were interviewed on lifestyle factors and main lifetime industry and occupation (Williams et al., 1977). Respondents and non-respondents were comparable in age, race, sex, marital status, method of diagnosis, country of birth and cancer site distributions. The proportions of specific main-lifetime industries and occupations among patients with cancer at one site were compared (separately for men and women) to those of patients having cancer at other sites combined, controlling for age, race, education, tobacco use and alcohol consumption. The total numbers of cancer cases in printing, publishing and allied products industries were 59 among men and 33 among women. The only significantly increased relative risk was found for cancer of the oral cavity in males (odds ratio, 4.5; $p < 0.05$; 7 cases). Nonsignificant excesses were observed for cancer of the pancreas both in men (odds ratio, 7.0; 3 cases) and women (odds ratio, 8.2; 2 cases). Among men, there were four urinary bladder cancer cases (odds ratio, 2.0) and two cases of melanoma (odds ratio, 3.0). [The Working Group noted the low response and the resultant possibility of bias.]

In a population-based case–control study of cancer at 19 sites (Siemiatycki, 1991), described in detail in the monograph on Carbon Black in this volume, pp. 188–189, analyses were carried out relating to the printing and publishing industry, employment as a printer and employment as a printing press worker. For the printing and publishing industry, the odds ratio for lung cancer was 2.0 ([95% CI, 1.2–3.5]; 35 cases); no significant excess was noted for any other site. In printers, significant excesses were found for lung cancer (odds ratio for any exposure, 2.1 [95% CI, 1.1–4.1]; 26 cases; odds ratio for over 10 years' exposure, 1.7 [95% CI, 0.7–4.1]; 13 cases) and pancreatic cancer (odds ratio for over 10 years' exposure, 3.7 [95% CI, 1.2–11.2]; 4 cases); for French printers, a significant excess for renal cancer was observed (odds ratio for any exposure, 3.4 [95% CI, 1.2–9.5]; 5 cases; odds ratio for over 10 years' exposure, 3.6 [95% CI, 1.0–10.6]; 3 cases). In the more narrowly defined category of printing press workers, the odds ratio for lung cancer was 3.1 ([95% CI, 1.1–8.7]; 15 cases), and, when this association was further analysed by histological type of lung cancer, the odds ratio for adenocarcinoma of the lung was 7.0 ([95% CI, 1.8–27.9]; 6 cases).

One of the substances assessed by the occupational hygienists in the Montréal (Canada) study (Siemiatycki, 1991) was inks, including printing inks. Nearly 4% of the total study sample exposure was attributed to inks in one or another of their jobs. The most common job titles in which it was attributed were: typesetters and printers, draftsmen and business machine repairmen. In analyses using cancer controls adjusting for age, smoking, ethnic groups and socioeconomic status, the following sites of cancer were not associated with exposure to inks: oesophagus (odds ratio, 0.9; 3 cases), stomach (odds ratio, 0.9; 7 cases), colon (odds ratio, 0.5; 9 cases), rectum (odds ratio, 1.2; 12 cases), pancreas (odds ratio, 1.3; 5 cases), prostate (odds ratio, 1.3; 13 cases), urinary bladder (odds ratio, 1.2 [95% CI, 0.7–2.1]; 18 cases), melanoma of the skin (odds ratio,

1.0; 5 cases) and lymphoma (odds ratio, 0.8; 6 cases). For lung cancer, there was an indication of excess among all those considered to be exposed (odds ratio, 1.6 [95% CI, 1.0–2.7]; 37 cases) but no greater risk among the subset considered to be substantially exposed (odds ratio, 1.5 [95% CI, 0.7–3.1]; 18 cases). For renal cancer, there was an indication of excess risk for those considered to be substantially exposed (odds ratio, 2.5 [95% CI, 1.0–5.4]; 7 cases).

In a reanalysis of the above data focusing on urinary bladder cancer (Siemiatycki *et al.* 1994), having worked in the printing and publishing industry for less than 10 years yielded an odds ratio of 0.3 (95% CI, 0.1–1.2; 2 cases), while for 10 or more years in the industry, the odds ratio was 1.9 (95% CI, 0.9–3.9; 11 cases). An elevated odds ratio of 3.0 (95% CI, 0.9–10.1; 4 cases) was found for substantial exposure to photographic products among photoengravers and photographic processors.

(*b*) *Urinary bladder cancer* (see Table 9)

Wynder *et al.* (1963) examined occupation and other risk factors associated with urinary bladder cancer risk in 300 male and 70 female cases and an equal number of controls from seven New York hospitals in 1957–61. Controls, matched by age and sex, were selected among patients of the same hospitals excluding subjects with cancers of the respiratory system and of the upper alimentary tract and those with myocardial infarction. Information on possible risk factors was collected through interviews. Three cases had ever worked as printers, engravers or lithographers [crude odds ratio, 2.2; 95% CI, 0.4–13.2].

A case–control study was conducted in eastern Massachusetts, United States (Cole *et al.*, 1972), on 510 male and female histologically confirmed cancers of the lower urinary tract (94% were urinary bladder cancer) diagnosed between January 1967 and June 1968 (Cole *et al.*, 1971). Each case was matched by sex and age to a general population control; 92% of the cases and 91% of the controls were interviewed and usable occupational history was collected for 461 cases (356 men and 105 women) and 485 controls (374 men and 111 women). The age-adjusted odds ratio for men whose 'usual' occupation was printing was 0.8 (95% CI, 0.3–1.9; 5 cases). Men ever employed in printing had an age- and smoking-adjusted odds ratio of 1.3 (95% CI, 0.7–2.5; 14 cases).

Sixty-five male and 10 female white patients with histologically confirmed urinary bladder cancer, admitted in 1978 to two community hospitals in northern New Jersey, United States, and 142 (123 men, 19 women) controls matched by age, race, place of birth, sex, hospital and residence were investigated by Najem *et al.* (1982). Controls were selected from patients treated for other conditions in the same hospitals, excluding those with a history of neoplasms or tobacco-related heart diseases. Lifetime occupational histories and a variety of other personal factors were collected through interviews. Among subjects employed in the printing industry for at least one year, the crude odds ratio was 2.7 (95% CI, 0.8–9.6; 7 cases). [The Working Group noted that these were prevalent cases and that this was a hospital-based series.]

Cartwright (1982) conducted a case–control study on urinary bladder cancer in three hospital districts of West Yorkshire, United Kingdom. A total of 625 prevalent cases were identified in October 1978 (472 men, 153 women) and 366 incident cases (272 men, 94 women) were identified between 1978 and 1980. Prevalent cases were matched to one hospital control and incident cases with two controls. Controls were selected randomly among patients without malignant diseases and matched by age, sex and health district. Information on a variety of personal habits and occupational history was derived from interviews. A significant odds ratio, adjusted by sex and type of case (incident or prevalent), of 3.1 (95% CI, 1.4–6.8; 18 cases) was detected for subjects working for at least six months as printers (defined as those exposed to ink-fly from high-speed presses).

Four papers are available giving results on urinary bladder cancer risks in printers based on various subsets of the United States National Bladder Cancer Study (Silverman *et al.*, 1983; Schoenberg *et al.*, 1984; Silverman *et al.*, 1989, 1990). This population-based case–control study was carried out in 10 areas of the United States during 1977–78. Age frequency-matched population controls were selected for each geographical area. Employment was considered as 'ever' or 'usual' in each occupation or industry; subjects never employed in the industry served as referents.

In the Detroit, MI, United States, area, out of 420 male subjects with carcinomas, or papillomas not specified as benign, of the lower urinary tract (95% being urinary bladder cancer) were identified and 339 (81%) were interviewed (Silverman *et al.*, 1983). Analysis was restricted to 303 white male cases and 296 white male population controls. Subjects ever employed in the printing industry did not show an increased risk (crude odds ratio, 1.1; 95% CI, 0.7–1.7; 50 cases). A nonsignificant elevation was seen for subjects who ever worked as printers (odds ratio, 3.0; 95% CI, 0.6–14.8; 6 cases).

In the New Jersey area, United States, 658 incident male cases of urinary bladder cancer and 1258 general population male controls were studied by Schoenberg *et al.* (1984). Age- and smoking-adjusted odds ratios were presented for each employment/material category in the multiple regression model. The findings suggested no increased risk for men ever employed in the printing industry (odds ratio, 0.9; 95% CI, 0.5–1.5; 20 cases). However, self-reported exposure to printing inks yielded an odds ratio of 1.6 (95% CI, 1.0–2.5; 42 cases).

In the latter two of this group of four papers, Silverman *et al.* (1989) reported on 2100 white male cases with urinary bladder cancer and 3874 population controls. The smoking-adjusted odds ratio was 0.8 for white men who had ever worked as printers (95% CI, 0.5–1.2; 37 cases). There were 652 cases of urinary bladder cancer and 1266 controls among white women (Silverman *et al.*, 1990). The smoking-adjusted odds ratio for women who had ever worked as printers was 0.2 (95% CI, < 0.1–1.4; 1 case).

A death certificate-based case–control study included 291 urinary bladder cancer deaths occurring in England and Wales in the period 1975–79 among males under the age of 50 years (Coggon *et al.*, 1984). For each case, two controls were selected among men who had died from any other causes, matched by sex, year of birth, year of death and residence. Based on occupation recorded on the death certificate, exposure to four known and five putative carcinogens was evaluated and graded (high, low, none), applying a job–exposure matrix. No increased risk was found for those occupations involving exposure to printing inks (odds ratio, 1.1; 95% CI, 0.5–2.3; 12 cases). However, the risk was elevated for occupations with high exposure to printing inks (odds ratio, 5.0; 95% CI, 1.0–25.8; 5 cases). [The Working Group noted the young age of the study subjects and that occupational status was derived from death certificates.]

Vineis and Magnani (1985) conducted a hospital-based study of 512 male cases of urinary bladder cancer diagnosed between 1978 and 1983, and 596 male hospital controls, matched by age, having benign urological (other than haematuria, cystitis and benign polyps of the urinary tract; Vineis *et al.*, 1984) and surgical conditions. All cases and controls lived in the province of Turin, Italy. Information on lifelong smoking habits and a complete occupational history were collected through interviews. The age-adjusted odds ratio for subjects ever employed in the broad category of the printing and publishing industry for at least six months was 1.8 (95% CI, 0.8–4.0; 17 cases). Adjustment for smoking (and not for age) resulted in a significant odds ratio of 1.7. When incident cases only were considered, the age-adjusted odds ratio was 1.2 (95% CI, 0.5–3.3) [number of incident cases not given].

Baxter and McDowall (1986) examined the association between occupation and urinary bladder cancer risk using death certificate information in the six London boroughs that had the highest mortality from this disease in England and Wales. Cases comprised 1080 deaths from urinary bladder cancer between 1968 and 1978 among male residents of the six boroughs. Two controls for each case were chosen: one selected randomly from male deaths from all other cancers and another from male deaths from all causes. Both were matched to the cases by residence, year of death and age. Odds ratios for printers were 1.5 (controls from all other cancers) and 1.2 (controls from all other causes including cancers) ($p > 0.05$; 21 cases). [The Working Group noted that the number of exposed controls was not given, that the method used to calculate the odds ratios was not clearly stated, and that occupational status was derived from death certificates.]

Using data derived from a large case–control study of smoking and cancer among hospitalized patients in six cities in the United States, Kabat *et al.* (1986) examined the potential risk factors for urinary bladder cancer for reported lifetime nonsmokers in a series of 76 male and 76 female cases and 238 male and 254 female controls. Controls, matched by age, sex, race, hospital and year of interview, were selected among patients with non-tobacco-related cancers (67% in men, 59% in women) or non-neoplastic diseases (33% in men, 41% in women). All subjects were interviewed between 1976 and 1983 and the usual (longest-held) occupation recorded. The study failed to detect an increased risk for male printers: two cases and six controls yielded a nonsignificant odds ratio of 1.1. [The Working Group noted the small number of exposed cases in this study.]

The Missouri Cancer Registry has been collecting data on cancer cases since 1972. Data on smoking, alcohol consumption and occupation (the longest-held job) are routinely derived from a standardized form administered to each cancer patient. Brownson *et al.* (1987) selected 823 histologically confirmed urinary bladder cancers diagnosed among white males between 1984 and 1986. For each case, three white male controls, frequency matched by age, were chosen among other patients in the registry, excluding smoking-related cancers. Analysis by occupational history was performed using low-risk employment categories as reference. The low-risk occupations included professionals, managers and sales and clerical workers. Printing machine operators had a statistically significant age-, smoking- and alcohol-adjusted odds ratio of 3.1 (95% CI, 1.1–8.9; 7 cases).

Iscovich et al. (1987) reported on a case–control study of 117 histologically confirmed urinary bladder cancer cases diagnosed between 1983 and 1985 in the La Plata area, Argentina. Cases were individually matched by age and sex to one neighbourhood control and one non-neoplastic hospital control. Information on smoking, demographic, socioeconomic and medical variables and occupational history for the three occupations of longest duration and the most recent occupation were collected from subjects by questionnaire. Three cases and two controls had ever been employed in printing industries, which yielded a nonsignificant age- and smoking-adjusted odds ratio of 2.7. Seven subjects (5 cases, 2 controls) working as printers either in the printing industry or elsewhere gave a statistically significant age-adjusted odds ratio of 5.4; after adjustment for smoking, the odds ratio was 5.6 but no longer significant.

As part of a methodological investigation of data sources for occupational studies, Steenland et al. (1987) conducted a case–control study of male urinary bladder cancer deaths in Hamilton County, OH, United States. A total of 731 urinary bladder cancer cases, deceased during 1960–82, were matched on age, sex, race, date of death and residence at death to two controls per case. Controls were chosen from among all other deaths with the exclusion of deaths from urinary tract tumours and pneumonia. Commercial city directories report yearly occupation and employer for all residents in the county over the age of 18, making it possible to reconstruct occupational lifetime histories. Occupational data from city directories were available for 648 cases and 1275 controls. Odds ratios, conditional on matched sets, were calculated by occupation, industry and employer. The odds ratio for having ever been employed as a press machine operator was 5.0 ([95% CI, 1.3–19]; 5 cases). For employment over 20 years in the printing industry, the odds ratio was 1.3 ($p > 0.05$; 12 cases).

A case–control study of occupation and urinary bladder cancer was carried out by González et al. (1989) in five provinces of Spain. Cases of histologically confirmed urinary bladder cancer were recruited from the registers of 12 hospitals between 1985 and 1986. A total of 497 (438 men and 59 women) cases, alive at the time of interview, resident in the province where the hospital was located and below the age of 79, were interviewed (71 were not traced or refused interview). The proportion of incident cases was 51%. For each case, two controls, matched by sex and age, were selected. The first was selected from the same hospital as the case, excluding patients with diagnoses possibly associated with the risk factors under study. Among hospital controls, 583 (62 women) were interviewed; 159 were not traced or refused interview. A second control was randomly selected from the same section of the census or municipality register as that of the case. Of the population controls, 530 (65 women) were interviewed; 200 were not traced or refused interview. A questionnaire on lifestyle and occupational history was administered to each subject. The age- and sex-adjusted odds ratio was 2.1 for male printers (95% CI, 1.0–4.3; 15 cases). The odds ratio was slightly lower (1.8; 95% CI, 0.8–4.1) after further adjustment for tobacco consumption and employment in other high-risk occupations. In the analysis by occupational subgroups, typesetters and linotypists were associated with an odds ratio of 2.0 (95% CI, 0.5–7.5; 9 cases) after adjusting for smoking and employment in other high-risk occupations.

Using data derived from an on-going case–control study conducted in the area of Milan, Italy, La Vecchia et al. (1990) examined the relationship between occupation and urinary bladder cancer. The case series included 263 (219 men, 44 women) histologically confirmed invasive urinary bladder cancers diagnosed between 1985 and 1988. The controls were 287 patients (210 men, 77 women) with acute medical and surgical conditions and non-neoplastic and non-urinary tract diseases. The proportion of cases ever employed in the printing industry (2.3%) was similar to that among controls (2.4%) [crude odds ratio, 0.9; 95% CI, 0.3–2.8].

Steineck et al. (1990) examined the relationship between urothelial cancer and exposure to benzene and exhaust in a population-based case–control study in Sweden. The study was based on men born between 1911 and 1945 and living in the county of Stockholm for all or part of the observation period of September 1985 to November 1987. Incident cases of urothelial cancers and/or squamous-cell carcinoma in the lower urinary tract were identified from the regional Cancer Registry and the urological departments (320 subjects). Controls (363 subjects) were selected by stratified (gender and year of birth) random sampling during the observation period from a computerized register covering the population of Stockholm. Information on exposure was collected by a postal questionnaire and all subjects were contacted at their homes. An industrial hygienist classified the subjects as having been exposed or unexposed to 38 agents and groups of substances. The adjusted odds ratio among men classified as having been exposed to carbon black (including printing inks) was 2.0 (95% CI, 0.8–4.9; 14 cases). The odds ratios were 3.2 (95% CI, 0.4–27.1) for low exposure to printing inks, 0.5 (95% CI, 0.1–2.1) for moderate exposure and 3.6 (95% CI, 0.8–12.1) for high exposure.

A population-based case–control study was conducted in the United States to examine the relationship between urinary bladder cancer, usual occupation and industry, and cigarette smoking (Burns & Swanson, 1991a). Cases and controls were selected through the Metropolitan Detroit Cancer Surveillance System, which had collected population-based cancer data for the three-county metropolitan Detroit area since 1973. Incident cases occurring among white and black subjects of each sex between the ages 40 and 84 were enrolled into the study. Subjects, or their surrogate respondents, were interviewed by telephone; the response rate for cases and controls was 94% and 95%, respectively. A total of 2160 urinary bladder cancer cases and 3979 controls with cancers of the colon and rectum, with complete histories of occupation and tobacco use, were included in the analysis. The analysis by usual occupation did not reveal an increased risk for printers (adjusted odds ratio, 0.9; 95% CI, 0.4–1.9; 12 cases). Nor was there an excess risk found for the printing industry (adjusted odds ratio, 0.7; 95% CI, 0.4–1.2; 22 cases). Case series in this study partly overlapped with the United States National Bladder Cancer Study (see Silverman et al., 1983, 1989, 1990). [The Working Group was concerned that only colonic and rectal cancers were used as controls.]

A hospital–based case–control study of 531 male matched pairs was conducted in South Lower Saxony, Germany (Claude et al., 1988; Kunze et al., 1992). Of the cases, 93% had histologically confirmed benign or malignant epithelial tumours of the urinary bladder; the remainder had tumours of the lower urinary tract. The cases were ascertained between 1977 and 1985 from three hospitals. Controls were patients with non-neoplastic diseases of the lower urinary tract from the same hospitals, individually matched by age to the cases. Among controls, 64% had hyperplasia of the prostate and 19% infection of the lower urinary tract. All subjects were interviewed on lifestyle, familial occurrence of urinary bladder cancer and occupational history. Cases and controls were compared according to whether they were ever employed in specific industries or occupations or ever exposed to selected agents. A significantly increased crude odds ratio of 5.0 (95% CI, 1.3–19.6; 11 cases) was observed among men ever employed in the printing industry. Analysis by occupation showed an elevated, although statistically nonsignificant, risk for printing workers (odds ratio, 3.0; 95% CI, 0.7–13.8; 7 cases). Analysis by length of employment was not performed due to the small numbers. [The Working Group noted that the definition of printing workers was unclear.]

Urinary bladder cancer and occupational exposures were investigated in a multicentre hospital-based case–control study in France between 1984 and 1987 (Cordier et al., 1993). Incident and prevalent cases, diagnosed after 1982, of histologically verified urinary bladder cancers in patients under 80 years of age were collected in urology departments of seven university hospitals in five regions of France. There was a total of 658 male cases. For each case, one hospital control was selected; controls were matched for age, ethnic origin and place of residence and had been admitted to other departments of the same hospital for diseases other than cancer, respiratory diseases or symptoms suggestive of urinary bladder cancer. Detailed interview data were collected for jobs held for at least six months. Occupational industrial titles were classified by a team of industrial hygiene experts. An unconditional logistic regression model was employed, controlling for age, hospital, place of residence and smoking status. The adjusted odds ratio for men ever employed in the printing and publishing industry for at least six months was 0.9 (95% CI, 0.5–1.5; 26 cases). The odds ratio for men who, based on job title, were classified as printers was 1.5 (95% CI, 0.6–3.5; 14 cases); the subgroup classified as printing pressmen had an odds ratio of 4.2 (95% CI, 0.8–20.8; 8 cases).

(c) *Lung cancer* (see Table 10)

Coggon et al. (1984), in a case–control study described on pp. 73–74, examined the relationship between lung cancer and occupation, based on information derived from death certificates. The study included 598 lung cancer deaths occurring in the period 1975–79 among males under the age of 40 in England and Wales. For each case, two controls were selected among men who had died from any other causes (matched by year of birth, year of death and residence). Information on occupation was obtained as described previously. A nonsignificant excess was found in occupations involving exposure to printing inks (odds ratio, 1.6; 95% CI, 0.9–2.7; 28 cases). The odds ratio for

Table 9. Case–control studies of urinary bladder cancer among workers in the printing industry

Reference, country	Type of controls	Exposure	Sex	No of exposed cases/controls	Odds ratio	95% CI	Comments
Wynder et al. (1963) USA	Hospital-based, excluding cancers of the respiratory system and upper alimentary tract, myocardial infarction	Printers, engravers, lithographers (ever)	M	3/2	[2.2]	[0.4–13.2]	—
Cole et al. (1972) USA	Population-based	Printing (usual) Printing (ever)	M M	5/7.7 (exp.) 14/15.1 (exp.)	0.8 1.3	0.3–1.9 0.7–2.5	Adjusted for age Adjusted for age and smoking
Williams et al. (1977) USA	Population-based, other cancers	Printing, publishing and allied products industry	M	4/NG	2.0	NG	Not significant. Adjusted for smoking
Najem et al. (1982) USA	Hospital-based, tobacco-related heart diseases and neoplasms excluded	Printing industry (≥ 1 year)	M+F	7/5	2.7	0.8–9.6	Crude odds ratio
Cartwright (1982) United Kingdom	Hospital-based, non-malignant diseases	Printers (exposed to ink-fly from high-speed presses)	M+F	18/NG	3.1	1.4–6.8	Adjusted for type of case (incident or prevalent) and sex
Silverman et al. (1983) USA	Population-based	Printing industry (ever) Printers (ever)	M M	50/45 6/2	1.1 3.0	0.7–1.7 0.6–14.8	Crude odds ratio
Coggon et al. (1984) United Kingdom	Deaths from other causes	Printing inks Printing inks (high exposure)	M M	12/21 5/2	1.1 5.0	0.5–2.3 1.0–25.8	Job exposure matrix applied to occupations recorded on death certificates; age < 50 years, cases and controls'
Schoenberg et al. (1984) USA	Population-based	Printing workers (ever) Printing ink (self-reporting)	M M	20/38 42/53	0.9 1.6	0.5–1.5 1.0–2.5	Adjusted for age, smoking and other employments

Table 9 (contd)

Reference, country	Type of controls	Exposure	Sex	No of exposed cases/controls	Odds ratio	95% CI	Comments
Vineis & Magnani (1985) Italy	Hospital-based, benign urological and surgical conditions	Printing and publishing industry (ever)	M	17/12	1.8 1.7	0.8–4.0 NG	Adjusted for age Adjusted for smoking; significant ($p < 0.05$) Only incident cases, adjusted for age
			M	NG	1.2	0.5–3.3	
Baxter & McDowall (1986) United Kingdom	Other cancers All causes of death	Printers (stated on death certificates)	M	21/NG	1.5 1.2	$p > 0.05$ $p > 0.05$	Against other cancers; matched on residence, year of death, age
Kabat et al. (1986) USA	Hospital-based, nonsmoking-related diseases	Printers	M	2/6	1.1	NG	Not significant. Nonsmoking cases and controls; blue-collar industrial occupations under-represented in the study population
Brownson et al. (1987) USA	Population-based, other nonsmoking-related cancers	Printing machine operators (longest-held job)	M M	7/8 7/6	3.1 2.3	1.1–8.9 0.8–7.2	All controls Prostate cancer excluded from controls. Adjusted for age, smoking and alcohol
Iscovich et al. (1987) Argentina	Neighbourhood and hospital-based, non-neoplastic diseases	Printing industry (ever) Printers (either in printing industry or elsewhere)	M+F M+F	3/2 5/2	2.7 5.6	NG NG	Not significant, adjusted for age and smoking

Table 9 (contd)

Reference, country	Type of controls	Exposure	Sex	No of exposed cases/controls	Odds ratio	95% CI	Comments
Steenland et al. (1987) USA	Other deaths, excluding urinary tract tumours and pneumonia	Pressing machine operator (ever)	M	5/2	5.0	[1.3–19]	Data from city directories
		Printing industry > 20 years	M	12/19	1.3	$p > 0.05$	Data from city directories. All matched on age, sex, race, date of death, residence at death
Gonzáles et al. (1989) Spain	Hospital- and population-based	Printers	M	15/19	1.8	0.8–4.1	Adjusted for smoking and exposure in other at-risk occupations
		Typesetters, linotypists	M	9/10	2.0	0.5–7.5	Adjusted for smoking and exposure in other at-risk occupations
Silverman et al. (1989, 1990) USA	Population-based	Printer (ever)	M (white)	37/77	0.8	0.5–1.2	Adjusted for smoking; frequency matching for age and geographic area
			F (white)	1/10	0.2	< 0.1–1.4	
La Vecchia et al. (1990) Italy	Hospital-based, other non-neoplastic, non-urinary diseases	Printing (ever)	M + F	[6/7]	[0.9]	[0.3–2.8]	Crude odds ratio
Steineck et al. (1990) Sweden	Population-based	Carbon blacks (including printing inks)	M	14/9	2.0	0.8–4.9	Adjusted for smoking; frequency matching for sex and year of birth
		Printing ink, low exposure	M	NG	3.2	0.4–27.1	
		Printing ink, moderate	M	NG	0.5	0.1–2.1	
		Printing ink, high	M	NG	3.6	0.8–12.1	

Table 9 (contd)

Reference, country	Type of controls	Exposure	Sex	No of exposed cases/controls	Odds ratio	95% CI	Comments
Burns & Swanson (1991a) USA	Colonic and rectal cancers	Printers (usual)	M + F	12/19	0.9	0.4–1.9	Adjusted for smoking, race, sex, age at diagnosis
		Printing industry	M + F	22/41	0.7	0.4–1.2	
Kunze et al. (1992) Germany	Hospital-based, non-neoplastic diseases of the lower urinary tract	Printing industry (ever)	M	11/3	5.0	1.3–19.6	Crude odds ratios
		Printing workers (ever)	M	7/3	3.0	0.7–13.8	
Cordier et al. (1993) France	Hospital-based, neoplastic, respiratory and urological conditions excluded	Printing and publishing industry (ever)	M	26/28	0.9	0.5–1.5	Adjusted for age, hospital, residence, smoking
		Printers	M	14/9	1.5	0.6–3.5	
		Printing pressmen	M	8/2	4.2	0.8–20.8	
Siemiatycki (1991) Canada	Hospital-based, other cancers	Inks (any)	M	18/NG	1.2	[0.7–2.1]	Adjusted for age, smoking
		Inks (substantial)		6/NG	0.9	[0.4–2.3]	
Siemiatycki et al. (1994) Canada	Population and hospital-based, other cancers, excluding lung and kidney sites	Printing and publishing industry	M				Adjusted for age, family income, smoking, coffee consumption, ethnicity, respondent status
		< 10 years		2/NG	0.3	0.1–1.2	
		≥ 10 years		11/NG	1.9	0.9–3.9	
		Photographic products (substantial exposure)		4/NG	3.0	0.9–10.1	

M, males; NG, not given; F, females

occupations with high exposure was 2.0 (95% CI, 0.8–5.0; 9 cases). [The Working Group noted the young age of the study subjects and that occupational status was derived from death certificates.]

Occupational risk factors for lung cancer were investigated in a case–control study in France (Benhamou et al., 1985, 1988). The data were collected during the period 1976–80 in 16 hospitals, 13 of which were in Paris. A total of 1260 male cases with histoogically confirmed lung cancer and 2084 matched (by age at diagnosis, hospital of admission, and interviewer) controls hospitalized for non-tobacco-related diseases were included. Complete occupational histories were recorded in personal interviews. The smoking-adjusted matched odds ratio for printers and related workers was 1.2 (95% CI, 0.7–1.9; 32 cases). There was no evidence of an increase in risk with duration of exposure.

A case–control study of lung cancer in Florence, Italy, was conducted to investigate occupational risk factors (Buiatti et al., 1985). All histologically confirmed cases of primary lung cancer admitted during 1981–83 to the regional general hospital (referral centre for all lung cancers) were selected for the study. For each case, one or two controls were selected from the same hospital by sex, age (plus or minus five years), date of admission and smoking status. All cases and controls were interviewed after their first admission to the hospital; in total, 592 cases (547 men, 45 women) and 1036 controls (955 men, 81 women) were interviewed. After restricting both series to residents of metropolitan Florence, 376 cases (340 men, 36 women) and 892 controls (817 men, 75 women) were available for the analysis. The odds ratio (adjusted for age, smoking and place of birth) for men having ever worked in paper or printing occupations was 1.6 (95% CI, 0.7–3.5; 11 cases). In women, two cases and four controls had ever worked in paper and printing occupations.

The association between occupation and lung cancer risk was examined in a population-based case–control study in New Mexico, United States (Lerchen et al., 1987). Cases included Hispanic white and other white residents of New Mexico, aged 25–84 years, with primary lung cancer, other than bronchiolar/alveolar carcinoma, diagnosed between January 1980 and December 1982. Cases were identified through the New Mexico Tumor Registry. Controls were selected by random-digit dialling for persons aged 65 and older from the Health Care Financing Administration's roster of Medicare participants. The controls were frequency matched to cases for sex, ethnicity and 10-year age category. A personal interview was used to obtain lifetime occupational and smoking histories and self-reported history of exposure to specific agents. The overall case interview rate was 89%, half of which were conducted with next-of-kin. The analysis was based on data from 333 male cases and 499 male controls. The odds ratio adjusted for age, ethnicity and smoking was 0.8 (95% CI, 0.4–1.8; 11 cases) for workers in the printing industry. Five cases had ever been employed as printers (odds ratio adjusted for age, ethnicity and smoking, 0.8; 95% CI, 0.3–2.6).

The association between lung cancer risk and occupation was examined in a case–control study in six areas of New Jersey, United States (Schoenberg et al., 1987). The study included 763 histologically confirmed incident lung cancers in white males and

900 general population white male controls between September 1980 and October 1981. The response rates were 70% among cases and 64% among controls. A total of 429 cases and 334 next-of-kin were interviewed. For self-respondent cases, controls were selected from State Drivers License files and frequency matched by race, age and residence. For deceased or incapacitated cases, controls, individually matched by race, age, residence and closest date of death or diagnosis, were selected from State mortality files, excluding deaths from lung cancer and respiratory diseases. Among controls, 564 were self-respondents. The smoking-adjusted odds ratio for subjects ever employed as printing workers (printing pressmen, compositors, typesetters, photoengravers, lithographers, printing engravers, printing trade apprentices) was 2.5 (95% CI, 1.0–6.1; 20 cases). After adjustment for additional variables (age, area, respondent type, education, vegetable consumption), the odds ratio was 2.3 (95% CI, 1.0–5.3). A significant crude odds ratio of 8.4 (7 cases) was found for duration of employment of 10 or more years as a printing worker.

The association between lung cancer risk and occupation was studied in a case–control study in Missouri, United States (Hoar Zahm *et al.*, 1989). Study subjects were identified through the Missouri Cancer Registry. Cases were all white male Missouri residents who had been diagnosed with histologically confirmed lung cancer from January 1980 through to November 1985; controls were white male Missouri residents diagnosed with cancer, excluding cancers of the lip, oral cavity, oesophagus, lung, urinary bladder, ill-defined sites and unknown sites. The study involved 4431 cases and 11 326 controls. Occupation at the time of diagnosis was abstracted from questionnaires administered to the cancer patients at the hospitals. Forty-eight percent of lung cancer cases and 55% of controls had unknown occupation on the records. The age- and smoking-adjusted odds ratio for printing occupations was 1.1 (95% CI, 0.6–1.9; 21 cases). An excess of adenocarcinoma was found for printing occupations (odds ratio, 1.8; 95% CI, 0.7–4.2; 7 cases). [The Working Group noted that many cases and controls had no known occupation.]

Occupational risk factors for lung cancer were assessed in a population-based case–control study in the Detroit Metropolitan Area, United States (Burns & Swanson, 1991b). Incident cancers occurring among white and black men and women between the ages of 40 and 84 were enrolled into the study through a rapid reporting system within two to six weeks of diagnosis. There were 5935 lung and bronchial cancer patients in the case group and 3956 colonic and rectal cancer patients in the control group. Subjects or their surrogate respondents were interviewed by telephone. Interviews with surrogates for deceased subjects were conducted with 39% of lung cancers, compared to 13% of the colonic and rectal cancer controls. The overall response rate was 93%. No elevated risk was found for printers (odds ratio adjusted for age at diagnosis, race, smoking and gender, 0.8; 95% CI, 0.4–1.7; 18 cases). The adjusted odds ratio for the printing industry was 0.7 (95% CI, 0.4–1.1; 54 cases). [The Working Group was concerned that only colonic and rectal cancers were used as controls and noted that the series of controls used in this study is probably the same as that in Burns & Swanson (1991a).]

Morabia *et al.* (1992) reported on a multicentre case–control study involving 1793 histologically confirmed male lung cancer cases at 24 hospitals in nine metropolitan

Table 10. Case–control studies of lung cancer among workers in the printing industry

Reference, country	Type of controls	Exposure	Sex	No of exposed cases/controls	Odds ratio	95% CI	Comments
Viadana et al. (1976) USA	Hospital-based, non-neoplastic diseases	Printing workers	M	7/NG	1.5	NG	Not significant
Coggon et al. (1984) United Kingdom	Deaths from other causes	Printing inks Printing inks (high exposure)	M M	28/36 9/9	1.6 2.0	0.9–2.7 0.8–5.0	Job exposure matrix applied to occupations recorded on death certificates; age < 40 years, cases and controls
Buiatti et al. (1985) Italy	Hospital-based	Paper or printing occupations (ever)	M F	11/16 2/4	1.6	0.7–3.5	Adjusted for age, smoking, place of birth; matched on sex, age, date of admission
Lerchen et al. (1987) USA	Population-based	Printing industry (ever) Printers	M M	11/17 5/7	0.8 0.8	0.4–1.8 0.3–2.6	Adjusted for age, ethnicity, smoking; matched on age, ethnicity and sex
Schoenberg et al. (1987) USA	Population-based	Printing workers ≥ 10 years Printing industry	M M M	20/11 7/1 37/31	2.5 8.4 1.3	1.0–6.1 NG 0.8–2.3	Adjusted for smoking ($p < 0.05$, crude) Adjusted for smoking
Benhamou et al. (1988) France	Hospital-based, non-tobacco-related diseases	Printers and related workers	M	32/51	1.2	0.7–1.9	Matched for sex, age at diagnosis, hospital, interviewer; adjusted for smoking
Hoar Zahm et al. (1989) USA	Selected cancer sites	Printing occupations	M	21/41 7/[4] (adeno-carcinoma)	1.1 1.8	0.6–1.9 0.7–4.2	Adjusted for age, smoking. Occupations unknown for about half of cases and controls
Burns & Swanson (1991) USA	Colonic and rectal cancers	Printers (usual) Printing industry	M+F	18/19 54/41	0.8 0.7	0.4–1.7 0.4–1.1	Adjusted for age at diagnosis, race, smoking, sex

Table 10 (contd)

Reference, country	Type of controls	Exposure	Sex	No of exposed cases/controls	Odds ratio	95% CI	Comments
Siemiatycki (1991) Canada	Hospital-based, other cancers	Printing and publishing industry	M	35/NG	2.0	[1.2–3.5]	Smoking-adjusted
		Printers	M	26/NG	2.1	[1.1–4.1]	Smoking-adjusted
		Printers (> 10 years)	M	13/NG	1.7	[0.7–4.1]	Smoking-adjusted
		Printing process workers	M	15/NG	3.1	[1.1–8.7]	Smoking-adjusted
			M	6/NG (adenocarcinoma)	7.0	[1.8–27.9]	Smoking-adjusted.
		Inks (any)	M	37/NG	1.6	[1.0–2.7]	Smoking-adjusted
		Inks (substantial)	M	18/NG	1.5	[0.7–3.1]	Smoking-adjusted
Morabia et al. (1992) USA	Hospital-based	Bookbinders and related printing trade workers	M	11/6	3.3	1.2–8.9	Adjusted for age, race, smoking, geographical area

M, male; NG, not given; F, female

areas in the United States during 1980–89. Information on sociodemographic characteristics, cigarette smoking and occupational history (usual job title and 44 specific exposures) were collected through a standardized questionnaire. For each case, one or two controls were selected from among patients without lung cancer, individually matched by race, age, hospital and cigarette smoking habits. Out of 3228 controls, 69% had a diagnosis of cancer other than of the lung and 31% had a diagnosis of non-neoplastic diseases. Unconditional logistic regression was used to compute odds ratios adjusted for matching variables. A statistically significantly increased risk was found for bookbinders and related printing trade workers (odds ratio, 3.3; 95% CI, 1.2–8.9; 11 cases).

(d) *Lymphohaematopoietic neoplasms* (see Table 11)

Occupational risks of leukaemia were investigated using the data from the Tri-State Leukemia Survey collected in New York, Baltimore and Minnesota, United States, during the period 1959–62 (Viadana & Bross, 1972). A subsample of 1345 white adult leukaemia cases and 1237 controls randomly selected from households in the same area as the cases were interviewed and included in the analysis. No relationship was found between leukaemia and occupation among women. Men reporting employment as printers in any of the five most recent jobs prior to interviews had a nonsignificant excess when compared to either non-printers (age-adjusted odds ratio, 1.5 [95% CI, 0.6–3.7]; 17 cases) or to clerks (age-adjusted odds ratio, 1.9 [95% CI, 0.7–5.0]).

Blair *et al.* (1993) examined the relationship between non-Hodgkin's lymphoma and occupation in a population-based case–control study in Iowa and Minnesota, United States. Cases were 715 white men with histologically confirmed non-Hodgkin's lymphoma reported to the Iowa State Health Registry during 1981–83 and patients identified through a surveillance network of hospitals in Minnesota during 1980–82. Controls were selected by random-digit dialling from the medical files of the Health Care Finance Administration and state vital records. All controls were frequency matched by state, age (15-year categories) and by year of death for deceased cases. A total of 622 cases and 1245 controls were interviewed on lifestyle and occupational history. The proportion of next-of-kin interviews was 30% for cases and 34% for controls. Subjects employed only as farmers were excluded, which left 546 cases and 1087 controls in the study. Odds ratios adjusted for age, state, smoking, type of respondent, family history of lymphoma and agricultural exposure to pesticides were computed. Six cases employed in printing press occupations yielded an odds ratio of 1.5 (95% CI, 0.4–5.1). A significant excess was found among workers with a duration of employment in the printing and publishing industry longer than 10 years (odds ratio, 2.5; 95% CI, 1.1–5.7); a decreased odds ratio of 0.5 was found for duration of employment less than 10 years (95% CI, 0.2–1.3).

(e) *Oropharyngeal cancer* (see Table 11)

Huebner *et al.* (1992) conducted a population-based case–control study of oropharyngeal cancers in four areas of the United States (Atlanta, New Jersey, Los Angeles and San Francisco). A total of 1114 black and white incident cases (762 males and 352 females) of primary histologically confirmed cancers of the oral cavity and pharynx (salivary glands and nasopharynx excluded) were identified from population-based cancer registries during 1984–85. A total of 1268 controls were obtained through random-digit dialling (aged 18–64 years) and from Health Care Financing Administration files (aged 65–79). These were frequency matched to the cases by race, sex, five-year age group and study area. Interviews on personal habits and occupation were conducted with the subjects or with next of kin for 75% of cases and 76% of controls. Surrogate respondents constituted 22% of cases and 2% of controls. The odds ratio adjusted for smoking, alcohol consumption, study location, age and race for men ever employed in the printing industry was 0.8 (95% CI, 0.5–1.5; 28 cases) and that for women was 2.9 (95% CI, 0.7–11.6; 8 cases). Analysis by job category revealed an odds ratio of 0.7 (95% CI, 0.3–1.4; 16 cases) for male printers/pressmen.

(f) *Testicular cancer* (see Table 11)

Coldman *et al.* (1982) examined the relationship of testicular seminoma to several risk factors in a case–control study of 128 histologically confirmed cases diagnosed between 1970 and 1977 in a regional treatment centre in Vancouver, Canada. One-hundred-and-twenty-eight controls were matched by age and year of diagnosis, and selected from other patients with skin cancers or Hodgkin's disease. A detailed occupational history was obtained for 89% of cases and 88% of controls. Seven cases had ever worked in the printing industry (odds ratio, 7.2; 95% CI, 0.9–162.3). No significant differences between cases and controls for exposure to inks were noted [actual results not reported].

The risk for testicular cancer associated with socioeconomic status and occupation was examined by Swerdlow *et al.* (1991) in England. A total of 259 cases of primary testicular cancer, incident between January 1977 and February 1981 and aged at least 10 years, were enrolled in six radiotherapy and oncology centres. Two control groups were selected from the same centres: 238 patients with other cancers (35% represented Hodgkin's disease, 13% non-Hodgkin's lymphoma, 10% brain cancer, 8% urinary bladder cancer) and, in other departments, 251 hospital inpatients with a wide range of non-malignant conditions. The subjects were interviewed on education, lifetime occupational history and father's occupation. Odds ratios adjusted for age and region of residence were estimated using conditional logistic regression models. When the results of the analyses for each control group were similar, the two sets of controls were combined. The risk was examined for occupation at age 20, at age 30, for the longest held occupation, for ever held occupation and for that most recently held. The only substantially increased odds ratio was reported for subjects ever employed in paper and printing (odds ratio, 2.1; 95% CI, 0.8–5.0; 10 cases).

(g) *Other cancer sites* (see Table 11)

The Missouri Cancer Registry identified 1993 white male cases of histologically confirmed colonic cancer and 9965 age-matched cancer controls, diagnosed between 1984 and 1987 (Brownson *et al.*, 1989). Two control groups were selected. The first contained all other cancers, the second excluded cancer sites known to be related to occupations (leukaemia, peritoneum, nasal cavity, lung, pleura, urinary bladder). Data on occupation and industry were collected routinely from hospital medical records at the time of diagnosis. Age-adjusted odds ratios were computed. Since odds ratios were generally of similar magnitude using each of the two control groups, all other cancers were used as controls. Excesses were found for printing machine operators (odds ratio, 1.9; 95% CI, 1.0–3.3; 18 cases) and workers employed in the printing and publishing industry (odds ratio, 1.8; 95% CI, 1.2–2.7; 33 cases).

Using the above-mentioned Missouri Cancer Registry, Brownson et al. (1990) evaluated the risk of brain cancer in relation to employment history in white males in a study of 312 histologically confirmed brain and other central nervous system cancers (ICD codes 191–192) and 1248 other cancer controls, frequency matched by age. Controls excluded cancers of ill-defined or unknown primary sites. The study covered the period 1984–88. Data on occupation and tobacco smoking habits were collected as in Brownson et al. (1989). Of the eligible subjects, 34% of cases and 38% of controls were excluded due to missing occupational information. [The Working Group was concerned by this high proportion of missing information.] The age- and smoking-adjusted odds ratio for workers in the printing and publishing industry was 2.8 (95% CI, 1.0–8.3; 7 cases). The elevated risk was equally distributed between astrocytic- and 'other'-cell type cancers [odds ratios not reported].

The role of occupational risk factors in the occurrence of pancreatic cancer was investigated in a case–control study conducted in France (Pietri et al., 1990), with 171 cases (105 men, 66 women) enrolled in seven public hospitals in Paris between 1982 and 1985. For each case, two controls, matched by sex, age at interview (within five years), hospital and interviewer, were selected: one among patients with cancers other than neoplasms of the biliary tract, liver, stomach and oesophagus and the other among patients with non-neoplastic diseases. The total number of controls was 317 (196 men, 121 women). All subjects agreed to be interviewed. Odds ratios were derived from unconditional logistic regression models adjusting for age, sex, foreign origin, education, coffee, cigarette and alcohol consumption using the two sets of controls combined. Cases and controls did not significantly differ in having been employed for at least one year in the printing industry (odds ratio, 1.3; 95% CI, 0.5–3.4; 8 cases). Among manual workers, three cases had ever been employed in printing occupations (odds ratio, 0.8; 95% CI, 0.2–3.4).

A case–control study of renal-cell cancer was conducted in Finland (Partanen et al., 1991). A total of 672 primary incident cases of renal-cell adenocarcinoma were identified through the Finnish Cancer Registry in 1977–78. Two population controls were matched to each case for year of birth, gender and survival status at the time of data collection; they were randomly selected from the population register, and the total number of controls was 1344 (280 alive, 1064 deceased). Data were collected by questionnaire, which had response rates for self-respondents of 76% for cases and 79% for controls; for next-of-kin respondents, the rates were 67% for cases and 65% for controls. The analyses were based on 338 matched sets of cases and controls, 192 cases having one control and 146 cases having two controls. An ad-hoc scoring system (three levels) was developed based on exposure level (background, low, high), the proportion of annual work time spent in the higher level of exposure and the period and duration of exposure for nine

Table 11. Case–control studies of other cancer sites among workers in the printing industry

Reference, country	Cancer type	Type of controls	Exposure	Sex	No of exposed cases/controls	Odds ratio	95% CI	Comments
Viadana & Bross (1972) USA	Leukaemia	Population-based	Printers (any of five most recent jobs)	M	17/814	1.5	[0.6–3.7]	In comparison with non-printers
				M	17/114	1.9	[0.7–5.0]	In comparison with clerks. Adjusted for age
Blair et al. (1993) USA	Non-Hodgkin's lymphoma	Population-based	Printing press occupations	M	6/5	1.5	0.4–5.1	Adjusted for age, state, smoking, family, history of malignant lymphoproliferative diseases, agricultural exposure to pesticides, use of hair dyes, direct or surrogate respondent
			Printing and publishing industry	M	NG NG	0.5 2.5	0.2–1.3 1.1–5.7	Duration < 10 years Duration > 10 years
Huebner et al. (1992) USA	Oral cavity and pharynx, excluding salivary glands and nasopharynx	Population-based	Printing industry (ever)	M F	28/36 8/4	0.8 2.9	0.5–1.5 0.7–11.6	Adjusted for age, race, smoking, alcohol, study location
			Printers/pressmen (ever)	M	16/27	0.7	0.3–1.4	
Coldman et al. (1982) Canada	Testicular seminoma	Hospital-based, patients with melanoma, other skin cancers and Hodgkin's diseases	Printing industry (ever)	M	7/1	7.2	0.9–162.3	Adjusted for age

Table 11 (contd)

Reference, country	Cancer type	Type of controls	Exposure	Sex	No of exposed cases/controls	Odds ratio	95% CI	Comments
Swerdlow et al. (1991) United Kingdom	Testis	Hospital-based, other cancers and non-neoplastic diseases	Paper and printing workers (ever)	M	10/12	2.1	0.8–5.0	Adjusted for age, residence
Brownson et al. (1989) USA	Colon	Population-based, other cancers	Printing and publishing industry	M	33/92	1.8	1.2–2.7	Adjusted for age
			Printing machine operators	M	18/49	1.9	1.0–3.3	
Brownson et al. (1990) USA	Brain and other central nervous system	Population-based, other cancers, excluding ill-defined or unknown primary site	Printing and publishing industry	M	7/10	2.8	1.0–8.3	Adjusted for age, smoking
Pietri et al. (1990) France	Pancreas	Hospital-based, non-neoplastic diseases and other cancers, excluding biliary tract, liver, stomach, oesophagus	Printing industry, all workers (at least 1 year)	M+F	8/12	1.3	0.5–3.4	Adjusted for age, sex, foreign origin, education, coffee, alcohol, smoking
			Printing industry, manual workers only (ever)	M+F	3/7	0.8	0.2–3.4	
Partanen et al. (1991) Finland	Renal-cell adenocarcinoma	Population-based	Printing and publishing industry, duration of employment ≥ 5 years	M	6/NG	4.6	0.9–23.5	Adjusted for obesity, smoking, coffee
				F	5/NG	8.0	0.9–69.8	
			Printers, duration of employment ≥ 5 years	M	4/NG	6.0	0.7–54.5	

Table 11 (contd)

Reference, country	Cancer type	Type of controls	Exposure	Sex	No of exposed cases/controls	Odds ratio	95% CI	Comments
Sinks et al. (1992) USA	Renal-cell cancer	Controls selected from paperboard printing plant workers (nested case–control study)	Duration of employment in finishing departments ≥ 5 years	M	3/3	16.6	1.7–453	Adjusted for age
Mandel et al. (1995) Australia, Denmark, Germany, Sweden, USA	Renal-cell cancer	Population-based	Printing or graphic industry (ever/never)	M	39/41	1.3	0.8–2.0	Adjusted for age, smoking, body mass index, education, study center
Multisite case–control studies								
Viadana et al. (1976) USA	Buccal cavity and pharynx	Hospital-based, non-neoplastic diseases	Printing workers	M	11/NG	2.6	NG	In comparison with clerical workers; significant ($p < 0.05$), adjusted for age. Not significant when adjusted for smoking (odds ratio 2.1)
Williams et al. (1977) USA	Oral cavity	Population-based, other cancers	Printing, publishing and allied products industry	M	7/NG	4.5	NG	Significant ($p < 0.05$)
	Pancreas			M	3/NG	7.0	NG	Not significant
	Pancreas			F	2/NG	8.2	NG	Not significant
	Melanoma			M	2/NG	3.0	NG	Not significant Adjusted for age, race, education, smoking, alcohol

Table 11 (contd)

Reference, country	Cancer type	Type of controls	Exposure	Sex	No of exposed cases/controls	Odds ratio	95% CI	Comments
Siemiatycki (1991) Canada	Pancreas	Hospital-based, other cancers	Printers (> 10 years exposure)	M	4/NG	3.7	[1.2–11.2]	Smoking-adjusted
	Kidney	Hospital-based, other cancers	Printers (any exposure)	M	5/NG	3.4	[1.2–9.5]	Smoking-adjusted
			Printers (> 10 years exposure)	M	3/NG	3.6	[1.0–10.6]	

M, male; F, female; NG, not given

selected exposures. For men and women, conditional logistic models were used to calculate separately odds ratio estimates adjusted for smoking status, coffee consumption and obesity. The odds ratio associated with employment for at least five years in the printing and publishing industry was 4.6 (95% CI, 0.9–23.5; 6 cases) in men and 8.0 (95% CI, 0.9–69.8; 5 cases) in women. Analysis by occupation revealed an increased risk among male printers who had worked for at least five years (odds ratio, 6.0, 95% CI, 0.7–54.5; 4 cases). Out of 10 cases with a high level of exposure to gasoline (odds ratio, 7.4; 95% CI, 1.6–35), three worked as operators of a relief-printing press and one as a rotogravure pressman who cleaned the press cylinders with gasoline; one of the two controls with high exposure was a press operator and foreman in a relief-printing shop.

A population-based multicentric international case–control study investigated the relationship between renal-cell adenocarcinoma and occupation (men and women) (Mandel *et al.*, 1995). Study centres in Australia, Denmark, Germany, Sweden and the United States interviewed 1732 incident cases and 2309 population controls (about 60% were men). The elicited occupational data were coded as ever/never in an occupation and industry. Ever employment in the printing or graphic industry was associated with an odds ratio of 1.3 (95% CI, 0.8–2.0; 39 cases) for men, adjusted for age, smoking status, body mass index, education and study centre.

2.2.4 *Cohort studies*

Cohort studies (industry- and community-based) are summarized in Table 12.

Mortality by occupation and industry among men who had served in the United States Armed Forces at some time between 1917 and 1940 was studied (Hrubec *et al.*, 1992). Information on tobacco use, occupation and industry was obtained through a mailed questionnaire sent in 1954 or 1957. The response rate was 84%. Included in the analysis were 248 046 respondents, who accumulated 4 530 604 person-years. A total of 164 785 deaths occurred during the period 1954–80. Mortality experience of this cohort during 1954–69 has been reported by Rogot and Murray (1980). Relative risks (RRs) for 1954–80 were calculated using a Poisson regression model adjusted for age at observation, calendar time, type of smoking and number of cigarettes smoked. Mortality from respiratory cancer was increased in printing pressmen and plate printers (adjusted RR, 1.6 [95% CI, 0.5–3.1]; 9 cases), in electrotypers and stereotypers (adjusted RR, 2.2; nonsignificant; 4 cases) and in bookbinders (adjusted RR, 2.5; nonsignificant; 3 cases; 1 case in nonsmokers). The risk of urinary bladder cancer was significantly increased in printing pressmen and plate printers (RR, 3.4 [95% CI, 1.3–9.0]; 4 cases). Two cases of renal cancer yielded a RR of 3.4 [95% CI, 0.6–18.7] in printing pressmen and plate printers. There were two cases of leukaemia in printing pressmen and plate printers (RR, 1.6), two cases in photoengravers and lithographers (RR, 1.7) and one case in other engravers (RR, 3.1).

2.3 Industry-based studies (see Table 12)

To evaluate lung cancer risk associated with occupations, Dunn and Weir (1968) carried out a prospective mortality study among 68 153 men, aged 35–64 years,

Table 12. Cohort studies (industry- and community-based) among workers in the printing industry

Reference, country	Study subjects	Period of follow-up	Occupation/exposure	Cancer site/cause of death	No. obs.	RR	95% CI	Comments
Dunn & Weir (1968) USA	68 153 male members of union organizations in 1954	1954–65	Printers	Lung	30	[0.8]	[0.5–1.1]	SMR, adjusted for age and smoking
Goldstein et al. (1970) USA	About 460 pressroom workers and 700 compositors in a newspaper plant	1947–62	Mineral oil mist Pressmen Compositors	Lung	3 6	[0.9]	[0.2–3.7]	Crude RR, pressmen versus compositors
Pasternack & Ehrlich (1972) USA	778 pressmen and 1207 compositors in a newspaper plant	1958–69	Pressmen Compositors	Mouth and respiratory system	6 8	[1.2]	[0.4–3.4]	Crude RR, pressmen versus compositors
Moss et al. (1972) United Kingdom	3485 deaths among male workers in London and Manchester newspaper printing companies	1952–66	Printing trade workers Machine-room men (London) Machine-room men (Manchester) Publishing-room men (London) Publishing-room men (Manchester)	Lung Lung Lung Lung Lung	365 71 38 91 22	1.3 1.2 2.0 1.4 1.6	[1.2–1.5] [1.0–1.6] [1.5–2.8] [1.1–1.7] [1.1–2.5]	PMR
Greenberg (1972) United Kingdom	670 deaths among newspaper printing workers in London	1954–66		All cancers Lung Stomach Leukaemia Urinary tract Tonsil	195 93 29 4 6 2	1.2 1.3 1.4 1.2 0.7 6.7	[1.0–1.4] [1.1–1.6] [1.0–2.0] [0.5–3.2] [0.3–1.5] [1.7–26.8]	PMR Based on national reference Based on national reference

Table 12 (contd)

Reference, country	Study subjects	Period of follow-up	Occupation/exposure	Cancer site/cause of death	No. obs.	RR	95% CI	Comments
Lloyd et al. (1977) USA	2604 deaths among members of the International Printing Pressmen and Assistants' Union	1966–68	Newspaper pressmen	All cancers	138	1.1	[0.9–1.3]	PMR
				Buccal cavity and pharynx	9	[2.4]	[1.2–4.6]	
				Lung	41	[1.1]	[0.8–1.5]	
				Leukaemia	8	[1.6]	[0.8–3.1]	
				Pancreas	9	[1.2]	[0.6–2.3]	
				Rectum	7	[1.6]	[0.7–3.3]	
			Commercial pressmen	All cancers	336	1.0	[0.9–1.1]	
				Lung	97	[1.0]	[0.8–1.3]	
				Buccal cavity and pharynx	13	[1.3]	[0.8–2.3]	
				Leukaemia	12	[0.9]	[0.5–1.5]	
Greene et al. (1979b) USA	347 cancer deaths at Government Printing Office in Washington DC	1948–77	Whole cohort	All haematopoietic	40	1.4	1.1–1.9	PCMR using age-, race- and time period-specific mortality proportions in Washington DC for 1950–69
				Multiple myeloma	10	2.2	1.2–4.0	
				Hodgkin's disease	7	2.3	1.1–4.7	
				Leukaemia	16	1.4	0.9–2.2	
				Colon	42	1.4	1.1–1.8	
				Liver/biliary tract	13	1.5	0.9–2.5	
				Urinary bladder	17	1.4	0.9–2.2	
				Melanoma	5	1.8	0.8–4.1	
				Lung	84	0.9	0.7–1.1	
				Kidney	8	1.1	0.6–2.2	
				Prostate	15	1.7	1.1–2.6	Non-white subjects
			Compositors	Multiple myeloma	8	5.3	2.7–10.1	White subjects
			Binders	Leukaemia	4	3.2	1.2–7.2	White subjects

Table 12 (contd)

Reference, country	Study subjects	Period of follow-up	Occupation/exposure	Cancer site/cause of death	No. obs.	RR	95% CI	Comments
Bertazzi & Zocchetti (1980) Italy	700 male workers employed for at least 5 years in a newspaper plant between 1940 and 1955	1956–75	Whole cohort	All causes	199	1.1	[0.9–1.2]	SMR
				All cancers	51	1.2	[0.9–1.6]	
				Lung	13	1.5	[0.8–2.5]	
				Larynx	3	2.0	[0.4–5.8]	
				Lymphohaemato-poietic	3	1.2	[0.2–3.4]	
			Packers and dispatch workers	All causes	66	1.5	[1.1–1.8]	
				All cancers	19	1.8	[1.1–2.8]	
				Lung	6	2.5	[0.9–5.4]	
Paganini-Hill et al. (1980) USA	1361 white newspaper pressmen, members of the Los Angeles Pressmen's Union for at least 1 year between 1949 and 1965	1949–78		All cancers	68	1.0	[0.8–1.2]	SMR
				Buccal cavity and pharynx	2	0.9	[0.1–3.3]	
				Lung	22	1.5	[0.9–2.3]	
				Kidney	5	3.0	[1.0–7.1]	
				Leukaemia	7	2.5	[1.0–5.1]	
Zoloth et al. (1986) USA	1401 deaths among white male commercial pressmen members of a labour union	1958–81		All cancers	315	1.3	1.1–1.4	PMR; 14% of this population overlaps with that of Lloyd et al. (1977)
				Colon	40	1.6	1.2–2.2	
				Rectum	19	2.1	1.3–3.2	
				Non-Hodgkin's lymphoma	11	2.1	1.2–3.8	
				Lung	79	1.2	0.9–1.5	
				Leukaemia	9	0.9	0.5–1.7	
				Urinary bladder	11	1.1	0.6–2.0	
				Kidney	8	1.5	0.7–2.9	
				Liver	9	2.2	[1.1–4.1]	Membership ≥ 20 years
				Pancreas	18	1.6	[1.0–2.6]	Membership ≥ 20 years

Table 12 (contd)

Reference, country	Study subjects	Period of follow-up	Occupation/exposure	Cancer site/cause of death	No. obs.	RR	95% CI	Comments
Svensson et al. (1990) Sweden	1020 male workers employed for at least 3 months in eight rotogravure plants	1952–86 (mortality) 1958–85 (morbidity)	Rotogravure printers	All causes	129	1.0	0.9–1.2	SMR; mortality
				Stomach	7	2.7	1.1–5.6	Mortality
					7	2.3	0.9–4.8	Morbidity
				Colon-rectum	7	2.2	0.9–4.5	Mortality
					9	1.5	0.7–2.8	Morbidity
				Respiratory system	11	1.4	0.7–2.5	Mortality
					16	1.8	1.0–2.9	Morbidity
					9	1.3	0.6–2.4	Morbidity, ≥ 5 yrs of exposure and > 10 yrs of latency
				Leukaemia, lymphoma	3	1.0	0.2–2.8	Mortality
				Leukaemia	3	1.7	0.3–4.9	Morbidity
				Urinary organs	1	0.5	0.01–2.5	Mortality
					4	0.6	0.2–1.7	Morbidity
Michaels et al. (1991) USA	1261 male members of the International Typographical Union employed by two newspaper plants in 1961, New York city	1961–84	Newspaper printers	All causes	498	0.7	0.7–0.8	SMR
				All cancers	123	0.8	0.7–1.0	
				Buccal cavity and pharynx	5	1.1	0.4–2.7	
				Lung	37	0.9	0.6–1.2	
				Urinary bladder	8	1.5	0.7–3.0	
				Leukaemia and aleukaemia	5	1.0	0.3–2.4	
Hrubec et al. (1992) USA	248 046 veterans	1954–80	Printing pressmen and plate printers	Respiratory system	9	1.6	[0.5–3.1]	Adjusted for smoking
				Kidney	2	3.4	[0.6–18.7]	
				Urinary bladder	4	3.4	[1.3–9.0]	

Table 12 (contd)

Reference, country	Study subjects	Period of follow-up	Occupation/exposure	Cancer site/cause of death	No. obs.	RR	95% CI	Comments
Sinks et al. (1992) USA	2050 white male workers in a paperboard printing plant	1957–89		All causes	141	1.0	0.9–1.2	SMR; mortality
				All cancers	16	0.6	0.3–0.9	
				Respiratory system	5	0.5	0.2–1.2	
				Renal-cell cancer	6	3.7	1.4–8.1	
				Urinary bladder	3	1.1	0.2–3.1	SIR; morbidity
Leon (1994) United Kingdom	9232 full-time male members of NGA and NATSOPA trade unions, between 1949 and 1963	1950–83	NATSOPA members (unskilled workers)	All cancers	509	1.2	1.1–1.3	SMR; national reference
						1.0	0.9–1.1	SMR; local reference
				Lung	242	1.4	1.2–1.6	National reference
						0.9	0.8–1.1	Local reference
				Urinary bladder	18	1.1	0.7–1.8	National reference
						1.0	0.6–1.6	Local reference
			NATSOPA newspaper machine assistants	Lung	94	1.8	1.4–2.2	National reference
						1.2	1.0–1.5	Local reference
			NATSOPA publishing room men	Brain and other CNS	4	3.7	1.0–9.5	National reference
						3.7	1.0–9.6	Local reference
			NATSOPA editorial workers	Buccal cavity and pharynx	2	10.5	1.3–38.0	National reference
						6.7	0.8–24.1	Local reference
			NATSOPA clerical staff	Buccal cavity and pharynx	3	6.4	1.3–18.6	National reference
						4.1	0.8–11.8	Local reference
			NGA members	Lung	150	0.8	0.7–1.0	National reference
						0.6	0.5–0.7	Local reference
				Urinary bladder	11	0.6	0.3–1.1	National reference
						0.6	0.3–1.0	Local reference

RR, relative risk; PMR, proportionate mortality ratio; PCMR, proportionate cancer mortality ratio; SIR, standardized incidence ratio; SMR, standardized mortality ratio; NGA, National Graphical Association; NATSOPA, National Society of Operative Printers, Graphical and Media Personnel; CNS, central nervous system identified through their 1954 membership in union organizations in California, United States. Data on occupational and smoking histories (daily consumption and years of smoking) were collected through questionnaires mailed during 1954 and 1957. Causes of death during 1954–65 were identified. The expected numbers of deaths were derived from age- and smoking-specific rates in the union population. A nonsignificant deficit of lung cancer ([SMR, 0.8; 95% CI, 0.5–1.1]; 30 observed cases) was reported for printers. No relationship was found with duration of employment.

Goldstein *et al.* (1970) compared the mortality from respiratory diseases during the years 1947–62 among about 460 actively employed and retired male pressroom workers in a large newspaper plant in New York City, United States, with that of about 700 compositors in the same plant. Environmental exposure to mineral oil mist in the pressroom ranged from 5 to 21 mg/m^3. The particle size of the mineral oil mist showed a mass median diameter of 'about' 15 μm with about 15% of the material in droplets generally considered to be 'respirable'. The particle size of carbon black in the ink used was about 0.1–0.2 μm. There were three deaths from pulmonary carcinoma out of 2797 person-years at risk among pressmen (crude rate, 1.1 per thousand) and six among compositors with 5127 person-years at risk (crude rate, 1.2 per thousand), giving a crude RR of [0.9; 95% CI, 0.2–3.7] for pressmen versus compositors. [The Working Group noted that the age distribution of the subjects in the exposed groups and the degree of completeness of death ascertainment were not given. The interpretation of this study is limited by the absence of an external comparison group]

Pasternack and Ehrlich (1972) expanded the study of Goldstein *et al.* (1970) examining the mortality of 778 pressmen and a comparison group of 1207 compositors employed in the same plant in the years 1958–69. Mortality data for active full-time employees and pensioners were available through health records maintained in the newspaper plant medical department. Observed and expected (on the basis of age- and period-specific mortality of compositors) deaths were compared. Cause-specific mortality was not analysed. Six cases of malignant neoplasms of the mouth and respiratory system among pressmen (5841 person-years) and eight cases among compositors (9189 person-years) yield a crude RR of [1.2; 95% CI, 0.4–3.4] for pressmen versus compositors. [The Working Group noted the possibility of under-ascertainment of deaths because of the type of information used, even when an internal comparison of cancer cases was performed.]

Moss *et al.* (1972) and Moss (1973) examined the mortality experience of 3485 full-time male workers from 16 newspaper printing companies in London and Manchester, United Kingdom, who died in 1952–66. Age- and calendar period-standardized PMRs were computed using regional statistics as the referent. Based on the occupational status supplied by the employers, workers were subdivided into five categories: compositors, readers and foundrymen; machine-room men; publishing-room men; other manual production workers; and office and white-collar workers. Significant excesses of deaths from lung cancer were observed for all manual printing trade workers (i.e. the first four categories above) (PMR, 1.3 [95% CI, 1.2–1.5]; 365 cases). Among machine-room men, a significant excess was found in Manchester (PMR, 2.0 [95% CI, 1.5–2.8]; 38 cases), whereas 71 lung cancer deaths were observed in London (PMR, 1.2 [95% CI, 1.0–1.6]).

Statistically significant excesses were also found among publishing-room men both in London (PMR, 1.4 [95% CI, 1.1–1.7]; 91 cases) and in Manchester (PMR, 1.6 [95% CI, 1.1–2.5]; 22 cases). Lung cancer mortality among non-manual workers did not differ from expectations. No information on smoking status was available. Observed deaths from bronchitis did not differ from the number expected among printing trade workers either in London or in Manchester.

Greenberg (1972) conducted a proportionate mortality study of male newspaper printing workers in London, United Kingdom, who died in the years 1954–66 based on 670 death certificates obtained from company death benefit fund records. Expected numbers of deaths by age group and calendar period were computed using regional statistics. Significant excesses of deaths from all neoplasms (PMR, 1.2 [95% CI, 1.0–1.4]; 195 cases) and cancer of the lung and bronchus (PMR, 1.3 [95% CI, 1.1–1.6]; 93 cases) were detected. There were 29 deaths from gastric cancer (PMR, 1.4 [95% CI, 1.0–2.0]) and four deaths from leukaemia (PMR, 1.2 [95% CI, 0.5–3.2]). Using national rates as reference, six deaths from cancer of the urinary organs (bladder and kidney) (PMR, 0.7 [95% CI, 0.3–1.5]) and two from cancer of the tonsil (PMR, 6.7 [95% CI, 1.7–26.8]) were observed. Low mortality from bronchitis and emphysema was observed, whereas the number of observed deaths from arteriosclerotic and degenerative heart disease did not differ from expectation. This study included workers from one of the factories studied by Moss *et al.* (1972).

A total of 2604 deaths occurring between 1966 and 1968 among white male members of a labour union representing printing pressmen in the United States were identified from death benefit fund records and examined for proportionate mortality (Lloyd *et al.*, 1977). Age-adjusted numbers of expected deaths by cause were computed using the mortality experience of white males in the United States in 1967. Among 676 deaths of newspaper pressmen, a significant excess of cancers of the buccal cavity and pharynx emerged ([PMR, 2.4; 95% CI, 1.2–4.6]; 9 deaths), which was mainly confined to men aged 20–54 years ([PMR, 10.0; 95% CI, 4.8–21.0]; 7 deaths). Mortality from lung cancer did not differ from expectation ([PMR, 1.1; 95% CI, 0.8–1.5]; 41 deaths). There were eight leukaemia deaths [PMR, 1.6; 95% CI, 0.8–3.1], nine deaths from pancreatic cancer [PMR, 1.2; 95% CI, 0.6–2.3] and seven rectal cancer deaths [PMR, 1.6; 95% CI, 0.7–3.3]. A total of 1840 deaths were observed among commercial pressmen; none of the specific cancer sites showed statistically significant differences between the observed and expected deaths in the whole group. Mortality from arteriosclerotic heart diseases was slightly elevated among commercial pressmen and newspaper pressmen.

Greene *et al.* (1979b) selected 347 cancer deaths occurring between 1948 and 1977 in former male United States Government Printing Office (GPO) employees in Washington DC whose last job was with the GPO and who had a survivor receiving death annuity payments. Proportionate cancer mortality ratios by specific cancer sites were computed using the age-, race- and time period-specific mortality proportions in the Washington DC area for the period 1950–69. A significant (44%) excess from all haematopoietic neoplasms (proportionate cancer mortality ratio (PCMR), 1.4; 95% CI, 1.1–1.9; 40 cases) was observed, with statistically significant increases for multiple myeloma (PCMR, 2.2; 95% CI, 1.2–4.0; 10 cases) and Hodgkin's disease (PCMR, 2.3; 95% CI, 1.1–4.7;

7 cases). Sixteen leukaemia deaths were found (PCMR, 1.4; 95% CI, 0.9–2.2). There was a statistically significant (36%) increase in the relative frequency of colonic cancer (95% CI, 1.1–1.8; 42 cases). Nonsignificantly elevated PCMRs related to cancers of the liver and biliary tract (PCMR, 1.5; 95% CI, 0.9–2.5; 13 cases) and urinary bladder (PCMR, 1.4; 95% CI, 0.9–2.2; 17 cases) and melanoma (PCMR, 1.8; 95% CI, 0.8–4.1; 5 cases) were observed. The observed number of lung cancers (84) did not exceed the expected number (PCMR, 0.9; 95% CI, 0.7–1.1), nor did that of renal cancers (PCMR, 1.1; 95% CI, 0.6–2.2; 8 cases). A significant increase in mortality from prostatic cancer was found among non-white subjects (PCMR, 1.7; 95% CI, 1.1–2.6; 15 cases). Analysis by occupation (composers, binders, pressmen, other) was performed only for multiple myeloma, leukaemia and colonic cancer. The excess of multiple myeloma was restricted to white compositors (PCMR, 5.3; 95% CI, 2.7–10.1; 8 cases); deaths from leukaemia occurred in excess among white bindery workers (PCMR, 3.2; 95% CI, 1.2–7.2; 4 cases), whereas the colonic cancer excess was not attributable to any specific occupation. Benzene had been used at the GPO on a limited basis for specialized processes, particularly in the bindery, while the main exposure of compositors was to inorganic lead.

Bertazzi et al. (1979) and Bertazzi and Zocchetti (1980) conducted a mortality study to investigate the reportedly high occurrence of cancer in a large newspaper plant in Milan, Italy. A total of 700 male workers employed between 1940 and 1955 with at least five years' service and alive on 31 December 1955 were admitted to the study. Mortality was studied in the period 1956–75. Expected deaths were computed from national mortality rates. Vital status ascertainment was 97% successful. Overall mortality was higher than expected (SMR, 1.1 [95% CI, 0.9–1.2]; 199 deaths). A total of 51 cancer deaths were observed in the whole cohort (SMR, 1.2 [95% CI, 0.9–1.6]). There were 13 deaths from lung cancer (SMR, 1.5 [95% CI, 0.8–2.5]), three deaths from laryngeal cancer (SMR, 2.0 [95% CI, 0.4–5.8]) and three from lymphatic and haematopoietic neoplasms (SMR, 1.2 [95% CI, 0.2–3.4]). None of the cancer site-specific SMRs was statistically significant. No clear-cut pattern of cancer mortality according to length of employment, duration of follow-up or time since first employment emerged. Analysis by subcategory showed a statistically significant excess among packers and dispatch workers for all causes (SMR, 1.5 [95% CI, 1.1–1.8]; 66 deaths) and all neoplasms (SMR, 1.8 [95% CI, 1.1–2.8]; 19 cases). Nonsignificant increases in lung cancer were found in packers and dispatch workers (SMR, 2.5 [95% CI, 0.9–5.4]; 6 cases) as well as in other workers (SMR, 1.6 [95% CI, 0.5–3.8]; 5 cases). With regard to non-malignant diseases, mortality from ischaemic heart diseases did not differ from expectation in the whole cohort, whereas packers and dispatch workers experienced a particularly elevated mortality from these causes (SMR, 2.1 [95% CI, 1.2–3.5]; 16 cases).

Paganini-Hill et al. (1980) carried out a mortality study of 1361 white newspaper web pressmen who were members of the Los Angeles Pressmen's Union for at least one year between 1949 and 1965. Of the subjects, 65% had worked for 20 or more years as pressmen, and most began employment before the age of 35. Vital status as of 1978 was determined for 1261 (91%), and death certificates were obtained for 344 of the 354 decedents. SMRs, adjusted for age and calendar period, were computed based on the general United States white male population. Overall mortality among web pressmen was 8% less than expected (SMR, 0.9 [95% CI, 0.8–1.0]; 344 cases); 68 deaths from all cancers were observed compared to 69 expected. Elevated SMRs were seen for renal cancer (SMR, 3.0 [95% CI, 1.0–7.1]; 5 cases) and leukaemia (SMR, 2.5 [95% CI, 1.0–5.1]; 7 cases). None of the other specific cancer sites showed statistically significant differences between observed and expected deaths, including lung cancer (SMR, 1.5 [95% CI, 0.9–2.3]; 22 deaths).

Following a report of a cluster of urinary bladder cancer, Zoloth *et al.* (1986) conducted a proportionate mortality study based on 1401 death certificates of white male members of a labour union of commercial printers in New York City, Long Island and New Jersey, United States, who died between 1958 and 1981. Expected numbers of deaths were calculated using age-, sex-, calendar- and time-specific United States population rates. Based on the amount of death benefit paid to the survivors of deceased members, the length of membership (< 20 years; > 20 years) was established for 1264 members (90% of the cohort). Only 164 members had length of membership less than 20 years. Mortality from all cancers was significantly elevated (PMR, 1.3; 95% CI, 1.1–1.4; 315 cases). Excesses at specific cancer sites were statistically significant for cancers of the colon (PMR, 1.6; 95% CI, 1.2–2.2; 40 cases) and rectum (PMR, 2.1; 95% CI, 1.3–3.2; 19 cases) and for non-Hodgkin's lymphoma (PMR, 2.1; 95% CI, 1.2–3.8; 11 cases). All deaths from liver (9 cases) and pancreatic cancer (18 cases) occurred among the longest-working population and these yielded significant excesses (PMRs, 2.2 [95% CI, 1.1–4.1] and 1.6 [95% CI, 1.0–2.6], respectively); there were 79 deaths from lung cancer (PMR, 1.2; 95% CI, 0.9–1.5). No excess was found for either deaths from leukaemia (PMR, 0.9; 95% CI, 0.5–1.7; 9 cases) or urinary bladder cancer (PMR, 1.1; 95% CI, 0.6–2.0; 11 cases). Three deaths due to myelofibrosis, a very rare myeloproliferative disease, were observed. With regard to nonmalignant diseases, the most notable excesses were found for arteriosclerotic and chronic rheumatic heart disease. [The data of this study are not fully independent from that of Lloyd *et al.* (1977); at most, 14% of this population potentially overlap with that study.]

Svensson *et al.* (1990) examined mortality (1952–86) and cancer incidence (1958–85) in a cohort of 1020 male workers employed for at least three months during 1925–85 in eight rotogravure plants in Sweden. Age-standardized mortality and incidence ratios were computed using mortality rates of the geographical area where the factories were located; cancer incidence rates for the same areas were obtained from the Swedish Tumour Registers. Vital status was ascertained for 99% of the cohort members. In 1983 and 1986, air concentrations of toluene were measured in three plants: the median concentrations were 33 ppm [124 mg/m^3] in two plants and 7 ppm [26 mg/m^3] in the third. The yearly average air concentrations of toluene were estimated in each plant; they reached a maximum of about 450 ppm [1665 mg/m^3] in the 1940s and 1950s, while later, there was a sharp fall. Cumulative doses (ppm-years) of toluene were calculated. Benzene was known to have been used in the 1940s, while its use decreased in the 1950s and 1960s. In one plant, the mean level of benzene in the air was 3 ppm [9.6 mg/m^3] ranging from 0.3 to 25 ppm [1–80 mg/m^3] in 1960, and from 0 to 61 ppm [0–195 mg/m^3] in 1962. Other aromatic and aliphatic hydrocarbons (such as 'naphtha') were used in decreasing proportions. After 1969, only toluene had been used in all plants, except one where substantial amounts of ethanol and ethyl acetate were used in addition to toluene. Mortality from all causes did not differ from that expected (SMR, 1.0; 95% CI, 0.9–1.2; 129 cases). A total of 41 cancer deaths were observed (SMR, 1.4; 95% CI, 1.0–1.9). Increases in deaths from cancer of the stomach (SMR, 2.7; 95% CI, 1.1–5.6; 7 cases) and colon/rectum (SMR, 2.2; 95% CI, 0.9–4.5; 7 cases) were detected; five out of seven cancers of the stomach and six out of seven cancers of the colon/rectum were found among the subcohort of workers with at least five years of exposure and more than 10 years of latency (SMR, 2.5; 95% CI, 0.8–5.9, for stomach; SMR, 2.4; 95% CI, 0.9–5.3, for colon and rectum). Eleven respiratory cancer deaths (7.9 expected) and three lympho-haematopoietic cancer deaths (3.1 expected) were observed. When cancer morbidity was analysed, only cancers of the respiratory tract were significantly increased (SMR, 1.8; 95% CI, 1.0–2.9; 16 cases); however, when a minimal employment period of five years and a latency period of 10 years were applied, the increase was no longer statistically significant (SMR, 1.3; 95% CI, 0.6–2.4; 9 cases). There were seven cancers of the stomach compared to three expected, nine cancers of the colon and rectum against six expected and three leukaemias versus 1.8 expected. Both mortality and morbidity for urinary tract cancers were nonsignificantly lower than expected. There were no associations between cumulative doses of toluene and SMRs for all tumour sites, gastrointestinal cancers or respiratory cancers. The authors cautioned about the overall accuracy of estimates of the exposure to toluene.

To investigate the effects of low-level exposure to lead, Michaels *et al.* (1991) examined the mortality experience (1961–84) of 1309 male members of the International Typographical Union employed in two New York City newspapers on 1 January 1961. The cohort was composed primarily of linotypists and slug makers; there was also a small group of mechanics and proof-readers. Although direct measurements of exposure were not available, historical industrial hygiene studies in the United States suggest that the typographers in this study had been exposed to airborne lead levels below the United States Occupational Safety and Health Administration permissible exposure level of 50 μg/m³. Forty-eight subjects (4%) were excluded since date of birth was unavailable. Out of 1261 workers, vital status was unknown for 39 cohort members (3%; assumed alive at the end of the study period). Age-, calendar time-, sex- and race-specific standardized mortality ratios were computed using the mortality rates of New York City as the comparison population. A surrogate measure of years of lead exposure was computed using the number of years between enrolment into the Union and last year of employment or until the end of 1976, which ever was earlier. The year 1976 was chosen since the use of hot lead in the plants ended during 1974–78. A significant deficit for overall mortality was observed (SMR, 0.7; 95% CI, 0.7–0.8; 498 cases). Mortality from all cancers was lower than expected (SMR, 0.8; 95% CI, 0.7–1.0; 123 cases). Five deaths from cancer of the buccal cavity and pharynx were observed (4.4 expected); 37 lung cancer deaths (SMR, 0.9; 95% CI, 0.6–1.2), eight urinary bladder cancers (SMR, 1.5; 95% CI, 0.7–3.0) and five leukaemias (SMR, 1.0; 95% CI, 0.3–2.4) occurred.

Following a physician's alert about one renal and one urinary bladder cancer in workers employed in the finishing department of a paperboard printing plant in Georgia, United States, Sinks *et al.* (1992) investigated mortality and cancer morbidity of 2050 subjects employed for more than one day between 1957 and 1988 in the facility. Vital status was successfully ascertained for 90% of the cohort. Out of 141 deceased subjects, death certificates were unavailable for 26 (18%). Expected numbers of deaths were computed based on United States mortality rates. The total number of person-years at risk accrued during the study period was 36 744; 71% had a duration of employment of less than four years and years-since-first-employment of less than 19 years. Age-standardized SIRs for urinary bladder and renal cancer were calculated using reference rates from the Atlanta-SEER (Surveillance, Epidemiology and End Results) Registry for the years 1973–77. Mortality from all causes did not differ from expectation (SMR, 1.0; 95% CI, 0.9–1.2; 141 cases), while a statistically significant deficit was observed for all cancer deaths (SMR, 0.6; 95% CI, 0.3–0.9; 16 cases). There were five deaths from cancer of the digestive organs (6.1 expected) and five from cancer of the respiratory system (9.4 expected), the remaining sites being represented by one or no cases. Six incident renal-cell cancers were identified (SIR, 3.7; 95% CI, 1.4–8.1). The excess risk persisted when the index case was excluded from the analysis (SIR, 3.1; 95% CI, 1.1–6.8). No increased risk for urinary bladder cancer was found (SIR, 1.1; 95% CI, 0.2–3.1; 3 cases). The review of the company safety data-sheets led to the identification of three potential carcinogens in the plant: methylene chloride (dichloromethane), formaldehyde (see IARC, 1995a) and trichloroethylene (see IARC, 1995b). The supplier information that several coloured inks had been manufactured from 3,3'-dichlorobenzidine and derivatives of toluidine was not confirmed by the analysis of 16 bulk pigment samples. The authors discussed possible limitations in the mortality study related to the high percentage of subjects lost to follow-up and of unknown causes of deaths. They also noted that the number of renal cancer cases could have been underestimated; moreover, a potential for selection bias exists due to the fact that four of the six renal cancer cases had been identified through employee interviews.

A nested case–control study was conducted for the six renal cancer cases identified in the above cohort of workers employed in a paperboard printing plant (Sinks *et al.*, 1992) (see Table 11). Forty-eight controls, matched by age and sex, were selected according to incidence density sampling. Controls were required to be of a younger age at first employment than the case's age at diagnosis. Conditional maximum likelihood odds ratio estimates were computed. Analysis by length of employment in each department (cut and crease, finishing, maintenance, office, press operators) was performed. Five out of six cases (83%) had worked in the plant for five years or more, whereas only 19% of the controls had a duration of employment greater then five years. Employment in the finishing department for five years or more was associated with a statistically significant increased risk (odds ratio, 16.6; 95% CI, 1.7–453.1; 3 cases). The odds ratio did not change when employment during the most recent five or 10 years was not included in the calculation. Time since first employment, date of hire and age at hire were added to the model; these covariates did not alter the odds ratio.

A retrospective cohort mortality study (Leon, 1994) of printing workers was undertaken after an anecdotal report of a cluster of urinary bladder cancers (three cases between 1971 and 1979 against 0.7 expected based on national rates) in a newspaper plant in Manchester, United Kingdom. The cohort included 9232 men born in 1890 or later who had been full-time members for at least six months in 1949–63 in two trade unions (National Graphical Association (NGA) and National Society of Operative Printers, Graphical and Media Personnel (NATSOPA)) in the Manchester area. The plant from which the cluster of urinary bladder cancers originated was not included. Mortality was examined from 1950 to 1983 and follow-up was 98% complete for NGA members and 97% complete for NATSOPA members. SMRs were computed. Expected numbers of deaths were obtained using age- and calendar-specific mortality figures for men in England and Wales. SMRs for Manchester County Borough in the period 1968–78 were applied as correction factors. Among NATSOPA members, significant elevations were found in SMRs for 'all causes of death' (SMR, 1.1; 95% CI, 1.1–1.2), all cancers (SMR, 1.2; 95% CI, 1.1–1.3), lung cancer (SMR, 1.4; 95% CI, 1.2–1.6) and ill-defined or secondary cancers (SMR, 1.6; 95% CI, 1.1–2.4). When adjustment for local rates was applied, all SMR values decreased and none of the above remained elevated; for lung cancer, the adjusted SMR was 0.9 (95% CI, 0.8–1.1). Mortality from urinary bladder cancer did not differ from expectation based on either national (SMR, 1.1; 95% CI, 0.7–1.8) or locally adjusted rates (SMR, 1.0; 95% CI, 0.6–1.6). For NGA members, mortality from all causes, from all cancers and from urinary bladder and lung cancer was lower than expected based on both national and locally adjusted rates. When workers were subdivided into four major occupational groups (NGA machine managers, NGA compositors, NATSOPA machine assistants and NATSOPA publishing-room men), the risk of lung cancer was elevated among NATSOPA machine assistants (SMR, 1.8; 95% CI, 1.4–2.2, based on national rates; SMR, 1.2; 95% CI, 1.0–1.5, based on locally adjusted rates; 94 cases). The only other statistically significant SMR was for malignant neoplasms of the brain and other sites in the central nervous system among NATSOPA publishing room men, although this was based on only four cases (SMR, 3.7; 95% CI, 1.0–9.5). Among NATSOPA non-production workers, two cancers of the buccal cavity and pharynx were found in the editorial group and three among clerical staff — the SMRs based on national rates were 10.5 (95% CI, 1.3–38.0) and 6.4 (95% CI, 1.3–18.6), respectively; adjustment for local rates reduced both mortality ratios (SMR, 6.7; 95% CI, 0.8–24.1, for editorial group; SMR, 4.1; 95% CI, 0.8–11.8, for clerical workers). RRs of NATSOPA versus NGA members were calculated via Poisson regression, controlling for age and calendar period. RRs were high for most of the causes. In particular, the RR for lung cancer was 1.7 (95% CI, 1.4–2.1). The authors noted that a strong socioeconomic component can explain the differences: NATSOPA members were mainly semiskilled or unskilled workers whereas NGA members had higher skills.

To evaluate the risk of lung cancer, a case–control study was nested within the above cohort of printing workers (Leon *et al.*, 1994) (see Table 10). A total of 110 lung cancer deaths were identified between 1949 and 1986 in machine assistants who were members of NATSOPA and who had entered the union after 1915. For each case, five controls were selected at random from all other machine assistants with a similar work record, born within 2.5 years of the corresponding case and alive at the date of the case's death. A complete occupational history was collected for all cases and 316 controls. Conditional logistic regression was used to compute odds ratio estimates. Duration of union membership was taken as an indicator of duration of work. Duration of work (< 10, 10–

19, 20–29, 30–39, > 40 years) in newspaper machine rooms was positively associated with the risk of lung cancer (odds ratios, 1; 1.2; 95% CI, 0.5–3.0; 1.4; 95% CI, 0.6–3.3; 2.0; 95% CI, 0.8–4.9; and 2.0; 95% CI, 0.7–5.8; p for trend = 0.07). Adjustment for period of first exposure gave similar results. When an analysis was carried out restricted to workers with a minimum of 15 years of latency, the duration–response relationship was not apparent; however, this analysis was hampered by small numbers and low precision. An analysis for period of first entry failed to show any clear pattern.

2.4 Childhood cancer in relation to parental exposure

Many case–control studies have been conducted to assess whether occupation of the father or of the mother at the time of conception constitutes a risk factor for malignant diseases in offspring. In the majority of the studies, the grouping of occupations is too broad to make it possible to evaluate the risk specific for printing-related occupations.

Kwa and Fine (1980) compared the parental occupation for 692 children who died of cancer before the age of 15 in Massachusetts, United States, and who were born during the periods 1947–57 and 1963–67 with that of 1384 population controls. For each case, two controls were selected among children whose birth registration immediately preceded or followed that of the case subject. Parental occupations were extracted from the birth certificates. Fifteen cancer cases had 'printer' as father's occupation at the time of birth [crude odds ratio, 1.8; 95% CI, 0.9–3.6]. The crude odds ratio for children with leukaemia and lymphoma and with printers as father's occupation was [1.7 (95% CI, 0.8–3.9; 9 cases)].

In a study designed to test hypotheses that paternal occupational exposure to lead and hydrocarbons is associated with Wilms' tumour in offspring, Wilkins and Sinks (1984) compared paternal occupation recorded on birth certificates of 62 cases identified through the Columbus Children's Hospital Tumor Registry in Ohio, United States, during the period 1950–81 with that of 124 controls matched on sex, year of birth and ethnic group. Half of the controls were also matched on the mother's county of residence when the child was born. None of the fathers of the cases was a printer; two of the fathers of the control children worked as printers at the time of birth of their children.

In the inter-regional epidemiological study of childhood cancer in the Northwest, West Midlands and Yorkshire regions of the United Kingdom during the period 1980–83, 555 cases of newly incident childhood cancer were identified in paediatric oncology centres and through cancer registries (Birch et al., 1985). Two groups of age- and sex-matched controls were recruited: 555 through the general practitioners or group practices with whom the cases were registered and 555 selected from among children in hospital for reasons other than neoplastic disease and who did not have either a genetic or other constitutional disease or malformation known to be associated with increased risk of cancer, or any other major malformation or chronic disease. Information on parental occupation was obtained by interview with the parents. No association was found between leukaemia or lymphoma (234 cases) and paternal employment classified as 'paper and printing workers' (McKinney et al., 1987). No association was apparent between central nervous system tumours (78 cases) and paternal employment in the year before pregnancy and during pregnancy in 'paper and printing' (Birch et al., 1990).

Cole Johnson et al. (1987) conducted a case–control study in Texas, United States, to evaluate paternal occupational exposure to hydrocarbons and the risk of childhood cancer of the nervous system. Cases included 499 children born in Texas, who had died of intracranial or spinal cord tumours with a birth year distribution from 1950 through 1979 at 0–14 years of age. A total of 998 controls, matched by sex, race and year of birth, were selected from live births in Texas during the same period. Paternal occupation was abstracted from the birth certificates. Nine cases had paternal occupation defined as 'printing workers' yielding a statistically significant odds ratio of 4.5 (95% CI, 1.4–14.7); the odds ratio for father's occupation in 'newspaper and printing industries' was 5.1 (95% CI, 1.6–16.3; 10 cases).

In a mortality-based case–control study in Ohio, United States, during the period 1959–78, information on paternal occupation was obtained from the birth certificates of the children (Wilkins & Koutras, 1988). A total of 491 deaths due to brain tumours at ages of 19 and under were ascertained and 491 controls were selected randomly after matching for sex, ethnic group and year of birth. The odds ratio associated with the father being employed in 'printing occupations' was 0.9 (95% CI, 0.3–2.8; 6 exposed cases), adjusted for parental age, birth order, birth weight, mother's residence at the time of birth of the index child, the sex of the child and the year of the child's birth.

In a study of the occupational exposure of parents of 204 children diagnosed with acute non-lymphoblastic leukaemia under 18 years of age during the period 1980–84, ascertained through the registration files of a cooperative clinical trial groups in the United States and Canada, and a similar number of controls selected by random-digit dialling, no association with reported exposure to printing inks of either parent was found (Buckley et al., 1989). [The Working Group noted that no details were given of the association other than the inclusion of this occupational exposure in the questionnaire]

Olsen et al. (1991) identified 1747 cases of childhood cancer from the Danish Cancer Registry and 8630 controls randomly selected from the Central Population Register after matching for sex and date of birth and survival without cancer until the date of diagnosis of the case. Employment histories of parents were established by record linkage with the files of the nationwide Supplementary Pension Fund. Eight fathers of cases with childhood cancer of any type were printers, and the odds ratio associated with this job title was 1.0; there was no significant association in any specific cancer sites.

Kristensen and Andersen (1992) investigated the occurrence of cancer in offspring of male members of the Oslo (Norway) printers' unions during the period 1930–74. Children born between 1950 and 1987 were traced for cancer in the Norway Cancer Registry. The study population comprised 12 440 children who were followed until 1987. Age-, gender- and calendar period-SIRs were computed based on incidence rates in Oslo. Using the employment description in the union records, each worker was categorized as exposed to lead (compositors, monotype casters and stereotype workers) or lead and solvents (rotary press printers and assistants) or solvents (all other printers). The total observed number of offspring cancers was 33 compared to 39.2 expected. After restricting the analysis to children whose father joined the union before the child's birth (10 829), seven cancer cases were identified in children younger than 14 years of age (SIR, 0.5; 95% CI, 0.2–1.0); 20 cancers occurred in those of older age (SIR, 1.0; 95% CI, 0.6–1.6). Among children aged 0–14 years, there were two acute leukaemia cases (4.5 expected cases) and two tumours of the central nervous system (4.1 expected cases); four cases of tumours of the central nervous system were observed among older children (3.0 expected). Since lead was abandoned in Oslo printing shops in the mid-1970s, all fathers were reclassified as lead-exposed only until 1975. Among 3221 children of these fathers, none developed cancer before the age of 15 (3.7 expected cases).

In a study of 163 cases of astrocytoma diagnosed under the age of 15 years during the period 1980–86 in eight hospitals in Pennsylvania, New Jersey and Delaware, United States, information on parental occupation was obtained by telephone interview (Kuijten *et al.*, 1992). Controls were selected by random-digit dialling. For the pre-conceptional period, the odds ratio associated with paternal employment as a printing worker was 4.0 (95% CI, 0.4–195.1, based on 5 discordant pairs); for paternal employment in this category during the index pregnancy, the odds ratio was 3.0 (95% CI, 0.2–158, based on 4 discordant pairs); and for the period between birth and one year prior to diagnosis, the odds ratio was 2.5 (95% CI, 0.4–26.2, based on 7 discordant pairs).

3. Studies of Cancer in Experimental Animals

Subcutaneous injection

Mouse: Groups of 20 male CB stock mice, 11 weeks of age, received 15–22 weekly injections into the right flank of unknown amounts of one of 22 different printing inks of unknown composition diluted or suspended in distilled water or arachis oil. Three groups of 40 mice each received 80 injections of distilled water or arachis oil alone or were untreated and served as controls. Survival at 18 months ranged from 7/20 to 14/20 in all groups with the exception of the group treated with Medium Sepia, which had no survivors by that time. No other toxic effect was reported. Local sarcomas developed at the injection site in 1/20 mice in each of five experimental groups: as early as at seven months in a mouse given Medium Sepia and from 18 to 19 months in mice receiving Process Black, Bronze Blue L/P, Concentrated Brown and Brown No. 3. The neoplasms that developed later were spindle-cell tumours, while the earlier tumour was more anaplastic. No tumour developed at the injection site in control mice (Carter *et al.*, 1969). [The Working Group noted many inadequacies in study design and reporting, and the small numbers of animals used.]

4. Other Data Relevant to an Evaluation of Carcinogenicity and its Mechanisms

4.1 Absorption, distribution, excretion and metabolism

In two press operators exposed to black offset ink during bulk material transfer, 1-hydroxypyrene levels were measured in pre- and post-shift urine. In one of the operators, post-shift urine 1-hydroxypyrene was elevated by about 60%. The other operator exhibited only a very slight increase (~ 5%) (Jongeneelen *et al.*, 1988).

4.2 Toxic effects

4.2.1 *Humans*

The first systematic study on the health effects of working in printing trades was carried out by Ramazzini at the end of the seventeenth century, who noted that there was a high prevalence of visual and respiratory symptoms (Ramazzini, 1933).

No excess mortality from all causes was observed among printing industry workers in the studies described in Section 2.3 (Paganini-Hill *et al.*, 1980; Svensson *et al.*, 1990; Michaels *et al.*, 1991; Sinks *et al.*, 1992; Leon, 1994).

No excess mortality from diseases of the circulatory system (Zoloth *et al.*, 1986; Svensson *et al.*, 1990) or from cardiovascular diseases (Greenberg, 1972; Paganini-Hill *et al.*, 1980; Sinks *et al.*, 1992; Leon, 1994) was observed among printing industry workers in the studies described in Section 2.3.

A small but significant excess mortality from arteriosclerotic heart disease (PMR, 1.09; $p < 0.01$) was observed among commercial pressmen, and a similar, but non-significant excess among newspaper pressmen (PMR, 1.05) and paper handlers (PMR, 1.04) (Lloyd *et al.*, 1977). An excess in mortality of ischemic heart disease (SMR, 3.0; $p < 0.01$) was reported among newspaper printing workers 25–54 years of age, which was associated with length of employment, age at the time of hiring and duration of follow-up (Bertazzi *et al.*, 1979; Bertazzi & Zocchetti, 1980). [No increase was observed in the 55–74-year and > 74-year age groups.]

Elevated mortality from cerebrovascular disease, which was of borderline statistical significance, was observed among typesetters in a study in the United States (Michaels *et al.*, 1991). The SMR for cerebrovascular disease mortality was 1.1 in an early British study (Greenberg, 1972) and 1.0 in a more recent study (Leon, 1994). The PMR for vascular lesions of the central nervous system was 0.76 ($p < 0.05$) among commercial pressmen in the United States (Zoloth *et al.*, 1986). A deficit of mortality from cerebrovascular lesions was also observed among commercial pressmen (PMR, 0.84; $p < 0.05$) and paper handlers (PMR, 0.41; not significant), but not among newspaper pressmen (PMR, 1.02) (Lloyd *et al.*, 1977).

No excess mortality from nonmalignant diseases of the respiratory tract was observed among printing industry workers in the studies described in Section 2.3 (Greenberg, 1972; Lloyd *et al.*, 1977; Zoloth *et al.*, 1986; Svensson *et al.*, 1990; Michaels *et al.*,

1991; Leon, 1994). Mortality from bronchitis was not elevated among men who worked full-time in the newspaper printing industry in London and Manchester (see Section 2.3) (Moss et al., 1972).

Five-year mortality from respiratory diseases among newspaper pressmen (11 deaths in 2797 man-years at risk, i.e. 3.9/1000 man-years) was not statistically significantly different from the mortality experience of compositors (17/5127; 3.3/1000 man-years) in the same company (Goldstein et al., 1970).

When occupational causes of asthma were sought in a case–control study in a community-based outpatient clinic in Singapore, an elevated odds ratio (2.2; 95% CI, 1.2–4.3) was observed for printers, after adjusting for age, sex, race, smoking and clinical atopy (Ng et al., 1994). The frequency of dyspnoea was not elevated among former printers in a prospective study of the relationship between main lifetime occupation and dyspnoea after the age of 65 (Nejjari et al., 1993). Several outbreaks of humidifier lung disease have been reported among printing workers (Hauck & Baur, 1990; Jost & Lehmann, 1990; Mamolen et al., 1993).

No excess mortality from gastrointestinal tract diseases was observed among printing industry workers in the studies described in Section 2.3 (Svensson et al., 1990; Michaels et al., 1991; Leon, 1994).

Excess mortality from liver cirrhosis was observed among newspaper web pressmen (Paganini-Hill et al., 1980) and newspaper pressmen (Lloyd et al., 1977). A nonsignificant increase in the mortality from liver cirrhosis was observed in a subcohort of British printing industry union members (Leon, 1994).

Hepatic function has been studied among printers, mostly with emphasis on exposure to toluene. Although a high prevalence (55/181) of abnormalities in biochemical tests of liver function was observed among workers at a rotogravure plant, this was mostly explained by alcohol abuse and overweight (Boewer et al., 1988). Similarly, in a four-year follow-up study of printers, no association was observed between occupational solvent exposure and biochemical markers of hepatic damage (Nasterlack et al., 1994). The prevalence of abnormal serum alanine, aspartate transaminase or alkaline phosphatase was not different among printing factory workers and its job applicants (Guzelian et al., 1988). Mild pericentral fatty change was observed in liver biopsies from six workers, in whom no cause other than work exposure could be identified for their elevated serum enzyme activities. An elevated urinary excretion of D-glucaric acid (the end-product of the glucuronic acid pathway and proposed as a marker substance for microsomal enzyme induction) was reported among printing plant workers, but was not correlated to levels of toluene exposure (Moretto & Lotti, 1990). In another printing plant, toluene-exposed workers had higher D-glucaric acid excretion than those not exposed to toluene [toluene concentrations not mentioned] (Sandstad et al., 1993).

No excess mortality from urinary tract diseases was observed among printing industry workers in studies described in Section 2.3, in which the indicators of exposure were toluene (Svensson et al., 1990) and lead (Michaels et al., 1991).

Creatinine clearance or urinary excretion rates of albumin or β2-microglobulin were not different among 43 flexoprint or rotogravure printers and 43 controls not exposed to organic solvents (Krusell et al., 1985). In the same subjects, urinary excretion of albumin and β2-microglobulin was not significantly elevated by 6.5-h exposure to toluene (382 mg/m^3) in an exposure chamber (Nielsen et al., 1985). A 31% prevalence of haematuria (dipstick) was observed among workers at a printing company compared to 25% in the control group from other industries. The authors noted that both the prevalences from exposed and control groups were higher than those reported in earlier studies (Sinclair et al., 1993). Hashimoto et al. (1991) reported that the prevalence of microscopic haematuria, leukocyturia and albuminuria was elevated among solvent-exposed newspaper pressworkers in comparison to compositors not exposed to organic solvents.

Prompted by a case of aplastic anaemia in a printer in a printing shop, where most of the work involved five-colour printing with UV-cured and air-cured offset ink, Cullen et al. (1983) investigated the blood-forming organs of seven workers exposed to a variety of solvents and resins. The solvents used were chemically analysed for benzene; none was detected. All workers examined were clinically healthy and none showed any abnormalities in the cellular elements of the peripheral blood. However, all workers had one or more findings in the bone-marrow aspirate that was considered to be pathological. No concurrent healthy controls were examined.

UV-cured inks and printing plates (often containing acrylates) are frequent causes of allergic contact dermatitis among printing workers (Magnusson & Mobacken, 1972; Wahlberg, 1974; Beurey et al., 1976; Pye & Peachey, 1976; Calas et al., 1977; Emmett, 1977; Malten, 1977); sensitized people have shown positive reactions on patch testing toward alkyl acrylates (Björkner & Dahlquist, 1979; Björkner, 1981), multifunctional acrylic monomers such as tripropylene glycol diacrylate (Smith, 1977; Whitfeld & Freeman, 1991), pentaerythritol triacrylate (Emmett, 1977; Emmett & Kominsky, 1977; Smith, 1977; Nethercott, 1978; Björkner & Dahlquist, 1979; Nethercott et al., 1983), trimethylol propane triacrylate (Emmett, 1977; Emmett & Kominsky, 1977; Smith, 1977; Nethercott, 1978; Björkner & Dahlquist, 1979; Björkner et al., 1980; Nethercott et al., 1983) or hexanediol diacrylate (Emmett & Kominsky, 1977), 2-hydroxyethyl methacrylate, different secondary acrylamides (Pedersen et al., 1982, 1983) as well as toward urethane, epoxy or polyester acrylate resins and oligomers (Emmett & Kominsky, 1977; Malten, 1977; Björkner & Dahlquist, 1979; Björkner et al., 1980; Nethercott, 1981; Nethercott et al., 1983). During a two-year period in Toronto, Canada, 21 printing tradesmen with contact dermatitis were evaluated. Of these, 14 (67%) had allergic contact dermatitis (positive patch testing) and seven (33%) had an irritant contact dermatitis (negative patch testing). The irritant contact dermatitis was attributed to solvent exposure. Six of the cases of allergic contact dermatitis and one of the cases of irritant contact dermatitis had exposure to UV-cured ink components (Nethercott & Nosal, 1986).

Single cases of allergic contact dermatitis, caused by ethoxylated phenol surfactant (Ashworth & White, 1991) and colophony (used for paper sizing) (Castelain et al., 1980; Bergmark & Meding, 1983), have also been described among printers.

Allergic contact dermatitis has been described in a female silk-screen printer in the manufacture of circuit boards; patch testing was positive toward diaminodiphenylmethane, 2-hydroxyethyl methacrylate and triglycidyl isocyanurate (Jolanki *et al.*, 1994).

No excess mortality from nervous system diseases was observed among printing industry workers in the studies described in Section 2.3 (Svensson *et al.*, 1990; Leon, 1994).

Mucous membrane irritation, neurological symptoms and neurophysiological and neuropsychological findings have been described among printers; these have usually been linked to exposure to different solvents or to lead (Baelum *et al.*, 1982; Ruijten *et al.*, 1991; Matsumoto *et al.*, 1993; Sinha *et al.*, 1993; Uchida *et al.*, 1993). The solvents most often incriminated have been hexane, toluene, xylene and chlorinated solvents, but often the exposure has been mixed.

n-Hexane-induced polyneuropathy has been observed among printing factory workers (Wang *et al.*, 1986; Chang, 1987; Chang *et al.*, 1993).

The frequency of neurasthenic symptoms was higher among rotogravure printers (mainly exposed to toluene) than among controls not exposed to solvents; the printers also had slightly lower scores in psychometric tests (Ørbaek & Nise, 1989). Changes in auditory, visual and somatosensory evoked potentials (Urban & Lukás, 1990; Abbate *et al.*, 1993; Stêtkárová *et al.*, 1993) and nerve conduction velocities (Stêtkárová *et al.*, 1993) have been suggested. However, in other studies, no relationship between exposure to toluene and reported central nervous system symptoms, clinical central nervous system signs, cardiovascular reflexes, psychological tests, electroencephalography or computerized tomography has been observed among rotogravure printers (Antti-Poika *et al.*, 1985; Juntunen *et al.*, 1985; Hänninen *et al.*, 1987) or in neuropsychological test performance among offset printers mainly exposed to either isopropanol, naphtha, acetone, xylene and hexane (Maizlish *et al.*, 1985) or acetone, isopropanol, toluene, xylene and 2-ethoxyethanol (Baird *et al.*, 1994). No deterioration of neuropsychological test performance or increase in subjective symptoms was observed in a two-year follow up study of screen printing workers exposed to a variety of solvents (toluene, methyl ethyl ketone, mineral spirits, dichloromethane, 'β-cellosolve' [probably 2-ethoxyethanol]) (White *et al.*, 1995). Decreases in the plasma concentrations of luteinizing and follicle-stimulating hormones and testosterone (Svensson *et al.*, 1992a,b) have also been reported among rotogravure printers.

Solvent-exposed printers were reported to have acquired colour vision deficit (Mergler *et al.*, 1988); however, these findings were not corroborated in another study comparing printers with binders (Baird *et al.*, 1994).

Noise-induced hearing loss was accentuated among rotogravure workers, who were exposed to toluene and high noise levels (88–98 dBA) in comparison to workers exposed to noise only (Morata *et al.*, 1993).

4.2.2 *Experimental systems*

In 30-day toxicity studies with Fischer 344/N rats and C3H mice, five males and five females of each species were given daily dermal applications (100 µL for mice, 250 µL for rats) of five different letterpress and five different offset newsprint inks on five days per week for a total of 21–22 applications. In female rats treated with either undiluted or a 3 : 1 diluted composite mixture of the letterpress inks, a decrease in body weight gain (12–14%) was observed. No such change was observed after treatment with the other inks. Scaliness at the site of application was observed in mice treated with letterpress inks (Mahler, 1992).

In a subsequent experiment, rats and mice (10 males and 10 females of each species) were given dermal applications (20 µL for mice, 50 µL for rats) twice a week for 13 weeks of the letterpress ink mixture, offset ink mixture, mineral oil used as printing ink extender, USP mineral oil (rats) or the four individual lots (mice only) of both letterpress and offset inks. Decreased body weight gain was observed in female rats treated with the printing ink mineral oil (5%) or letterpress ink mixture (9%), and increased relative liver and kidney weights (12–15%) in both male and female rats treated with USP mineral oil. In mice, body weight gain was not affected, but liver weight was increased in most groups treated with either inks or mineral oils. No histological changes were observed in the liver. Dermal scaliness and/or irritation were observed in male and female mice treated with USP mineral oil or one of the letterpress inks (Mahler, 1992). Microscopically, skin lesions were observed at the site of application in all treated groups of mice.

Potent sensitizers identified in tests with guinea pigs were trimethylol propane triacrylate, pentaerythritol triacrylate, hexanediol diacrylate, tripropylene glycol diacrylate and the prepolymers, epoxy and urethane diacrylate (Nethercott, 1978; Björkner, 1980, 1981; Nethercott *et al.*, 1983; Björkner *et al.*, 1984). Oligomeric methacrylated polyester was a moderate sensitizer in a similar experiment (Björkner, 1982).

4.3 Reproductive and developmental effects

4.3.1 *Humans*

(a) *Cohort study*

Kristensen *et al.* (1993) investigated the offspring of a cohort of 10 992 men who had been members of the Oslo (Norway) unions of compositors, lithographers and bookbinders between 1 January 1930 and 31 December 1974, and who were alive at the time of the 1960 census. A total of 6251 infants born during the period 1967–86 was identified by linking records from the printers' unions and the Norwegian Medical Births Registry. The reference group was all 118 403 births to married Oslo couples during the period 1967–86. The morbidity ratio, standardized for maternal age, year of birth, sex and birth order for total birth defects (ICD-8 codes 740–759.9) was 0.9 (95% CI, 0.8–1.0). When specific congenital anomalies were considered, the highest morbidity ratio was 1.6 (95% CI, 1.0–2.5) for cleft lip and/or palate. Further analyses in this study related to exposure to lead and solvents on the basis of job classification. On the basis of 69 job codes in the union records, each worker was categorized *a priori* into one of four exposure groups: lead only, solvents only, lead and solvents or 'other' exposure. The father's job category one year prior to the child's birth was considered in the allocation of exposures, as the a-priori hypothesis was that exposure acts on the developing paternal germ cells, that is up to three months prior to conception. None of the paternal exposure categories was associated with the risk of total birth defects. However, boys whose fathers had been exposed to lead were at increased risk of cleft lip (using the reference population of total births to married couples in Oslo, the SMR was 4.1; 95% CI, 1.8–8.1). The authors noted that they did not have an a-priori hypothesis about cleft lip, and the possibility of a chance cluster could not be excluded.

In addition to congenital anomalies, the authors considered fetal growth and birth weight, gestational age and perinatal death. A higher proportion (67/1000 births) of printers' children than that of total births in Oslo (54/1000 births) was small for gestational age; there was no consistent association with exposure category between internal analysis and analysis using the external reference group. There was no association with birth weight, and the overall distribution of gestational age was similar between printers' children and total births in Oslo. The proportion of printers' children that were delivered as early pre-term births was 6.5/1000, as compared to 8/1000 for total births to married couples in Oslo. Compared with 'other' exposures, the RR for early pre-term births (16–27 weeks) was 5.4 (95% CI, 1.7–17.4) for infants with paternal exposure to solvents and 8.6 (95% CI, 2.7–27.3) for infants with paternal exposure to lead and solvents; these RRs were adjusted for fathers' occupational status, year of birth, twin births, maternal age, birth order and sex. Compared with the external reference group, the elevated risk associated with external exposure to lead and solvents was still apparent (SMR, 2.1; 95% CI, 1.2–3.3). The corresponding SMR for exposure to solvents only was 1.3 (95% CI, 0.6–2.3). The rate of late abortion (still-born infants with gestational age of less than 28 completed weeks) was 2.7/1000 births to printers, compared with 5.3/1000 for the reference population. The rate of still births and early neonatal deaths was 16.0/1000 infants born to printers, as compared to 12.6/1000 for the reference population. The RR for total perinatal deaths was 2.4 (95% CI, 1.2–4.9) for children with fathers exposed to lead and 1.9 (95% CI, 0.96–3.7) for children with fathers exposed to lead and solvents. The association with lead only was not apparent in the comparison with the external reference group, but the SMR for paternal exposure to lead and solvents was 1.6 (95% CI, 1.2–2.2) (Kristensen et al., 1993). [The Working Group noted that the parents' smoking behaviour was unknown.]

(b) Case–control studies of congenital anomalies

In the Metropolitan Atlanta Congenital Defects Programme (United States), parents of infants with a variety of anomalies have been interviewed since 1970n about a range of exposures three to six months after the birth of their children (Edmonds et al., 1981). The occupational distribution of parents of babies with one malformation was compared with that of parents with babies with all other malformations (Erickson et al., 1979). [The period of study was not specified]. In general, the mother was the sole informant; some 85% of eligible mothers were interviewed. Information adequate for coding the industry of paternal employment at the time of conception was obtained from 76% of the total 989 women interviewed. Eleven (1.6%) of the 705 fathers for whom this information was available worked in the printing and publishing industry at the time of conception; this proportion is similar to that of all employed persons aged 16 or more as determined in the 1970 census. Analyses were made by industry, occupation and occupation–industry cross classification. Results were presented if the RR was greater than or equal to 2, associated with a p value of less than 0.05 and the exposure of at least two fathers. The only category involving paternal employment in printing for which a result was presented was for cleft palate; two fathers of affected births were clerical personnel in the printing industry. Both were mailers and handled freshly printed material. None of the fathers of infants with other types of malformations had this category of employment. In regard to maternal occupation, the reference period during which exposure was considered was the first trimester and during the period up to and including the time of conception. Two mothers of infants with Down's syndrome were managers in printing firms; none of the mothers of infants with other types of malformations were so employed. Three mothers of babies with omphalocele and gastroschisis were also employed in the printing industry (RR, 7.7; $p < 0.05$); two operated printing presses while one worked both as a binder and a press operator. Another mother employed in the printing industry operated a press and her baby was affected by microcephaly, cleft palate and limb defects.

In a large study based on the national congenital malformation system in England and Wales during the period 1974–79, information on the occupations of both mothers and fathers of births with congenital anomalies was obtained from the notification form which is completed by Area Health Authority staff (Office of Population Consensus and Surveys, 1982). About 60% of births with malformations notified had a father's and mother's occupation stated. In a total of 47 352 births with notified malformations, the father's occupation was stated; 20 778 births with malformations had a mother's occupation stated. In the analysis of paternal occupation, the analysis was performed for two separate periods — 1974–76 and 1977–79 — and births with malformations notified in those periods were compared with total births in the middle year of each period (i.e. 1975 and 1978, respectively) to produce prevalence rates at birth by father's occupation. The distribution of paternal occupation for total births was obtained from birth certificate data, so the sources of information for cases and total births differ. The malformation ratio (observed cases/expected cases) associated with paternal employment in the paper and printing industries was 0.92 in 1974–76 and 0.86 in 1977–79. Data were presented on three groups of malformations — neural tube defects, cleft lip and/or palate and Down's syndrome. No increase in the malformation ratio was apparent for any of these groups. When analysis was made according to a smaller subdivision of types of occupations, the malformation ratio associated with the father being a printer was 1.6 in the first period and 1.4 in the second period. These excesses were not statistically significant. This information was not presented for the three selected groups of malformation. As noted by the authors, the differences in sources of information for cases and total births are a potentially serious source of bias as, at birth registration, details of occupation are collected by the Registrar with the aid of several questions whereas, at notification, details of occupation are derived either from a record obtained by health workers during pregnancy or from an enquiry made to the mother by a health visitor. Thus, the information obtained from birth registration tends to be more precise than that obtained from notification, and this may cause certain occupations to be classified differently between the sources, with a differential effect on numerator and denominator. This bias may account for the nonsignificant excess of malformations in the offspring of printers. As there are no comparable data on maternal occupation from birth registration, the analysis by maternal occupation was limited to a study of the proportions of different malformations in births delivered to mothers of certain occupational groups, compared to the overall distribution of total births with malformations. For neural tube defects, the proportionate malformation ratio associated with the mother being a paper or printing worker was 1.4 during the period 1974–76, and 1.2 during the period 1977–79. For cleft lip and/or palate, the corresponding ratios were 1.2 and 1.7; for Down's syndrome, the ratios were 1.0 and 1.7, respectively. These ratios were not statistically significant.

A study of similar design was made for births with malformations notified in the period 1980–82 (McDowall, 1985). Notification may be made of malformations recorded in still-births and in live-births during the first week of life. Of a total of 40 346 births with malformation notified in England and Wales during this period, 62% (24 922) had a father's occupation stated and 28% (11 115) a mother's occupation. Standardized congenital malformation ratios for paternal occupations were calculated as the ratio of observed notifications to expected notifications, the former being adjusted for each specific type of malformation to account for the under-reporting of paternal occupation at notification. The effect of this adjustment was to make the standardized congenital malformation ratio equal to 1 rather than about 0.6 for a paternal occupation that had the same malformation notification rate as all cases for which an occupation was recorded. The number of expected notifications was derived by applying maternal age-specific rates of notified malformations to the number of births in each paternal occupation unit as determined from live- and still-birth registration. A significant positive association (standardized congenital malformation ratio, 1.4 based on 159 exposed cases) between total malformations and paternal employment as a printer was found, but the author noted that this would have been affected by the bias arising from the use of different sources of information for the numerator and the denominator. With regard to anomalies of specific types, a significant excess was found for rectal and anal atresia and stenosis (standardized congenital malformation ratio, 3.5 based on five exposed cases). No excess was found for related occupations; the standardized congenital malformation ratio was 0.4 for 'compositors', based on 11 exposed cases, 0.4 for 'electrotypers, stereotypes, printing plate and cylinder preparers' based on four exposed cases, 0.6 for 'printing machine minders and assistants' based on 26 exposed cases and 0.8 for 'screen and block printers' based on six exposed cases. As in the earlier analysis (Office of Population Censuses and Surveys, 1982), only standardized proportionate congenital malformation ratios could be calculated for maternal occupation. A significant excess was found for reduction deformities associated with maternal employment as a printer (standardized proportionate congenital malformation ratio, 10.1 based on three exposed cases). No excess was apparent for related occupations.

A large survey assessing the effect of occupational factors on pregnancy outcome in employed workers was conducted in Montréal, Canada, during the period 1982–84. A sample of 56 067 women who had just given birth or been treated for a spontaneous abortion in 11 hospitals, in which about 90% of deliveries in the city take place, were interviewed about their index pregnancies and all previous pregnancies (McDonald et al., 1987, 1988). In the main analyses relating to the women's occupation, 60 occupational groups were considered; none of the groups was related specifically to printing and related occupations. However, noting the findings of Erickson et al. (1979) relating to omphalocele or gastroschisis in employees in the printing industry, McDonald et al. (1988) noted that there was no evidence of such an association in the Montréal data. The only malformations in printing workers were two cases of Fallot's tetralogy. The women in the Montréal study were also asked about the husband's or partner's employment at the time of the first missed menstrual period for both current and past pregnancies (McDonald et al., 1989). For spontaneous abortion, the analysis was limited to those in which the women were employed for 30 h a week or more at the time of conception; there were 47 326 such pregnancies — 24 711 index and 22 615 previous. The RR of spontaneous abortion associated with the father working in 'printing operations' was 0.9 ([95% CI, 0.7–1.1]; 81 exposed cases). The analysis relating to congenital anomalies was based on 47 822 pregnancies (27 472 index and 30 350 prior) of women employed 15 h a week or more at the time of conception. This total excluded spontaneous abortions of less than 20 weeks gestation but included therapeutic abortions following prenatal diagnosis of fetal abnormality. These RRs were adjusted for maternal age, gravidity, previous miscarriage, ethnic group, educational level, smoking and alcohol consumption. Two pregnancies associated with chromosomal anomalies were observed where the father was employed in printing operations (RR, 1.8 [95% CI, 0.3–9.9]). The RR of developmental defects, defined to include neural tube defects, cleft lip and/or palate and anomalies of the heart and of the respiratory, digestive and urinary tracts, was 0.6 ([95% CI, 0.2–1.8]; 4 exposed cases). In the analysis relating to congenital anomalies, no adjustment was made for confounding variables. No excess was apparent for specific types of congenital anomaly considered and, in particular, none of the fathers of cases with cleft lip and/or palate was a printer.

Olshan et al. (1991) reported a study based on 14 415 live births with birth defects recorded in the population-based registry in British Columbia, Canada, in the period 1952–73. Two controls were matched to each case on month, year and hospital of birth. Information on paternal occupation was obtained from the birth-registration record. Potential confounding by parental age, ethnic group and outcome of previous pregnancy was considered. Twenty categories of congenital anomalies were considered in relation to 58 occupational categories. Of the fathers of infants with congenital anomalies, 1.7% (245/14 415) had been employed as printers at the time of birth of the child. This occupation was associated with RRs greater than or equal to 1.5 for five groups of anomalies, and RRs of 0.7 or less for six groups of anomaly. RRs of 2 or more were observed for obstructive renal defects (RR, 2.0; 95% CI, 0.4–9.9; 3 exposed cases), atresia of urethra (RR, 4.5; 95% CI, 1.0–20.8; 7 exposed cases) and clubfoot (RR, 2.2; 95% CI, 1.2–4.1; 19 exposed cases). The RR for cleft palate was 1.6 (95% CI, 0.3–7.1; 3 exposed cases), that for cleft lip was 1.0 (95% CI, 0.1–11.1; 1 exposed case) and that for cleft palate with cleft lip was 1.2 (95% CI, 0.3–4.5; 3 exposed cases).

Fedrick (1976) determined the RRs, defined as the rate in specific categories relative to the rate in a total population, for anencephalus associated with specific paternal occupational orders at the time of birth in a population-based study in Oxfordshire and West Berkshire, United Kingdom, during the period 1965–72. Paternal occupation could be classified to an 'order' for 88% (92 083/104 854) of births in 1965–71 (data on total births in 1972 were not included) and 85% (151/177) of the births with anencephalus. There was a statistically significant increase in risk associated with work in the paper and printing industries (RR, 1.9; 5 exposed cases). All five case fathers employed in the paper and printing industries were printers; the RR for printers was 6.7. [The Working Group noted that although a p value of less than 0.001 was reported, detailed subgroup analysis was not reported for the other occupational groups.]

Using data from a case–control study of neural tube defects in upstate New York, United States, during the period 1968–74 which was aimed primarily at assessing a possible association with oral contraceptive use and included 201 mothers of cases, Polednak and Janerich (1983) analysed parental occupations as reported on the birth records of 171 case–control pairs. The RR associated with paternal employment in the printing industry (other than as a salesman) was 0.5 (not statistically significant, based on three discordant pairs).

Ericson et al. (1988) examined the association between neural tube defects and maternal occupation in Sweden, using three sets of data. In all three data sets, the defects were ascertained from the records of the Swedish Medical Birth Registry. In the first data set, the maternal occupation recorded in the 1975 census was determined for 158 infants with neural tube defects identified during the period 1976–77 and 316 controls matched on year of birth, maternal age and parity. In the second data set, the maternal occupation recorded at the 1980 census was determined for 103 infants with neural tube defects born in 1980 and 1981 and 206 randomly selected controls; the age and parity distributions did not differ significantly between cases and controls. In the third data set, maternal occupation recorded at the first antenatal clinic visit was determined for 87 infants with spina bifida born during the period 1982–84 and 174 controls matched on year of birth, maternal age and parity. Anencephalus was not considered in this last period since few cases were born as a consequence of intensive prenatal screening. No statistically significant association with any type of maternal occupation was found in any of the three data sets or in a pooled analysis of these. The RR associated with maternal employment in the 'graphic industry' was 1.0 (1 exposed case and 2 exposed controls in pooled analysis).

Brender and Suarez (1990) examined the association between paternal occupations thought to involve exposures to solvents and anencephalus in Texas, United States, during the period 1981–86. A total of 727 cases were identified by death certificates, live-birth records and records of fetal death. Of these, 528 (72.6%) were included in the analysis, exclusions having to be made either because the record from which the case was ascertained could not be matched with a live birth record from which the information of paternal occupation was obtained, or because the recorded occupation could not be classified. Similarly, only 1160 (79.2%) of 1464 randomly selected live-born controls, frequency matched on ethnic group and year of birth, were included. The RR associated with the father being employed as a printer was 1.6 (95% CI, 0.4–5.5; 4 exposed cases).

Thus, the only group of congenital anomalies associated with printing-related work in more than one study was cleft lip and/or cleft palate (Erickson *et al.*, 1979; Kristensen *et al.*, 1993). [The Working Group noted that cleft lip with or without associated cleft palate and cleft palate without associated cleft lip are considered to be etiologically distinct.]

(c) *Case–control study of infertility*

Rachootin and Olsen (1983) examined the association between infertility and occupation in a case–control study using data collected from medical records and mailed questionnaires. A total of 1069 infertile case couples were identified from records of the Odense University Hospital in Denmark during the period 1977–80 and were compared with 4305 fertile control couples who had a healthy child born at the same hospital during the period 1977–79. Information on occupation was obtained from 927 case and 3728 control couples, representing a response rate of about 87% for each group. The analyses focused on three subgroups: (1) 258 couples in which the male partner was diagnosed as having abnormalities of sperm density, mortality or morphology; (2) 305 couples in which the female partner was diagnosed as having amenorrhoea, anovulation, luteal insufficiency or other endocrine malfunction; and (3) 129 couples with idiopathic infertility. The first two subgroups were not mutually exclusive as 48 couples were diagnosed as having both of these reproductive disorders. Analysis in relation to occupation was made restricting the control group to couples who reported that the index child had been conceived within a year of the decision to have the child and where the female partner was 20 years of age or older at the time the questionnaire was sent out. Statistically significant positive associations were found between sperm abnormalities and the male partner being employed as a typesetter (a) in the year prior to hospital admission of spouse for infertility investigation and (b) as the longest-held occupation prior to hospital admission. Odds ratios were not presented as there were fewer than 20 controls in this occupational category. There was no association with idiopathic infertility or with delayed conception, defined as conceiving more than a year after the decision was taken to have a child. In addition, there were significant positive associations between the female member of the pair having been employed as a typesetter both in the year prior to hospital admission and for whom this was the longest-held occupation prior to hospital admission. There was no association between female employment in typesetting and idiopathic infertility or delayed conception. (Again no odds ratio was presented as fewer than 20 controls were employed in this category.) A total of 55 categories of male employment and 43 categories of female employment were analysed.

(d) *Case–control study of mental retardation*

A case–control study of parental employment and mental retardation of unknown etiology was carried out in children referred to the paediatric or child neurology departments of the Nijmegen University Hospital (The Netherlands) or to rehabilitation centres in the vicinity during the period 1979–87 (Roeleveld *et al.*, 1993). A total of 340 cases with mental and psychomotor retardation (ICD codes 317–319) with an IQ of less than 80 according to the Dutch School Criteria for the Mentally Retarded were compared with 362 children with other congenital handicaps. The main diagnoses of the control children were familial neuromuscular and metabolic disorders and cerebral palsy. Data on parental occupation were obtained by interview with 306 parents of cases (participation rate, 90%) and 322 parents of controls (participation rate, 89%). Using a job–exposure matrix developed for use in England and Wales, the RR, adjusted for primigravidity, prematurity, alcohol consumption and leisure-time activities, associated with maternal occupational exposure to printing inks during the last trimester of pregnancy was 1.6 [95% CI, 0.8–3.0]. The corresponding RR associated with paternal exposure during the three months prior to conception was 0.7 [95% CI, 0.4-1.3]. When the father was asked directly about exposure to printing inks, the associated RR was 0.7 [95% CI, 0.2–2.8]. For 57% of the cases and 47% of the controls, the father provided information about his own exposure; for the remaining subjects, the mother supplied information about the father and the appropriate exposure checklists were left for him to complete. None of the mothers reported exposure to printing inks during the third trimester in response to a direct question about this. The lag time between the critical period and the interview varied from two to 25 years, with an average lag time of 10.7 years. A number of exposures that could not be identified by the job–exposure matrix approach were studied. Among these, maternal exposure in the last trimester to copying machines was associated with a RR of 3.0 [95% CI, 1.1–8.3] and exposure to correction fluid was associated with a RR of 1.7 [95% CI, 0.7–4.1]; these RRs were adjusted for primigravidity, prematurity, alcohol consumption and leisure time activities. Paternal preconceptional exposure to copying machines was associated with a RR of 1.5 [95% CI, 0.8–2.8].

4.3.2 *Experimental systems*

Groups of 10 male and 10 female Fischer 344/N rats, seven to eight weeks of age, were given dermal applications to hair-clipped skin of 50 µL neat letterpress or offset newsprint inks, or of mineral oil vehicles at twice-weekly intervals for 13 weeks (Mahler, 1992). Control rats were hair-clipped but untreated. The letterpress and offset ink mixtures were prepared by mixing four separate lots of each type, each lot representing a batch from each of the four major manufacturers in the United States. There were no effects of any of the applications on epididymal sperm motility, sperm density, testicular spermatid head count, vaginal cytology or oestrus cycle length.

4.4 Genetic and related effects

The genotoxicity of various printing ink components, e.g. solvents, such as toluene and benzene, and dyes, has been reviewed (Dean, 1985; WHO, 1986; IARC, 1987l, 1989b,c, 1993b,c,d; Snyder *et al.*, 1993; WHO, 1993; McGregor, 1994; Snyder & Kalf, 1994).

Azo dyes are used extensively in the printing industry. A review on the mutagenicity of 84 azo dyes has been given by Chung and Cerniglia (1992). Magenta and CI Basic

Red 9 (preparation of printing inks) and CI Pigment Red 53:1 and CI Pigment Red 3 (preparation and in printing ink) have been evaluated by IARC (1993b,c,d).

4.4.1 *Humans*

(a) *Urinary mutagenicity*

Crebelli *et al.* (1985) studied urinary mutagenicity in workers employed in the use of an IBM Printing System 3800, which implies exposure to small amounts of 2,4,7-trinitro-9-fluorenone. Urine samples were collected in the morning after at least three workdays from 22 exposed white men (11 smokers, 11 nonsmokers) and 18 white men (7 smokers, 11 nonsmokers) working in the same bank, but only occasionally in contact with the printing system. Analytical measurements showed the presence of 2,4,7-trinitro-9-fluorenone residues in both printed and photocopied sheets (0.4–4 µg in 21 × 27 cm sheets) and in the surrounding air (0–0.4 µg/m^3). Urine samples were concentrated by adsorption onto XAD-2 resin and elution with acetone. Mutagenicity was studied both by the plate incorporation assay and the microtitre fluctuation assay using *Salmonella typhimurium* TA98. Pure 2,4,7-trinitro-9-fluorenone was highly mutagenic, but urine concentrates of both exposed and presumed unexposed workers were mostly negative. A few concentrates were significantly mutagenic but with no relationship to exposure (one among exposed and four among controls). In the presence of an exogenous metabolic activation system, five of eight concentrates from both unexposed and exposed smokers were mutagenic, whereas the urine of nonsmokers was completely inactive.

(b) *Cytogenetic damage in lymphocytes*

Forni *et al.* (1971) studied chromosomal aberrations in peripheral lymphocytes in 34 male workers employed in a rotogravure plant, 10 of whom were exposed to both benzene (for 1–22 years) and toluene (for 12–14 years) and 24 to toluene only (for 3–15 years). Each worker was matched with a healthy control of the same sex and approximately the same age with no history of exposure to benzene or its homologues or to therapeutic or occupational X-irradiation. There was no control for tobacco smoking or alcohol consumption, but people having had a recent viral disease or vaccination were excluded. The concentrations of benzene and homologues of benzene expressed as concentrations of benzene were 131 ppm [419 mg/m^3] in the centre of the rotary machine room, 125 ppm [400 mg/m^3] near the windows, 363 ppm [1162 mg/m^3] near a folding machine and 532 ppm [1702 mg/m^3] in the auxiliary room. After a benzene-poisoning epidemic, only toluene was used as an ink diluent. Concentrations of toluene in different parts of the working department varied between 56 and 824 ppm [211 and 3106 mg/m^3] up to 1966, when the plant was moved and better ventilation was obtained (156 ppm [588 mg/m^3] near folding machines and 265 ppm [999 mg/m^3] between machine elements). In the benzene–toluene group, a statistically significant increase in the incidence of unstable chromosomal aberrations was found (1.66% compared with 0.61% in the controls). This was not found in the toluene-exposed group.

Funes-Cravioto *et al.* (1977) examined eight rotoprinting factory workers exposed to benzene during the 1940s and to toluene from 1950 onwards for periods ranging from two to 26 years. Air concentrations of benzene were not measured. Air concentrations of toluene showed time-weighted average values of 100–200 ppm [377–754 mg/m^3] with occasional rises up to 500–700 ppm [1885–2639 mg/m^3]. The frequency of chromosome breaks was 11.9% (5.6% in controls). No increase in sister chromatid exchange frequency was found in the four workers examined in this group. In 14 other workers exposed to toluene only (1.5–26 years), the frequency of chromosome breaks was 8.9%. [The Working Group noted that statistical evaluations were not made on the separate groups.]

Mäki-Paakkanen et al. (1980) studied 32 men working in two different rotoprinting factories and 15 control men from a research institute. Toluene was the only solvent used in the two factories. Concentrations of benzene, which is a contaminant, had been checked regularly since 1962 and had always been below 0.05% (average, 0.006%) as measured by gas chromatography. The personal toluene exposures of the printers were calculated as time-weighted averages for an 8-h work day and were between 7 and 112 ppm [26–420 mg/m^3]. The cytogenetic study showed little difference between the workers and the controls (chromatid type, 1.0 versus 0.7%; chromosome type, 0.5 versus 0.9%; sister chromatid exchange, 8.5 versus 8.9 per cell). A higher frequency of sister chromatid exchange was observed among smokers both in the group of workers and in the controls. No effect of smoking was observed as regards chromosomal aberrations.

Bauchinger et al. (1982) examined 20 men working for more than 16 years on a rotogravure machine. The continuously measured toluene concentration in the air of the rotary machine was between 200 and 300 ppm [754 and 1131 mg/m^3]. The toluene used contained < 0.3% benzene. From the same plant, a group of 24 unexposed workers [sex unspecified] was used as controls. There were 12 smokers and eight nonsmokers in the exposed group, and nine smokers and 15 nonsmokers in the control group. In the exposed group, a significantly larger number of peripheral lymphocytes with structural chromosomal changes were found: chromatid breaks (0.0036 versus 0.0019 per cell) and chromatid exchanges (0.0015 versus 0.0004 per cell), but not acentrics or dicentrics. The yield of gaps was also significantly increased (0.0248 versus 0.019 per cell). Significantly higher levels of sister chromatid exchange were observed in the nonsmoking rotogravure workers compared with nonsmoking controls (8.55 versus 7.75 per cell). This was also the case for toluene-exposed smokers compared with control smokers (10.33 versus 8.89 per cell). In both groups, smokers had significantly higher sister chromatid exchange values than nonsmokers.

A second study of chromosomal changes in peripheral lymphocytes was made on 27 workers from the same factory who had not been exposed to toluene for four months to five years (Schmid et al., 1985). These 27 workers were subdivided into two groups: those without exposure for up to two years, and those without exposure for more than two years. They were compared with 26 controls. The exposure conditions had been the same as in the first study. The group of 13 workers without exposure for up to two years still showed a significantly larger number of cells with structural chromosomal aberrations (S-cells; 0.69 versus 0.49%) and chromatid-type aberrations (but not gaps) than the controls (0.39 versus 0.20/100 cells). No significant difference was observed in the group of 14 workers without exposure for more than two years compared with the controls. In all groups, smokers showed a significantly higher incidence of sister chromatid exchange than nonsmokers, but the two post-exposure groups did not differ from the controls as regards incidence of sister chromatid exchange.

Chromosomal aberrations in peripheral lymphocytes of rotogravure printing plant workers [sex unspecified] were studied in two groups of exposed persons by Pelclová *et al.* (1990): 42 persons (37 smokers, 5 nonsmokers) who had been exposed to highly purified toluene (104–1170 ppm; 390–4380 mg/m^3) and to printing dyes for an average of 12 years; 28 office and technical employees (17 smokers, 11 nonsmokers) of the same printing plant exposed to low levels of toluene (2.1–4.3 ppm; 8-16 mg/m^3) and working on average 2 h per day in the rotary machine room. A control group of 32 persons (17 smokers, 15 nonsmokers) was selected from offices in a nearby brewery and dairy. Exposure was assessed by measuring toluene in blood at the end of the working shift (124.0 and 10.3 μmol/L in the two exposed groups, respectively; not measured in the control group). The percentages of aberrant cells were 3.6 and 3.3 in the exposed groups and 2.1 in the control group. There were significantly higher proportions of chromatid breaks per cell in both exposed groups and gaps and chromosomal exchanges in the low toluene-exposed group only, whereas the proportions of chromosomal breaks per cell were not different between the three groups. A small difference in the percentage of cells with aberrations was seen between smokers and nonsmokers (3.4 versus 2.4%; $p < 0.05$). Smoking printers had 3.8% aberrant cells, smoking office and technical employees had 3.4% and smoking controls had 2.5%. The difference between smoking printers and smoking controls was significant ($p < 0.05$).

Micronuclei and chromosomal aberrations in lymphocytes have been studied in 21 men employed as rotogravure printers by Nise *et al.* (1991). As controls, 21 workers were selected from a margarine factory with no exposure to solvents or other components of the printing trade. The median time of exposure was 25 years; median weekly individual exposure measured in 1986 was 150 mg/m^3 when the median blood toluene content was 1.6 μmol/L (measured between 11.00–14.00 h in the middle of a working week). For lymphocytes stimulated with pokeweed mitogen, there was a significantly increased incidence of micronuclei in the printers compared with the controls, when allowance was made for age and smoking (2.8 versus 1.5/1000 cells). There was also a significant difference in the incidence of small micronuclei for which the size ratio micronucleus:main nucleus was ≤ 0.03 (1.0 versus 0.3/1000 cells) and, in the exposed group, an association between blood toluene and small micronuclei (0.17/1000 cells/μmol toluene/L blood; $p = 0.0005$). No difference was found with lymphocytes stimulated with phytohaemagglutinin (2.8 versus 2.5/1000 cells). There was no significant difference in chromosome breaks between the exposed and the control group, but there was an association between chromosome breaks and earlier benzene exposure. [The Working Group noted that there is considerable variation in micronucleus frequency in pokeweed mitogen-stimulated lymphocytes. The relevant historical database from the same laboratory gives frequencies of 1.5 (this study), 3.4, 6.0, 7.0 and 2/1000 pokeweed mitogen-stimulated cells (Högstedt *et al.*, 1988; Hagmar *et al.*, 1989; Högstedt *et al.*, 1991; Nise *et al.*, 1991; Holmén *et al.*, 1994).]

Lead concentrations in urine and the frequency of chromosomal aberrations and sister chromatid exchange in lymphocytes were studied in 84 printers exposed to lead (Chen et al., 1992). A linear correlation was found between urinary lead concentrations and the frequency of chromosomal aberrations and sister chromatid exchange.

Richer et al. (1993) exposed five male volunteers to 50 ppm [187.5 mg/m^3] toluene, which is a prominent solvent in the printing industry, in a controlled exposure chamber for 7 h per day for three days on three occasions at two-weekly intervals. Blood samples were taken before and after exposure. No effect upon sister chromatid exchange frequencies was observed in peripheral lymphocytes.

4.4.2 *Experimental systems*

Toluene also shows little mutagenic potential *in vitro*, while *in vivo* there have been some reports of the induction of both micronuclei and chromosomal aberrations in rodent bone-marrow cells. These responses may be largely due to contamination of the toluene with benzene, and it has been suggested that no definitive conclusion can be reached at present regarding the genetic activity of toluene in rodents (McGregor, 1994). Most studies have failed to show any mutagenic potential of benzene *in vitro*. Positive responses have been reported in *Salmonella typhimurium* TA100 in a microfluctuation test (Glatt et al., 1989) and in three studies with the Syrian hamster embryo cell transformation assay (IARC, 1987l). The mutagenicity of the benzene metabolites *trans*-1,2-benzene dihydrodiol, benzene diol epoxides, 1,2,4-trihydroxybenzene, hydroquinone, catechol and *trans,trans*-muconaldehyde has been demonstrated (reviewed by Snyder et al., 1993). Many in-vivo studies have demonstrated that benzene is clastogenic in assays for the induction of rodent bone-marrow cell micronuclei and chromosomal aberrations (IARC, 1987l). However, covalent association of benzene with cellular DNA occurs at a very low level, which has been difficult to detect *in vivo* (Snyder et al., 1993).

Co-exposure of BDF1 mice to toluene and benzene has been shown to reduce the DNA damage in peripheral leukocytes, bone-marrow cells and liver cells induced by benzene alone (Plappert et al., 1994). An interaction between benzene and toluene resulting in a reduced response to benzene has also been described with cytogenetic endpoints in mice and rats (reviewed by McGregor, 1994).

Milvy and Kay (1978) studied the mutagenicity of 19 dyes used in the graphic arts and printing industry with the *S. typhimurium*/mammalian microsome reversion test system. Two of these were found to be mutagenic. Para red [1-(4-nitrophenyl)azo-2-naphthol] was mutagenic in TA98 and TA1538 when activated with S9 homogenate. Dinitroaniline orange was a direct mutagen with TA98. No mutagenicity was found with lithol red (CI 15630), alkali blue (CI 42750), cadmium red (CI 77196), naphtol red (CI 12315), phthalocyanine green (CI 74260), elemental aluminium, Red Lake C (CI 15585) (see IARC, 1987o), rhodamine (CI 45160) (see IARC, 1987p), diarylide orange (CI 21110), red 2B (CI 15865), Hansa yellow (CI 11680), phthalocyanine blue (CI 74160), diarylide yellow (CI 21090), fire red (CI 12085), molybdate orange (CI 77605), lithol rubine (CI 15850) or iron blue (CI 77510).

Møller et al. (1983) found that extracts of typewriter ribbons and carbon papers were mutagenic in the *Salmonella*/microsome assay with strain TA98. A sample of black ink used for a word-processing system was also shown to be mutagenic. Fractionation of the ribbon extracts indicates that at least two to three different classes of mutagenic components were present in the extracts.

The mutagenicity of letterpress and offset black newsprint ink (see section 1.3) has been tested under the United States National Toxicology Program (Mahler, 1992). The inks were tested in a preincubation protocol at concentrations of 100–10 000 μg/plate in *S. typhimurium* strains TA100 and TA98 with and without rat and Syrian hamster liver S9. In the absence of S9, none of the inks was mutagenic. In the presence of 30% hamster liver S9, both inks were mutagenic in both *Salmonella* strains. In the presence of rat liver S9, both inks were non-mutagenic with strain TA100, but induced mutations in strain TA98.

The azo dye D & C Red No. 9 is used for manufacturing printing inks. It has previously been tested in several in-vitro and in-vivo test systems and has been shown to be non-mutagenic (see IARC, 1993b). Dillon et al. (1994) tested the dye in a specific *Salmonella* mutagenicity assay, incubating the test compound with a rat caecal preparation under anaerobic conditions to reduce the azo bond. Ethyl acetate extracts of this incubation mixture gave dose-related mutagenic responses in TA100 and a weak response in TA98 when incubated with rat liver S9. The presumed major reduction product, 1-amino-2-naphtol, was mutagenic to TA100, but not TA98, with S9. The authors suggest that the previous non-mutagenicity of this dye may have been due to insufficient reductive cleavage.

5. Summary of Data Reported and Evaluation

5.1 Exposure data

Printing inks are mixtures of three main types of ingredients: pigments, vehicles and additives. Pigments used in printing inks include both inorganic pigments such as carbon black and titanium dioxide and organic pigments, which are frequently dyes rendered insoluble by complexing with a metal ion. Most organic pigments are prepared from azo, anthraquinone and triarylmethane dyes, and phthalocyanines.

There are five main printing processes, and inks are designed for the specific process. Lithography and letterpress are collectively known as the 'paste ink' processes and use inks that are essentially non-volatile at normal temperatures. Flexography and gravure are known as the 'liquid ink' processes and are based upon volatile solvents that evaporate readily at room temperatures. Screen printing uses inks that fall between the other two groups.

Choice of the vehicle (solvent with resins) for a printing ink depends on the printing process, how the ink will be dried, and the substrate on which the image is to be printed. In lithography and letterpress, where inks are dried by absorption and oxidation, vehicles are generally mixtures of mineral and vegetable oils and resins. Flexographic inks, which are designed to dry quickly by evaporation, can be either water-based or based on organic solvents (such as ethanol, ethyl acetate, n-propanol or isopropanol) with a wide variety of resins. Vehicles for gravure inks, which also dry by evaporation, may also contain aromatic or aliphatic hydrocarbons and ketones as solvents. Inks for screen printing use organic solvents that are somewhat less volatile than those used for flexography or gravure (higher glycol ethers and aromatic and aliphatic hydrocarbons). Additives in inks include driers, waxes and plasticizers.

Ultraviolet radiation-cured inks, commonly based on acrylates, are used in all of the printing processes to varying degrees.

The manufacture of inks consists of dissolving or dispersing resins in organic solvents or oils to produce the vehicle (varnish), mixing and dispersing the pigment or dye into the vehicle, introduction of any additives and packaging. Some or all of these stages may be done manually or automatically in a batch process or as a continuous process.

During the manufacture of printing inks, exposure to pigments, vehicles and additives can occur through inhalation or skin contact during mixing and dispersion and during clean-up of mixers. Exposures are higher with liquid inks than with paste inks. During newspaper printing by letterpress or lithography, the major exposure is to ink mist. Rotary letterpress was the dominant process for the production of newspapers until the 1970s. It has now been largely replaced by web offset litho, in which exposures to ink mist are considerably lower than for letterpress. In other lithographic and letterpress printing, the major exposure is to hydrocarbon-based cleaning solvents and isopropanol from damping solutions. In flexographic, gravure and screen printing, exposures are mainly to organic solvents. Historically, some workers in both ink manufacture and printing were exposed to much higher levels of lead, polycyclic aromatic hydrocarbons and benzene than today, and the development and use of modern control technologies have made possible the marked reduction in solvent and ink mist exposures.

5.2 Human carcinogenicity data

A large volume of epidemiological data deals with potential cancer risks in printing processes. Because of the presence of a fairly large number of adequate cohort and case–control studies, it was considered that there was no marginal benefit in considering further the descriptive studies based on simple tabulations of death certificate causes of death and mentions of occupation. In any case, these latter studies did not provide clear patterns of results.

The evaluation of results of case–control and cohort studies, in particular those regarding relatively rare neoplasms, was hampered by the possibility of reporting or publication bias. A second problem was the poor specificity of exposure information. While most studies were based on crude designations of the exposure variable, a few, most notably some of the cohort studies, did describe risks for subgroups of the printing industry that are more homogenous in exposure circumstances. The Working Group tried to identify such subgroup studies with presumably more well-defined common exposure circumstances. In a small number of studies, there was an explicit attempt to identify a group of workers exposed to printing inks. This, like the designation of exposure on the basis of the job or industry title, is of poor specificity. Further, most of these were in the context of community-based case–control studies, and the attribution of exposure was based on job–exposure matrices, which do not discriminate among subsectors of the printing industry.

A third problem was that most of the cohort and record-linkage studies had no information on some important counfounders, notably cigarette smoking. It has previously been shown that confounding by smoking is unlikely to distort the relative risk estimate between occupation titles and lung cancer by more than 30%. For other sites that are affected by smoking, the maximal bias is likely to be even lower. The Working Group considered these possible biases when interpreting results.

Apart from cancers of the lung, oropharynx, urinary bladder and kidney and leukaemia, which are presented below, the Working Group considered that the findings are not strong or consistent enough to be evaluated.

Lung cancer

Ten community-based case–control studies examined the relationship between lung cancer and occupation and reported results regarding printing industry and/or printing related occupations. Increased relative risks were found in eight studies; smoking was controlled for in six of them and the smoking-adjusted relative risks for 'printing occupations' ranged from 1.1 to 3.3. Two studies reported findings for exposure to printing inks: both found a positive association. A Canadian study found that a small excess of lung cancer detected in printers as a whole was concentrated in printing process workers and was very high for adenocarcinoma of the lung in particular.

Six census-based record-linkage studies reported results for lung cancer. The Swedish study found a statistically significant 60% excess of lung cancer in blue-collar workers in printing enterprises. One study in Denmark showed an increased risk only for women employed in printing, publishing and allied industries. A subsequent Danish study revealed a slight, statistically significantly increased risk in printing and bookbinding industry workers; the risk was higher and still significant in workers employed in newspaper and magazine production. The Finnish study and the two Italian studies did not detect a statistically significant increased risk for lung cancer related to printing occupations.

Among the industry-based studies, five proportionate mortality studies evaluated the risk for lung cancer. In one study, statistically significantly increased risks were found among printing trade workers in two different areas of the United Kingdom. A subsequent, partially overlapping study in London, United Kingdom, reported a statistically significant 30% excess of lung cancer among newspaper printing workers. A third study examined United States newspaper and commercial pressmen separately; neither newspaper nor commercial pressmen showed an increased mortality from lung cancer. The other two proportionate mortality studies in printing workers in the United States failed to show an increased risk for lung cancer.

Among the industry-based studies, seven standardized mortality ratio studies examined lung cancer risk in printing trade workers. Respiratory cancer was elevated in four studies; in none was smoking controlled. An increased risk for lung cancer was found in the Italian cohort of newspaper workers. The historical cohort of trade union members in the United Kingdom printing industry had a statistically significantly increased risk for lung cancer among mainly unskilled workers in newspaper letterpress printing (machine assistants). Newspaper web pressmen in Los Angeles, United States, showed a nonsignificantly increased lung cancer risk. A Swedish study of rotogravure printers revealed increased risk from respiratory cancers.

In addition, in a cohort of United States Army veterans, an increased smoking-adjusted relative risk for respiratory cancer was found in 'printing pressmen and plate printers'.

In some studies, it was possible to separate newspaper printing workers from other less-well defined employment in the printing industries. Seven cohort studies examined lung cancer risk in workers employed in the newspaper printing industry, mainly during the period 1945–1970. Three were proportionate mortality studies, two of United Kingdom newspaper printing companies and one that examined United States newspaper pressmen and commercial pressmen separately. Four were standardized mortality ratio studies of newspaper web pressmen in Los Angeles, United States, of newspaper printers in two plants in New York City, United States, of newspaper workers in one Italian plant and of newspaper machine assistants who were trade union members in the printing industry in the United Kingdom. Five of the seven cohort studies reported increased relative risk estimates ranging from 1.2 to 1.5, of which three were statistically significant. In none of them was smoking taken into account. In addition, a Danish record-linkage study found a two-fold, statistically significantly increased risk for workers in newspaper and magazine production whose typical job was the operation of rotary letterpress machines.

Cancer of the oropharynx

The risk for cancer of the buccal cavity and pharynx was examined in three case–control studies in the United States. One study on multiple cancer sites showed a smoking-adjusted elevated risk in printing workers. Another similar study detected a significantly high smoking- and alcohol-adjusted risk for cancer of the oral cavity in workers in the printing and publishing industry. A third case–control study on oropharyngeal cancer did not find an increased risk among men employed in the printing industries, whereas a nonstatistically significantly increased risk was found for women. A nonstatistically significantly increased risk was found in male workers in printing and bookbinding industries in a Danish record-linkage study.

Four cohort studies reported results for cancer of the buccal cavity and pharynx. The United States study that examined separately newspaper pressmen and commercial pressmen found a higher than two-fold statistically significantly increased risk in newspaper pressmen only. Increased risks were not found in a standardized mortality ratio study of newspaper pressmen in Los Angeles nor in a cohort of newspaper printers in two plants in New York City. The cohort of trade union members in the United Kingdom printing industry found increased risks in non-production workers (editorial and clerical staff) only.

Urinary bladder cancer

Thirty-five studies have reported findings for urinary bladder cancer and employment in the printing industry. A positive association between urinary bladder cancer and either a printing occupation or employment in the printing industry was reported in 14 of the 23 case–control studies. The range of relative risk estimates derived from these studies was from 1.1 to 5.6. These associations were statistically significant in only three of the case–control studies. Generally, the interpretation of the case–control studies was limited by their use of broad job and industry categories such as printers or the printing industry. One study in Spain reported a nonsignificantly increased relative risk for typesetters and linotypists.

Six cohort studies and six record-linkage studies have also been reported. Increased rate ratios were reported in five of the cohort studies of workers in the printing industry. However, the rate ratios in two of them were close to 1.0 and a statistically significant increase was found only for printing pressmen and plate printers in the cohort of United States Army veterans. Similarly, the relative risk estimates derived from the record-linkage studies were close to unity and the only ones that achieved statistical significance were from the two Danish studies: one study found an increased risk among men employed in printing and bookbinding industries and the other among men employed in printing, publishing and allied industries. Five case–control studies reported results for exposure to printing inks and urinary bladder cancer risk was observed to be elevated in all five studies.

Cancer of the kidney

Slight to moderate excesses of cancer of the kidney have been reported in the printing industry in five industry-based studies in different cohorts in the United States, and in two record-linkage studies in Italy and Sweden. Ten cohort studies did not report results for cancer of the kidney at all. Four case–control studies, one nested in a cohort of paperboard printing workers and three representing different populations in three continents reported odds ratios ranging from 1.3 to 16.6. Most of these were not statistically significant. By far the most powerful case–control study, a multicentric study conducted in Australia, Denmark, Germany, Sweden and the United States, reported a 30% non-significant excess associated with employment in printing and graphic industry.

Leukaemia

Results regarding leukaemia risk in printing workers have been reported in one case–control study and seven cohort studies. The case–control study found a nonstatistically significantly increased risk for printers. Significantly increased risks were found in two cohort studies.

The proportionate mortality study that examined newspaper pressmen and commercial pressmen separately found a 60% excess of leukaemia risk only in newspaper pressmen. In another proportionate mortality study in printing workers in the United States, a statistically significantly increased risk for leukaemia was detected primarily among bindery workers who may have had exposure to benzene. Newspaper web pressmen in Los Angeles, United States, also showed a higher than two-fold increased risk. A Swedish study of rotogravure printers revealed an increased risk for leukaemia, although this was based on a very small number of cases. Both newspaper web pressmen and rotogravure printers may have been exposed to benzene and other organic solvents in the past. Three other cohort studies in newspaper printing workers in London, United Kingdom, commercial pressmen in the United States and newspaper printers in New York City failed to show an increased risk.

Overall, notwithstanding the variability in the results, there are indications of excess risks among printing process workers for some types of cancer. In its evaluation of these data, the Working Group considered the likelihood of publication bias, the possibility of confounding by cigarette smoking, and the imprecision and inconsistency of the designation of exposure groups. Based on these considerations, the Working Group concluded that there is weak evidence of an increased risk of lung and urinary bladder cancers among workers in the printing industry.

While there was a suggestion of an increased risk of lung and urinary bladder cancers in relation to exposure to printing inks, the quality of the data was weak.

The Working Group noted that the vast majority of epidemiological studies covered workers who were in the printing industry in North America or Europe during the middle of the twentieth century. Very few of the studies included workers whose employment was after 1980. Given the rapid technological changes that have gone on in this industry in North America and Europe in the past decade, it is questionable whether the exposure circumstances that were prevalent in the past are still prevalent. However, there may be areas of the world in which the older processes are still prevalent. Where the technologies have substantially changed from those of the past and insofar as this has changed the exposure conditions, the present evaluation may not be relevant.

5.3 Animal carcinogenicity data

Twenty-two different printing inks were tested for carcinogenicity in one study in mice by subcutaneous injection. The study was inadequate for evaluation.

5.4 Other relevant data

No consistent association between employment in printing trades and morbidity from non-malignant diseases has been observed. Solvent-induced central nervous system damage has been observed in several but not all studies on employees in printing trades. Ultraviolet radiation-cured printing inks are a frequent cause of allergic contact dermatitis.

One study suggested that occupational exposures may induce hepatic damage in printers, but several other studies failed to confirm this finding.

An early report of an increased risk of anencephalus associated with paternal employment in printing has not been confirmed in subsequent studies of neural tube defects. In two studies, an association between this exposure and cleft lip and/or palate has been observed. However, in one of these, the association was apparent only for cleft palate, and in the other only for cleft lip, and no noteworthy association has been observed in a further three studies. In a single study in rats, dermal exposure to newspaper inks had no effect on sperm numbers or motility, vaginal cytology or oestrus cycle length.

Several pigments and dyes used in printing inks are mutagenic in *Salmonella typhimurium*: para red, dinitroaniline orange, azo dye D & C Red No. 9.

An increased frequency of chromosomal aberrations in peripheral lymphocytes in printing workers exposed to *inter alia* toluene was found in two studies, but not in two other studies. In one study, an increased frequency of chromosomal aberrations was found in workers exposed to toluene and benzene. In one study of a group exposed to toluene, an increased frequency of sister chromatid exchange was found, but not in two other studies. In one study of printers exposed to lead, increased frequencies of chromosomal aberrations and sister chromatid exchange were found. In one study, an increased frequency of micronuclei was observed in printing workers exposed to toluene. In one study of volunteers exposed to toluene, no increase in sister chromatid exchange was observed in lymphocytes.

5.5 Evaluation[1]

There is *limited evidence* that occupational exposures in printing processes are carcinogenic.

There is *inadequate evidence* for the carcinogenicity in humans of printing inks.

There is *inadequate evidence* for the carcinogenicity in experimental animals of printing inks.

Overall evaluation

Occupational exposures in printing processes are *possibly carcinogenic to humans (Group 2B)*.

Printing inks are *not classifiable as to their carcinogenicity to humans (Group 3)*.

---

[1] For definition of the italicized terms, see Preamble, pp. 24–27.

6. References

Abbate, C., Giorgianni, C., Munaò, F. & Brecciaroli, R. (1993) Neurotoxicity induced by exposure to toluene. An electrophysiologic study. *Int. Arch. occup. environ. Health*, **64**, 389–392

American Conference of Governmental Industrial Hygienists (1995) *1995–1996 Threshold Limit Values (TLVs) for Chemical Substances and Physical Agents and Biological Exposure Indices (BEIs)*, Cincinnati, OH

American Newspaper Publishers Association (1988) *Newspaper Inks*, Washington DC

Antti-Poika, M., Juntunen, J., Matikainen, E., Suoranta, H., Hänninen, H., Seppäläinen, A.M. & Liira, J. (1985) Occupational exposure to toluene: neurotoxic effects with special emphasis on drinking habits. *Int. Arch. occup. environ. Health*, **56**, 31–40

Aronson, K.J. & Howe, G.R. (1994) Utility of a surveillance system to detect associations between work and cancer among women in Canada, 1965–1991. *J. occup. Med.*, **36**, 1174–1179

Ashworth, J. & White, I.R. (1991) Contact allergy to ethoxylated phenol (Short communication). *Contact Derm.*, **24**, 133–134

Atkinson, D. (1995) Environmental compliance using cationic UV ink systems. *FlexoTech*, **April/May**, 26–28

Baelum, J. (1990) *Human Solvent Exposure: Factors Influencing the Pharmacokinetics and Acute Toxicity* (Institute of Environmental and Occupational Medicine Report), Aarhus, Denmark, University of Aarhus

Baelum, J., Andersen, I. & Mølhave, L. (1982) Acute and subacute symptoms among workers in the printing industry. *Br. J. ind. Med.*, **39**, 70–75

Baird, B., Camp, J., Daniell, W. & Antonelli, J. (1994) Solvents and color discrimination ability. Nonreplication of previous findings. *J. occup. Med.*, **36**, 747–751

Bassemir, R.W., Bean, A., Wasilewski, O., Kline, D., Hillis, W., Su, C., Steel, I.R. & Rusterholz, W.E. (1995) Inks. In: Kroschwitz, J.I. & Howe-Grant, M., eds, *Kirk-Othmer Encyclopedia of Chemical Technology*, 4th Ed., Vol. 14, New York, John Wiley & Sons, pp. 482–503

Bauchinger, M., Schmid, E., Dresp., J., Kolin-Gerresheim, J., Hauf, R. & Suhr, E. (1982) Chromosome changes in lymphocytes after occupational exposure to toluene. *Mutat. Res.*, **102**, 439–445

Baxter, P.J. & McDowall, M.E. (1986) Occupation and cancer in London: an investigation into nasal and bladder cancer using the Cancer Atlas. *Br. J. ind. Med.*, **43**, 44–49

Beaulieu, H.J. & Anderson, D.O. (1978) Newspaper production is plagued by excessive noise and aerosols. *Occup. Health Saf.*, **Sept/Oct**, 62–70

Benhamou, S., Benhamou, E., Tirmarche, M. & Flamant, R. (1985) Lung cancer and use of cigarettes: a French case–control study. *J. natl Cancer Inst.*, **74**, 1169–1175

Benhamou, S., Benhamou, E. & Flamant R. (1988) Occupational risk factors for lung cancer in a French case–control study. *Br. J. ind. Med.*, **45**, 231–233

Bergmark, G. & Meding, B. (1983) Allergic contact dermatitis from newspaper print (Short communication). *Contact Derm.*, **9**, 330

Bertazzi, P.A. & Zocchetti, C. (1980) A mortality study of newspaper printing workers. *Am. J. ind. Med.*, **1**, 85–97

Bertazzi, P.A., Zocchetti, C., Della Foglia, M., Guercilena, S. & Riboldi, L. (1979) Mortality experience among newspaper workers. *Med. Lav.*, **6**, 421–437 (in Italian)

Beurey, J., Mougeolle, J.-M. & Weber, M. (1976) Cutaneous manifestations associated with resins used in printing. *Ann. Dermatol. Syphiligr.*, **103**, 423–430 (in French)

Birch, J.M., Mann, J.R., Cartwright, R.A., Draper, G.J., Waterhouse, J.A.H., Hartley, A.L. Johnston, H.E., McKinney, P.A., Stiller, C.A. & Hopton, P.A. (1985) The inter-regional epidemiological study of childhood cancer (IRESCO). Study design, control selection and data collection. *Br. J. Cancer*, **52**, 915–922

Birch, J.M., Hartley, A.L., Teare, M.D., Blair, V., McKinney, P.A., Mann, J.R., Stiller, C.A., Draper, G.J., Johnston, H.E., Cartwright, R.A. & Waterhouse, J.A.H. (1990) The inter-regional epidemiological study of childhood cancer (IRESCC): case–control study of children with central nervous system tumours. *Br. J. Neurosurg.*, **4**,17–26

Björkner, B. (1980) Allergenicity of trimethylol propane triacrylate in ultraviolet curing inks in the guinea pig (Short report). *Acta derm.-venereol.*, **60**, 528–531

Björkner, B. (1981) Sensitization capacity of acrylated prepolymers in ultraviolet curing inks tested in the guinea pig. *Acta derm.-venereol.*, **61**, 7–10

Björkner, B. (1982) Sensitization capacity of polyester methacrylate in ultraviolet curing inks tested in the guinea pig. *Acta derm.-venereol.*, **62**, 153–182

Björkner, B. & Dahlquist, I. (1979) Contact allergy caused by UV-cured acrylates (Short communication). *Contact Derm.*, **5**, 403–404

Björkner, B., Dahlquist, I. & Fregert, S. (1980) Allergic contact dermatitis from acrylates in ultraviolet curing inks. *Contact Derm.*, **6**, 405–409

Björkner, B., Niklasson, B. & Persson, K. (1984) The sensitizing potential of di-(meth)acrylates based on bisphenol A or epoxy resin in the guinea pig. *Contact Derm.*, **10**, 286–304

Blair, A., Linos, A., Stewart, P.A., Burmeister, L.F., Gibson, R., Everett, G., Schuman, L. & Cantor, K. P. (1993) Evaluation of risks for non-Hodgkin's lymphoma by occupation and industry exposures from a case–control study. *Am. J. ind. Med.*, **23**, 301–312

Blot, W.J. & Fraumeni, J.F., Jr (1978) Geographic patterns of bladder cancer in the United States. *J. natl Cancer Inst.*, **61**, 1017–1023

Bœwer, C., Enderlein, G., Wollgast, U., Nawka, S., Palowski, H. & Bleiber, R. (1988) Epidemiological study on the hepatotoxicity of occupational toluene exposure. *Int. Arch. occup. environ. Health*, **60**, 181–186

Bowles, R.F. (1939) Developments in ink drying. *Penrose Annual*, **41**, 162–165

Brender, J.D. & Suarez, L. (1990) A brief original contribution. Paternal occupation and anencephaly. *Am. J. Epidemiol.*, **131**, 517–521

Brownson, R.C., Chang, J.C. & Davis, J.R. (1987) Occupation, smoking, and alcohol in the epidemiology of bladder cancer. *Am. J. public Health*, **77**, 1298–1300

Brownson, R.C., Hoar Zahm, S., Chang, J.C. & Blair, A. (1989) Occupational risk of colon cancer. An analysis by anatomic subsite. *Am. J. Epidemiol.*, **130**, 675–687

Brownson, R.C., Reif, J.S., Chang, J.C. & Davis, J. R. (1990) An analysis of occupational risks for brain cancer. *Am. J. public Health*, **80**, 169–172

Brugnone, F., Perbellini, L., Apostoli, P., Bellomi, M. & Caretta, D. (1983) Isopropanol exposure: environmental and biological monitoring in a printing works. *Br. J. ind. Med.*, **40**, 160–168

Bruno, M.H. (1982) Printing processes. In: Mark, H.F., Othmer, D.F., Overberger, C.G., Seaborg, G.T. & Grayson, N., eds, *Kirk-Othmer Encyclopedia of Chemical Technology*, 3rd Ed., Vol. 19, New York, John Wiley & Sons, pp. 110–163

Buckley, J.D., Robison, L.L., Swotinsky, R., Garabrant, D.H., LeBeau, M., Manchester, P., Nesbit, M.E., Odom, L., Peters, J.M., Woods, W.G. & Hammond, G.D. (1989) Occupational exposures of parents of children with acute nonlymphocytic leukemia: a report from the Childrens Cancer Study Group. *Cancer Res.*, **49**, 4030–4037

Buiatti, E., Kriebel, D., Geddes, M., Santucci, M. & Pucci, N. (1985) A case–control study of lung cancer in Florence, Italy. I. Occupational risk factors. *J. Epidemiol. Community Health*, **39**, 244–250

Burns, P.B. & Swanson, G.M. (1991a) Risk of urinary bladder cancer among blacks and whites: the role of cigarette use and occupation. *Cancer Causes Contr.*, **2**, 371–379

Burns, P.B. & Swanson, G.M. (1991b) The Occupational Cancer Incidence Surveillance Study (OCISS): risk of lung cancer by usual occupation and industry in the Detroit Metropolitan area. *Am. J. ind. Med.*, **19**, 655–671

Calas, E., Castelain, P.Y., Raulot-Lapointe, H., Ducos, P., Cavelier, C., Duprat, P. & Poitou, P. (1977) Allergic contact dermatitis to a photopolymerizable resin used in printing. *Contact Derm.*, **3**, 186–194

Carlick, D.J. (1971) The Suncure system. *Penrose Annual*, **64**, 168–170

Carter, R.L., Mitchley, B.C.V. & Roe, F.J.C. (1969) Preliminary survey of 22 printing inks for carcinogenic activity by the subcutaneous route in mice. *Food Cosmet. Toxicol.*, **7**, 53–58

Cartwright, R. (1982) Occupational bladder cancer and cigarette smoking in West Yorkshire. *Scand. J. Work Environ. Health*, **8** (Suppl. 1), 79–82

Casey, P., Hagger, R. & Harper, P. (1983) A collaborative study of 'ink mist' in U.K. newspaper press-rooms. *Ann. occup. Hyg.*, **27**, 127–135

Castelain, P.-Y., Pirious, A., Raulot-Lapointe, H. & Robaglia, J.-L. (1980) Sensitization to abieto-formo-phenolic resin in printing ink (Short communication). *Contact Derm.*, **6**, 145-146

Chang, Y.-C. (1987) Neurotoxic effects of *n*-hexane on the human central nervous system: evoked potential abnormalities in *n*-hexane polyneuropathy. *J. Neurol. Neurosurg. Psych.*, **50**, 269–274

Chang, C.M., Yu, C.W., Fong, K.Y., Leung, S.Y., Tsin, T.W., Yu, Y.L., Cheung, T.F. & Chan, S.Y. (1993) *n*-Hexane neuropathy in offset printers. *J. Neurol. Neurosurg. Psych.*, **56**, 538–542

Chen, Q., Kan, X., Ma, X., Li, Z., Wang, Z., Ren, H. & Yang, B. (1992) Lead concentrations in urine correlated with cytogenetic damages in workers exposed to lead. *Chin. J. prev. Med.*, **26**, 334–335

Chung, K.-T. & Cerniglia, C. E. (1992) Mutagenicity of azo dyes: structure–activity relationships. *Mutat. Res.*, **277**, 201–220

Claude, J., Frentzel-Beyme, R.R. & Kunze, E. (1988) Occupation and risk of cancer of the lower urinary tract among men. A case–control study. *Int. J. Cancer*, **41**, 371–379

Coggon, D., Pannett, B. & Acheson, E.D. (1984) Use of job–exposure matrix in an occupational analysis of lung and bladder cancers on the basis of death certificates. *J. natl Cancer Inst.*, **72**, 61–65

Coldman, A.J., Elwood, J.M. & Gallagher, R.P. (1982) Sports activities and risk of testicular cancer. *Br. J. Cancer*, **46**, 749–756

Cole, P., Monson, R.R., Haning, H. & Friedell, G.H. (1971) Smoking and cancer of the lower urinary tract. *New Engl. J. Med.*, **284**, 129–134

Cole, P., Hoover, R. & Friedell, G.H. (1972) Occupation and cancer of the lower urinary tract. *Cancer*, **29**, 1250–1260

Cole Johnson, C., Annegers, J.F., Frankowski, R.F., Spitz, M.R. & Buffler, P.A. (1987) Childhood nervous system tumors — an evaluation of the association with paternal occupational exposure to hydrocarbons. *Am. J. Epidemiol.*, **126**, 605–613

Cordier, S., Clavel, J., Limasset, J.C., Boccon-Gibod, L., Le Moual, N., Mandereau, L. & Hemon, D. (1993) Occupational risks of bladder cancer in France: a multicentre case–control study. *Int. J. Epidemiol.*, **22**, 403–411

Costa, G., Faggiano, F. & Lagorio, S., eds (1995) *Mortality by Occupation in Italy in the 1980s*, Rome, Istituto Superiore per la Prevenzione e la Sicurezza del Lavoro (in Italian)

Crebelli, R., Aquilina, G., Falcone, E., Carere, A., Caperle, M., Crespi, M. & Zito, R. (1985) Monitoring of urinary mutagenicity in workers exposed to low doses of 2,4,7-trinitro-9-fluorenone. *Scand. J. Work Environ. Health*, **11**, 295–300

Cullen, M.R., Rado, T., Waldron, J.A., Sparer, J. & Welch, L.S. (1983) Bone marrow injury in lithographers exposed to glycol ethers and organic solvents used in multicolor offset and ultraviolet curing printing processes. *Arch. environ. Health*, **38**, 347–354

Daniels, W.J. (1982) *Reporter Printing Company, Fond Du Lac, Wisconsin* (Health Hazard Evaluation Report, No. HETA-81-311-1139), Cincinnati, OH, United States National Institute for Occupational Safety and Health

Dean, B.J. (1985) Recent findings on the genetic toxicology of benzene, toluene, xylenes and phenols. *Mutat. Res.*, **154**, 153–181

Deng, J.-F., Wang, J.-D., Shih, T.-S. & Lan, F.-L. (1987) Outbreak of carbon tetrachloride poisoning in a color printing factory related to the use of isopropyl alcohol and an air conditioning system in Taiwan. *Am. J. ind. Med.*, **12**, 11–19

De Rosa, E., Bartolucci, G.B., Sigon, M., Corona, P.C., Perbellini, L. & Brugnone, F. (1986) Environmental and biological monitoring of workers exposed to low levels of toluene. *Appl. ind. Hyg.*, **1**, 132–137

Dillon, D., Combes, R. & Zeiger, E. (1994) Activation by caecal reduction of the azo dye D & C Red No. 9 to a bacterial mutagen. *Mutagenesis*, **9**, 295–299

Dubrow, R. (1986) Malignant melanoma in the printing industry. *Am. J. ind. Med.*, **10**, 119–126

Dubrow, R. & Wegman, D.H. (1984) Cancer and occupation in Massachusetts: a death certificate study. *Am. J. ind. Med.*, **6**, 207–230

Dunn, J.E., Jr & Weir, J.M. (1968) A perspective study of mortality of several occupational groups. Special Emphasis on lung cancer. *Arch. environ. Health*, **17**, 71–76

Edmonds, L.D., Layde, P.M., James, L.M., Flynt, J.W., Erickson, J.D. & Oakley, G.P., Jr (1981) Congenital malformations surveillance: two American systems. *Int. J. Epidemiol.*, **10**, 247–252

Emmett, E.A. (1977) Contact dermatitis from polyfunctionmal acrylic monomers. *Contact Derm.*, **3**, 245–248

Emmett, E.A. & Kominsky, J.R. (1977) Allergic contact dermatitis from ultraviolet cured inks. Allergic contact sensitization to acrylates. *J. occup. Med.*, **19**, 113–115

Erickson, J.D., Cochran, W.M., & Anderson, C.E. (1979) Parental occupation and birth defects. A preliminary report. *Contr. Epidem. Biostatist.*, **1**, 107–117

Ericson, A., Källén, B. & Löfkvist, E. (1988) Environmental factors in the etiology of neural tube defects: a negative study. *Environ. Res.*, **45**, 38–47

European Confederation of Paint, Printing Ink and Artists' Colours Manufacturers' Associations (CEPE) (1995) *Printing and Printing Ink Industries*, Market Information, Brussels

Fedrick, J. (1976) Anencephalus in the Oxford record linkage study area. *Develop. Med. Child Neurol.*, **18**, 643–656

FIOH (Finnish Institute of Occupational Health) (1995) *Industrial Hygiene Measurements, 1950–70* (Data Base), Helsinki

Forni, A., Pacifico, E. & Limonta, A. (1971) Chromosome studies in workers exposed to benzene or toluene or both. *Arch. environ. Health*, **22**, 373–378

Foussereau, J., Benezra, C., Maibach, H.I. & Hjorth, N. (1982) *Occupational Contact Dermatitis. Clinical and Chemical Aspects*, Philadelphia, W.B. Saunders Company

Funes-Cravioto, F., Zapata-Gayon, C., Kolmodin-Hedman, B., Lambert, B., Lindsten, J., Norberg, E., Nordenskjöld, M., Olin, R. & Swensson, Å. (1977) Chromosome aberrations and sister-chromatid exchange in workers in chemical laboratories and a rotoprinting factory and in children of women laboratory workers. *Lancet*, **ii**, 322–325

Gallagher, R.P., Threlfall, W.J., Band, P.R. & Spinelli, J.J. (1989) *Occupational Mortality in British Columbia, 1950–1984*, Vancouver, BC, Canada, Cancer Control Agency of British Columbia

Glatt, H., Padykula, R., Berchtold, G.A., Ludewig, G., Platt, K.L., Klein, J. & Oesch, F. (1989) Multiple activation pathways of benzene leading to products with varying genotoxic characteristics. *Environ. Health Perspectives*, **82**, 81–89

Goldstein, D.H., Benoit, J.N. & Tyroler, H.A. (1970) An epidemiologic study of an oil mist exposure. *Arch. environ. Health*, **21**, 600–603

González, C., López-Abente, G., Errezola, M., Escolar, A., Riboli, E., Izarzugaza, I. & Nebot, M. (1989) Occupation and bladder cancer in Spain: a multi-centre case–control study. *Int. J. Epidemiol.*, **18**, 569–577

Greenberg, M. (1972) A proportional mortality study of a group of newspaper workers. *Br. J. ind. Med.*, **29**, 15–20

Greenburg, L., Mayers, M.R., Goldwater, L. & Smith, A.R. (1939) Benzene (benzol) poisoning in the rotogravure printing industry in New York City. *J. ind. Hyg.*, **21**, 395–420

Greene, M.H., Dalager, N.A., Lamberg, S.I., Argyropoulos, C.E. & Fraumeni, J.F., Jr (1979a) Mycosis fungoides: epidemiologic observations. *Cancer Treat. Rep.*, **63**, 597–606

Greene, M.H., Hoover, R.N., Eck, R.L. & Fraumeni, J.F., Jr (1979b) Cancer mortality among printing plant worker. *Environ. Res.*, **20**, 66–73

Guzelian, P., Mills, S. & Fallon, H.J. (1988) Liver structure and function in print workers exposed to toluene. *J. occup. Med.*, **30**, 791–796

Hagmar, L., Högstedt, B., Welinder, H., Karlsson, A. & Rassner, F. (1989) Cytogenetic and hematological effects in plastics workers exposed to styrene. *Scand. J. Work Environ. Health*, **15**, 136–141

Hamilton, A. (1925) *Industrial Poisons in the United States*, New York, The Macmillan Company

Hänninen, H., Antti-Poika, M. & Savolainen, P. (1987) Psychological performance, toluene exposure and alcohol consumption in rotogravure printers. *Int. Arch. occup. environ. Health*, **59**, 475–483

Hargreaves, I. (1995) Energy cured inks offer many advantages. *Folding Carton Ind.*, **22**, 30, 32

Hashimoto, D.M., Kelsey, K.T., Seitz, T., Feldman, H.A., Yakes, B. & Christiani, D.C. (1991) The presence of urinary cellular sediment and albuminuria in newspaper pressworkers exposed to solvents. *J. occup. Med.*, **33**, 516–526

Hauck, R. & Baur, X. (1990) Forms of humidifier lung. *Klin. Wochenschr.*, **68**, 512–517 (in German)

Hoar Zahm, S., Brownson, R.C., Chang, J.C. & Davis, J.R. (1989) Study of lung cancer histologic types, occupation, and smoking in Missouri. *Am. J. ind. Med.*, **15**, 565–578

Högstedt, B, Bratt, I., Holmén, A., Hagmar, L. & Skerfving, S. (1988) Frequency and size distribution of micronuclei in lymphocytes stimulated with phytohaemagglutinin and pokeweed mitogen in workers exposed to piperazine. *Hereditas*, **109**, 139–142

Högstedt, B., Holmén, A., Karlsson, A., Raihle, G., Nillius, K. & Vestlund, K. (1991) Gasoline pump mechanics had increased frequencies and sizes of micronuclei in lymphocytes stimulated by pokeweed mitogen. *Mutat. Res.*, **263**, 51–55

Holmén, A., Karlsson, A., Bratt, I., Raihle, G. & Högstedt, B. (1994) Increased frequency of micronuclei in lymphocytes of Swedish chimney sweeps. *Int. Arch. occup. environ. Health*, **66**, 185–187

Hrubec, Z., Blair, A.E., Rogot, E. & Vaught, J. (1992) *Mortality Risks by Occupation among U.S. Veterans of Known Smoking Status 1954–1980* (NIH Publication No. 92-3407), Bethesda, MD, National Institutes of Health

Huebner, W.W., Schoenberg, J.B., Kelsey, J.L., Wilcox, H.B., McLaughlin, J.K., Greenberg, R.S., Preston-Martin, S., Austin, D.F., Stemhagen, A., Blot, W.J., Winn, D.M. & Fraumeni, J.F., Jr (1992) Oral and pharyngeal cancer and occupation: a case–control study. *Epidemiology*, **3**, 300–309

IARC (1983) *IARC Monographs on the Evaluation of the Carcinogenic Risk of Chemicals to Humans*, Vol. 32, *Polynuclear Aromatic Compounds, Part 1, Chemical, Environmental and Experimental Data*, Lyon

IARC (1987a) *IARC Monographs on the Evaluation of Carcinogenic Risks to Humans*, Suppl. 7, *Overall Evaluations of Carcinogenicity: An Updating of* IARC Monographs *Volumes 1 to 42*, Lyon, pp. 194-195

IARC (1987b) *IARC Monographs on the Evaluation of Carcinogenic Risks to Humans*, Suppl. 7, *Overall Evaluations of Carcinogenicity: An Updating of* IARC Monographs *Volumes 1 to 42*, Lyon, pp. 230-232

IARC (1987c) *IARC Monographs on the Evaluation of Carcinogenic Risks to Humans*, Suppl. 7, *Overall Evaluations of Carcinogenicity: An Updating of* IARC Monographs *Volumes 1 to 42*, Lyon, pp. 216-219

IARC (1987d) *IARC Monographs on the Evaluation of Carcinogenic Risks to Humans*, Suppl. 7, *Overall Evaluations of Carcinogenicity: An Updating of* IARC Monographs *Volumes 1 to 42*, Lyon, pp. 193-194

IARC (1987e) *IARC Monographs on the Evaluation of Carcinogenic Risks to Humans*, Suppl. 7, *Overall Evaluations of Carcinogenicity: An Updating of* IARC Monographs *Volumes 1 to 42*, Lyon, pp. 133-134

IARC (1987f) *IARC Monographs on the Evaluation of Carcinogenic Risks to Humans*, Suppl. 7, *Overall Evaluations of Carcinogenicity: An Updating of* IARC Monographs *Volumes 1 to 42*, Lyon, pp. 252-254

IARC (1987g) *IARC Monographs on the Evaluation of Carcinogenic Risks to Humans*, Suppl. 7, *Overall Evaluations of Carcinogenicity: An Updating of* IARC Monographs *Volumes 1 to 42*, Lyon, p. 229

IARC (1987h) *IARC Monographs on the Evaluation of Carcinogenic Risks to Humans*, Suppl. 7, *Overall Evaluations of Carcinogenicity: An Updating of* IARC Monographs *Volumes 1 to 42*, Lyon, p. 70

IARC (1987i) *IARC Monographs on the Evaluation of Carcinogenic Risks to Humans*, Suppl. 7, *Overall Evaluations of Carcinogenicity: An Updating of* IARC Monographs *Volumes 1 to 42*, Lyon, p. 70

IARC (1987j) *IARC Monographs on the Evaluation of Carcinogenic Risks to Humans*, Suppl. 7, *Overall Evaluations of Carcinogenicity: An Updating of* IARC Monographs *Volumes 1 to 42*, Lyon, p. 70

IARC (1987k) *IARC Monographs on the Evaluation of Carcinogenic Risks to Humans*, Suppl. 7, *Overall Evaluations of Carcinogenicity: An Updating of* IARC Monographs *Volumes 1 to 42*, Lyon, p. 73

IARC (1987l) *IARC Monographs on the Evaluation of Carcinogenic Risks to Humans*, Suppl. 7, *Overall Evaluations of Carcinogenicity: An Updating of* IARC Monographs *Volumes 1 to 42*, Lyon, pp. 120-122

IARC (1987m) *IARC Monographs on the Evaluation of Carcinogenic Risks to Humans*, Suppl. 7, *Overall Evaluations of Carcinogenicity: An Updating of* IARC Monographs *Volumes 1 to 42*, Lyon, pp. 143-144

IARC (1987n) *IARC Monographs on the Evaluation of Carcinogenic Risks to Humans*, Suppl. 7, *Overall Evaluations of Carcinogenicity: An Updating of* IARC Monographs *Volumes 1 to 42*, Lyon, pp. 362-363

IARC (1987o) *IARC Monographs on the Evaluation of Carcinogenic Risks to Humans*, Suppl. 7, *Overall Evaluations of Carcinogenicity: An Updating of* IARC Monographs *Volumes 1 to 42*, Lyon, p. 61

IARC (1987p) *IARC Monographs on the Evaluation of Carcinogenic Risks to Humans*, Suppl. 7, *Overall Evaluations of Carcinogenicity: An Updating of* IARC Monographs *Volumes 1 to 42*, Lyon, p. 71

IARC (1989a) *IARC Monographs on the Evaluation of Carcinogenic Risks to Humans*, Vol. 47, *Some Organic Solvents, Resin Monomers and Related Compounds, Pigments and Occupational Exposures in Paint Manufacture and Painting*, Lyon, pp. 125-156

IARC (1989b) *IARC Monographs on the Evaluation of Carcinogenic Risks to Humans*, Vol. 47, *Some Organic Solvents, Resin Monomers and Related Compounds, Pigments and Occupational Exposures in Paint Manufacture and Painting*, Lyon, pp. 307–326

IARC (1989c) *IARC Monographs on the Evaluation of Carcinogenic Risks to Humans*, Vol. 47, *Some Organic Solvents, Resin Monomers and Related Compounds, Pigments and Occupational Exposures in Paint Manufacture and Painting*, Lyon, pp. 79-123

IARC (1989d) *IARC Monographs on the Evaluation of Carcinogenic Risks to Humans*, Vol. 45, *Occupational Exposures in Petroleum Refining; Crude Oil and Major Petroleum Fuels*, Lyon, pp. 159-201

IARC (1990) *IARC Monographs on the Evaluation of Carcinogenic Risks to Humans*, Vol. 49, *Chromium, Nickel and Welding*, Lyon, pp. 49-256

IARC (1992) *IARC Monographs on the Evaluation of Carcinogenic Risks to Humans*, Vol. 55, *Solar and Ultraviolet Radiation*, Lyon

IARC (1993a) *IARC Monographs on the Evaluation of Carcinogenic Risks to Humans*, Vol. 58, *Beryllium, Cadmium, Mercury, and Exposures in the Glass Manufacturing Industry*, Lyon, pp. 119-237

IARC (1993b) *IARC Monographs on the Evaluation of Carcinogenic Risks to Humans*, Vol. 57, *Occupational Exposures of Hairdressers and Barbers and Personal Use of Hair Colourants; Some Hair Dyes, Cosmetic Colourants, Industrial Dyestuffs and Aromatic Amines*, Lyon, pp. 203-212

IARC (1993c) *IARC Monographs on the Evaluation of Carcinogenic Risks to Humans*, Vol. 57, *Occupational Exposures of Hairdressers and Barbers and Personal Use of Hair Colourants; Some Hair Dyes, Cosmetic Colourants, Industrial Dyestuffs and Aromatic Amines*, Lyon, pp. 215-234

IARC (1993d) *IARC Monographs on the Evaluation of Carcinogenic Risks to Humans*, Vol. 57, *Occupational Exposures of Hairdressers and Barbers and Personal Use of Hair Colourants; Some Hair Dyes, Cosmetic Colourants, Industrial Dyestuffs and Aromatic Amines*, Lyon, pp. 259-267

IARC (1995a) *IARC Monographs on the Evaluation of Carcinogenic Risks to Humans*, Vol. 62, *Wood Dust and Formaldehyde*, Lyon, pp. 217–362

IARC (1995b) *IARC Monographs on the Evaluation of Carcinogenic Risks to Humans*, Vol. 63, *Dry Cleaning, Some Chlorinated Solvents and Other Industrial Chemicals*, Lyon, pp. 75–158

International Labour Office (1991) *Occupational Exposure Limits for Airborne Toxic Substances: Values of Selected Countries* (Occupational Safety and Health Series No. 37), 3rd Ed., Geneva

Iscovich, J., Castelletto, R., Estève, J., Muñoz, N., Colanzi, R., Coronel, A., Deamezola, I., Tassi, V. & Arslan, A. (1987) Tobacco smoking, occupational exposure and bladder cancer in Argentina. *Int. J. Cancer*, **40**, 734–740

Jolanki, R., Kanerva, L., Estlander, T. & Tarvainen, K. (1994) Concomitant sensitization to triglycidyl isocyanurate, diaminodiphenylmethane and 2-hydroxyethyl methacrylate from silk-screen printing coatings in the manufacture of circuit boards. *Contact Derm.*, **30**, 12–15

Jongeneelen, F.J., Anzion, R.B.M., Scheepers, P.T.J., Bos, R.P., Henderson, P.T., Nijenhuis, E.H., Veenstra, S.J., Brouns, R.M.E. & Winkes, A. (1988) 1-Hydroxypyrene in urine as a biological indicator of exposure to polycyclic aromatic hydrocarbons in several work environments. *Ann. occup. Hyg.*, **32**, 35–43

Jost, M. & Lehmann, M. (1990) Humidifier syndrome as occupational disease in Switzerland. *Schweiz. Runds. Med. (Praxis)*, **79**, 797–803 (in German)

Juntunen, J., Matikainen, E., Antti-Poika, M., Suoranta, H. & Valle, M. (1985) Nervous system effects of long-term occupational exposure to toluene. *Acta neurol. scand.*, **72**, 512–517

Kabat, G.C., Dieck, G.S. & Wynder, E.L. (1986) Bladder cancer in nonsmokers. *Cancer*, **57**, 362–367

Kay, K. (1976) Toxicologic and cancerogenic evaluation of chemicals used in the graphic arts industries. *Clin. Toxicol.*, **9**, 359–390

Kennaway, E.L. & Kennaway, N.M. (1947) A further study of the incidence of cancer of the lung and larynx. *Br. J. Cancer*, **1**, 260–298

Krishnan, E.R., Goodman, R.J. & McCammon, C.S. (1987) *Industrial Hygiene Walk-Through Survey Report of American-National Can Company, Philadelphia Plant, Philadelphia, Pennsylvania* (Report No. 157.25), Cincinnati, OH, United States National Institute for Occupational Safety and Health

Kristensen, P. & Andersen, A. (1992) A cohort study on cancer incidence in offspring of male printing workers. *Epidemiology*, **3**, 6–10

Kristensen, P., Irgens, L.M., Daltveit, A.K. & Andersen, A. (1993) Perinatal outcome among children of men exposed to lead and organic solvents in the printing industry. *Am. J. Epidemiol.*, **137**, 134–144

Kronoveter, K.J. & Gill, J. (1977) *Herald-Times, Inc., Bloomington, Indiana* (Report No. HHE-76-96-390), Cincinnati, OH, United States National Institute for Occupational Safety and Health

Krusell, L., Nielsen, H.K., Baelum, J., Lundqvist, G., Omland, Ø., Vaeth, M., Husted, S.E., Mogensen, C.E. & Geday, E. (1985) Renal effects of chronic exposure to organic solvents. A clinical controlled trial. *Acta med. scand.*, **218**, 323–327

Kübler, R. (1993) Printing inks. In: Elvers, B., Hawkins, S., Russey, W. & Schulz, G., eds, *Ullmann's Encyclopedia of Industrial Chemistry*, 5th rev. Ed., Vol. A22, New York, VCH Publishers, pp. 143–156

Kuijten, R.R., Bunin, G.R., Nass, C.C. & Meadows, A.T. (1992) Parental occupation and childhood astrocytoma: results of a case–control study. *Cancer Res.*, **52**, 782–786

Kunze, E., Chang-Claude, J. & Frentzel-Beyme, R. (1992) Life style and occupational risk factors for bladder cancer in Germany. *Cancer*, **69**, 1776–1790

Kwa, S.-L. & Fine, L.J. (1980) The association between parental occupation and childhood malignancy. *J. occup. Med.*, **22**, 792–794

La Vecchia, C., Negri, E., D'Avanzo, B. & Franceschi, S. (1990) Occupation and the risk of bladder cancer. *Int. J. Epidemiol.*, **19**, 264–268

Leach, R.H. & Pierce, R.J. (1993) *The Printing Ink Manual*, 5th Ed., London, Chapman & Hall

Leon, D.A. (1994) Mortality in the british printing industry: a historical cohort study of trade union members in Manchester. *Occup. environ. Med.*, **51**, 79–86

Leon, D.A., Thomas, P. & Hutchings, S. (1994) Lung cancer among newspaper printers exposed to ink mist: a study of trade union members in Manchester, England. *Occup. environ. Med.*, **51**, 87–94

Lerchen, M.L., Wiggins, C.L. & Samet, J.M. (1987) Lung cancer and occupation in New Mexico. *J. natl Cancer Inst.*, **79**, 639–645

Lewis, R.J., Sr (1993a) *Hawley's Condensed Chemical Dictionary*, 12th Ed., New York, Van Nostrand Reinhold Co., p. 585

Lewis, R.J., Sr (1993b) *Hawley's Condensed Chemical Dictionary*, 12th Ed., New York, Van Nostrand Reinhold Co., p. 876

Lewis, P. (1994) Report of airborne solvent concentrations in a printing ink manufacturing plant in Cape Town, Republic of South Africa. *Appl. occup. environ. Hyg.*, **9**, 147–151

Lloyd, J.W., Decoufle, P. & Salvin, L.G. (1977) Unusual mortality experience of printing pressmen. *J. occup. Med.*, **19**, 543–550

Lynge, E., Andreassen Rix, B., Villadsen, E., Andersen, I., Hink, M., Olsen, E., Lucht Møller, U. & Silfverberg, E. (1995) Cancer in printing workers in Denmark. *Occup. environ. Med.*, **52**, 738–744

Magnusson, B. & Mobacken, H. (1972) Allergic contact dermatitis from acrylate printing plates in a printing plant. *Berufsdermatosen*, **20**, 138–142

Mahler, J.F. (1992) *NTP Technical Report on Toxicity Studies of Black Newsprint Inks Administered Topically to F344/N Rats and C3H Mice* (NIH Publication 92-3340; Toxicity Report Series No. 17), Research Triangle Park, NC, United States National Toxicology Program

Maizlish, N.A., Langolf, G.D., Whitehead, L.W., Fine, L.J., Albers, J.W., Goldberg, J. & Smith, P. (1985) Behavioural evaluation of workers exposed to mixtures of organic solvents. *Br. J. ind. Med.*, **42**, 579–590

Mäki-Paakkanen, J., Husgafvel-Pursiainen, K., Kalliomäki, P.-L., Tuominen, J. & Sorsa, M. (1980) Toluene-exposed workers and chromosome aberrations. *J. Toxicol. environ. Health*, **6**, 775–781

Malker, H.S.R. & Gemne, G. (1987) A register-epidemiology study on cancer among swedish printing industry workers. *Arch. environ. Health*, **42**, 73–82

Malten, K.E. (1977) Letterflex photoprepolymer sensitization in newspaper printers due to penta erythritol tetrakis 3 mercaptopropionate and 3 mercaptopropionic acid. *Contact Derm.*, **3**, 257–262

Malten, K.E. (1982) Old and new, mainly occupational dermatological problems in the production and processing of plastics. In: Maibach, H.I. & Gellin, G.A., eds, *Occupational and Industrial Dermatology*, Chicago, IL, Year Book Medical Publishers, pp. 237–283

Mamolen, M., Lewis, D.M., Blanchet, M.A., Satink, F.J. & Vogt, R.L. (1993) Investigation of an outbreak of 'humidifier fever' in a print shop. *Am. J. ind. Med.*, **23**, 483–490

Mandel, J.S., McLaughlin, J.K., Schlehofer, B., Mellemgaard, A., Helmert, U., Lindblad, P., McCredie, M. & Adami, H.-O. (1995) International renal-cell cancer study. IV. Occupation. *Int. J. Cancer*, **61**, 601–605

Matsumoto, T., Fukaya, Y., Yoshitomi, S., Arafuka, M., Kubo, N. & Ohno, Y. (1993) Relations between lead exposure and peripheral neuromuscular functions of lead-exposed workers — results of tapping test. *Environ. Res.*, **61**, 299–307

McCammon, C.S., Krishnan, E.R. & Goodman, R.J. (1987a) *Industrial Hygiene Walk-Through Survey Report of Tetra Pak Incorporated, Denton, USA Plant, Denton, Texas* (Report No. 157.18), Cincinnati, OH, United States National Institute for Occupational Safety and Health

McCammon, C.S., Krishnan, E.R. & Goodman, R.J. (1987b) *Industrial Hygiene Walk-Through Survey Report of Metallized Products, Inc., Winchester, Massachusetts* (Report No. 157.21), Cincinnati, OH, United States National Institute for Occupational Safety and Health

McDonald, A.D., McDonald, J.C., Armstrong, B., Cherry, N.M., Delorme, C., Nolin, A.D. & Robert, D. (1987) Occupation and pregnancy outcome. *Br. J. ind. Med.*, **44**, 521–526

McDonald, A.D., McDonald, J.C., Armstrong, B., Cherry, N.M., Côté, R., Lavoie, J., Nolin, A.D. & Robert, D. (1988) Congenital defects and work in pregnancy. *Br. J. ind. Med.*, **45**, 581–588

McDonald, A.D., McDonald, J.C., Armstrong, B., Cherry, N.M., Nolin, A.D. & Robert, D. (1989) Fathers' occupation and pregnancy outcome. *Br. J. ind. Med.*, **46**, 329–333

McDowall, M.E. (1985) *Occupational Reproductive Epidemiology. The Use of Routinely Collected Statistics in England and Wales 1980–1982*, London, Her Majesty's Stationery Office

McGregor, D.B. (1994) The genetic toxicology of toluene. *Mutat. Res.*, **317**, 213–228

McKinney, P.A., Cartwright, R.A., Saiu, J.M.T., Mann, J.R., Stiller, C.A., Draper, G.J., Hartley, A.L., Hopton, P.A., Birch, J.M., Waterhouse, J.A.H. & Johnston, H.E. (1987) The interregional epidemiological study of childhood cancer (IRESCC): a case–control study of aetiological factors in leukaemia and lymphoma. *Arch. Dis. Child.*, **62**, 279–287

McLaughlin, J.K., Malker, H.S.R., Blot, W.J., Ericsson, J.L.E., Gemne, G. & Fraumeni, J.F., Jr (1988) Malignant melanoma in the printing industry. *Am. J. ind. Med.*, **13**, 301–304

Menck, H.R. & Henderson, B.E. (1976) Occupational differences in rates of lung cancer. *J. occup. Med.*, **18**, 797–801

Mergler, D., Bélanger, S., De Grosbois, S. & Vachon, N. (1988) Chromal focus of acquired chromatid discrimination loss and solvent exposure among printshop workers. *Toxicology*, **49**, 341–348

Michaels, D., Zoloth, S.R. & Stern, F.B. (1991) Does low-level lead exposure increase risk of death? A mortality study of newspaper printers. *Int. J. Epidemiol.*, **20**, 978–983

Milham, S., Jr (1992) *Occupational Mortality in Washington State 1950–1989*, Cincinnati, OH, United States National Institute for Occupational Safety and Health

Milvy, P. & Kay, K. (1978) Mutagenicity of 19 major graphic arts and printing dyes. *J. Toxicol. environ. Health*, **4**, 31–36

Møller, M., Alfheim, I., Löfroth, G. & Agurell, E. (1983) Mutagenicity of extracts from typewriter ribbons and related items. *Mutat. Res.*, **119**, 239–249

Monster, A.C., Kezic, S., van de Gevel, I. & de Wolff, F.A. (1993) Evaluation of biological monitoring parameters for occupational exposure to toluene. *Int. Arch. occup. environ. Health*, **65**, S159–S162

Morabia, A., Markowitz, S., Garibaldi, K. & Wynder, E.L. (1992) Lung cancer and occupation: results of a multicentre case–control study. *Br. J. ind. Med.*, **49**, 721–727

Morata, T.C., Dunn, D.E., Kretschmer, L.W., Lemasters, G.K. & Keith, R.W. (1993) Effects of occupational exposure to organic solvents and noise on hearing. *Scand. J. Work Environ. Health*, **19**, 245–254

Moretto, A. & Lotti, M. (1990) Exposure to toluene increases the urinary excretion of D-glucaric acid. *Br. J. ind. Med.*, **47**, 58–61

Moss, E. (1973) A mortality survey in the newspaper industry. *Ann. occup. Hyg.*, **16**, 195–196

Moss, E., Scott, T.S. & Atherley, G.R.C. (1972) Mortality of newspaper workers from lung cancer and bronchitis 1952-66. *Br. J. ind. Med.*, **29**, 1-14

Najem, G.R., Louria, D.B., Seebode, J.J., Thind, I.S., Prusakowski, J.M., Ambrose, R.B. & Fernicola, A.R. (1982) Life time occupation, smoking, caffeine, saccharine, hair dyes and bladder carcinogenesis. *Int. J. Epidemiol.*, **11**, 212–217

Nasterlack, M., Triebig, G. & Stelzer, O. (1994) Hepatotoxic effects of solvent exposure around permissible limits and alcohol consumption in printers over a 4-year period. *Int. Arch. occup. environ. Health*, **66**, 161–165

National Association of Printing Ink Manufacturers (1988) *Printing Ink Handbook*, Harrison, NY

Nejjari, C., Tessier, J.F., Dartigues, J.F., Barberger-Gateau, P., Letenneur, L. & Salamon, R. (1993) The relationship between dyspnoea and main lifetime occupation in the elderly. *Int. J. Epidemiol.*, **22**, 848–854

Nethercott, J.R. (1978) Skin problems associated with multifunctional acrylic monomers in ultraviolet curing inks. *Br. J. Dermatol.*, **98**, 541–552

Nethercott, J.R. (1981) Allergic contact dermatitis due to an epoxy acrylate. *Br. J. Dermatol.*, **104**, 697–703

Nethercott, J.R. & Nosal, R. (1986) Contact dermatitis in printing tradesmen. *Contact Derm.*, **14**, 280–287

Nethercott, J.R., Jakubovic, H.R., Pilger, C. & Smith, J.W. (1983) Allergic contact dermatitis due to urethane acrylate in ultraviolet curing inks. *Br. J. ind. Med.*, **40**, 241–250

Ng, T.P., Hong, C.Y., Goh, L.G., Wong, M.L., Koh, K.T.C. & Ling, S.L. (1994) Risks of asthma associated with occupations in a community-based case–control study. *Am. J. ind. Med.*, **25**, 709–718

Nielsen, H.K., Krusell, L., Baelum, J., Lundqvist, G., Omland, Ø., Vaeth, M., Husted, S.E., Mogensen, C.E. & Geday, E. (1985) Renal effects of acute exposure to toluene. A controlled clinical trial. *Acta med. scand.*, **218**, 317–321

Nise, G., Högstedt, B., Bratt, I. & Skerfving, S. (1991) Cytogenetic effects in rotogravure printers exposed to toluene (and benzene). *Mutat. Res.*, **261**, 217–223

Office of Population Censuses and Surveys (1982) Congenital malformations and parents' occupation. *OPCS Monitor*, **MB3 82/1**, 1–10

Olsen, J.H. & Jensen, O.M. (1987) Occupation and risk of cancer in Denmark. An analysis of 93 810 cancer cases, 1970–1979. *Scand. J. Work Environ. Health*, **13** (Suppl. 1)

Olsen, J.H., de Nully Brown, P., Schulgen, G. & Møller Jensen, O. (1991) Parental employement at time of conception and risk of cancer in offspring. *Eur. J. Cancer*, **27**, 958–965

Olshan, A.F., Teschke, K. & Baird, P.A. (1991) Paternal occupation and congenital anomalies in offspring. *Am. J. ind. Med.*, **20**, 447–475

Ørbaek, P. & Nise, G. (1989) Neurasthenic complaints and psychometric function of toluene-exposed rotogravure printers. *Am. J. ind. Med.*, **16**, 67–77

Paganini-Hill, A., Glazer, E., Henderson, B.E. & Ross, R.K. (1980) Cause-specific mortality among newspaper web pressmen. *J. occup. Med.*, **22**, 542–544

Pardoen, H. (1995) Future of offset resins. *Polymers Paint Colour J.*, **185**, S10–S11

Partanen, T., Heikkilä, P., Hernberg, S., Kauppinen, T., Moneta, G. & Ojajärvi, A. (1991) Renal cell cancer and occupational exposure to chemical agents. *Scand. J. Work Environ. Health*, **17**, 231–239

Pasternack, B. & Ehrlich, L. (1972) Occupational exposure to an oil mist atmosphere. A 12-year mortality study. *Arch. environ. Health*, **25**, 286–294

Pedersen, N.B., Chevallier, M.-A. & Senning, A. (1982) Secondary acrylamides in nyloprint® printing plate as a source of contact dermatitis. *Contact Derm.*, **8**, 256–262

Pedersen, N.B., Senning, A. & Nielsen, A.O. (1983) Different sensitising acrylic monomers in Napp® printing plate. *Contact Derm.*, **9**, 459–464

Pelclová, D., Rössner, P. & Picková, J. (1990) Chromosome aberrations in rotogravure printing plant workers. *Mutat. Res.*, **245**, 299–303

Petersen, G.R. & Milham, S., Jr (1980) *Occupational Mortality in the State of California 1959–61 (DHEW (NIOSH) Publication No. 80-104; PB80 176423)*, Cincinnati, OH, United States National Institute for Occupational Safety and Health

Pfirrmann, W. (1994) Four-color process with UV and water-based inks: perfect prints require perfect prepress. *Screen Printing*, **84**, 54–61

Pietri, F., Clavel, F., Auquier, A. & Flamant, R. (1990) Occupational risk factors for cancer of the pancreas: a case–control study. *Br. J. ind. Med.*, **47**, 425–428

Plappert, U., Barthel, E. & Seidel, H.J. (1994) Reduction of benzene toxicity by toluene. *Environ. mol. Mutag.*, **24**, 283–292

Polednak, A.P. & Janerich, D.T. (1983) Uses of available record systems in epidemiologic studies of reproductive toxicology. *Am. J. ind. Med.*, **4**, 329–348

Pukkala, E. (1995) *Cancer Risk by Social Class and Occupation. A Survey of 109 000 Cancer Cases among Finns of Working Age* (Contributions to Epidemiology and Biostatistics, Vol. 7), Basel, Karger

Purdham, J.T., Bozek, P.R. & Sass-Kortsak, A. (1993) The evaluation of exposure of printing trade employees to polycyclic aromatic hydrocarbons. *Ann. occup. Hyg.*, **37**, 35–44

Pye, R.J. & Peachey, R.D.G. (1976) Contact dermatitis due to Nyloprint. *Contact Derm.*, **2**, 144–146

Rachootin, P. & Olsen, J. (1983) The risk of infertility and delayed conception associated with exposures in the Danish workplace. *J. occup. Med.*, **25**, 394–402

Ramazzini, B. (1933) *De Morbis Artificum Diatriba* [Discussion on unnatural death], Turin, Minerva Medica, p. 114

Richer, C.-L., Chakrabarti, S., Senécal-Quevillon, M., Duhr, M.A., Zhang, X.X. & Tardif, R. (1993) Cytogenetic effects of low-level exposure to toluene, xylene, and their mixture on human blood lymphocytes. *Int. Arch. occup. environ. Health*, **64**, 581–585

Roeleveld, N., Zielhuis, G.A. & Gabreëls, F. (1993) Mental retardation and parental occupation: a study on the applicability of job exposure matrices. *Br. J. ind. Med.*, **50**, 945–954

Rogot, E. & Murray, J. (1980) Cancer mortality among nonsmokers in an insured group of US veterans. *J. natl Cancer Inst.*, **65**, 1163–1168

Ruijten, M.W.M.M., Verberk, M.M. & Sallé, H.J.A. (1991) Nerve function in workers with long term exposure to trichloroethene. *Br. J. ind. Med.*, **48**, 87–92

Sakurai, H. (1982) Monitoring health effects due to hazardous working environment organic solvents. *J. Jpn med. Assoc.*, **88**, 1193–1208

Samimi, B. (1982) Exposure to isophorone and other organic solvents in a screen printing plant. *Am. ind. Hyg. Assoc. J.*, **43**, 43–48

Sandstad, O., Osnes, T., Skar, V. & Osnes, M. (1993) Urinary D-glucaric acid, a marker substance for microsomal enzyme induction. Methodological aspects, responses to alcohol and findings in workers exposed to toluene. *Scand. J. clin. Lab. Invest.*, **53**, 327–333

Schmid, E., Bauchinger, M. & Hauf, R. (1985) Chromosome changes with time in lymphocytes after occupational exposure to toluene. *Mutat. Res.*, **142**, 37–39

Schoenberg, J.B., Stemhagen, A., Mogielnicki, A.P., Altman, R., Abe, T. & Mason, T.J. (1984) Case–control study of bladder cancer in New Jersey. I. Occupational exposures in white males. *J. natl Cancer Inst.*, **72**, 973–981

Schoenberg, J.B., Stemhagen, A., Mason, T.J., Patterson, J., Bill, J. & Altman, R. (1987) Occupation and lung cancer risk among New Jersey white males. *J. natl Cancer Inst.*, **79**, 13–21

Searle, C. (1993) Choosing the right green solvent. *Br. Printer*, **October**, 42, 44

Siemiatycki, J., ed. (1991) *Risk Factors for Cancer in the Workplace*, Boca Raton, CRC Press

Siemiatycki, J., Dewar, R., Nadon, L. & Gérin, M. (1994) Occupational risk factors for bladder cancer: results from a case–control study in Montréal, Quebec, Canada. *Am. J. Epidemiol.*, **140**, 1061–1080

Silverman, D.T., Hoover, R.N., Albert, S. & Graff, K.M. (1983) Occupation and cancer of the lower urinary tract in Detroit. *J. natl Cancer Inst.*, **70**, 237–245

Silverman, D. T., Levin, L.I., Hoover, R.N. & Hartge, P. (1989) Occupational risk factors of bladder cancer in the United States: I. White men. *J. natl Cancer Inst.*, **81**, 1472–1480

Silverman, D.T., Levin, L.I. & Hoover, R.N. (1990) Occupational risk factors of bladder cancer among white women in the United States. *Am. J. Epidemiol.*, **132**, 453–461

Sinclair, M.I., McNeil, J.J., Atkins, R.C., Turnidge, J.D., Wood, C.J. & Matthews, B.J. (1993) Investigation of hematuria at a printing company. *J. occup. Med.*, **35**, 1055–1061

Sinha, S.P., Sharma, V., Srivastava, S. & Srivastava, M.M. (1993) Neurotoxic effects of lead exposure among printing press workers. *Bull. environ. Contam. Toxicol.*, **51**, 490–493

Sinks, T., Lushniak, B., Haussler, B.J., Sniezek, J., Deng, J.-F., Roper, P., Dill, P. & Coates, R. (1992) Renal cell cancer among paperboard printing workers. *Epidemiology*, **3**, 483–489

Smith, W.D.L. (1977) Allergic dermatitis due to a triacrylate in ultraviolet cured inks. *Contact Derm.*, **3**, 312–314

Snyder, R. & Kalf, G.F. (1994) A perspective on benzene leukemogenesis. *Crit. Rev. Toxicol.*, **24**, 177–209

Snyder, R., Witz, G. & Goldstein, B.D. (1993) The toxicology of benzene. *Environ. Health Perspectives*, **100**, 293–306

Steenland, K., Burnett, C. & Osorio, A.M. (1987) A case–control study of bladder cancer using city directories as a source of occupational data. *Am. J. Epidemiol.*, **126**, 247–257

Steineck, G., Plato, N., Gerhardsson, M., Norell, S. & Hogstedt, C. (1990) Increased risk of urothelial cancer in Stockholm during 1985–87 after exposure to benzene and exhausts. *Int. J. Cancer*, **45**, 1012–1017

Stětkárová, I., Urban, P., Procházka, B. & Lukáš, E. (1993) Somatosensory evoked potentials in workers exposed to toluene and styrene. *Br. J. ind. Med.*, **50**, 520–527

Svensson, B.-G., Nise, G., Englander, V., Attewell, R., Skerfving, S. & Möller, T. (1990) Deaths and tumours among rotogravure printers exposed to toluene. *Br. J. ind. Med.*, **47**, 372–379

Svensson, B.-G., Nise, G., Erfurth, E.-M., Nilsson, A. & Skerfving, S. (1992a) Hormone status in occupational toluene exposure. *Am. J. ind. Med.*, **22**, 99–107

Svensson, B.-G., Nise, G., Erfurth, E.-M. & Olsson, H. (1992b) Neuroendocrine effects in printing workers exposed to toluene. *Br. J. ind. Med.*, **49**, 402–408

Swerdlow, A.J., Douglas, A.J., Huttly, S.R.A. & Smith, P.G. (1991) Cancer of the testis, socioeconomic status, and occupation. *Br. J. ind. Med.*, **48**, 670–674

Taggi, A.J. & Walker, P. (1996) Printing processes. In: Kroschwitz, J.I. & Howe-Grant, M., eds, *Kirk-Othmer Encyclopedia of Chemical Technology*, 4th Ed., Vol. 19 (in press)

The Freedonia Group (1995) *Printing Inks to 2000*, Cleveland, OH

Uchida, Y., Nakatsuka, H., Ukai, H., Watanabe, T., Liu, Y.-T., Huang, M.-Y., Wang, Y.-L., Zhu, F.-Z., Yin, H. & Ikeda, M. (1993) Symptoms and signs in workers exposed predominantly to xylenes. *Int. Arch. occup. environ. Health*, **64**, 597–605

Ukai, H., Takada, S., Inui, S. & Ikeda, M. (1986) Relationship between exposure and environmental concentrations in organic solvent workplaces. *Tohoku J. exp. Med.*, **149**, 251–260

Ukai, H., Takada, S., Inui, S., Imai, Y., Kawai, T., Shimbo, S.-I. & Ikeda, M. (1994) Occupational exposure to solvent mixtures: effects on health and metabolism. *Occup. environ. Med.*, **51**, 523–529

United States National Institute for Occupational Safety and Health (1995) *National Occupational Exposure Survey (1981–1983)*, Cincinnati, OH

United States Occupational Safety and Health Administration (1994) Air contaminants. *US Code fed. Regul.*, **Title 29**, Part 1910.1000

Urban, P. & Lukáš, E. (1990) Visual evoked potentials in rotogravure printers exposed to toluene. *Br. J. ind. Med.*, **47**, 819–823

Veulemans, H., Groeseneken, D., Masschelein, R. & Van Vlem, E. (1987) Survey of ethylene glycol ether exposures in Belgian industries and workshops. *Am. ind. Hyg. Assoc. J.*, **48**, 671–676

Viadana, E. & Bross, I.D.J. (1972) Leukemia and occupations. *Prev. Med.*, **1**, 513–521

Viadana, E., Bross, I.D.J. & Houten, L. (1976) Cancer experience of men exposed to inhalation of chemicals or to combustion products. *J. occup. Med.*, **18**, 787–792

Vineis, P. & Magnani, C. (1985) Occupation and bladder cancer in males: a case–control study. *Int. J. Cancer*, **35**, 599–606

Vineis, P., Estève, J. & Terracini, B. (1984) Bladder cancer and smoking in males: types of cigarettes, age at start, effect of stopping and interaction with occupation. *Int. J. Cancer*, **34**, 165–170

Wahlberg, J.E. (1974) Contact sensitivity to Nyloprint[R] printing plates. *Contact Derm. Newslett.*, **16**, 510–511

Wang, J.-D., Chang, Y.-C., Kao, K.-P., Huang, C.-C., Lin, C.-C. & Yeh, W.-Y. (1986) An outbreak of *n*-hexane induced polyneuropathy among press proofing workers in Taipei. *Am. J. ind. Med.*, **10**, 111–118

Wang, Q.-S., Boffetta, P., Parkin, D.M. & Kogevinas, M. (1995) Occupational risk facktors for lung cancer in Tianjin, China. *Am. J. ind. Med.*, **28**, 353–362

White, R.F., Proctor, S.P., Echeverria, D., Schweickert, J. & Feldman, R.G. (1995) Neurobehavioral effects of acute and chronic mixed-solvent exposure in the screen printing industry. *Am. J. ind. Med.*, **28**, 221–231

Whitfeld, M. & Freeman, S. (1991) Allergic contact dermatitis to ultra violet cured inks. *Australas J. Dermatol.*, **32**, 65–68

WHO (1986) *Toluene* (Environmental Health Criteria 52), Geneva

WHO (1993) *Benzene* (Environmental Health Criteria 150), Geneva

Wilkins, J.R., III & Koutras, R.A. (1988) Paternal occupation and brain cancer in offspring: a mortalitiy-based case control study. *Am. J. ind. Med.*, **14**, 299–318

Wilkins, J.R., III & Sinks, T.H., Jr (1984) Paternal occupation and Wilms' tumour in offspring. *J. Epidemiol. Community Health*, **38**, 7–11

Williams, C.H. (1992) *The Printer's Ink Handbook*, Barnet, Hestfordshire, UK, Maclean Hunter

Williams, C.H. (1994a) Water based inks for flexo. *Paper Focus*, **November**, 36

Williams, C.H. (1994b) Flexo inks — where are they now? *FlexoTech*, **November**, 24–25

Williams, R.R., Stegens, N.L. & Goldsmith, J.R. (1977) Association of cancer site and type with occupation and industry from the third national cancer survey interview. *J. natl Cancer Inst.*, **59**, 1147–1185

Winchester, R.V. (1985) Solvent exposure of workers during printing ink manufacture. *Ann. occup. Hyg.*, **29**, 517–519

Wood, S. (1994) The history of printing inks. *Prof. Printer*, **38**, 12–17

Wynder, E.L., Onderdonk, J. & Mantel, N. (1963) An epidemiological investigation of cancer of the bladder. *Cancer*, **16**, 1388–1407

Zoloth, S.R., Michaels, D.M., Villalbi, J.R. & Lacher, M. (1986) Patterns of mortality among commercial pressmen. *J. natl Cancer Inst.*, **76**, 1047–1051

CARBON BLACK

1. Exposure Data

Carbon black was considered by previous Working Groups (IARC, 1984, 1987a). New data have since become available, and these are included in the present monograph and have been taken into consideration in the evaluation.

1.1 Chemical and physical data

1.1.1 *Nomenclature*

The Chem. Abstr. Serv. Reg. No. for all carbon blacks is 1333-86-4.

Acetylene black

Chem. Abstr. Name: Carbon black, acetylene
IUPAC Systematic Name: Carbon black, acetylene
Synonyms: CI 77266; CI Pigment Black 7; explosion acetylene black; explosion black
Trade Names: P68; P1250; Shawinigan Acetylene Black; Ucet

Channel black

Chem. Abstr. Name: Carbon black, channel
IUPAC Systematic Name: Carbon black, channel
Synonym: CI 77266; CI Pigment Black 7; Impingement black
Trade Names: Aroflow; Arrow; Atlantic; Black Pearls; Carbolac; Carbomet; CK3; Collocarb; Conductex, Continental; Croflex; Crolac; Degussa; Dixie; Dixiecell; Dixiedensed; Elf; Excelsior; Farbruss; Fecto; Huber; Kosmink; Kosmobil; Kosmolak; Kosmos; Kosmovar; Micronex; Mogul; Monarch; Neo-Spectra; Peerless; Printex; Raven; Regent; Royal Spectra; Special Black IV & V; Spheron; Superba; Super-Carbovar; Super-Spectra; Texas; Triangle; United; Witco; Wyex

Furnace black

Chem. Abstr. Name: Carbon black, furnace
IUPAC Systematic Name: Carbon black, furnace
Synonyms: CI 77266; CI Pigment Black 7; gas-furnace black; oil-furnace black
Trade Names: Aro; Arogen; Aromex; Arovel; Arotone; Atlantic; Black Pearls; Carbodis; Collocarb; Conductex, Continex; Corax; Croflex; Dixie; Durex; Elftex; Essex; Furnal; Furnex; Gastex; Huber; Humenegro; Kosmos; Metanex; Modulex;

Mogul; Molacco; Monarch; Neotex; Opal; Peerless; Pelletex; Philblack; Printex; Rebonex; Regal; Special Schwarz; Statex; Sterling; Texas; Ukarb; United; Vulcan

Lampblack

Chem. Abstr. Name: Carbon black, lamp

IUPAC Systematic Name: Carbon black, lamp

Synonyms: CI 77266; CI Pigment Black 6

Trade Names: Carbon Black BV and V; Durex; Eagle Germantown; Flamruss; Magecol; Tinolite; Torch Brand

Thermal black

Chem. Abstr. Name: Carbon black, thermal

IUPAC Systematic Name: Carbon black, thermal

Synonyms: CI 77266; CI Pigment Black 7; therma-atomic black

Trade Names: Atlantic; Cancarb; Croflex; Dixitherm; Huber; Kosmotherm; Miike 20; P-33; Sevacarb; Shell Carbon; Statex; Sterling; Thermatomic; Thermax; Thermblack; Velvetex

1.1.2 General description

Carbon black is sometimes confused with soot but it is a very different material (Medalia *et al.*, 1981). Carbon black is a powdered form of elemental carbon manufactured by the controlled vapour-phase pyrolysis of hydrocarbons. Different types of carbon black have a wide range of particle sizes, high surface areas per unit mass, quite low contents of ash and toluene-extractable materials and varying degrees of particle aggregation. A carbon black with a high degree of aggregation is said to have a high 'structure'. Structure is determined by the size and shape of the aggregated particles, the number of primary particles per aggregate and their average mass.

The fundamental unit of a carbon black is the aggregate. This is a chain of roughly spherical carbon particles that are permanently fused together in a random branching structure. The aggregate may consist of a few or hundreds of spherical particles (or, as in thermal black, primarily single spheres rather than chains). The chains are open structures and are used to absorb fluids and reinforce materials such as rubber. The aggregates can bind together by van der Waals forces in more loosely associated agglomerates, or they may be compressed in pellets (up to 0.5 cm) held together by means of binders (molasses/lignosulfonates) (Dannenberg *et al.*, 1992; Gardiner *et al.*, 1992a).

To describe a carbon black aggregate, two dimensions are necessary:

(1) *Mean diameter of the component spheres in the chain*. This is a measure of the chain 'thickness'. This is called the primary particle size and generally is inversely proportional to the surface area of the black.

(2) *Extent of the branched chain aggregate*. This is called the aggregate size and is the dimension of the rigid framework that is the aggregate.

In addition to these two dimensions, there is a property or 'structure' which is the volume of space that is 'reinforced' by the aggregate — essentially, the amount of fluid it can absorb internally. A standard method of measuring this property is by the dibutyl phthalate absorption of a black, in units of mL/100 g.

Also of importance for human exposure is the behaviour of carbon black in air and its deposition in the respiratory tract upon inhalation. This is determined by the aerodynamic diameter of the particles. The aerodynamic diameter can be measured by impactors and is dependent upon the geometric diameter, material density and shape factor of the aggregates. Most commonly, the size distribution of airborne particles is expressed as its mass median aerodynamic diameter (MMAD) with its geometric standard deviation.

Carbon black is variously known as acetylene black, channel black, furnace black, lampblack or thermal black, depending on the specific process by which it is manufactured (see Section 1.2.1). The properties of typical types of carbon black are presented in Tables 1–6.

In contrast to carbon black, soot is a material of varying and often unknown composition, which is an unwanted by-product of the incomplete combustion of all kinds of carbon-containing materials, such as waste oil, coal, paper, rubber, plastic, garbage and also fuel oils or gasoline. Soots have a low available carbon surface area owing to their large particle size and small carbon component. They typically contain large quantities of dichloromethane- and toluene-extractable materials, and their ash content can be 50% or more (European Committee for Biological Effects of Carbon Black, 1982).

Two other commercial carbonaceous products are activated carbon and bone black. Activated carbon is a collective name for a group of porous carbons. These are manufactured either by the treatment of carbon with gases or by the carbonization of carbonaceous materials with simultaneous activation by chemical treatment. Activated carbon possesses a porous structure, usually with small amounts of chemically bonded oxygen and hydrogen, and can contain up to 20% of mineral matter, which is usually indicated as ash or residue as a result of ignition. The nature of this mineral material depends on the raw materials used, and can consist of silica and compounds of alkali and alkaline-earth metals, for example. X-Ray investigations show that the carbon is mainly in the form of very small crystallites with a graphite-like structure (Vohler *et al.*, 1986).

Bone black is a by-product of the bone char industry. Bone char is made by carbonizing bones and is used principally in sugar refining. Bone black is a pigment derived from bone char, used primarily as a colourant in artists' paint and for tinting vinyl fabrics for upholstery and automotive interiors. The carbon content of bone black is usually approximately 10% (Lewis, 1988, 1993).

1.1.3 *Chemical and physical properties of the technical products*

All commercially available types of carbon black are insoluble in water and organic solvents, but various types differ in other chemical and physical properties. The ranges of properties of each of the four types of carbon black are summarized in Table 1. Analyses of samples of carbon black produced commercially in the United States of America and

Europe are given in Table 2, and those of the carbon black types produced in Japan are given in Table 3.

Table 1. Typical ranges of properties for four types of carbon black pigment

Property	Acetylene black	Furnace black	Lampblack	Thermal black
Average aggregate diameter (nm)	NR	80–500	NR	300–810[a]
Average primary particle diameter (nm)	35–50	17–70	50–100	150–500
Surface area[a], N_2 (m^2/g)	60–70	20–200	20–95[b] (17–25)[c]	6–15
Oil absorption (mL/g)	3.0–3.5	0.67–1.95	1.05–1.65	0.30–0.46
pH	5–7	5–9.5	3–7	7–8
Volatile matter (%)	0.4	0.3–2.8	0.4–9.0[b] (0.5–1.5)[c]	0.10–0.50
Hydrogen (%)	0.05–0.10	0.45–0.71	NR	0.3–0.5
Oxygen (%)	0.10–0.15	0.19–1.2	NR	0.00–0.12
Benzene extract (%)	0.1	0.01–0.18	0.00–1.4	0.02–1.7
Ash (%)	0.00	0.1–1.0	0.00–0.16	0.02–0.38
Sulfur (%)	0.02	0.05–1.5	NR	0.00–0.25
Density (g/mL)	NR	1.80	1.77[b]	NR

From Garret (1973), Weast (1981) and Hess & Herd (1993)
NR, not reported
[a] Surface area calculated by the nitrogen adsorption method
[b] Value in the United States
[c] Value in Europe (data provided by European carbon black manufacturers (IARC, 1984))

Extractable polycyclic aromatic hydrocarbons (PAHs), nitro-PAHs, and sulfur-containing aromatics from carbon black are discussed in Section 1.1.4.

Acetylene black

Acetylene black is characterized by its high purity, low oxygen content and an extremely high degree of aggregation or structure. X-Ray analysis indicates that acetylene black is the most crystalline or graphitic of the commercial blacks. The levels of ash content and benzene-extractable materials are very low, and acetylene black is not readily wetted by water, since its surface is saturated with hydrogen atoms.

Channel black

Channel black is no longer produced in the United States but, in Europe, its manufacture still continues and, in some operations, is expanding. In the past, channel black was characterized by its small particle size, low degree of aggregation or structure, relatively high content of oxygen complexes on its surface and acidic pH. Natural gas was used as the feedstock and water was not required for the quenching of the reaction,

Table 2. Analyses of samples of several carbon blacks produced commercially in the United States and Europe

Property	Acetylene		Furnace (HAF/N330)	Furnace (FEF/N550)	Lampblack		Thermal (FT/N880)	Thermal (MT/N990)
	USA	Germany			USA	Europe		
Average aggregate diameter (nm)	NR	NR	260	470	NR	NR	300	400
Average particle diameter (nm)	40	35	28	42	65	95	200	400
Surface area (BET) (m^2/g)	65	70	80	42	22	20	12	7
Toluene extract (%)	0.1	0.05^a	0.08	NR	0.2^a	0.1^a	0.8	0.3
pH	4.8	7.0	7.5	NR	3.0	7.0	9.0	8.5
Volatile matter (%)	0.3	0.05	1.0	1.0	1.5	1.0	0.5	0.5
Ash (%)	0.0	0.05	0.3	0.2	0.02	0.05	0.1	0.3
Composition (%)								
Carbon	99.7	99.8	97.9	98.4	98	98	99.2	99.3
Hydrogen	0.1	0.05	0.3	0.4	0.2	0.4	0.5	0.3
Sulfur	0.02	0.005	0.6	0.7	0.8	0.6	0.01	0.01
Oxygen	0.2	0.05	0.8	0.4	0.8	0.4	0.3	0.1

From Hoechst Aktiengesellschaft (1979); Dannenberg (1978); IARC (1984); Robertson & Smith (1994)
NR, not reported; HAF, high-abrasion furnace; FEF, fast extrusion furnace; FT, fine thermal; MT, medium thermal; BET, the Brunauer, Emmett and Teller procedure for calculating surface area
^{a} Benzene extract (%)

and, therefore, channel black had a very low ash content. The volatile content of channel black was about 5% but could be increased to as much as 18% by after-treatments. The surface of this channel black reportedly contained hydroxyl, carbonyl and carboxylic acid groups (Garret, 1973; Claassen, 1978).

Table 3. Typical analyses of three types of carbon black commercially available in Japan

Property	Acetylene black	Furnace black	Thermal black
Average particle diameter (nm)	40	21	90
Iodine adsorption (mg/g)[a]	105	119	26
DBPA[b] (mL/100g)	125	115	27
pH	7.5	7.7	8.5
Volatile matter (%)	0.2	1.4	0.5
Ash (%)	0.1	0.25	0.4

From IARC (1984)
[a] Iodine adsorption is reported as the milligrams of iodine adsorbed per gram of carbon black under specified conditions established by the American Society for Testing and Materials (American Society for Testing and Materials, 1995a) and has been used to approximate the surface area of carbon blacks (Johnson & Eberline, 1978)
[b] DBPA, the dibutyl phthalate absorption method, a standard procedure for measuring void volume, a characteristic related to structure

Channel black, when manufactured, was available as a dry chemical in either powder or pelleted form. Three types were available for reinforcing rubber — easy, medium and hard processing. These varied slightly in particle diameter, that with the largest particle diameter (approximately 29 nm) being known as easy-processing channel and that with the smallest (approximately 22 nm) as hard-processing channel (IARC, 1984).

The average diameters of the channel blacks used for colour and ink applications are shown in Table 4. Medium-flow channel blacks and long-flow channel blacks received an after-treatment with hot air to increase their volatile contents and thereby increase the 'flow' of the lithographic inks in which they were used.

Carbon black made in Germany by an impingement roller process from aromatic hydrocarbon-containing coal-tar residues and coke-oven gases is said to have similar properties to those of older channel black. It has an acidic pH, a volatile content of about 5%, a surface area of about 100 m^2/g and an average particle diameter of 10–30 nm (Claassen, 1978; Dannenberg et al., 1992).

Furnace black

Furnace black consists of irregularly shaped aggregate structures of spherical particles. Originally produced by the gas-furnace process, it is now produced almost entirely by the oil-furnace process. Gas-furnace black was characterized by a low degree

Table 4. Average diameters of channel blacks used in colours and inks

Channel black	Symbol	Average diameter (nm)
High-colour channel	HCC	12
Medium-colour channel	MCC	16
Regular colour channel	RCC	25
Medium-flow channel	MFC	25
Long-flow channel	LFC	25

From IARC (1984)

of structure and had properties that led to a low-to-medium reinforcing performance. Oil-furnace blacks have a substantially higher degree of aggregation and structure and, consequently, furnace black is now available with a wide range of characteristics depending on the desired product performance. The quality of furnace black is controlled by variations of the raw materials, operating temperatures, atmospheric turbulence and by alteration of furnace design. A very small percentage of the total quantity of furnace blacks is subjected to after-treatment by various oxidation processes for particular applications.

Furnace black is available in several grades. Those used in rubber products have been classified by the American Society for Testing and Materials (ASTM) according to a standard four-character nomenclature system. In this system, the letter N indicates that the product gives a normal curing rate, while the letter S indicates that it reduces the rate of cure. The first digit is used to designate the typical average particle size (e.g. 1 indicates 11–19 nm and 9 indicates 201–500 nm) and the last two digits are assigned arbitrarily. The N100 series are super-abrasion furnace blacks; the N200 series are intermediate super-abrasion furnace blacks; the N300 series are high-abrasion furnace blacks; the N500 series are fast-extrusion furnace blacks; and the N700 series are semi-reinforcing furnace blacks. The N900 series are thermal blacks (American Society of Testing and Materials, 1995a).

Table 5 provides the ASTM designations for furnace blacks used in rubber, a description of their types, the symbols used to designate the types, as well as three typical measures of surface area (iodine adsorption, cetyl trimethyl ammonium bromide adsorption and nitrogen adsorption), one measure of the degree of aggregation (dibutyl phthalate absorption) and one rough measure of particle size (tinting strength). Table 6 provides similar information for furnace blacks used in inks, paints and plastics.

Lampblack

Lampblack is considered to be the forerunner of all carbon blacks. Essential and typical properties of lampblack are its high degree of aggregation or structure and low surface area. Formerly, lampblack was an oily product sold in the fluffy state or partially

Table 5. Typical properties of currently available furnace blacks for rubber[a]

ASTM designation	Type of black	Symbol	Iodine adsorption[b] (mg/g)	CTAB[c] (m^2/g)	Surface area, N_2[d] (m^2/g)	DBPA[e] (mL/100 g)	Tinting strength[f] (%)
N110	Super-abrasion furnace	SAF	145	126	143	113	124
N115	Super-abrasion furnace	SAF	160	128	143	113	123
N121	Super-abrasion furnace, high structure	SAF-HS	121	121	124	132	121
N125	Super-abrasion furnace	SAF	117	126	122	104	123
N134	Super-abrasion furnace	SAF	142	134	145	127	132
N135	Super-abrasion furnace	SAF	151	127	141	135	119
S212	Intermediate super-abrasion furnace, low structure, slow cure	—	90	119	120	85	115
N220	Intermediate super-abrasion furnace	ISAF	121	111	115	114	115
N231	Intermediate super-abrasion furnace, low modulus	ISAF-LM	121	108	111	92	117
N234	Improved intermediate super-abrasion furnace, high structure	ISAF-HS	120	119	126	125	124
N293	Conductive furnace	CF	145	114	130	100	117
N299	Intermediate super-abrasion furnace, high structure	ISAF-HS	108	104	103	124	113
S315	High-abrasion furnace, low structure, slow curing	HAF-LS-SC	60	95	91	79	—
N326	High-abrasion furnace, low structure	HAF-LS	82	83	78	72	112
N330	High-abrasion furnace	HAF	82	82	79	102	103
N335	High-abrasion furnace	HAF	92	88	85	110	110
N339	Improved high-abrasion furnace, high structure	HAF-HS	90	93	96	120	110
N343	High-abrasion furnace	HAF	92	95	97	130	114
N347	High-abrasion furnace, high structure	HAF-HS	90	87	85	124	103
N351	High-abrasion furnace, high structure	HAF-HS	68	73	71	120	100
N356	Super-processing furnace, high structure	SPF-HS	92	93	91	154	105
N358	High-abrasion furnace, high structure	HAF-HS	84	88	82	150	99
N375	Improved high-abrasion furnace, high structure	HAF-HS	90	96	93	114	115
N539	Fast-extruding furnace, low structure	FEF-LS	43	41	40	111	—
N550	Fast-extruding furnace	FEF	43	42	42	121	—
N582	Acetylene black	FEF	100	76	80	180	67
N630	General-purpose furnace, low structure	GPF-LS	36	35	34	78	—

Table 5 (contd)

ASTM designation	Type of black	Symbol	Iodine adsorption[b] (mg/g)	CTAB[c] (m²/g)	Surface area, N₂[d] (m²/g)	DBPA[e] (mL/100 g)	Tinting strength[f] (%)
N642	General-purpose furnace, very low structure	–	36	34	39	64	–
N650	General-purpose furnace, high structure	GPF-HS	36	38	37	122	–
N660	General-purpose furnace	GPF	36	36	35	90	–
N683	General-purpose furnace, high structure	GPF-HS	35	39	37	133	–
N754	Semi-reinforcing furnace, low structure	SRF-LS	24	29	25	58	–
N762	Semi-reinforcing furnace, low modulus	SRF-LM	27	29	28	65	–
N765	Semi-reinforcing furnace, high structure	SRF-HS	31	33	36	115	–
N772	Semi-reinforcing furnace	SRF	30	33	31	65	–
N774	Semi-reinforcing furnace, high modulus	SRF-HM	29	29	29	72	–
N787	Semi-reinforcing furnace, high modulus	SRF-HM	30	31	30	80	–
N907	Medium thermal, non-staining, free flowing	MT-NS-FF	–	–	10	34	–
N908	Medium thermal, non-staining	MT-NS	–	–	10	34	–
N990	Medium thermal, free flowing	MT-FF	0	9	9	43	–
N991	Medium thermal	MT	–	8	9	35	–

From Ford & Lyon (1973), Garret (1973), Dannenberg (1978), Smith (1982), Dannenberg et al. (1992) and American Society for Testing and Materials (1995a)

[a] The values given are often averages of typical values supplied by several manufacturers.
[b] ASTM standard method No. D1510
[c] CTAB, cetyl trimethyl ammonium bromide measurement of surface area; ASTM standard method No. D3765
[d] ASTM standard method No. 3037/4820, surface area calculated by the nitrogen adsorption method
[e] DBPA, dibutyl phthalate absorption; ASTM standard method No. D2414
[f] ASTM standard method No. 3265

Table 6. Typical properties of furnace blacks for inks, paints, paper and plastics

Furnace black	Surface area[a] (m²/g)	Particle size (nm)	Aggregate size (nm)	DBPA (mL/100 g) Fluffy	DBPA (mL/100 g) Pellets	Bulk density (g/L) Fluffy	Bulk density (g/L) Pellets	Nigrometer index[b]	Tinting strength (%)	Volatile content (%)
Normal furnace grades										
High colour	250–300	14–15	60	70–75	60–65	50–300	400–550	65–76	117–124	1.2–2.0
Medium colour	150–220	16–24	50–160	47–122	46–117	130–300	390–550	74–78	118–124	1.0–1.5
Regular colour	45–140	20–37	60–220	42–125	42–124	176–420	350–600	84–93	73–119	0.9–1.5
Low colour	24–45	41–75	280–430	71	64–120	256	352–512	94–99	48–69	0.6–0.9
Surface oxidized grades										
High colour	400–600	10–20	–	121	105	–	–	64	100–135	8.0–9.5
Medium colour (long flow)	100–138	23–24	–	49–60	55	240–360	530	83–84	112–135	3.5–5.0
Medium colour (medium flow)	96–110	25	–	49–72	70	225–360	480	84	112–114	2.5–3.5
Low colour	30–40	50–56	–	48–93	–	260–500	–	92–100	64	3.5

From Dannenberg et al. (1992)
DBPA, dibutyl phthalate absorption
[a] As calculated by the Brunauer, Emmett and Teller (BET) procedures
[b] A method for measuring the diffuse reflectance from a black paste with a black tile standard. The low numbers represent the 'jettest', or most intense, black grades

compressed; however, recent grades are essentially free of residual oil and sold as a dry pigment or as a suspension in linseed oil.

The ASTM (American Society for Testing and Materials, 1995b) specifications for the dry pigment are: moisture and other volatile matter, 3.0% max.; acetone extract, 1.0% max.; ash, 0.5% max.; and coarse particles (residue on a No. 325 sieve), 0.5% max. Specifications for the paste in oil are: pigment, 25% min.; linseed oil, 75% max.; moisture and other volatile matter, 0.7% max.; and coarse particles (residue on a No. 325 sieve), 1.0% max.

Thermal black

Thermal black exhibits the largest particle size and the lowest surface area of the commercial carbon blacks. Consisting of discrete spherical particles, it also has the lowest degree of aggregation or structure, and is also characterized by low oxygen content.

Thermal blacks are available in several grades. Table 5 provides information on the ASTM classification of these products, according to the system described above for furnace black.

1.1.4 Extractable impurities in carbon black

Because of their source materials, the methods of their production and their large surface areas and surface characteristics, commercial carbon blacks typically contain varying quantities of adsorbed by-products from the production process, particularly aromatic compounds. A number of methods have been developed and used to extract and characterize these adsorbed chemicals (see Section 1.1.5(*b*)). The classes of chemicals most commonly identified in these extracts are PAHs, nitro derivatives of PAHs (nitro-PAHs) and sulfur-containing PAHs. Examples of these three classes of chemicals identified in carbon black extracts are given in Table 7.

The specific chemicals detected in carbon black extracts and their relative quantities vary widely from sample to sample. Extraction method, type and grade of carbon black and after-treatments all appear to be factors that affect the types and quantities of impurities obtained. However, substantial batch-to-batch variation is typical.

Among the PAHs frequently found at the highest levels in carbon black extracts are benzo[*ghi*]perylene, coronene, cyclopenta[*cd*]pyrene, fluoranthene and pyrene. For example, in a study of five types of furnace black used in tyre manufacture, extraction with hot benzene after 250 h yielded means of 252–1417 mg extract per kg carbon black. The quantities of various PAHs found in the extracts were as follows (mg/kg): anthanthrene, < 0.5–108; benzacridine derivative, < 0.5; benzo[*def*]dibenzothiophene and benzo[*e*]acenaphthylene, < 0.5; benzofluoranthenes (total), < 0.5–17; benzo[*ghi*]fluoranthene, 20–161; benzo[*ghi*]perylene, 23–336; benzopyrenes (total), 2–40; cyclopenta-[*cd*]pyrene, < 0.5–264; coronene and isomer, 13–366; dimethylcyclopentapyrene and/or dimethylbenzofluoranthene, 2–57; fluoranthene, 10–100; indeno[1,2,3-*cd*]pyrene, 1–59; phenanthrene and/or anthracene, < 0.5–5; and pyrene, 46–432 (Locati *et al.*, 1979). The

Table 7. Some compounds identified in carbon black extracts

Polycyclic aromatic hydrocarbons (PAHs) (see also IARC, 1983)
    Acenaphthylene
    Anthanthrene
    Anthracene
    Benz[*a*]acenaphthylene
    Benz[*a*]anthracene
    Benzo[*b*]fluoranthene
    Benzo[*ghi*]fluoranthene
    Benzo[*j*]fluoranthene
    Benzo[*k*]fluoranthene
    Benzo[*a*]pyrene
    Benzo[*e*]pyrene
    Benzo[*ghi*]perylene
    Chrysene
    Coronene
    4*H*-Cyclopenta[*def*]phenanthrene
    Cyclopenta[*cd*]pyrene
    Dibenz[*a,h*]anthracene
    Fluoranthene
    Indeno[1,2,3-*cd*]pyrene
    Naphthalene
    Perylene
    Phenanthrene
    Pyrene

Nitro derivatives of PAHs (nitro-PAHs) (see also IARC, 1987b, 1989)
    1,3-Dinitropyrene
    1,6-Dinitropyrene
    1,8-Dinitropyrene
    9-Nitroanthracene
    3-Nitro-9-fluorenone
    1-Nitronaphthalene
    1-Nitropyrene
    1,3,6-Trinitropyrene

Sulfur-containing PAHs
    Benzo[*def*]dibenzothiophene
    Dibenzothiophene
    Phenanthro[4,5-*bcd*]thiophene
    Triphenyleno[4,5-*bcd*]thiophene

From Falk & Steiner (1952); Gabor *et al.* (1969); Gold (1975); Qazi & Nau (1975); Renes (1975); Wallcave *et al.* (1975); Lee & Hites (1976); Fitch *et al.* (1978); Nakajima *et al.* (1978); Fitch & Smith (1979); Locati *et al.* (1979); De Wiest (1980); Rosenkranz *et al.* (1980); Taylor *et al.* (1980); Sanders (1981); Ramdahl *et al.* (1982); Rivin & Smith (1982); Butler *et al.* (1983); Novrocik *et al.* (1983); Nishioka *et al.* (1986); Jin *et al.* (1987); Agurell & Löfroth (1993) results of two similar studies, which used benzene to extract adsorbates from a number of oil-furnace blacks and one thermal black are shown in Table 8 (Taylor et al., 1980; Zoccolillo et al., 1984).

Table 8. Benzo[a]pyrene concentrations in the benzene extracts of 10 carbon blacks

ASTM designation[a]	Surface area (m^2/g)	Total extract (mg/kg) (no. of samples)	Benzo[a]pyrene concentration (mg/kg) (no. of samples)
N220	118	250 (2)	0.29 (4)
N234	128	630 (2)	1.08 (5)
N326	80	225 (1)	0.18 (1)
N339	90	510 (4)	1.46 (2)
N347	90	343 (1)	0.50 (1)
N351	70	780 (3)	5.47 (5)
N375	101	1020 (5)	3.81 (2)
N550	42	610 (1)	0.14 (1)
N660	36	653 (6)	4.8 (6)
N990	10	8020 (1)	35.00 (1)

From Taylor et al. (1980) and Zoccolillo et al. (1984)
[a] For American Society for Testing and Materials designations of types, see Table 5

PAH fractions from six different batches of the same furnace black (ASTM designation N660) were analysed and ranged from 200 to 736 mg/kg; benzo[a]pyrene concentrations ranged from 1.2 to 9.7 mg/kg in benzene extracts (Zoccolillo et al., 1984).

Seven types of carbon black used in tyre production in Poland (domestic: JAS-220, JAS-330, JAS-530; imported: HAF-N-326, HAF-N-330, SRF-N-762 and Durex-0) were analysed. The toluene-soluble extractable compounds, including PAHs, were determined by the gravimetric method, and benzo[a]pyrene by high-performance liquid chromatography (HPLC) with a spectrometric detector. Toluene-soluble compounds were found to amount to 0.12–0.25% (by weight). Benzo[a]pyrene, at a range of 1.44–3.07 ppm [mg/kg], was detected in five of the seven carbon blacks examined (Rogaczewska et al., 1989).

Agurell and Löfroth (1993) studied the variation of impurities in a furnace carbon black (N330) manufactured in Sweden over a three-year period. The following PAHs were determined at the following ranges of concentration in benzene extracts (mg/kg carbon black): phenanthrene, 0.9–15; fluoranthene, 4.5–72; pyrene, 26–240; benzo[ghi]fluoranthene, 7.2–72; cyclopenta[cd]pyrene, 6.6–188; chrysene, 0.1–1.3; benzo[b]fluoranthene, benzo[j]fluoranthene and benzo[k]fluoranthene, 0.4–18; benzo[e]pyrene, 0.9–19; benzo[a]pyrene, 0.9–28; perylene, 0.1–3.5; indeno[1,2,3-cd]pyrene, 2–43; benzo[ghi]perylene, 14–169; and coronene, 14–169.

Somewhat higher total levels of PAH were found in extracts of thermal blacks. A 24-h benzene extract of an N990-type thermal black yielded approximately 4000 mg extract per kg carbon black. Individual PAHs included (mg/kg): benzo[*ghi*]perylene, 1217; coronene, 800; pyrene, 603; anthanthrene, 299; fluoranthene, 197; benzo[*a*]pyrene, 186; and benzo[*e*]pyrene, 145 (De Wiest, 1980). The total level of PAH in the benzene extract of another N990 thermal black sample was 2140 mg/kg, which included 35 mg/kg benzo[*a*]pyrene (Zoccolillo *et al.*, 1984).

Nitro-PAHs were identified in extracts of some samples of channel black and furnace black that had been subjected to an oxidative treatment using nitric acid. Discovery of these by-products in a photocopy toner in the late 1970s led to modifications in this oxidative treatment process, and these steps have reportedly eliminated nitro-PAHs from commercial furnace black produced since 1980 (Fitch *et al.*, 1978; Fitch & Smith, 1979; Rosenkranz *et al.*, 1980; Sanders, 1981; Ramdahl *et al.*, 1982; Butler *et al.*, 1983).

A number of oxidized PAHs (e.g. ketones, quinones, anhydrides, carboxylic acids) were also identified in carbon black samples that had undergone oxidative treatment (Fitch *et al.*, 1978; Fitch & Smith, 1979; Rivin & Smith, 1982), and one study reported that 3-nitro-9-fluorenone was detected in a nitric acid-treated carbon black used for making carbon ink in China (Jin *et al.*, 1987).

Carbon black made from high-sulfur feedstocks frequently contains detectable quantities of extractable sulfur-containing aromatic compounds such as benzothiophene derivatives (Lee & Hites, 1976; Nishioka *et al.*, 1986).

Trace amounts of a variety of inorganic elements (e.g. calcium, iron, potassium, lead, arsenic, chromium, selenium) also have been identified in some analyses of carbon black samples (Collyer, 1975; Sokhi *et al.*, 1990).

1.1.5 *Analysis*

This section briefly reviews methods of analysis to detect the presence of carbon black in various matrices, as well as methods used to isolate and analyse surface contaminants (see Section 1.1.4).

(*a*) *Carbon black in various matrices*

Because of the difficulty of separating carbon black from other airborne particulates in the workplace, total dust is usually measured as a surrogate for airborne concentration of carbon black in facilities producing or using carbon black. Both personal membrane-filter and static high-volume sampling techniques are used to collect carbon black in the work environment, followed by gravimetric analysis to arrive at the total dust concentration. Free carbon has been determined by predigestion of a sample with nitric acid to destroy organic matter followed by weighing of the residue and ignition between 140 and 700 °C. The amount of free carbon is determined by the loss of weight upon ignition (United States National Institute for Occupational Safety and Health, 1978). The gravimetric method [Method 5000] of the United States National Institute for Occupational Safety and Health for determining total dust has a working range of 1.5–10 mg/m^3

for a 200-L sample of air and an estimated limit of detection of 0.03 mg/sample (Eller, 1994).

The American Society for Testing and Materials (1995a) has published similar methods for the analysis of carbon black in several natural and synthetic rubbers. It has been reported that thermogravimetric analysis is accurate for determining the carbon black content of rubbers in the range of 0.1 to 30% by weight (Charsley & Dunn, 1981).

(b)   Adsorbates on carbon blacks

Several methods have been reported for the extraction and analysis of adsorbates on carbon black. Soxhlet extraction with various organic solvents has been the primary method used to remove the adsorbed chemicals from the carbon black samples, but vacuum sublimation or extraction combined with sonification have also been used (Zoccolillo *et al.*, 1984). The efficiency of the Soxhlet extraction depends on the extraction time and solvent, the type of carbon black, the relationship between sample weight/solvent volume and the amount of extractable material. Some solvents can react with the surface groups of carbon black and form artifacts during the extraction (Fitch *et al.*, 1978).

Locati *et al.* (1979) found that in five furnace blacks a Soxhlet extraction time of 150 h was necessary to remove 95% of the benzene-extractable matter and 250 h for exhaustive extraction. They also observed that the lower the relative molecular mass was, the shorter the time necessary to obtain extraction was.

Taylor *et al.* (1980) examined the solvent efficiency of three solvents (24-h Soxhlet) as measured by benzo[*a*]pyrene extractability from five furnace blacks. They found that toluene and benzene had quite similar efficiencies, but that cyclohexane could not remove more than 10% of the benzene-extractable benzo[*a*]pyrene from any of the furnace blacks. Toluene was, however, clearly the best extractant when the adsorbate content of the carbon black was low (less than 1 mg/kg).

Giammarise *et al.* (1982) found that benzene, toluene, monochlorobenzene and *ortho*-dichlorobenzene were all effective extraction solvents for nitropyrenes from an old carbon black sample with a high level of nitropyrene impurities (approximately 70 mg/kg). Monochlorobenzene was the best extractant. When a current carbon black with only traces of nitropyrene impurities (approximately 0.5 mg/kg) was extracted, monochlorobenzene removed more than 90% of the nitropyrenes within 24 h, while toluene extracted only 60% in that time.

Analytical methods used to determine the components of the carbon black extracts produced by Soxhlet extraction of carbon black with various solvents have been summarized (Jacob & Grimmer, 1979). Common methods include gas chromatography (GC) with packed and capillary columns and HPLC with spectrophotometric and spectrofluorimetric detection.

Zoccolillo *et al.* (1984) reported the determination of PAHs in carbon black by Soxhlet extraction with benzene, purification by silica gel thin-layer chromatography and analysis by GC and/or HPLC.

Colmsjö and Östman (1988) reported a method to isolate and fingerprint some high-molecular-weight PAHs in carbon black. The PAH fraction of a carbon black extract (Soxhlet-extracted with dichloromethane) was isolated with a backflush technique and applied to an amino-bonded stationary phase for HPLC. This fraction was further separated by reverse-phase HPLC and each subfraction was analysed by low-temperature fluorescence.

Sigvardson and Birks (1984) reported that selective detection of nitro-PAHs in carbon black was achieved by Soxhlet extraction with toluene, evaporation to dryness, dissolution in dichloromethane and direct injection into the HPLC column. The nitro-PAHs were reduced online to the corresponding amino-PAHs and detected by peroxyoxalate chemiluminescence.

Jin et al. (1987) described a method for the analysis of nitroarenes in carbon black. The method involved the Soxhlet extraction of the sample with organic solvents (the use of chlorobenzene resulted in the highest overall yield), pre-separation by column chromatography on silica gel and separation and determination by reverse-phase HPLC with ultraviolet detection.

The bioavailability of the PAHs adsorbed onto the surface of the carbon black has been assessed by means of quantifying the concentration of the major adsorbed PAH, pyrene, by use of its urinary metabolite, 1-hydroxypyrene. The urine was adjusted to pH 5.0 and incubated with 50 µL β-glucuronidase/aryl sulfatase for 4 h at 37 °C. After extraction and washing, the hydrolysed urine was injected into a HPLC unit with a fluorescence detector. The limit of detection was approximately 0.075 nmol/L [16 ng/L] (Gardiner et al., 1992b).

1.2 Production and use

1.2.1 Production

The early Chinese and Hindus produced carbon black for their inks and lacquers by a simple lampblack process. Lampblack supplied the needs of the pigment industry until the opening of the natural gas fields in the United States in 1872 and the establishment of the channel process. At that time, annual world consumption of carbon black was less than 1000 tonnes. Consumption increased rapidly following the discovery in 1904 of carbon black's usefulness in the reinforcement of rubber; the needs of the growing world rubber industry were met by supplies of channel black and lampblack from the United States. In 1922, the gas-furnace process was introduced in the United States (Garret, 1973). The increasing cost of natural gas led to a switch to the oil-furnace process in the early 1940s and to the final closure of channel black manufacture in the United States in 1976. Since oil feedstock is readily transported, the oil-furnace process can be located close to the user industries, and, following the end of the Second World War, carbon black manufacture was established in many industrialized countries (Dannenberg et al., 1992).

Worldwide production of carbon black in 1993 was about six million tonnes (estimate based on demand for carbon black; see Table 11). Production data on carbon black in several countries from 1987 to 1994 are presented in Table 9, and production data on carbon black by grade in the United States, western Europe and Japan in 1988 are presented in Table 10.

Table 9. Production of carbon black in several countries from 1987 to 1994 (thousand tonnes)

Country	1987	1988	1989	1990	1991	1992	1993	1994
Canada	172	181	190	178	157	161	166	167
China	290	290	320	327	334	NA	NA	NA
France	232	241	270	252	224	232	204	235
Germany	361	379	401	393	380	375	334	297
Italy	190	200	202	180	185	182	171	187
Japan	635	740	779	783	793	771	702	698
Republic of Korea	NA	NA	NA	NA	232	248	300	311
USA	1355	1324	1320	1302	1234	1370	1462	1503

From Anon. (1989, 1991a); Dannenberg et al. (1992); Anon. (1993); China National Chemical Information Centre (1993); Anon. (1995)
NA, not available

Table 10. Production of carbon blacks for rubber in 1988 in the United States, western Europe and Japan (thousand tonnes)

Grade	USA	Western Europe	Japan
Tread grades			
N100	35	28	37
N200	158	161	118
N300	555	528	300
Total	748 (55.2%)	717 (63.8%)	455 (61.5%)[a]
Non-tread grades			
N500	120	153	136
N600	326	137	87
N700	129	103	29
N900 (thermal)	23	–	9
Total	598 (44.1%)	393 (35.0%)	261 (35.3%)[a]
Other grades			
Acetylene	9	14	24
Total carbon black	1355	1124	740[a]

Adapted from Dannenberg et al. (1992)
[a] Calculated by the Working Group

Carbon black is produced by many companies in Russia, 12 companies in China, six companies each in India, Japan and the United States, three companies each in Canada, France, Italy and the Netherlands, two companies each in Australia, Brazil, Germany, the Republic of Korea, Poland, Spain, Thailand and the United Kingdom, and one company each in Chile, Colombia, the Czech Republic, Estonia, Hungary, Iran, Malaysia, Mexico, the Philippines, Portugal, Romania, South Africa, Sweden, Taiwan and Turkey (Chemical Information Services, 1994; Cabot Co., 1995).

Acetylene black

Acetylene black was first made commercially in Germany in 1928, in Canada in the 1930s, in Japan in 1942 and in the United States in 1964. It is currently estimated to comprise substantially less than 1% of total carbon black production (Dannenberg et al., 1992).

The dissociation of acetylene into carbon and hydrogen was achieved as early as 1861, and the first commercial process was based on the partial combustion of acetylene (Bean, 1964). Subsequently, a process based on the explosion of an electric arc was developed in Germany. The process currently in use — continuous thermal decomposition — is covered by a series of patents going back to 1938 and was reported in 1964 to have been in commercial use for some years in continental Europe, Asia and Canada (Bean, 1964; Claassen, 1978).

In the continuous thermal decomposition process for acetylene black, the reaction is initiated by burning the acetylene feedstock with a controlled amount of air. When the reaction temperature is sufficiently high (e.g. 800 °C), the air supply is shut off, oxidation ceases and an exothermic self-sustained dissociation of acetylene to form hydrogen and acetylene black occurs at temperatures of up to 1000 °C (Dannenberg et al., 1992).

Channel black

The channel process for making carbon black was first used commercially in the United States in 1872 (Garret, 1973). From the First World War to the Second World War, the channel black process accounted for most of the carbon black used worldwide for rubber and pigment applications. However, rising prices of natural gas, smoke pollution, low yield and the rapid development of the furnace-process grades of carbon black have been given as reasons for this process being abandoned in 1976 in the United States; operations still exist and are being expanded in Europe (Dannenberg, 1978; Dannenberg et al., 1992).

In the channel or impingement process, small natural-gas flames were impinged on channel irons that collected the deposited carbon black (Garret, 1973). This process gave only very low yields (5%); however, in Germany, a plant making carbon black by an impingement process, sometimes called 'gas black', is reported to give yields of 60% using coal-tar residues containing naphthalene or anthracene as the carbon feedstock. In this process, the molten material is evaporated by a stream of hot coke-oven gas and heated to about 370 °C prior to reaching the burners. The flames are directed onto revolving water-cooled pipes and the carbon black, which forms on impingement, is continuously scraped from the pipes. For the production of finer-particle black for use as pigments, the amount of oil carried by the gas is decreased and the vapours to the burner are diluted with air (Claassen, 1978).

Production of channel black in the United States reached a peak of 307 000 tonnes in 1948, but it had fallen to 132 000 tonnes by 1960 and showed a steady decline until production stopped in 1976. The quantity of carbon black made by the manufacturer in Germany using the impingement process is believed to constitute less than 1% of total world production of carbon black.

Furnace black

The gas-furnace process for making carbon black was first introduced in the United States in 1922, and the oil-furnace process in 1943 (Garret, 1973). Furnace black was first produced commercially in Japan around 1950. The gas-furnace process, which is based on the partial combustion of natural gas, was carried out using refractory-lined retorts or furnaces at a temperature of 1200–1500 °C. The process achieved only low yields of carbon black and has not been used in the United States since the 1960s (Dannenberg, 1978). In the oil-furnace process, which is now used to produce over 95% of total output of all carbon black, a heavy aromatic feedstock from a petroleum refinery or petrochemical operation is injected by atomization into a high-velocity stream of combustion gases produced by the complete burning of an auxiliary fuel (such as natural gas) with excess air. Although some of the feedstock is burned at 1200–1850 °C, most is converted to hydrogen and carbon black with high yields. Downstream, the reaction gases are cooled by spraying with water. The carbon black particles are then separated from the gases and pelletized. A very small percentage of furnace black is subjected to after-treatment by various oxidation processes, some of which have involved nitration. In the United States, about 95% of feedstocks are decant oils (clarified heavy distillates from the catalytic cracking of gas oils); European feedstock sources are 50% decant oils and 50% ethylene tars and creosote oils (Dannenberg *et al.*, 1992).

Lampblack

Lampblack was first produced commercially in the United States in the 1840s (Patterson, 1980). It is made principally by burning aromatic petroleum oils and coal-tar products, such as creosote and anthracene oils, in open, shallow pans using a restricted air supply (Smith, 1964). This is carried out at temperatures lower than those for other carbon black processes. The lampblack is separated from the gases and pelletized. Currently, only a few plants located mostly in western and eastern Europe still produce these rather coarse blacks (mean particle diameter, approximately 100 nm) which have special properties (Vohler *et al.*, 1986; Dannenberg *et al.*, 1992). Lampblack is not produced in Japan.

Lampblack is produced on a small scale; total production is believed to constitute less than 1% of total world production of carbon black.

Thermal black

In the thermal process, which dates back to 1922, a chamber filled with checkered brickwork is heated to about 1300 °C by injecting a burning mixture of gas and air. When the required temperature has been reached, the flow of burning gas is stopped and the hydrocarbon feedstock (usually gas) is injected. Contact with the hot bricks causes the feedstock to crack, forming carbon black and hydrogen. This process is run cyclically using two chambers, one being heated while the other produces carbon black. In one plant in the United Kingdom, medium thermal black is produced from oil rather than using natural gas as the raw material (Johnson & Eberline, 1978; Dannenberg et al., 1992).

Total production of thermal black is estimated to be about 2% of total carbon black production of North America, western Europe and Japan.

1.2.2 Use

The primary use of carbon black (for example, approximately 90% in the United States) is in rubber products (see IARC, 1982), including tyres (69% in the United States), tubes, treads and other automotive products (about 10% in the United States) and other industrial rubber products (11% in the United States). Miscellaneous non-rubber uses (approximately 10%) include applications as pigments in paints, plastics, paper, inks and ceramics. These levels of use have been steady for the past 10 years (Anon., 1985, 1988, 1991b, 1994). World demand for carbon black by region is presented in Table 11. Western Europe consumes 74% in tyres and other automotive products and almost 20% in other industrial rubber products. Applications as pigments in western Europe and Japan account for 5–6% of consumption (Dannenberg et al., 1992).

Table 11. World carbon black demand by region (thousand tonnes)

Region	1983	1993
North America	1387	1664
Central and South America	221	336
Western Europe	1058	1102
Eastern Europe	1022	556
Africa and the Middle East	107	147
Asia and Oceania	1139	2173
World	4934	5978

From The Freedonia Group (1994a)

Information on the quantities of carbon black used in various applications is very seldom presented in a form that provides separate data on the individual types of carbon black. However, it can be inferred that the major carbon black used is furnace black, since this is the predominant item in commerce, and that thermal black follows as a distant second place; minor quantities of the other carbon blacks are used in highly specialized applications. Most carbon black is supplied as wet or dry pellets, but very small amounts are still shipped in bags as fluffy black.

Carbon black is an intense black pigment, but its principal industrial use today is based on its ability to reinforce natural and synthetic rubbers (for a description of the rubber manufacturing processes in which carbon blacks are used, see IARC, 1982). Addition of carbon black in quantities in the range 10–150 parts per 100 parts by weight of rubber polymer results in very marked improvements in the properties of vulcanized rubbers, particularly in terms of resistance to abrasion, tear strength, tensile strength, stiffness and hardness. Addition of carbon black also changes the properties of rubbers in the unvulcanized condition, so improving handling and shaping in the manufacture of all types of rubber products. Carbon black is particularly useful in its ability to reinforce rubber. The world rubber industry is thus dependent on the use of carbon black.

The most important product of the rubber industry is the pneumatic tyre and this represents the single largest application of carbon black. For every 100 parts by weight of rubber used in the manufacture of a tyre, there are about 60 parts by weight of carbon black. Since tyres also contain steel and textile materials, carbon black represents about 25% of the total weight of a finished pneumatic tyre. The consumption of the various grades of carbon black can be divided into 'tread grades' for tyre reinforcement and 'non-tread' grades for non-tread tyre use and other applications. In the United States, 55% of carbon black produced for rubber is for tread grades; tread-grade production is 64% in western Europe and 60% in Japan (Dannenberg et al., 1992).

Many of the non-tyre applications of carbon black in rubber are also for the automotive industry — for example, in hoses, weatherstrips, sponge seals and engine mountings. Overall, about 80% of total carbon black consumption is for automotive applications.

For more than a thousand years before the discovery in 1904 of its reinforcing effect in rubber, carbon black was used as a pigment. Today, although pigments represent less than 10% of the total usage, it is used in inks, paints, lacquers, cements, paper, coatings and in plastics, where it is also used as an ultraviolet absorber.

Acetylene black

Acetylene black is used primarily for speciality applications because of its relatively high cost. Approximately 95% of worldwide use of acetylene black is in the manufacture of dry-cell batteries. Because of its ability to absorb large quantities of electrolyte, acetylene black imparts greater capacity, longer shelf-life and lower resistance to dry cells than any other filler. Its high thermal and electrical conductivity also imparts desirable properties to certain rubber and plastic products, such as thermal insulators, belt drives, cable sheathing, hoses and shoe soles (Bean, 1964; Union Carbide Corp., 1964; Claassen, 1978; Gulf Oil Chemicals Company, 1982; Vohler et al., 1986).

Channel black

Channel black has been used for rubber reinforcement and as a pigment. In rubber reinforcement, it was reported to yield products with high tensile strength, high elongation and high tear resistance. With the smallest particle size, channel black gave high colour intensity when used as a pigment in paint, ink and plastics. The carbon black made by the impingement process in Germany also reportedly finds limited use in rubber reinforcement and more extensive use in pigment applications in printing inks, plastics, lacquers and coatings. High-quality oxidized black from this process is particularly useful in deep black lacquers and coatings (Claassen, 1978; Vohler *et al.*, 1986).

Furnace black

Over 95% of the carbon black in the rubber industry is produced by the furnace process (Garret, 1973), and furnace black is also used in printing inks, plastics and paints. For these different purposes, a wide variety of specially tailored grades possessing the necessary properties are available (for example, different grades are used in tyre sidewalls).

The furnace black process has the advantage of being continuous and carried out in closed reactors, which means that all parameters and inputs can be controlled. Properties of carbon black, such as surface area, particle size, structure, absorptivity, abrasion resistance, tint strength and others, can systematically be varied in the furnace black process by adjusting the operating parameters. Thus, most semi-reinforcing rubber blacks (SRF, GPF, FEF) with specific surface areas of 20–60 m^2/g, and active reinforcing blacks (HAF, SAF, ISAF) (see Table 5 for definition of these terms) with specific surface areas of 65–150 m^2/g are manufactured by this process, as are, to an increasing extent, pigment-grade types of carbon black with much larger surface areas and smaller particle sizes (Vohler *et al.*, 1986).

The major application of furnace black in the rubber industry is in the manufacture of tyres, retread rubber and inner tubes. Other automotive uses in elastomers include belts, hose, motor mounts, O-rings and wire and cable covers. Non-automotive uses in elastomers include coated fabrics, conveyor belts, floor mats, footwear, gaskets, gloves, hard rubber products, hose, packaging, pontoons, toys and wire and cable covers.

The majority of the furnace black used in printing inks is for newspaper inks, with the remainder divided among lithographic/offset, gravure, letterpress, flexographic and other inks. Furnace black is also used as a colourant in alkyd and acrylic enamels, industrial finishes, lacquers and a variety of other paints.

Furnace black is used in plastics principally for the following purposes: as an antistatic agent, colourant, filler (sometimes to impart strength) and ultraviolet radiation stabilizer and as an additive to increase or decrease electrical conductivity. End-uses include appliances, automotive accessories, extrusion and calender coatings, film, housewares, phonograph records, pipe and conduit, and wire and cable (Dannenberg, 1978; Dannenberg *et al.*, 1992).

Lampblack

Most of the lampblacks produced currently are coarse particulates with special properties (Vohler *et al.*, 1986). They are used mainly as non-reinforcing or semi-reinforcing blacks in rubber goods. Lampblacks are used to a lesser extent as pigments for tinting and shading cosmetics, enamels, inks, lacquers, paints and plastics. Lampblack pigments are readily dispersible and have little tendency to float in paint or ink formulations (Claassen, 1978). The principal use of lampblack in the pigmentation of artists' paints is in water colours; it has a more minor role in oil colours (Levison, 1973).

The physical and electrical properties of lampblack make it useful in the production of arc carbons, brushes and resistors. Its high tinting strength and hiding power has led to its use in blackboards, cement, crayons and leather (Garret, 1973).

Thermal black

Thermal black is used principally in mechanical rubber goods with high filler content (Vohler *et al.*, 1986). Thermal black is used in non-tyre rubber when low reinforcement is required, and it is used in speciality polymers and in neoprene, nitrile and ethylene–propylene elastomers (Patterson, 1980). End-product applications include belts, footwear, gaskets, hose, mechanical goods, V-belts, O-rings and seals, tyre innerliners and wire insulation (Dannenberg, 1978).

1.3 Occurrence

1.3.1 *Natural occurrence*

Carbon black is not known to occur as natural product.

1.3.2 *Occupational exposure*

Occupational exposure by any route to carbon black has been reduced markedly in the last 30–35 years, mainly by technological improvements, increases in the proportion of the product that is bulk loaded (by trucks and trains) and legislative enforcement. Until recently, very few reliable data on occupational exposure to carbon black were available, but two major prospective cross-sectional studies in the United States and Europe (France, Germany, Italy, the Netherlands, Spain, Sweden and the United Kingdom) have characterized exposure accurately in these workforces. These data and those from the previous studies have been reviewed (Gardiner, 1995a) but a further summary of the more important data is given below.

The majority of the studies of occupational exposure do not report adequately the sampling strategies with which the data were collected. Only in one study were the measurements taken specifically for the purpose of an epidemiological study and hence used techniques such as person/day randomization. Rarely were the type (personal/-static), duration or rationale for the number of samples stated. It is possible to measure a variety of aerosol fractions, but usually respirable and total inhalable dust are measured. Respirable dust is that fraction of an aerosol with an aerodynamic diameter suitable for penetration into the alveoli/gas exchange region of the lung (typically < 10 μm). Total inhalable dust is that fraction of an aerosol with an aerodynamic diameter suitable for inspiration into the respiratory system (typically < 100 μm). Differences in definitions of these fractions and in the methodologies by which they are measured require that inter-study comparison should be undertaken with care.

Kollo (1960) took 160 measurements in a Russian channel black plant where airborne dust levels ranged from 44 to 407 mg/m^3 in the factory area, 25.3 to 278.6 mg/m^3 in the working aisles, 9.3 to 972 mg/m^3 in the pelletizing area and 26.7 to 208.6 mg/m^3 in the packing area.

Sands and Benitez (1961) used a static high-volume sampler to measure total dust in various areas of three factories processing rubber (one in Uruguay and two in the United States). The range in the three factories was 0.14–4.59 mg/m^3, 1.06–17.66 mg/m^3 and 1.06–38.84 mg/m^3, with the range being 1.77–38.84 mg/m^3 by the Banbury mixer loading area, 1.41–13.42 mg/m^3 during milling and 0.14–4.24 mg/m^3 in the general air of the milling rooms. It was suggested by the authors that 3.5 mg/m^3 represented a safe and achievable air standard which prompted the American Conference of Governmental Industrial Hygienists (ACGIH) to propose a threshold limit value (TLV) of 3.5 mg/m^3 in 1965, which was adopted in 1967 (American Conference of Governmental Industrial Hygienists, 1993).

Komarova (1965) measured exposure to carbon black in the packaging department of two Russian factories manufacturing lampblack and furnace black. The number of measurements was not specified, but the ranges were 166–1000 mg/m^3 (lampblack) and 60–78 mg/m^3 (furnace black). Slepicka et al. (1970) found exposures ranging from 8.4 to 29.0 mg/m^3 in two Czechoslovakian channel black factories between 1960 and 1968, although neither the number of samples nor their location was quoted.

A survey found a range of concentrations of 90–196 mg/m^3 (from an unspecified number of samples) in a Russian furnace black factory (Spodin, 1973). The lowest and highest average concentrations recorded by another Russian study were 1.53 ± 0.4 mg/m^3 for workers by the hatches of the electrostatic filter and 34.5 ± 8.9 mg/m^3 for workers involved in cleaning the production areas; in total, 109 samples were taken. It was noted that throughout the 1960s and 1970s, workers packing carbon black were exposed to two to seven times the maximal permissible concentration (10 mg/m^3 in 1975) for 60–70% of their working shifts (Troitskaya et al., 1975, 1980).

Between July 1972 and January 1977, the United States Occupational Safety and Health Administration conducted 85 workplace investigations to determine compliance with the occupational exposure limit for carbon black of both manufacturers and users. Approximately 20% of the workplaces inspected were in violation of the total inhalable exposure limit of 3.5 mg/m^3, and about 60% of these were one to two times above the limit (United States Occupational Safety and Health Administration, 1977).

A number of Health Hazard Evaluations have been conducted by the United States National Institute of Occupational Safety and Health in facilities either producing or using carbon black (Belanger & Elesh, 1979; Hollett, 1980; Salisbury, 1980; Boiano & Donohue, 1981). In general, these measurements were less than 3.5 mg/m^3, although these studies involved a limited number of samples and a limited number of days over which the measurements were taken.

In the rubber industry, employees are exposed to carbon black mainly in the compounding and Banbury mixing areas. It has been reported that for total dust (in which carbon black was one component) the median levels in 14 tyre and tube manufacturing plants in the United States were 1.7 mg/m^3 for the compounding area samples (individual plant means ranged up to 3.9 mg/m^3) and 1.3 mg/m^3 for the Banbury mixing-area samples (for which the highest plant mean was 4.2 mg/m^3). The values for personal samples were 3.1 mg/m^3 for the compounding area (highest plant mean, 5.0 mg/m^3) and 1.9 mg/m^3 for the Banbury area (highest plant mean, 5.8 mg/m^3) (Williams et al., 1980). A United States National Institute for Occupational Safety and Health study (Heitbrink & McKinnery, 1986) evaluated the effect of control measures at Banbury mixers and the mills beneath the mixers in tyre factories and found lower exposures than those found by Williams et al. (1980). The geometric means of mixer operators' exposures at five factories ranged from 0.08 to 1.54 mg/m^3 and the geometric means of milling operators' exposures at three factories ranged from 0.20 to 1.22 mg/m^3.

Over a six-month period, beginning in October 1979, a total of 1951 personal samples (1564 total dust, 387 respirable dust) were collected from 24 carbon black production facilities in the United States (Smith & Musch, 1982). A summary of the results are provided in Table 12. Workers involved in filling and stacking bags of carbon black (material handling) had the highest mean total dust exposures up to 2.2 mg/m^3. Samples were not taken from all employment areas in every factory and the numbers of samples taken differed from area to area. These data were subsequently used in the first cross-sectional analyses of the American respiratory morbidity study (Robertson et al., 1988) and the update of the American cohort study examining circulatory, respiratory and malignant diseases (Robertson & Ingalls, 1989).

Table 12. Summary of average dust exposure by employment area in United States carbon black production facilities (1979–80)

Area of employment	Total dust			Respirable dust		
	No. of plants	No. of samples	Geometric mean (mg/m^3)	No. of plants	No. of samples	Geometric mean (mg/m^3)
Administration	8	72	0.01	2	28	0.00
Laboratory	17	133	0.04	10	35	0.01
Production	22	480	0.44	14	111	0.13
Maintenance	19	386	0.59	11	89	0.12
Material handling	20	493	1.45	13	124	0.35

From Smith & Musch (1982)

In a mortality study conducted in the United Kingdom (Hodgson & Jones, 1985), the authors used a limited amount of exposure data collected by Her Majesty's Factory

Inspectorate in 1976. Personal samples were taken of 47 people in five carbon black factories, 24 (51%) of these being above 3.5 mg/m^3. The highest exposure recorded for routine work was 79 mg/m^3, but workers engaged in filter-bag replacement may have been exposed to even higher levels, although exposure measurements were not reported.

The United States particulate sampling survey of 1979–80 (Smith & Musch, 1982) was repeated twice, once in 1980–82 and then again in 1987 (Musch & Smith, 1990). The number of participating companies decreased from seven to six and the number of plants decreased from 24 to 17. In 1980–82, 973 total dust samples were taken; the number fell to 577 in 1987. The data are summarized in Table 13. A drop of approximately 50% in exposure was evident in maintenance and material handling parts of the factories. Of the job categories in the maintenance section, the following reduction was seen between the second and third surveys: utility, 0.89 to 0.55 mg/m^3; in plant, 0.79 to 0.52 mg/m^3; shop, 1.00 to 0.07 mg/m^3; instrument, 0.47 to 0.17 mg/m^3; and foreman, 0.35 to 0.18 mg/m^3. Of the job categories in the material handling section, the following reduction was seen between the second and third surveys: stack and bag, 1.92 to 0.77 mg/m^3; bagger, 2.67 to 0.85 mg/m^3; bulk loader, 2.07 to 0.82 mg/m^3; stacker, 1.15 to 0.70 mg/m^3; fork-lift truck driver, 0.53 to 0.34 mg/m^3; and foreman, 0.18 to 0.02 mg/m^3.

Table 13. Summary of average total dust exposure by employment area in United States carbon black production facilities in 1980–82 and 1987

Area of employment	1980–82		1987	
	No. of samples	Geometric mean (mg/m^3)	No. of samples	Geometric mean (mg/m^3)
Administration	4	0.06	2	0.02
Laboratory	85	0.51	23	0.20
Production	273	0.45	164	0.45
Maintenance	363	0.71	181	0.36
Material handling	248	1.63	207	0.71

From Musch & Smith (1990)

The most recent and comprehensive data come from the exposure assessment element of the trans-European respiratory morbidity study (Gardiner et al., 1993). The first data published from this study were from a pilot study assessing the bioavailability of the adsorbed PAHs. Five individuals packing carbon black into 25-kg bags were assessed over the period of a week; their weekly personal mean dust exposures were 1.53, 5.30, 9.56, 9.99 and 13.21 mg/m^3 (Gardiner et al., 1992b).

The first cross-sectional phase of the prospective European study was conducted between 1987 and 1989 with the second extending from 1990 to 1992. A fully randomized, epidemiologically based sampling strategy was used to minimize the inherent biases of worker selection, and a statistically relevant proportion of all job categories in all 18 plants was taken. In addition, both respirable and total inhalable dust fractions were measured. In the first phase, 1278 respirable and 1288 total inhalable dust samples were taken (Gardiner *et al.*, 1992a). The use of a unique multiplication factor derived from the variability of the phase I data (Gardiner, 1995b) means that, in phase II, significantly more samples were taken — 2941 respirable and 3433 total inhalable dust samples (Gardiner *et al.*, 1996). The respirable dust data for both phases are presented in Table 14 and the data for total inhalable dust are summarized in Table 15.

As with the data presented by Musch and Smith (1990), it is evident that in the three years between the two surveys exposure had decreased by approximately 50% (total inhalable dust, 49.9%; respirable dust, 42%) (Gardiner *et al.*, 1996).

The National Occupational Exposure Survey conducted by the United States National Institute for Occupational Safety and Health (1995) between 1981 and 1983 indicated that about 1 729 000 employees in the United States were potentially exposed to carbon black. The estimate is based on a survey of companies and did not involve measurements of actual exposure, and might, for many workers, involve very low levels and/or incidental exposure to carbon black.

No data were available on exposure to carbon black in the non-automotive rubber, paint, printing or printing ink (i.e. 'user') industries. Operators in these user industries who handle the fluffy or pelletized carbon black during rubber, paint and ink production are expected to have significantly lower exposures to carbon black than workers in carbon black production. Other workers in user industries have little opportunity for exposure. End users of these products (rubber, ink or paint) are not exposed to carbon black *per se*, since it is bound within the product matrix.

1.3.3 *Ambient air*

In 1978, it was estimated that 1240 tonnes of carbon black were emitted during carbon black manufacture in the United States (Rawlings & Hughes, 1979). Table 16 summarizes typical particulate carbon black emissions into the air during various stages of manufacture by the oil-furnace process prior to 1979. The particulate matter was reported to comprise carbon black (McBath, 1979).

Rivin and Smith (1982) reviewed the literature on emissions of carbon black into the atmosphere during its manufacture. Modern carbon black plants generally employ bag filters to reduce emissions; discharge from a bag filter in good condition during this process (under normal conditions) reportedly contains carbon black (wet basis) at less than 50 mg/m^3, a concentration that is not visible (Johnson & Eberline, 1978).

Tyre dust, of which carbon black is a component, was estimated, in a 1969 study, to account for approximately 0.8% of the aerosol above an urban area in California. An estimated 0.2% of the particulates in the aerosol consisted of elemental carbon contributed by tyre dust (Friedlander, 1973).

Carbon black was not detected in the atmosphere around a factory in Germany where it was manufactured (Deimel & Dulson, 1980).

Table 14. Respirable dust data from phase I and II of the European respiratory morbidity study

Job title	Phase I			Phase II		
	No. of samples	Geometric mean (mg/m^3)	Maximum (mg/m^3)	No. of samples	Geometric mean (mg/m^3)	Maximum (mg/m^3)
Administrative staff (office bound, phase I)	278	0.15	1.09	302	0.09	2.07
Administrative staff (non-office bound, phase II)	–	–	–	163	0.12	1.16
Laboratory assistant	134	0.20	16.36	320	0.12	2.98
Process control room operator	44	0.18	1.33	159	0.13	5.28
Instrument mechanic	61	0.20	1.94	181	0.18	24.65
Electrician	62	0.27	7.60	134	0.15	1.59
Process foreman	79	0.21	0.94	253	0.18	3.36
Furnace operator	48	0.27	5.59	144	0.22	4.16
Fitter	107	0.35	4.77	238	0.19	3.18
Welder	41	0.27	1.10	66	0.28	7.71
Process operator	177	0.29	19.36	310	0.16	3.35
Conveyor operator	53	0.20	1.74	79	0.30	3.10
Warehouse/packer	159	0.50	13.92	408	0.35	19.00
Cleaner	35	0.32	8.60	183	0.27	20.70

From Gardiner et al. (1996)

Table 15. Total inhalable dust data from phase I and II of the European respiratory morbidity study

Job title	Phase I				Phase II			
	No. of samples	Geometric mean (mg/m^3)	% > 3.5 mg/m^3	Maximum (mg/m^3)	No. of samples	Geometric mean (mg/m^3)	% > 3.5 mg/m^3	Maximum (mg/m^3)
Administrative staff (office bound, phase I)	289	0.18	0.0	2.77	302	0.12	0.0	1.71
Administrative staff (non-office bound, phase II)	–	–	–	–	186	0.19	0.0	2.56
Laboratory assistant	143	0.38	2.1	9.66	326	0.24	0.0	2.44
Process control room operator	50	0.29	0.0	2.19	157	0.17	0.0	1.40
Instrument mechanic	59	0.55	8.5	22.67	257	0.40	1.6	8.61
Electrician	58	0.65	10.3	31.12	193	0.33	0.0	3.33
Process foreman	70	0.37	0.0	3.42	296	0.25	0.7	5.03
Furnace operator	56	0.54	5.4	9.21	156	0.44	1.9	9.48
Fitter	95	1.28	11.6	22.77	270	0.53	2.6	6.87
Welder	41	1.18	7.3	9.56	75	1.06	10.7	8.75
Process operator	179	0.96	16.8	30.75	420	0.44	1.4	16.92
Conveyor operator	59	0.53	5.1	4.20	127	0.64	4.7	8.97
Warehouse/packer	151	1.96	35.1	41.11	490	0.96	12.0	19.95
Cleaner	38	1.24	21.1	21.17	178	0.58	8.4	18.04

From Gardiner *et al.* (1996)

Table 16. Typical particulate emissions during the manufacture of carbon black by the oil-furnace process

Source	Range (kg/tonne)	Average (kg/tonne)
Main process vent (uncontrolled)	0.1–5	3.27
Flare	1.2–1.5	1.35
Carbon monoxide boiler and incinerator	–	1.04
Dryer vent:		
Uncontrolled	0.05–0.40	0.23
Bag filter	0.01–0.40	0.12
Scrubber	0.01–0.70	0.36
Pneumatic system vent:		
Bag filter	0.06–0.70	0.29
Vacuum clean-up system vent:		
Bag filter	0.01–0.05	0.03
Fugitive emissions	–	0.10
Solid waste incinerator (where used)	–	0.12

From McBath (1979)

Because the production of carbon black often involves the combustion of aromatic residual oils from petroleum refining and coal tar, the carbon black industry must also consider potential ambient air emissions of sulfur and nitrogen oxides, hydrogen sulfide, volatile hydrocarbons and carbon monoxide (The Freedonia Group, 1994b).

1.4 Regulations and guidelines

Occupational exposure limits and guidelines for carbon black are presented in Table 17. The use of carbon black from hydrocarbon sources in food contact materials is provisionally accepted by FAO (Food and Agriculture Organization of the United Nations)/WHO (World Health Organization) (United Nations Environment Programme, 1995).

The United States Food and Drug Administration (1976, 1994a) has banned the use of carbon black (prepared by the impingement or channel process) for direct use in food, drugs and cosmetics (21 CFR 81.10). The United States Food and Drug Administration (1994b) has approved the following uses of carbon black:

– Carbon black (channel process) is permitted as an indirect food additive as a component of adhesives that come in contact with food (21 CFR 175.105);

– Carbon black (channel process, prepared by the impingement process from stripped natural gas) is permitted as a colourant (21 CFR 178.3297) in resinous and polymeric coatings that come in contact with food (21 CFR 175.300) and in rubber articles intended for repeated use that come in contact with food (21 CFR 177.2600);

Table 17. Occupational exposure limits and guidelines for carbon black

Country	Year	Concentration (mg/m^3)	Interpretation
Argentina	1991	3.5	TWA
Australia	1993	3	TWA
Belgium	1993	3.5	TWA
Bulgaria[a]	1995	3.5	TWA
Canada	1991	3.5	TWA
Colombia[a]	1995	3.5	TWA
Denmark	1993	3.5	TWA
Finland	1993	3.5	TWA
		7	STEL (15 min)
France	1993	3.5	TWA
Germany	1995	None	
Jordan[a]	1995	3.5	TWA
Mexico	1991	3.5	TWA
Netherlands	1994	None	
New Zealand[a]	1995	3.5	TWA
Philippines	1993	3.5	TWA
Republic of Korea[a]	1995	3.5	TWA
Russia	1993	4	MAC
Singapore[a]	1995	3.5	TWA
Sweden	1993	3	TWA
United Kingdom	1995	3.5	TWA
		7	STEL (15-min)
USA			
ACGIH (TLV)	1995	3.5[b]	TWA
NIOSH (REL)	1994	3.5[c] (Ca)	TWA
OSHA (PEL)	1994	3.5	TWA
Viet Nam[a]	1995	3.5	TWA

From US National Institute for Occupational Safety and Health (NIOSH) (1994a,b); US Occupational Safety and Health Administration (OSHA) (1994); American Conference of Governmental Industrial Hygienists (ACGIH) (1995); Deutsche Forschungsgemeinschaft (1995); Health and Safety Executive (1995); United Nations Environment Programme (1995)

TWA, time-weighted average; STEL, short-term exposure limit; MAC, maximal allowable concentration; TLV, threshold limit value; REL, recommended exposure limit; Ca, potential occupational carcinogen; PEL, permissible exposure limit

[a] Follows ACGIH TLVs

[b] Substance identified by other sources as a suspected or confirmed human carcinogen

[c] In the presence of polynuclear aromatic hydrocarbons (PAHs), the limit for PAHs is 0.1 mg/m^3 TWA, determined as cyclohexane-extractable fraction

- Carbon black (channel process) is permitted as a component of polysulfide polymer-polyepoxy resins that come in contact with dry food (21 CFR 177.1650);
- Carbon black (channel process or furnace combustion process) is permitted as an optional adjuvant substance: in perfluorocarbon-cured elastomers that come in contact with non-acid food (pH above 5.0) at concentrations not to exceed 15 parts per 100 parts of the terpolymer (21 CFR 177.2400) and in phenolic resins in molded articles that come in contact with non-acid food (pH above 5.0) (21 CFR 177.2410).

Because of the possibility that traces of oil feedstocks can appear in the wastes, the United States Environmental Protection Agency (1994a) has banned the discharge of process waste-water pollutants into navigable waters by carbon black manufacturers utilizing any of the following processes: furnace, thermal, channel, or lamp. The United States Environmental Protection Agency (1994b) has exempted carbon black from the requirement of a tolerance when used as a colourant/pigment in animal tags (40 CFR 180.1001).

2. Studies of Cancer in Humans

Industrial exposure to carbon black has occurred in the carbon black production industry and in a number of user industries, including the rubber, paint and printing industries. The cancer risks associated with these three exposure circumstances have been evaluated within the *IARC Monographs* programme (see IARC, 1982, 1989b; see also this volume).

The Working Group felt that epidemiological evidence concerning cancer risk in user industries, where there has been no attempt to identify which of the workers may have been exposed to carbon black, carries little weight in the present evaluation. Consequently, in this monograph, attention was restricted to those studies that explicitly attempted to identify carbon black-exposed workers. Some studies based on carbon black production workers and some studies of workers in user industries satisfied this criterion.

From the point of view of exposure patterns, the greatest potential for elucidating the carcinogenicity of carbon black would seem to be in the carbon black production industry. Also, it appears that concentration of exposure was substantially higher in the past in the carbon black production workers than in the user industries. A further advantage of studies among producers is the fact that, in this industry, carbon black was the dominant exposure in the industrial environment, whereas workers in user industries were often exposed to complex mixtures of substances, of which carbon black may have been a relatively minor component. Potential confounding of results by concomitant occupational exposures is therefore a significant, potential problem in studies among users. The potential for such confounding bias is greatest when the study population is concentrated in an industry in which there is a single, dominant exposure (or a small number of dominant exposures) other than carbon black. To the extent that the study population includes many different types of carbon black users, each with distinct profiles of concomitant exposures, the likelihood of a strong confounding effect would be diluted. From this point of view, the relative risk estimates due to carbon black exposure derived from studies that include a variety of carbon black user industries provide less opportunity for confounding bias than do studies in a user industry in which most workers have the same profile of concomitant exposures.

2.1 Industry-based studies

Table 18 summarizes cohort studies and Table 19 case–control studies of workers exposed to carbon black.

2.1.1 *Studies in the carbon black production industry*

The cancer occurrence among employees at carbon black production facilities in the United States has been followed for different periods since 1935 and is described in five reports (Ingalls, 1950; Ingalls & Risquez-Iribarren, 1961; Robertson & Ingalls, 1980, 1989; Robertson & Inman, 1996).

Mortality results from the first two reports were subsumed by Robertson and Ingalls (1980), which represents the most complete report on the mortality experience of workers employed by any of four major carbon black manufacturing companies, with production plants located in Texas, Oklahoma and Louisiana, United States (Ingalls, 1950). Eligible study subjects were male employees aged 15 years and over with 12 months or more service during the years 1935–74 at any of the carbon black plants. Between 1950 and 1975, the average number of active workers in all participating companies was about 1250 per year. In the early years, there were equal numbers of workers in the two major processes — channel black production and furnace black production; by 1960, channel black production was dwindling and most workers were involved in the furnace black plants. The mortality experience of the cohort was traced via insurance company records. This company provided death benefit insurance to participating members of the cohort between 1935 and 1974. Over the entire period, there was a total of 34 739 person-years of observation in the mortality follow-up. [It is not stated how many distinct individuals were in the study.] Only 2% of these person-years at risk were in the age group 65 and over. Expected numbers of deaths were calculated from state vital statistics (death rates for white men at five-year intervals starting in 1937 applied to annual employee censuses by age, in five-year groups). There were 29 observed cancer deaths (standardized mortality ratio (SMR) [0.7 (95% confidence interval [CI], 0.5–1.0)]) (Table 18). There were six observed deaths due to cancers of the digestive organs and the peritoneum (SMR [0.6; 95% CI, 0.2–1.4]). There were 13 observed deaths due to cancers of the respiratory system (SMR [0.9; 95% CI, 0.5–1.5]). For all malignancies combined, there was no evidence of increasing mortality with increasing years of service, no significant excess cancers in any of the eight five-year periods and no trend in relative risk of cancer over time.

In a short communication, Robertson and Inman (1996) made a preliminary report of an extension of the follow-up of this cohort. Cohort members from two of the original four companies and those from an additional company were traced for an additional

Table 18. Cohort studies among workers exposed to carbon black

Reference location	Study subjects	Period of follow-up	Occupation/exposure	Cancer site/cause of death	No. obs.	RR	95% CI	Comments
Robertson & Ingalls (1980) Southern USA	Male employees of US carbon black producers	1935–74	Employees with carbon black exposure (34 729 person-years)	All causes	190	[0.8]		Age- and race-adjusted comparison with state populations. No smoking information
				All cancers	29	[0.7]	[0.5–1.0]	
				Gastrointestinal	6	[0.6]	[0.2–1.4]	
				Respiratory	13	[0.9]	[0.5–1.5]	
			Subset with ≥ 20 years service	All cancers	15	[0.8]		Few person-years over age 65
Hodgson & Jones (1985) United Kingdom	Male employees of 5 United Kingdom carbon black producers	1947–80	Employees with carbon black exposure (19 266 person-years)	All causes	129	0.8	[0.7–1.0]	Age-adjusted comparison with local populations. No smoking information. Few person-years over age 65
				All cancers	42	1.0	[0.7–1.3]	
				Lung	25	1.5	[1.0–2.2]	
				Urinary bladder	3	2.5	[0.5–7.3]	
Blair et al. (1990) USA	Male employees at plants with formaldehyde exposure	Not given	Subset with carbon black exposure	Lung	20	1.3	[0.8–2.0]	Age-adjusted comparison with US population. No smoking information
			≥ 20 years duration	Lung	6	2.4	[0.9–5.2]	
			≥ 20 years latency	Lung	11	1.4	[0.7–2.5]	
Robertson & Inman (1996) Southern USA	Male employees of US carbon black producers	1935–94	Employees with carbon black exposure (55 784 person-years)	All causes	377	0.7	0.6–0.8	Age- and race-adjusted comparison with state populations. No smoking information. Short communication. Incompletely documented. This cohort largely overlaps with that of Robertson & Ingalls (1980).
				All cancers	79	0.7	0.6–0.9	
				Gastrointestinal	12	0.5	0.3–0.8	
				Respiratory	34	0.8	0.6–1.1	

RR, relative risk estimates — all SMRs (standardized mortality ratios)

Table 19. Case–control studies of workers exposed to carbon black

Reference location	Study base	Cases	Controls	Exposure	Cancer site	No. of exposed cases/controls	RR	95% CI	Comments
Robertson & Ingalls (1989) USA	Male employees of carbon black producers active in 1980	Previously diagnosed with skin cancer (n = 24)	One matched for age and one matched for age and duration of employment (n = 48)	Cumulative exposure index for carbon black	Skin	24/NG	0.9	0.3–3.2	Prevalent cases may well be unrepresentative. Not clear how exposure was dichotomized.
Bourguet et al. (1987) USA	Male employees of rubber manufacturing industry active in 1964 or earlier	Subsequently diagnosed with skin cancer in local hospitals (n = 65)	Four matched to each case on company, year of birth, year of hire (n = 254)	Intensity of exposure to carbon black: Low Medium High	Skin	14/47 14/49 8/48	0.7 1.2 0.7	NG NG NG	'Intensity of exposure' reflected concentration and frequency of exposure. The findings did not indicate any exposure–response relationship nor any trend by duration of exposure.
Steineck et al. (1990) Sweden	Male general population of Stockholm	Urothelial cancers 1985–87 (n = 254)	Population control frequency matched on sex and year of birth (n = 287)	Ever exposed to carbon black including printing inks	Urothelial	14/9	2.0	0.8–4.9	Adjusted for year of birth and smoking
Siemiatycki (1991) Canada	Male general population of Montréal	Incident cases from 1979 to 1985, with any of 19 types of cancer	Cancer controls, not matched	Ever exposed to carbon black (i.e. 'any' exposure)					Adjusted for age, social class, ethnicity and smoking
		Oesophagus (99)	2546		Oesophagus	11/NG	2.2	[1.1–4.4]	
		Stomach (251)	2397		Stomach	9/NG	0.8	[0.4–1.4]	
		Colon (497)	2056		Colon	17/NG	0.7	[0.5–1.1]	
		Rectum (257)	1299		Rectum	10/NG	0.7	[0.4–1.3]	
		Pancreas (116)	2454		Pancreas	3/NG	0.7	[0.2–2.2]	
		Lung (857)	1360		Lung	52/NG	1.6	[1.1–2.3]	
		Prostate (449)	1550		Prostate	25/NG	1.2	[0.7–1.9]	
		Urinary bladder (484)	1879		Urinary bladder	26/NG	1.2	[0.7–1.9]	
		Kidney (177)	2481		Kidney	14/NG	1.9	[1.1–3.3]	
		Skin melanoma (103)	2525		Skin melanoma	2/NG	0.4	[0.1–1.8]	
		Non-Hodgkin's lymphoma (215)	2357		Non-Hodgkin's lymphoma	9/NG	0.9	[0.5–1.8]	

Table 19 (contd)

Reference location	Study base	Cases	Controls	Exposure	Cancer site	No. of exposed cases/ controls	RR	95% CI	Comments
Parent et al. (1996) Canada	Male general population of Montréal	As in Siemyatycki (1991) lung (n = 857)	Cancer control (n = 1360)	Higher exposure to carbon black	All lung cancers		2.2	1.0–4.9	Adjusted for age, social class, ethnicity, smoking, asbestos and chromium compounds. 'Higher' exposure reflects concentration, frequency, confidence in attribution and duration of exposure.
					Oat-cell carcinomas		5.1	1.7–14.9	
					Squamous-cell carcinomas		0.8	0.2–2.8	
					Adenocarcinomas		2.1	0.5–8.2	
			Population control (n = 533)		All lung cancers		1.5	0.6–4.0	
					Oat-cell carcinomas		4.8	1.4–17.0	
					Squamous-cell carcinomas		0.4	0.1–1.6	
					Adenocarcinomas		1.8	0.4–7.7	

RR, relative risk — all ORs (odds ratios); NG, not given

20 years to 1994, bringing the total person-years of observation up to 54 784. [Of these, 7% were in age groups over 65.] Expected numbers of deaths were based on white male, age-, calendar year- and state-specific death rates. The overall SMR was 0.7 and the SMR for all cancers was 0.7 (95% CI, 0.6–0.9) (Table 18). The SMR for respiratory cancer, based on 34 observed cases, was 0.8 (95% CI, 0.6–1.1).

[The Working Group noted that the system used for ascertainment of vital status, based on records of the insurance company, may well have led to under-ascertainment of deaths in this cohort of workers. Such under-reporting might not lead to bias in estimates of risk if departures from the company were unrelated to mortality risk and if the person-years of observation have been appropriately adjusted for 'loss to view'. The published reports do not provide sufficient detail to reassure the Working Group that these problems have been given adequate consideration. The fairly low overall SMR of 0.7 lends some weight to the hypothesis of a systematic downward bias.]

Information on cancer morbidity was presented in the first, second and fourth reports from this group. The incidence of cancer for the earlier periods (1944–49 and 1949–57) was ascertained from annual examinations and insurance claims (Ingalls, 1950; Ingalls & Risquez-Iribarren, 1961). These cancer incidences were compared with the cancer incidence reported from New York State in 1950 and with that of workers in the carbon black industry whose jobs did not involve exposure to carbon black. Among carbon black workers, only a handful of cases of cancer were ascertained (three in the first report and six in the second); of these nine cases in the 'exposed' cohort, five were skin cancers of which two were melanomas. There were three more skin cancers in the presumed 'unexposed' cohort, of which two were melanomas. Based on the absence of an excess of observed over expected deaths in the entire cohort and the absence of any apparent difference between the 'exposed' and 'unexposed' subcohorts, the authors concluded that there was no evidence of excess cancer incidence. [The Working Group noted that the data presented in the two early reports are obscure as to case-ascertainment procedures and the definition of comparison groups, which are based on very small numbers.]

The fourth report from this group described a study examining morbidity among the same cohort of carbon black industry workers in the United States (Robertson & Ingalls, 1989). The study base consisted of workers aged 15 and over employed in 1980 on site at any of seven carbon black producers. A nested case–control design was used (see Table 19). A case was defined as a member of the study population who had filed a health insurance claim with a physician's diagnosis of either a malignant neoplasm, a disease of the circulatory system or a disease of the respiratory system. For each case, two controls were selected from the workers in the study base who had none of these diagnoses, one matched for age alone and one matched for age and duration of employment. Exposure to carbon black was estimated by attributing to each job the exposure levels found in a previous hygiene survey of jobs and combining it with duration of employment to produce a cumulative exposure index. In total, 36 cases of malignant neoplasm were ascertained, of which 24 were skin cancers (12 basal-cell and 12 squamous-cell), three were lung cancers and the remainder were spread among other sites. The odds ratio for all cancers was 1.1 (95% CI, 0.4–2.7) and that for skin cancers was 0.9 (95% CI, 0.3–3.2).

[The Working Group noted that the use of a study base of workers active at a single point in time led to small numbers. Further, the distribution of cancer sites among prevalent cases would not reflect the site distribution of incident cases; for example, cancers with long survival (e.g. skin cancer) would be over-represented and cancers with short survival (e.g. lung cancer) would be under-represented. The exposure contrast used to compute the odds ratio was unclear.]

A historical cohort study was carried out among carbon black production workers in the United Kingdom (Hodgson & Jones, 1985). Data were collected on 1422 male manual workers with at least one year of service between 1947 and 1974 in any of five major carbon black production factories. For three of the companies, the investigators believed that they had compiled virtually complete rosters of all eligible workers; for the other two, they did not have available those workers who had left the industry before the late 1960s. The subjects identified were traced via national vital statistics registers. For the three 'completely enumerated' subcohorts, the follow-up period began one year after initial exposure to carbon black (i.e. after beginning employment) and for the two 'incompletely enumerated' subcohorts, the follow-up period began after the late 1960s. For members of all subcohorts, the follow-up ended on 31 December 1980, unless truncated by death or emigration. There were a total of 19 266 person-years of observation in the mortality follow-up, of which only 4% were in the age group over 65 years. Overall, 9% of the study subjects were known to have died during the period of observation. Irrespective of whether the comparison was with national mortality rates or with regionally specific mortality rates, there was a deficit in this cohort of deaths from all causes (of approximately 15%) (this was made up mainly of deficits of circulatory and non-malignant respiratory deaths). For specific cancer sites, there was a deficit of deaths from stomach cancer (0 observed, 4.1 expected), parity in deaths from colorectal cancer (3 observed, 4.0 expected), an excess of deaths from urinary bladder cancer (3 observed, 1.3 expected) and most notably an excess of deaths from respiratory cancer (SMR, 1.5 [95% CI, 1.0–2.3]; 25 observed cases) (Table 18). [The deficit of deaths from all causes was probably due largely to the 'healthy worker effect' but also partly to a survivor bias in the two 'incompletely enumerated' factories as those inception subcohorts included workers who had remained (i.e. survived) in the industry for up to 20 years.] The excess of deaths from lung cancer was most evident in one of the five factories (10 observed, 4.8 expected), although there were slight excesses in the other four factories combined as well (15 observed, 11.7 expected). [The Working Group noted that this factory was one that had incomplete early data, but it is unlikely that this could have biased the results in the observed direction.] Industrial hygiene measurements had been taken in 1976 in the various factories. About one-half of all personal samples were above the benchmark TLV for carbon black of 3.5 mg/m^3. There was no indication that the factory with the greatest excess risk had higher exposure levels than the other factories. On the contrary, the measured levels were somewhat lower on average, although they were based on small numbers of measurements. A nested case–control study was carried out, comparing the cases of lung cancer with controls chosen from the unaffected members of the cohort, matched for factory and date of birth. The duration of employment did not differ between cases and controls. [The Working Group noted that the report contained too little information on this study to evaluate the findings adequately.] While no data were presented on smoking habits in this population, the authors refer to an early study in this industry in which smoking habits were not found to have differed greatly from those of the general population.

A general excess risk of cancer was reported in workers in one carbon black producing plant in the former USSR (Troitskaya *et al.*, 1980). [The Working Group noted that neither absolute figures nor the method of calculating observed to expected ratios were given.]

2.1.2 Studies in carbon black user industries

A nested case–control study was conducted in the tyre and rubber manufacturing industry to examine the association of squamous-cell carcinoma of the skin with rubber manufacturing materials presumed to be contaminated by PAHs (Bourguet *et al.* 1987). Cases of skin cancer were identified from the records of four hospitals located in Akron, OH, United States, and these were cross-checked against a list of past and present employees of two local rubber companies who had been enumerated in 1964 for historical cohort studies conducted previously in this industry. Sixty-five cases of squamous-cell skin cancer in white men were thereby ascertained in this cohort. The authors acknowledge that their case-ascertainment system may not have identified all cases in the cohort. Controls were selected from remaining cohort members and were matched to cases on company, year of birth and year of hire, and were required to have been employed in the industry until the corresponding case's date of diagnosis or date of leaving the industry. A total of 254 matched controls were identified, with approximately four matched controls selected for each case. Two experienced industrial hygienists assessed each study subject's exposure to five substances: carbon black, extender oils, lubricating oils, rubber solvents and rubber stocks. Conditional logistic regression analyses were carried out with all five substances included in the models, and each one categorized into three exposure subgroups reflecting concentration and frequency of exposure. For carbon black, the odds ratios in these three exposure subgroups were 0.7, 1.2 and 0.7, respectively, indicating the lack of an exposure–response relationship (see Table 19). There was also no evidence for any trend by duration of exposure.

A historical cohort of 26 561 workers employed in 10 facilities was assembled to evaluate cancer risks associated with exposure to formaldehyde (see IARC, 1995) (Blair *et al.* 1990). The plants were drawn from a variety of industries in which exposure to formaldehyde can be substantial and were located across the United States. The project was characterized by a very extensive assessment of exposure to formaldehyde. About 85% of the workers were thought to have been exposed to formaldehyde at levels above 0.1 ppm [0.123 mg/m^3]. In order to assess possible confounding and modification of effect due to other occupational substances, an assessment was made of each worker's exposure to a number of other substances, one of which was carbon black. The exposure status of subjects was inferred from their recorded work histories, linked to estimates of exposure in different jobs in these plants. The latter estimates were derived by industrial hygienists who carried out site visits, discussed exposure conditions with workers and plant managers and consulted available hygiene monitoring data. Although this study was not designed primarily to assess risk in relation to carbon black exposure, the data could be used for that purpose, and, in one report focusing primarily on exposure to formaldehyde and lung cancer risk, results were presented showing the associations between each of the other substances collected and lung cancer. Expected numbers of deaths were computed using national rates. For all levels and durations of exposure to carbon black combined, there was a slight excess risk for lung cancer (SMR, 1.3 [95% CI, 0.8-2.0]; 20 observed cases) (Table 18). Based on 142 observed cases, the SMR for formaldehyde was 1.4 [95% CI, 1.2–1.6] for ≥ 20 years after first exposure. There was no clear trend by duration of exposure and the pattern of results was similar when restricted to 20 years or more since first exposure. [The Working Group noted that the description of methods of exposure assessment and analysis for carbon black was limited. It was not clear whether all workers exposed to carbon black were also exposed to formaldehyde.]

2.2 Community-based case–control studies

In a Swedish case–control study of urothelial cancer during 1985–87 (Steineck *et al.*, 1990), described in the monograph on printing trades and printing inks in this volume, p. 76, the odds ratio for men having been exposed to carbon black was 2.0 (95% CI, 0.8–4.9; 14 cases), adjusted for year of birth and for smoking (Table 19). However, some of these workers had been exposed to printing inks and other substances.

A population-based case–control study of cancer among male residents of Montréal, Canada, aged 35–70, included histologically confirmed cases of cancer at 11 major sites, newly diagnosed between 1979 and 1985, in 19 major hospitals (Siemiatycki, 1991). With a response rate of 82%, 3730 cancer patients were successfully interviewed. For each site of cancer analysed, two control groups were used, giving rise to two separate sets of analyses and results: one control group was selected from among cases of cancer at the other sites studied (cancer controls; see Table 19) and the other group consisted of 533 age-stratified population controls from the general population (response rate, 72%). The interview was designed to obtain detailed lifetime job histories and information on potential confounders. Each job was reviewed by a trained team of chemists and industrial hygienists who translated jobs into occupational exposures, using a checklist of 293 common occupational substances. Cumulative exposure indices were created for each substance, on the basis of duration, concentration, frequency and the degree of certainty in the exposure assessment itself, and these were analysed at two levels: 'any' and 'substantial' exposure; the latter is a subset of 'any'. Five percent of the entire study population had been exposed to carbon black at some time (i.e. lifetime exposure prevalence). Among the main occupations in which carbon black was attributed in this study were painters (26%), printing industry workers (17%), motor vehicle mechanics (8%) and occupations in rubber and plastics products (6%) (Parent *et al.*, 1996). The presentation of published results was based mainly on the cancer control group. For the following cancer sites, there was no indication of excess risk in relation to any exposure to carbon black, after adjustment for age, ethnic group, social class and smoking (number of exposed cases; odds ratio): stomach (9; 0.8), colon (17; 0.7), rectum (10; 0.7), pancreas (3; 0.7), prostate (25; 1.2), urinary bladder (26; 1.2), skin melanoma (2; 0.4) and non-Hodgkin's lymphoma (9; 0.9). For the following sites there was indication of excess risk (number of exposed cases; odds ratio [95% CI]): oesophagus (11; 2.2 [1.1–4.4]), kidney (14; 1.9 [1.1–3.3]) and lung (52; 1.6 [1.1–2.3]).

To investigate further the possible link between carbon black and lung cancer, an additional analysis of the Montréal data set was carried out. A synthetic exposure index was created, composed of the indices deduced for each exposed subject (concentration, frequency, confidence in the attribution of exposure, duration), and this index was used to designate a lower and a higher cumulative exposure subgroup. Logistic regression analyses were carried out, adjusting for the same covariates as in the above analyses, as well as for two recognized lung carcinogens, asbestos and chromium compounds. Using cancer controls, the odds ratios for lower and higher exposure were 1.1 (95% CI, 0.7–1.8) and 2.2 (95% CI, 1.0–4.9); using population controls, the odds ratios for lower and higher exposure were 0.9 (95% CI, 0.5–1.6) and 1.5 (95% CI, 0.6–4.0), respectively. The excess among highly exposed workers was most pronounced for oat-cell tumours of the lung: odds ratios, 5.1 (95% CI, 1.7–14.9) using cancer controls and 4.8 (95% CI, 1.4–17.0) using population controls (Table 19) (Parent *et al.*, 1996).

3. Studies of Cancer in Experimental Animals

The studies described in the following sections include those investigating the potential carcinogenicity of carbon black, solvent-extracted carbon black and the materials extracted from carbon black (carbon black extracts). However, a detailed review of individual materials extracted from various carbon blacks is not part of this monograph. Some of these individual components (e.g. nitroaromatic compounds) have been evaluated in previous monographs (IARC, 1989a).

Several early studies compared the carcinogenicity of carbon black or carbon black extracts when administered orally or by skin or subcutaneous application. More recent studies have examined the carcinogenicity of inhaled or intratracheally administered carbon black or solvent-extracted carbon black. Many of these studies were part of large studies carried out to investigate the carcinogenicity of diesel exhaust (see also IARC, 1989c).

The Working Group also considered some issues relating to the interpretation of several of the inhalation and intratracheal instillation studies of carbon black. A lesion frequently seen in treated rats has been described variously as 'proliferating squamous cyst', 'proliferative keratin cyst', 'proliferating squamous epithelioma', 'benign cystic keratinizing squamous-cell tumour' or 'cystic keratinizing squamous-cell (CKSC) tumour'. Various authors have included this lesion in tumour counts, but the neoplastic nature of this lesion has been debated (Vainio *et al.*, 1992; Carlton, 1994; Dungworth *et al.*, 1994; Mauderly *et al.*, 1994); its relationship to pulmonary neoplasia is uncertain. Therefore, where possible, the Working Group has listed incidences of this lesion separately from those of other pulmonary neoplasms.

The Working Group considered reports by von Haam and Mallette (1952), von Haam *et al.* (1958), Nau *et al.* (1958a, 1960, 1962), Shabad *et al.* (1972) and Davis *et al.* (1975) in their evaluation, but, because of deficiencies in detail of design, performance and/or reporting, did not use this information in reaching its conclusion.

3.1 Oral administration

3.1.1 *Mouse*

After two weeks of acclimatization, two groups of 31 and 28 female weanling CF1 mice received a diet for two years that did or did not (controls) include furnace black (ASTM N-375; 2.05 g/kg diet). At necropsy, all tissues were examined for gross pathology. Only tissues with macroscopically diagnosed alterations were examined histologically. Survival at two years was similar in treated mice (84%) and in controls (71%). No increase in tumour incidence was observed (Pence & Buddingh, 1985). [The Working Group noted the small numbers of animals and the incomplete histopathological examination.]

3.1.2 *Rat*

After two weeks of acclimatization, two groups of 29 female weanling Sprague-Dawley rats received a diet for two years that did or did not (controls) include furnace black (ASTM N-375; 2.05 g/kg diet). At necropsy, all tissues were examined for gross pathology. Only tissues with macroscopically diagnosed lesions were examined histologically. Survival at two years was similar in controls (38%) and treated animals (45%). No increase in tumour incidence was observed (Pence & Buddingh, 1985). [The Working Group noted the small numbers of animals and the incomplete histopathological examination.]

3.2 Inhalation

3.2.1 *Mouse*

Groups of 80 female Crl: NMRI BR mice, seven weeks old, were exposed to high purity furnace black (Printex 90; primary particle size, 14 nm; specific surface area, 227 ± 18.8 m^2/g; MMAD of particles in the exposure chambers, 0.64 µm). The extractable organic mass of the carbon black was 0.04%; the content of benzo[*a*]pyrene was 0.6 pg/mg and that of 1-nitropyrene was < 0.5 ng/mg particle mass. For 18 h per day on five days per week, the animals were exposed in whole-body exposure chambers to 7.4 mg/m^3 carbon black for four months followed by 12.2 mg/m^3 for 9.5 months. After exposure, the mice were kept in clean air for another 9.5 months. A control group was exposed to clean air throughout. Histopathology was performed on the nasal and paranasal cavities, larynx, trachea and lung. After 11 months and up to 17 months, body weights were significantly lower (5–7%) in the carbon black-exposed group compared with the control group. During the last months, no difference between the groups was observed. After 13.5 months, mortality was 20% in the carbon black-exposed group and

10% in the control group; 50% mortality was reached after 19 months in the carbon black-exposed group and after 20 months in the control group. In exposed mice, the lung particle burden was 0.8, 2.3 and 7.4 mg carbon black per lung after three, six and 12 months, respectively; and, at 12 months, this corresponded to a lung particle burden of 37 mg/g clean air control lung. Among tissues examined, tumours were only observed in the lung. However, no statistical difference was observed between experimental and control animals: carbon black-exposed mice, 11.3% [9/80] adenomas and 10% [8/80] adenocarcinomas; controls, 25% [20/80] adenomas and 15.4% [12/80] adenocarcinomas (Heinrich et al., 1995).

3.2.2 Rat

Two groups of 72 female Wistar Crl:(WI)BR rats, seven weeks old, were exposed by inhalation for 17 h per day on five days per week to 6 mg/m^3 furnace black (Printex 90; 0.04% extractable mass of organics (content of benzo[a]pyrene, 0.6 pg/mg and that of 1-nitropyrene, < 0.5 pg/mg carbon black; primary particle size, 15 nm; MMAD of particles in the exposure chamber, 1.1 μm; specific surface area, 230 m^2/g). One group was exposed for 43 weeks and kept for an additional 86 weeks in clean air and the other group was exposed for 86 weeks and housed in clean air for an additional 43 weeks. Two clean air control groups were kept for 129 weeks. The respiratory tract of all animals was examined histopathologically. The 43-week exposure group had a lung tumour rate of 18% [13/72] (2 bronchiolar/alveolar adenomas, 7 benign CKSC tumours, 4 bronchiolar/alveolar adenocarcinomas and 1 squamous-cell carcinoma). The 86-week exposure group had a lung tumour rate of only 8% [6/72] (1 bronchiolar/alveolar adenoma, 4 benign CKSC tumours and 1 squamous-cell carcinoma). In addition to the six tumours of the latter group, six rats showed lung lesions in the borderline between non-neoplastic and neoplastic (described as marked hyperplasia or marked squamous-cell proliferation). [The difference in the tumour rates of the two exposure groups was not statistically significant.] No tumour was observed in the clean air controls (Heinrich et al., 1994).

A group of 100 female Wistar Crl:(WI)BR rats, seven weeks old, was exposed to high purity furnace black (Printex 90; particle size 14 nm; specific surface area, 227 ± 18.8 m^2/g; MMD of particles in the exposure chamber, 0.64 μm). The extractable organic mass of the furnace black was 0.04%; the content of benzo[a]pyrene was 0.6 pg/mg and that of 1-nitropyrene was < 0.5 ng/mg particle mass. Rats were exposed for 18 h per day on five days per week in whole-body exposure chambers to 7.4 mg/m^3 carbon black for four months followed by 12.2 mg/m^3 for 20 months (average 11.6 mg/m^3). After exposure, the rats were kept in clean air for another six months. Controls were exposed to clean air throughout. Additional groups of 9–20 rats were also exposed to carbon black and were killed at six, 12, 18 and 24 months. Histopathology was performed on the nasal and paranasal cavities, larynx, trachea and lung. Mortality in the carbon black-exposed group was 56% after 24 months of exposure and 92% after 30 months. In the clean air group, mortality was 42% after 24 months and 85% after 30 months. Compared to the controls, the mean lifespan of the treated rats was significantly reduced (Kaplan-Meier method using the SAS-lifetest programme). Mean body weights were significantly lower from day 300 to the end of exposure (carbon black-exposed, 325 g; control, 417 g). The lung burden of carbon black at 24 months was 43.9 ± 4.3 mg/lung [equivalent to 31.3 mg/g clean air control lung] and 6.7 mg/animal in the lung-associated lymph nodes (determined after 22 months of exposure). Benign and malignant lung tumours were increased in the treated groups. The numbers of rats with lung tumours are summarized in Table 20 (Heinrich et al., 1995).

Table 20. Number of female rats with lung tumours after carbon black exposure

Exposure period	Clean air control	Carbon black-exposed (average concentration of carbon black, 11.6 mg/m^3)
6 months	0/21	0/20
12 months	0/21	0/18
18 months	0/18	0/16
24 months	0/10	1/9[a]
30 months	1/217[b]	20/100[a]
		13/100[b]
		4/100[c]
		13/100[d]
No. of animals with tumours[e]	1/217	39/100 (28/100)[f]

From Heinrich et al. (1995)
[a] Benign CKSC tumours
[b] Adenocarcinomas
[c] Squamous-cell carcinomas
[d] Adenomas
[e] Some animals had two lung tumours
[f] Excluding 11 animals that had only benign cystic keratinizing squamous-cell tumours

Groups of 135–136 female and 138–139 male Fischer 344/N specific pathogen-free rats, seven to nine weeks old, were exposed to 0, 2.5 or 6.5 mg/m^3 furnace black (Elftex-12) for 16 h per day on five days per week for up to 24 months in whole-body exposure chambers. The carbon black aerosol was produced by an air jet dust generator and was diluted with filtered air. The carbon black particle size distribution in the chamber was bimodal with 67% in the large-size mode (MMAD, 2.0 µm) and 33% in the small-size mode (MMAD, 0.1 µm). The level of extractable organic material was 0.04–0.29% (mean value during the course of exposure was 0.12%). Observations throughout the complete lifespan were made for the majority of rats in each group (that is, for approximately 100 males and 100 females in total). From these data, body weight, survival and carcinogenicity were evaluated. After exposure for 24 months, surviving rats were kept in clean air until mortality reached 90%. Three female and three male rats selected randomly from each group were killed after three, six, 12, 18 or 23 months for multiple evaluations, including particle burden and histopathology. The high-dose exposure to carbon black reduced the median lifespan of both females and males significantly. The survival of males was also significantly reduced by the low-dose exposure to carbon black (Kaplan-Meier method for determining survival curves; statistical method, log-rank test of Harrington and Flemming). A significant reduction in the body weights of female and male rats exposed to the high-dose carbon black first occurred on days 309 and 449, respectively. For the low-dose exposure to carbon black, this effect was seen only after day 509 of exposure for both males and females. After about 22 months of exposure to the high-dose carbon black, the mean reduction in body-weight was 16% for females and 14% for males. For the low-dose exposure to carbon black, these figures were below 10%. The exposure caused progressive, dose-related accumulation of carbon black particles in the lungs. After 23 months, the mean lung burden reached 12.4 mg/g of clean air control lung in low-dose males, 13.9 mg/g of clean air control lung in low-dose females, 20.2 mg/g of clean air control lung in high-dose males and 30 mg/g of clean air control lung in high-dose females. Full necropsies were performed on all animals and lungs and suspected lung tumours were examined microscopically. The incidences of the various types of lung tumours are shown in Table 21. Statistical comparisons were performed using logistic regression modelling. The incidences of adenomas and adenocarcinomas were significantly increased in females, particularly at the high-dose level. The percentages of male and female rats with lung tumours are given in Table 22. Exposure-related squamous cysts in the lung were classified as non-neoplastic lesions. In animals dying later than 18 months after the start of the exposure, squamous cysts (1 or more per animal) were observed in 0/86 male controls, 1/73 low-dose males and 4/74 high-dose males and in 0/91, 8/90 and 13/87 control, low-dose and high-dose females, respectively (Mauderly *et al.*, 1994; Nikula *et al.*, 1995).

3.3 Intratracheal administration

Rat

A group of 37 female Wistar rats, 15 weeks old, was instilled intratracheally under CO_2 anaesthesia with furnace black (Printex 90) of a high specific surface area (270 m^2/g). The carbon black was suspended in 0.9% sodium chloride using ultrasonication and 3 mg/rat were instilled once a week for 15 weeks. A control group of 39 female rats was instilled with 0.4 ml 0.9% saline once a week for 15 weeks. The animals died spontaneously or were killed when moribund or after 131 weeks at the latest. More than 50% of rats in the treated and control groups survived to 100 weeks. The respiratory tract was evaluated microscopically. No primary lung tumour was found in the control group. In the treated animals, 65% [24] of the rats had primary lung tumours — three rats had adenomas, six rats had adenocarcinomas and one additional rat had an adenocarcinoma and a CKSC tumour, four rats had CKSC tumours and one additional rat had a CKSC tumour and an adenoma, three rats had squamous-cell carcinomas and six rats had squamous-cell carcinomas and additional lung tumours (1 adenoma, 1 adenocarcinoma, 3 adenocarcinomas and CKSC tumours, 1 CKSC tumour) (Pott & Roller, 1994; Pott *et al.*, 1994).

Table 21. Numbers of different types of lung neoplasms observed and numbers of rats with each type of neoplasm[a]

Type of tumour	Control			Low-dose carbon black (2.5 mg/m³)			High-dose carbon black (6.5 mg/m³)		
	Female	Male	Total	Female	Male	Total	Female	Male	Total
No. of animals examined[b]	114	118	232	116	115	231	114	115	229
Adenoma									
No. of neoplasms	0	1	1	2	1	3	17	0	17
No. of rats with neoplasms	0	1	1	2	1	3	13	0	13
Adenocarcinoma									
No. of neoplasms	0	1	1	6	1	7	23	1	24
No. of rats with neoplasms	0	1	1	6	1	7	20	1	21
Squamous-cell carcinoma									
No. of neoplasms	0	1	1	0	0	0	1	2	3
No. of rats with neoplasms	0	1	1	0	0	0	1	2	3
Adenosquamous carcinoma									
No. of neoplasms	0	0	0	0	0	0	1	1	2
No. of rats with neoplasms	0	0	0	0	0	0	1	1	2
Malignant tumour not otherwise specified[c]									
No. of neoplasms	0	0	0	1	0	1	0	0	0
No. of rats with neoplasms	0	0	0	1	0	1	0	0	0

From Mauderley et al. (1994)

[a] Several rats had multiple types of tumours or multiple tumours of a single type or both; thus, these rats (or their tumours) are counted more than once in this table
[b] Including all rats that underwent gross necropsy and microscopic examinations of the lung whether the rats died spontaneously, were euthanized or were killed
[c] This tumour was of a mixed mesenchymal and epithelial type

Table 22. Summary of numbers and percentages of rats examined for lung neoplasms that had one or more neoplasms[a]

Group	Sex	No. of rats at risk for neoplasms[b]	Rats with malignant neoplasms		Rats with malignant or benign neoplasms	
			No.	Percentage	No.	Percentage
Control	Female	105	0	0	0	0
	Male	109	2	1.8	3	2.8
	Combined	214	2	0.9	3	1.4
Low-dose carbon black (2.5 mg/m^3)	Female	107	7	6.5	8	7.5
	Male	106	1	0.9	2	1.9
	Combined	213	8	3.8	10	4.7
High-dose carbon black (6.5 mg/m^3)	Female	105	21	20	28	26.7
	Male	106	4	3.8	4	3.8
	Combined	211	25	11.8	32	15.2

From Mauderly et al. (1994)

[a] Each rat with one or more neoplasm was counted only once in each neoplasm category.

[b] Values include all rats examined by gross necropsy and microscopy except rats killed at three, six and 12 months. The first lung neoplasm was observed between 12 and 18 months of exposure; thus all rats that died spontaneously or were euthanized in moribund condition plus those killed at 18 months or later were considered to be at risk for lung neoplasms. The total number of rats examined, including those killed at three, six and 12 months, is listed in Table 21.

Groups of 48 female Wistar Crl:(WI)BR rats, seven weeks of age, were treated by intratracheal injection once a week for 16–17 weeks with approximately 1 mg of two types of extracted carbon black (furnace Black Printex 90 or Lampblack 101). Resultant total particle doses were 15 mg/animal. A control group of 47 rats was treated with the vehicle (0.9% sodium chloride + 0.25% Tween 80 solution). Although the amount of organic material that could be extracted from the two carbon blacks was small (< 0.1%), the particles were re-extracted with heated toluene for 4 h before they were used in this experiment. The specific surface areas (extracted) and primary particle sizes of Printex 90 and Lamp Black 101 were 270 m^2/g and 14 nm, and 22 m^2/g and 95 nm, respectively. Satellite groups of two to four animals were used to determine the lung particle load one day after the last treatment. Both groups showed a lung particle load of 11 mg [8.1 mg/g of clean air control lung]. Fifty percent of the animals in both groups were alive at 18 months. After an experimental time of 27 months, the respiratory tract of the 48 treated animals per group was investigated histopathologically. In the Printex 90 carbon black-treated rats, 10 had lung tumours ($p < 0.001$, Fisher's exact test) (9 benign (CK) squamous-cell tumours, 1 bronchiolar/alveolar adenoma and 4 bronchiolar/alveolar carcinomas). In the lampblack-treated animals, four rats had benign (CK) squamous-cell tumours. No lung tumour was observed in the 47 vehicle-treated controls (Heinrich, 1994; Dasenbrock et al., 1996).

3.4 Skin application

Mouse: In a series of experiments by Nau *et al.* (1958b), groups of CFW white and C3H brown mice [sex unspecified], six to 10 weeks old, received thrice-weekly skin applications by brush of 10% or 20% carbon black (of several types) suspended in cottonseed oil or mineral oil (oil suspension) or in 1% carboxymethyl cellulose (water suspension). Tests were also carried out with 20% extracted carbon black or benzene extracts of various carbon blacks. The types of carbon black were said to be representative of materials used at that time (see Table 23).

A total of 240 CFW white and C3H brown mice [sex unspecified], six to ten weeks old, received thrice-weekly skin applications by brush of 10% or 20% of three carbon blacks (product No. 5, furnace black; product No. 8, thermal black; and product No. 13, channel black; see Table 23), which contained 0–1% benzene-extractable material and were suspended in cottonseed, mineral oil or 1% carboxymethyl cellulose. These were applied on the shaved back of the mice for 12–18 months (estimated total dose, 3.6–12.8 g/mouse). No skin tumour was reported, but five tumours occurred in other organs in channel black-painted mice. Another 130 animals received treatment with a benzene-extracted carbon black (product No. 5, furnace black; see Table 23) (estimated total dose, 6.3–23.4 g/mouse); no skin tumour was observed, but two lymphosarcomas were reported (Nau *et al.*, 1958b).

Groups of male C3H and CFW mice [numbers and age unspecified] received thrice-weekly skin applications of carbon black extracts obtained by hot benzene extraction from eight different carbon blacks for up to 12 months. All but one of the extracts was reported to show moderate to strong carcinogenicity for the skin (see Table 24). Groups of positive controls (162 mice) received thrice-weekly skin applications of 3-methylcholanthrene (estimated total dose, 7–24 mg/mouse) or benzo[*a*]pyrene (26–27 mg/mouse) in water, oil or benzene (32 mg/mouse) for six to 18 months. [The Working Group noted several deficiencies in these experiments, namely inadequacies in experimental design with the use of 1% benzene as a vehicle for some extracts, and the limited reporting.] (Nau *et al.*, 1958b).

3.5 Subcutaneous and/or intramuscular administration

Mouse: Groups of 50 male and female C57Bl mice, 5–5.5 months of age, received subcutaneous injections of the following: 300 mg of a furnace black (surface area, 15 m^2/g; average particle diameter of about 80 nm) containing 300 mg/kg (ppm) benzo[*a*]pyrene, either suspended in 1 ml tricaprylin or as a pellet; 300 mg of a channel black (surface area, 380 m^2/g; average particle diameter of about 17 nm) from which no aromatic hydrocarbons were detected after extraction with benzene ('non-benzo[*a*]pyrene extractable') either in 1.5 mL tricaprylin or as a pellet; 300 mg channel black plus 0.09 mg benzo[*a*]pyrene either in tricaprylin or as a pellet; benzene extract from 300 mg furnace black in 1 mL tricaprylin; the extracted carbon black from the 300 mg furnace black after benzene extraction in 1 mL tricaprylin; 300 mg furnace black treated for 3 h with hot chromic acid and suspended in 1 mL tricaprylin; and 600 mg of a mixture of

Table 23. Types and properties of carbon blacks used in Nau et al. studies

Product No.	Supplier No.	Type	Parent material	Iodine surface area, average[a] (m²/g)	Benzene extract, average (%)	pH, average	Volatile, average (%)	Grade of black
2	1	Oil furnace	Oil residue	52.7	0.374	9.09	3.53	HAF
1	1	Oil furnace	Oil residue	184.5	0.15	8.86	3.17	CF
9	2	Gas furnace	Gas	21.5	0.06	9.69	0.70	HMF
10	2	Furnace	Gas-oil	28.4	0.07	8.85	1.42	FEF or MAF
11	2	Furnace	Gas-oil	61.7	0.05	9.22	1.29	HAF
12	2	Furnace	Gas-oil	99.3	0.05	9.02	2.34	ISAF
3	3	Oil furnace	Oils	76.0	0.053	9.58	4.51	HAF
4	3	Oil furnace	Oils	35.5	0.125	9.53	2.05	FEF
13	4	Channel	Gas	108.0	0.00	4.99	5.08	MPC
14	4	Channel special	Gas	126.0	0.02	9.00	2.50	STC
5	5	Oil furnace	Oils	35.5	0.17	9.08	1.96	FEF
8	5	Thermal combustion	Gas	11.5	1.04	7.47	1.04	MT
6	5	Oil furnace	Heavy aromatic tar plus natural gas	67.7	0.246	9.70	2.46	HAF
7	5	Gas furnace	Gas	21.7	0.118	9.83	0.69	SRF
15	5	Oil furnace	Oil	62.4	0.26	6.72	3.01	HAF
16	5	Gas furnace	Gas	111.0	0.08	5.81	5.15	Ink black

HAF, high-abrasion furnace; CF, conducting furnace; HMF, high modules furnace; FEF, fast-extruding furnace; MAF, medium-abrasion furnace; ISAF, intermediate super-abrasion furnace; MPC, medium-processing channel; STC, special thermal channel; MT, medium thermal; SRF, semi-reinforcing furnace

[a] Iodine surface area levels are lower than nitrogen surface area levels.

Table 24. Induction of skin tumours with various carbon black extracts in mice treated for at least 12 months

Type of material used[a]	Total dose of benzene extract (mg)	Final tumour index[b] (%)
High-abrasion furnace black		
Product No. 2	12.7	0
	45.0	15.0
	51.4	16.0
	201.7	82.0
Product No. 3	12.6	0
	20.0	0
	24.3	0
	147.4	33.0
Product No. 6	12.6	0
	16.5	0
	32.5	24.0
	36.5	52.0
	170.0	85.0
Fast-extruding furnace black		
Product No. 4	10.8	0
	21.8	7.0
	26.6	9.0
	135.0	44.0
Product No. 5	27.0	33.0
	32.9	25.0
	129.0	25.0
	201.0	74.0
Semi-reinforcing furnace black		
Product No. 7	6.3	0
	7.9	14.0
	8.9	0
	15.4	0
	18.0	0
	24.0	0
	132.6	73.0
Conducting furnace black		
Product No. 1	17.9	0
	21.5	0
	136.6	0
Medium thermal black		
Product No. 8	117.1	85.0

From Nau et al. (1958b)
[a] See also Table 23
[b] Defined by the authors as the percentage of animals, not dying from other cases, developing skin tumours during the treatment period

furnace black and channel black in 1.5 mL tricaprylin. Further groups of 50 mice received injections of 1.0 mL tricaprylin (vehicle controls) or 0.09 mg benzo[a]pyrene per mouse in 1 mL tricaprylin (positive controls). The experiment was terminated at 20 months after injection of the test materials. All questionable tumours found post-mortem were examined microscopically. Tumour incidence was calculated as a percentage and was based on the number of animals alive five months after the start of the study, which was the time at which the first deaths from tumours occurred. In nine of the 12 groups (treated and controls), few or no sarcomas were induced and 70% of the animals were still alive 12 months after the start of the experiment; in the other three, 52–66% of the mice were still alive at this time. In animals treated with the carbon black containing benzo[a]pyrene (a furnace black), a significantly increased incidence of subcutaneous sarcomas (a few of which metastasized) was observed when compared with tricaprylin controls. The incidence of subcutaneous sarcomas induced by the injection of test materials and the 'average fatal time' (i.e. time to death from subcutaneous sarcoma) are summarized in Table 25. High incidences of sarcoma (18/46) were observed in mice receiving furnace black with extractable benzo[a]pyrene administered in tricaprylin, in those receiving the carbon black extract from furnace black containing benzo[a]pyrene (22/45) and in positive controls (39/41). It should be noted that administration of furnace black containing benzo[a]pyrene in pellet form in the absence of tricaprylin induced an incidence of sarcomas of only 2/47 and that of non-benzo[a]pyrene-extractable channel black in pellet form induced an incidence of only 1/47. Extracted furnace black, that is furnace black following benzene treatment, induced one sarcoma in 37 animals; no subcutaneous sarcoma developed in any of the other groups. It was found that mixing non-benzo[a]pyrene-extractable carbon black with benzo[a]pyrene-extractable carbon black (furnace and channel blacks) resulted in a loss of carcinogenicity of the latter (Steiner, 1954).

A series of 21 groups of 10–20 male or female C3H brown or CFW white mice (total number, 344), eight to 10 weeks old, received a total dose of 17–300 mg of different carbon blacks suspended in cooking oil, tricaprylin or 1% carboxymethyl cellulose in water as one or two subcutaneous injections and were observed for 20 months. The authors reported an 8–13% tumour index in three groups receiving subcutaneous injections of carbon black (product Nos 4, 7 and 8 (two furnace blacks and one thermal black); see Table 23) in cooking oil. The tumour index was defined by the author as the percentage of tumours occurring in 'animals excluding those found dead of causes unknown'. The tumours were described as 'subcutaneous mixed tumours' (Nau et al., 1960).

Three groups of 20 C3H, 20 C3H and CFW or 10 C3H male and female mice received two subcutaneous injections of an extracted furnace black (product No. 5; see Table 23) (after extraction in hot benzene for 24 h) (total dose, 0.14–150 mg) in cooking oil or in 1% carboxymethyl cellulose in water and were observed for 20 months. No tumour occurred at the injection site among 19 mice killed at the end of the experiment or in animals dying during the course of the experiment (Nau et al., 1960).

Table 25. Carcinogenicity of carbon blacks

Materials tested	Tumours/ survivors at 5 months	Tumour yield (%)	Average fatal time (days)
Benzo[a]pyrene-containing furnace black[a], tricaprylin	18/46	39.1	363
Benzo[a]pyrene-containing furnace black[a], pellets	2/47	4.3	411
Non-benzo[a]pyrene-extractable channel black, tricaprylin	0/48	0.0	–
Non-benzo[a]pyrene-extractable channel black, pellets	1/47	2.1	524
Non-benzo[a]pyrene-extractable channel black plus benzo[a]pyrene, tricaprylin	0/43	0.0	–
Non-benzo[a]pyrene-extractable channel black plus benzo[a]pyrene, pellets	0/48	0.0	–
Benzene extract of benzo[a]pyrene-containing furnace black, tricaprylin	22/45	48.9	295
Furnace black[a] residue, tricaprylin	1/37	2.7	405
Benzo[a]pyrene-containing furnace black[a] treated with chromic acid, tricaprylin	0/47	0.0	–
Benzo[a]pyrene-containing furnace black[a] plus non-benzo[a]pyrene-extractable channel black, tricaprylin	0/41	0.0	–
Tricaprylin, 1.0 mL	0/43	0.0	–
Benzo[a]pyrene (0.09 mg), tricaprylin	39/41	95.1	233

From Steiner (1954)
[a]Furnace black from which benzo[a]pyrene and six other PAHs can be extracted with benzene

Groups of 10–30 male and female C3H and CFW mice, eight to 10 weeks old, received one or two subcutaneous injections of benzene extracts of different carbon blacks in cooking oil (total dose, 0.01–6.5 mg) (product Nos 1–10 and 15–17; see Table 23). In 31/36 groups, tumour indices of 15–100% were reported, 22 of which had an index of ≥ 50%. No subcutaneous tumour was observed in five groups. Further groups given subcutaneous injections of 0.2–3.25 mg benzene extracts of carbon black (product No. 5; see Table 23) in 1% carboxymethyl cellulose in water or methanol extracts of carbon blacks (product Nos. 1, 5–7; see Table 23) in water were reported to have a nil tumour index. Similar groups of 9–20 C3H mice received one subcutaneous injection of 0.1 or two subcutaneous injections of 0.2 mg extracted material readsorbed onto carbon black (product No. 5; see Table 23) in cooking oil or in carboxymethyl cellulose in water; a nil tumour index was again reported. Further groups of 19–20 C3H mice received as one or two subcutaneous injections 0.5–1.0 mL of cooking oil, which had been incubated with carbon black (product No. 5; see Table 23) for one to six months then centrifuged to remove the carbon black; the subcutaneous tumour index in these animals was 17–92% (Nau et al., 1960).

As a positive control, 14 groups of 10–20 CFW and/or C3H mice, eight to 10 weeks of age, received as one or two subcutaneous injections 0.002–1.0 mg 3-methylcholanthrene (MCA) in cooking oil; the reported tumour indices ranged from 25–100%. When four groups of 20 C3H mice received as one or two subcutaneous injections 0.05–

0.25 mg MCA in carboxymethyl cellulose in water, reported tumour indices ranged from 12 to 95%, whereas a nil tumour index was reported for five groups of 12 female C3H mice receiving one subcutaneous injection of 0.002-0.01 mg MCA in carboxymethyl cellulose in water. Eleven further groups of 15–21 male and female C3H mice received single subcutaneous injections of 0.01–0.2 mg MCA adsorbed onto different carbon blacks (product Nos 1, 5–7, 13 and 18; see Table 23) in cooking oil. For three of these groups (product Nos 1, 13 and 7), the reported tumour indices were 5, 5 and 7%, respectively; for the other products, the tumour index was nil. When four groups of 10–22 C3H mice received the same amount of MCA adsorbed onto carbon blacks (product Nos 5 and 14; see Table 23) in carboxymethyl cellulose in water, the reported tumour index was nil. Two groups of 20 C3H mice injected with 0.1 mg benzo[*a*]pyrene alone or adsorbed onto carbon black (product No. 5; see Table 23) in carboxymethyl cellulose in water were reported to have tumour indices of 56 and 0%, respectively. Four groups of 20–31 C3H mice were injected with 0.5–1.0 mL tricaprylin or cooking oil, and the tumour indices ranged from 0 to 5%. Of a total of 943 untreated CFW and C3H controls, six were reported to have malignant skin neoplasms, one a malignant neoplasm of the liver and one a malignant neoplasm of the spleen. [No detail as to the histology of these tumours was available.] No animal developed a subcutaneous sarcoma (Nau *et al.*, 1960). [The Working Group noted deficiencies in experimental design and reporting in the above experiments; in particular, difficulty was experienced in interpreting the data presented in tabular form.]

3.6 Intraperitoneal administration

Rat: A group of 36 female Wistar rats was injected intraperitoneally once per week for four weeks with 20 mg furnace black 'Corax L' suspended in 2 ml saline [1.2% volatiles; toluene extract, < 0.1%; primary particle size, 23 nm; surface area, 150 m^2/g]. Fifty percent of the rats lived longer than 119 weeks, and after 132 weeks 20% of the animals were still alive. One out of 35 animals examined histopathologically at the end of the experiment had a sarcoma in the abdominal cavity (tumours of the uterus were excluded). Other groups treated in the same way with total doses of 80 mg diesel soot (no local tumour in 34 rats), 20 mg titanium dioxide (0/47), 250 mg non-fibrous silicon carbide (1/22), 250 mg activated carbon (1/25), 160 mg magnetite (2/34) or 160 mg iron (III) oxide (0/33) did not show an increased tumour incidence either. Fibrous dust administered in the same way induced significantly increased incidences of mesotheliomas/sarcomas in the abdominal cavity with total doses sometimes as low as 1–2 mg; for example, 13/21 rats injected intraperitoneally with 1.25 mg fibrous silicon carbide developed mesotheliomas/sarcomas (Pott *et al.*, 1991). [The Working Group noted the apparent low power of this assay to detect carcinogenesis arising from exposure to non-fibrous particles.]

3.7 Combined administration with known carcinogens

3.7.1 Mouse

After two weeks of acclimatization, two groups of 30 and 33 female weanling CF1 mice received a diet for 52 weeks that did or did not (controls) include furnace black (ASTM N-375, 2.05 g/kg diet). Both groups of mice received six weekly intraperitoneal injections of 20 mg/kg bw 1,2-dimethylhydrazine at the start of the study. At necropsy, all tissues were examined for gross pathology. Only tissues with macroscopically diagnosed lesions were examined histologically. Survival was similar in treated and control animals. Carbon black did not enhance the incidence of colonic tumours induced by 1,2-dimethylhydrazine (Pence & Buddingh, 1985).

3.7.2 Rat

After two weeks of acclimatization, two groups of 44 and 45 female weanling Sprague-Dawley rats received a diet for 52 weeks that did or did not (controls) include furnace black (ASTM N-375, 2.05 g/kg diet). Both groups of rats received 16 weekly intraperitoneal injections of 10 mg/kg bw 1,2-dimethylhydrazine at the start of the experiment. At necropsy, all tissues were examined for gross pathology. Only tissues with macroscopically diagnosed lesions were examined histologically. Carbon black did not enhance the incidence of colonic tumours induced by 1,2-dimethyl-hydrazine (Pence & Buddingh, 1985).

Two groups of 72 female Crl:(WI)BR rats, seven weeks old, were exposed by inhalation for 17 h per day on five days per week to a PAH-rich hard coal-tar pitch condensation aerosol (T/P aerosol) that contained no carbon particles. The exposure concentration of 2.6 mg/m^3 T/P aerosol contained 50 µg/m^3 benzo[*a*]pyrene among other PAHs. The MMAD of this aerosol was 0.5 µm. Four other groups of 72 female rats each were exposed to two mixtures of furnace black (Printex 90) and T/P vapour resulting in benzo[*a*]pyrene concentrations in these exposure atmospheres of 50 µg/m^3. Two of these four groups were exposed to the T/P aerosol containing 2 mg/m^3 carbon black and the other two groups were exposed to the T/P aerosol containing 6 mg/m^3 carbon black. The T/P vapour condensed onto the surface of the carbon black particles. One group of each of the three exposure atmospheres was exposed for 43 weeks and kept in clean air for 86 weeks. The other was exposed for 86 weeks and kept in clean air for 43 weeks. The two control groups of 72 rats each were kept in clean air for 129 weeks. No lung tumour was observed in clean air groups. Comparing the three 43-week exposure groups, the lung tumour rates of the groups combining T/P aerosol with carbon black showed an approximately two-fold higher increase compared to the groups exposed to T/P aerosol only. There was no difference in the lung tumour rates between the three exposure groups exposed for 86 weeks (Heinrich *et al.*, 1994).

3.7.3 Hamster

Syrian golden hamsters from the TNO/Holland breeding farm [sex unspecified] were treated intratracheally with a total dose of 60 mg/animal carbon black [not further specified] together with 3 and 9 mg benzo[a]pyrene. Benzo[a]pyrene was dissolved in acetone with the carbon black added to obtain smaller benzo[a]pyrene particles and to provide condensation of benzo[a]pyrene on carbon black after acetone vaporization. Two other groups of hamsters were treated with 3 and 9 mg benzo[a]pyrene without carbon black. The total dose was administered by 40 instillations once per week in 0.1 mL saline solution containing 0.5% Tween 80. Between 40 and 43 hamsters per group were examined histopathologically at the end of the experiment. Malignant and benign tumour incidences in the larynx, trachea and lung were reported. The authors stated that carbon black did not enhance the carcinogenic effect of benzo[a]pyrene (Pott & Stöber, 1983) [The Working Group noted the inadequate reporting of many experimental details in relation to mortality and duration of the study.]

4. Other Data Relevant to an Evaluation of Carcinogenicity and its Mechanisms

4.1 Absorption, distribution, metabolism and excretion

4.1.1 *Humans*

Studies of lung tissue from workers in carbon black factories have shown widespread deposits of large amounts of carbon black (Rosmanith *et al.*, 1969; Beck *et al.*, 1985). [No quantification of data were given.]

Five nonsmoking warehouse packers in a furnace black manufacturing plant were examined for exposure to dust, the level of which was assessed by air sampling and from the urinary excretion of 1-hydroxypyrene (derived from pyrene) in post-shift urine for five consecutive days during one work week. The average dust concentrations over the five days ranged from 1.53 to 13.21 mg/m^3. The average urinary excretion of 1-hydroxy-pyrene was 0.103–0.475 μmol/mol creatinine. The urinary excretion was lower on Mondays than on the other days. The authors concluded that urinary excretion was affected by dust exposure, and that the pyrene on the dust was bioavailable (Gardiner *et al.*, 1992b). [The Working Group noted the pyrene content of the carbon black was not measured.]

4.1.2 *Experimental systems*

(a) *Kinetics*

Several review articles, mostly focusing on particulate toxicity and carcinogenicity have also described the retention kinetics of particles (including carbon black) after their deposition in the lungs of experimental animals (Morrow, 1988; Snipes, 1989; Kreyling, 1990; Morrow, 1992; Muhle *et al.*, 1994; Oberdörster, 1995).

A number of studies, summarized in Tables 26–33, using intratracheal instillation and inhalation in mice and rats evaluated the retention kinetics of different carbon black materials after deposition into the lung. Bowden and Adamson (1984) instilled 4 mg of colloidal carbon (primary particle size, 30 nm diameter) into the trachea of Swiss mice and followed its clearance in groups of three mice killed at intervals over a six-month period. They reported that most of the carbon black was cleared via the mucociliary escalator, but some transepithelial passage via type I cells occurred as well. Heavily laden alveolar macrophages stayed in the lung for the whole observation period and there was some, although low, clearance via the lymphatic system. No quantitation of the results was reported.

Several groups evaluated the retention kinetics of inhaled carbon black in the lungs of rats. Lee et al. (1987) and Strom et al. (1989) used two different furnace blacks (RCF [regular colour furnace]-7 and Elftex 12), which were inhaled in whole-body exposure chambers for 20 h per day on seven days a week for one to 11 weeks (for details, see Table 26). The MMADs were 0.22 and 0.24 µm, respectively. Both studies found a significantly prolonged retention half-life with increasing lung burdens. Lee et al. (1987) determined the retention kinetics of subsequently administered ^{14}C-diesel particles and found that pulmonary half-life ($t_{1/2}$) increased with increasing lung burden. Strom et al. (1989) determined carbon black lung burdens and lymph nodal burdens of carbon black. Lung burdens of 1.1, 3.5 and 5.9 mg carbon black were achieved after one, three and six weeks' exposure, respectively; the one-year retention fractions were 8, 46 and 61% of the lung burden at the end of the exposure periods, respectively. Additionally, carbon black was retained in the regional thoracic lymph nodes at 1, 21 and 27% of the initial lung burden, respectively. The authors concluded from their results that a carbon black lung (macrophage compartment) burden in the rat of ~0.8 mg results in a doubling of the normal retention half-time of about 50 days.

Muhle et al. (1990) confirmed these results in an inhalation study using Wistar rats. These inhaled furnace black (Printex 90) of a particle size of 0.64 µm MMAD. The inhalation was for 95 h per week at a concentration of 7.4 ± 1.5 mg/m^3 for a total of 4.5 months. The retained carbon black at the end of exposure amounted to 13.7 ± 2.0 mg. The retention $t_{1/2}$ of subsequently inhaled ^{85}Sr-labelled polystyrene test particles was 472 days in these rats compared with 61 days in air controls. The carbon black data fitted well into the retention kinetic data obtained with other low-toxicity, low-solubility particles and the authors concluded that a rat lung burden of ~0.5 mg of such particles, including carbon black, resulted in a prolongation of retention half-time. This quantitative relationship is similar to that observed with other particulate materials.

Another study by the same group (Creutzenberg et al., 1990; Muhle et al., 1994) used the same carbon black material (Printex 90); in a whole-body exposure system, female Wistar rats inhaled an average exposure concentration of 7.5 mg/m^3 for up to four months, then 12 mg/m^3 for 19 h per day on five days a week for up to 24 months. At three, six, 12, 18 and 24 months of exposure, the retention of carbon black was determined as well as the influence of exposure to carbon black on the alveolar macrophage-mediated clearance function in the lung. The carbon black lung burden reached a level of 50.2 ± 10.9 mg at 18 months and the respective retention $t_{1/2}$ for carbon black was determined to be 550 days (95% CI, 322–1868 days) following termination of exposure. Test particle clearance of ^{59}Fe$_2$O$_3$ (0.35 µm diameter) was significantly prolonged with increasing carbon black exposure duration, with $t_{1/2}$ ranging

Table 26. Kinetics of carbon black in experimental animals

Particle type	Particle diameter and surface area	Species (age and sex)	Route of exposure and dose/exposure concentration	Duration of study	Findings	Comments	Reference
Colloidal carbon	30 nm	Swiss mouse	Intratracheal instillation, 4 mg	6 months	Most CB cleared via MC escalator; some transepithelial passage, very low lymphatic clearance; heavily laden AM remained for months in lung	No quantitative results; findings based on qualitative histological data	Bowden & Adamson (1984)
^{7}Be-labelled carbon particles (Elftex 8; furnace black)	0.01–1 μm (primary 37 nm)	Swiss mouse (4 weeks and 18 months; female)	Gavage, 7 mg	14 days	^{7}Be activity was mainly confined to the gastrointestinal tract; retained dose at 14 days: young — 3.3×10^{-5} %; old — 8.4×10^{-5} %; some activity in non-intestinal tissue	Very small fraction of CB may penetrate via Peyer's patches	LeFevre & Joel (1986)
RCF-7 (furnace black)	0.22 μm MMAD (primary 37 nm)	Fischer 344 rat	Inhalation, 20 h/day, 7 days/week, 6.6 mg/m^3	1–11 weeks; followed by ^{14}C-diesel exposure for 45 min + 1 year observation	Linear increase of CB lung burden with exposure duration; lung burden ~30 mg; increased CB and ^{14}C-diesel pulmonary $t_{1/2}$ with increasing lung burden	Methodology for determining $t_{1/2}$ is not described adequately	Lee et al. (1987)
Elftex 12 (furnace black)	0.24 μm MMAD	Fischer 344 rat	Inhalation, 20 h/day, 7 days/week, 7 mg/m^3	1, 3, 6 weeks exposure, followed by up to 1 year observation	Lung burdens: 1.1, 3.5, 5.9 mg CB; 1-year retention: 8, 46, 61%; LN burden: 1, 21, 27% of initial lung CB; doubling of normal $t_{1/2}$ of ~50 days occurs at CB lung burden of ~0.8 mg	Authors propose AM sequestration model to explain retarded CB clearance at higher CB burden	Strom et al. (1989)
Printex 90 (furnace black)	0.64 μm MMAD (primary 14 nm)	Wistar rat	Inhalation 95 h/week, 7.4 mg/m^3	4.5 months	Retained CB: 13.7 mg; $t_{1/2}$ of ^{85}Sr-labelled test particles: 472 days; prolonged $t_{1/2}$ at rat lung burden of ~0.5 mg	Quantitative relationship observed is similar to those of other low-toxicity low-solubility particles	Muhle et al. (1990)

Table 26 (contd)

Particle type	Particle diameter and surface area	Species (age and sex)	Route of exposure and dose/exposure concentration	Duration of study	Findings	Comments	Reference
Printex 90 (furnace black)	0.64 μm MMAD	Wistar rat (female)	Inhalation 19 h/day, 5 days/week, 12 mg/m³	24 months for CB; 3, 12, 18 months for ⁵⁹Fe₃O₄ and ⁸⁵Sr-polystyrene particles	CB lung burden: 50.2 mg; CB $t_{1/2}$ 550 days; ⁵⁹Fe₃O₄ $t_{1/2}$ 244–591 days; ⁸⁵Sr $t_{1/2}$ 472 days at 3 months then back to normal $t_{1/2}$ of 50–60 days	No data are provided to demonstrate lower alveolar deposition of ⁸⁵Sr particles at high lung burdens	Creutzenberg et al. (1990); Muhle et al. (1994)
Elftex 12 (furnace black)	2–2.4 μm MMAD (large mode) 0.02–0.1 μm DED (small mode, 10–30%)	Fischer 344 rat	Inhalation 3.5 mg/m³ 98 mg/m³	16 h/day, 7 days/week; 6 h/day, 5 days/week; 4 h/day, 1 day/week; 12 weeks exposure + 24 weeks post-exposure	Pulmonary retention half-life $t_{1/2}$ not different for different exposure rates; average $t_{1/2}$ ~520 days (95% CI, 350-950)		Henderson et al. (1992)
Elftex 12 (furnace black)	20 μm MMAD (large mode) 0.1 μm MMDD (small mode, 33%) 43 m²/g	Fischer 344/N rat	2.5 mg/m³ 6.5 mg/m³	16 h/day, 5 days/week 24 months	Double exponential clearance; slow phase, no clearance in CB-exposed group compared to $t_{1/2}$ in controls of 113–135 days		Mauderly et al. (1994)
Printex 90 (furnace black)	0.64 μm MMAD 227 m²/g	Wistar rat NMRI mouse	1.6 mg/m³ (average)	18 h/day, 5 days/week, 24 months (rat) + 6 months post-exposure 13.5 months (mouse) + 9.5 months post-exposure	CB accumulation kinetics test particle clearance; CB accumulation in rat and mouse lung similar (at 1 year of exposure)		Heinrich et al. (1995)

CB, carbon black; MC, mucociliary; LN, lymph node; AM, alveolar macrophages; MMAD, mass median aerodynamic diameter; $t_{1/2}$, retention half-life; DED, diffusion equivalent diameter from 244 to 591 days, compared with 61–96 days in air controls. In contrast, ^{85}Sr-labelled polystyrene microsphere (3.5 μm diameter) clearance showed only a prolonged retention half-time after three months of exposure to carbon black with a retention $t_{1/2}$ of 472 days, whereas at the 12- and 18-month exposure time points, the test particle clearance returned to control values of about 50–60 days. The authors explained this by a change in the deposition site of the larger ^{85}Sr-labelled polystyrene microspheres due to altered lung architecture (in response to carbon black-induced inflammation and other changes) and breathing pattern. [The Working Group noted that this seems to be a reasonable explanation, especially in view of the increase in lung weight due to inflammatory responses after the heavy exposure to carbon black particles of 1.7-fold and six-fold at three and 12 months, respectively. However, no data were provided to illustrate the lower alveolar deposition of ^{85}Sr-labelled test particles.]

In a study of chronic inhalation in Wistar rats and NMRI mice exposed to furnace black (Printex 90; 11.6 mg/m^3), the pulmonary particulate accumulation was measured (Heinrich et al., 1995). The rats were exposed for 24 months and the mice for 13.5 months. Both rats and mice showed similar accumulation kinetics over the exposure time; at one year of exposure, the normalized lung burden (mg/g of control lung) was 32 mg (rats) and 37 mg (mice). In addition, rats showed significantly prolonged retention of tracer particles compared to controls as early as at three months of exposure and which persisted through 12 and 18 months of exposure and until three months after the 18 months' exposure (see Creutzenberg et al., 1990).

A study by Henderson et al. (1992) evaluated the pulmonary retention of furnace black (Elftex 12) inhaled at three different dose rates such that the product of concentration × time was very similar (392 mg × h/m^3 per week). Lung burdens ranged between 3 and 4 mg. The retention $t_{1/2}$ determined over a 24-week post-exposure period was not statistically significant between the different groups and was reported as ~520 days with a 95% CI of 350–950 days.

Mauderley et al. (1994) studied the retention of tracer doses of [^{7}Be]-furnace black (Elftex 12) in rats at three and 18 months of chronic exposure to two concentrations of unlabelled carbon black (2.5 mg/m^3 and 6.5 mg/m^3). Clearance of the labelled carbon black followed a two-exponential model with about 50% of it cleared in control animals with retention $t_{1/2}$ of 14 and 19 days and in the exposed rats between 14 and 40% cleared with retention $t_{1/2}$ ranging between four and 10 days. The most striking differences were found in the slow-phase clearance components, which showed little or no clearance over a time period of 126 days for the low- and high-dose groups compared to retention $t_{1/2}$ of 113 and 135 days for control rats.

In an attempt to determine the translocation of carbon black particles after gastric administration, LeFevre and Joel (1986) gavaged four-week-old and 18-month-old female Swiss mice with 7 mg ^{7}Be-labelled furnace black particles (Elftex 8). They then determined the isotope distribution at 4 h and one, two, five and 14 days later; they concluded from their findings that there is uptake and distribution from the gut and that transit is more rapid in young mice. Peyer's patches of older mice take up more carbon than those of younger mice.

Overall, these studies demonstrate that exposures to carbon black that achieve lung burdens exceeding about 0.5–1 mg/g rat lung result in significant prolongation of the retention $t_{\frac{1}{2}}$ of carbon black in the lung and, moreover, affect the clearance function of alveolar macrophages for other particulate material. This is in agreement with the concept of particle lung overload (see review, Morrow 1988), which is defined as a significant impairment of the alveolar macrophage-mediated particle clearance function due to high loading of alveolar macrophages with low-toxicity, low-solubility particles. The only study comparing the kinetics of inhaled carbon black in rats and mice found that the pulmonary accumulation was similar in both species leading to a condition of particle overload. There are no kinetic data on sex differences in rats or differences in other species.

(b)   Kinetics of carbon black adsorbed material

Concern had been raised in the past that material, including carcinogenic compounds, adsorbed onto carbon black particles will be retained longer in the lung upon inhalation and will subsequently lead to a greater availability of carcinogens to target cells in the lung. In particular, this would be of importance for materials such as diesel exhaust particles which are known to contain PAHs adsorbed onto a carbon core and which have been thought to contribute to a carcinogenic response of inhaled diesel exhaust. The studies considered are summarized in Table 27.

Pylev *et al.* (1970a,b) used intratracheal instillation of [^{3}H]-benzo[*a*]pyrene adsorbed onto furnace black particles (26–160 nm) and followed retention of radioactivity for 21 days in Syrian hamsters. Compared to [^{3}H]-benzo[*a*]pyrene suspended in aminosol vitrum, retention of [^{3}H]-benzo[*a*]pyrene was longer when adsorbed onto carbon black.

Male Fischer 344/Crl rats were exposed by inhalation to Elftex 12 (furnace black; primary particle size, 37 nm; surface area, 43 m^2/g) for 30 days with an adsorbed [3,4,9,10-^{14}C]-1-nitropyrene (Wolff *et al.*, 1989) or [7-^{14}C]benzo[*a*]pyrene (Sun *et al.*, 1989). A total concentration of 100 mg/m^3 was used with either 2 mg/m^3 1-nitropyrene added or 0.2, 2 or 20% benzo[*a*]pyrene. The investigators found that the long-term retention of radioactivity from both 1-nitropyrene and benzo[*a*]pyrene increased when adsorbed onto carbon black. For both adsorbed compounds, a biphasic clearance was found with most being cleared from the lungs within one to two days. At all time points, 16–60 times more radioactivity was retained after dosing the adsorbed compounds compared to administration of the pure compound. Covalent interaction of these compounds with lung macromolecules was also higher when they were co-administered with carbon black particles.

These three studies demonstrate that carbon black administered to rat and hamster lung either by inhalation or instillation can act as a carrier of adsorbed material which subsequently is cleared much more slowly than the compound given without carbon black adsorption.

Table 27. Kinetics and effects of carbon black adsorbed compounds

Carbon black characteristics	Adsorbed compound	Test system	Duration	End-points	Findings	Reference
Furnace black 26–160 nm	[^{3}H]BaP	Intratracheal instillation, Syrian hamster	21 days	Macrophage response and BaP retention	CB + BaP elicited more macrophages; longer BaP retention with CB than without	Pylev et al. (1970a,b)
Elftex 12 (furnace black) 37 nm; 43 m^2/g	^{14}[C]BaP	Inhalation; Fischer 344/N rat; 100 mg/m^3 mass with 0.2, 2 or 20% BaP; BaP alone, 2, 20 mg/m^3; intratracheal instillation of 500 μg CB ± 10 or 100 μg BaP	2 h exposure (nose only) + 30 days	BaP lung retention	Biphasic lung retention; long-term retention of BaP increased 16–60 times when coated onto CB; more pronounced after instillation compared to inhalation; covalent interactions of BaP with lung macromolecules increased when administered as CB coating	Sun et al. (1989)
Elftex 12 (furnace black) 37 nm; 43 m^2/g	^{14}[C]-1-Nitropyrene	Inhalation; Fischer 344/N rat; 98 mg/m^3 CB + 2 mg/m^3 nitropyrene; nitropyrene alone	2 h exposure (nose only) + 30 days	Nitropyrene lung retention	Biphasic nitropyrene retention increased when adsorbed to CB; nitropyrene covalently bound to macromolecules was 10-fold greater at 30 days when inhaled adsorbed compared to nitropyrene alone	Wolff et al. (1989)

BaP, benzo[a]pyrene; CB, carbon black

4.1.3 Comparison between animals and humans

Although carbon black has been identified in humans, no quantitative data are available on retention of carbon black in humans. However, based on studies with other highly insoluble particulate materials, it can be assumed that the normal retention $t_{1/2}$ in humans is longer than that measured in rats by a factor of approximately 10. Thus, Bailey et al. (1985) found that retention $t_{1/2}$ of inhaled monodisperse 1 and 4 μm diameter fused aluminosilicate particles in humans ranged from ~200–700 days, depending on the time after exposure, with an average of ~500 days for most of the particles to be cleared, in comparison to rat data (Muhle et al., 1990) which demonstrate a retention $t_{1/2}$ of 61–96 days. This presumes that the retention kinetics of different particles of low solubility and low toxicity are the same, as has been demonstrated in rats. Normal pulmonary retention $t_{1/2}$ for low-solubility, low-toxicity particles in mice have been reported as ~55 days (Kreyling, 1990).

Heavy exposure to carbon black in occupational settings may lead to high carbon black burdens in the human lung. In analogy to the rat, if this lung burden exceeds ~0.5 mg/g lung, it would be expected that the normal retention half-life may be prolonged. Indeed, there is some evidence showing that occupations leading to heavy particulate loads of the lung (e.g. coal mining) show a prolonged clearance of the dust from the alveolar space (Stober et al., 1965; Freedman & Robinson, 1988; Freedman et al., 1988)

4.2 Toxic effects

4.2.1 Humans

Comprehensive reviews of the toxicity of carbon black to humans are available (United States National Institute for Occupational Safety & Health, 1978; Rivin & Smith, 1982; IARC, 1984; Gardiner, 1995a).

Gärtner and Brauss (1951) first described radiological changes analogous to pneumoconiosis in 31 workers in a carbon black factory. However, these individuals had no functional lung abnormality. Since that time, a series of other reports have been published on pneumoconiosis in carbon black workers.

In 56 workers in two German carbon black factories (one produced carbon black from oil that was burned with lightgas, the other from acetylene), two (both belonging to the 16 workers who had an exposure time exceeding 10 years) had chest X-ray changes, compared to none among 52 controls (who had had radiographs taken without suspicion of lung disease) (Mai, 1966). [The selection of workers is not clear, neither are the criteria for diagnosis.]

In a study by Valic et al. (1975), respiratory function was measured in a group of 35 carbon black workers (average age, 38.8 years) and 35 controls matched for age, body weight and smoking habits examined in 1964 and again in 1971. Measurements of carbon black concentrations in the work environment showed that the respirable concentration (mean, < 1 μm) was 7.2 mg/m^3 in 1964 and 7.9 mg/m^3 in 1971. By 1971, the average duration of exposure was 12.9 years and was more than 10 years for 26/35 workers. In 1971, carbon black workers who smoked exhibited a reduced forced vital capacity and a reduced forced expiratory volume compared with smoking controls. For carbon black workers who did not smoke, there was no significant difference compared with nonsmoking controls. However, for both smoking and nonsmoking carbon black workers, the annual declines in these parameters were three- to four fold greater than expected. Radiological lung changes, characterized by slight interstitial fibrosis, were found in 17.1% of the workers. These lung changes progressed between 1964 and 1971. [The selection of exposed and unexposed populations is unclear.]

In a study on carbon black (furnace) workers, Cocarla et al. (1976) found that 29 (20.3%) of 143 workers exposed for a mean of 19 years had pneumoconiosis, often accompanied by generalized or local emphysema and disturbances of lung perfusion. In the same subpopulation, an increase in levels of serum immunoglobulin (Ig)A and a decrease in IgM were observed. Measurements of air levels of carbon black were not reported.

In 125 carbon black-exposed workers in dry-cell battery and tyre manufacture (mean exposure time of about five years), a reduction in pulmonary function and an increase in respiratory symptoms were observed in comparison to 145 controls (healthy non-industrially exposed men with a [sic] history of respiratory disease). The largest reduction in lung function was observed in the group that had exposure to the highest mean dust level (31 mg/m^3) at the dry-cell battery factory. The most common respiratory symptoms were cough with phlegm (28 and 22% in the battery and tyre factories, respectively). Radiographs were reported to show no abnormality (Oleru et al., 1983). [Only a fraction of all workers was examined; the selection criteria are not clear. No adjustment was made for smoking; however, only 12.8% of workers in the study group were smokers.]

In a Czech factory producing carbon black from anthracene, out of 12 workers exposed for ≥ 15 years, one had chest radiographical findings consistent with pneumoconiosis. One additional worker, who had been exposed for 27 years and who had no symptoms of respiratory disease or lung function disturbances, had fine, diffuse changes in the chest radiogram; an open lung biopsy was taken from this worker. There was heavy pigmentation of the lung surface. Histological examination revealed heavy deposits of carbon black particles and slight, mainly reticular fibrosis with associated emphysema (Rosmanith et al., 1969). [No information on exposure levels; selection of the workers for examination is not clear, neither are the smoking habits.]

Beck et al. (1985) studied X-radiographs of various thicknesses of carbon black deposits. They concluded that radiographical changes in workers were due to tissue reaction rather than to simple deposition. Further, the authors examined lung tissue from two subjects [smoking not defined] with radiographical changes and found fibrous tissue and emphysema around the carbon deposits. On the other hand, a biopsy from a worker with generalized nodulation did not reveal fibrosis (Slepicka et al., 1970).

In a study of 83 currently exposed carbon black workers (at least two years' exposure) and 144 controls, the current workers had statistically significantly higher prevalences than the controls of the following: chronic bronchitis (60% versus 19%), 'obstructive disturbance of ventilation' at spirometric examination (24% versus 6%) and nonspecific bronchial hyper-reactivity on histamine test (28% versus 3%). In a total of 83 currently and 46 formerly exposed workers, 2.3% (average exposure time, 23.3 years) displayed findings in chest radiograms suggesting pneumoconiosis and an additional 6.9% had slighter changes. Data on dust exposure were not given (Kandt, 1985). [The type of carbon black factory and the selection of workers and controls are not well described.]

In 3027 carbon black workers (92.3% men, 7.7% women) employed in 19 plants (mean employment time, 10.9 years), slight associations were found between prevalences of chronic cough and sputum production and dust exposure, as assessed from job titles in nonsmokers (39–49% in different exposure intensity categories). In smokers, there was no association. There were no data on dust levels. Further, spirometry indicated minor exposure-associated reductions in the forced vital capacity and forced expiratory volume. Moreover, routine chest radiographs, taken during the previous two years, of 935 of the workers from 11 of the 19 plants where they could be obtained revealed six cases of a simple type of pneumoconiosis [smoking status not stated], all belonging to the group of 396 workers who had been employed for more than 10 years (Crosbie, 1986). [A large proportion of the radiographs was taken by the mass miniature technique, which is not optimal for detection of minor changes.]

In a multicentre European study, a population of 1742 employees (92% men, 8% women in the original study base) in 15 carbon black factories were examined (mean duration of employment 14.2 years). [Most of them were identical to those studied earlier by Crosbie (1986).] In a total of 1317 samples of total inhalable dust, the geometric mean level was 0.57 (geometric standard deviation, 4.0) mg/m^3 and in 1298 respirable dust samples the geometric mean level was 0.21 (2.7) mg/m^3. Associations were found between frequencies of cough, sputum and symptoms of chronic bronchitis (mean prevalence, 10%) and current exposure. Forced vital capacity and forced expiratory volume showed slight decreases with increasing dust exposure in both smokers and nonsmokers. Chest radiographs could only be taken in 10 of the 15 plants, and they were taken for 1096 workers. Of these, 24.4% showed small opacities (International Labour Organization category 0/1 or greater), with a strong association with rising cumulated exposure (five categories). Preliminary analyses indicated associations between radiographical findings and lung function (Gardiner et al., 1993).

In 913 employees of six carbon black producers, there was no consistent association between forced vital capacity or forced expiratory volume (fractions of predicted values) and cumulative dust exposure (range < 50–≥ 200 mg/m^3 × months), when age and smoking habits were taken into consideration. Information on exposure was obtained from measurements made one year earlier in 24 plants [relation to the population studied for health effects unclear]. Some 1500 total dust samples obtained by personal monitoring showed geometric mean time-weighted average levels ranging from 0 to 2.0 mg/m^3, with 76% below 1.0 mg/m^3 (Robertson et al., 1988). [Because of the lack of occupational histories from several of the factories, there was a major loss of workers (804) from the original study base (1717), and this might have affected the outcome. The Working Group noted that the spirometry values had already been standardized by comparison with referents to calculate their predicted percentage before being used again in the age-specific two-way analysis of variance.]

Three years later, 697 (76%) of these workers were retested. The total group showed small, statistically nonsignificant increases in forced vital capacity and forced expiratory volume. The author states that the most likely explanation for those findings is the variability of lung function testing (Robertson, 1996).

In a cross-sectional study, chest radiograms were taken for 507 predominantly male workers in a German carbon black factory. The average duration of employment was given as approximately 25 years. Of the radiograms, 75.5% were classified as International Labour Organization (ILO) category 0/0, 13.0% as 0/1, 8.9% as 1/0, 2.4% as 1/1 and 0.2% as 2/1. The authors ascribed the findings in the latter two categories to two cases of silicosis unrelated to exposure to carbon black and to prevalent smoking (70% smokers or ex-smokers) (Küpper et al., 1994a,b).

In the same factory, lung function (body plethysmography, including measurements after inhalation of a single concentration of methacholine) was investigated in 578 carbon black-exposed and 99 unexposed workers. The fine dust concentration was 0.01–9.14 mg/m^3 (total dust, 1.08–19.95 mg/m^3). The duration of exposure was not stated, but the average 'dust-years' (dust exposure times duration of exposure) were 11.3 in non-smokers, 21.5 in ex-smokers and 16.3 in smokers. In multiple linear regression analysis (allowing for age), exposure to carbon black had a significant effect on deterioration of lung function in the nonsmoking group only. Thus, there was a significant decrease in airway resistance (which, however, was dependent on two extreme values) and expiratory flow. However, in the smokers' group, pathological findings (as compared to expected values in reference populations) in airway resistance were more frequent among the carbon black-exposed workers than in the unexposed controls. There was no significant association between bronchial hyper-reactivity and exposure to carbon black (Küpper et al., 1996).

In a case–control study of employees of seven carbon black producers in the United States, workers who had submitted medical insurance claims with diagnoses of selected diseases of the respiratory and circulatory systems were individually matched for age and year of service with undiseased co-workers. Individual cumulative total dust exposures for cases and controls were estimated. Cases with a disease of the respiratory system (27 pairs) had a nonsignificantly higher cumulative total dust exposure than did the controls; cases with diseases of the circulatory system (48 pairs) had significantly lower cumulative total dust exposure than the controls (Robertson & Ingalls, 1989).

Among other effects of exposure to carbon black (lampblack), dermatological lesions, such as the presence of carbon black 'tattoos' on hands and forearms, as well as follicular blackheads containing carbon black on uncovered skin surfaces, have been reported (Capusan & Mauksch, 1969). A brownish discoloration of the conjunctiva has occurred after long-term use of eye-liner containing carbon black, with accumulation of pigment in the tissues (Sugar & Kobernick, 1966; Haddad & Zehetbauer, 1980).

In a study of 58 current and 35 former male workers (at least two years of exposure) in a carbon black factory (channel black) and 60 controls, the currently exposed workers had a significantly higher prevalence of nasal complaints (including hyposmia) than either former workers or controls. Moreover, rhinoscopy revealed mucosal findings in 27% of the currently exposed workers versus only 9% of the controls. However, X-rays of the paranasal sinuses and cytology of nasal secretions did not reveal differences. The total dust concentration ranged from 9.0 to 13.35 mg/m^3 in the machine room and from 4.4 to 8.2 mg/m^3 in the operators' room (Kandt & Biendara, 1985).

4.2.2 *Experimental systems*

(*a*) *Inhalation studies*

A number of inhalation studies discussed in the following paragraphs are summarized in Table 28.

An early study by Snow (1970) showed that inhalation of thermal black with a particle diameter of ~0.15–0.2 μm and a surface area of between 10 and 15 m^2/g resulted in oedema of the laryngeal folds in Syrian hamsters exposed for 6 h per day on five days per week for a total of 53 and 172 days to a concentration of 105 mg/m^3 and for 236 days to a concentration of 56 mg/m^3 (see Table 28). Evaluation was limited to the larynx and trachea and consequently no effects on the lower respiratory tract were reported.

In another early study by Rhoades (1972), Long-Evans rats were exposed to gas channel black with a mass median diameter of 2.2 μm at a concentration of 4 mg/m^3 continuously for 16 days. The focus here was on the surfactant system, and findings of alveolar thickening and atelectasis were reported but no further effect related to surfactant. [The Working Group noted the lack of detail in the reporting of lung damage.]

Nau *et al.* (1976) exposed rhesus monkeys, Syrian hamsters and guinea-pigs to thermal black at a concentration of 53 mg/m^3 for 6 h per day on six days per week; the rodents were exposed for a total of 236 days and the monkeys for three years. No significant change in the lung was found in guinea-pigs despite the high concentration and long duration of exposure and no finding was given for the hamster lung. In monkeys, particles were adsorbed in regional lymph nodes and a moderate to severe emphysematous response with massive particle accumulation was observed in the lung. The investigators also reported right hypertrophy of the ventricular septum and, to a degree, of the left ventricle in the exposed monkeys. [The reporting of the findings was inadequate and no determination of the lung dose was made.]

Fenters *et al.* (1979) exposed CD-1 mice to a concentration of 1.5 mg/m^3 (0.3 μm MMAD) Sterling MT CT-6729 medium thermal black by inhalation for 3 h per day on five days per week for a total of four, 12 and 20 weeks. A challenge with virus and bacteria was included and it was found that there was decreased resistance to these infections, as seen by a reduced bactericidal capacity in the lung. They also observed morphological changes in the conducting airways and alveoli by scanning electron microscopy. [Alterations in immune responses using a plaque-forming assay were inconsistent, and no lung dose was reported.]

Wright (1986) found no increased type II cell proliferation after exposure of Fischer 344 rats to a concentration of 6.1 mg/m^3 of an unspecified carbon black (MMAD, 0.22 μm) for 20 h per day on seven days a week for 14 days [no lung burden data given].

Table 28. Non-neoplastic effects of carbon black in inhalation studies in experimental animals

Particle type	Particle diameter and surface area	Species (age and sex)	Exposure concentration	Duration of study	End-points	Findings	Comments	Reference
Thermal black	0.15–0.2 µm, 10–15 m^2/g	Syrian hamster	~56 mg/m^3 ~105 mg/m^3	6 h/day, 5 days/week, 53 and 172 days (high), 236 days (low)	Histology of larynx and trachea	Oedema of laryngeal folds; retention of tracheal and subglottic glands of amorphous eosinophilic material; no morphological changes	Study limited to larynx and trachea	Snow (1970)
Gas channel black	2.2 µm MMD	Long-Evans rats	4 mg/m^3	16 days (continuous)	Surfactant properties	Alveolar thickening, atelectasis; no surfactant effects	No lung dose; poor reporting of lung damage	Rhoades (1972)
Thermal black		Rhesus monkey, C3H mouse, hamster, guinea-pig	53 mg/m^3	6 h/day, 6 days/week, 236 days (rodents), 3 years (monkey)	Histology of lung and heart; pulmonary function (monkey)	Monkey: particles in the lymph; moderate to severe emphysematous lesions; massive particle accumulation; ventricular hypertrophy; guinea-pig: no significant lung changes	Inadequate reporting of findings; high concentration	Nau et al. (1976)
Sterling MT CT-6729 (medium thermal black)	0.3 µm MMAD	CD-1 mouse	1.5 mg/m^3	3 h/day, 5 days/week, 4, 12, 20 weeks	Immune response; infectivity challenge (virus and bacteria)	Decreased resistance to infection; bactericidal capacity in lung reduced; SEM changes in conducting airways and alveoli	Changes in immune response were inconsistent (plaque-forming assay); no lung dose reported	Fenters et al. (1979)
Carbon black [unspecified]	0.22 µm MMAD	Fischer 344 rat	6.1 mg/m^3	20 h/day, 7 days/week up to 14 days	Cell proliferation	Type II cell proliferation increased after diesel exhaust or NO_2 but not after CB	Lung burden not determined	Wright (1986)

Table 28 (contd)

Particle type	Particle diameter and surface area	Species (age and sex)	Exposure concentration	Duration of study	End-points	Findings	Comments	Reference
Elftex 12 (furnace black)	2.0 μm MMAD (large mode) 0.02–0.12 μm MMDD (small mode) 43 m²/g	Fischer 344 rat (14–15 weeks; male)	10 mg/m³	7 h/day, 5 days/week, 12 weeks	Lung inflammation, histology	12% PMN in BAL; 2.2 mg CB lung burden; macrophage aggregates engorged with CB in terminal bronchioles and alveolar ducts; type II cell hyperplasia	Effects of CB and concurrently tested diesel exhaust were very similar	Wolff et al. (1990)
Elftex 12 (furnace black)	2–2.4 μm MMAD (large mode) 0.02–0.12 μm DED (small mode, 10–30%)	Fischer 344/N rat (11–15 weeks; female)	3.5 mg/m³ 13 πg/m³ 98 πg/m³	16 h/day, 7 days/week; 6 h/day, 5 days/week; 4 h/day, 1 day/week; 12 weeks exposure + 24 weeks post-exposure	BAL analysis and histopathology at different exposure rates and constant c × t	Lung burden 3–4 mg; cellular (PMN) and biochemical (protein, LDH, β-glucuronidase) parameters significantly increased at all exposure rates to same degree; slight thickening of alveolar septa with hypertrophic epithelial cells; increased lung weights	Exposure rate does not influence toxic effects of inhaled CB	Henderson et al. (1992)
Printex 90 (furnace black) (± pyrolyzed pitch ± irritant gases)	0.64 μm MMAD 227 m²/g	Wistar rat	6 mg/m³	18 h/day, 5 days/week 10 months + 20 months post-exposure	Histology/ morphology	Squamous metaplasia; squamous differentiation of type II cells as precursor stages of squamous metaplasia in rats	Results specific to exposure to CB only not reported; pitch-related effects and BaP + CB cannot be distinguished	Nolte et al. (1993)

CARBON BLACK

Table 28 (contd)

Particle type	Particle diameter and surface area	Species (age and sex)	Exposure concentration	Duration of study	End-points	Findings	Comments	Reference
Printex 90 (furnace black) (± pyrolyzed pitch)	0.64 µm MMAD 227 m^2/g	Wistar rat	6 mg/m^3	18 h/day, 5 days/week, 10 months + 20 months post-exposure	Histopathology; focus on hyperplasia and metaplasia	Hyperplastic bronchiolar epithelium, metaplasia, inflammation, alveolar histiocytosis, alveolar lipoproteinosis	Focus is on cellular changes rather than CB specific effects; no clear distinction was made between effects of CB and pitch	Nolte et al. (1994)
Regal GR (furnace black) (96.9% carbon)	2.4 µm MMAD, 88 m^2/g	Swiss mouse (5 weeks; female)	10 mg/m^3	4 h/day, 4 days + 7 days observation	Infectivity model, bacterial and viral challange the day following exposure	No effect of CB alone on integrity of lung defences against bacterial and viral challenge; no pulmonary inflammation by BAL; co-exposure to 2.5 ppm acrolein impaired all defences	Short exposure duration; no lung burden data	Jakab (1993)
Regal GR (furnace black)	2.4 µm MMAD, 88 m^2/g	Swiss mouse (5 weeks; female)	10 mg/m^3	4 h, 24 h post-exposure	BAL analysis and AM phagocytosis	No change from controls; co-exposure to 1.5 ppm O$_3$ showed significant inflammation and suppression of phagocytosis	Exposure duration short	Jakab & Hemenway (1994)

Table 28 (contd)

Particle type	Particle diameter and surface area	Species (age and sex)	Exposure concentration	Duration of study	End-points	Findings	Comments	Reference
Elftex 12 (furnace black)	2.0 μm MMAD (large mode) 0.1 μm MMDD (small mode) (33%) 43 m²/g	Fischer 344/N rat (male and female)	2.5 mg/m³ 6.5 mg/m³	16 h/day, 5 days/week, 24 months	Histo-pathology; particle clearance; BAL parameters	Reduction in body weight; lung burden 21.0 and 38.5 mg CB; LN accumulation; dose-related increase in lung weight; BAL parameters (PMN, protein, LDH, β-glucuronidase); histology: chronic active inflammation, bronchiolar/alveolar metaplasia, alveolar proteinosis, septal and focal fibrosis, alveolar hyperplasia, squamous cysts; impaired clearance of ⁷Be CB particles at 3 and 18 months; accumulation of fluorescent microspheres in macrophage aggregates; female rats show greater response than males	Results nearly identical to those of studies performed concurrently using diesel exhaust	Mauderly et al. (1994); Nikula et al. (1995)
Printex 90 (furnace black)	0.64 μm MMAD 227 m²/g	Wistar rat (female)	6 mg/m³ 11.3 mg/m³	17 h/day, 5 days/week, 10 months + 20 months post-exposure or 20 months + 10 months post-exposure 18 h-day, 5 days/week, 24 months + 6 months post-exposure	Histo-pathology; hyperplasia; metaplasia	Inflammation: bronchiolar/alveolar hyperplasia, alveolar histiocytosis, lipoproteinosis; squamous metaplasia (probably from type II cells)		Dungworth et al. (1994)

Table 28 (contd)

Particle type	Particle diameter and surface area	Species (age and sex)	Exposure concentration	Duration of study	End-points	Findings	Comments	Reference
Monarch 880 (furnace black)	0.88 μm MMAD (16 nm primary particle) 220 m²/g	Fischer 344 rat	1.1, 7.1, 52.8 mg/m³	6 h/day, 5 days/week, for 13 weeks exposure; 3 months and 8 months post-exposure	BAL parameters; cell proliferation; histopathology; dosimetry	Lung burdens, 0.35, 1.8 and 7.8 mg; no change in any parameter at lowest concentration; dose-related increase at middle and high dose in cellular and biochemical lavage parameters; dose-related increase in alveolar cell proliferation		Driscoll et al. (1996)
Printex 90 (furnace black)	0.64 μm MMAD 227 m²/g	Wistar rat (female) NMRI mouse (female)	11.6 mg/m³ (average)	*Rats*: 18 h/day, 5 days/week 24 months + 6 months post-exposure *Mice*: 18 h/day, 5 days/week 13.5 months + 9.5 months post-exposure	Histopathology; lung clearance; dosimetry	*Rats*: body weights decreased, lung weight increased; lung burden, 44 mg; LN, 6.7 mg; impaired lung clearance; BAL: cellular and biochemical parameters increased; moderate to high-grade bronchiolar/alveolar hyperplasia; slight to moderate interstitial fibrosis *Mice*: body weights decreased, lung weights increased; lung burden, 7.4 mg (37 mg/g control lung, similar to rat)	Responses were very similar in both species compared to concurrently run diesel exhaust and ultrafine titanium dioxide	Heinrich et al. (1995)

MMD, mass median diameter; MMAD, mass median aerodynamic diameter; SEM, scanning electron microscopic; CB, carbon black; MMDD, mass median diffusion diameter; PMN, polymorphonuclear neutrophils; BAL, bronchoalveolar; c × t, concentration × time; LDH, lactate dehydrogenase; BaP, benzo[*a*]pyrene; AM, alveolar macrophages; LN, lymph node

More detailed characterizations of pulmonary responses to carbon black were reported by Wolff et al. (1990) and Henderson et al. (1992). Elftex 12 (furnace black) was aerosolized to give a large and small mode particle size distribution with a MMAD of 2 μm and mass median diffusive diameter (MMDD) of 0.02–0.12 μm. In the study by Wolff et al. (1990) an exposure concentration of 10 mg/m^3 was delivered to 14–15-week-old male Fischer 344/N rats for 7 h per day on five days per week for a total of 12 weeks. At a carbon black lung burden of 2.2 mg, it was found that 12% of the lavageable cells consisted of polymorphonuclear neutrophils. Aggregates of macrophages engorged with carbon black were found in the terminal bronchioles and alveolar ducts, and type II cell hyperplasia was also observed. Concurrently tested diesel exhaust particles showed very similar effects.

In the study by Henderson et al. (1992), the aim was to expose the rats to a different dose rate such that the product of concentration × time was very similar. Thus, 11–15-week-old female Fischer 344/N rats were exposed to concentrations of 3.5 mg/m^3 for 16 h per day on seven days a week, 13 mg/m^3 for 6 h per day on five days a week and 98 mg/m^3 for 4 h on one day per week. Each exposure lasted for a total period of 12 weeks, followed by a 24-week post-exposure period. Resulting lung burdens ranged between 3 and 4 mg. In spite of the different exposure rates, the cellular and biochemical parameters of bronchoalveolar lavage fluid increased to the same degree. In all exposure groups, a slight thickening of alveolar septa with hypertrophic epithelial cells was found and lung weights were significantly increased. The authors concluded that the exposure rate over this range does not influence resulting toxic effects of inhaled carbon black in the lungs.

Histological and morphological end-points were evaluated in a study by Nolte et al. (1993, 1994) following inhalation of Printex 90 (furnace black) by Wistar rats with or without the addition of pyrolyzed pitch and irritant gases such as nitrogen dioxide. The exposure concentration was 6 mg/m^3 for 18 h per day on five days per week for a total of 10 months with a 20-month post-exposure observation period. Findings were squamous metaplasia and squamous differentiation of type II cells, which was considered to be a precursor stage of the squamous metaplasia in this rat model of particle inhalation. However, it was difficult to differentiate the results due to exposure to carbon black only from those of changes induced by the combination of carbon black and pitch. [The Working Group noted the poor reporting of exposure concentration, exposure details and results specific to carbon black.]

Jakab (1993) determined the influence of carbon black inhalation on the resistance of mice to a bacterial and viral challenge one day after exposure. Five-week-old female Swiss mice were exposed to 10 mg/m^3 Regal GR (furnace black) (particle size, 2.4 ± 2.75 μm MMAD) for 4 h per day for a total of four days followed by a seven-day observation period. Carbon black alone had no effect on the integrity of lung defences against the bacterial and viral challenges and no pulmonary inflammatory response was detected in parameters of bronchoalveolar lavage fluid. In contrast, co-exposure to 2.5 ppm [5.7 mg/m^3] acrolein resulted in an impairment of the measured lung defence parameters (total and differential cell counts and albumin levels). This change in biological effect was not observed with acrolein alone. [The Working Group noted the relatively short duration of exposure to carbon black by inhalation. No lung burden data were reported.]

In an effort to determine the effect of exposure to carbon black in combination with ozone on alveolar macrophage phagocytosis, Jakab and Hemenway (1994) exposed five-week-old female Swiss mice to 10 mg/m^3 Regal GR (furnace black) (MMAD, 2.4 ± 2.75 µm) for 4 h. Analysis of bronchoalveolar lavage fluid 24 h later showed that there was no change in comparison to controls in animals exposed to carbon black only, and alveolar macrophage phagocytosis was not affected. Exposure to ozone alone caused significant changes in these parameters. [The Working Group noted that the exposure duration was rather short to see a significant effect of carbon black on the end-points studied.]

A detailed study comparing exposures to carbon black and diesel exhaust particles was reported by Mauderly *et al.* (1994) and Nikula *et al.* (1995). Male and female Fischer 344 rats were exposed to 2.5 mg/m^3 and 6.5 mg/m^3 Elftex 12 (furnace black) for 16 h per day on five days per week for a total of 24 months. Three males and three females each were sacrificed at three, six, 12, 18 and 23 months of exposure for evaluation of lung burden and lymph node burdens, histopathology, bronchoalveolar parameters and clearance of test particles. Particle size measurements resulted in a bimodal distribution with 2.0 µm MMAD for the large fraction, measured by a cascade impactor, and 0.1 µm MMDD for the small fraction, measured by parallel-flow diffusion battery (33%). Results showed a reduction in body weight at a final lung burden of 21.0 and 38.5 mg carbon black in the two exposure groups. Significant accumulation of carbon black with time occurred in the lymph nodes, and there was a dose-related increase in lung weight. Polymorphonuclear neutrophils and biochemical (protein, lactate dehydrogenase, β-glucuronidase) parameters of bronchoalveolar lavage fluid showed significant dose-dependent increases. Table 29 lists the occurrence of these and other non-neoplastic lung lesions found in the carbon black-exposed rats before and after 18 months of exposure. For most of the end-points examined, no significant difference in severity or incidence was observed between female and male rats. However, chronic active inflammation, alveolar proteinosis and bronchiolar/alveolar metaplasia occurred consistently with greater incidences and somewhat greater severity in females compared to males. Alveolar hyperplasia and squamous cysts were also found. Even at the low-exposure level, alveolar epithelial hyperplasia was observed in 4/6 rats after three months and 6/6 rats after six months of exposure. Bronchiolar/alveolar metaplasia was first observed in the high-dose group after 12 months of exposure (2/6 rats) and was present in 18/18 rats of this group and 22/24 rats of the low-dose group killed after either 23 months of exposure or after an additional six weeks without exposure. A total of 12 rats with squamous cysts were observed, 10 of which were in the group kept the additional six weeks without exposure. None of 19 squamous cysts increased in size after implantation into athymic mice, compared with 1/2 squamous-cell carcinomas and 2/8 lung adenocarcinomas. When ^{7}Be carbon black particles were administered at three and 18 months of exposure, significant retardation of their respective lung clearance was observed which increased with duration of exposure. Moreover, inhalation of fluorescent-labelled microspheres at three and 18 months of exposure caused significant

Table 29. Percentages (and severity scores) of rats dying, euthanized or sacrificed that had non-neoplastic lung lesions during a two-year exposure to carbon black (CB)[a,b]

Lesion	Female rats						Male rats					
	Control		Low CB (2.5 mg/m³)		High CB (6 mg/m³)		Control		Low CB (2.5 mg/m³)		High CB (6 mg/m³)	
	<18 months	>18 months	<18 months	>18 months	<18 months	>18 months	<18 months	>18 months	<18 months	>18 months	<18 months	>18 months
Alveolar macrophage hyperplasia	0 (0–0)	4 (0–1)	100 (1–3)	100 (2–4)	96 (0–4)	100 (3–4)	30 (0–2)	0 (0–0)	100 (1–3)	100 (1–3)	100 (2–4)	100 (2–4)
Alveolar epithelial hyperplasia	0 (0–0)	9 (0–2)	90 (0–3)	100 (1–4)	93 (0–4)	100 (2–4)	9 (0–0)	2 (0–2)	98 (0–4)	100 (1–3)	100 (1–4)	100 (1–4)
Chronic active inflammation	0 (0–0)	5 (0–1)	24 (0–2)	34 (0–2)	37 (0–3)	63 (0–2)	0 (0–0)	1 (0–3)	9 (0–2)	14 (0–2)	20 (0–2)	34 (0–3)
Septal fibrosis	0 (0–0)	2 (0–1)	52 (0–3)	96 (0–3)	78 (0–3)	100 (1–4)	3 (0–1)	1 (0–1)	61 (0–3)	92 (0–3)	75 (0–4)	99 (0–4)
Alveolar proteinosis	0 (0–0)	1 (0–2)	29 (0–2)	66 (0–3)	52 (0–3)	97 (0–4)	0 (0–0)	0 (0–0)	2 (0–1)	15 (0–1)	25 (0–2)	66 (0–4)
Focal fibrosis with epithelial hyperplasia	0 (0–0)	0 (0–0)	0 (0–0)	17 (0–4)	7 (0–2)	31 (0–3)	0 (0–0)	0 (0–0)	0 (0–0)	6 (0–2)	2 (0–2)	25 (0–3)
Squamous metaplasia	0 (0–0)	0 (0–0)	0 (0–0)	6 (0–2)	0 (0–0)	24 (0–3)	0 (0–0)	0 (0–0)	2 (0–1)	1 (0–1)	2 (0–3)	3 (0–2)
Number of rats	23	91	21	95	27	87	32	86	44	71	44	71

From Nikula et al. (1995)

[a] Animals dying, euthanized or sacrificed before and after 18 months of exposure were pooled
[b] Severity scores range from 1 to 4, with 4 being the most severe effect/lesion

sequestration of these particles in aggregated alveolar macrophages in the alveolar space as well as in the pulmonary interstitium, and this increased with exposure time and exposure concentration. Thus, at 18 months of exposure, both the low-dose and high-dose carbon black-exposed groups exhibited ~50% of the fluorescent microspheres in aggregated macrophages in the alveolar parenchyma. All of these results were nearly identical to those of comparison groups exposed concurrently to the same concentration of diesel exhaust.

A number of studies performed at the Fraunhofer Institute (Hannover, Germany) focused on histopathological events after exposure to carbon black, as well as inflammatory responses and parameters of lung clearance. The carbon black material used was Printex 90 (furnace black) with a surface area of 227 m^2/g; the aerosolized particles had an MMAD of 0.64 µm. Heinrich et al. (1995) reported results in female Wistar rats and female NMRI mice exposed to an average concentration of 11.6 ± 1.9 mg/m^3 Printex 90 for 18 h per day on five days per week for a total of 24 months (in rats) with an additional six-month post-exposure period, or 13.5 months (in mice) with an additional 9.5-month post-exposure time. Findings included decreased body weight and progressively increasing lung weight in rats, first measured at three months, with a lung burden (not contributing to lung weight) of 44 mg carbon black per lung at the 24-month time point. Carbon black (6.7 mg) was found in the lung-associated lymph nodes, and impaired lung clearance of test particles was found during this study (see Muhle et al., 1994). Significant increases in cellular and biochemical parameters of lung lavage were also observed. Histologically, moderate to high-grade bronchiolar/alveolar hyperplasia and slight to moderate interstitial fibrosis were observed in addition to significant findings of lung tumours which are reported in the previous section.

In mice, there was also a decrease in body weight and an increase in lung weight at a lung burden of 7.4 mg at 12 months of exposure. This lung burden, when normalized to a control lung, corresponded to 37 mg carbon black/g control lung which was similar to the normalized rat lung burden of carbon black of 32 mg/g control lung. No detailed description of nonneoplastic histological changes in the lungs of mice was provided, although there was elevated mortality during exposure in this and, more so, in other groups so that the planned 18-month exposure duration had to be shortened to 13.5 months. Overall, responses found in rats and mice exposed to carbon black were very similar to the responses in the respective groups of animals concurrently exposed to diesel exhaust or ultrafine titanium dioxide particles (Heinrich et al., 1995).

Dungworth et al. (1994) gave a more detailed description of effects induced in female Wistar rats by inhalation of 6 mg/m^3 carbon black for 17 h per day on five days a week for 10 months with a subsequent 20-month post-exposure period (72 rats) or inhalation for 20 months with a subsequent 10-month post-exposure period (72 rats). Another group (100 rats) was exposed to 11.3 mg/m^3 for 18 h per day on five days a week for a total of 24 months with up to a six-month post-exposure period. Findings were chronic active inflammation, including bronchiolar/alveolar hyperplasia, alveolar histiocytosis, lipoproteinosis and squamous metaplasia (probably originating from type II cells).

Nolte *et al.* (1994) also reported results after exposure of Wistar rats to 6 mg/m^3 of this same material for 18 h per day on five days a week for a total of 10 months with an additional 20-month post-exposure period. As in the study described previously (Nolte *et al.*, 1993), a clear distinction in their reporting was not drawn between those effects found in animals exposed to carbon black only and those found in animals that were co-exposed to carbon black and pyrolyzed pitch — the focus of this study was on cellular changes and cell dynamics. Thus, they reported hyperplastic bronchiolar epithelium and metaplasia as well as inflammatory responses including alveolar histiocytosis and alveolar lipoproteinosis.

A subchronic inhalation study with Monarch 880 (furnace black) was performed in Fischer 344 rats exposed to 1.1, 7.1 and 52.8 mg/m^3 for 6 h per day on five days a week for a total of 13 weeks with an eight-month post-exposure period (Driscoll *et al.*, 1996). The MMAD of the carbon black aerosol was 0.88 µm and the primary particle size was 16 nm with a specific surface area of 220 m^2/g. Lung burdens at the end of exposure were 0.35, 1.8 and 7.8 mg, respectively; there was continued accumulation of carbon black in the tracheobronchial lymph nodes during the post-exposure period in the mid- and high-dose groups only. Inflammatory cellular and biochemical parameters of lung lavage (total and differential cell counts; protein and lysosomal and cytoplasmic enzymes) and cell proliferative responses as well as histopathological evaluation showed no change in the low-exposure group as compared to controls at any time point. The groups exposed to the higher concentrations had dose-related increases in cellular and biochemical broncho-alveolar parameters as well as in alveolar cell proliferative responses at the end of exposure which remained elevated throughout the post-exposure period in the highest exposure group. Cellular inflammatory parameters also remained elevated during post-exposure in the mid-exposure group whereas lavage protein and enzyme levels returned to control values. Table 30 summarizes the findings of non-neoplastic end-points determined in this study at the end of the subchronic exposure and up to eight months post-exposure.

In summary, these studies show that once a certain lung burden has been achieved, inhalation of carbon black in rats results in significant inflammatory responses in the lung. The reported inflammatory pulmonary responses may be mechanistically related to subsequent fibrotic as well as neoplastic effects observed in long-term chronic inhalation studies at high-exposure concentrations. The effects of carbon black appear to be more severe in rats than in mice, based on the single mouse study. In addition, female rats seemed to respond with higher incidence and greater severity than males with respect to chronic, active inflammation, alveolar proteinosis and bronchiolar/alveolar metaplasia.

(b) Instillation studies

A number of studies evaluating non-neoplastic effects of carbon black were performed by intratracheal instillation (see Table 31). Bowden and Adamson in a number of studies from 1978 through 1982 evaluated cellular responses in mice, specifically focusing on the kinetics of macrophages and polymorphonuclear neutrophils. They used India ink (Pelikan Co.) in most of their studies which has a primary particle diameter of

Table 30. Three-month multi-exposure inhalation study of carbon black in rats

End-point	Exposure concentration (mg/m³)		
	1.1	7.1	52.8
Continued post-exposure accumulation in lymph nodes	–	+	+
No. of AM increased:			
End of exposure	–	+	+
Post-exposure	–	–	+
No. of PMN increased:			
End of exposure	–	+	+
Post-exposure	–	+	+
Cell proliferation:			
End of exposure	–	+	+
Post-exposure	–	–	(+)
Fibrotic response	–	(+)	+

Compiled from Driscoll *et al.* (1996)
–, no response; +, marked response; (+), mild response; AM, alveolar macrophages; PMN, polymorphonuclear macrophages

30–40 nm and was instilled mostly at doses of 4 mg into Swiss mice [sex not always specified]. Subsequent evaluation showed that there was generally a biphasic macrophage response in which the first phase occurred without mitotic activity whereas the second phase showed mitosis of primarily interstitial macrophages. There was also a high initial response in terms of elicitation of polymorphonuclear neutrophils, which reached twice the number of alveolar macrophages (Adamson & Bowden, 1978; Bowden & Adamson, 1978). They also observed that this very high dose and extremely high dose rate (4 mg instilled) resulted in rapid migration of blood monocytes to pulmonary alveoli and rapid production of monocytes in the bone marrow (Adamson & Bowden, 1980). In a study with male Swiss mice using doses as low as 0.1 mg and as high as 8 mg (dose–response study), these authors observed (Adamson & Bowden, 1981) that the number of alveolar macrophages elicited was correlated to the dose delivered and that this response was very similar to that to other different types of particles administered. In further studies (Adamson & Bowden, 1982a), they confirmed that the responses are not unique to carbon black but also occur after different particles, including latex, are instilled into the mouse. Further, they showed that chemotactic factors are elicited in the alveolar space and detectable in bronchoalveolar lavage fluid. They further determined the importance of the macrophage to elicit the initial response; when whole-body irradiated mice were instilled with carbon black, a limited macrophage response occurred which was followed by proliferation of interstitial macrophages still present in the lungs of these irradiated mice (Adamson & Bowden, 1982b; Bowden & Adamson, 1982).

Table 31. Non-neoplastic effects of carbon black by instillation studies in experimental animals

Particle type	Particle diameter and surface area	Species (age and sex)	Dose	Observation period	End-points	Findings	Comments	Reference
India ink (Pelikan Co.)	30–40 nm	Swiss mouse	4 mg	Up to 28 days	Macrophage proliferation and CB transport in lung	Biphasic macrophage response; first phase without mitotic activity; second-phase mitosis of interstitial macrophages; initial large PMN increase in BAL (twice AM numbers); some CB particles crossed epithelium to reach peribronchial and perivascular sites	Very high dose and dose rate	Adamson & Bowden (1978); Bowden & Adamson (1978)
India ink (Pelikan Co.)	30–40 nm	Swiss mouse	4 mg	Up to 7 weeks	Macrophage and blood monocyte response	Rapid migration of blood monocytes to pulmonary alveoli and rapid production of monocytes in bone marrow after CB dosing	Very high dose and dose rate	Adamson & Bowden (1980)
India ink	30 nm	Swiss mouse (male)	0.1, 1, 2, 4, 8 mg	Up to 14 days	Macrophage response	Confirming findings (Adamson & Bowden (1978); number of AM elicited is related to dose delivered; lowest dose showed very little response	Focus of study was on macrophage kinetics after different particle types	Adamson & Bowden (1981)
Colloidal carbon	30 nm	Swiss mouse	4 mg (donor)[a] BAL supernatant (recipient)[a]	2 days Up to 14 days	PMN and macrophage response	Chemotactic factors in BAL after CB cause PMN and AM influx	Similar response also after latex particles, not unique to CB	Adamson & Bowden (1982a)
Colloidal carbon	30 nm	Swiss mouse	4 rr g, whole-body irradiated and controls	Up to 20 weeks	Macrophage response after monocyte depletion	Limited initial macrophage response after monocyte depletion, followed by interstitial macrophage proliferation; increased translocation of CB to interstitium in depleted mice with decreased AM output	Response not specific to CB; importance of AM for containment of particles in alveolar space	Adamson & Bowden (1982b); Bowden & Adamson (1982)

Table 31 (contd)

Particle type	Particle diameter and surface area	Species (age and sex)	Dose	Observation period	End-points	Findings	Comments	Reference
Furnace black Regal 660	~20 nm	Fischer 344 rat (male)	0.5 mg	24 h	BAL response, interstitial passage	Inflammatory response (PMN and protein in BAL) is correlated with particle surface area; CB passage to interstitium is less than for ultrafine TiO_2; interstitial access can influence alveolar inflammation	Surface area correlation applies to both ultrafine and larger-sized TiO_2 particles	Oberdörster et al. (1992)

CB, carbon black; PMN, polymorphonuclear macrophages; BAL, bronchoalveolar lavage; AM, alveolar macrophages; TiO_2, titanium dioxide
ᵃ 4 mg carbon black instilled into donor mice, from which supernatant of BAL was obtained and administered to recipient mice

Macrophage depletion also resulted in an increased translocation of administered carbon black to interstitial sites. Overall, the studies of these investigators confirmed the importance of the alveolar macrophage in pulmonary defences to particles including carbon black. Both the alveolar macrophage and the interstitial macrophage systems are of importance in this response as part of the pulmonary defence system against particles.

Oberdörster et al. (1992) administered 0.5 mg furnace black (Regal 660) with a primary particle size of ~20 nm to male Fischer 344 rats by intratracheal instillation. In addition, other particle types (ultrafine and larger-sized titanium dioxide) were administered at different dose levels, and the bronchoalveolar lavage response and interstitial access of particles was determined 24 h later. They found that the inflammatory response as measured by lavaged polymorphonuclear macrophages and lavage protein was correlated best with the particle surface area of different particle types of titanium dioxide; the inflammatory response induced by carbon black also fitted the same regression. Interstitial access of ultrafine titanium dioxide particles was greater than that for carbon black, with higher doses resulting in a diminished inflammatory response in the alveolar space (larger polymorphonuclear macrophage) and a shift of the inflammation towards the interstitium.

In a study of the effects of benzo[a]pyrene on rat lung, Davis et al. (1975) dosed intratracheally groups of 18 female Wistar rats, 12–16 weeks of age at the beginning of the experiment, with 0.5, 1.0 or 2.0 mg benzo[a]pyrene with or without 0.5 mg carbon black on 18 occasions at biweekly intervals. One group received carbon black alone and one received no treatment. Group mean survival times ranged from 73 to 109 weeks. At autopsy, all rats given carbon black had black deposits in their lungs, mainly in alveolar macrophages. These animals also showed significantly more severe columnar and cuboidal metaplasia of the alveolar epithelium, whereas rats receiving benzo[a]pyrene alone showed no increased severity of metaplasia. However, squamous metaplasia of the alveolar epithelium was increased in rats receiving carbon black alone as well as carbon black with benzo[a]pyrene.

Overall, these results from intratracheal instillation studies with carbon black show that high acute doses of carbon black elicit a significant pulmonary inflammatory response which is possibly related to the large specific surface area of the particles.

(c) *Ex-vivo and in-vitro studies*

A number of studies have been performed with carbon black using either in-vivo exposure to the particulate compound with subsequent isolation of cells and specific in-vitro investigations or primary in-vitro exposure of cell systems to evaluate effects. These studies are summarized in Table 32.

Miller and Zarkower (1974) exposed Balb/c mice by inhalation to 5.4 mg/m^3 carbon black [unspecified] with a mass median diameter of 1.8 μm continuously for seven to 28 days. The aim was to investigate effects on the immune system. After exposure, spleen and lung lymph node T and B cell lymphocytes were isolated and an in-vitro lymphocyte assay, including a transformation test, was performed. They reported significant changes in the responsiveness to mitogens of both B and T lymphocytes as well as changes in

Table 32. In-vitro toxicity studies of carbon black

Particle type	Particle diameter and surface area	Test system	Dose/exposure concentration	Findings	Comments	Reference
Not specified	1.8 μm	Balb/c mouse: inhalation exposure; isolation of spleen and lung lymph node T and B cells; in-vitro transformation of lymphocyte and migration of macrophage	5.4 mg/m^3, 7–28 days continuously	Significant changes in responsiveness to both B- and T-lymphocyte-specific mitogens; changes in lymphocyte populations	Lung dose not determined; altered immune response, possibly due to high CB lung burden and inflammation	Miller & Zarkower (1974)
Fisher carbon black (thermal black)	25 μm 31 m^2/g	Effect of adsorption of BaP; uptake into 1 ml of rat liver microsomes	16.7 mg CB + 5 μg BaP; 30 min incubation	BaP was not released from CB and there was no uptake into microsomes, in contrast to other particles (Fe$_3$O$_4$, SiO$_2$, asbestos)	Relevancy of test system (high doses) is questionable	Lakowicz & Bevan (1979, 1980)
Fisher carbon black (thermal black)	25 μm 31 m^2/g	Effect of adsorption of BaP on particles for uptake rate into model membranes	16.7 mg CB + 5 μg BaP; 30 min incubation	BaP was not released from CB, no uptake into membranes in contrast to BaP adsorbed on other particles	Relevancy of test system (high doses) is questionable	Lakowicz et al. (1980)
3 Oil furnace blacks	96, 168, 220 nm 128, 101, 70 m^2/g	In-vitro elution of BaP from CB using plasma, serum, lung lavage fluid; in-vivo feeding study with CB in mice to study AHH induction	In-vitro: 5–20 g CB with biological medium, 24 h In-vivo: 0.08, 2 and 20 g/kg diet, 30–180 days	Less than 0.005% of adsorbed BaP can be eluted by biological media; in-vivo feeding at high dose does not induce AHH in mouse lung or liver	Results are in contrast to other in-vivo bioavailability studies of BaP; test system may not be relevant	Buddingh et al. (1981)
4 Oil furnace blacks	96, 168, 175, 220 nm 128, 101, 90, 70 m^2/g	Elution of adsorbed BaP from CB into phospholipid vesicles	100 μg CB	Elution of BaP from CB depends on amount of BaP present; rate and extent of elution is lowered with less adsorbed BaP	Study aims at bioavailability of PAH adsorbed onto CB particles; high-dose study not relevant for in vivo	Bevan & Worrell (1985)

Table 32 (contd)

Particle type	Particle diameter and surface area	Test system	Dose/exposure concentration	Findings	Comments	Reference
4 Oil furnace blacks	96, 168, 175, 220 nm 128, 101, 90, 70 m²/g	Elution of adsorbed BaP from CB into phospholipid vesicles from rat lung homogenate and simulated lung fluid	Up to 200 mg CB	0.2–0.6% of BaP can be eluted; more elution from low surface area CB and high BaP content	Very low elution in vitro using biological systems	Bevan & Yonda (1985)
Regal 660 (furnace black)	10 m²/g	In-vitro activation of rat serum with particles to determine chemotactic activity for AM	5–25 mg CB/ml serum; 10% serum for AM chemotaxis	CB has highest activity for different particles (SiO₂, asbestos, PVC, TiO₂) to activate serum chemotactic factors	High doses	Oberdörster et al. (1989)
2 Oil furnace carbon blacks (N339 & Black Pearl 2000)	Not given. Surface area of N 339 is 15-fold less than that of Black Pearl 2000	Phagocytosis assay with AM from Wistar rats; CB adsorbed with polar and semi-polar compounds	1.5 ml of 0.04 mg/ml suspension for 2×10^5 AM, 45 min	AM phagocytosis of CB + adsorbate suppressed only for low surface CB; Fc-receptor mediated phagocytosis of sheep red blood cells was impaired after previous uptake only of low surface area CB + adsorbate	Surface properties are important for fate of particle-pollutant complexes	Jakab et al. (1990)

CB, carbon black; BaP, benzo[a]pyrene; AHH, arylhydrocarbon hydroxylase; PAH, polycyclic aromatic hydrocarbons; AM, alveolar macrophage lymphocyte populations. Spleen lymphocytes of carbon black-exposed mice also exhibited significantly enhanced ratios of transformation after sensitization of the mice with tuberculosis antigen.

Lakowicz and Bevan (1979, 1980) and Lakowicz et al. (1980) investigated the effect of adsorption of benzo[a]pyrene onto Fisher carbon black (thermal black) (25 µm particle size) on uptake into either rat liver microsomes or into model membranes. 5 µg benzo[a]pyrene had been adsorbed onto 16.7 mg carbon black which was incubated with model systems for 30 min. Release of benzo[a]pyrene from the carbon black was not detected, and no uptake into liver microsomes or model membranes occurred. This is in contrast to the enhanced uptake of benzo[a]pyrene absorbed onto certain other particles (haematite, silica and asbestos) that were also tested by these authors.

Additional studies were performed to evaluate the in-vitro elution of benzo[a]pyrene from carbon black using either plasma, serum or lung lavage fluid from rats, phospholipid vesicles, or phospholipid vesicles, rat lung homogenates and simulated lung fluids (Buddingh et al., 1981; Bevan & Worrell, 1985; Bevan & Yonda, 1985). These studies are listed in Table 32 and show that very little of the benzo[a]pyrene adsorbed onto carbon black particles can be eluted by biological media. The eluted amount depends on the total amount of benzo[a]pyrene present on the carbon black. Together with their in-vitro studies, Buddingh et al. (1981) also reported results of a feeding study with carbon black in mice designed to examine the induction of arylhydrocarbon hydroxylase in the lung and liver; they did not find induction of this enzyme even at the highest dose level of 20 g/kg of diet carbon black for up to 180 days.

Oberdörster et al. (1989) investigated activation of chemotactic serum factor *in vitro* by furnace black (Regal 660) and compared it to other particle types such as titanium dioxide, polyvinyl chloride (PVC), asbestos and silicon dioxide. They found that, on a mass basis, carbon black showed the greatest ability to induce chemotactic factors for alveolar macrophages in rat serum. The authors suggested that surface area may play an important role in the inducibility of such factors, since titanium dioxide, with a very low surface area, showed the lowest response.

Jakab et al. (1990) studied alveolar macrophage phagocytosis of two different oil furnace carbon blacks with high and low surface areas and with adsorbed polar and semi-polar compounds. The carbon blacks (1.5 ml of a 0.04 mg/ml suspension) were incubated with 2×10^5 alveolar macrophages from Wistar rats for 45 min. For the low surface area carbon black particles only, they observed depressed alveolar macrophage phagocytosis of the carbon black and the adsorbates as well as for sheep red blood cells that had previously been incubated with the carbon black–adsorbate complex. The authors suggested that surface properties are important parameters to determine the fate of particle–pollutant complexes in the lung.

No firm conclusions about the in-vitro toxicity or effects on cell systems can be drawn from these limited in-vitro studies reported in the literature. With respect to elution of adsorbed benzo[a]pyrene, it appears that desorption from carbon black particles occurs at a very low rate and to a very low degree. However, in-vivo studies (see section 4.1.2(*b*)) have demonstrated clearly that adsorbed material can be eluted readily.

(d) Other studies

A number of studies have been performed with carbon black that may not be directly relevant for an evaluation of carcinogenicity — the route of exposure was in most cases rather unusual. These studies consist of experiments of intracardiac, intravenous, intrabladder-wall and intraventricular (central nervous system) injections of carbon black into experimental animals. The material used was mostly India ink (Pelikan Co.) and the major conclusion from these studies is that systemically administered carbon black particles can be trapped in the pulmonary circulation and be transported to pulmonary interstitial sites and alveolar macrophages. This is a suggested route of elimination of foreign bodies from the systemic circulation according to the authors (Blau & Veall, 1967; Vales *et al.*, 1967; Bertheussen & Nissen, 1976; Bertheussen *et al.*, 1978 (see Table 33)).

4.3 Reproductive and developmental effects

No data were available to the Working Group.

4.4 Genetic and related effects

4.4.1 *Humans*

No data were available to the Working Group.

4.4.2 *Experimental systems* (see also Table 34 and Appendices 1 and 2)

The activity of carbon black particles and of their corresponding solvent extracts in short-term assays must be considered separately. When carbon black particles are tested, results may be influenced by such experimental conditions as the presence of serum, the concentration of dimethyl sulfoxide or other solvents, or the duration of exposure. In addition, these assays may underestimate in-vivo exposure owing to the short duration of the experiments. Conversely, the amount of chemicals eluted by solvent extracts of carbon blacks may be greater than that which would be eluted by biological fluids (Buddingh *et al.*, 1981). Additionally, the nature of the solvent and the temperature and duration of the Soxhlet extraction influence the final biological response (Sanders, 1981; Giammarise *et al.*, 1982; Butler *et al.*, 1983).

Several different carbon blacks have been assayed in short-term tests. These include a rubber-grade furnace black (N339), a nitric acid after-treated black (Black Pearls), a third carbon black of unspecified type, and several unspecified carbon black pastes. In addition, extracts of three of the above and 20 other carbon black samples have been tested. The data are summarized in Table 34.

Table 33. Other studies of carbon black related to non-neoplastic end-points

Particle type	Particle diameter and surface area	Test system	Dose	Findings	Comments	Reference
India ink (Pelikan) (+ 4.3% fish glue, 1% phenol)	20–50 nm	Guinea-pig: intracardiac injection (animals turned slaty grey for a few min); study of thymus uptake	0.1–0.15 ml/100 g bw of 10% suspension	CB found in macrophages throughout thymus. Foreign bodies can reach thymus	Not relevant for CB toxicity	Blau & Veall (1967)
Not given	16 and 70 nm	Intravenous injection into rabbits; reaction of pulmonary vessels	5 mg, up to 13 weeks follow-up	Formation of emboli in pulmonary vessels; endothelial hyperplasia; passage of CB to alveolar space and elimination	Not relevant for CB toxicity	Vales et al. (1967)
India ink (Pelikan)	Not given	Injection into urinary bladder wall of rat; sacrifice after 1–24 h	0.1–0.2 ml of 25% suspension	CB found in AM at 24 h, suggesting pulmonary excretion of foreign bodies	Not relevant for CB toxicity	Bertheussen & Nissen (1976)
India ink (Pelikan)	Not given	Injection into ventricular system of Wistar rat	3 : 1 mixture of saline : CB	CB found in macrophages in alveolar septae of lung as elimination pathway	Not relevant for CB toxicity	Bertheussen et al. (1978)

CB, carbon black; AM, alveolar macrophages

Table 34. Genetic and related effects of carbon blacks or their formulations

Test system	Result[a] Without exogenous metabolic system	Result[a] With exogenous metabolic system	Dose[b] (LED/HID)	Reference
SA0, *Salmonella typhimurium* TA100, reverse mutation	–[c]	–[c]	3750	Kirwin et al. (1981)
SA0, *Salmonella typhimurium* TA100, reverse mutation	+[d]	+[d]	NR	Agurell & Löfroth (1983)
SA0, *Salmonella typhimurium* TA100, reverse mutation	–[e]	–[e]	50	Venier et al. (1987)
SA5, *Salmonella typhimurium* TA1535, reverse mutation	–[c]	–[c]	3750	Kirwin et al. (1981)
SA7, *Salmonella typhimurium* TA1537, reverse mutation	–[c]	–[c]	3750	Kirwin et al. (1981)
SA8, *Salmonella typhimurium* TA1538, reverse mutation	–[c]	–[c]	3750	Kirwin et al. (1981)
SA9, *Salmonella typhimurium* TA98 reverse mutation	–[f]	0	250	Rosenkranz et al. (1980)
SA9, *Salmonella typhimurium* TA98. reverse mutation	(+)[g]	0	500	Rosenkranz et al. (1980)
SA9, *Salmonella typhimurium* TA98, reverse mutation	+[h]	0	5.0	Rosenkranz et al. (1980)
SA9, *Salmonella typhimurium* TA98, reverse mutation	–[c]	–[c]	3750	Kirwin et al. (1981)
SA9, *Salmonella typhimurium* TA98, reverse mutation	+[d]	+[d]	NR	Agurell & Löfroth (1983)
SA9, *Salmonella typhimurium* TA98, reverse mutation	(+)[e]	+[e]	2.5	Venier et al. (1987)
SAS, *Salmonella typhimurium* TA98NR, reverse mutation	+[d]	0	NR	Agurell & Löfroth (1983)
SAS, *Salmonella typhimurium* TA98/1,8DNP, reverse mutation	+[d]	0	NR	Agurell & Löfroth (1983)
DMM, *Drosophila melanogaster*, somatic mutation (mosaics)	–[c]		10 000 larval feeding	Kirwin et al. (1981)

Table 34 (contd)

Test system	Result[a]		Dose[b] (LED/HID)	Reference
	Without exogenous metabolic system	With exogenous metabolic system		
DMX, *Drosophila melanogaster*, sex-linked recessive mutation	−[c]		10 000 larval feeding	Kirwin et al. (1981)
DML, *Drosophila melanogaster*, dominant lethal test	−[c]		10 000 larval feeding	Kirwin et al. (1981)
DMN, *Drosophila melanogaster*, aneuploidy (sex-chromosome loss)	−[c]		10 000 larval feeding	Kirwin et al. (1981)
G5T, Gene mutation, mouse lymphoma L5178Y cells, *tk* locus *in vitro*	−[c]	−[c]	40 000	Kirwin et al. (1981)
SIC, Sister chromatid exchange, Chinese hamster ovary CHO cells *in vitro*	−[c]	−[c]	1000	Kirwin et al. (1981)
MIA, Micronucleus induction, M3E3/C3 hamster epithelial cells	+[i]	0	1	Riebe-Imre et al. (1995)
TCM, Cell transformation, C3H/10T½ mouse fibroblasts *in vitro*	−[c]	0	16 000	Kirwin et al. (1981)
TCL, Anchorage independent growth, M3E3/C3 hamster epithelial cells *in vitro* (undifferentiated and small mucus granule cell stage)	+[j]	0	100	Riebe-Imre et al. (1995)
GVA, *p53*, K-*ras* in pulmonary carcinomas in F344/N rats	−[k]		3.5 inh 16 h/d × 5 d/wk × 24 mo	Swafford et al. (1995)
GVA, *hprt* Mutation analysis in type II alveolar cells isolated from rats after exposure	+[l]		1.5 inh. 6 h/d × 5 d/wk × 13 wk	Driscoll et al. (1995)
BVD, Binding to DNA (^{32}P-postlabelling) in F344/N rat alveolar type II cells *in vivo*	+[k]		3.5 inh 16 h/d × 5 d/wk × 12 wk	Bond et al. (1990)

Table 34 (contd)

Test system	Result[a]		Dose[b] (LED/HID)	Reference
	Without exogenous metabolic system	With exogenous metabolic system		
BVD, Binding to DNA (^{32}P-postlabelling) in Wistar rat lung *in vivo*	–[m]		7.3 inh 18 h/d × 5 d/wk × 2 yr	Gallagher *et al.* (1994)

[a] +, positive; (+), weak positive; –, negative; 0 not tested; ?, inconclusive

[b] LED, lowest effective dose; HID, highest ineffective dose. In-vitro tests, μg/mL; in-vivo tests, mg/kg bw; NR, dose not reported; MMAD, mass median aerodynamic diameter; MMDD, mass median diffusion diameter

[c] Rubber-grade furnace black N339, surface area 100 m^2/g, 48-h toluene extractables 0.15%; particles suspended in DMSO (Ames test + SCE assay), acetone (cell transformation test) or culture media (mouse lymphoma test).

[d] Carbon blacks from various manufacturers (20 samples). Soxhlet extraction of 1-g samples with 200 mL benzene for 16 h and solvent exchange into DMSO

[e] Carbon black used for refining tanned skins (7 samples). (a) Sonication of 2 g samples in 40 mL benzene for 0.5 h; (b) Soxhlet extraction of 4-g samples with 50 mL toluene for 48 h. Solvent exchange into DMSO (1 g extract/ml).

[f] Black Pearls L (furnace black, manufacture of which involves a nitration-oxidation step). Suspension in DMSO at 5 mg/mL for 5 h before testing.

[g] Raven 5750 (furnace black, oxidative aftertreated). Soxhlet extraction of 10-g samples with toluene for 48 h. Low-temperature concentration and solvent exchange into 1 mL DMSO.

[h] Black Pearls L (furnace black, manufacture of which involves a nitration-oxidation step). Soxhlet extraction of 10 g samples with toluene for 48 h. Low-temperature concentration and solvent exchange into 1 mL DMSO.

[i] Carbon black [not otherwise characterised]; carbon black suspended in the culture medium containing undifferentiated cells for 72 h.

[j] Carbon black [not otherwise characterised]; carbon black suspended in the culture medium containing undifferentiated or differentiated cells for 72 h.

[k] Elftex-12 (furnace black); 2 μm MMAD (large mode); 0.1 μm MMDD (small mode); surface area, 43 m^2/g. Whole-body exposure

[l] Monarch 880 (furnace black); 0.8 μm MMAD (16 nm primary particle); surface area, 220 m^2/g. Whole-body exposure

[m] Printex 90 (furnace black); MMAD, 0.65 m; surface area, 270 m^2/g. Carbon black in air at 2 yr mean of 11.3 mg/m^3. Whole-body exposure

In an extensive study, Kirwin et al. (1981) tested a rubber-grade furnace black (N339; surface area, 100 m^2/g; toluene extractables (48-h), 0.15 wt%) in the following five short-term assays:

(1) Mutagenicity in *Salmonella typhimurium*: no mutagenic activity was observed in *S. typhimurium* strains TA1535, TA1537, TA1538, TA98 or TA100 at concentrations of carbon black of up to 7.5 mg/plate in the presence or absence of an Aroclor-induced rat-liver homogenate supernatant fraction (S9); cellular toxicity (TA100) was assessed and viable count was reduced by 27% at 7497 μg/plate of carbon black. [It was not reported whether, or for how long, the carbon black particles were suspended in dimethyl sulfoxide (DMSO) prior to testing.]

(2) Sister chromatid exchange in Chinese hamster ovary cells: the carbon black was suspended in DMSO at 100 mg/mL [time and temperature unspecified] and then diluted in culture medium to give a final concentration range of 0.00032–1 mg/mL; cells were exposed for 2 h both in the presence and absence of S9; very small increases in the frequency of sister chromatid exchange as compared to the control value with and without S9 were observed for several concentrations, but these were not dose related.

(3) L5178Y tk$^{+/-}$ mouse lymphoma mutagenicity assay: cells were exposed for 4 h [time extended for an unspecified time owing to difficulty in separating carbon black from cells] to concentrations of carbon black of 10–40 mg/mL in the absence of S9 and of 5–15 mg/mL in the presence of S9; cell survival was < 1% at the highest concentration; no mutagenicity was observed.

(4) C3H/10T½ CL8 mouse embryo morphological cell transformation assay: carbon black suspended in acetone was tested at four concentrations ranging from 2–16 mg/mL; no transformed focus was observed.

(5) Genetic activity in *Drosophila melanogaster*: larvae were fed diets containing 1% carbon black until pupation; flies were scored for mosaics, Y-chromosome loss, chromosomal aberrations and dominant lethal and sex-linked lethal mutations; no genetic effect was observed.

A nitric acid-treated furnace black (Black Pearls; surface area, 115 m^2/g; toluene extractables (48-h), 0.3 wt%) (Sanders, 1981) was tested for mutagenicity in *S. typhimurium*. Particles were first suspended in DMSO for 5 h and an aliquot containing 500 μg carbon black was then tested in strain TA98. No mutagenic activity was observed (Rosenkranz et al., 1980).

An aliquot of a 48-h Soxhlet toluene extract (solvent exchanged into DMSO) equivalent to 10 μg of the above carbon black was, however, mutagenic in the same strain. This carbon black contained nitrated pyrenes at a level of 67 mg/kg (Sanders, 1981). More recent production lots of this grade of carbon black had a 200-fold reduction in nitrated pyrene content; extracts had a mutagenicity that was reduced by the same order of magnitude (Rosenkranz et al., 1980; Agurell & Löfroth, 1983; Butler et al., 1983).

Benzene or acetone extracts of 20 commercial carbon blacks were tested in the *Salmonella* mutagenicity assay. Of the 20 extracts (some of which required activation with rat-liver S9), 15 were mutagenic to strains TA98 and/or TA100 and five were inactive (Agurell & Löfroth, 1983).

Venier *et al.* (1987) tested the mutagenicity in *S. typhimurium* strains TA98 and TA100 of seven carbon black pastes that are used as commercial leather dyes. Samples were assayed for mutagenicity either directly or after extraction with benzene. The compounds that were tested were in the form of thick pastes and the carbon black content ranged from 5 to 8%. In all compounds but one, carbon black was dispersed in 10–15% casein solution in water containing small amounts of sulfonated castor oil and cresols. Different extraction procedures were used for the pastes. No mutagenicity was observed either directly or after sonication with benzene in any of the carbon black samples tested. After a 48-h extraction of carbon blacks with boiling toluene, four carbon black samples were mutagenic in strain TA98 in the presence of S9. The activity ranged from 1.3 to 9.7 induced revertants/mg equivalent of extract. A weak direct mutagenic activity in strain TA98 was shown by one extract. The presence of PAHs in the toluene extracts was reported by the authors to explain the mutagenicity of only one carbon black sample. Low or undetectable levels of PAHs were found in other mutagenic extracts.

Two studies analysed the extent to which exposure of rats to carbon black induced DNA adducts in lung tissue (Bond *et al.*, 1990; Gallagher *et al.*, 1994). Both studies employed the ^{32}P-postlabelling assay to measure DNA adducts.

Gallagher *et al.* (1994) exposed female Wistar rats (Crl:(WI)BR) to furnace black (Printex 90) particles. The carbon black exposure was 7.5 mg/m^3 for the first four months and 12 mg/m^3 for the last 20 months. Exposures were for 18 h per day on five days a week for two years using whole-body exposure chambers. The carbon black surface area was 270 m^2/g. The MMAD of the carbon black particles for the exposures was 0.65 μm. The extractable organic matter, as determined by solvent extraction with dichloromethane, was 0.039%. After two years of exposure, animals were killed and the distal tip of the peripheral left lung lobe was removed for analysis of DNA adducts using the nuclease P1 or butanol extraction versions of the ^{32}P-postlabelling assay. ^{32}P-Postlabelling analysis detected one major radiolabelled spot that was referred to as adduct 1. DNA adduct levels for adduct 1 after two years of exposure to carbon black were about 9 adducts/10^9 bases and about 17 for the filtered-air controls. Adduct 1 was found to increase in an age-related fashion and was presumed by the authors to be a ^{32}P-labelled I-compound. Adduct levels were determined for the diagonal radioactive zone; however, no significant elevation in adduct levels in this zone was observed in lung DNA isolated from the carbon black-exposed animals.

Bond *et al.* (1990) exposed male and female Fischer 344/N rats to filtered air or carbon black (6.2 mg/m^3) for 12 weeks. The carbon black was Elftex-12 (furnace black) of which 59% of the mass had a 1.9 μm MMAD and 41% of the mass had a 0.10 μm geometric mean diameter and surface area characteristics similar to those of eluted diesel soot but negligible amounts of extractable organic chemicals and no measurable mutagenic activity. Rats were exposed for 16 h per day on five days a week for 12 weeks. DNA adducts in alveolar type II cells were assessed at the end of the exposure. The authors report the presence of several adducts as assessed by the nuclease P1 version of the ^{32}P-postlabelling assay. The level of carbon black-induced adducts was significantly elevated above that seen in controls. Total levels of DNA adducts in type II cells from control and carbon black-exposed rats were approximately 5 and 25 adducts/10^9 bases, respectively. The authors could not determine whether exposure to carbon black induced an increase in the level of adducts already present in cells from sham-exposed rats or if carbon black induced the formation of new adducts with chromatographic characteristics of I-spots.

Riebe-Imre et al. (1994) assessed the ability of carbon black particles [the Working Group was aware that this was Printex 90 (furnace black)] to induce cytotoxicity, cell transformation and micronuclei formation in a fetal Syrian hamster lung epithelial cell line. The cytotoxicity of the carbon black particles was negligible. Carbon black particles induced in-vitro transformation in small mucus granule cell-differentiated M3E3/C3 cells and in undifferentiated M3E3/C3 cells. Carbon black particles showed a much weaker activity in undifferentiated cells compared to differentiated cells. They also reported cytoskeletal changes. Concentrations of carbon black used in the transformation studies ranged from 100 to 300 µg/mL. Peak responses in the differentiated cells occurred at 200 µg/mL and were approximately four-fold over those of controls. In the undifferentiated cells, peak concentrations were observed at 300 µg/mL and were approximately eight-fold over those of controls. A dose-related increase was observed in the frequency of micronuclei over the dose range 0.1–2.0 µg/mL. However, maximal responses were only approximately 50% greater than the control frequency of about 4.5%; there was no indication of variation, and the possibility of different responses at higher doses (that were tolerated in the cell transformation test) were not reported.

Swafford et al. (1995) analysed pulmonary carcinomas from rats exposed to diesel exhaust, furnace black (Elftex 12) and air for alterations in K-ras and p53 to determine if mutations were similar. Details of the exposure conditions are described in Nikula et al. (1995). Briefly, male and female Fischer 344/N rats (seven to nine weeks of age) were exposed for 16 h per day on five days per week for 24 months to diesel exhaust or carbon black at concentrations of 2.44 and 6.33 mg/m^3 for diesel exhaust and 2.46 and 6.55 mg/m^3 for carbon black. Controls were exposed to air only. The number of carcinomas analysed were 28 for diesel exhaust, 18 for carbon black and five for air only. K-ras exon 1 mutations were found in two neoplasms, one each from diesel exhaust and carbon black exposure groups. No mutations in the K-ras gene were observed in lung neoplasms from control rats. Immunohistochemical staining revealed evidence of p53 inactivation in 2/4 squamous-cell carcinomas and adenocarcinomas from carbon black-exposed rats. p53 Mutational analyses revealed the presence of one mutation in a diesel exhaust-induced squamous-cell carcinoma. This mutation was reported by the authors to be a silent mutation. [The Working Group noted that it is not clear whether the only (silent) mutation was in a diesel exhaust or carbon black-exposed rat.]

A subchronic inhalation study with Monarch 880 (furnace black) was performed in Fischer 344 rats exposed to 1.1, 7.1 and 52.8 mg/m^3 for 6 h per day on five days a week for a total of 13 weeks with an eight-month post-exposure period (Driscoll et al., 1996). The MMAD of the carbon black aerosol was 0.88 µm and the primary particle size was 16 nm with a specific surface area of 220 m^2/g. The rat alveolar type II cell isolation and the *hprt* clonal selection assay were used. Mutant frequencies ranged from 8.2 to 5.2 mutants/10^6 epithelial cells in the air control animals. Exposure to 52.8 mg/m^3 carbon black resulted in *hprt* mutant frequencies which were 4.3-, 3.2- and 2.7-fold greater than the air control group, immediately, three and eight months after exposure, respectively. A significant increase in the frequency of *hprt* mutants was detected immediately after 13 weeks of exposure to 7.1 mg/m^3 carbon black but not after three or eight months of recovery. No significant changes in the *hprt* mutant frequency were observed for alveolar epithelial cells from rats exposed to 1.1 mg/m^3 carbon black. This mutagenic response occurred at exposures that also resulted in significant pulmonary inflammation, epithelial hyperplasia and fibrosis.

4.5 Mechanistic considerations related to carcinogenicity

Figure 1 describes a hypothesized mechanistic model for effects of particle exposure in the lung. More specifically, the model, derived from inhalation studies in rats, pertains to exposure to low-toxicity low-solubility particles. This topic has also been discussed at a recent symposium (Mauderly & McCunney, 1996). Phagocytosis of such particles by alveolar macrophages leads to activation by alveolar macrophage and the subsequent release of inflammatory cytokines, growth factors, chemokines, enzymes and reactive oxygen species. This in turn recruits polymorphonucleocytes from the circulatory system into the alveolar space, thus amplifying the inflammatory response including the release of additional reactive oxygen species. This inflammatory response is dependent on the dose of deposited and phagocytized particles. Particularly, impairment of alveolar macrophage-mediated particle clearance due to lung particle overload results in further particle accumulation and amplifies this process, leading to chronic inflammation, including fibrotic changes. The continuous release of reactive oxygen species can result in increased mutation frequencies in specific target cells, which become manifest during increased cell proliferative responses. Subsequent responses include metaplastic changes, which finally result in tumour formation.

This specific mechanism involving reactive oxygen species is based on studies by Driscoll *et al.* (1996), which showed that carbon black in rats led in a dose-dependent manner to increased mutation frequency of the type II cells in those cases where significantly increased numbers of inflammatory cells were present. Earlier studies by Driscoll (1996) have shown that co-incubation of rat lung epithelial cells with inflammatory polymorphonucleocytes also resulted in increased mutation frequencies and that this response could be significantly decreased in the presence of antioxidants. Thus, oxidative damage to DNA due to released reactive oxygen species from inflammatory cells appears to be a plausible mechanism underlying the particle-induced rat tumour response. In further support of this hypothesis are the studies of Bond *et al.* (1990) who reported elevated levels of DNA damage in alveolar type II cells of rats exposed to concentrations of carbon black known to induce an inflammatory response.

An alternative mechanism relates to physical phenomena due to particles taken up by target cells. As pointed out in Figure 1, high pulmonary particle burdens result in

Figure 1. Mechanistic chain of events for pulmonary effects of low-toxicity, low-solubility particles assumed to be operative in rats

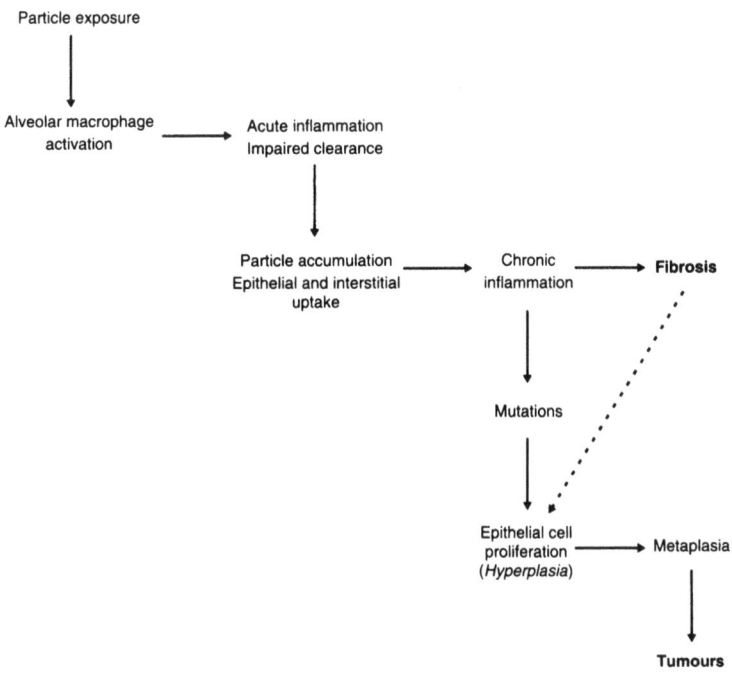

Adapted from Oberdörster (1995)

increased uptake by epithelial cells and access of particles into the pulmonary interstitium. A study by Riebe-Imre et al. (1994) reported that a fetal Syrian hamster lung epithelial cell line, when incubated with different doses of carbon black, showed increased transformation, particularly when these cells were already differentiated. Carbon black also induced a dose-dependent enhancement of micronucleus formation in these cells, mostly due to clastogenic effects. Since these effects were observed in an epithelial cell line derived from hamsters after in-vitro exposures, the relevance to the in-vivo situation needs to be addressed. Particle-induced lung tumours have not been observed in hamsters, and target cells in the only species that shows particle-induced tumours (the rat) are most likely of alveolar origin, including the type II cells. This mechanistic hypothesis, based on physically induced DNA alterations, may be less plausible and relevant than the one of particle-induced oxidative damage.

The particle-associated rat lung tumours (e.g. diesel exhaust) cannot be extrapolated to mice or hamsters; these species do not, at comparable lung burdens of particles, develop lung tumours. For carbon black specifically, this species difference has been demonstrated between rats and mice. Several inhalation studies with low-solubility, low-toxicity particles, one of them with carbon black, have shown that no lung tumour was induced in mice at exposure concentrations and lung burden that exceeded the capacity of the lung to clear the particles and induced significant toxicity.

A central question is whether the toxic and defensive mechanisms suggested to operate in rats also operate in humans. Very little is known about the relationship between overload and lung cancer risk in humans, although it may be assumed that this overload-related mechanism could occur in humans exposed to sufficient levels (or doses). Limited indirect inferences regarding this issue may be derived from the available epidemiological studies of workers exposed to carbonaceous particles. The epidemiological studies of carbon black workers are not very informative in this regard. It is interesting to note that studies of coal miners have generally failed to detect an increased risk for lung cancer (Merchant et al., 1986; Harrington & Levy, 1994). This evidence has been interpreted by some scientists as suggesting that a lung overload-related mechanism does not induce cancer in humans. However, there are other plausible interpretations for these observations that should be considered. It is difficult, as with nearly all epidemiological studies, to rule out that limitations in sample size, bias and other study design issues might explain these negative findings. The Working Group considered that the fact that coal miners are not permitted to smoke while working may have introduced a negative confounding bias for lung cancer in these studies. This bias is consistent with the observation of a deficit in lung cancer risk in these studies. Furthermore, it is important to consider the surface area characteristics of the inhaled low-toxicity, low-solubility particle (including coal dust) and its impact on dose.

Lung particle burden in the lungs of these workers by mass were on average ~15 mg/g lung. However, although pulmonary particle accumulation by mass is very high in coal miners, particle mass may not be the most relevant dose parameter for a correlation with specific long-term effects. It has been suggested, based on results of a number of studies, that surface area of retained particles may be a better parameter for correlation with pulmonary inflammation and neoplastic events. Thus, it may well be that coal miners, in spite of the high mass loading of particles in the lungs, did not reach a sufficient surface area of the retained particles. Also in rats, it has been found in some studies that lung particle mass burdens in this range, and even higher, did not result in increased tumour incidences. For example, chronic studies in rats with toner particles and pigment-grade titanium dioxide at exposure concentrations of 16 mg/m^3 and 50 mg/m^3, respectively, resulted in typical findings of particle overload (e.g. impaired particle clearance at lung burdens of ~12 and 60 mg/g lung, respectively) but with less inflammatory response and without induction of lung tumours. When these pulmonary mass particle burdens are expressed in terms of their surface area, the retained dose (by particle surface area) is lower than the dose (by surface area) observed to induce lung tumours in rats in other studies.

Epidemiological studies of diesel-exposed workers may also contribute something to this issue. A recent review of the epidemiological literature in this area concluded that these studies suggest a small-to-moderate increase in lung cancer risk, and that these findings do not appear to be fully explicable by confounding or other sources of bias (Cohen & Higgins, 1995). These studies are pertinent because it has been suggested, based on the recent experimental studies of rats exposed to high concentrations of diesel exhaust and carbon black (Nikula et al., 1994; Heinrich et al., 1995), that the increased risk for lung cancer associated with diesel exhaust in rats might be explained by the carbonaceous core rather than the organic fraction of diesel soot. Workers in these studies were generally exposed to diesel exhaust levels below 200 $\mu g/m^3$, which is below the level at which overload of the lung in humans is believed to occur (Cohen & Higgins, 1995). These findings might be interpreted as suggesting either that lung overload occurs at lower levels than expected among humans, or that an overload-related mechanism may play the dominant role only at the high-exposure levels used in experimental studies.

In addition to these dose considerations, it is also of importance to consider differences in specific defence mechanisms between rats, mice and humans. The Working Group is not aware of studies which have evaluated specific pulmonary defences, including antioxidant levels, in humans under particle load conditions; such data, in contrast, are available for rats and mice. Thus, whether humans respond to chronic inhalation of particles, including carbon black, more like a rat or more like a mouse cannot be decided at present. It should be emphasized that the dose plays a most important role in the chain of mechanistic events outlined in Figure 1.

5. Summary of Data Reported and Evaluation

5.1 Exposure data

Carbon black is a powdered form of elemental carbon manufactured by the vapour-phase pyrolysis of hydrocarbon mixtures, such as heavy petroleum distillates and residual oils, coal-tar products, natural gas and acetylene. Worldwide production of carbon black in 1993 was approximately 6 million tonnes.

Carbon blacks are categorized as acetylene black, channel black, furnace black, lampblack or thermal black, according to the process by which they are manufactured. Lampblack is the oldest type of carbon black, having been used as a pigment for centuries. Channel black, produced from natural gas, was introduced in the late nineteenth century and was the major carbon black used worldwide in the early twentieth century for rubber and pigment applications; with the exception of a special product made in Germany, it is no longer produced. Acetylene, furnace and thermal blacks have been produced since the early twentieth century. Over 90% of all carbon black produced today is furnace black.

The primary use of carbon black is in rubber products, mainly tyres and other automotive products, but also in many other rubber products such as hoses, gaskets and coated fabrics. Much smaller amounts of carbon black are used in inks and paints, in plastics and in the manufacture of dry-cell batteries.

Types of carbon black are characterized by the size distribution of the primary particles, the degree of their aggregation and agglomeration and the various chemicals adsorbed onto the surfaces. Average primary particle diameters in several commercially produced carbon blacks range from 10 to 400 nm, while average aggregate diameters range from 100 to 800 nm. Typical classes of chemicals adsorbed onto the carbon black surface are polycyclic aromatic hydrocarbons (PAHs), nitro derivatives of PAHs and sulfur-containing PAHs. Examples of PAHs extracted most frequently from carbon black using a variety of extraction methods (e.g. prolonged Soxhlet extraction with benzene or toluene) include benzopyrenes, benzo[ghi]perylene, coronene, fluoranthene and pyrene.

Exposures to carbon black vary markedly within any production facility. The highest levels of exposure are experienced by those who interact with the process the most, including fitters/welders, warehouse packers and site cleaners. Exposures can vary greatly among factories and regionally.

Several studies in the 1960s found very high levels of exposure, even up to 1000 mg/m^3 in furnace, lamp- and channel black plants. Later studies in some countries have found lower levels, although many of these were in excess of the existant occupational exposure limits. In the late 1980s and early 1990s, more extensive studies in western Europe and the United States have found (geometric mean) personal exposure to total inhalable carbon black to be on average less than 1 mg/m^3. Even lower exposures may occur among some workers in industries using carbon black, such as rubber, printing ink and paint manufacture, and exposures to carbon black in the use of rubber, printing ink or paint are negligible.

5.2 Human carcinogenicity data

The greatest potential for elucidating the carcinogenicity of carbon black is in the carbon black production industry where carbon black has been the prime industrial exposure and where exposure levels have been high. Cohort studies of carbon black production workers have been conducted in the United States and in the United Kingdom. Interpretation of the study in the United States is hampered by problems of uncertainty in the completeness of the cohort and in the definition and completeness of follow-up. The study in the United Kingdom also had some problems in completeness of the cohort, but the follow-up was probably complete. In both cohorts, fewer observed than expected deaths due to all causes occurred and, in the study in the United States, this may in part have been attributable to under-ascertainment of deaths or to inflation of person-years of follow-up. The study in the United States found no excess mortality due to any type of cancer when compared to state vital statistics rates; in fact there were deficits for some types of cancer. The study in the United Kingdom found an excess of respiratory cancer deaths (standardized mortality ratio, 1.5; 95% confidence interval, 1.0–2.2).

A nested case–control study within the United States cohort was hampered by very small numbers and problems of interpretation. Most cases were of non-melanoma skin cancer. Neither for all cancers combined nor for skin cancers alone was there evidence that cases had higher cumulative exposure to carbon black than controls.

A cohort study was carried out among workers in the United States to assess cancer risks due to exposure to formaldehyde. Ten participating plants were spread across several industries in which workers may have experienced exposure to formaldehyde. To control for confounding and modification of effect by other exposures, workers' exposures to various other chemicals, including carbon black, were assessed by industrial hygienists. For all assessed levels and durations of exposure to carbon black combined, there was a slight nonsignificant excess of lung cancer. There was no clear trend by duration of exposure. Carbon black-exposed workers in this cohort may also have been exposed to formaldehyde and other substances.

Another industry-based study was a nested case–control study conducted in the tyre and rubber manufacturing industry to examine the association of squamous-cell carcinoma of the skin with rubber manufacturing materials. For each study subject, industrial hygienists assessed exposure to five substances, including carbon black, based on evaluations of each subject's job history. The results of this study indicated no effect of carbon black on skin cancer.

In a community-based case–control study in Canada, interviews were designed to obtain detailed lifetime job histories and information on potential confounders. Potential occupational exposures were identified for each job description, and among the exposures assessed was carbon black. In this study population, potential exposure to carbon black occurred in some individuals in user industries, notably among painters and in the printing and rubber industries. For the following cancer sites, there was no indication of excess risk in relation to carbon black: stomach, colon, rectum, pancreas, prostate, urinary bladder, skin melanoma and non-Hodgkin's lymphoma. For the following sites there was indication of excess risk: oesophagus, kidney and lung. The lung cancer excess was particularly concentrated among oat-cell cancers.

A Swedish case–control study reported a nonstatistically significantly increased risk for urothelial cancer for men exposed to carbon black.

In assessing all the available data, there is no evidence of an effect of carbon black for most cancer sites. For cancers of the urinary bladder, kidney and oesophagus, isolated results indicate excess risks, but these are not sufficient to support an evaluation of human carcinogenicity.

Two studies were informative for non-melanoma skin cancer (a nested case–control study among the United States carbon black production cohort and a nested case–control study among rubber workers); neither demonstrated any excess risk for skin cancer due to carbon black.

Of the studies listed above, four were considered informative for lung cancer. Of those, two indicated excess risk among carbon black-exposed workers at borderline statistical significance (the carbon black production cohort in the United Kingdom and the Canadian community-based study), one indicated excess risk but was not significant (the United States formaldehyde cohort) and the other indicated no excess (the United States carbon black production cohort).

Each of the available studies has limitations for the specific purpose of assessing the carcinogenicity of carbon black. The Working Group considered the study of carbon black producers in the United Kingdom to be the most informative for this purpose. That study indicated an excess risk of borderline significance. Confounding by smoking could not be excluded, although some information was presented indicating that it was unlikely. The formaldehyde cohort study indicated a slight excess of lung cancer among the subgroup exposed to carbon black, but this could easily have been due to chance or confounding by formaldehyde or other occupational substances. The community-based study in Montréal of exposure in a variety of user industries showed an elevated risk in the subgroup categorized as having high exposure to carbon black; the result was of borderline statistical significance using a cancer series control group and not significant using a population control group. It is not clear which control group provides the most valid estimates. Even the high-exposure subgroup of this study was unlikely to have experienced exposure levels of the same order of magnitude as did workers in the carbon black production industry. Although the United States carbon black worker study, which was negative, was large, its methodological limitations detracted from its value. The Working Group therefore considered the whole body of evidence rather weak and the results conflicting.

5.3 Animal carcinogenicity data

No adequate study of the carcinogenicity of carbon black administered by the oral route was available.

In one study in female mice by inhalation exposure, carbon black did not increase the incidence of respiratory tract tumours.

Two different carbon black products were tested in two inhalation studies in female rats and in one study using rats of each sex. Significant increases in the incidence of malignant lung tumours and the incidence of benign and malignant lung tumours combined were observed in female rats in all three studies. In addition, increased incidences of lesions described as benign cystic keratinizing squamous-cell tumours or squamous cysts were observed.

In two studies in female rats by intratracheal administration, using one type of carbon black, both extracted and non-extracted material increased the incidence of benign and malignant lung tumours. In one of the studies, a different type of extracted carbon black with a larger primary particle size increased the incidence of lesions described as benign cystic keratinizing squamous-cell tumours.

In several skin-painting studies in mice using various carbon blacks, no carcinogenic effect on the skin was observed; the painting of several carbon black extracts (benzene extracts) resulted in skin tumours.

In a series of studies in male and female mice by subcutaneous injection, a carbon black containing demonstrable quantities of carcinogenic PAHs produced local sarcomas, whereas a carbon black from which no PAH was detected did not produce such sarcomas. In several studies in mice, solvent extracts of carbon black produced sarcomas following subcutaneous injection.

5.4 Other relevant data

Upon inhalation exposure of humans to carbon black, these particles are deposited in the lung. The exposure may cause slight radiological changes. The prevalence of radiological findings has varied considerably among different studies, probably because of varying radiological techniques and possibly also due to different exposure circumstances and possible concomitant exposures to other compounds. Further, workers may develop chronic bronchitis and a slight reduction in lung function. These findings may be interpreted mainly as a slight nonspecific irritant effect of heavy dust exposure. On the other hand, some data indicate a fibrous tissue reaction in the area surrounding the carbon deposits in the lung parenchyma.

Studies on the pulmonary retention of inhaled carbon blacks in rats and mice have shown that these particles behave very similarly to other low-solubility, low-toxicity particles. Carbon black displayed normal retention characteristics in rats at lung burdens not exceeding a certain level which is approximately in the range of 0.5–1 mg/g of lung. At higher lung burdens, a prolonged clearance is found. Impaired particle clearance due to high loading of carbon black in experiments with rats results in increased accumulation of particles. Subsequent inflammatory responses occur which develop into chronic active inflammation. Increased collagen deposition from proliferating fibroblasts, increased epithelial cell proliferation and metaplasia have been found at high lung burdens of carbon black. It appears that the high specific surface area of most carbon blacks may be an important parameter in the induction of inflammatory and subsequent other responses in the lung. One study with carbon black in rats confirmed findings with other particles that females are more sensitive than males.

Most assays for mutagenicity are negative for carbon black. In rats exposed to carbon black by inhalation, *hprt* mutant frequency was elevated in type II cells following a 12-week exposure. Carbon black did not induce a significant increase in DNA adducts in peripheral lung tissue of rats after two years of inhalation exposure. In another study, exposure of rats by inhalation to carbon black increased DNA adduct levels in type II cells. K-*ras* mutations were found in one out of 18 neoplasms analysed from a carbon black-exposed rat. No exposure-related *p53* mutation was found.

Some mechanistic considerations on particle-induced lung neoplasms are presented.

5.5 Evaluation[1]

There is *inadequate evidence* in humans for the carcinogenicity of carbon black.

There is *sufficient evidence* in experimental animals for the carcinogenicity of carbon black.

There is *sufficient evidence* in experimental animals for the carcinogenicity of carbon black extracts.

Overall evaluation

Carbon black is *possibly carcinogenic to humans (Group 2B)*.

---

[1]For definition of the italicized terms, see Preamble, pp. 24–27.

6. References

Adamson, I.Y.R. & Bowden, D.H. (1978) Adaptive responses of the pulmonary macrophagic system to carbon. II. Morphologic studies. *Lab. Invest.*, **38**, 430–438

Adamson, I.Y.R. & Bowden, D.H. (1980) Role of monocytes and interstitial cells in the generation of alveolar macrophages. II. Kinetic studies after carbon loading. *Lab. Invest.*, **42**, 518–524

Adamson, I.Y.R. & Bowden, D.H. (1981) Dose response of the pulmonary macrophagic system to various particulates and its relationship to transepithelial passage of free particles. *Exp. Lung Res.*, **2**, 165–175

Adamson, I.Y.R. & Bowden, D.H. (1982a) Chemotactic and mitogenic components of the alveolar macrophage response to particles and neutrophil chemoattractant. *Am. J. Pathol.*, **109**, 71–77

Adamson, I.Y.R. & Bowden, D.H. (1982b) Effects of irradiation on macrophagic response and transport of particles across the alveolar epithelium. *Am. J. Pathol.*, **106**, 40–46

Agurell, E. & Löfröth, G. (1983) Presence of various types of mutagenic impurities in carbon black detected by the Salmonella assay. In: Waters, M.D., ed., *Short-term Bioassays in the Analysis of Complex Environmental Mixtures, III*, New York, Plenum Press, pp. 297–306

Agurell, E. & Löfroth, G. (1993) Impurity variations in a carbon black: characterization by the Ames *Salmonella* mutagenicity assay and polycyclic aromatic hydrocarbon analysis. *Environ. Toxicol. Chem.*, **12**, 219–223

American Conference of Governmental Industrial Hygienists (1993) *Documentation of the Threshold Limit Values*, Cincinnati, OH, pp. 220–221

American Conference of Governmental Industrial Hygienists (1995) *1995–1996 Threshold Limit Values (TLVs) for Chemical Substances and Physical Agents and Biological Exposure Indices (BEIs)*, Cincinnati, OH, p. 15

American Society for Testing and Materials (1995a) *1995 Annual of ASTM Standards*, Section 9, *Rubber*, Vol. 09.01, *Rubber, Natural and Synthetic — General Test Methods: Carbon Black*, Philadelphia, PA, pp. 1–46, 335–338, 445–446

American Society for Testing and Materials (1995b) *1995 Annual of ASTM Standards*, Section 6, *Paints, Related Coatings, and Aromatics*, Volume 06.03, *Paint — Pigments, Drying Oils, Polymers, Resins, Naval Stores, Cellulosic Esters, and Ink Vehicles*, Philadelphia, PA, pp. 35

Anon. (1985) Chemical profile: carbon black. *Chem. Mark. Rep.*, **228**, 53–54

Anon. (1988) Chemical profile: carbon black. *Chem. Mark. Rep.*, **234**, 41–41

Anon. (1989) Facts & figures for the chemical industry. *Chem. Eng. News*, **67**, 36–90

Anon. (1991a) Facts & figures for the chemical industry. *Chem. Eng. News*, **69**, 28–69

Anon. (1991b) Chemical profile: carbon black. *Chem. Mark. Rep.*, **240**, 25, 54

Anon. (1993) Facts & figures for the chemical industry. *Chem. Eng. News*, **71**, 38–83

Anon. (1994) Chemical profile: carbon black. *Chem. Mark. Rep.*, **246**, 48–49

Anon. (1995) Facts & figures for the chemical industry. *Chem. Eng. News*, **73**, 36–79

Arbeidsinspectie [Labour Inspection] (1994) *De Nationale MAC-Lijst 1994* [National MAC list 1994], The Hague, Netherlands

Bailey, M.R., Fry, F.A. & James, A.C. (1985) Long-term retention of particles in the human respiratory tract. *J. Aerosol Sci.*, **46**, 295–305

Bean, D.C. (1964) Acetylene black. In: Mark, H.E., McKetta, J.J., Jr & Othmer, D.F., eds, *Encyclopedia of Chemical Technology*, 2nd Ed., Vol. 4, New York, Interscience, p. 278

Beck, B., Gohlke, R., Sturm, W., Bergmann, L. & Wolf., E. (1985) Carbon black lung as an occupational disease. *Z. Erkrank. Atm.-Org.*, **164**, 254–266 (in German)

Belanger, P.L. & Elesh, E. (1979) *Health Hazard Evaluation and Technical Assistance HHE 78-72-618, Kentile Floors, Inc., South Plainfield, New Jersey*, Cincinnati, OH, United States National Institute for Occupational Safety and Health

Bertheussen, K.J. & Nissen, H.M. (1976) Pulmonary excretion of foreign bodies and necrotic material from the bladder wall. *Eur. Urol.*, **2**, 34–36

Bertheussen, K.J., Diemer, N.H., Praestholm, J. & Klinken, L. (1978) Pulmonary excretion of carbon black injected into the cerebral ventricles of the rat. *Acta pathol. microbiol. scand., Sect. A.*, **86**, 90–92

Bevan, D.R. & Worrell, W.J. (1985) Elution of benzo[a]pyrene from carbon blacks into biomembranes *in vitro*. *J. Toxicol. environ. Health*, **15**, 697–710

Bevan, D.R. & Yonda, N.T. (1985) Elution of polycyclic aromatic hydrocarbons from carbon blacks into biomembranes *in vitro*. *Toxicol. ind. Health*, **1**, 57–67

Blair, A., Stewart, P.A. & Hoover, R.N. (1990) Mortality from lung cancer among workers employed in formaldehyde industries. *Am. J. ind. Med.*, **17**, 683–699

Blau, J.N. & Veall, N. (1967) The uptake and localization of proteins, Evans blue and carbon black in the normal and pathological thymus of the guinea-pig. *Immunology*, **12**, 363–372

Boiano, J. & Donohue, M. (1981) *Health Hazard Evaluation HHE 80-203-960, Phillips Chemical Company, Toledo, Ohio*, Cincinnati, OH, United States National Institute for Occupational Safety and Health

Bond, J.A., Johnson, N.F., Snipes, M.B. & Mauderly, J.L. (1990) DNA adduct formation in rat alveolar type II cells: cells potentially at risk for inhaled diesel exhaust. *Environ. mol. Mutag.*, **16**, 64–69

Bourguet, C.C., Checkoway, H. & Hulka, B.S. (1987) A case–control study of skin cancer in the tire and rubber manufacturing industry. *Am. J. ind. Med.*, **11**, 461–473

Bowden, D.H. & Adamson, I.Y.R. (1978) Adaptive responses of the pulmonary macrophagic system to carbon. I. Kinetic studies. *Lab. Invest.*, **38**, 422–429

Bowden, D.H. & Adamson, I.Y.R. (1982) Alveolar macrophage response to carbon in monocyte-depleted mice. *Am. Rev. respir. Dis.*, **126**, 708–711

Bowden, D.H. & Adamson, I.Y.R. (1984) Pathways of cellular efflux and particulate clearance after carbon instillation to the lung. *J. Pathol.*, **143**, 117–125

Buddingh, F., Bailey, M.J., Wells, B. & Haesemeyer, J. (1981) Physiological significance of benzo[a]pyrene adsorbed to carbon blacks: elution studies, AHH determinations. *Am. ind. Hyg. Assoc. J.*, **42**, 503–509

Butler, M.A., Evans, D.L., Giammarise, A.T., Kiriazides, D.K., Marsh, D., McCoy, E.C., Mermelstein, R., Murphy, C.B. & Rosenkranz, H.S. (1983) Application of the *Salmonella* mutagenicity assay to carbon blacks and toners. In: Cooke, M.W. & Dennis, A.J., eds, *Polycyclic Aromatic Hydrocarbons, 7th International Symposium*, Columbus, OH, Battelle Press, pp. 225–232

Cabot Co. (1995) *Companies Manufacturing Carbon Black*, Boston, MA

Capusan, I. & Mauksch, J. (1969) Skin diseases in workers at a carbon black producing factory (with special reference to the lampblack process). *Berufsdermatosen*, **17**, 28–37 (in German)

Carlton, W.W. (1994) 'Proliferative keratin cyst', a lesion in the lungs of rats following chronic exposure to para-aramid fibrils. *Fundam. appl. Toxicol.*, **23**, 304–307

Charsley, E.L. & Dunn, J.G. (1981) The application of thermogravimetry (TG) to the characterization and quantitative determination of carbon blacks. *Plast. Rubber Process. Appl.*, **1**, 3–7

Chemical Information Services (1994) *Directory of World Chemical Producers 1995/96 Standard Edition*, Dallas, TX, p. 151

China National Chemical Information Centre (1993) *China Chemical Industry (English Edition), World Chemical Industry Yearbook, 1993*, Beijing, China, p. 169

Claassen, E.J. (1978) Carbon black, nonfurnace types. In: McKetta, J.J. & Cunningham, W.A., eds, *Encyclopedia of Chemical Processing and Design*, Vol. 6, New York, Marcel Dekker, pp. 265, 266, 273, 274, 275, 279

Cocarla, A., Cornea, G., Dengel, H., Gabor, S., Milea, M. & Pappilian, V.V. (1976) Carbon black pneumoconiosis. *Int. Arch. occup. Environ. Health*, **36**, 217–228 (in French)

Cohen, A.J. & Higgins, M.W.P. (1995) Health effects of diesel exhaust: epidemiology. In: *Diesel Exhaust: A Critical Analysis of Emissions, Exposure, and Health Effects*, Cambridge, MA, Health Effects Institute, pp. 251–292

Collyer, H.J. (1975) Carbon black and ecology. In: Ayer, F.A., ed., *Environmental Aspects of Chemical Use in Rubber Processing Operations, March 1975, Akron, OH* (Conference Proceedings, EPA-560/1-75-002; PB244 172), Washington DC, United States Environmental Protection Agency, pp. 130–136

Colmsjö, A.L. & Östman, C.E. (1988) Isolation and fingerprinting of some high-molecular-weight polynuclear aromatic compounds. *Anal. chim. Acta*, **208**, 183–193

Creutzenberg, O., Bellmann, B., Heinrich, U., Fuhst, R., Koch, W. & Muhle, H. (1990) Clearance and retention of inhaled diesel exhaust particles, carbon black, and titanium dioxide in rats at lung overload conditions. *J. Aerosol Sci.*, **21** (Suppl. 1), S455–S458

Crosbie, W.A. (1986) The respiratory health of carbon black workers. *Arch. environ. Health*, **41**, 346–353

Dannenberg, E.M. (1978) Carbon black. In: Mark, H.F., Othmer, D.F., Overberger, C.G. & Seaborg, G.T., eds, *Kirk-Othmer Encyclopedia of Chemical Technology*, 3rd Ed., Vol. 4, New York, John Wiley & Sons, pp. 631–666

Dannenberg, E.M., Paquin, L. & Gwinnell, H. (1992) Carbon black. In: Kroschwitz, J.I. & Howe-Grant, M., eds, *Kirk-Othmer Encyclopedia of Chemical Technology*, 4th Ed., Vol. 4, New York, John Wiley & Sons, pp. 1037–1074

Dasenbrock, C., Peters, L., Crentzenberg, O. & Heinrich, U. (1996) The carcinogenic potency of carbon particles with and without PAH after repeated intratracheal administration in the rat. *Toxicol. Lett.* (in press)

Davis, B.R., Whitehead, J.K., Gill, M.E., Lee, P.N., Butterworth, A.D. & Roe, F.J.R. (1975) Response of rat lung to 3,4-benzpyrene administered by intratracheal instillation in infusine with or without carbon black. *Br. J. Cancer*, **31**, 443–452

Deutsche Forschungsgemeinschaft (1995) *MAK and BAT Values 1995* (Report No. 31), Weinheim, VCH Verlagsgesellschaft

De Wiest, F. (1980) Experimental study of the blood elution process of the polycyclic aromatic hydrocarbons adsorbed on carbon blacks and soots. I. Physico-chemical characterization of carbon particles. *J. Pharm. belg.*, **35**, 253–265 (in French)

Deimel, M. & Dulson, W. (1980) PAH emission measurements in the environment of a carbon black factory. *VDI-Bericht*, **358**, 139–145 [*Chem. Abstr.*, **93**; 244515h] (in German)

Driscoll, K.E. (1996) The role of inflammation in the development of rat lung tumors in response to chronic particle exposure. *Inhal. Toxicol.*, **8** (Suppl.)

Driscoll, K.E., Carter, J.M., Howard, B.W., Hassenbein, D., Pepelko, W., Baggs, R.B. & Oberdörster, G. (1996) Pulmonary inflammatory, chemokine and mutagenic responses in rats after subchronic inhalation of carbon black. *Toxicol. appl. Pharmacol.* (in press)

Dungworth, D.L., Mohr, U., Heinrich, U., Ernst, H. & Kittel, B. (1994) Pathologic effects of inhaled particles in rat lungs: associations between inflammatory and neoplastic processes. In: Mohr, U., Dungworth, D.L., Mauderly, J.L. & Oberdörster, G., eds, *Toxic and Carcinogenic Effects of Solid Particles in the Respiratory Tract*, Washington DC, ILSI Press, pp. 75–98

Eller, P.M., ed. (1994) Carbon black — Method 5000. In: *United States NIOSH Manual of Analytical Methods*, 4th Ed. (DHHS (NIOSH) Publ. No. 94-113), Washington DC, United States Government Printing Office

European Committee for Biological Effects of Carbon Black (1982) *A Comparative Study of Soot and Carbon Black* (Bulletin No. 2, January), Boston, MA, Cabot Corp.

Falk, H.L. & Steiner, P.E. (1952) The identification of aromatic polycyclic hydrocarbons in carbon blacks. *Cancer Res.*, **12**, 30–39

Fenters, J.D., Bradof, J.N., Aranyi, C., Ketels, K., Ehrlich, R. & Gardner, D.E. (1979) Health effects of long-term inhalation of sulfuric acid mist-carbon particle mixtures. *Environ. Res.*, **19**, 244–257

Fitch, W.L. & Smith, D.H. (1979) Analysis of adsorption properties and adsorbed species on commercial polymeric carbons. *Environ. Sci. Technol.*, **13**, 341–346

Fitch, W.L., Everhart, E.T. & Smith, D.H. (1978) Characterization of carbon black adsorbates and artifacts formed during extraction. *Anal. Chem.*, **50**, 2122–2126

Ford, F.P. & Lyon, F. (1973) The morphology and chemistry of carbon black. In: Patton, T.C., ed., *Pigment Handbook*, Vol. 3, New York, John Wiley & Sons, pp. 194–195

Freedman, A.P. & Robinson, S.E. (1988) Noninvasive magnetopneumographic studies of lung dust retention and clearance in coal miners. In: Frantz, R.L. & Ramani, R.V., eds, *Respirable Dust in the Mineral Industries: Health Effects, Characterization and Control*, University Park, PA, Penn State University Press, pp. 181–186

Freedman, A.P., Robinson, S.E. & Street, M.R. (1988) Magnetopneumographic study of human alveolar clearance in health and disease. *Ann. occup. Hyg.*, **32** (Suppl. 1), 809

Friedlander, S.K. (1973) Chemical element balances and identification of air pollution sources. *Environ. Sci. Technol.*, **7**, 235–240

Gabor, S., Raucher, G., Stefanescu, A. & Ossian, A. (1969) Prevention of occupational hazards in the carbon black industry. *Igiena*, **18**, 57–62 (in Romanian)

Gallagher, J., Heinrich, U., George, M., Hendee, L., Phillips, D.H. & Lewtas, J. (1994) Formation of DNA adducts in rat lung following chronic inhalation of diesel emissions, carbon black and titanium dioxide particles. *Carcinogenesis*, **15**, 1291–1299

Gardiner, K. (1995a) Effects on respiratory morbidity of occupational exposure to carbon black: a review. *Arch. environ. Health*, **50**, 44–60

Gardiner, K. (1995b) The methodological problems of multinational epidemiological studies with particular reference to carbon black studies. *Occup. Med.*, **45**, 247–255

Gardiner, K., Trethowan, W.N., Harrington, J.M., Calvert, I.A. & Glass, D.C. (1992a) Occupational exposure to carbon black in its manufacture. *Ann. occup. Hyg.*, **36**, 477–496

Gardiner, K., Hale, K.A., Calvert, I.A., Rice, C. & Harrington, J.M. (1992b) The suitablity of the urinary metabolite 1-hydroxypyrene as an index of polynuclear aromatic hydrocarbon bioavailability from workers exposed to carbon black. *Ann. occup. Hyg.*, **36**, 681–688

Gardiner, K., Trethowan, N.W., Harrington, J.M., Rossiter, C.E. & Calvert, I.A. (1993) Respiratory health effects of carbon black: a survey of European carbon black workers. *Br. J. ind. Med.*, **50**, 1082–1096

Gardiner, K., Calvert, I.A., van Tongeren, M.J.A. & Harrington, J.M. (1996) Occupational exposure to carbon black in its manufacture: data from 1987–1992. *Ann. occup. Hyg.*, **40**, 65–77

Garret, M.D. (1973) Carbon black, bone black, lampblack. In: Patton, T.C., ed., *Pigment Handbook*, Vol. 1, New York, John Wiley & Sons, pp. 709–712, 714, 716, 721, 726, 727, 740, 741, 743

Gärtner, H. & Brauss, F.W. (1951) Studies on the question of carbon black lung and on the harmfulness of carbon dust part in dust of carbon factories. *Med. Welt*, **8**, 253–255 (in German)

Giammarise, A.T., Evans, D.L., Butler, M.A., Murphy, C.B., Kiriazides, D.K., Marsh, D. & Mermelstein, R. (1982) Improved methodology for carbon black extraction. In: Cooke, M., Dennis, A.J. & Fisher, G.L., eds, *Polycyclic Aromatic Hydrocarbons: Physical and Biological Chemistry, 6th International Symposium*, Columbus, OH, Battelle Press, pp. 325–345

Gold, A. (1975) Carbon black adsorbates: separation and identification of a carcinogen and some oxygenated polyaromatics. *Anal. Chem.*, **47**, 1469–1472

Gulf Oil Chemicals Company (1982) Discover this unique acetylene carbon black. In: Sweum, A., ed., *1982 Rubber Red Book*, 34th Ed., Atlanta, GA, Communication Channels, p. 496

von Haam, E. & Mallette, F.S. (1952) Studies on the toxicity and skin effects of compounds used in the rubber and platics industries. III. Carcinogenicity of carbon black extracts. *Arch. ind. Hyg. occup. Med.*, **6**, 237–242

von Haam, E., Titus, H.L., Caplan, I. & Shinowara, G.Y. (1958) Effect of carbon blacks on carcinogenic compounds. *Proc. Soc. exp. Biol. Med.*, **98**, 95–98

Haddad, R. & Zehetbauer, G. (1980) Problems arising from the use of cosmetics on the eye margin. *Klin. Mbl. Augenheilk.*, **177**, 829–831 (in German)

Harrington, J.M. & Levy, L.S. (1994) Lung cancer. In: Parkes, W.R., ed., *Occupational Lung Disorders*, 3rd Ed., London, Butterworth-Heinmann, pp. 644–666

Health and Safety Executive (1995) *Occupational Exposure Limits 1995* (Guidance Note EH 40/95), Sudbury, Suffolk, HSE Books, p. 30

Heinrich, U. (1994) Carcinogenic effects of solid particles. In: Mohr, U., Dungworth, D.L., Mauderly, J.L. & Oberdörster, G., eds, *Toxic and Carcinogenic Effects of Solid Particles in the Respiratory Tract*, Washington DC, ILSI Press, pp. 57–73

Heinrich, U., Peters, L., Creutzenberg, O., Dasenbrock, C. & Hoymann, H.-G. (1994) Inhalation exposure of rats to tar/pitch condensation aerosol or carbon black alone or in combination with irritant gases. In: Mohr, U., Dungworth, D.L., Mauderly, J.L. & Oberdörster, G., eds, *Toxic and Carcinogenic Effects of Solid Particles in the Respiratory Tract*, Washington DC, ILSI Press, pp. 433–441

Heinrich, U., Fuhst, R., Rittinghausen, R., Creutzenberg, O., Bellmann, B., Koch, W. & Levsen, K. (1995) Chronic inhalation exposure of Wistar rats and two different strains of mice to diesel exhaust, carbon black, and titanium dioxide. *Inhal. Toxicol.*, **7**, 533–556

Heitbrink, W.A. & McKinney, W.N., Jr (1986) Control of air contaminants at mixers and mills used in tyre manufacturing. *Am. ind. Hyg. Assoc. J.*, **47**, 312–321

Henderson, R.F., Barr, E.B., Cheng, Y.S., Griffith, W.C. & Hahn, F.F. (1992) The effect of exposure pattern on the accumulation of particles and the response of the lung to inhaled particles. *Fundam. appl. Toxicol.*, **19**, 367–374

Hess, W.M. & Herd, C.R. (1993) Microstructure, morphology, and general physical properties. In: Donnet, J.-B., Bansal, R.C. & Wang, M.-J., eds, *Carbon Black, Science and Technology*, 2nd Ed., New York, NY, Marcel Decker, pp. 89–151

Hodgson, J.T. & Jones, R.D. (1985) A mortality study of carbon black workers employed at five United Kingdom factories between 1947 and 1980. *Arch. environ. Health*, **40**, 261–268

Hoechst Aktiengesellschaft (1979) *Carbide and Basic Materials for Batteries: Acetogen® — Carbon Black*, Hürth, Germany

Hollett, B.A. (1980) *Health Hazard Evaluation Determination Report HHE 78-7-666, Kawecki Berylco Industries, Inc., Boyerton, Pennsylvania*, Cincinnati, OH, United States National Institute for Occupational Safety and Health

IARC (1982) *IARC Monographs on the Evaluation of the Carcinogenic Risk of Chemicals to Humans*, Vol. 28, *The Rubber Industry*, Lyon, pp. 47–88, 121–147, 183–230

IARC (1983) *IARC Monographs on the Evaluation of the Carcinogenic Risk of Chemicals to Humans*, Vol. 32, *Polynuclear Aromatic Compounds, Part 1, Chemical, Environmental and Experimental Data*, Lyon

IARC (1984) *IARC Monographs on the Evaluation of the Carcinogenic Risk of Chemicals to Humans*, Vol. 33, *Polynuclear Aromatic Hydrocarbons, Part 2, Carbon Blacks, Mineral Oils and Some Nitroarenes*, Lyon, pp. 35–85

IARC (1987a) *IARC Monographs on the Evaluation of Carcinogenic Risks to Humans*, Suppl. 7, *Overall Evaluations of Carcinogenicity: An Updating of* IARC Monographs *Volumes 1 to 42*, Lyon, pp. 142–143

IARC (1987b) *IARC Monographs on the Evaluation of Carcinogenic Risks to Humans*, Suppl. 7, *Overall Evaluations of Carcinogenicity: An Updating of* IARC Monographs *Volumes 1 to 42*, Lyon, p. 67

IARC (1989a) *IARC Monographs on the Evaluation of Carcinogenic Risks of Chemicals to Humans*, Vol. 46, *Diesel and Gasoline Engine Exhausts and Some Nitroarenes*, Lyon, pp. 201, 215, 231, 291, 321

IARC (1989b) *IARC Monographs on the Evaluation of Carcinogenic Risks of Chemicals to Humans*, Vol. 47, *Some Organic Solvents, Resin Monomers and Related Compounds, Pigments and Occupational Exposures in Paint Manufacture and Painting*, Lyon, pp. 329–442

IARC (1989c) *IARC Monographs on the Evaluation of Carcinogenic Risks of Chemicals to Humans*, Vol. 46, *Diesel and Gasoline Engine Exhausts and Some Nitroarenes*, Lyon, pp. 41–185

IARC (1995) *IARC Monographs on the Evaluation of the Carcinogenic Risk of Chemicals to Humans*, Vol. 62, *Wood Dust and Formaldehyde*, Lyon, pp. 217–362

Ingalls, T.H. (1950) Incidence of cancer in the carbon black industry. *Arch. ind. Hyg. occup. Med.*, **1**, 662–676

Ingalls, T.H. & Risquez-Iribarren, R. (1961) Periodic search for cancer in the carbon black industry. *Arch. environ. Health*, **2**, 429–433

Jacob, J. & Grimmer, G. (1979) Extraction and enrichment of polycyclic aromatic hydrocarbons (PAH) from environmental matter. In: Egan, H., Castegnaro, M., Bogovski, P.R., Kunte, H., & Walker, E.A., eds, *Environmental Carcinogens. Selected Methods of Analysis*, Vol. 3, *Analysis of Polycyclic Aromatic Hydrocarbons in Environmental Samples* (IARC Scientific Publications No. 29), Lyon, IARC, pp. 79–89

Jakab, G.J. (1993) The toxicologic interactions resulting from inhalation of carbon black and acrolein on pulmonary antibacterial and antiviral defenses. *Toxicol. appl. Pharmacol.*, **121**, 167–175

Jakab, G.J. & Hemenway, D.R. (1994) Concomitant exposure to carbon black particulates enhanced ozone-induced lung inflammation and suppression of alveolar macrophage phagocytosis. *J. Toxicol. environ. Health*, **41**, 221–231

Jakab, G.J., Risby, T.H., Sehnert, S.S., Hmieleski, R.R. & Gilmour, M.I. (1990) Suppression of alveolar macrophage membrane-receptor-mediated phagocytosis by model particle-adsorbate complexes: physicochemical moderators of uptake. *Environ. Health Perspectives*, **89**, 169–174

Jin, Z.-L., Dong, S.-P., Xu, W.-B., Li, Y.-Q. & Xu, X.-B. (1987) Analysis of mutagenic nitroarenes in carbon black by high-performance liquid chromatography. *J. Chromatogr.*, **386**, 185–190

Johnson, P.H. & Eberline, C.R. (1978) Carbon black, furnace black. In: McKetta, J.J. & Cunningham, W.A., eds, *Encyclopedia of Chemical Processing and Design*, Vol. 6, New York, Marcel Dekker, pp. 228–231, 236–237, 246–252, 255–257

Kandt, D. (1985) Clinical symptoms and findings of the lower airways in workers exposed to carbon black. *Z. Erkrank. Atm. Org.*, **165**, 25–41 (in German)

Kandt, D. & Biendara, E. (1985) Clinical symptoms and findings of the upper airways in workers exposed to carbon black. *Z. Erkrank. Atm. Org.*, **165**, 13–24 (in German)

Kirwin, C.J., LeBlanc, J.V., Thomas, W.C., Haworth, S.R., Kirby, P.E., Thilager, A., Bowman, J.T. & Brusick, D.J. (1981) Evaluation of the genetic activity of industrially produced carbon black. *J. Toxicol. environ. Health*, **7**, 973–989

Kollo, R.M. (1960) A health evaluation of working conditions in a channel black plant. *Trud. Leningrads. Sanit.-Gig. med. Inst.*, **62**, 128–131 (in Russian)

Komarova, L.T. (1965) The effect of air pollution on the morbidity and health of workers in carbon black production. *Nauchn. Trud. Omsk. med. Inst.*, **61**, 115–121 (in Russian)

Kreyling, W.G. (1990) Interspecies comparison of lung clearance of 'insoluble' particles. *J. Aerosol Med.*, **3** (Suppl. 1), S-93–S-110

Küpper, H.U,. Breitstadt, R., Wiebe, V., Mohry, P. & Ulmer, W.T. (1994a) Lung function tests and radiological examination of employees in the German carbon black industry. In: *Carbon Black World 94, Houston TX, February 23–24, 1994*

Küpper, H.U,. Breitstadt, R., Wiebe, V., Mohry, P. & Ulmer, W.T. (1994b) Carbon black dust exposition for workers of the chemical industry. Radiological and lung function studies for the estimation of the effects on the bronchopulmonary system. In: Kessel, R., ed., *Proceedings of the German Society of Occupational Medicine and Environmental Medicine, 34th Annual Meeting, Wiesbaden, May 16–19, 1994*, Stuttgart, Gentner Verlag, pp. 255–262 (in German)

Küpper, H.U., Breitstadt, R. & Ulmer, W.T. (1996) Effects on the lung function of exposure to carbon black dusts — Results of a study carried out on 677 members of staff of the Degussa factory in Kalscheuren/Germany. *Arbeitsmed. Sozialmed. Umweltmed.* (in press) (in German)

Lakowicz, J.R. & Bevan, D.R. (1979) Effects of asbestos, iron oxide, silica, and carbon black on the microsomal availability of benzo[*a*]pyrene. *Biochemistry*, **18**, 5170–5176

Lakowicz, J.R. & Bevan, D.R. (1980) Benzo[*a*]pyrene uptake into rat liver microsomes: effects of adsorption of benzo[*a*]pyrene to asbestos and non-fibrous mineral particulates. *Chem.-biol. Interactions*, **29**, 129–138

Lakowicz, J.R., Bevan, D.R. & Riemer, S.C. (1980) Transport of a carcinogen, benzo[*a*]pyrene, from particulates to lipid bilayers. A model for the fate of particle-adsorbed polynuclear aromatic hydrocarbons which are retained in the lungs. *Biochim. biophys. Acta*, **629**, 243–258

Lee, M.L. & Hites, R.A. (1976) Characterization of sulfur-containing polycyclic aromatic compounds in carbon blacks. *Anal. Chem.*, **48**, 1890–1893

Lee, P.S., Gorski, R.A., Hering, W.E. & Chan, T.L. (1987) Lung clearance of inhaled particles after exposure to carbon black generated from a resuspension system. *Environ. Res.*, **43**, 364–373

LeFevre, M.E. & Joel, D.D. (1986) Distribution of label after intragastric administration of ^{7}Be-labeled carbon to weaning and aged mice. *Proc. Soc. exp. Biol. Med.*, **182**, 112–119

Levison, H.W. (1973) Pigmentation of artists' colors. In: Patton, T.C., ed., *Pigment Handbook*, Vol. II, New York, John Wiley & Sons, p. 428

Lewis, P.A. (1988) Carbon Black. In: Lewis, P.A., ed., *Pigment Handbook*, Vol. I, *Properties and Economics*, 2nd Ed., New York, John Wiley & Sons, pp. 743–758

Lewis, R.J., Sr (1993) *Hawley's Condensed Chemical Dictionary*, 12th Ed., New York, Van Nostrand Reinhold Co., p. 160

Locati, G., Fantuzzi, A., Consonni, G., Li Gotti, I. & Bonomi, G. (1979) Identification of polycyclic aromatic hydrocarbons in carbon black with reference to carcinogenic risk in tire production. *Am. ind. Hyg. Assoc. J.*, **40**, 644–652

Mai, O. (1966) Carbon black lungs. *Z. ges. Hyg.*, **12**, 421–425 (in German)

Mauderly, J.L. (1994) Noncancer pulmonary effects of chronic inhalation exposure of animals to solid particles. In: Mohr, U., Dungworth, D.L., Mauderly, J.L. & Oberdörster, G., eds, *Toxic and Carcinogenic Effects of Solid Particles in the Respiratory Tract*, Washington DC, ILSI Press, pp. 43–55

Mauderly, J.L. (1996) Lung overload: The dilemma and opportunities for resolution. *Inhal. Toxicol.*, **8** (Suppl.), 1–20

Mauderly, J.L. & McCunney, R.J. (1996) Particle overload in the rat lung and lung cancer: relevance for human risk assessment. *Inhal. Toxicol.*, **8** (Suppl.)

Mauderly, J.L., Snipes, M.B., Barr, E.B., Belinsky, S.A., Bond, J.A., Brooks, A.L., Chang, I.-Y., Cheng, Y.S., Gillett, N.A., Griffith, W.C., Henderson, R.F., Mitchell, C.E., Nikula, K.J. & Thomassen, D.G. (1994) *Pulmonary Toxicity of Inhaled Diesel Exhaust and Carbon Black in Chronically Exposed Rats. Part I: Neoplastic and Nonneoplastic Lesions* (HEI Research Report Number 68), Cambridge, MA, Health Effects Institute

McBath, A. (1979) Carbon black. In: *Compilation of Air Pollutant Emission Factors*, 3rd Ed., Suppl. 9 (EPA Report AP-42; PB81-244097), Research Triangle Park, NC, Office of Air Quality Planning and Standards, United States Enviromental Protection Agency, pp. 49–56

Medalia, A.I., Rivin, D. & Sanders, D.R. (1981) *A Comparison of Carbon Black with Soot* (Cabot Research Paper, No. 148-230), Boston, MA, Cabot Corp.

Merchant, J.A. Taylor, G. & Hodous, T.K. (1986) Coal-workers pneumoconiosis and exposure to other carbonaceous dusts. In: Merchant, J.A., ed., *Occupational Respiratory Diseases* (DHHS Publications Series NIOSH 86-102), Washington DC, United States Government Printing Office, pp. 329–384

Miller, S.D. & Zarkower, A. (1974) Effects of carbon dust inhalation on the cell-mediated immune response in mice. *Infect. Immun.*, **9**, 534–539

Mohr, U., Dungworth, D.L., Mauderly, J.L. & Oberdörster, G., eds (1994) *Toxic and Carcinogenic Effects of Solid Particles in the Respiratory Tract*, Washington DC, ILSI Press

Morrow, P.E. (1988) Possible mechanisms to explain dust overloading of the lungs. *Fundam. appl. Toxicol.*, **10**, 369–384

Morrow, P.E. (1992) Contemporary issues in toxicology. Dust overloading of the lungs: update and appraisal. *Toxicol. appl. Pharmacol.*, **113**, 1–12

Muhle, H., Creutzenberg, O., Bellmann, B., Heinrich, U. & Mermelstein, R. (1990) Dust overloading of lungs: investigations of various materials, species differences, and irreversibility of effects. *J. Aerosol Med.*, **3** (Suppl. 1), S-111–S-128

Muhle, H., Bellmann, B. & Creutzenberg, O. (1994) Toxicokinetics of solid particles in chronic rat studies using diesel soot, carbon black, toner, titanium dioxide, and quartz. In: Mohr, U., Dungworth, D.L., Mauderly, J.L. & Oberdörster, G., eds, *Toxic and Carcinogenic Effects of Solid Particles in the Respiratory Tract*, Washington, DC, ILSI Press, pp. 29–41

Musch, D.C. & Smith, R.G. (1990) Characterizing occupational exposure to carbon black; results from three particulate sampling studies. In: *Proceedings of Annual Meeting of the American Industrial Hygiene Association, St Louis, MO, June 1989*, Ann Arbor, MI, University of Michigan

Nakajima, I., Hirokado, M., Usami, H., Mizoiri, S. & Endo, F. (1978) Polynuclear aromatic hydrocarbon in food additives. IV. Analysis of benzo(a)pyrene in activated carbon and carbon black. *Tokyotoritsu Eisei Kenkyusho Kenkyo Nempo*, **29**, 203–205 (in Japanese)

Nau, C.A., Neal, J. & Stembridge, V.A. (1958a) A study of the physiological effects of carbon black. I. Ingestion. *Arch. ind. Health*, **17**, 21–28

Nau, C.A., Neal, J. & Stembridge, V.A. (1958b) A study of the physiological effects of carbon black. II. Skin contact. *Arch. ind. Health*, **18**, 511–520

Nau, C.A., Neal, J. & Stembridge, V.A. (1960) A study of the physiological effect of carbon black. III. Adsorption and elution potentials; subcutaneous injections. *Arch. environ. Health*, **1**, 512–533

Nau, C.A., Neal, J. & Stembridge, V.A. (1962) Physiological effect of carbon black. IV. Inhalation. *Arch. environ. Health*, **4**, 415 – 431

Nau, C.A., Taylor, G.T. & Lawrence, C.H. (1976) Properties and physiological effects of thermal carbon black. *J. occup. Med.*, **18**, 732–734

Nikula, K.J., Snipes, M.B., Barr, E.B., Griffith, W.C., Henderson, R.F. & Mauderly, J.L. (1994) Influence of particle-associated organic compounds on the carcinogenicity of diesel exhausts. In: Mohr, U., Dungworth, D.L., Mauderly, J.L. & Oberdörster, G., eds, *Toxic and Carcinogenic Effects of Solid Particles in the Respiratory Tract*, Washington DC, ILSI Press, pp. 566–568

Nikula, K.J., Snipes, M.B., Barr, E.B., Griffith, W.C., Henderson, R.F. & Mauderly, J.L. (1995) Comparative pulmonary toxicities and carcinogenicities of chronically inhaled diesel exhaust and carbon black in F344 rats. *Fundam. appl. Toxicol.*, **25**, 80–94

Nishioka, M., Chang, H.-C. & Lee, M.L. (1986) Structural characteristics of polycyclic aromatic hydrocarbon isomers in coal tars and combustion products. *Environ. Sci. Technol.*, **20**, 1023–1027

Nolte, T., Thiedemann, K.-U., Dungworth, D.L., Ernst, H., Paulini, I., Heinrich, U., Dasenbrock, C., Peters, L., Ueberschär, S. & Mohr, U. (1993) Morphology and histogenesis of squamous cell metaplasia of the rat lung after chronic exposure to a pyrolyzed pitch condensate and/or carbon black, or to combinations of pyrolyzed pitch condensate, carbon black and irritant gases. *Exp. toxicol. Pathol.*, **45**, 135–144

Nolte, T., Thiedemann, K.-U., Dungworth, D.L., Ernst, H., Paulini, I., Heinrich, U., Dasenbrock, C., Peters, L., Ueberschär, S. & Mohr, U. (1994) Histological and ultrastructural alterations of the bronchioloalveolar region in the rat lung after chronic exposure to a pyrolyzed pitch condensate or carbon black, alone or in combination. *Inhal. Toxicol.*, **6**, 459–483

Novrocik, J., Novrocikova, M. & Frycka, J. (1983) Condensed aromatic hydrocarbons in rubber carbon blacks. *Plasty a Kaucuk*, **20**, 173–177 (in Czech)

Oberdörster, G. (1995) Lung particle overload: implications for occupational exposures to particles. *Regul. Toxicol. Pharmacol.*, **27**, 123–135

Oberdörster, G., Ferin, J., Corson, N., Gavett, S. & Hemenway, D. (1989) Particle induced in vitro chemotaxis of alveolar macrophages (AM) and neutrophils and in vivo effects. In: Mossman, B.T. & Bégin, R.O., eds, *Effects of Mineral Dusts on Cells* (NATO ASI Series, Vol. H30), Berlin, Springer-Verlag, pp. 313–320

Oberdörster, G., Ferin, J., Gelein, R., Soderholm, S.C. & Finkelstein, J. (1992) Role of the alveolar macrophage in lung injury: studies with ultrafine particles. *Environ. Health Perspectives*, **97**, 193–199

Oleru, U.G., Elegbeleye, O.O., Enu, C.C. & Olumide, Y.M. (1983) Pulmonary function and symptoms of Nigerian workers exposed to carbon black in dry cell battery and tire factories. *Environ. Res.*, **30**, 161–168

Parent, M.-E., Siemiatycki, J. & Renaud, G. (1996) Case–control study of exposure to carbon black in the occupational setting and risk of lung cancer. *Am. J. ind. Med.* (in press)

Patterson, W.J. (1980) An introduction to carbon black. In: *17th Akron Rubber Group Lecture Series-1980*, Boston, MA, Cabot Corp.

Pence, B.C. & Buddingh, F. (1985) The effect of carbon black ingestion on 1,2-dimethylhydrazine-induced colon carcinogenesis in rats and mice. *Toxicol. Lett.*, **25**, 273–277

Pott, F. & Roller, M. (1994) Relevance of nonphysiological exposure routes for carcinogenicity of solid particles. In: Mohr, U., Dungworth, D.L., Mauderly, J.L. & Oberdörster, G., eds, *Toxic and Carcinogenic Effects of Solid Particles in the Respiratory Tract*, Washington DC, ILSI Press, pp. 109–125

Pott, F. & Stöber, W. (1983) Carcinogenicity of airborne combustion products observed in subcutaneous tissue and lungs of laboratory rodents. *Environ. Health Perspectives*, **47**, 293–303

Pott, F., Roller, M., Rippe, R.M., Germann, P.-G. & Bellmann, B. (1991) Tumours by the intraperitoneal and intrapleural routes and their significance for the classification of mineral fibres. In: Brown, R.C., Hoskins, J.A. & Johnson, N.F., eds, *Mechanisms in Fibre Carcinogenesis*, New York, Plenum Press, pp. 547–565

Pott, F., Dungworth, D.L., Heinrich, U., Muhle, H., Kamino, K., Germann, P.-G., Roller, M., Rippe, R.M. & Mohr, U. (1994) Lung tumours in rats after intratracheal instillation of dusts. *Ann. occup. Hyg.*, **38** (Suppl. 1), 357–363

Pylev, L.N., Roe, F. & Vorvik, D. (1970a) A study of macrophage reaction of the pulmonary tissue of hamsters in response to the intratracheal insufflation of tritium labelled (^{3}H)-benz[*a*]pyrene mixed with carbon black and asbestos. *Patol. Fiziol. eksp.*, **14**, 47–51 (in Russian)

Pylev, L.N., Roe, F. & Vorvik, D. (1970b) The distribution and elimination of ^{3}H-benz[*a*]pyrene in the animals organism after its intratracheal injection with asbestos and carbon black. *Vop. Onkol.*, **16**, 61–69

Qazi, A.H. & Nau, C.A. (1975) Identification of polycyclic aromatic hydrocarbons in semi-reinforcing furnace carbon black. *Am. ind. Hyg. Assoc. J.*, **36**, 187–192

Ramdahl, T., Kveseth, K. & Becher, G. (1982) Analysis of nitrated polycyclic aromatic hydrocarbons by glass capillary gas chromatography using different detectors. *J. high Resol. Chromatogr. & Chromatogr. Comm.*, 5, 19–26

Rawlings, G.D. & Hughes, T.W. (1979) Emission inventory data for acrylonitrile, phthalic anhydride, carbon black, synthetic ammonia, and ammonium nitrate. In: Frederick, E.R., ed., *Proceedings of the Specialty Conference of Emission Factors and Inventories, Anaheim. CA, November 1978*, Pittsburgh, PA, Air Pollution Control Association, pp. 173–183

Renes, L.E. (1975) Toxicology and physiological effects of Phillips oil furnace carbon blacks. In: Ayer, F.A., ed., *Environmental Aspects of Commercial Use in Rubber Processing Operations Conference Proceedings, March 1975, Akron, OH* (EPA-560/1-75-002, PB-244 172), Washington DC, United States Environmental Protection Agency, pp. 437–441

Rhoades, R.A. (1972) Effect of inhaled carbon on surface properties of rat lung. *Life Sci.*, **11**, 33–42

Riebe-Imre, M., Aufderheide, M., Gärtner-Hübsch, S., Peraud, A. & Straub, M. (1994) Cytotoxic and genotoxic effects of insoluble particles *in vitro*. In: Mohr, U., Dungworth, D.L., Mauderly, J.L. & Oberdörster, G., eds, *Toxic and Carcinogenic Effects of Solid Particles in the Respiratory Tract*, Washington DC, ILSI Press, pp. 519–523

Rivin, D. & Smith, R.G. (1982) Environmental health aspects of carbon black. *Rubber Chem. Technol.*, **55**, 707–761

Robertson, J.McD. (1996) Epidemiologic studies in North American carbon black workers. *Inhal. Toxicol.*, **8** (Suppl.), 41–50

Robertson, J.McD. & Ingalls, T.H. (1980) A mortality study of carbon black workers in the United States from 1935 to 1974. *Arch. environ. Health*, **35**, 181–186

Robertson, J.McD. & Ingalls, T.H. (1989) A case–control study of circulatory, malignant, and respiratory morbidity in carbon black workers in the United States. *Am. ind. Hyg. Assoc. J.*, **50**, 510–515

Robertson, J.McD. & Inman, K.J. (1996) Mortality in carbon black workers in the United States: a preliminary report. *J. occup. environ. Med.* (in press)

Robertson, J.McD. & Smith, R.G. (1994) Carbon black. In: Clayton, G.D. & Clayton, F.E., eds, *Patty's Industrial Hygiene and Toxicology*, 4th Ed., Vol. 2, Part D, New York, John Wiley & Sons, pp. 2395–2423

Robertson, J.McD., Diaz, J.F., Fyfe, I.M. & Ingalls, T.H. (1988) A cross-sectional study of pulmonary function in carbon black workers in the United States. *Am. ind. Hyg. Assoc. J.*, **49**, 161–166

Rogaczewska, T., Ligocka, D. & Nowicka, K. (1989) Hygienic characteristics of carbon black used in tyre production. *Pol. J. occup. Med.*, **2**, 367–375

Rosenkranz, H.S., McCoy, E.C., Sanders, D.R., Butler, M., Kiriazides, D.K. & Mermelstein, R. (1980) Nitropyrenes: isolation, identification, and reduction of mutagenic impurities in carbon black and toners. *Science*, **209**, 1039–1043

Rosmanith, J., Kandus, J. & Holuša, R. (1969) Anthracofibrosis during the production of carbon black (carbon black lung). *Int. Arch. Gewerbepathol. Gewerbehyg.*, **25**, 292–298 (in German)

Salisbury, S. (1980) *Health Hazard Evaluation HHE 79-075-784, St. Clair Rubber Co., Marysville, Michigan*, Cincinnati, OH, United States National Institute for Occupational Safety and Health

Sanders, D.R. (1981) Nitropyrenes: the isolation of trace mutagenic impurities from the toluene extract of an after treated carbon black. In: Cooke, M. & Dennis, A.J., eds, *Chemical Analysis and Biological Fate: Polycyclic Aromatic Hydrocarbons, 5th International Symposium*, Columbus, OH, Battelle Press, pp. 145–158

Sands, F.W. & Benitez, J.C. (1961) Evaluation of the dust exposure in the use of carbon black in a rubber factory in Uruguay. In: *Proceedings of the Thirteenth International Congress on Occupational Health, New York, 28 July 1960*, New York, Book Craftsmen Associates, pp. 531–532

Shabad, L.M., Pylev, L.N. & Nasyrov, R.L. (1972) The significance of certain physical properties of various industrial types of carbon black for the liberation of the carcinogenic hydrocarbon benzo[a]pyrene from them. *Gig. Tr. prof. Zabol.*, **1**, 9–12 (in Russian)

Siemiatycki, J. (1991) *Risk Factors for Cancer in the Workplace*, Boca Raton, FL, CRC Press

Sigvardson, K.W. & Birks, J.W. (1984) Detection of nitro-polycyclic aromatic hydrocarbons in liquid chromatography by zinc reduction and peroxyoxalate chemiluminescence. *J. Chromatogr.*, **316**, 507–518

Slepicka, J., Eisler, L., Mirejowský, P. & Šimecek, R. (1970) Pulmonary changes in workers long-term exposed in soot production. *Pracov. Lék.*, **22**, 276–281 (in Czech)

Smith, W.R. (1964) Carbon (carbon black). In: Mark, H.F., McKetta, J.J., Jr & Othmer, D.F., eds, *Encyclopedia of Chemical Technology*, 2nd Ed., Vol. 4, New York, Interscience, pp. 254, 255

Smith, D.R., ed. (1982) *Blue Book 1982*, New York, Bill Communications, pp. 193–195, 197–203

Smith, R.G. & Musch, D.C. (1982) Occupational exposure to carbon black: a particulate sampling study. *Am. ind. Hyg. Assoc. J.*, **43**, 925–930

Snipes, M.B. (1989) Long-term retention and clearance of particles inhaled by mammalian species. *Crit. Rev. Toxicol.*, **20**, 175–211

Snow, J.B., Jr (1970) Carbon black inhalation into the larynx and trachea. *Laryngoscope*, **80**, 267–287

Sokhi, R.S., Gray, C., Gardiner, K. & Earwaker, L.G. (1990) PIXE (particle-induced X-ray emission) analysis of carbon black for elemental impurities. *Nucl. Instr. Meth. Phys. Res.*, **B49**, 414–417

Spodin, Y.N. (1973) Preventive sanitary inspection of the Kremenchug carbon black plant. *Gig. Trud.*, **9**, 22–26 (in Russian)

Steineck, G., Plato, N., Gerhardsson, M., Norell, S.E. & Hogstedt, C. (1990) Increased risk of urothelial cancer in Stockholm during 1985–87 after exposure to benzene and exhausts. *Int. J. Cancer*, **45**, 1012–1017

Steiner, P.E. (1954) The conditional biological activity of the carcinogens in carbon blacks and its elimination. *Cancer Res.*, **14**, 103–110

Stenbäck, F., Rowland, J. & Sellakumar, A. (1976) Carcinogenicity of benzo[*a*]pyrene and dusts in the hamster lung (instilled intratracheally with titanium oxide, aluminum oxide, carbon and ferric oxide). *Oncology*, **33**, 29–34

Stöber, W., Einbrodt, H.J. & Klosterkötter, W. (1965) Quantitative studies of dust retention in animal and human lungs after chronic inhalation. In: Davies, C.N., ed., *Inhaled Particles and Vapours II*, Edinburgh, Pergamon Press, pp. 409–418

Strom, K.A., Johnson, J.T. & Chan, T.L. (1989) Retention and clearance of inhaled submicron carbon black particles. *J. Toxicol. environ. Health*, **26**, 183–202

Sugar, H.S. & Kobernick, S. (1966) Subconjunctival pigmentation associated with the use of eye cosmetics containing carbon black. *Am. J. Ophthalmol.*, **62**, 146–149

Sun, J.D., Wolff, R.K., Maio, S.M. & Barr, E.B. (1989) Influence of adsorption to carbon black particles on the retention and metabolic activation of benzo[*a*]pyrene in rat lungs following inhalation exposure or intratracheal instillation. *Inhal. Toxicol.*, **1**, 1–19

Swafford, D.S., Nikula, K.J., Mitchell, C.E. & Belinsky, S.A. (1995) Low frequency of alterations in *p53*, K-*ras*, and *mdm2* in rat lung neoplasms induced by diesel exhaust or carbon black. *Carcinogenesis*, **16**, 1215–1221

Taylor, G.T., Redington, T.E., Bailey, M.J., Buddingh, F. & Nau, C.A. (1980) Solvent extracts of carbon black — Determination of total extractables and analysis for benzo(*a*)pyrene. *Am. ind. Hyg. Assoc. J.*, **41**, 819–825

The Freedonia Group (1994a) *Executive Summary: World Carbon Black Demand by Region (thousand metric tons)*, Cleveland, OH

The Freedonia Group (1994b) *Market Environment: Regulatory and Environmental Considerations*, Cleveland, OH

Troitskaya, N.A., Velichkovsky, B.T., Bikmullina, S.K., Sazhina, T.G., Gorodnova, N.V. & Andreeva, T.D. (1975) Substantiation of the maximum permissible concentration of industrial carbon black in the air of workrooms. *Gig. Tr. prof. Zabol.*, **17**, 32–36 (in Russian)

Troitskaya, N.A., Velichkovsky, B.T., Kogan, F.M. & O'Kuz'minukh, A.I. (1980) On carcinogenic hazards in the carbon black industry. *Vopr. Oncol.*, **26**, 63–67 (in Russian)

Union Carbide Corp. (1964) *Union Carbide Olefins Brings First US Acetylene Black Plant on Stream*, New York, Union Carbide Olefins Company

United Nations Environment Programme (1995) *International Register of Potentially Toxic Chemicals, Legal File, Carbon Black*, Geneva

United States Environmental Protection Agency (1994a) Carbon black manufacturing point source category. *US Code fed. Regul.*, **Title 40**, Part 458, pp. 368–373

United States Environmental Protection Agency (1994b) Exemptions from the requirement of a tolerance. *US Code fed. Regul.*, **Title 40**, Part 180.1001, pp. 439, 457–463

United States Food and Drug Administration (1976) Color additives. Termination of provisional listing of carbon black. *Fed. Regist.*, **41**, 41857–41859

United States Food and Drug Administration (1994a) Termination of provisional listings of color additives. *US Code fed. Regul.*, **Title 21**, Part 81.10, pp. 369–370

United States Food and Drug Administration (1994b) Food and drugs. *US Code fed. Regul.*, **Title 21**, Parts 175.105, 175.300, 177.1650, 177.2400, 177.2410, 177.2600, 178.3297, pp. 127–142, 145–160, 272, 292–294, 303–308, 341–344

United States National Institute for Occupational Safety and Health (1978) *Criteria for a Recommended Standard ... Occupational Exposure to Carbon Black* (DHEW Publication No. (NIOSH) 78-204), Washington DC, Department of Health, Education and Welfare

United States National Institute for Occupational Safety and Health (1994a) *Pocket Guide to Chemical Hazards* (DHHS (NIOSH) Publ. No. 94-116), Cincinnati, OH, pp. 52–53

United States National Institute for Occupational Safety and Health (1994b) *RTECs Chem.-Bank*, Cincinnati, OH

United States National Institute for Occupational Safety and Health (1995) *National Occupational Exposure Survey (1981–1983)*, Cincinnati, OH, pp. 52–53, 346

United States Occupational Safety and Health Administration (1977) *Test for Hazardous Substance 527 (Carbon Black) from OSHA Inception through January 1977*, Washington DC

United States Occupational Safety and Health Administration (1994) Air contaminants. *US Code fed. Regul.*, **Title 29**, Part 1910.1000, p. 9

Vainio, H., Magee, P.N., McGregor, D.B. & McMichael, A.J., eds (1992) *Mechanisms of Carcinogenesis in Risk Identification* (IARC Scientific Publications No. 116), Lyon, IARC

Vales, O., Gherdol, A. & Bit Chakoch, G. (1967) Reaction of the intima of the pulmonary vessels of the rabbit to the injection of hydrophilic carbon black. *Rev. Soc. argent. Biol.* **43**, 40–45 (in Spanish)

Valic, F., Beritic-Stahuljak, D. & Mark, B. (1975) A follow-up study of functional and radiological lung changes in carbon-black exposure. *Int. Arch. Arbeitsmed.*, **34**, 51–63

Venier, P., Tecchio, G., Clonfero, E. & Levis, A.G. (1987) Mutagenic activity of carbon black dyes used in the leather industry. *Mutagenesis*, **2**, 19–22

Vohler, O., von Sturm, F., Wege, E., von Kienle, H., Voll, M. & Kleinschmit, P. (1986) Carbon. In: Gerhartz, W., Yamamoto, Y.S., Campbell, F.T., Pfefferkorn, R. & Rounsaville, J.F., eds, *Ullmann's Encyclopedia of Industrial Chemistry*, 5th rev. Ed., Vol. A5, New York, NY, VCH Publishers, pp. 124–163

Wallcave, L., Nagel, D.L., Smith, J.W. & Waniska, R.D. (1975) Two pyrene derivatives of widespread environmental distribution: cyclopenta[cd]pyrene and acepyrene. *Environ. Sci. Technol.*, **9**, 143–145

Weast, R.C. (1981) *CRC Handbook of Chemistry and Physics*, 62nd Ed., Boça Raton, FL, CRC Press, p. F-74

Williams, T.M., Harris, R.L., Arp, E.W., Symons, M.J. & Van Ert, M.D. (1980) Worker exposure to chemical agents in the manufacture of rubber tires and tubes: particulates. *Am. ind. Hyg. Assoc. J.*, **41**, 204–211

Wolff, R.K., Sun, J.D., Barr, E.B., Rothenberg, S.J. & Yeh, H.C. (1989) Lung retention and binding of [^{14}C]-1-nitropyrene when inhaled by F344 rats as a pure aerosol or adsorbed to carbon black particles. *J. Toxicol. environ. Health*, **26**, 309–325

Wolff, R.K., Bond, J.A., Henderson, R.F., Harkema, J.R. & Mauderly, J.L. (1990) Pulmonary inflammation and DNA adducts in rats inhaling diesel exhaust or carbon black. *Inhal. Toxicol.*, **2**, 241–254

Wright, E.S. (1986) Effects of short-term exposure to diesel exhaust on lung cell proliferation and phospholipid metabolism. *Exp. Lung Res.*, **10**, 39–55

Zoccolillo, L., Liberti, A., Coccioli, F. & Ronchetti, M. (1984) Routine determination of polycyclic aromatic hydrocarbons in carbon black by chromatographic techniques. *J. Chromatogr.*, **288**, 347–355

2-CHLORONITROBENZENE, 3-CHLORONITROBENZENE AND 4-CHLORONITROBENZENE

1. Exposure Data

1.1 Chemical and physical data

1.1.1 *Nomenclature*

2-Chloronitrobenzene

Chem. Abstr. Serv. Reg. No.: 88-73-3

Chem. Abstr. Name: 1-Chloro-2-nitrobenzene

IUPAC Systematic Name: 1-Chloro-2-nitrobenzene

Synonyms: *ortho*-Chloronitrobenzene; 2-chloro-1-nitrobenzene; 2-CNB; 2-nitrochlorobenzene; *ortho*-nitrochlorobenzene; 1-nitro-2-chlorobenzene

3-Chloronitrobenzene

Chem. Abstr. Serv. Reg. No.: 121-73-3

Chem. Abstr. Name: 1-Chloro-3-nitrobenzene

IUPAC Systematic Name: 1-Chloro-3-nitrobenzene

Synonyms: *meta*-Chloronitrobenzene; 3-chloro-1-nitrobenzene; 3-CNB; 3-nitrochlorobenzene; *meta*-nitrochlorobenzene; 1-nitro-3-chlorobenzene

4-Chloronitrobenzene

Chem. Abstr. Serv. Reg. No.: 100-00-5

Chem. Abstr. Name: 1-Chloro-4-nitrobenzene

IUPAC Systematic Name: 1-Chloro-4-nitrobenzene

Synonyms: *para*-Chloronitrobenzene; 4-chloro-1-nitrobenzene; 4-CNB; 4-nitrochlorobenzene; *para*-nitrochlorobenzene; 1-nitro-4-chlorobenzene; 4-nitro-1-chlorobenzene; *para*-nitrophenyl chloride

1.1.2 Structural and molecular formulae and relative molecular mass

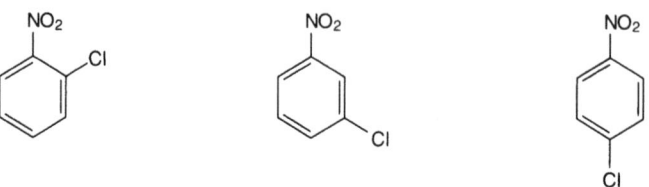

2-Chloronitrobenzene   3-Chloronitrobenzene   4-Chloronitrobenzene

$C_6H_4ClNO_2$ Relative molecular mass: 157.56

1.1.3 Chemical and physical properties of the pure substance

2-Chloronitrobenzene

(a) *Description*: Monoclinic needles (Lide, 1993)

(b) *Boiling-point*: 246 °C (Lide, 1993)

(c) *Melting-point*: 34–35 °C (Lide, 1993)

(d) *Spectroscopy data*: Infrared (prism [3231, 9802], grating [29726]), ultraviolet (UV) [21893] and nuclear magnetic resonance (proton [10548], C-13 [4218]) spectral data have been reported (Sadtler Research Laboratories, 1980).

(e) *Solubility*: Slightly soluble in water (590 mg/L at 20 °C) (BUA, 1992a); soluble in acetone, benzene, diethyl ether and ethanol (Lide, 1993)

(f) *Volatility*: Vapour pressure, 0.4 mm Hg [53.3 Pa] at 25 °C; relative vapour density (air = 1), 5.44 (Monsanto Co., 1993a)

(g) *Octanol/water partition coefficient (P)*: log P, 2.24 (Hansch et al., 1995)

(h) *Conversion factor*: $mg/m^3 = 6.44 \times ppm$[1]

3-Chloronitrobenzene

(a) *Description*: Pale yellow, rhombic prisms (from ethanol) (Lide, 1993)

(b) *Boiling-point*: 235–236 °C (Lide, 1993)

(c) *Melting-point*: 46 °C (Lide, 1993)

(d) *Spectroscopy data*: Infrared (prism [3232], grating [15443]), UV [958] and nuclear magnetic resonance (proton [6923], C-13 [6221]) spectral data have been reported (Sadtler Research Laboratories, 1980).

(e) *Solubility*: Slightly soluble in water (390 mg/L at 20 °C) (BUA, 1992b); sparingly soluble in cold ethanol; soluble in benzene, carbon disulfide, chloroform, diethyl ether, glacial acetic acid and hot ethanol (Budavari, 1989; Lide, 1993)

(f) *Volatility*: Vapour pressure, 1.85 Pa at 25 °C (BUA, 1992b)

(g) *Octanol/water partition coefficient (P)*: log P, 2.46 (Hansch et al., 1995)

---

[1] Calculated from: $mg/m^3 = (\text{relative molecular mass}/24.45) \times ppm$, assuming temperature (25 °C) and pressure (101 kPa)

(h) *Conversion factor*: mg/m^3 = 6.44 × ppm[1]

4-Chloronitrobenzene

(a) *Description*: Monoclinic prisms (Lide, 1993)

(b) *Boiling-point*: 242 °C (Lide, 1993)

(c) *Melting-point*: 83.6 °C (Lide, 1993)

(d) *Spectroscopy data*: Infrared (prism [4683], grating [435]), UV [1290] and nuclear magnetic resonance (proton [10629, V122], C-13 [628]) spectral data have been reported (Sadtler Research Laboratories, 1980).

(e) *Solubility*: Slightly soluble in water (243 mg/L at 20 °C) (BUA, 1992b); soluble in acetone, boiling ethanol, diethyl ether and carbon disulfide; sparingly soluble in cold ethanol (Budavari, 1989; Lide, 1993)

(f) *Volatility*: Vapour pressure, 0.15 mm Hg [20 Pa] at 30 °C; relative vapour density (air = 1), 5.44 (Monsanto Co., 1993b)

(g) *Octanol/water partition coefficient (P)*: log P, 2.39 (Hansch et al., 1995)

(h) *Conversion factor*: mg/m^3 = 6.44 × ppm[1]

1.1.4 *Technical products and impurities*

2-Chloronitrobenzene is available commercially at purities of > 99–99.85% (DuPont Chemicals, 1993; Monsanto Co., 1993a; Aldrich Chemical Co., 1994).

3-Chloronitrobenzene is available commercially at purities of 95–98% (Aldrich Chemical Co., 1994).

4-Chloronitrobenzene is available commercially at purities of 99–99.3% (Monsanto Co., 1993b; Aldrich Chemical Co., 1994).

1.1.5 *Analysis*

In evaluating exposure to chloronitrobenzenes, zinc reduction of the nitro group in acidified urine has been used for the routine determination of urinary chloronitrobenzenes at concentrations of 5–50 mg/L. After diazotization and coupling to 1-amino-8-naphthol-2,4-disulfonic acid (Chicago acid), spectrophotometric determination of the primary aromatic amine as a red azo dye was carried out. An alternative method has been described, in which the aromatic nitro compounds are reduced under alkaline conditions using formamidine sulfinic acid (thiourea dioxide) (Koniecki & Linch, 1958).

Mitchell and Deveraux (1978) reported the role of various detectors in gas chromatography in the determination of traces of organic compounds in the atmosphere. For the monochloronitrobenzenes, electron capture detection was approximately 4000 times more sensitive than flame ionization detection. Other selective detectors include thermionic ionization, Hall electrolytic conductivity and mass spectrometry (BUA, 1992b).

---

[1] Calculated from: mg/m^3 = (relative molecular mass/24.45) × ppm, assuming temperature (25 °C) and pressure (101 kPa)

Simple and accurate high-performance liquid chromatographic (HPLC) methods have been developed for determining 4-chloronitrobenzene in plasma (Yoshida, 1991) and 4-chloronitrobenzene and its metabolites in rat urine (Yoshida, 1993a). Samples are diluted with methanol, centrifuged and the methanol eluate analysed by isocratic reverse-phase HPLC with UV detection. Detection limits for 4-chloronitrobenzene are 0.01 µg/mL in plasma and 0.1 µg/mL in urine.

Selected methods for the analysis of chloronitrobenzenes in various media are presented in Table 1.

Table 1. Methods for the analysis of 2-, 3- and 4-chloronitrobenzenes

Sample matrix	Sample preparation	Assay procedure	Limit of detection	Reference
Air	Draw sample through solid sorbent tube; desorb with methanol	GC/FID	0.1 µg/sample[a]	Eller (1994) [Method 2005]
	Draw sample through Tenax TA; desorb thermally	GC/FID	< 0.04 µg/L[a,b,c]	Patil & Lonkar (1994)
	Trap in ethanol or isopropanol; reduce to aniline form with zinc/hydrochloric acid; couple with 1,2-naphthoquinone-4-sulfonic acid, disodium salt; extract with carbon tetrachloride; read at 450 nm	Colorimetry	10 µg/L[a]	Dangwal & Jethani (1980)
Water	Degas sample; adsorb on Tenax; desorb thermally	GC/MS	NR[a,b,c]	Braunstein et al. (1989)
	Extract sample with dichloromethane or adsorb on Amberlite XAD resin and elute with dichloromethane	GC/ECD	NR[a,b,c]	Feltes et al. (1990)
	Liquid–liquid extraction with dichloromethane; dry with anhydrous sodium sulfate; evaporate to dryness; redissolve in methanol	SFC/FID	10 ppm (mg/L)[c] 5 ppm (mg/L)[a]	Ong et al. (1992)
Industrial wastewater	Add sodium oxalate/EDTA/perchloric acid solutions; filter; adjust to pH 3.0 with perchloric acid	LC/UV	NR[a]	Nielen et al. (1985)
Soil	Extract with methanol; clean-up with solid phase extraction	HPLC/UV	NR[b,c]	Grob & Cao (1990)
		CGC	NR[b,c]	Greco & Grob (1990)
Urine	Reduce to aniline form with zinc/hydrochloric acid; couple with 1,2-naphthoquinone-4-sulfonic acid, disodium salt; extract with carbon tetrachloride; read at 450 nm	Colorimetry	0.6 mg/L[a]	Dangwal & Jethani (1980)

Table 1 (contd)

Sample matrix	Sample preparation	Assay procedure	Limit of detection	Reference
Blood	Extract from separated plasma and concentrate simultaneously using 2,2,4-trimethylpentane	GC/ECD	1.0 μg/L[a]	Lewalter & Ellrich (1991)
Fish	Extract sample with acetonitrile; transfer to petroleum ether; clean-up by Florisil column chromatography	GC/ECD	5–25 ppb[b,c] (μg/kg)	Yurawecz & Puma (1983)

GC, gas chromatography; FID, flame ionization detection; MS, mass spectrometry; NR, not reported; ECD, electron capture detection; SFC, capillary supercritical fluid chromatography; EDTA, ethylenediamine tetraacetic acid; LC, liquid chromatography; UV, ultraviolet detection; HPLC, high-performance liquid chromatography; CGC, capillary gas chromatography

[a] 4-Chloronitrobenzene
[b] 3-Chloronitrobenzene
[c] 2-Chloronitrobenzene

1.2 Production and use

1.2.1 *Production*

Continuous or batch nitration of chlorobenzene at 40–70 °C with mixed acids (30% HNO_3 : 56% H_2SO_4 : 14% H_2O) typically gives a 98% yield of an isomer mix consisting of 34–36% 2-chloronitrobenzene, 63–65% 4-chloronitrobenzene and about 1% 3-chloronitrobenzene. The isomers can be separated by a combination of fractional crystallization and distillation (Dunlap, 1981; Booth, 1991).

Chlorination of nitrobenzene at 35–45 °C in the presence of iron [III] chloride gives an isomer mixture containing 86% 3-chloronitrobenzene, 10% 2-chloronitrobenzene and 4% 4-chloronitrobenzene. A continuous process has been described that uses a series of reactors operating at 35–55 °C, with a residence time of 5 h. Purification is achieved by a combination of distillation and crystallization. Final purification of 3-chloronitrobenzene may be achieved chemically by caustic hydrolysis of the residual 2- and 4-chloronitrobenzenes and washing them out as nitrophenols (Booth, 1991).

Production of monochloronitrobenzenes in the United States of America in the 1970s was approximately 70 000 tonnes per year, of which about 25 000 tonnes were the 2-isomer and 45 000 tonnes were the 4-isomer (Dunlap, 1981). Estimated annual production figures in 1985 for 2- and 4-chloronitrobenzenes were 60 000 tonnes in Germany, 40 000 tonnes in the United States and 30 000 tonnes in Japan (Booth, 1991). In Germany, an estimated 1000–3000 tonnes of 3-chloronitrobenzene are produced annually (BUA, 1992b).

2-Chloronitrobenzene is produced by 17 companies in China, four companies in India, three companies each in Germany and Japan, two companies each in France, Italy and the United States, and one company each in Brazil, Poland, the Republic of Korea and the United Kingdom. 4-Chloronitrobenzene is produced by 17 companies in China, four companies each in India and Japan, three companies in Germany, two companies each in France, Italy and the United States and one company each in Brazil, Poland, the Republic of Korea and Romania. 3-Chloronitrobenzene is apparently produced in small quantities in several countries, including, at least, Germany, India and the United Kingdom (BUA, 1992b; Chemical Information Services, 1994).

1.2.2 *Use*

The chloronitrobenzenes are used almost exclusively as chemical intermediates. 2-Chloronitrobenzene is an important intermediate in the synthesis of colourants. Reduction with iron produces 2-chloroaniline (Fast Yellow G Base) and electrolytic reduction followed by rearrangement of the resulting hydrazo derivative leads to 3,3'-dichlorobenzidine (see IARC, 1982a, 1987a), both of which are important diazo components. Treatment of 2-chloronitrobenzene with aqueous sodium hydroxide at 130 °C produces 2-nitrophenol; treatment with methanol–sodium hydroxide gives 2-nitroanisole (see this volume); and treatment with ethanol–sodium hydroxide gives 2-nitrophenetole. All of these are used as precursors of the derived amines and many other products. Treatment with aqueous ammonia at 175 °C under pressure yields 2-nitroaniline (Fast Orange GR Base). Sulfonation and chlorosulfonation also give important sulfonic acid and sulfonyl chloride derivatives (Booth, 1991).

2-Chloronitrobenzene is also used in the preparation of *ortho*-anisidine (see IARC, 1982b, 1987b) (Fast Red BB Base), *ortho*-phenetidine, 3-amino-4-hydroxybenzenesulfonamide, picric acid, lumber preservatives, diaminophenol hydrochloride (a photographic developer), corrosion inhibitors, pigments and agricultural chemicals (Farris, 1978; Dunlap, 1981; Hathaway *et al.*, 1991; Lewis, 1993).

Most 3-chloronitrobenzene is reduced to 3-chloroaniline (Orange GC Base), a dye intermediate, with minor quantities finding other applications. Crude 3-chloronitrobenzene can be chlorinated exhaustively to give pentachloronitrobenzene, the use of which as a fungicide (Terrachlor) has led to several series of nitro-containing agrochemicals (Booth, 1991).

4-Chloronitrobenzene and its derivatives are also used in many synthetic processes. Common chemical intermediates produced from 4-chloronitrobenzene include 4-chloroaniline (see IARC, 1993), 4-nitrophenol, 4-nitroanisole, *para*-anisidine (IARC, 1982b, 1987b), 4-nitroaniline (Fast Red GC base), 6-chloro-3-nitrobenzenesulfonic acid, 2,4-dinitrochlorobenzene and 3,4-dichloronitrobenzene. 4,4'-Diaminodiphenyl sulfone (Dapsone; see IARC, 1980, 1987c), a drug used to treat leprosy, is synthesized from 4-chloronitrobenzene by one of two methods. In addition, condensation of 4-chloronitrobenzene with 2,4-dichlorophenol gives 2,4-dichloro-4'-nitrodiphenyl ether, which is used as a herbicide (nitrofen; see IARC, 1983, 1987d); 4-chloronitrobenzene can also be reacted with aniline to give 4-nitrodiphenylamine, which on reductive *N*-alkylation gives important antioxidants for rubber (see IARC, 1982c, 1987e; Dunlap, 1981; Budavari, 1989; Booth, 1991; Lewis, 1993).

1.3 Occurrence

1.3.1 Natural occurrence

Chloronitrobenzenes are not known to occur as natural products.

1.3.2 Occupational exposure

Occupational exposures to 4-chloronitrobenzene were measured for 10 workers in a dye factory in Japan producing 4-nitrophenol and *para*-anisidine (Yoshida *et al.*, 1988). The workers packed 4-chloronitrobenzene into bags and also transferred the raw material into a reaction vessel. Personal exposures were monitored for three seasons and averaged 0.31 mg/m^3 in autumn (range, 0.12–0.80 mg/m^3), 0.16 mg/m^3 in winter (range, 0.04–0.66 mg/m^3) and 0.38 mg/m^3 in summer (range, 0.16–0.88 mg/m^3). Workers wore simple dust respirators.

Long-term occupational exposure to aromatic nitro and amino compounds, including 2- and 4-chloronitrobenzenes, has been studied in Japan in 35 male workers involved in the production of dyes or pharmaceuticals. Urinary excretion of diazo metabolites was significantly higher ($p < 0.01$) in exposed workers compared to office workers. The main duties of the exposed workers included loading aromatic nitro and amino compounds from paper bags into reaction vessels, removing the reaction products from the reaction vessels and bagging the reaction products in paper bags. These jobs were carried out in separate rooms in the chemical factories. A few exposed workers carried the bags containing aromatic nitro and amino compounds or reaction products by fork lift truck to the warehouse for storage. The workers wore protective fabric clothing and cotton gloves. When the workers were engaged in the bagging operations, they also wore particulate-filter respirators and polyvinyl chloride-coated cotton gloves. The authors concluded that the workers were exposed to aromatic nitro-amino compounds routinely at .more than 0.3 mg/m^3 (as 4-chloronitrobenzene) and that exposure occurred through skin contact and inhalation (Yoshida *et al.*, 1989a).

Patil and Lonkar (1994) described the development of a method for monitoring chloronitrobenzenes in the presence of chlorobenzene in the workplace environment. In order to validate the method, air samples were collected in India at a chemical plant manufacturing chloronitrobenzenes using chlorobenzene as raw material. The relative humidity and temperature of the workplace during the period of investigation were found to vary in the ranges of 50–90% and 20–40 °C, respectively. Concentrations ranged from 0.03 to 0.52 mg/m^3 for 2-chloronitrobenzene, from 0.03 to 0.17 mg/m^3 for 3-chloronitrobenzene and from 0.16 to 0.96 mg/m^3 for 4-chloronitrobenzene.

The National Occupational Exposure Survey conducted between 1981 and 1983 indicated that 2900 and 2950 employees in the United States were potentially exposed to 2-chloronitrobenzene and 4-chloronitrobenzene, respectively. The estimate is based on a survey of companies and did not involve measurements of actual exposure (United States National Institute for Occupational Safety and Health, 1995).

1.3.3 *Environmental occurrence*

The major sources of environmental release of nitro aromatic compounds, such as chloronitrobenzenes, appear to be from chemical plants where they are produced and/or used as intermediates. Minor sources of release into the environment may occur during transport, storage or land burial, and chloronitrobenzenes may form in the environment through the oxidation of man-made aromatic amines or the reaction of nitrogen oxides in highly polluted air with chlorinated aromatic hydrocarbons (Howard, 1989).

Water

In the United States, 2-chloronitrobenzene was detected in river water samples from the Mississippi at Cape Girardeau, MO, and New Orleans, LA, at concentrations of 4–37 µg/L (0.004–0.037 ppm) and 1–2 µg/L (0.001–0.002 ppm), respectively, from September 1957 to April 1959 (Middleton & Lichtenberg, 1960). 2- and 4-Chloronitrobenzenes have been identified but not quantified in drinking-water from New Orleans, LA (Lucas, 1984).

From 1977 to 1982, 2- and 4-chloronitrobenzenes were detected in the River Rhine at levels ranging from less than 0.1 µg/L to 1.0 µg/L and from less than 0.1 µg/L to 0.11 µg/L, respectively. In a related study, 4-chloronitrobenzene was monitored in the River Rhine at a concentration of 1 µg/L, but was not detected in related tap-water (Piet & Morra, 1983). Concentrations of 3-chloronitrobenzene in the River Rhine have generally been less than 0.1 µg/L, although occasional samples have contained up to 1.2 µg/L. Higher concentrations of 2- and 3-chloronitrobenzenes were reported in 1978–79 in some samples from the Dutch section of the River Rhine (BUA, 1992a,b).

Chloronitrobenzenes have been identified in water samples from three sampling points along the River Elbe in Germany. In samples collected near Lauenberg, the concentrations of 2-, 3- and 4-chloronitrobenzenes were 0.2, 0.2 and 0.3 µg/L, respectively; in samples collected near Brokdorf, they were 0.1, 0.04 and 0.1 µg/L, respectively; and in samples collected near Brunsbüttel, they were 0.04, < 0.02 and 0.04 µg/L, respectively (Feltes *et al.*, 1990).

Concentrations of chloronitrobenzenes were measured at five sampling sites on the River Bormida in north-western Italy between September 1989 and February 1990. Twelve samples were taken at each site, and the mean concentrations of 2, 3 and 4-chloronitrobenzenes were highest near Bistagno (0.48, 0.12 and 0.19 µg/L, respectively) and lowest near the estuary at Alessandria (0.03, 0.01 and 0.01 µg/L, respectively) (Trova *et al.*, 1991)

In Dutch coastal waters, mean concentrations of 11 ng/L (maximum concentration, 50 ng/L) 2-chloronitrobenzene, 4.1 ng/L (maximum concentration, 14 ng/L) 3-chloronitrobenzene and 6.9 ng/L (maximum concentration, 31 ng/L) 4-chloronitrobenzene were detected (van de Meent *et al.*, 1986). Chloronitrobenzenes were also identified as micropollutants in water samples collected in 1986 from the Scheldt Estuary, located in the south-west of the Netherlands and the north-west of Belgium. Dissolved concentrations in water ranged from 0.5 to 2.1 ng/L (median, 1.3 ng/L) for 2-chloronitrobenzene, 0.1 to 1.3 ng/L (median, 0.3 ng/L) for 3-chloronitrobenzene and 0.5 to 2.5 ng/L (median, 1.4 ng/L) for 4-chloronitrobenzene. Concentrations of suspended particulate matter reached a maximum of 1.9 ng/g for 2-chloronitrobenzene, 1.5 ng/g (median, 0.2 ng/g) for 3-chloronitrobenzene and 1.9 ng/g (median, 0.3 ng/g) for 4-chloronitrobenzene (van Zoest & van Eck, 1991).

Chloronitrobenzenes were identified as pollutants in groundwater samples collected from January to March 1987 in Degrémont, France. 2-Chloronitrobenzene was the primary pollutant, accounting for 70% of the pollution; concentrations ranged from 970 to 1828 µg/L, with an average of 1500 µg/L. Concentrations of 3- and 4-chloronitrobenzenes ranged from 72 to 179 µg/L and 5 to 123 µg/L, respectively (Duguet et al., 1988).

2-Chloronitrobenzene has been identified in treated effluents from an advanced wastewater treatment plant in Orange County, CA, United States (Lucas, 1984). 2-, 3- and 4-Chloronitrobenzenes have also been identified in the effluent from 3-chloronitrobenzene production at concentrations ranging from 1500 to 1800 mg/L (Howard et al., 1976).

Wastewaters discharged from a chloronitrobenzene manufacturing plant in India were found to contain 24–93 mg/L (mean, 71 mg/L) 2-chloronitrobenzene, 1.3–3.0 mg/L (mean, 2.0 mg/L) 3-chloronitrobenzene and 112–227 mg/L (mean, 166 mg/L) 4-chloronitrobenzene (Swaminathan et al., 1987).

1.3.4 Food

In the United States, 2-chloronitrobenzene (at concentrations of 0.006–0.24 ppm (mg/kg)), 3-chloronitrobenzene (at 0.057 ppm) and 4-chloronitrobenzene (at 0.008–0.63 ppm) were found in the edible portion of various species of fish taken from the Mississippi River at sampling points 0, 60 and 150 miles south of St Louis, MO. Chloronitrobenzenes were not detected (< 0.005 ppm) in fish taken from the Mississippi River 100 miles north of St Louis (2 samples) or 260–400 miles south of St Louis (3 samples) or in those taken from the Missouri River (6 samples) (Yurawecz & Puma, 1983).

Chlorinated nitrobenzenes, including 2-, 3- and 4-chloronitrobenzenes, have been identified in fish from the River Main in Germany (Steinwandter, 1987)

1.4 Regulations and guidelines

Occupational exposure limits and guidelines for 4-chloronitrobenzene in several countries are given in Table 2.

2. Studies of Cancer In Humans

No data were available to the Working Group.

Table 2. Occupational exposure limits and guidelines for 4-chloronitrobenzene

Country	Year	Concentration (mg/m³)	Interpretation
Argentina	1991	0.6 (Sk)	TWA
Australia	1993	0.6 (Sk)	TWA
Austria	1993	1 (Sk)	TWA
Belgium	1993	0.64 (Sk)	TWA
Bulgaria[a,b]	1995	0.64 (Sk)	TWA
Canada	1991	3 (Sk)	TWA
Columbia[a,b]	1995	0.64 (Sk)	TWA
Czech Republic[b]	1993	1	TWA
		2	STEL
Denmark	1993	1 (Sk)	TWA
Finland	1993	1 (Sk)	TWA
		3	STEL (15 min)
Germany[b]	1995	None (Sk) IIIB	–
Hungary	1993	1 (Sk)	TWA
		2	STEL
Japan	1993	0.64 (Sk)	TWA
Jordan[a,b]	1995	0.64 (Sk)	TWA
Mexico	1991	1 (Sk)	TWA
		2	STEL (15-min)
Netherlands	1994	1 (Sk)	TWA
New Zealand[a,b]	1995	0.64 (Sk)	TWA
Philippines[a,b]	1995	0.64 (Sk)	TWA
Poland[b]	1993	1	TWA
Republic of Korea[a,b]	1995	0.64 (Sk)	TWA
Russia	1993	0.64 (Sk)	TWA
		1[b]	STEL
Singapore[a,b]	1995	0.64 (Sk)	TWA
Switzerland	1993	1 (Sk)	TWA
		2	STEL
United Kingdom	1995	1 (Sk)	TWA
		2	STEL (15-min)
USA			
ACGIH (TLV)	1995	0.64 (Sk)[c,d]	TWA
OSHA (PEL)	1994	1 (Sk)	TWA
NIOSH (REL)	1994	(Sk,Ca)	LFC

Table 2 (contd)

Country	Year	Concentration (mg/m^3)	Interpretation
Vietnam[a,b]	1995	0.64 (Sk)	TWA

From Arbeidsinspectie (1994); United States National Institute for Occupational Safety and Health (NIOSH) (1994a,b); United States Occupational Safety and Health Administration (OSHA) (1994); American Conference of Governmental Industrial Hygienists (ACGIH) (1995); Deutsche Forschungsgemeinschaft (1995); Health and Safety Executive (1995); United Nations Environment Programme (1995)

Sk, absorption through the skin may be a significant source of exposure; TWA, time-weighted average; STEL, short-term exposure limit; IIIB, suspected of having carcinogenic potential; TLV, threshold limit value; PEL, permissible exposure level; REL, recommended exposure level; LFC, lowest feasible concentration; Ca, potential occupational carcinogen

[a] Follows ACGIH values

[b] Same occupational exposure limit for 2-chloronitrobenzene and/or 3-chloronitrobenzene

[c] Substance identified in the BEI (Biological Exposure Indices) documentation for inducers of methaemoglobin

[d] Substance identified by other sources as a suspected or confirmed carcinogen

3. Studies of Cancer in Experimental Animals

2-Chloronitrobenzene

Oral administration

Mouse: Groups of 25 male and 25 female CD-1 mice (derived from HaM/ICR mice), six to eight weeks of age, were fed diets containing 0 (control), 3000 or 6000 mg/kg (ppm) 2-chloronitrobenzene (97–99% pure) for eight months. After that time, the dietary concentrations were lowered to 0, 1500 and 3000 mg/kg (ppm) for 10 months and mice were then held for three months on control diet prior to terminal examination. Mice that died within the first six months of the study were discarded without necropsy. Complete gross necropsy was carried out on all other animals; certain tissues were examined histopathologically, including those of the lung, liver, spleen, kidney, adrenal glands, heart, urinary bladder, stomach, intestines and reproductive organs; gross lesions and tissue masses were also examined histopathologically. Information on survival or body-weight gain was not reported. The incidence of hepatocellular carcinomas in males (3/18 controls, 7/17 at the low dose and 3/16 at the high dose) was increased at the low dose ($p < 0.025$ versus a pooled control incidence of 7/99, Fisher's exact test) and in females

(0/20 controls, 5/22 at the low dose and 5/19 at the high dose) was increased at the low and high doses ($p < 0.025$ versus concurrent and pooled control incidence of 1/102) (Weisburger et al., 1978). [The Working Group noted the small number of animals and the limited histopathological evaluation and reporting.]

Rat: Groups of 25 male Charles River CD rats (derived from Sprague Dawley), six to eight weeks of age, were fed diets containing 0 (control), 1000 or 2000 mg/kg (ppm) 2-chloronitrobenzene (97–99% pure) for six months. After that time, the dietary concentrations were lowered to 0, 500 and 1000 mg/kg (ppm) for 12 months and rats were then held for six months on control diet prior to terminal examination. Rats that died within the first six months of the study were discarded without necropsy. Complete gross necropsy was carried out on all other animals; certain tissues were examined histopathologically, including those of the lung, liver, spleen, kidney, adrenal glands, heart, urinary bladder, stomach, intestines and pituitaries; gross lesions and tissue masses were also examined histopathologically. Information on survival or body-weight gain and individual tumour incidence was not reported. The total number of animals bearing multiple tumours was increased in low-dose rats (incidence of multiple tumours: controls, 1/22; low-dose rats, 7/22; high-dose rats, 1/19; $p < 0.025$ versus concurrent and pooled control incidence of 14/111, Fisher's exact test) (Weisburger et al., 1978). [The Working Group noted the small number of animals, the short duration of treatment, the limited histopathological evaluation and the inadequate reporting.]

4-Chloronitrobenzene

Oral administration

Mouse: Groups of 25 male and 25 female CD-1 mice (derived from HaM/ICR mice), six to eight weeks of age, were fed diets containing 0, 3000 or 6000 mg/kg (ppm) 4-chloronitrobenzene (97–99% pure) for 18 months. The mice were then held for three months on control diet prior to terminal examination. Mice that died within the first six months of the study were discarded without necropsy. Complete gross necropsy was carried out on all other animals; certain tissues were examined histopathologically including those of the lung, liver, spleen, kidney, adrenal glands, heart, urinary bladder, stomach, intestines and reproductive organs; gross lesions and tissue masses were also examined histopathologically. Information on survival or body-weight gain was not reported. The incidence of hepatocellular carcinomas was increased in low-dose males (controls, 1/14; low-dose mice, 4/14; $p < 0.025$ versus pooled control incidence of 7/99, Fisher's exact test; high-dose mice, 0/14). The incidence of vascular tumours was increased in high-dose males (controls, 0/14; low-dose mice, 2/14; high-dose mice, 4/14; $p < 0.025$ versus concurrent and pooled control incidence of 5/99) and in high-dose females (controls, 0/15; low-dose mice, 2/20; high-dose mice, 7/18; $p < 0.025$ versus concurrent and pooled control incidence of 9/102) (Weisburger et al., 1978). [The Working Group noted the small number of animals and the limited histopathological evaluation and reporting.]

Rat: Groups of 25 male Charles River CD rats (derived from Sprague Dawley rats), six to eight weeks of age, were fed diets containing 0, 2000 or 4000 mg/kg (ppm) 4-chloronitrobenzene (97.99% pure) for three months. After that time, the dietary concentrations were lowered to 0, 250 and 500 mg/kg (ppm) for two months, then raised to 0, 500 and 1000 mg/kg (ppm) for 13 months; rats were then held for six months on control diet prior to terminal examination. Rats that died within the first six months of the study were discarded without necropsy. Complete gross necropsy was carried out on all other animals; certain tissues were examined histopathologically, including the lung, liver, spleen, kidney, adrenal glands, heart, urinary bladder, stomach, intestines and pituitaries; gross lesions and tissue masses were also examined histopathologically. Information on survival or body-weight gain was not reported. No increase in tumour incidences was reported (Weisburger *et al.*, 1978). [The Working Group noted the small number of animals, the short duration of dosing and the limited histopathological evaluation and reporting.]

3-Chloronitrobenzene

No data were available to the Working Group.

4. Other Data Relevant for an Evaluation of Carcinogenicity and its Mechanisms

4.1 Absorption, distribution, metabolism and excretion

4.1.1 *Humans*

The measurement of urinary excretion of diazo-positive metabolites has been employed for biological monitoring of 4-chloronitrobenzene-exposed workers. The time-weighted average air concentrations in the breathing zones of 10 workers in a dye-producing factory that used 4-chloronitrobenzene were 0.31, 0.16 and 0.38 mg/m^3 in the autumn, winter and summer, respectively. In unexposed controls (laboratory staff), the excretion of diazo-positive metabolites (4-aminophenol equivalents) pre- and post-shift was 0.261 and 0.274 [mg/g] creatinine, respectively. In the exposed workers, the average values ranged from 0.375 to 0.543 [mg/g] creatinine pre-shift and from 0.677 to 1.086 [mg/g] creatinine post-shift. Pre-shift values rose over the working week, which indicates slow elimination and consequent gradual accumulation (Yoshida *et al.*, 1988). [The authors use mg/mg creatinine, but the Working Group considered mg/g creatinine to be the most likely units.]

The excretion of urinary metabolites of 4-chloronitrobenzene has been studied in connection with an outbreak of accidental poisoning with 4-chloronitrobenzene (Yoshida *et al.*, 1992, 1993). In the more extensive study of the outbreak, Yoshida *et al.* (1993) analysed urinary elimination of metabolites for up to 29 days in eight workers. The parent compound was not excreted. However, metabolites were eliminated for the total observation period (total metabolites, 179–1076 mg). The elimination of each metabolite seemed to fit into a one-compartment model with an elimination half-life of about one week. There was a complicated metabolic pattern of considerable interindividual variation, which had three major pathways (see Figure 1):

(1) The most important route of metabolism was glutathione conjugation, which resulted in the excretion of the mercapturic acid, N-acetyl-S-(4-nitrophenyl)-L-cysteine (48% of the total metabolites).

(2) There was also a slow reduction of the nitro group, resulting in excretion of 4-chloroaniline (29.9%), which was rapidly metabolized further by (a) fast N-acetylation to 4-chloroacetanilide and further to 4-chloro-oxanilic acid, (b) ring-hydroxylation to give 2-amino-5-chlorophenol (8.7%) — which was N-acetylated into 4-chloro-2-hydroxyacetanilide — and (c) chlorination to give 2,4-dichloroaniline (chlorination was a novel metabolic pathway in humans), which was readily excreted (1.2%).

(3) Finally, there was a slow ring-hydroxylation of the parent compound to give 2-chloro-5-nitrophenol (12.2%).

Figure 1. Metabolic pathway of 4-chloronitrobenzene in humans

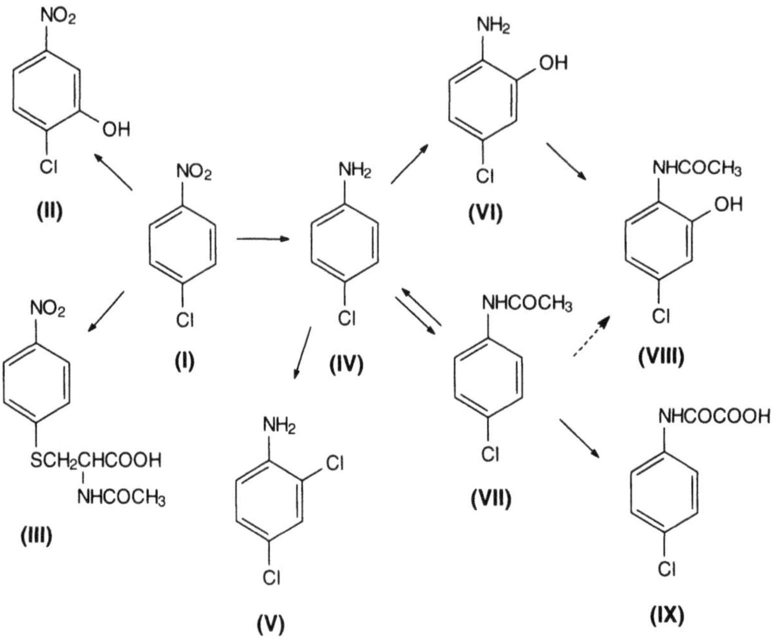

From Yoshida et al. (1993)

I, 4-Chloronitrobenzene; II, 2-chloro-5-nitrophenol; III, N-acetyl-S-(4-nitrophenyl)-L-cysteine; IV, 4-chloroaniline; V, 2,4-dichloroaniline; VI, 2-amino-5-chlorophenol; VII, 4-chloroacetanilide; VIII, 4-chloro-2-hydroxyacetanilide; and IX, 4-chloro-oxanilic acid.

Dotted line: It was not clear whether VIII was formed by hydroxylation of VII.

No data concerning the absorption, distribution, metabolism and excretion of 2- and 3-chloronitrobenzenes in humans were available to the Working Group.

4.1.2 *Experimental systems*

Dermal absorption studies with 2- and 4-chloronitro[^{14}C]benzenes were conducted in male Fischer 344 rats using dermal applications of [0.0325, 0.32, and 3.25 mg/cm^2] (0.65, 6.5 and 65 mg/kg bw, respectively) (United States National Toxicology Program, 1993). At all dose levels, 33–40% of 2-chloronitrobenzene and 51–62% of 4-chloronitrobenzene were absorbed from the skin within 72 h, based upon measurements of eliminated radioactivity. The absorbed radioactivity was excreted in urine and faeces. Urinary excretion of 4-chloronitrobenzene-derived radioactivity was substantially greater than that of 2-chloronitrobenzene-derived radioactivity over 72 h (43–45% versus 21–28% of the dose). Approximately 5–12% of 4-chloronitrobenzene-derived radioactivity were excreted in faeces, while 11–15% of 2-chloronitrobenzene-derived radioactivity were excreted via this route.

In studies by oral administration, rats were given 2, 20 or 200 mg/kg bw 2-chloronitrobenzene or 4-chloronitrobenzene, and urine and faeces were collected for up to 72 h. At all doses, 61–77% of the 2-chloronitrobenzene and 73–78% of the 4-chloronitrobenzene were absorbed. Both isomers were rapidly metabolized and excreted primarily in the urine. At the lower doses, about 6% and 3% of the administered doses of 2-chloronitrobenzene and about 23% and 5% of the administered doses of 4-chloronitrobenzene were found in tissues after 24 and 72 h, respectively. For 2-chloronitrobenzene, most tissues contained less than 0.1% of the dose administered. Most of the radioactivity was in the liver, followed by fat, muscle and kidney. For 4-nitrochlorobenzene, the highest concentrations of ^{14}C at 24 h were found in fat, followed by blood cells, kidney, liver and spleen. HPLC analysis of urine revealed the presence of up to 23 metabolites from 2-chloronitrobenzene and 25 metabolites from 4-chloronitrobenzene [metabolites unspecified] (United States National Toxicology Program, 1993).

In repeated-dose studies, groups of four young adult (10–12 weeks of age) or geriatric (19–20 months of age) male Fischer 344 rats received 65 mg/kg bw 2-chloronitro[^{14}C]-benzene or 4-chloronitro[^{14}C]benzene by gavage on days 1, 5 and 9. Unlabelled parent compound was administered in corn oil at the same dose on days 2, 3, 4, 6, 7, 8, 10 and 11. In young adult rats after 0, 4 or 8 days of pretreatment with 2-chloronitrobenzene, ^{14}C was excreted primarily in urine for the first 24 h. Radioactivity was excreted rapidly in both urine and faeces following pretreatment. Approximately 5% of the administered radioactivity was found in the tissues 72 h after the day 9 dose, and most was in the liver. Similar results were reported for 4-chloronitrobenzene. However, with 4-chloronitrobenzene, most of the radioactivity was found in blood cells and fat. The absorption, metabolism and excretion of 2-chloronitrobenzene and 4-chloronitrobenzene were not greatly affected by age. Geriatric rats absorbed, metabolized and cleared these compounds as readily as young adults at the 65-mg/kg dose. Only very minor variations in rates of excretion in urine and faeces were noted (United States National Toxicology Program, 1993).

Bray et al. (1956) examined the metabolism of the three isomers of chloronitrobenzenes in rabbits. Doe rabbits were given [route unspecified] 0.1 g/kg [whether diet or body weight not specified] 2-chloronitrobenzene and 0.2 g/kg 3- and 4-chloronitrobenzenes. Urine was collected over 24-h periods until metabolites were no longer excreted (usually after 48 h). The main metabolic processes were reduction and hydroxylation. For each isomer, nearly the entire dose was excreted in urine as the chloroaniline or derivatives of phenolic metabolites. For all three isomers, about 10% of the administered dose was excreted as chloroaniline. The phenols formed were excreted mainly as conjugates with sulfuric and glucuronic acids (66% of the administered dose for 2-chloronitrobenzene, 51% for 3-chloronitrobenzene and 40% for 4-chloronitrobenzene).

Yoshida et al. (1991) identified the urinary metabolites of 4-chloronitrobenzene in rats by gas chromatography–mass spectrometry. Male Sprague-Dawley rats were given a single intraperitoneal injection of 100 mg/kg bw 4-chloronitrobenzene and urine was collected from 8 to 24 h after dosage. Rats excreted eight urinary metabolites: 4-chloroaniline, 2,4-dichloroaniline, 4-nitrothiophenol, 2-chloro-5-nitrophenol, 2-amino-5-chlorophenol, 4-chloroformanilide, 4-chloro-2-hydroxyacetanilide and a small amount of 4-chloroacetanilide. Only trace amounts of unchanged 4-chloronitrobenzene were detected.

Rickert and Held (1990) studied the metabolism of the radiolabelled isomers of chloronitrobenzenes by isolated male Fischer 344 rat hepatocytes. Incubation of rat hepatocytes with 2-chloronitro[^{14}C]benzene resulted in the detection of 2-chloroaniline (19.2% of total radioactivity), 2-chloroaniline-N-glucuronide (14.2% of total radioactivity and S-(2-nitrophenyl)glutathione (13.3% of total radioactivity). Metabolism of 3-chloronitro[^{14}C]benzene by isolated male Fischer 344 rat hepatocytes produced 3-chloroaniline as the major metabolite (30.9% of total radioactivity). Much smaller quantities of 3-chloroacetanilide (16.7%) and 3-chloroaniline-N-glucuronide (6.7%) were also produced by this isomer. No glutathione conjugate was detected. 4-Chloronitro[^{14}C]benzene was metabolized to 4-chloroaniline (15.4% of total radioactivity), S-(4-nitrophenyl) glutathione (10.4%) and 4-chloroacetaniline (16.3%). Studies of metabolism with hepatic microsomes demonstrated that the formation of the chloroanilines from all isomers of the chloronitrobenzenes was mediated by cytochrome P450-dependent metabolism; formation was inhibited by SKF 525-A, metyrapone and carbon monoxide.

4.2 Toxic effects

4.2.1 *Humans*

Eleven longshoremen were poisoned with 4-chloronitrobenzene due to the accidental opening of bags. It was assumed that both inhalation and skin absorption occurred. The symptoms included headache, palpitation, dizziness, nausea, vomiting and poor appetite. On physical examination they had cyanosis. Laboratory tests revealed methaemoglobinaemia, anaemia, reticulocytosis and Heinz bodies (Yoshida et al., 1992, 1993).

No data concerning the toxic effects of 2-chloronitrobenzene (or 3-chloronitrobenzene) in humans were available to the Working Group.

4.2.2 Experimental systems

Vernot et al. (1977) determined the acute toxicity of 2-chloronitrobenzene and 4-chloronitrobenzene. In male Sprague-Dawley rats, single oral LD_{50} values for 2-chloronitrobenzene and 4-chloronitrobenzene were 270 and 810 mg/kg bw, respectively. For male CF-1 mice, acute oral LD_{50} values were reported as 140 mg/kg bw for 2-chloronitrobenzene and 1410 mg/kg bw for 4-chloronitrobenzene.

In a teratogenicity study with New Zealand white rabbits, 8/18 died during treatment with 4-chloronitrobenzene at 40 mg/kg bw per day on gestation days 7–19 (Nair et al., 1985).

Nair et al. (1986a) exposed male and female Sprague Dawley rats by inhalation to 2-chloronitrobenzene atmospheres at concentrations of 0, 10, 30 and 60 mg/m^3 (measured values, 0, 9.9, 30 and 59 mg/m^3) for 6 h per day on five days per week for four weeks. A dose-related increase in blood methaemoglobin levels was observed after two weeks of exposure. The increase was statistically significant at the mid- and high-exposure concentrations compared to control animals. Four weeks of exposure to 30 and 60 mg/m^3 produced a significant increase in blood methaemoglobin levels and a decrease in haemoglobin, haematocrit and red blood cell count values in both sexes. Increases in liver, kidney and spleen weights were also observed at 30 and 60 mg/m^3. Only the spleen was found to have microscopic changes, such as an increase in extramedullary haematopoiesis, at exposures of 30 and 60 mg/m^3. In addition, there was an increase in haemosiderosis in the spleen at all doses of 2-chloronitrobenzene.

In a similar study, Nair et al. (1986b) also examined the inhalation toxicity of 4-chloronitrobenzene in male and female Sprague-Dawley rats. Animals were exposed to concentrations of 0, 5, 15 and 45 mg/m^3 (measured values, 6, 19 and 46 mg/m^3) 4-chloronitrobenzene in air for 6 h per day on five days per week for four weeks. 4-Chloronitrobenzene caused dose-dependent cyanosis. Cyanosis of the conjunctival and nasal area was observed at all concentration levels, while rats in the mid- and high-exposure groups were cyanotic over the entire body. 4-Chloronitrobenzene produced significant methaemoglobinaemia at the two-week interval in mid-exposure females (11.5% ± 5.1 versus 2.7 ± 2.8 for controls) and in high-exposure males (14.0 % ± 9.2 versus 3.2 % ± 2.7) and females (14.9 % ± 1.6 versus 2.7 % ± 2.8). After four weeks of exposure, 4-chloronitrobenzene produced a significant increase in methaemoglobin levels in female rats exposed to mid and high doses and in male rats exposed to low, mid and high doses. This isomer also produced statistically significant reductions in red blood cell counts, haematocrit and haemoglobin at both 15 and 45 mg/m^3 in both males and females exposed for two and four weeks. A significant increase in white blood cell counts was seen in mid- and high-dose females and high-dose males after two weeks of exposure. At four weeks, white blood counts were elevated in high-dose males and females. Splenic enlargement, accompanied by an increase in absolute and relative spleen weights, was observed in male rats exposed to the high dose of 4-chloronitrobenzene. Mean and relative liver weights were also significantly increased in high-dose male rats. Microscopic examination revealed that high-dose exposure also increased the incidence of splenic congestion and extramedullary haematopoiesis and haemosiderosis.

Nishida et al. (1982) observed increased methaemoglobin levels, increased osmotic fragility of erythrocytes and increased formation of Heinz bodies in rabbits [sex unspecified] following a single subcutaneous dose of 50–200 mg/kg bw 4-chloronitrobenzene.

Watanabe et al. (1976) examined the toxicity of the three isomers of chloronitrobenzene in male Wistar rats injected intraperitoneally with 100 μmol/kg bw [15.8 mg/kg bw] 2-, 3- or 4-chloronitrobenzenes. Animals were killed 5 h later to determine methaemoglobin levels. 3-Chloronitrobenzene was the most potent of the isomers in the production of methaemoglobinaemia (31.9 ± 5.8% versus 20.6 ± 6.9% for 2-chloronitrobenzene and 16.3 ± 7.5% for 4-chloronitrobenzene). Methaemoglobin formation was also studied *in vitro*. Both 3- and 4-chloronitrobenzenes produced levels of methaemoglobin that were significantly greater than control levels.

In a study conducted by Yoshida et al. (1989b) male Fischer 344 rats were injected intraperitoneally with 1.0 mmol [158 mg]/kg bw 4-chloronitrobenzene. Urine was collected for 24 h and rats were killed 48 h after the injection to remove kidneys and collect blood. 4-Chloronitrobenzene had no effect on the urea nitrogen content of blood but produced a significant elevation in urine output and a significant increase in urinary N-acetyl-β-D-glucosaminidase activity. 4-Chloronitrobenzene caused no significant histopathological change in renal tubular epithelial cells. The compound caused a significant elevation in methaemoglobin levels as compared to controls (5.4% ± 0.5 versus 1.1% ± 0.2). 4-Chloronitrobenzene caused no significant change in serum aspartate aminotransferase or alanine aminotransferase activities.

Two-week and 13-week whole-body exposure studies of 2-chloronitrobenzene and 4-chloronitrobenzene were conducted in male and female Fischer 344 rats and in male and female B6C3F1 mice (United States National Toxicology Program, 1993). During the studies, rats and mice were exposed to 0, 1.1, 2.3, 4.5, 9 or 18 ppm [0, 6.4, 14.8, 29, 58 or 116 mg/m^3] 2-chloronitrobenzene vapour or 0, 1.5, 3, 6, 12 or 24 ppm [9.6, 19.3, 38.6, 77 or 155 mg/m^3] 4-chloronitrobenzene vapour for 6 h per day on five days per week, excluding weekends and holidays, for either 12 exposure days or 13 weeks.

Male and female rats exposed to 2-chloronitrobenzene for two weeks had exposure-related increases in absolute and relative liver weights. Absolute and relative spleen weights were increased in males and females in the 18-ppm groups, while relative kidney weights were slightly increased in males exposed to 18 ppm. In this exposure group, haemosiderin deposition was observed in the spleens and in the portal and central areas of the livers of all rats. Mice also demonstrated exposure-related increases in liver weights and increases in spleen weights at 18 ppm after two weeks' exposure to 2-chloronitrobenzene. Mice, especially in the 18 ppm-exposure group, had liver changes such as coagulative necrosis and granulomatous inflammation. Mice also had increased haematopoietic cell proliferation and haemosiderin deposition in the spleen at exposure concentrations as low as 4.5 ppm (United States National Toxicology Program, 1993)

In the rats exposed to 2-chloronitrobenzene by inhalation for 13 weeks, elevated methaemoglobin concentrations were detected in all treated rats; indeed, elevated methaemoglobin levels were detected in all treated male rats by day 23. Histopathological changes in male and female rats included increased basophilia of centrilobular hepatocytes, pigmentation and regeneration of the proximal convoluted tubules of the kidney and hyperplasia of the nasal cavity epithelia. Exposure of mice to 2-chloronitrobenzene for 13 weeks produced hepatocellular necrosis, cytomegaly, mineralization and chronic inflammation of the liver after exposure to 18 ppm. Mice also demonstrated increased haematopoietic activity (United States National Toxicology Program, 1993)

Exposure to 4-chloronitrobenzene for two weeks induced exposure-related increases in absolute and relative spleen and liver weights in both male and female rats. Exposure to this isomer also produced splenic enlargement and dark discoloration in all males and females exposed to 12 or 24 ppm and microscopic changes in the spleen, such as an increase in haematopoiesis, in male and female rats exposed to 6 ppm and above. Microscopic changes in the proximal convoluted tubules of the kidneys of males and females in the 12 and 24 ppm groups — hyaline droplet nephropathy in male rats and an increase in haemosiderin deposition in female rats — were also noted (United States National Toxicology Program, 1993).

Mice exposed to 4-chloronitrobenzene for two weeks showed a dose-related increase in absolute and relative liver and spleen weights. Microscopic changes in the spleen, including increased haematopoietic cell proliferation and pigmentation, occurred in all animals exposed to 12 and 24 ppm (United States National Toxicology Program, 1993).

More severe methaemoglobinaemia occurred in rats exposed to 4-chloronitrobenzene for 13 weeks as compared to those exposed to 2-chloronitrobenzene. Spleen weights were also increased in exposed rats. Lesions in the spleen, liver and kidney were similar to those described in the two-week study (United States National Toxicology Program, 1993).

Mice exposed to 4-chloronitrobenzene for 13 weeks showed microscopic changes in liver and spleen similar to those observed after two weeks of inhalation exposure. In addition, male and female mice showed increased haematopoiesis and haemosiderin deposition in bone marrow while, in female mice, squamous-cell hyperplasia of the forestomach epithelium was noted (United States National Toxicology Program, 1993).

4.3 Reproductive and developmental effects

4.3.1 *Humans*

No data were available to the Working Group.

4.3.2 *Experimental systems*

Groups of 24 Sprague-Dawley rats were administered 0, 5, 15 or 45 mg/kg bw per day 4-chloronitrobenzene by gavage on gestation days 6–19. The dams receiving the highest dose had reduced weight gain and increased spleen weights. There was no significant difference in the mean number of implantations or viable fetuses after administration of 5 and 15 mg/kg bw per day when compared with controls. The highest dose was associated with an increased number of resorptions per dam (5.6 ± 5.8, compared with 0.5 ± 0.7 in controls). The frequency of skeletal malformations was increased in the group receiving the highest dose (29 skeletal malformations in 10 litters) compared with controls (two malformations in two litters). The predominant skeletal malformation was angulated ribs occurring either alone or in association with misshapen forelimb bones. The frequency of external and soft-tissue malformations was similar between treated and untreated groups, and the types of these malformations were considered to be representative of sporadic isolated occurrences. Groups of 18 New Zealand rabbits were also administered 0, 5, 15 or 40 mg/kg bw per day by gavage on gestation days 7–19. However, eight of the 18 rabbits dosed with 40 mg/kg bw per day chloronitrobenzene died and therefore this treatment was abandoned. No effect of the lower doses on maternal or fetal toxicity or on teratogenesis was observed (Nair et al., 1985).

In a reproduction study of 4-chloronitrobenzene, 0.1, 0.7 or 5.0 mg/kg bw per day were administered by gavage in corn oil to groups of 15 male and 30 female Sprague-Dawley rats for about 14 weeks prior to mating, during mating and during gestation and lactation (Nair et al., 1989). F_1 parental rats were dosed similarly from prior to mating to throughout lactation. Other than a slight decrease in the pregnancy rate and male fertility index in the F_0 generation receiving 5 mg/kg bw per day, no reproductive effect was observed in the F_0 or the F_1 generation. [This study is described only in the form of an abstract.]

The United States National Toxicology Program (1993) studied the effects of inhalation of 2-chloronitrobenzene or 4-chloronitrobenzene for 13 weeks on sperm morphology and vaginal cytology in Fischer 344/N rats and B6C3F1 mice. Groups of 10 male and 10 female rats and 10 male and 10 female mice were exposed to 0, 4.5, 9 and 18 ppm [29, 58 and 116 mg/m^3] 2-chloronitrobenzene or 0, 6, 12 and 24 ppm [39, 77 and 155 mg/m^3] 4-chloronitrobenzene vapour through whole-body exposure for 6 h plus T_{90} per day on five days per week. (T_{90} signifies the time following the start of exposure for the vapour concentration to read 90% of the final stable concentration in the chamber. T_{90} was 20–25 min for 2-chloronitrobenzene and 15–18 min for 4-chloronitrobenzene.) In male rats exposed to the highest dose of 2-chloronitrobenzene, the left cauda epididymal weight, the number of spermatid heads per testes and the spermatid count were significantly lower than those of controls. No significant change occurred in exposed females. In mice, sperm motility was significantly decreased in all groups of exposed males, but again no significant change occurred in female mice. In male rats exposed to 4-chloronitrobenzene at the highest concentration, atrophy of the seminiferous tubules was observed, together with reductions in spermatid count, spermatozoa concentration and sperm motility. The oestrus cycle length was decreased in all groups of female rats exposed to the compound. In mice, no significant change occurred in males exposed to 4-chloronitrobenzene, but the oestrus cycle length of females exposed to the highest concentration was significantly increased.

Continuous breeding studies in CD-1 Swiss mice were performed to determine whether the relatively mild changes seen in these assays would be indicative of significant deficits in fertility in more exhaustive studies (United States National Toxicology Program, 1993). Groups of 20 breeding pairs received 40, 80 or 160 mg/kg bw per day 2-chloronitrobenzene or 62.5, 125 or 250 mg/kg bw per day 4-chloronitrobenzene in corn oil by gavage for seven days prior to cohousing and for 98 days of continuous breeding.

Forty breeding pairs received the corn oil vehicle only. These doses were up to approximately one-third to one-half of those that caused mortality in two-week range-finding studies. No noteworthy effect of 2-chloronitrobenzene on reproductive performance or outcome was observed. 4-Chloronitrobenzene produced significant and progressive deficits in fertility in the F_0 generation; the number of high-dose breeding pairs delivering litters declined by the second litter, and the difference was significant for the third and fourth litters. The average number of litters per pair decreased slightly with increasing dose. Weight gains of F_1 and F_2 pups were lower than those of the control pups.

4.4 Genetic and related effects

4.4.1 *Humans*

No data were available to the Working Group.

4.4.2 *Experimental systems* (see also Table 3 and Appendices 1 and 2)

2-Chloronitrobenzene gave negative results in the *Escherichia coli* SOS-chromotest. Mutagenic activity was reported for strain TA100 of *Salmonella typhimurium* when tested in the presence of induced hamster or rat-liver S9. Conflicting responses were obtained with TA98. 2-Chloronitrobenzene did not induce sex-linked recessive lethal mutations in germ cells of male *Drosophila melanogaster* when administered to adults either by feeding or by injection or to larvae by feeding. 2-Chloronitrobenzene injected intraperitoneally into Swiss CD-1 mice produced DNA single-strand breaks in the liver, kidney and brain. 2-Chloronitrobenzene induced sister chromatid exchange and chromosomal aberrations in Chinese hamster ovary CHO cells.

Studies with 3-chloronitrobenzene were negative in the *E. coli* SOS-chromotest and in *S. typhimurium* mutagenicity assays. Negative results were also obtained in Chinese hamster ovary cell tests for the induction of chromosomal aberrations and sister chromatid exchange.

4-Chloronitrobenzene gave negative results in the *E. coli* SOS-chromotest, but was mutagenic in *S. typhimurium* when tested in the presence and absence of induced liver S9. 4-Chloronitrobenzene did not induce sex-linked recessive lethal mutations in germ cells of male *Drosophila melanogaster* when administered to adults either by feeding or by injection or to larvae by feeding. 4-Chloronitrobenzene injected intraperitoneally into male Swiss CD-1 mice induced DNA single-strand breaks in liver, kidney and brain. In non-proliferating cultured rat hepatocytes, 4-chloronitrobenzene induced DNA single-strand breaks at 5 mM which were almost completely repaired within 24 h. At 50 mM only about one-half of the induced breaks were repaired within 48 h and most repair occurred during the second day (Cesarone *et al.*, 1984). In Chinese hamster ovary cells, 4-chloronitrobenzene induced sister chromatid exchange in the presence of S9 and chromosomal aberrations with and without S9, but the positive response for chromosomal aberrations occurred only at doses that were severely toxic.

Table 3. Genetic and related effects of chloronitrobenzenes

Test system	Result[a]		Dose[b] (LED/HID)	Reference
	Without exogenous metabolic system	With exogenous metabolic system		
2-Chloronitrobenzene				
PRB, Prophage induction, SOS repair test, DNA strand breaks, cross links or related damage	–	–	NR	von der Hude et al. (1988)
SA0, Salmonella typhimurium TA100, reverse mutation	–	–	NR	Gilbert et al. (1980)
SA0, Salmonella typhimurium TA100, reverse mutation	–	+	38	Haworth et al. (1983)
SA0, Salmonella typhimurium TA100, reverse mutation	–	0	50	Suzuki et al. (1983)
SA0, Salmonella typhimurium TA100, reverse mutation	–	0	630	Shimizu et al. (1983)
SA0, Salmonella typhimurium TA100, reverse mutation	–	+	38	US National Toxicology Program (1993)
SA3, Salmonella typhimurium TA1530, reverse mutation	–	–	NR	Gilbert et al. (1980)
SA5, Salmonella typhimurium TA1535, reverse mutation	–	–	NR	Gilbert et al. (1980)
SA5, Salmonella typhimurium TA1535, reverse mutation	–	–	128	Haworth et al. (1983)
SA5, Salmonella typhimurium TA1535, reverse mutation	–	0	630	Shimizu et al. (1983)
SA7, Salmonella typhimurium TA1557, reverse mutation	–	–	NR	Gilbert et al. (1980)
SA7, Salmonella typhimurium TA1557, reverse mutation	–	–	128	Haworth et al. (1983)
SA7, Salmonella typhimurium TA1557, reverse mutation	–	0	630	Shimizu et al. (1983)
SA8, Salmonella typhimurium TA1538, reverse mutation	–	–	NR	Gilbert et al. (1980)
SA8, Salmonella typhimurium TA1538, reverse mutation	+	0	80	Shimizu et al. (1983)
SA9, Salmonella typhimurium TA98, reverse mutation	–	–	NR	Gilbert et al. (1980)
SA9, Salmonella typhimurium TA98, reverse mutation	–	(+)	77	Haworth et al. (1983)
SA9, Salmonella typhimurium TA98, reverse mutation	–	–	50	Suzuki et al. (1983)
SA9, Salmonella typhimurium TA98, reverse mutation	+	0	80	Shimizu et al. (1983)
SA9, Salmonella typhimurium TA98, reverse mutation	–	(+)	50	US National Toxicology Program (1993)
SA9, Salmonella typhimurium TA98, reverse mutation	0	–	2.5	Suzuki et al. (1987)

Table 3 (contd)

Test system	Result[a]		Dose[b] (LED/HID)	Reference
	Without exogenous metabolic system	With exogenous metabolic system		
2-Chloronitrobenzene (contd)				
SAS, *Salmonella typhimurium* TA1532, TA1950, TA1975, TA1978 or G46, reverse mutation	–		NR	Gilbert et al. (1980)
SAS, *Salmonella typhimurium* TA98NR, reverse mutation	0	–	5	Suzuki et al. (1987)
SAS, *Salmonella typhimurium* TA98NR/1,8-DNP₆, reverse mutation	0	–	2.5	Suzuki et al. (1987)
DMX, *Drosophila melanogaster*, sex-linked recessive lethal mutation	–		125 (adult feed)	Zimmering et al. (1985)
DMX, *Drosophila melanogaster*, sex-linked recessive lethal mutation	–		10 000 (adult inj.)	Zimmering et al. (1985)
DMX, *Drosophila melanogaster*, sex-linked recessive lethal mutation	–		60 (larval feed)	Zimmering et al. (1989)
SIC, Sister chromatic exchange, Chinese hamster cells *in vitro*	(+)[d]	(+)[d]	50	US National Toxicology Program (1993)
CIC, Chromosomal aberrations, Chinese hamster cells *in vitro*	?	(+)[d]	465	US National Toxicology Program (1993)
DVA, DNA strand breaks, cross-links or related damage in Swiss mouse brain cells *in vivo*	+		60	Cesarone et al. (1980)
DVA, DNA strand breaks, cross-links or related damage in Swiss mouse liver and kidney cells *in vivo*	+		60	Cesarone et al. (1982)
3-Chloronitrobenzene				
PRB, Prophage induction, SOS repair test, DNA strand breaks, cross links or related damage	–	–	NR	von der Hude et al. (1988)
SA0, *Salmonella typhimurium* TA100, reverse mutation	–	–	128	Haworth et al. (1983)
SA0, *Salmonella typhimurium* TA100, reverse mutation	–	–	50	Suzuki et al. (1983)
SA0, *Salmonella typhimurium* TA100, reverse mutation	–	–	630	Shimizu et al. (1983)

286   IARC MONOGRAPHS VOLUME 65

Table 3 (contd)

Test system	Result[a]		Dose[b] (LED/HID)	Reference
	Without exogenous metabolic system	With exogenous metabolic system		
3-Chloronitrobenzene (contd)				
SA5, *Salmonella typhimurium* TA1535, reverse mutation	–	–	128	Haworth et al. (1983)
SA5, *Salmonella typhimurium* TA1535, reverse mutation	–	–	630	Shimizu et al. (1983)
SA7, *Salmonella typhimurium* TA1537, reverse mutation	–	–	128	Haworth et al. (1983)
SA7, *Salmonella typhimurium* TA1537, reverse mutation	–	–	630	Shimizu et al. (1983)
SA8, *Salmonella typhimurium* TA1538, reverse mutation	–	–	630	Shimizu et al. (1983)
SA9, *Salmonella typhimurium* TA98, reverse mutation	–	–	128	Haworth et al. (1983)
SA9, *Salmonella typhimurium* TA98, reverse mutation	–	–	50	Suzuki et al. (1983)
SA9, *Salmonella typhimurium* TA98, reverse mutation	–	–	630	Shimizu et al. (1983)
SIC, Sister chromatid exchange, Chinese hamster cells *in vitro*	–	–	160	Galloway et al. (1987)
CIC, Chromosomal aberrations, Chinese hamster cells *in vitro*	–	–	500	Galloway et al. (1987)
4-Chloronitrobenzene				
PRB, Prophage induction, SOS repair test, DNA strand breaks, cross links or related damage	–	–	NR	von der Hude et al. (1988)
SA0, *Salmonella typhimurium* TA100, reverse mutation	–	–	NR	Gilbert et al. (1980)
SA0, *Salmonella typhimurium* TA100, reverse mutation	–	+	128	Haworth et al. (1983)
SA0, *Salmonella typhimurium* TA100, reverse mutation	–	–	50	Suzuki et al. (1983)
SA0, *Salmonella typhimurium* TA100, reverse mutation	(+)	0	630	Shimizu et al. (1983)
SA0, *Salmonella typhimurium* TA100, reverse mutation	–	+	192	US National Toxicology Program (1993)
SA3, *Salmonella typhimurium* TA1530, reverse mutation	–	–	NR	Gilbert et al. (1980)
SA5, *Salmonella typhimurium* TA1535, reverse mutation	–	–	NR	Gilbert et al. (1980)
SA5, *Salmonella typhimurium* TA1535, reverse mutation	–	(+)	256	Haworth et al. (1983)
SA5, *Salmonella typhimurium* TA1535, reverse mutation	+	0	315	Shimizu et al. (1983)

Table 3 (contd)

Test system	Result[a]		Dose[b] (LED/HID)	Reference
	Without exogenous metabolic system	With exogenous metabolic system		
4-Chloronitrobenzene (contd)				
SA5, *Salmonella typhimurium* TA1535, reverse mutation	(+)	+	256	US National Toxicology Program (1993)
SA7, *Salmonella typhimurium* TA1537, reverse mutation	–	–	NR	Gilbert *et al.* (1980)
SA7, *Salmonella typhimurium* TA1537, reverse mutation	–	–	384	Haworth *et al.* (1983)
SA7, *Salmonella typhimurium* TA1537, reverse mutation	–	0	630	Shimizu *et al.* (1983)
SA8, *Salmonella typhimurium* TA1538, reverse mutation	–	0	630	Shimizu *et al.* (1983)
SA9, *Salmonella typhimurium* TA98, reverse mutation	–	–	NR	Gilbert *et al.* (1980)
SA9, *Salmonella typhimurium* TA98, reverse mutation	–	–	384	Haworth *et al.* (1983)
SA9, *Salmonella typhimurium* TA98, reverse mutation	–	–[c]	50	Suzuki *et al.* (1983)
SA9, *Salmonella typhimurium* TA98, reverse mutation	–	0	630	Shimizu *et al.* (1983)
SA9, *Salmonella typhimurium* TA98, reverse mutation	0	–[c]	50	Suzuki *et al.* (1987)
SA9, *Salmonella typhimurium* TA98, reverse mutation	–	–	385	US National Toxicology Program (1993)
SAS, *Salmonella typhimurium* TA1532, TA1950, TA1975, TA1978 or G46, reverse mutation	–	–	NR	Gilbert *et al.* (1980)
SAS, *Salmonella typhimurium* TA98NR, reverse mutation	0	–[c]	50	Suzuki *et al.* (1987)
SAS, *Salmonella typhimurium* TA98NR/1,8-DNP₆, reverse mutation	0	–[c]	50	Suzuki *et al.* (1987)
DMX, *Drosophila melanogaster*, sex-linked recessive lethal mutation	–		100 ppm (adult feed)	Zimmering *et al.* (1985)
DMX, *Drosophila melanogaster*, sex-linked recessive lethal mutation	–		100 (adult inj.)	Zimmering *et al.* (1985)
DMX, *Drosophila melanogaster*, sex-linked recessive lethal mutation	–		80 (larval feed)	Zimmering *et al.* (1989)

Table 3 (contd)

Test system	Result[a]		Dose[b] (LED/HID)	Reference
	Without exogenous metabolic system	With exogenous metabolic system		
4-Chloronitrobenzene (contd)				
DIA, DNA strand breaks, cross-links or related damage, animal cells *in vitro*	+	0	5	Cesarone *et al.* (1984)
RIA, DNA repair exclusive of unscheduled DNA synthesis, animal cells *in vitro*	+	0	5	Cesarone *et al.* (1984)
SIC, Sister chromatid exchange, Chinese hamster cells *in vitro*	−	+	250	Galloway *et al.* (1987)
CIC, Chromosomal aberrations, Chinese hamster cells *in vitro*	(+)	+	600	Galloway *et al.* (1987)
CIC, Chromosomal aberrations, Chinese hamster cells *in vitro*	(+)	+	600	US National Toxicology Program (1993)
DVA, DNA strand breaks, cross-links or related damage, animal cells *in vivo*	+		30	Cesarone *et al.* (1983)

[a] +, positive; (+), weak positive; −, negative; 0 not tested; ?, inconclusive
[b] LED, lowest effective dose; HID, highest ineffective dose. In-vitro tests, μg/ml; in-vivo tests, mg/kg bw; NR, dose not reported
[c] Positive in the presence of 200 μg/plate norharman (CAS 244-63-3)
[d] Weak positive in one laboratory, negative in another

5. Summary of Data Reported and Evaluation

5.1 Exposure data

2-, 3- and 4-Chloronitrobenzenes are produced as a mixture by nitration of chlorobenzene. After separation, the three isomers are used as important chemical intermediates in the production of colourants, pharmaceuticals and a variety of other products. Human exposure to chloronitrobenzenes may occur during the production and use of these intermediates.

5.2 Human carcinogenicity data

No data on the carcinogenicity of 2-, 3- or 4-chloronitrobenzene to humans were available to the Working Group.

5.3 Animal carcinogenicity data

2-Chloronitrobenzene was tested for carcinogenicity by oral administration in the diet in one study in mice and in one study in rats. The studies were inadequate for an evaluation.

4-Chloronitrobenzene was tested for carcinogenicity by oral administration in the diet in one study in mice and in one study in rats. Although the study in mice reported an increased incidence of vascular tumours in exposed males and females, neither study was considered adequate for an evaluation.

3-Chloronitrobenzene has not been tested for carcinogenicity in experimental animals.

5.4 Other relevant data

4-Chloronitrobenzene is absorbed through inhalation and/or via the skin upon human exposure after which urinary metabolites of 4-chloronitrobenzene appear, which are the results of N-acetylation, nitro-group reduction and — to a lesser extent — ring-hydroxylation. Metabolism is slow, with elimination of metabolites occurring over many days. Considerable interindividual variation occurs in this metabolism.

The urinary metabolites of 4-chloronitrobenzene are qualitatively similar in humans and rats.

No data concerning the absorption, distribution, metabolism and excretion or toxic effects of 2- or 3-chloronitrobenzene in humans were available to the Working Group.

The disposition of 2-chloronitrobenzene in rats is similar to that of 4-chloronitrobenzene.

In humans, exposure to 4-chloronitrobenzene is associated with such symptoms as headache, palpitation, dizziness, nausea, vomiting and poor appetite. Cyanosis, methaemoglobinaemia and anaemia also occur. Methaemoglobinaemia and anaemia occur in rats exposed to 4-chloronitrobenzene, 3- chloronitrobenzene or 2- chloronitrobenzene.

In a single study in rats, a maternally toxic dose of 4-chloronitrobenzene increased the resorption rate and the frequency of skeletal malformations. In female rats and mice, inhalation exposure to 4-chloronitrobenzene increased the oestrus cycle length. In rats, but not mice, inhalation exposure to the compound decreased spermatogenesis with atrophy of the seminiferous tubules. In a continuous breeding study, a progressive decrease in fertility was noted in mice receiving 4-chloronitrobenzene.

In rats and mice exposed to 2-chloronitrobenzene by inhalation, decreased spermatogenesis was observed. No significant change was observed in exposed females. In a continuous breeding study, fertility was not affected in mice receiving 2-chloronitrobenzene.

2-Chloronitrobenzene induced reverse mutations but not primary DNA damage in bacteria. It was not mutagenic to insects. In mammalian cells *in vitro*, it induced sister chromatid exchange and chromosomal aberrations. Intraperitoneal injection into mice resulted in DNA damage in the liver, kidney and brain.

3-Chloronitrobenzene gave negative results in bacterial mutagenicity assays and in cultured mammalian cell chromosomal assays.

4-Chloronitrobenzene induced reverse mutations but not primary DNA damage in bacteria. It was not mutagenic to insects. At toxic doses, it induced chromosomal aberrations, sister chromatid exchange and repairable DNA breaks in cultured mammalian cells. Intraperitoneal injection into mice induced DNA damage in the liver, kidney and brain.

5.5 Evaluation[1]

There is *inadequate evidence* in humans for the carcinogenicity of chloronitrobenzenes.

There is *inadequate evidence* in experimental animals for the carcinogenicity of chloronitrobenzenes.

Overall evaluation

Chloronitrobenzenes are *not classifiable as to their carcinogenicity to humans (Group 3)*.

6. References

Aldrich Chemical Co. (1994) *Aldrich Catalog/Handbook of Fine Chemicals 1994–1995*, Milwaukee, WI, p. 340

---

[1] For definition of the italicized terms, see Preamble, pp. 24–27.

American Conference of Governmental Industrial Hygienists (1995) *1995-1996 Threshold Limit Values for Chemical Substances and Physical Agents and Biological Exposure Indices*, Cincinnati, OH, p. 28

Arbeidsinspectie [Labour Inspection] (1994) *De Nationale MAC-Lijst 1994* [National MAC list 1994], The Hague, p. 35

Booth, G. (1991) Nitro compounds, aromatic. In: Elvers, B., Hawkins, S. & Schulz, G., eds, *Ullmann's Encyclopedia of Industrial Chemistry*, 5th rev. Ed., Vol. A17, New York, NY, VCH Publishers, pp. 411-455

Braunstein, L., Hochmüller, K. & Spengler, K. (1989) Analysis of water-carried substances amenable to gas chromatography by coupling the gas chromatograph to a mass-selective detector. *Vom Wasser*, **73**, 167-189 (in German)

Bray, H.G., James, S.P. & Thorpe, W.V. (1956) The metabolism of the monochloronitrobenzenes in the rabbit. *Biochem. J.*, **64**, 38-44

BUA (1992a) *GDCh-Advisory Committee on Existing Chemicals of Environmental Relevance (BUA): o-Chloronitrobenzene (BUA Report 2) (October 1985)*, Weinheim, VCH Verlagsgesellschaft mbH

BUA (1992b) *GDCh-Advisory Committee on Existing Chemicals of Environmental Relevance (BUA): m-Chloronitrobenzene, p-Chloronitrobenzene (BUA Report 11) (February 1988)*, Weinheim, VCH Verlagsgesellschaft mbH

Budavari, S., ed. (1989) *The Merck Index*, 11th Ed., Rahway, NJ, Merck & Co., p. 332

Cesarone, C.F., Bolognesi, C. & Santi, L. (1980) Function of the structure and of the chemico-physico property of the chemical agent on the induction of DNA alterations: nitrochlorobenzenes. *Boll. Soc. Ital. Biol. sper.*, **54**, 1680-1686 (in Italian)

Cesarone, C.F., Bolognesi, C. & Santi, L. (1982) Evaluation of damage to DNA after in vivo exposure to different classes of chemicals. *Arch. Toxicol.*, **Suppl. 5**, 355-359

Cesarone, C.F., Bolognesi, C. & Santi, L. (1983) DNA damage induced *in vivo* in various tissues by nitrobenzene derivatives. *Mutat. Res.*, **116**, 239-246

Cesarone, C.F., Fugassa, E., Gallo, G., Voci, A. & Orunesu, M. (1984) Influence of the culture time on DNA damage and repair in isolated rat hepatocytes exposed to nitrochlorobenzene derivatives. *Mutat. Res.*, **131**, 215-222

Chemical Information Services (1994) *Directory of World Chemical Producers 1995/96*, Oceanside, NY, p. 180

D'Addario, A.P. & Jagannath, D.R. (1981) Structure and resonance effects on the observed mutagenic response of mononitro halobenzenes (Abstract P58). *Environ. Mutag.*, **3**, 325

Dangwal, S.K. & Jethani, B.M. (1980) A simple method of determination of nitrobenzene and nitrochlorobenzene in air and urine. *Am. ind. Hyg. Assoc. J.*, **41**, 847-850

Deutsche Forschungsgemeinschaft (1995) *MAK and BAT Values 1995* (Report No. 31), Weinheim, VCH Verlagsgesellschaft, p. 35

Duguet, J.P., Anselme, C., Mazounie, P. & Mallevialle, J. (1988) Application of the ozone-hydrogen peroxide combination for the removal of toxic compounds from a groundwater. In: Angeletti, G. & Bjørseth, A., eds, *Organic Micropollutants in the Aquatic Environment, Proceedings of the Fifth European Symposium, Rome, October 20-22, 1987*, Dordrecht, Kluwer Academic Publishers, pp. 299-309

Dunlap, K.L. (1981) Nitrobenzene and nitrotoluenes. In: Mark, H.F., Othmer, D.F., Overberger, C.G., Seaborg, G.T. & Grayson, N., eds, *Kirk-Othmer Encyclopedia of Chemical Technology*, 3rd Ed., Vol. 15, New York, NY, John Wiley & Sons, pp. 916–932

DuPont Chemicals (1993) *Material Safety Data Sheet: o-Nitrochlorobenzene*, Wilmington, DE

Eller, P.M., ed. (1994) Nitrobenzenes — Method 2005. In: *NIOSH Manual of Analytical Methods*, 4th Ed., Vol 2 (E-N), (DHHS (NIOSH) Publ. No. 94-113), Washington DC, United States Government Printing Office

Farris, R.E. (1978) Aminophenols. In: Mark, H.F., Othmer, D.F., Overberger, C.G., Seaborg, G.T. & Grayson, N., eds, *Kirk-Othmer Encyclopedia of Chemical Technology*, 3rd Ed., Vol. 2, New York, NY, John Wiley & Sons, pp. 422–440

Feltes, J., Levsen, K., Volmer, D. & Spiekermann, M. (1990) Gas chromatographic and mass spectrometric determination of nitroaromatics in water. *J. Chromatogr.*, **518**, 21–40

Galloway, S.M., Armstrong, M.J., Reuben, C., Colman, S., Brown, B., Cannon, C., Bloom, A.D., Nakamura, F., Ahmed, M., Duk, S., Rimpo, J., Margolin, B.H., Resnick, M.A., Anderson, B. & Zeiger, E. (1987) Chromosome aberrations and sister chromatid exchanges in Chinese hamster ovary cells: evaluation of 108 chemicals. *Environ. mol. Mutag.*, **10** (Suppl. 10), 1–175

Gilbert, P., Saint-Ruf, G., Poncelet, F. & Mercier, M. (1980) Genetic effects of chlorinated anilines and azobenzenes on *Salmonella typhimurium*. *Arch. environm. Contam. Toxicol.*, **9**, 533–541

Greco, T.G. & Grob, R.L. (1990) Determination of aromatic organic compounds following liquid extraction from soil, solid-phase extraction concentration, and analysis by capillary gas chromatography. *J. environ. Sci. Health*, **A25**, 185–202

Grob, R.L. & Cao, K.B. (1990) High performance liquid chromatographic study of the recovery of aromatic amine and nitro compounds from soil. *J. environ. Sci. Health*, **A25**, 117–136

Hansch, C., Leo, A. & Hoekman, D.H. (1995) *Exploring QSAR*, Washington DC, American Chemical Society, p. 17

Hathaway, G.J., Proctor, N.H., Hughes, J.P. & Fischman, M.L., eds (1991) *Proctor and Hughes' Chemical Hazards of the Workplace*, 3rd Ed., New York, Van Nostrand Reinhold, pp. 428–431

Haworth, S., Lawlor, T., Mortelmans, K., Speck, W. & Zeiger, E. (1983) *Salmonella* mutagenicity test results for 250 chemicals. *Environ. Mutag.*, **Suppl. 1**, 3–142

Health and Safety Executive (1995) *Occupational Exposure Limits 1995* (Guidance Note EH 40/95), Sudbury, Suffolk, HSE Books, p. 30

Howard, P.H. (1989) *Handbook of Environmental Fate and Exposure Data for Organic Chemicals*, Vol. 1, Chelsea, MI, Lewis Publishers, pp. 146–160

Howard, P.H., Santodonato, J., Saxena, J., Malling, J. & Greninger, D. (1976) *Investigation of Selected Environmental Contaminants: Nitroaromatics* (US EPA Report No. EPA-560/2-76-010; US NTIS PB-275078), Washington DC, United States Environmental Protection Agency, Office of Toxic Substances

von der Hude, W., Behm, C., Gürtler, R. & Basler, A. (1988) Evaluation of the SOS chromotest. *Mutat. Res.*, **203**, 81–94

IARC (1980) *IARC Monographs on the Evaluation of the Carcinogenic Risk of Chemicals to Humans*, Vol. 24, *Some Pharmaceutical Drugs*, Lyon, pp. 59–36

IARC (1982a) *IARC Monographs on the Evaluation of the Carcinogenic Risk of Chemicals to Humans*, Vol. 29, *Some Industrial Chemicals and Dyestuffs*, Lyon, pp. 239–256

IARC (1982b) *IARC Monographs on the Evaluation of the Carcinogenic Risk of Chemicals to Humans*, Vol. 27, *Some Aromatic Amines, Anthraquinones and Nitroso Compounds, and Inorganic Fluorides Used in Drinking-water and Dental Preparations*, Lyon, pp. 63–80

IARC (1982c) *IARC Monographs on the Evaluation of the Carcinogenic Risk of Chemicals to Humans*, Vol. 28, *The Rubber Industry*, Lyon

IARC (1983) *IARC Monographs on the Evaluation of the Carcinogenic Risk of Chemicals to Humans*, Vol. 30, *Miscellaneous Pesticides*, Lyon, pp. 271–282

IARC (1987a) *IARC Monographs on the Evaluation of Carcinogenic Risks to Humans*, Suppl. 7, *Overall Evaluations of Carcinogenicity: An Updating of* IARC Monographs *Volumes 1 to 42*, Lyon, pp. 193–194

IARC (1987b) *IARC Monographs on the Evaluation of Carcinogenic Risks to Humans*, Suppl. 7, *Overall Evaluations of Carcinogenicity: An Updating of* IARC Monographs *Volumes 1 to 42*, Lyon, p. 57

IARC (1987c) *IARC Monographs on the Evaluation of Carcinogenic Risks to Humans*, Suppl. 7, *Overall Evaluations of Carcinogenicity: An Updating of* IARC Monographs *Volumes 1 to 42*, Lyon, pp. 185–186

IARC (1987d) *IARC Monographs on the Evaluation of Carcinogenic Risks to Humans*, Suppl. 7, *Overall Evaluations of Carcinogenicity: An Updating of* IARC Monographs *Volumes 1 to 42*, Lyon, p. 67

IARC (1987e) *IARC Monographs on the Evaluation of Carcinogenic Risks to Humans*, Suppl. 7, *Overall Evaluations of Carcinogenicity: An Updating of* IARC Monographs *Volumes 1 to 42*, Lyon, pp. 332–334

IARC (1993) *IARC Monographs on the Evaluation of Carcinogenic Risks of Chemicals to Humans*, Vol. 57, *Occupational Exposures of Hairdressers and Barbers and Personal Use of Hair Colourants; Some Hair Dyes, Cosmetic Colourants, Industrial Dyestuffs and Aromatic Amines*, Lyon, pp. 305–321

Koniecki, W.B. & Linch, A.L. (1958) Determination of aromatic nitro compounds. *Anal. Chem.*, **30**, 1134–1137

Lewalter, J. & Ellrich, D. (1991) Nitroaromatic compounds (nitrobenzene, *p*-nitrotoluene, *p*-nitrochlorobenzene, 2,6-dinitrotoluene, *o*-dinitrobenzene, 1-nitronaphthalene, 2-nitronaphthalene, 4-nitrobiphenyl). In: Angerer, J. & Schaller, K.H., eds, *Analyses of Hazardous Substances in Biological Materials*, Vol. 3, New York, VCH Publishers, pp. 207–229

Lewis, R.J., Sr (1993) *Hawley's Condensed Chemical Dictionary*, 12th Ed., New York, Van Nostrand Reinhold Co., p. 269

Lide, D.R., ed. (1993) *CRC Handbook of Chemistry and Physics*, 74th Ed., Boca Raton, FL, CRC Press, p. 3–81

Lucas, S.V. (1984) *GC/MS Analysis of Organics in Drinking Water Concentrates and Advanced Waste Treatment Concentrates. 2: Computer-printed Tabulations of Compound Identification Results from Large Volume Concentrates* (US EPA Report No. EPA-600/1–84-020b; US NTIS PB85-128239), Research Triangle Park, NC, United States Environmental Protection Agency van de Meent, D., den Hollander, H.A., Pool, W.G., Vrendenbregt, M.J., van Oers, H.A.M., de Greef, E. & Luijten, J.A. (1986) Organic micropollutants in Dutch coastal waters. *Water Sci. Technol.*, **18**, 73–81

Middleton, F.M. & Lichtenberg, J.J. (1960) Measurements of organic contaminants in the nation's rivers. A beginning survey, using the carbon filter technique, shows variability of contaminant concentrations and reveals important specific contaminants in five major rivers. *Ind. Eng. Chem.*, **52**, 73A–76A

Mitchell, J., Jr & Deveraux, H.D. (1978) Determination of traces of organic compounds in the atmosphere: role of detectors in gas chromatography. *Anal. chim. Acta*, **100**, 45–52

Monsanto Co. (1993a) *Material Safety Data Sheet: ortho-Nitrochlorobenzene*, St Louis, MO

Monsanto Co. (1993b) *Material Safety Data Sheet: para-Nitrochlorobenzene*, St Louis, MO

Nair, R.S., Johannsen, F.R. & Schroeder, R.E. (1985) Evaluation of teratogenic potential of *para*-nitroaniline and *para*-nitrochlorobenzene in rats and rabbits. In: Rickert, D.B., ed., *Toxicity of Nitroaromatic Compounds*, Washington DC, Hemisphere, pp. 61–85

Nair, R.S., Johannsen, F.R., Levinskas, G.J. & Terrill, J.B. (1986a) Assessment of toxicity of *o*-nitrochlorobenzene in rats following a 4-week inhalation exposure. *Fundam. appl. Toxicol.*, **7**, 609–614

Nair, R.S., Johannsen, F.R., Levinskas, G.J. & Terrill, J.B. (1986b) Subchronic inhalation toxicity of *p*-nitroaniline and *p*-nitrochlorobenzene in rats. *Fundam. appl. Toxicol.*, **6**, 618–627

Nair, R.S., Johannsen, F.R., Schroeder, R.E. & Auletta, C.S. (1989) Evaluation of chronic and reproductive effects of *p*-nitrochlorobenzene (PNCB) in rats (Abstract no. 846). *Toxicologist*, **9**, 212

Nielen, M.W.F., Brinkman, U.A.T. & Frei, R.W. (1985) Industrial wastewater analysis by liquid chromatography with precolumn technology and diode-array detection. *Anal. Chem.*, **57**, 806–810

Nishida, N., Nakamura, I., Kudo, Y. & Kagami, M. (1982) Effects of aromatic nitro and amino compounds on the osmotic fragility of red cells. *Jpn. J. ind. Health*, **24**, 172–180

Ong, C.P., Chin, K.P., Lee, H.K. & Li, S.F.Y. (1992) Analysis of nitroaromatics in aqueous samples by capillary supercritical fluid chromatography. *Int. J. environ. Stud.*, **41**, 17–25

Patil, S.F. & Lonkar, S.T. (1994) Evaluation of Tenax TA for the determination of chlorobenzene and chloronitrobenzenes in air using capillary gas chromatography and thermal desorption. *J. Chromatogr.*, **684**, 133–142

Piet, G.J. & Morra, C.F. (1983) *Artificial Groundwater Recharge* (Water Resources Eng. Series), Pitman Publ., pp. 31–42

Rickert, D.E. & Held, S.D. (1990) Metabolism of chloronitrobenzenes by isolated rat hepatocytes. *Drug Metab. Dispos.*, **18**, 5–9

Sadtler Research Laboratories (1980) *Sadtler Standard Spectra 1980. Cumulative Index*, Philadelphia, PA, p. 87

Shimizu, M., Yasui, Y. & Matsumoto, N. (1983) Structural specificity of aromatic compounds with special reference to mutagenic activity in *Salmonella typhimurium* — A series of chloro- or fluoro-nitrobenzene derivatives. *Mutat. Res.*, **116**, 217–238

Steinwandter, H. (1987) Research in environmental pollution. II. Determination of polychlorinated nitrobenzenes (PCNB's) in Main river fish. *Fresnius' Z. anal. Chem.*, **326**, 139–141

Suzuki, J., Koyama, T. & Suzuki, S. (1983) Mutagenicities of mono-nitrobenzene derivatives in the presence of norharman. *Mutat. Res.*, **120**, 105–110

Suzuki, J., Takahashi, N., Kobayashi, Y., Miyamae, R., Ohsawa, M. & Suzuki, S. (1987) Dependence of *Salmonella typhimurium* enzymes of mutagenicities of nitrobenzene and its derivatives in the presence of rat-liver S9 and norharman. *Mutat. Res.*, **178**, 187–193

Swaminathan, K., Kondawar, V.K., Chakrabarti, T. & Subrahmanyam, P.V.R. (1987) Identification and quantification of organics in nitro aromatic manufacturing wastewaters. *Indian J. environ. Health*, **29**, 32–38

Trova, C., Cossa, G. & Gandolfo, G. (1991) Behavior and fate of chloronitrobenzenes in a fluvial environment. *Bull. environ. Contam. Toxicol.*, **47**, 580–585

United Nations Environment Programme (1995) *International Register of Potentially Toxic Chemicals, Legal File, Chloronitrobenzenes*, Geneva

United States National Institute for Occupational Safety and Health (1994a) *Pocket Guide to Chemical Hazards* (DHHS (NIOSH) Publ. No. 94-116), Cincinnati, OH, pp. 226–227, 342

United States National Institute for Occupational Safety and Health (1994b) *RTECs Chem.-Bank*, Cincinnati, OH

United States National Institute for Occupational Safety and Health (1995) *National Occupational Exposure Survey (1981–1983)*, Cincinnati, OH

United States National Toxicology Program (1993) *NTP Technical Report on Toxicity Studies of 2-Chloronitrobenzene and 4-Chloronitrobenzene (CAS Nos. 88-73-3 and 100-00-5) Administered by Inhalation to F344/N Rats and B6C3F1 Mice* (NTP Toxicity Report Series No. 33; NIH Publication 93-3382), Research Triangle Park, NC

United States Occupational Safety and Health Administration (1994) Air contaminants. *US Code fed. Regul.*, **Title 29**, Part 1910.1000, p. 14

Vernot, E.H., MacEwen, J.D., Haun, C.C. & Kinkead, E.R. (1977) Acute toxicity and skin corrosion data for some organic and inorganic compounds and aqueous solutions. *Toxicol. appl. Pharmacol.*, **42**, 417–423

Watanabe, T., Ishihara, N. & Ikeda, M. (1976) Toxicity of and biological monitoring for 1,3-diamino-2,4,6-trinitrobenzene and other nitro-amino derivatives of benzene and chlorobenzene. *Int. Arch. occup. environ. Health*, **37**, 157–168

Weisburger, E.K., Russfield, A.B., Homburger, F., Weisburger, J.H., Boger, E., Van Dongen, C.G. & Chu, K.C. (1978) Testing of twenty-one environmental aromatic amines or derivatives for long-term toxicity or carcinogenicity. *J. environ. Pathol. Toxicol.*, **2**, 325–356

Yoshida, T. (1991) Determination of *p*-chloronitrobenzene in plasma by reversed-phase high-performance liquid chromatography. *J. Chromatogr.*, **570**, 321–328

Yoshida, T. (1993a) Determination of *p*-chloronitrobenzene and its metabolites in urine by reversed-phase high-performance liquid chromatography. *J. Chromatogr.*, **613**, 79–88

Yoshida, T. (1993b) Detection of *p*-chloro-oxalinic acid in urine of rats and human subjects exposed to *p*-chloronitrobenzene. *Chem. Express*, **8**, 17–20

Yoshida, M., Sunaga, M. & Hara, I. (1988) Urinary diazo-positive metabolites levels of workers handling *p*-nitrochlorobenzene in a dye producing factory. *Ind. Health*, **26**, 87–91

Yoshida, M., Sunaga, M., Hara, I., Katsumata, M. & Minami, M. (1989a) Elevation of urinary N-acetyl-β-D-glucosaminidase· and β-galactosidase activities in workers with long-term exposure to aromatic nitro-amino compounds. *Bull. environ. Contam. Toxicol.*, **43**, 1–8

Yoshida, M., Yoshikawa, H., Goto, H. & Hara, I. (1989b) Evaluation of the nephrotoxicity of aromatic nitro-amino compounds by urinary enzyme activities. *J. toxicol. Sci.*, **14**, 257–268

Yoshida, T., Andoh, K. & Tabuchi, T. (1991) Identification of urinary metabolites in rats treated with p-chloronitrobenzene. *Arch. Toxicol.*, **65**, 52–58

Yoshida, T., Tabuchi, T. & Andoh, K. (1992) Identification of urinary metabolites of human subjects acutely poisoned by *p*-chloronitrobenzene. *Xenobiotica*, **22**, 1459–1470

Yoshida, T., Tabuchi, T. & Andoh, K. (1993) Pharmacokinetic study of *p*-chloronitrobenzene in humans suffering from acute poisoning. *Drug Metab. Dispos.*, **21**, 1142–1146

Yurawecz, M.P. & Puma, B.J. (1983) Identification of chlorinated nitrobenzene residues in Mississippi River fish. *J. Assoc. off. anal. Chem.*, **66**, 1345–1352

van Zoest, R. & van Eck, G.T.M. (1991) Occurrence and behaviour of several groups of organic micropollutants in the Scheldt estuary. *Sci. total Environ.*, **103**, 57–71

Zimmering, S., Mason, J.M., Valencia, R. & Woodruff, R.C. (1985) Chemical mutagenesis testing in *Drosophila*. II. Results of 20 coded compounds tested for the National Toxicology Program. *Environ. Mutag.*, **7**, 87–100

Zimmering, S., Mason, J.M. & Valencia, R. (1989) Chemical mutagenesis testing in *Drosophila*. VII. Results of 22 coded compounds tested in larval feeding experiments. *Environ. mol. Mutag.*, **14**, 245–251

3,7-DINITROFLUORANTHENE AND 3,9-DINITROFLUORANTHENE

3,7- and 3,9-Dinitrofluoranthenes were considered by a previous Working Group in June 1988 (IARC, 1989). New data have since become available, and these are included in the present monograph and have been taken into consideration in the evaluation.

1. Exposure Data

1.1 Chemical and physical data

1.1.1 *Nomenclature*

3,7-Dinitrofluoranthene

Chem. Abstr. Serv. Reg. No.: 105735-71-5
Chem. Abstr. Name: 3,7-Dinitrofluoranthene
IUPAC Systematic Name: 3,7-Dinitrofluoranthene

3,9-Dinitrofluoranthene

Chem. Abstr. Serv. Reg. No.: 22506-53-2
Chem. Abstr. Name: 3,9-Dinitrofluoranthene
IUPAC Systematic Name: 3,9-Dinitrofluoranthene

1.1.2 *Structural and molecular formulae and relative molecular mass*

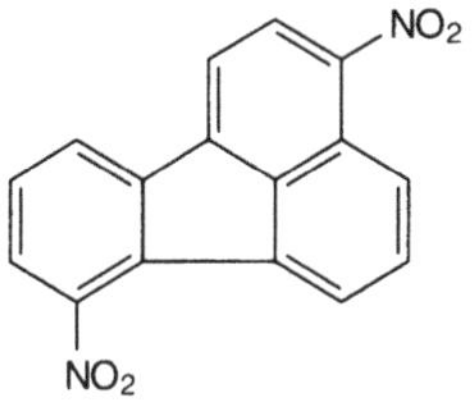

3,7-Dinitrofluoranthene

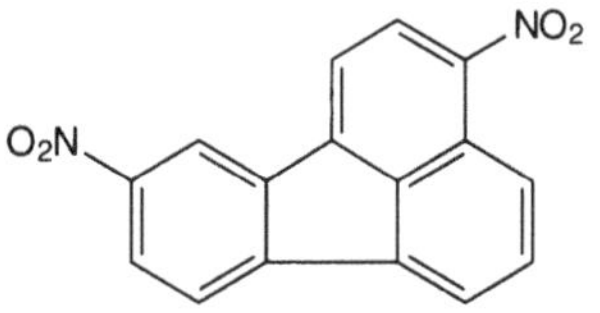

3,9-Dinitrofluoranthene

$C_{16}H_8N_2O_4$ Relative molecular mass: 292.3

1.1.3 *Chemical and physical properties of the pure substance*

From Nakagawa *et al.* (1987), unless otherwise specified

3,7-Dinitrofluoranthene

(a) *Description*: Yellow needles
(b) *Melting-point*: 203–204 °C
(c) *Spectroscopy data*: Nuclear magnetic resonance, ultraviolet and mass spectral data have been reported (Ramdahl *et al.*, 1988)
(d) *Octanol/water partition coefficient (P)*: log P, 4.44
(e) *Conversion factor*: mg/m^3 = 11.96 × ppm[1]

3,9-Dinitrofluoranthene

(a) *Description*: Yellow needles
(b) *Melting-point*: 222–224 °C
(c) *Spectroscopy data*: Nuclear magnetic resonance, ultraviolet and mass spectral data have been reported (Ramdahl *et al.*, 1988)
(d) *Octanol/water partition coefficient (P)*: log P, 4.44
(e) *Conversion factor*: mg/m^3 = 11.96 × ppm[1]

1.1.4 *Technical products and impurities*

No data were available to the Working Group

1.1.5 *Analysis*

Tokiwa *et al.* (1990) reported a method to separate and identify dinitrofluoranthenes in airborne particulates. The particulate matter was collected on a silica fibre filter and extracted with dichloromethane. The crude extracts were applied to a column filled with silica gel and were eluted step by step with hexane, hexane : benzene (1 : 1, v/v), benzene, benzene : methanol (1 : 1, v/v) and methanol. The components were fractionated and identified by high-performance liquid chromatography and gas chromatography with mass spectrometry.

1.2 Production and use

No evidence was found that either 3,7- or 3,9-dinitrofluoranthene has been produced in commercial quantities or used for anything other than laboratory applications. 3,7- or 3,9-Dinitrofluoranthene can be synthesized by nitration of fluoranthene or 3-nitrofluoranthene in the presence of fuming nitric acid, with subsequent fractionation and purification by recrystallization (Nakagawa *et al.*, 1987; Horikawa *et al.*, 1991; Matsuoka *et al.*, 1993).

---

[1] Calculated from: mg/m^3 = (relative molecular mass/24.45) × ppm, assuming temperature (25 °C) and pressure (101 kPa)

1.3 Occurrence

1.3.1 *Natural occurrence*

Neither 3,7- nor 3,9-dinitrofluoranthene is known to occur as a natural product.

1.3.2 *Occupational exposure*

No data were available to the Working Group

1.3.3 *Environmental occurrence*

3,7- and 3,9-Dinitrofluoranthenes were detected at a concentration of 0.028 mg/kg and 0.013 mg/kg, respectively, in particulates emitted from a diesel engine (Tokiwa *et al.*, 1986) and dinitrofluoranthenes have also been found in the incomplete combustion products of liquefied petroleum gas (Horikawa *et al.*, 1987).

In Sapporo, Hokkaido, Japan, in 1989, 3,7- and 3,9-dinitrofluoranthenes were detected in airborne particulates at a level of 0.01 mg/kg of particulate. 3,7-Dinitrofluoranthene has also been found in particulate emissions from a kerosene heater at 0.14 mg/kg of particulate (Tokiwa *et al.*, 1990, 1993).

1.4 Regulations and guidelines

No data were available to the Working Group.

2. Studies of Cancer in Humans

No data were available to the Working Group.

3. Studies of Cancer in Experimental Animals

3.1 Subcutaneous injection

Rat: Two groups of 21 and 11 male Fischer 344/DuCrj rats, six weeks of age, received twice-weekly subcutaneous injections of 0.05 mg 3,7-dinitrofluoranthene (> 99% pure) or 0.05 mg 3,9-dinitrofluoranthene (> 99% pure), respectively, in dimethyl sulfoxide (DMSO) for 10 weeks (total dose, 1 mg/rat). A vehicle-control group of 21 rats received injections of DMSO only. The rats were observed for 50 weeks, after which time surviving animals were killed. The mean survival time of 3,9-dinitrofluoranthene-treated rats was shorter than that of 3,7-dinitrofluoranthene-treated animals, both of which were reduced in comparison to controls. All organs were examined grossly and histopathological examination was carried out on all tumours and major organs. All rats treated with 3,7-dinitrofluoranthene and 10/11 rats treated with 3,9-dinitrofluoranthene developed subcutaneous tumours. Of the tumours that developed at the site of injection,

20/21 that were induced by 3,7-dinitrofluoranthene and 7/10 that were induced by 3,9-dinitrofluoranthene were classified as malignant fibrous histiocytomas, whereas the remainder showed typical features of rhabdomyosarcomas (1/21 and 3/10, respectively). The time of appearance of the subcutaneous tumours was earlier in the 3,9- than in the 3,7-dinitrofluoranthene-treated group: the first tumour in the 3,9-dinitrofluoranthene-treated group appeared on day 88 (average, 117), 10 weeks earlier than in the 3,7-dinitrofluoranthene-treated group (115; average, 186). No subcutaneous tumour was found in the control group (Tokiwa et al., 1987).

3.2 Intrapulmonary implantation

Rat: A total of 84 male Fischer 344/DuCrj rats, 11 weeks of age, were divided into five groups that received pulmonary implants of 50, 100 or 200 µg 3,9-dinitrofluoranthene, 200 µg 3,7-dinitrofluoranthene in a mixture of equal volumes of beeswax and tricaprylin or the beeswax–tricaprylin mixture alone. The rats were anaesthetized, a left lateral thoracotomy was performed and 0.5 ml of the beeswax-tricaprylin mixture containing the chemicals was injected directly into the lower third of the left lung. After injection, the mixture formed a solid, defined pellet in the lung. All rats were observed for up to 100 weeks. The mean body weights of all treated groups were significantly decreased compared to those of controls. In 3,9-dinitrofluoranthene-treated rats, the earliest deaths caused by lung cancer were observed at weeks 99, 53 and 37 in the low-, mid- and high-dose groups, respectively. The earliest death in the 3,7-dinitrofluoranthene-treated group occurred at week 50. Lungs, liver, spleen and kidneys were examined grossly, and grossly apparent lesions and tissues from all major organs were examined microscopically. The incidence of lung tumours was 1/10 (10%), 7/10 (70%) and 19/21 (90.5%) in the low-, mid- and high-dose 3,9-dinitrofluoranthene-treated groups and 12/22 (54.5%) in the 3,7-dinitrofluoranthene-treated group. Most of the lung tumours induced by both chemicals were invasive squamous-cell carcinomas. No lung tumour was found in control rats (Horikawa et al., 1991).

4. Other Data Relevant to an Evaluation of Carcinogenicity and its Mechanisms

4.1 Absorption, distribution, metabolism and excretion

4.1.1 *Humans*

No data were available to the Working Group.

4.1.2 *Experimental systems*

3,9-Dinitrofluoranthene was metabolized to an aminonitrofluoranthene (the isomeric structure of which could not be identified) by lung cytosol and microsomes from seven-to-nine-week-old male Fischer 344/N rats under anaerobic, but not under aerobic, conditions. Pretreatment of the animals with 3-methylcholanthrene increased the microsomal but not the cytosolic reduction of 3,9-dinitrofluoranthene (Mitchell *et al.*, 1993).

4.2 Toxic effects

No data were available to the Working Group.

4.3 Reproductive and developmental effects

No data were available to the Working Group.

4.4 Genetic and related effects

4.4.1 *Humans*

No data were available to the Working Group.

4.4.2 *Experimental systems* (see also Table 1 and Appendices 1 and 2)

The genetic and related effects of nitroarenes and of their metabolites have been reviewed (Rosenkranz & Mermelstein, 1983; Beland *et al.*, 1985; Rosenkranz & Mermelstein, 1985; Tokiwa & Ohnishi, 1986; Tokiwa *et al.*, 1993).

3,7-Dinitrofluoranthene was mutagenic at extremely low doses in *Salmonella typhimurium* strains and preferentially inhibited the growth of DNA repair-deficient *Bacillus subtilis*. In the *umu* test, *S. typhimurium* strains NM1011, which has increased nitrofurazone-reductase activity, and NM3009, which has high *O*-acetyltransferase and nitroreductase activities, were both particularly highly sensitive to the genetic activity of 3,7-dinitrofluoranthene.

3,7-Dinitrofluoranthene did not induce mutations to 6-thioguanine resistance in Chinese hamster V79 cells. In a Chinese hamster cell line (CHL), 3,7-dinitrofluoranthene induced chromosomal aberrations in the absence, but not in the presence, of rat-liver S9 mix.

In vivo, 3,7-dinitrofluoranthene induced micronuclei in mouse bone marrow.

3,9-Dinitrofluoranthene was mutagenic to *S. typhimurium* strains and preferentially inhibited the growth of DNA repair-deficient *B. subtilis*. The *umu* test with *S. typhimurium* NM1011, a strain with increased nitrofurazone-reductase activity, was particularly highly sensitive to 3,9-dinitrofluoranthene.

3,9-Dinitrofluoranthene did not induce mutations to 6-thioguanine resistance in Chinese hamster V79 lung cells. In a Chinese hamster cell line (CHL), 3,9-dinitrofluoranthene induced chromosomal aberrations in the absence, but not in the presence, of rat-liver S9 mix.

In vivo, 3,9-dinitrofluoranthene induced micronuclei in mouse bone marrow.

Table 1. Genetic and related effects of 3,7- and 3,9-dinitrofluoranthenes

Test system	Result[a] Without exogenous metabolic system	Result[a] With exogenous metabolic system	Dose[b] (LED/HID)	Reference
3,7-Dinitrofluoranthene				
PRB, Prophage induction, SOS repair test, DNA strand breaks, cross-links or related damage in *Salmonella typhimurium* TA1535/pSK1002	+	0	0.03	Oda et al. (1992)
PRB, Prophage induction, SOS repair test, DNA strand breaks, cross-links or related damage in *Salmonella typhimurium* TA1535/pSK1002	+	0	0.001	Oda et al. (1993)
PRB, Prophage induction, SOS repair test, DNA strand breaks, cross-links or related damage in *Salmonella typhimurium* NM1000	(+)	0	0.003	Oda et al. (1993)
PRB, Prophage induction, SOS repair test, DNA strand breaks, cross-links or related damage in *Salmonella typhimurium* NM1011	+	0	0.003	Oda et al. (1992)
PRB, Prophage induction, SOS repair test, DNA strand breaks, cross-links or related damage in *Salmonella typhimurium* NM1011	+	0	0.0001	Oda et al. (1993)
PRB, Prophage induction, SOS repair test, DNA strand breaks, cross-links or related damage in *Salmonella typhimurium* NM1000	–	0	0.03	Oda et al. (1992)
PRB, Prophage induction, SOS repair test, DNA strand breaks, cross-links or related damage in *Salmonella typhimurium* NM2000	+	0	0.0003	Oda et al. (1993)
PRB, Prophage induction, SOS repair test, DNA strand breaks, cross-links or related damage in *Salmonella typhimurium* NM2009	+	0	0.00001	Oda et al. (1993)
PRB, Prophage induction, SOS repair test, DNA strand breaks, cross-links or related damage in *Salmonella typhimurium* NM3009	+	0	0.00001	Oda et al. (1993)
BSD, *Bacillus subtilis* rec strains, differential toxicity	+	0	0.005	Nakagawa et al. (1987)
SA0, *Salmonella typhimurium* TA100, reverse mutation	+	+	0.0005	Nakagawa et al. (1987)
SA5, *Salmonella typhimurium* TA1535, reverse mutation	–	–	0.008	Nakagawa et al. (1987)
SA7, *Salmonella typhimurium* TA1537, reverse mutation	+	(+)	0.001	Nakagawa et al. (1987)
SA8, *Salmonella typhimurium* TA1538, reverse mutation	+	+	0.0005	Nakagawa et al. (1987)
SA9, *Salmonella typhimurium* TA98, reverse mutation	+	+	0.00013	Nakagawa et al. (1987)

Table 1 (contd)

Test system	Result[a] Without exogenous metabolic system	Result[a] With exogenous metabolic system	Dose[b] (LED/HID)	Reference
3,7-Dinitrofluoranthene (contd)				
SAS, *Salmonella typhimurium* TA97, reverse mutation	+	0	0.00025	Nakagawa *et al.* (1987)
SAS, *Salmonella typhimurium* TA98NR, reverse mutation	+	0	0.0005	Nakagawa *et al.* (1987)
SAS, *Salmonella typhimurium* TA98/1,8-DNP$_6$, reverse mutation	+	0	0.0005	Nakagawa *et al.* (1987)
SAS, *Salmonella typhimurium* TA1978, reverse mutation	–	0	0.5	Nakagawa *et al.* (1987)
G9H, Gene mutation, Chinese hamster lung V79 cells *hprt* locus	0	–	100	Tokiwa *et al.* (1988)
CIC, Chromosomal aberrations, Chinese hamster cells *in vitro*	+	–	2.5	Matsuoka *et al.* (1993)
MVM, Micronucleus test, mice *in vivo*	+		20 ip	Tokiwa *et al.* (1988)
3,9-Dinitrofluoranthene				
PRB, Prophage induction, SOS repair test, DNA strand breaks, cross-links or related damage in *Salmonella typhimurium* TA1535/pSK1002	+	0	0.003	Oda *et al.* (1992)
PRB, Prophage induction, SOS repair test, DNA strand breaks, cross-links or related damage in *Salmonella typhimurium* TA1535/pSK1002	+	0	0.0018	Oda *et al.* (1993)
PRB, Prophage induction, SOS repair test, DNA strand breaks, cross-links or related damage in *Salmonella typhimurium* NM1000	+	0	0.006	Oda *et al.* (1993)
PRB, Prophage induction, SOS repair test, DNA strand breaks, cross-links or related damage in *Salmonella typhimurium* NM1011	+	0	0.001	Oda *et al.* (1992)
PRB, Prophage induction, SOS repair test, DNA strand breaks, cross-links or related damage in *Salmonella typhimurium* NM1011	+	0	0.00056	Oda *et al.* (1993)
PRB, Prophage induction, SOS repair test, DNA strand breaks, cross-links or related damage in *Salmonella typhimurium* NM1000	–	0	0.01	Oda *et al.* (1992)
PRB, Prophage induction, SOS repair test, DNA strand breaks, cross-links or related damage in *Salmonella typhimurium* NM2000	+	0	0.0069	Oda *et al.* (1993)
PRB, Prophage induction, SOS repair test, DNA strand breaks, cross-links or related damage in *Salmonella typhimurium* NM2009	+	0	0.0001	Oda *et al.* (1993)

Table 1 (contd)

Test system	Result[a]		Dose[b] (LED/HID)	Reference
	Without exogenous metabolic system	With exogenous metabolic system		
3,9-Dinitrofluoranthene (contd)				
PRB, Prophage induction, SOS repair test, DNA strand breaks, cross-links or related damage in *Salmonella typhimurium* NM3009	+	0	0.00006	Oda et al. (1993)
BSD, *Bacillus subtilis* rec strains, differential toxicity	+	0	0.01	Nakagawa et al. (1987)
SA0, *Salmonella typhimurium* TA100, reverse mutation	+	+	0.001	Nakagawa et al. (1987)
SA5, *Salmonella typhimurium* TA1535, reverse mutation	–	–	0.008	Nakagawa et al. (1987)
SA7, *Salmonella typhimurium* TA1537, reverse mutation	+	+	0.001	Nakagawa et al. (1987)
SA8, *Salmonella typhimurium* TA1538, reverse mutation	+	+	0.00013	Nakagawa et al. (1987)
SA9, *Salmonella typhimurium* TA98, reverse mutation	+	+	0.00013	Nakagawa et al. (1987)
SA9, *Salmonella typhimurium* TA98, reverse mutation	+	0	0.001	Horikawa et al. (1994)
SAS, *Salmonella typhimurium* TA97, reverse mutation	+	0	0.00025	Nakagawa et al. (1987)
SAS, *Salmonella typhimurium* TA98NR, reverse mutation	+	0	0.0005	Nakagawa et al. (1987)
SAS, *Salmonella typhimurium* TA98/1,8-DNP$_6$, reverse mutation	+	0	0.0005	Nakagawa et al. (1987)
SAS, *Salmonella typhimurium* TA1978, reverse mutation	+	0	0.25	Nakagawa et al. (1987)
G9H, Gene mutation, Chinese hamster lung V79 cells *hprt* locus	0	–	100	Tokiwa et al. (1988)
CIC, Chromosomal aberrations, Chinese hamster cells *in vitro*	+	–	2.5	Matsuoka et al. (1993)
MVM, Micronucleus test, mice *in vivo*	+		10 ip	Tokiwa et al. (1988)

[a] +, positive; (+), weak positive; –, negative; 0, not tested
[b] LED, lowest effective dose; HID, highest ineffective dose. In-vitro tests, µg/mL; in-vivo tests, mg/kg bw

5. Summary of Data Reported and Evaluation

5.1 Exposure data

3,7- and 3,9-Dinitrofluoranthenes are produced for laboratory use by nitration of fluoranthene. 3,7- and 3,9-Dinitrofluoranthenes have been detected at low levels in emissions from diesel engines, kerosene heaters and other combustion sources.

5.2 Human carcinogenicity data

No data were available to the Working Group.

5.3 Animal carcinogenicity data

3,7- and 3,9-Dinitrofluoranthenes were tested for carcinogenicity in rats by subcutaneous injection in one study and by pulmonary implantation in another study. Subcutaneous injection of 3,7- and 3,9-dinitrofluoranthenes induced a high incidence of subcutaneous tumours at the site of injection, most of which were malignant fibrous histiocytomas. Pulmonary implantation of 3,7- and 3,9-dinitrofluoranthenes induced a high incidence of lung tumours, most of which were squamous-cell carcinomas.

5.4 Other relevant data

3,7- and 3,9-Dinitrofluoranthenes are highly mutagenic to bacteria, particularly in the absence of an exogenous metabolic system. In mammalian cells *in vitro*, these compounds induced chromosomal aberrations but not gene mutations. *In vivo*, they induced micronuclei in mouse bone marrow.

5.5 Evaluation[1]

There is *inadequate evidence* in humans for the carcinogenicity of 3,7- and 3,9-dinitrofluoranthenes.

There is *sufficient evidence* in experimental animals for the carcinogenicity of 3,7- and 3,9-dinitrofluoranthenes.

Overall evaluation

3,7- and 3,9-Dinitrofluoranthenes are *possibly carcinogenic to humans (Group 2B)*.

---

[1] For definition of the italicized terms, see Preamble, pp. 24–27.

6. References

Beland, F.A., Heflich, R.H., Howard, P.C. & Fu, P.P. (1985) The in vitro metabolic activation of nitro polycyclic aromatic hydrocarbons. In: Harvey, R.G., ed., *Polycyclic Hydrocarbons and Carcinogenesis* (ACS Symposium Series, No. 283), Washington DC, American Chemical Society, pp. 371–396

Horikawa, K., Otofuji, T., Nakagawa, R., Sera, N., Kuroda, Y., Otsuka, H. & Tokiwa, H. (1987) Mutagenicity of dinitrofluoranthenes (DNFs) and their carcinogenicity in F344 rats (Abstract no. 11). *Mutat. Res.*, **182**, 360–361

Horikawa, K., Sera, N., Otofuji, T., Murakami, K., Tokiwa, H., Iwagawa, M., Izumi, K. & Otsuka, H. (1991) Pulmonary carcinogenicity of 3,9- and 3,7-dinitrofluoranthene, 3-nitrofluoranthene and benzo[a]pyrene in F344 rats. *Carcinogenesis*, **12**, 1003–1007

Horikawa, K., Mohri, T., Tanaka, Y. & Tokiwa, H. (1994) Moderate inhibition of mutagenicity and carcinogenicity of benzo[a]pyrene, 1,6-dinitropyrene and 3,9-dinitrofluoranthene by Chinese medicinal herbs. *Mutagenesis*, **9**, 523–526

IARC (1989) *IARC Monographs on the Evaluation of Carcinogenic Risks to Humans*, Vol. 46, *Diesel and Gasoline Engine Exhausts and Some Nitroarenes*, Lyon, pp. 189–200

Matsuoka, A., Horikawa, K., Yamazaki, N., Sera, N., Sofuni, T. & Tokiwa, H. (1993) Chromosomal aberrations induced *in vitro* by 3,7- and 3,9-dinitrofluoranthene. *Mutat. Res.*, **298**, 255–259

Mitchell, C.E., Bechtold, W.E. & Belinsky, S.A. (1993) Metabolism of nitrofluoranthenes by rat lung subcellular fractions. *Carcinogenesis*, **14**, 1161–1166

Nakagawa, R., Horikawa, K., Sera, N., Kodera, Y. & Tokiwa, H. (1987) Dinitrofluoranthene: induction, identification and gene mutation. *Mutat. Res.*, **191**, 85–91

Oda, Y., Shimada T., Watanabe, M., Ishidate, M., Jr & Nohmi, T. (1992) A sensitive *umu* test system for the detection of mutagenic nitroarenes in *Salmonella typhimurium* NM1011 having a high nitroreductase activity. *Mutat. Res.*, **272**, 91–99

Oda, Y., Yamazaki, H., Watanabe, M., Nohmi, T. & Shimada, T. (1993) Highly sensitive *umu* test system for the detection of mutagenic nitroarenes in *Salmonella typhimurium* NM3009 having high O-acetyltransferase and nitroreductase activities. *Environ. mol. Mutag.*, **21**, 357–364

Ramdahl, T., Zielinska, B., Arey, J. & Kondrat, R.W. (1988) The electron impact mass spectra of di- and trinitrofluoranthenes. *Biomed. environ. Mass Spectrom.*, **17**, 55–62

Rosenkranz, H.S. & Mermelstein, R. (1983) Mutagenicity and genotoxicity of nitroarenes: all nitro-containing chemicals were not created equal. *Mutat. Res.*, **114**, 217–267

Rosenkranz, H.S. & Mermelstein, R. (1985) The genotoxicity, metabolism and carcinogenicity of nitrated polycyclic aromatic hydrocarbons. *J. environ. Sci. Health*, **C3**, 221–272

Tokiwa, H. & Ohnishi, Y. (1986) Mutagenicity and carcinogenicity of nitroarenes and their sources in the environment. *CRC crit. Rev. Toxicol.*, **17**, 23–60

Tokiwa, H., Otofuji, T., Nakagawa, R., Horikawa, K., Maeda, T., Sano, N., Izumi, K. & Otsuka, H. (1986) Dinitro derivatives of pyrene and fluoranthene in diesel emission particulates and their tumorigenicity in mice and rats. In: Ishinishi, N., Koizumi, A., McClellan, R.O. & Stöber, W., eds, *Carcinogenic and Mutagenic Effects of Diesel Engine Exhaust*, Amsterdam, Elsevier, pp. 253–270

Tokiwa, H., Otofuji, T., Horikawa, K., Sera, N., Nakagawa, R., Maeda, T., Sano, N., Izumi, K. & Otsuka, H. (1987) Induction of subcutaneous tumors in rats by 3,7- and 3,9-dinitrofluoranthene. *Carcinogenesis*, **8**, 1919–1922

Tokiwa, H., Horikawa, K., Omura, H. & Kuroda, Y. (1988) Mutagenicity in Chinese hamster V79 cells and induction of micronuclei in mice by nitrated fluoranthenes. *Exp. Oncol.*, **7**, 33–37

Tokiwa, H., Sera, N., Kai, M., Horikawa, K. & Ohnishi, Y. (1990) The role of nitroarenes in the mutagenicity of airborne particles indoors and outdoors. In: Waters, M.D., Daniel, F.B., Lewtas, J., Moore, M.M. & Nesnow, S., eds, *Genetic Toxicology of Complex Mixtures*, New York, Plenum Press, pp. 165–172

Tokiwa, H., Horikawa, K. & Ohnishi, Y. (1993) Genetic toxicology and carcinogenicity of mono- and dinitrofluoranthenes. *Mutat. Res.*, **297**, 181–195

2,4-DINITROTOLUENE, 2,6-DINITROTOLUENE AND 3,5-DINITROTOLUENE

1. Exposure Data

1.1 Chemical and physical data

1.1.1 *Nomenclature*

The CAS Reg. No. for dinitrotoluene, not otherwise specified, is 25321-14-6.

2,4-Dinitrotoluene

Chem. Abstr. Serv. Reg. No.: 121-14-2

Chem. Abstr. Name: 1-Methyl-2,4-dinitrobenzene

IUPAC Systematic Name: 2,4-Dinitrotoluene

Synonyms: Dinitrotoluene; 2,4-dinitrotoluol; DNT; 2,4-DNT; 4-methyl-1,3-dinitrobenzene

2,6-Dinitrotoluene

Chem. Abstr. Serv. Reg. No.: 606-20-2

Chem. Abstr. Name: 2-Methyl-1,3-dinitrobenzene

IUPAC Systematic Name: 2,6-Dinitrotoluene

Synonyms: 2,6-DNT; 1-methyl-2,6-dinitrobenzene

3,5-Dinitrotoluene

Chem. Abstr. Serv. Reg. No.: 618-85-9

Chem. Abstr. Name: 1-Methyl-3,5-dinitrobenzene

IUPAC Systematic Name: 3,5-Dinitrotoluene

Synonyms: 3,5-DNT

1.1.2 Structural and molecular formulae and relative molecular mass

2,4-Dinitrotoluene   2,6-Dinitrotoluene   3,5-Dinitrotoluene

$C_7H_6N_2O_4$ Relative molecular mass: 182.15

1.1.3 Chemical and physical properties of the pure substance

2,4-Dinitrotoluene

(a) *Description*: Yellow needles from ethanol or carbon disulfide (Booth, 1991; Lide, 1993)

(b) *Boiling-point*: 300 °C (decomposes) (Lide, 1993)

(c) *Melting-point*: 71 °C (Lide, 1993)

(d) *Spectroscopy data*: Infrared (prism [175, 258], grating [8017]), ultraviolet (UV) [2550], nuclear magnetic resonance (proton [3229], C-13 [4627]) and mass spectral data have been reported (Sadtler Research Laboratories, 1980)

(e) *Solubility*: Slightly soluble in water (270 mg/L at 22 °C); soluble in acetone, benzene, diethyl ether and ethanol (Mabey et al., 1982; Lide, 1993)

(f) *Volatility*: Vapour pressure, 0.00011 mm Hg [0.015 Pa] at 20 °C; relative vapour density (air = 1), 6.27 (Howard, 1989; Booth, 1991)

(g) *Stability*: Combustible when exposed to heat or flame; can react with oxidizing materials. Decomposes when heated at ≥ 250 °C. Mixture with nitric acid is a high explosive. Mixture with sodium carbonate (or other alkalies) can decompose with significant increase in pressure at 210 °C (Sax & Lewis, 1989).

(h) *Octanol/water partition coefficient (P)*: log P, 1.98 (Hansch et al., 1995)

(i) *Conversion factor*: $mg/m^3 = 7.45 \times ppm$[1]

2,6-Dinitrotoluene

(a) *Description*: Rhombic needles from ethanol (Lide, 1993)

(b) *Boiling-point*: 285 °C (Howard, 1989)

(c) *Melting-point*: 66 °C (Lide, 1993)

---

[1] Calculated from: mg/m^3 = (molecular weight/24.45) × ppm, assuming temperature (25 °C) and pressure (101 kPa)

(d) *Spectroscopy data*: Infrared (prism [17378], grating [676]), UV [5514], nuclear magnetic resonance (proton [895]) and mass spectral data have been reported (Sadtler Research Laboratories, 1980)

(e) *Solubility*: Slightly soluble in water (180 mg/L at 20 °C); soluble in ethanol (Mabey *et al.*, 1982; Lide, 1993)

(f) *Volatility*: Vapour pressure, 0.018 mm Hg [2.4 Pa] at 20 °C (Mabey *et al.*, 1982)

(g) *Octanol/water partition coefficient (P)*: log P, 2.10 (Hansch *et al.*, 1995)

(h) *Conversion factor*: $mg/m^3 = 7.45 \times ppm^1$

3,5-Dinitrotoluene

(a) *Description*: Yellow rhombic needles from acetic acid (Lide, 1993)

(b) *Boiling-point*: Sublimes (Lide, 1993)

(c) *Melting-point*: 93 °C (Lide, 1993)

(d) *Density*: 1.2772 at 11 °C/4 °C (Lide, 1993)

(e) *Solubility*: Soluble in benzene, chloroform, diethyl ether and ethanol (Lide, 1993)

(f) *Stability*: Combustible when exposed to heat or flame; can react with oxidizing materials. A moderate explosion hazard when exposed to heat (Sax & Lewis, 1989)

(g) *Octanol/water partition coefficient (P)*: log P, 2.28 (United States National Library of Medicine, 1995)

(h) *Conversion factor*: $mg/m^3 = 7.45 \times ppm^1$

1.1.4 *Technical products and impurities*

2,4-Dinitrotoluene is available commercially at purities ranging from 95% (with content of 2,6-isomer < 1.5%) to > 99% in high-melt and super high-melt grades. 2,6-Dinitrotoluene is available commercially at purities ranging from 97% to > 99%, sometimes with 10% water added. Mixtures of 95 : 5, 80 : 20 and 65 : 35 2,4-dinitrotoluene and 2,6-dinitrotoluene, respectively, are also available commercially (Girundus Corp., 1994; Miles, 1994; TCI America, 1994; Acros Organics, 1995; Air Products and Chemicals, 1995; Lancaster Synthesis, 1995).

1.1.5 *Analysis*

Spectrophotometric determination of 2,4-dinitrotoluene in acidified urine has been used routinely in the range of 5–50 mg/L. After diazotization and coupling to 1-amino-8-naphthol-2,4-disulfonic acid (Chicago acid), the primary aromatic amine was determined as a red azo dye. An alternative method uses formamidine sulfinic acid (thiourea dioxide) to reduce the aromatic nitro compounds under alkaline conditions (Konieck & Linch, 1958).

---

[1] Calculated from: mg/m^3 = (molecular weight/24.45) × ppm, assuming temperature (25 °C) and pressure (101 kPa)

Various instrumental techniques have been used for the determination of nitro-aromatic compounds, nitramines and explosives, with gas chromatography being perhaps the most widely used. Because some of these compounds may undergo thermal degradation during analysis by gas chromatography, high-performance liquid chromatographic (HPLC) methods have been developed, which use reverse-phase chromatography and ultraviolet detection in most instances. Spectrophotometry and polarography have also been used for the detection of explosives. Thin-layer chromatography has been used to identify nitroaromatic compounds and explosives (Steuckart et al., 1994).

Selected methods for the analysis of dinitrotoluenes in various media are presented in Table 1.

Table 1. Methods for the analysis of dinitrotoluenes

Sample matrix	Sample preparation	Assay procedure	Limit of detection	Reference
Air	Draw air through modified Tenax-GC tube; desorb with acetone	GC/TEA	20 µg/m^{3a}	US Occupational Safety and Health Administration (1990) [Method 44]
	Draw air through a filter/bubbler; extract with ethylene glycol	HPLC/UV	NR^a	Taylor (1978) [Method S215]
Water	Extract sample with dichloromethane or adsorb on Amberlite XAD resin and elute with dichloromethane	GC/ECD	NR^{a,b}	Feltes et al. (1990)
	Liquid–liquid extraction with dichloromethane; dry with anhydrous sodium sulfate; evaporate to dryness; redissolve in methanol	SFC/FID	30 ppm (mg/L)^a	Ong et al. (1992)
	Extract with toluene	GC/EC	3 ng/L^b 40 ng/L^a	Hable et al. (1991)
	Liquid–liquid extraction with dichloromethane	HPTLC	20 ng^{a,b}	Steuckart et al. (1994)
	Solid-phase microextraction with a polydimethylsiloxane-coated fibre; thermally desorb directly into GC	GC/FID	15 µg/L^{a,b}	Horng & Huang (1994)
Wastewater	Dilute sample with methanol and acetonitrile	HPLC/UV	10 µg/L^a	Jenkins et al. (1986)
	Extract with diethyl ether; dry over anhydrous magnesium sulfate; filter	GC/FID	NR^{a,b,c}	Spanggord et al. (1982a)

Table 1 (contd)

Sample matrix	Sample preparation	Assay procedure	Limit of detection	Reference
Municipal and industrial discharges	Extract with dichloromethane; dry; exchange to hexane	GC/ECD	0.01 µg/L^b 0.02 µg/L^a	US Environmental Protection Agency (1986a, 1994) [Methods 8090 & 609]
	Extract with dichloromethane at pH > 11 and at pH < 2; dry (packed column method)	GC/MS	1.9 µg/L^b 5.7 µg/L^a	US Environmental Protection Agency (1986b, 1994) [Methods 8250 & 625]
	Add isotopically labelled analogue to sample; extract with methylene chloride at pH 12–13 and at pH < 2; dry (capillary column method)	GC/MS	10 µg/L^{a,b}	US Environmental Protection Agency (1994) [Method 1625B]
Industrial effluents and seawater	Extract with benzene; inject into glass capillary GC column	GC/ECD	0.059 µg/L^b 0.13 µg/L^a 0.17 µg/L^c	Hashimoto et al. (1980)
Water, soil, sediment, waste	Extract with dichloromethane (capillary column method)	GC/MS	PQL^{a,b,d}	US Environmental Protection Agency (1986c) [Method 8270]
Soil	Extract with acetone in ultrasonic bath; extract with toluene, remove acetone; dry; add calcium chloride solution; extract with dichloromethane	HPTLC	20 ng^{a,b}	Steuckart et al. (1994)
	Extract with acetone; react supernatant with potassium hydroxide/sodium sulfite; read absorbance at 570 nm	Colorimetry	2 µg/g^a	Jenkins & Walsh (1992)
Freshwater and marine sediments	Extract with acetonitrile; clean-up with solid-phase extraction	GC/MS	NR^{a,b}	Davis et al. (1993)
Urine	Hydrolyse metabolites; extract; derivatize	GC/MS	NR^{a,b}	Turner et al. (1985)
	Extract with ethyl acetate	GC/ECD	NR^{a,b}	Woollen et al. (1985)
Blood	Extract with toluene	GC/ECD	NR^{a,b}	Woollen et al. (1985)
	Extract from separated plasma and concentrate simultaneously using 2,2,4-trimethylpentane	GC/ECD	2.0 µg/L^b	Lewalter & Ellrich (1991)

Table 1 (contd)

Sample matrix	Sample preparation	Assay procedure	Limit of detection	Reference
Materials exposed to explosives	Draw air through ethanol-soaked handswab; adsorb on alumina/octadecylsilylsilica; desorb with methanol/water (100 : 35 v/v)	HPLC/ED	12 pg/sample[a]	Lloyd (1983a,b)

GC, gas chromatography; TEA, thermal energy analyser; HPLC, high-performance liquid chromatography; UV, ultraviolet detection; NR, not reported; ECD, electrolytic conductivity detection; SFC, capillary supercritical fluid chromatography; FID, flame ionization detection; EC, electron capture detection; HPTLC, high-performance thin-layer chromatography

[a] 2,4-Dinitrotoluene
[b] 2,6-Dinitrotoluene
[c] 3,5-Dinitrotoluene
[d] PQL, practical quantitation limit: groundwater, 10 µg/L; low soil/sediment, 660 µg/kg; medium-level soil and sludges by sonicator, 4.95 mg/kg; non-water-miscible waste, 49.5 kg

1.2 Production and use

1.2.1 Production

A 96% yield of 2,4-dinitrotoluene can be produced by continuous nitration of 4-nitrotoluene with 'mixed acid' (which contains equimolar quantities of nitric and sulfuric acids) under controlled conditions. Alternative processes yield mixed isomer products — when toluene is nitrated directly under similar conditions with 2.1 equivalents of nitric acid, the product is approximately an 80 : 20 mixture of 2,4-dinitrotoluene and 2,6-dinitrotoluene, and nitration of 2-nitrotoluene, which is sometimes present in excess, gives an approximate 67 : 33 mixture of 2,4- and 2,6-dinitrotoluenes (Levine et al., 1985; Booth, 1991).

3,5-Dinitrotoluene can be prepared by the nitration of nitrotoluene with mixed acid (Lewis, 1993).

Mixtures of 2,4- and 2,6-dinitrotoluenes were produced in the United States of America at a rate of over 227 thousand tonnes per year in 1982 (Howard, 1989).

2,4-Dinitrotoluene is produced by three companies each in Japan and the United States and one company each in China, the Czech Republic, Egypt, Germany, Italy, Portugal, Turkey and the United Kingdom. 2,6-Dinitrotoluene is produced by one company each in Egypt, Germany and Italy. The mixture of 2,4- and 2,6-dinitrotoluenes is produced by two companies in the United States and by one company each in China, the Czech Republic, Germany, Poland, the Republic of Korea, Romania and Turkey (Chemical Information Services, 1994).

1.2.2 Use

Most of the 2,4-dinitrotoluene produced is hydrogenated (nickel catalyst) to 2,4-diaminotoluene (see IARC, 1978, 1987a) for conversion to toluene diisocyanate (see IARC, 1986, 1987b), which is a monomer in the production of polyurethane (see IARC, 1979). A much smaller amount is used in explosives and for further nitration to trinitrotoluene (see this volume). Crude mixtures of 2,4- and 2,6-dinitrotoluenes are also used to produce mixed toluene diamines (80 : 20 or 67 : 33, depending on the nitration process used), which in turn are converted to mixed toluene diisocyanates for polyurethane production. The use of these mixed isomers in polyurethane production has considerable cost benefits (Booth, 1991).

Dinitrotoluenes, including the 2,4-, 2,6- and 3,5-isomers, are used in organic synthesis in the production of toluidines and dyes. In the production of explosives, dinitrotoluenes are used to manufacture trinitrotoluene and gelatin explosives, to plasticize cellulose nitrate, to moderate the burning rate of propellants and to waterproof some smokeless powders (Howard, 1989; Lewis, 1993).

1.3 Occurrence

1.3.1 Natural occurrence

Dinitrotoluenes are not known to occur as natural products.

1.3.2 Occupational exposure

Exposure to dinitrotoluenes may occur from their use in the manufacture of toluene diisocyanate, in the production of explosives, in the manufacture of azo dye intermediates and in organic synthesis in the preparation of toluidines and dyes (Howard, 1989).

However, few reports of occupational exposures to dinitrotoluene exist. Levine *et al.* (1985) monitored in 1983 7-h time-weighted average (TWA) personal exposure to dinitrotoluene and urinary metabolites of dinitrotoluene in a dinitrotoluene manufacturing plant constructed in 1973. Exposures of production unit operators to both 2,4- and 2,6- dinitrotoluenes averaged 0.26 mg/m^3 (range, 0.05–0.59 mg/m^3). Exposures of loaders, who load storage tanks, collect samples and perform cleaning tasks, averaged 0.32 mg/m^3 (range, 0.14–0.49 mg/m^3). Exposures of maintenance mechanics averaged 0.12 mg/m^3 (range, 0.08–0.15 mg/m^3) and the exposure of acid stripper operators was 0.06 mg/m^3. The highest personal air concentrations and levels of urinary metabolites were found to be for loaders, followed by process operators. The levels of urinary metabolites of dinitrotoluene in loaders and operators exceeded those that would have resulted from the inhaled concentrations (as indicated by personal air monitoring), although the workers wore gloves for operations in which dermal exposures were possible.

A study of explosives manufacture in the United Kingdom also compared personal airborne exposures to dinitrotoluene with levels of urinary metabolites of dinitrotoluene (Woollen *et al.*, 1985). Personal exposures ranged from 'not detected' to 0.1 mg/m^3 dinitrotoluene. Area samples positioned near dusty parts of the process ranged from 0.02

to 2.68 mg/m³. However, atmospheric concentrations could not account for the observed excretion levels of the metabolite 2,4-dinitrobenzoic acid, indicating probable dermal uptake.

Quantitative occupational exposure data were not available for a mortality study of munitions workers producing rocket propellants in the United States (Stayner et al., 1993). Workers were classified into three exposure categories based on their opportunity for exposure to dinitrotoluenes.

1.3.3 Environmental occurrence

(a) Water

2,4-Dinitrotoluene has been detected in seawater, river water and in wastewater from 2,4,6-trinitrotoluene production. 2,4-Dinitrotoluene was found in Dokai Bay, Japan, at concentrations of up to 206 µg/L (Hashimoto et al., 1982), in water from the River Rhine in the Netherlands at 0.3 µg/L (Zoeteman et al., 1980), in water samples obtained from the River Potomac near Quantico, VA, United States, at a concentration of < 10 µg/L (Hall et al., 1987) and in Waconda Bay, Lake Chickamauga, TN, United States (range of means, < 0.10–22.1 µg/L) (Putnam et al., 1981). It has also been detected in groundwater at levels ranging from 2 to 90 500 µg/L (0.002 to 90.5 ppm) near a nitroaromatic manufacturing facility in Pasadena, TX, United States (Matson, 1986), and was detected but not quantified, in groundwater (one sample) at the Hawthorne Naval Ammunition Depot, NV, United States (Pereira et al., 1979).

2,4-Dinitrotoluene has also been found in condensate wastewater from 2,4,6-trinitrotoluene manufacture at an unspecified concentration (Liu et al., 1984) and in wastewater from 2,4,6-trinitrotoluene production at an average concentration of 9700 µg/L in 29/54 samples (Spanggord & Suta, 1982). It has been detected in effluents from coal mining (18 µg/L in 1/49 samples), iron and steel manufacture (530 µg/L in 1/5 samples), aluminium forming (77 µg/L in 1/2 samples), foundries (mean, 26 µg/L; range, 7–50 µg/L in 4/4 samples) and organic chemical manufacture (mean, 14 000 µg/L in 4 samples) (United States Environmental Protection Agency, 1980; Howard, 1989).

2,4-Dinitrotoluene concentrations of 3.1 and 13.0 µg/L were detected in surface-water samples collected from two brooks near Hischagen/Waldhof, Germany, which were close to sites used for Second World War munitions manufacture; the river into which the brooks fed (River Losse) was found to have a concentration of 0.5 µg/L. Two ponds in the Clausthal-Zellerfeld region of Germany, again near sites of previous munitions manufacture, had levels of 1.2 and 0.8 µg/L; the ponds feed into the River Oder, which was found to have a level of 0.02 µg/L. Concentrations at three locations (Brunsbüttel, Brokdorf, Lauenburg) of the River Elbe ranged from 0.1 to 1.3 µg/L (Feltes et al., 1990).

Concentrations of 2,4-dinitrotoluene ranged from 700 to 1180 µg/L in groundwater samples collected near a former explosives factory in Elsnig, Germany (Steuckart et al., 1994).

In wastewaters generated in the manufacture of 2,4,6-trinitrotoluene, 2,4-dinitrotoluene was found in all 54 samples at concentrations ranging from 40 to 48 600 µg/L over a 12-month sampling period (Spanggord et al., 1982a).

In 1993, estimated quantities of 2,4-dinitrotoluene released into the environment by industrial facilities in the United States were 850 kg into air and 150 kg into water (United States National Library of Medicine, 1995).

2,6-Dinitrotoluene has also been detected in seawater, in raw wastewater from a textile plant and in wastewater from 2,4,6-trinitrotoluene production. It was found in Dokai Bay, Japan, at concentrations ranging from 'not detected' to 14.8 µg/L (Hashimoto et al., 1982). It was found at concentrations ranging from 1.3 to 38.7 µg/L (mean, 19.4 µg/L) at Waconda Bay, Lake Chichamauga, TN, United States (Putnam et al., 1981) and was detected at concentrations ranging from not detected to 76 800 µg/L (mean, 16 763 µg/L) in groundwater near a nitroaromatic plant in Pasadena, TX, United States (Matson et al., 1986). The concentration of 2,6-dinitrotoluene in raw wastewater from a textile plant was 50 µg/L in one sample (Rawlings & Samfield, 1979) and averaged 4300 µg/L in 12/54 samples in wastewater from 2,4,6-trinitrotoluene production (Spanggord & Suta, 1982). It was also detected (5 µg/L) in wastewater from a nitrobenzene plant (Shafer, 1982). It was detected in effluents from coal mining (30 µg/L in 1/49 samples), iron and steel manufacture (range, 47–140 µg/L in 2/8 samples), nonferrous metals manufacture (max., 16 µg/L), foundries (mean, 20 µg/L in 6/6 samples; range 4–50 g/L), organic chemical manufacture (mean, 3800 µg/L in 4 samples), paint and ink formulations (max., 10 µg/L) and textile mills (54 µg/L in 1 sample) (United States Environmental Protection Agency, 1980; Howard, 1989).

2,6-Dinitrotoluene concentrations of 4.1 and 7.6 µg/L were detected in surface-water samples collected from two brooks near Hischagen/Waldhof, Germany, in the vicinity of munitions manufacture during the Second World War; the river into which the brooks fed (River Losse) had a concentration of 0.1 µg/L. Two ponds in the Clausthal-Zellerfeld region of Germany, again near previous munitions manufacture, had levels of 0.07 and 0.3 µg/L; the ponds feed into the River Oder which had a level of 0.02 µg/L. Concentrations at three locations (Brunsbüttel, Brokforf, Lauenburg) of the River Elbe ranged from 0.04 to 0.5 µg/L.

2,6-Dinitrotoluene was found in all 54 samples taken from wastewaters generated in the manufacture of 2,4,6-trinitrotoluene at a concentration ranging from 60 to 14 900 µg/L over a 12-month sampling period (Spanggord et al., 1982a).

The concentration of 2,6-dinitrotoluene ranged from not detected to 510 µg/L in groundwater samples collected near a former explosives factory in Elsnig, Germany (Steuckart et al., 1994).

In 1993, estimated quantities of 2,6-dinitrotoluene released into the environment by industrial facilities in the United States were 210 kg into air and 100 kg into water (United States National Library of Medicine, 1995).

3,5-Dinitrotoluene has been detected at concentrations ranging from 162 to 339 µg/L in condensate effluent resulting from the production of 2,4,6-trinitrotoluene (Spanggord & Suta, 1982).

In wastewaters generated in the manufacture of 2,4,6-trinitrotoluene, 3,5-dinitrotoluene was found in 51/54 samples with a range of 140–6480 µg/L over a 12-month sampling period (Spanggord et al., 1982a).

(b) Soil and sediments

In the United States, both 2,4- and 2,6-dinitrotoluenes were detected in one of two soil samples taken near the Buffalo River at the former site of a dye manufacturing plant (Nelson & Hites, 1980). 2,4- and 2,6-Dinitrotoluenes were detected in sediment from Waconda Bay, Lake Chickamauga, TN, at concentrations ranging from < 2.5 to ≤ 7.9 µg/kg and < 1.3–17 µg/kg, respectively (Putnam et al., 1981).

Concentrations of 2,4- and 2,6-dinitrotoluenes ranged from 3.4 to 226 mg/kg and 0.2 to 12.1 mg/g, respectively, in soil samples collected from two explosive ordnance sites in Mississippi and Alaska, United States (Jenkins & Walsh, 1992).

1.4 Regulations and guidelines

Occupational exposure limits and guidelines in several countries are presented in Table 2.

Table 2. Occupational exposure limits and guidelines for dinitrotoluenes (all isomers unless otherwise noted)

Country	Year	Concentration (mg/m^3)	Interpretation
Australia	1993	1.5 (Sk)	TWA
Belgium	1993	1.5 (Sk)	TWA
Bulgaria[a]	1995	0.15 (Sk)	TWA
Colombia[a]	1995	0.15 (Sk)	TWA
Czech Republic	1993	1	TWA
		2	STEL
Denmark	1993	1.5 (Sk)	TWA
Egypt	1993	1.5 (Sk)	TWA
Finland	1993	1.5 (Sk)	TWA
		3 (Sk)	STEL
Germany	1995	None (III A2) (Sk)	
India	1993	1.5 (Sk)	TWA
		5 (Sk)	STEL
Jordan[a]	1995	0.15 (Sk)	TWA
Netherlands	1994	1.5 (Sk)[b]	TWA
New Zealand[a]	1995	0.15 (Sk)	TWA
Philippines	1993	1.5 (Sk)	TWA
Poland	1991	1 (Sk)	TWA
Republic of Korea[a]	1995	0.15 (Sk)	TWA
Singapore[a]	1995	0.15 (Sk)	TWA
Switzerland	1993	1.5 (Sk,C)	TWA
Turkey	1993	1.5 (Sk)	TWA
United Kingdom	1995	None	
USA			
ACGIH (TLV)	1995	0.15 (Sk, A2)[c,d]	TWA
OSHA (PEL)	1994	1.5 (Sk)	TWA
NIOSH (REL)	1994	1.5 (Sk,Ca)	TWA

Table 2 (contd)

Country	Year	Concentration (mg/m³)	Interpretation
Viet Nam[a]	1995	0.15 (Sk)	TWA

From International Labour Office (1991); Arbeidsinspectie (1994); United States National Institute for Occupational Safety and Health (NIOSH) (1994a,b); United States Occupational Safety and Health Administration (OSHA) (1994); American Conference of Governmental Industrial Hygienists (ACGIH) (1995); Deutsche Forschungsgemeinschaft (1995); Health and Safety Executive (1995); United Nations Environmental Programme (1995)

Sk, absorption through the skin may be a significant source of exposure; TWA, time-weighted average; STEL, short-term exposure limit; III A2, substances shown to be clearly carcinogenic only in animal studies but under conditions indicative of carcinogenic potential at the workplace; C, suspected of being a carcinogen; TLV, threshold limit value; A2, suspected human carcinogen; PEL, permissible exposure level; REL, recommended exposure level; Ca, potential occupational carcinogen

[a] Follows ACGIH values
[b] 2,4-Dinitrotoluene only
[c] Substance identified in the BEI (Biological Exposure Indices) documentations for inducers of methaemoglobin
[d] Substance identified by other sources as a suspected or confirmed human carcinogen

2. Studies of Cancer in Humans

A retrospective cohort mortality study conducted by Levine *et al.* (1986) included workers exposed to dinitrotoluene at one of two munitions facilities in Joliet, IL, and Radford, VA, United States. The aim of this study was to investigate a possible interaction between exposure to dinitrotoluenes and incidence of liver cancer, as had been demonstrated in studies on animals (see Section 3). The study included 156 men from the Joliet facility and 301 men from the Radford facility with at least one month of exposure to dinitrotoluene during the 1940s and 1950s. The vital status of this cohort was ascertained to the end of 1980. Expected deaths and standardized mortality ratios (SMRs) were estimated using United States mortality rates for white men. The study failed to detect an increased risk for any cancer site. No death from liver or gall-bladder cancer was observed (0.5 expected). [The Working Group noted the low statistical power of this study.]

A study by Stayner *et al.* (1993) was conducted in one of the above two facilities (Radford, VA), although it differed in its identification of dinitrotoluene-exposed jobs and in the time period studied. This investigation included 4989 white male workers who were exposed to dinitrotoluene for at least one day and had worked for at least five months at the Radford facility between 1949 and 1980. Workers in this study were exposed to a mixture of approximately 98% 2,4-dinitrotoluene and 2% 2,6-dinitrotoluene. The vital status of this cohort was ascertained until the end of 1982. SMRs were estimated using United States mortality rates and standardized rate ratios (SRRs) were estimated using mortality rates from an internal unexposed cohort of 7436 workers identified at the Radford facility. An excess of cancer of the 'biliary passages, liver and gall-bladder' (six cases) was observed in this study based upon comparisons with both the United States population (SMR, 2.7; 95% confidence interval (CI), 1.0–5.8) and the internal unexposed cohort (SRR, 3.9; 95% CI, 1.0–14.4). No other cancer site was at increased risk. Hospital or pathology records were available for five of the six cases and all were validated. It was not possible to conduct informative duration-response analysis because very few workers had more than five years of exposure to dinitrotoluene.

3. Studies of Cancer in Experimental Animals

2,4-Dinitrotoluene

3.1 Oral administration

3.1.1 *Mouse*

Groups of 50 male and 50 female B6C3F1 mice, approximately six weeks of age, were fed diets containing 0 (control), 0.008 or 0.04% [0, 80 or 400 mg/kg diet (ppm)] 'practical-grade' 2,4-dinitrotoluene [purity stated to be > 95% by the supplier] for 78 weeks. This was followed by 13 weeks of control diet for all groups, after which time the mice were killed and subjected to complete histopathological examination. Studies with the low and high doses of 2,4-dinitrotoluene were run at different times, and, therefore, each had its own control group. Mean body weights of treated mice were reduced compared to controls during the study; final body-weight reductions were 9% and 18% in low- and high-dose males and 11% and 24% in low- and high-dose females, respectively. Survival was 78% in high-dose males compared to 74% in concurrent controls and 90% in low-dose males compared to 82% in concurrent controls. The respective figures in females were 72% at the high dose (concurrent controls, 70%) and 84% at the low dose (concurrent controls, 78%). No increase in tumour incidence was reported for any site in treated mice (United States National Cancer Institute, 1978).

In a screening assay based on increased multiplicity and incidence of lung tumours in a strain of mice highly susceptible to the development of this neoplasm, groups of 26 male and 26 female strain A mice, six to eight weeks old, were given 2,4-dinitrotoluene (at a purity of 92–95%, with the major impurity being 2,6-dinitrotoluene) in tricaprylin by gavage twice a week for 12 weeks (total doses, 0, 1200, 3000 and 6000 (maximal tolerated dose, MTD) mg/kg bw). Surviving mice were killed 18 weeks after the last treatment and examined for the gross appearance of lung tumours. Survival was 45/50 in controls, 47/52 at the low dose, 48/52 at the mid dose and 44/52 at the high dose. There was no increase in lung tumour incidence or in the number of lung tumours per mouse when compared to controls. The incidence of lung tumours in survivors was 27% in controls, 28% at the low dose, 31% at the mid dose and 23% at the high dose (Stoner et al., 1984).

After two weeks of acclimatization, groups of 38 male and 38 female weanling Charles River CD-1 mice received diets containing 0, 0.01%, 0.07% and 0.5% [100, 700 and 5000 mg/kg diet (ppm)] 2,4-dinitrotoluene (98.5–99% 2,4-dinitrotoluene, 1–1.5% 2,6-dinitrotoluene; Lee et al., 1985) for 24 months. After 12 months, eight males and eight females from each group were killed and necropsied. The remaining mice were killed at 24 months. Estimates of 2,4-dinitrotoluene intake for controls and low-, mid- and high-dose mice were 0, 14, 95 and 898 mg/kg bw per day, respectively. Body weights of low- and mid-dose females did not differ from those of controls. After three months, weight gains of mid-dose males were lower than those of controls [approximately 10%]; high-dose males and females were also lighter than controls [by approximately 22–35%]. Survival of high-dose mice began to decrease after month 8; median survival was reached in 21, 21, 19 and 9 months for control, low-, mid- and high-dose males and 20.5, 19, 20.5 and 10 months for control, low-, mid- and high-dose females, respectively. Tumours, described as cystic adenoma, cystic papillary adenoma, cystic papillary carcinoma and solid carcinoma, were observed in the renal tubular epithelium. The incidence of these tumours in males was 0/24 in controls, 6/22 at the low dose, 16/19 at the mid dose and 10/29 at the high dose; 2/8 high-dose males examined at one year also had a renal tumour. One of 23 high-dose females developed a renal tumour at 23 months (Hong et al., 1985).

3.1.2 Rat

Groups of 50 male and 50 female Fischer 344 rats, approximately six weeks of age, were fed diets containing 0, 0.008 (0.0075% for the first 19 weeks) or 0.02% [0, 80 (75) or 200 mg/kg diet (ppm)] 'practical-grade' 2,4-dinitrotoluene [purity stated to be > 95% by the supplier] for 78 weeks followed by control diet for 26 weeks. After this time they were killed and subjected to complete histopathological examination. The studies with low- and high-dose rats were run at different times and, therefore, each had its own control group. Mean body weights of treated rats were reduced compared to controls during the study; compared to their concurrent controls, body-weight reductions were very slight in low-dose males, 20–25% in high-dose males, variable in low-dose females and up to 18% in high-dose females. Survival was 58% in high-dose males compared to 52% in concurrent controls and 58% in low-dose males compared to 64% in concurrent controls. The respective figures in females were 52% at the high dose (concurrent controls, 48%) and 62% at the low dose (concurrent controls, 62%). No integumentary tumours were seen in controls, but the incidences of these tumours were increased in low- and high-dose males — fibromas of the skin/subcutaneous tissue occurred in 7/49 low-dose and 13/49 high-dose males; lipomas occurred in 3/49 high-dose males; fibrosarcomas occurred in 1/49 low-dose and 2/49 high-dose males, and squamous-cell papillomas and basal-cell carcinomas each occurred in single low-dose males. In treated females, the incidence of fibroadenomas of the mammary gland was increased (23/50 at the high dose versus 4/23 in controls; $p < 0.016$, Fisher's exact test) (United States National Cancer Institute, 1978).

After two weeks acclimatization, groups of 38 male and 38 female weanling Charles River CD rats received diets containing 0%, 0.0015%, 0.01% and 0.07% [15, 100 and 700 mg/kg diet (ppm)] 2,4-dinitrotoluene (98.5–99% 2,4-dinitrotoluene, 1-1.5% 2,6-dinitrotoluene) for up to 24 months. After 12 months, eight males and eight females from each group were killed and necropsied. The remaining rats were killed at 24 months. Estimates of 2,4-dinitrotoluene intake for control, low-, mid- and high-dose males were 0, 0.57 ± 0.02, 3.92 ± 0.15 and 34.5 ± 0.8 mg/kg bw per day, respectively, and those for females were 0, 0.71 ± 0.02, 5.14 ± 0.18 and 45.3 ± 1.4 mg/kg bw per day, respectively. Body weights of low-dose rats did not differ from those of controls. Weight gains of mid-dose rats were lower after month 9 [by as much as approximately 12% in males and females]. [High-dose rats were up to approximately 33–38% lighter than controls.] Survival of low- and mid-dose rats did not differ from that of controls, whereas all high-dose males and all but one high-dose females died before the end of the study, with 50% mortality being reached between 19 and 20 months for each sex. [Survival was approximately 40–45% for the controls.] At 12 months, neoplastic nodules of the liver were diagnosed in 1/8 low-dose males, 6/7 high-dose males and 7/8 high-dose females with hepatocellular carcinoma in 1/8 high-dose females. None was found in the other groups. In rats surviving longer than 12 months, the incidences of hepatocellular neoplastic nodules and hepatocellular carcinomas were as follows: males — hepatocellular neoplastic nodules: 1/25 in controls, 2/27 at the low dose, 1/19 at the mid dose, 2/27 at the high dose; hepatocellular carcinomas: 1/25, 0/27, 1/19, 6/27, respectively [$p = 0.06$; Fisher's exact test]; females — hepatocellular neoplastic nodules: 0/23 in controls, 2/28 at the low dose, 2/26 at the mid dose, 5/25 at the high dose; hepatocellular carcinomas: 0/23, 0/28, 1/26, 10/25, respectively [$p = 0.03$; Fisher's exact test]. The incidences of mammary gland tumours (predominantly fibroadenomas) in females were 11/23 in controls, 1/28 at the low dose, 16/26 at the mid dose and 21/25 at the high dose [$p < 0.05$; Dunnett's test]; the incidences of skin tumours (mostly fibromas) in males were 2/25 in controls, 4/27 at the low dose, 3/19 at the mid dose and 15/27 at the high dose [$p < 0.05$; Dunnett's test] (Lee et al., 1985).

Groups of 28 male CDF (Fischer 344)/CrlBR rats, weighing 130–150 g and having been in quarantine for four weeks, received a control diet or a diet containing a sufficient quantity of 2,4-dinitrotoluene ['purified', but purity unspecified] to provide a dose of 27 mg/kg bw per day for 52 weeks. The rats were then killed and the livers and lungs evaluated histopathologically. Body-weight gain of the treated group was about 25% less than that of the controls at the end of the study. No early death was indicated. Hepatocyte degeneration and vacuolization were apparent in the majority of treated animals; acidophilic and basophilic foci were reported in 70% and 10% of the livers of treated rats, respectively. One treated rat had a hepatic neoplastic nodule and no liver tumour was noted in the control group (Leonard et al., 1987). [The Working Group noted the small numbers of animals used, the short duration of the study and the incomplete histopathology.]

3.2 Intraperitoneal injection

Mouse: In a screening assay based on increased multiplicity and incidence of lung tumours in a strain of mice highly susceptible to the development of this neoplasm, groups of 52 or 53 male and female strain A mice, six to eight weeks old, were given thrice weekly intraperitoneal injections of 2,4-dinitrotoluene (at a purity of 92–95%, with the major impurity being 2,6-dinitrotoluene) in tricaprylin for eight weeks (total doses, 0, 600, 1500 and 3000 (MTD) mg/kg bw). Surviving mice were killed 22 weeks after the last injection and lung tumours appearing grossly were counted. Survival was 52/52 in controls, 52/53 at the low dose, 52/52 at the mid dose and 50/52 at the high dose. There was no increase in lung tumour incidence or in the number of lung tumours per mouse when compared to controls. The incidence of lung tumours in surviving mice was 29% in controls, 44% at the low dose, 19% at the mid dose and 26% at the high dose (Schut *et al.*, 1982; Stoner *et al.*, 1984).

2,6-Dinitrotoluene

3.1 Oral administration

3.1.1 *Mouse*

In a screening assay based on increased multiplicity and incidence of lung tumours in a strain of mice highly susceptible to the development of this neoplasm, groups of 26 male and 26 female strain A mice, six to eight weeks old, were given 2,6-dinitrotoluene (at a purity of 98%) in tricaprylin twice a week by gavage for 12 weeks (total doses, 0, 1200, 3000 and 6000 (MTD) mg/kg bw). Surviving mice were killed 18 weeks after the last dose and examined for the gross appearance of lung tumours. Survival was 45/50 in controls, 49/52 at the low dose, 50/52 at the mid dose and 38/52 at the high dose. There was no increase in lung tumour incidence or in the number of lung tumours per animal when compared to controls. The incidence of lung tumours in surviving mice was 27% in controls, 18% at the low dose, 22% at the mid dose and 34% at the high dose (Stoner *et al.*, 1984).

3.1.2 *Rat*

In a study designed to evaluate the influence of pectin-induced changes in gut microflora on the hepatocarcinogenicity of 2,6-dinitrotoluene, groups of 30 male CDF (Fischer 344)/CrlBR rats, weighing 130–150 g and having been in quarantine for two weeks, were placed on one of three diets containing sufficient quantities of 2,6-dinitrotoluene (at a purity of 99.9%) to produce daily doses of 0, 0.6–0.7 or 3–3.5 mg/kg bw. Ten animals from each treatment group were killed at three, six and 12 months and livers were evaluated histopathologically. The diets used were open formula, cereal-based NIH-07, purified AIN-76A and AIN-76A with 5% added pectin. Groups receiving the high dose of 2,6-dinitrotoluene gained about 10% less weight than the respective control groups. No early death was reported. The number and size of γ-glutamyl transpeptidase (γ-GT)-staining foci increased in a time- and dose-dependent manner in animals given

2,6-dinitrotoluene in the NIH-07 diet. The group on the NIH-07 diet receiving the high dose of 2,6-dinitrotoluene exhibited a 100% incidence of liver tumours (6/10 with hepatocellular carcinomas and/or 6/10 with neoplastic nodules) at 12 months. Three of 10 rats receiving the low dose in NIH-07 diet had neoplastic nodules at 12 months. No tumour was observed in rats receiving the control diets or 2,6-dinitrotoluene in the AIN-76A diet with or without added pectin (Goldsworthy *et al.*, 1986). [The Working Group noted that pectin did not influence the tumour outcome in this experiment].

Groups of 28 male CDF (Fischer 344)/CrlBR rats, weighing 130–150 g and having been in quarantine for four weeks, received a control diet or a diet containing a sufficient quantity of 2,6-dinitrotoluene ['purified', but purity unspecified] to provide doses of 7 or 14 mg/kg bw per day for 52 weeks. After this time, the rats were killed and the livers and lungs evaluated histopathologically. Body-weight gains of the low-dose and high-dose groups were about 18% and 32% less than those of the controls at the end of the study. No early death was indicated. Hepatocyte degeneration and vacuolization were apparent in the majority of treated animals; acidophilic and basophilic foci were reported in over 90% of treated rats. No liver tumour was noted in the control group. Neoplastic nodules were found in 18/20 rats at the low dose and 15/19 at the high dose. Hepatocellular carcinomas, described as trabecular, occurred in 17/20 at the low dose and 19/19 at the high dose, and one tumour described as an adenocarcinoma was found in a low-dose rat. Cholangiocarcinomas occurred in 2/20 low-dose rats. Liver tumours metastasized to the lung in 3/20 rats at the low dose and 11/19 at the high dose (Leonard *et al.*, 1987).

3.2 Intraperitoneal injection

Mouse: In a screening assay based on increased multiplicity and incidence of lung tumours in a strain of mice highly susceptible to the development of this neoplasm, groups of 26 male and 26 female strain A mice, six to eight weeks old, were given thrice weekly intraperitoneal injections of 2,6-dinitrotoluene (at a purity of 98%) in tricaprylin for eight weeks (total doses, 0, 600, 1500 and 3000 (MTD) mg/kg bw). Surviving mice were killed 22 weeks after the last injection and were examined for the gross appearance of lung tumours. Survival was 52/52 in controls, 50/52 at the low dose, 51/52 at the mid dose and 47/52 at the high dose. There was no increase in lung tumour incidence or in the number of lung tumours per mouse when compared to controls. The incidences of lung tumours in surviving mice were 29% in controls, 34% at the low dose, 45% at the mid dose and 30% at the high dose (Stoner *et al.*, 1984).

3,5-Dinitrotoluene

No adequate data were available to the Working Group.

Technical grades of dinitrotoluene

3.1 Oral administration

3.1.1 Mouse

In a screening assay based on increased multiplicity and incidence of lung tumours in a strain of mice highly susceptible to the development of this neoplasm, groups of 26 male and 26 female strain A mice, six to eight weeks old, were given 2 : 1 mixtures of 2,4-dinitrotoluene (at a purity of 92–95%, with the major impurity being 2,6-dinitrotoluene) and 2,6-dinitrotoluene (at a purity of 98%) in tricaprylin twice a week by gavage for 12 weeks (total doses, 0, 1200, 3000 and 6000 (MTD) mg/kg bw). Surviving mice were killed 18 weeks after the last dose and were examined for the gross appeareance of lung tumours. Survival was 45/50 in controls, 48/52 at the low dose, 48/52 at the mid dose and 48/52 at the high dose. There was no increase in lung tumour incidence or in the number of lung tumours per mouse when compared to controls. The incidence of lung tumours in survivors was 27% in controls, 35% at the low dose, 35% at the mid dose and 33% at the high dose (Stoner et al., 1984).

3.1.2 Rat

Popp and Leonard (1982), Rickert et al. (1984) and Leonard et al. (1987) reported various aspects of a study on the carcinogenicity of technical-grade dinitrotoluene (76.5% 2,4-dinitrotoluene, 18.8% 2,6-dinitrotoluene, 2.43% 3,4-dinitrotoluene, 1.54% 2,3- dinitrotoluene, 0.69% 2,5-dinitrotoluene and 0.04% 3,5-dinitrotoluene) in Fischer 344 rats. In Rickert et al. (1984), data that outlined the incidence of hepatic neoplastic lesions in Fischer 344 rats at all sample intervals were presented in tabular form. The compound was given in the diet at doses of 0, 3.5, 14 and 35 mg/kg bw per day. Due to the mortality in animals at the high-dose level, the final kill of these animals was carried out at 55 weeks. At the high-dose level, the incidences of hepatocellular carcinomas in males were 2/10 at 26 weeks, 10/10 at 52 weeks and 20/20 at 55 weeks, and in females were 0/10, 4/10 and 11/20, respectively; the incidences of neoplastic nodules in males were 0/10 at 26 weeks, 3/10 at 52 weeks and 5/20 at 55 weeks, and in females were 0/10, 8/10 and 12/20, respectively. In the mid-dose group sampled at 26, 52, 78 and 104 weeks, the incidences of hepatocellular carcinomas in males were 0/10, 3/10, 19/20 and 22/23, and in females were 0/10, 0/10, 10/20 and 41/63, respectively; similarly, the incidences of neoplastic nodules were 0/10, 4/10, 11/20 and 16/23 in males, and 0/10, 0/10, 0/20 and 53/69 in females. In the low-dose group, no liver tumour was observed in any animal at 26 or 52 weeks. At 104 weeks, the incidences of hepatocellular carcinomas were 9/70 and 12/61 in males and females, respectively (neoplastic nodules, 11/70 and 12/61, respectively). In control animals, a single hepatocellular carcinoma was observed in males (1/61) at 104 weeks, at which time neoplastic nodules were also seen in 9/61 males and 5/57 females.

Groups of 28 male CDF (Fischer 344)/CrlBR rats, weighing 130–150 g and having been in quarantine for four weeks, received a control diet or a diet containing a sufficient quantity of technical-grade dinitrotoluene (76.5% 2,4-dinitrotoluene, 18.8% 2,6-dinitrotoluene, 2.43% 3,4-dinitrotoluene, 1.54% 2,3-dinitrotoluene, 0.69% 2,5- dinitrotoluene and 0.04% 3,5-dinitrotoluene) to provide a dose of 35 mg/kg bw per day for 52 weeks. After this time the rats were killed and the livers and lungs were evaluated histopathologically. Body-weight gain of the treated rats was about 26% less than that of the controls at the end of the study. Hepatocyte degeneration and vacuolization were apparent in the majority of treated animals; acidophilic and basophilic foci were reported in over 90% of treated animals. No liver tumour was noted in the control group. Neoplastic nodules were found in 10/19 treated rats, hepatocellular carcinomas (trabecular) in 9/19 and cholangiocarcinomas in 2/19 (Leonard *et al.*, 1987).

3.2 Intraperitoneal injection

Mouse: In a screening assay based on increased multiplicity and incidence of lung tumours in a strain of mice highly susceptible to the development of this neoplasm, groups of 26 male and 26 female strain A mice, six to eight weeks old, were given thrice weekly intraperitoneal injections of a 2 : 1 mixture of 2,4-dinitrotoluene (92–95% pure with the major impurity being 2,6-dinitrotoluene) and 2,6 dinitrotoluene (with a purity of 98%) in tricaprylin for eight weeks (total doses, 0, 960, 2400 and 4800 (MTD) mg/kg bw). Surviving mice were killed 22 weeks after the last injection and were examined for the gross appearance of lung tumours. Survival was 52/52 in controls, 48/52 at the low dose, 50/52 at the mid dose and 40/52 at the high dose. There was no increase in lung tumour incidence or in the number of lung tumours per mouse when compared to controls. The incidence of lung tumours in surviving mice was 29% in controls, 33% at the low dose, 28% at the mid dose and 23% at the high dose (Stoner *et al.*, 1984).

Initiation–promotion experiments with various dinitrotoluenes

Oral administration

Rat: Popp and Leonard (1982) and Leonard *et al.* (1983) summarized results with several standard initiation–promotion assays in CDF (Fischer 344)/CrlBR rats, using technical-grade dinitrotoluene (76% 2,4-dinitrotoluene, 18% 2,6-dinitrotoluene and less than 3% of each 2,3-dinitrotoluene, 2,5-dinitrotoluene, 3,4-dinitrotoluene and 3,5-dinitrotoluene) and purified 2,6-, 2,3-, 2,4-, 2,5-, 3,4- and 3,5-dinitrotoluene isomers [of unspecified purity]. They reported weak hepatocyte-initiating activity (scored as γ-GT-positive foci) of technical-grade dinitrotoluene and 2,6-dinitrotoluene. No initiating activity was demonstrated with the other isomers.

The liver foci-promoting activity of technical-grade dinitrotoluene (76.5% 2,4-dinitrotoluene, 18.8% 2,6-dinitrotoluene, 1.5% 2,3-dinitrotoluene, 0.7% 2,5-dinitrotoluene, 2.4% 3,4-dinitrotoluene and 0.1% 3,5-dinitrotoluene) was compared to that of purified 2,4-dinitrotoluene and 2,6-dinitrotoluene (purity > 99.4% for both isomers) in a male CDF (Fischer 344)/CrlBR rat hepatic initiation–promotion protocol. Rats weighing 130–150 g were given a single dose of 150 mg/kg bw *N*-nitrosodiethylamine by intraperitoneal injection. Two weeks later, the animals were placed on diets containing 0.47% 2,4-dinitrotoluene, 0.06, 0.12 or 0.24% 2,6-dinitrotoluene or 0.55 or 0.2% technical-grade dinitrotoluene. Rats receiving technical-grade dinitrotoluene were killed after three or six weeks of feeding and those receiving the purified isomers after six or 12 weeks of feeding. Liver foci scored as γ-GT-positive were increased in all treatments suggesting that 2,4- and 2,6-dinitrotoluenes and technical-grade dinitrotoluene all have promoting potential (Leonard et al., 1986).

4. Other Data Relevant for an Evaluation of Carcinogenicity and its Mechanisms

4.1 Absorption, distribution, metabolism and excretion

4.1.1 *Humans*

The metabolism of dinitrotoluenes has been reviewed (Rickert et al., 1984; Rickert, 1987).

Metabolites of 2,4- and 2,6-dinitrotoluenes have been identified in the urine of workers exposed to technical-grade dinitrotoluene (about 80% 2,4-dinitrotoluene and 20% 2,6-dinitrotoluene). Metabolites of 2,4-dinitrotoluene were 2,4-dinitrobenzoic acid, 2-amino-4-nitrobenzoic acid, 2,4-dinitrobenzyl glucuronide and 2-(*N*-acetyl)amino-4-nitrobenzoic acid; and metabolites of 2,6-dinitrotoluene were 2,6-dinitrobenzoic acid and 2,6-dinitrobenzyl glucuronide (2-amino-6-nitrobenzoic acid was not detected). In addition, the urine contained unchanged dinitrotoluenes. 2,4-Dinitrobenzoic acid and 2-amino-4-nitrobenzoic acid accounted collectively for 74–86% of the dinitrotoluene metabolites detected. The elimination half-life of total dinitrotoluene-related material detected in urine ranged from 1.0 to 2.7 h (Turner et al., 1985).

In 17 workers exposed to technical-grade dinitrotoluene in a dinitrotoluene production plant, personal air sampling revealed levels ranging from 0.6 to 5.9 mg/10 m^3 (0.06–0.59 mg/m^3) 2,4- and 2,6-dinitrotoluenes. The wiping of skin suspected of being contaminated (mainly hands and forehead) showed levels of 'not detected' (< 2 μg) to 179.5†μg 2,4-dinitrotoluene [area unspecified]. 2,4- and 2,6-Dinitrobenzoic acids, 2,4- and 2,6-dinitrobenzyl glucuronides, 2-amino-4-nitrobenzoic acid and 2-(*N*-acetyl)amino-4-nitrobenzoic acid (but not 2-amino-6-nitrobenzoic acid) were found in the urine. Possibly, there was a sex difference as regards the pattern of metabolites; three women appeared to excrete relatively more dinitrobenzyl glucuronides than 14 men (33.3 versus 9.5% of all metabolites). The urine contained more metabolites than would have resulted from the dinitrotoluene present in the inhaled air, which strongly indicated dermal absorption. A rough estimate of the maximal absorbed daily dose encountered indicated an exposure of 0.24–1.00 mg/kg bw in one worker (Levine et al., 1985).

Urinary metabolites have been used for biomonitoring of workers exposed to technical grade dinitrotoluene. 2,4-Dinitrobenzoic acid was the major metabolite found in urine, and appeared within hours of the onset of exposure. In the 28 workers studied (Woolen et al., 1985), the highest concentrations were found in post-shift samples, which contained 17 mg/L 2,4-dinitrobenzoic acid on average. The levels varied considerably between workers and in individual workers from day to day. The atmospheric levels of dinitrotoluene were so low (range, 0.03–0.1 mg/m^3 personal sampling) that the authors concluded that the skin may be a major route of absorption. Pre-shift urinary samples contained very low levels of metabolites (< 1 mg/L), which shows that a large fraction has a rapid turn over — the elimination half-life of 2,4-dinitrobenzoic acid was 2–5 h. However, traces could still be detected after several days, which indicates the presence of a slow compartment. In addition to the metabolites described by Turner *et al.* (1985) and Levine *et al.* (1985), the urine also contained 2-amino-6-nitrobenzoic acid. 2,4- and 2,6-Dinitrotoluenes were found in blood samples.

Guest *et al.* (1982) showed that the faecal material and ileal contents from a patient with ileostomy were capable of metabolizing 2,4-dinitrotoluene, producing nitrosonitrotoluenes and aminonitrotoluenes. Mori *et al.* (1984a) found that all six dinitrotoluene isomers were metabolized to aminonitrotoluenes by *Escherichia coli* isolated from human intestinal contents.

No data on 2,3-dinitrotoluene were available to the Working Group.

4.1.2 *Experimental systems*

(*a*) *Whole animals*

A scheme for the metabolism of 2,6-dinitrotoluene is presented in Figure 1.

Male Wistar rats (200 g) were given an oral dose of 22.2 mg/kg bw 2,4-dinitro-[^{3}H]-toluene in salad oil [not identified] (Mori *et al.*, 1978). Radioactivity in blood and liver reached a peak at 6 h after administration with an elimination half-life from blood of 22 h [liver elimination half-life unspecified]. Approximately 10% of the radioactivity administered was excreted in the bile within 24 h. After seven days, about 46% of the radioactivity was excreted in urine or faeces (Mori *et al.*, 1977).

The distribution of uniformly [^{14}C]-ring-labelled 2,4-dinitrotoluene after oral gavage in corn oil was investigated in male Fischer 344 rats (80–90 days old) at three doses (10, 35 and 100 mg/kg bw) and in females at a dose of 100 mg/kg bw (Rickert & Long, 1980). In male rats, terminal elimination half-lives of total radioactivity ranged from 61 (for the two lowest doses) to 27 h (for the high dose) in plasma and from 51 to 36 h in livers, respectively. Terminal elimination half-lives of radioactivity were similar in male and female rats; however, in female livers, concentrations of radioactivity were only one-half of those found in males. The tissues of both sexes contained unmetabolized 2,4-dinitrotoluene, 2,4-dinitrobenzoic acid and 2-amino-4-nitrobenzoic acid. No evidence of the presence of 2,4-diaminotoluene was found in any tissue examined.

Rickert and Long (1981) also investigated the metabolism and excretion of 2,4-dinitrotoluene in a separate study of male and female Fischer 344 rats given 10, 35 or 100 mg/kg bw uniformly [^{14}C]-ring-labelled 2,4-dinitrotoluene by oral gavage in corn oil. In both males and females, urine was the major route of excretion of 2,4-dinitrotoluene metabolites at all doses: 4-(*N*-acetyl)amino-2-nitrobenzoic acid, 2,4-dinitrobenzoic acid, 2-amino-4-nitrobenzoic acid and 2,4-dinitrobenzyl alcohol glucuronide were the major metabolites, comprising > 85% of the radioactivity excreted in rat urine. Both male and

Figure 1. Metabolism of 2,6-dinitrotoluene

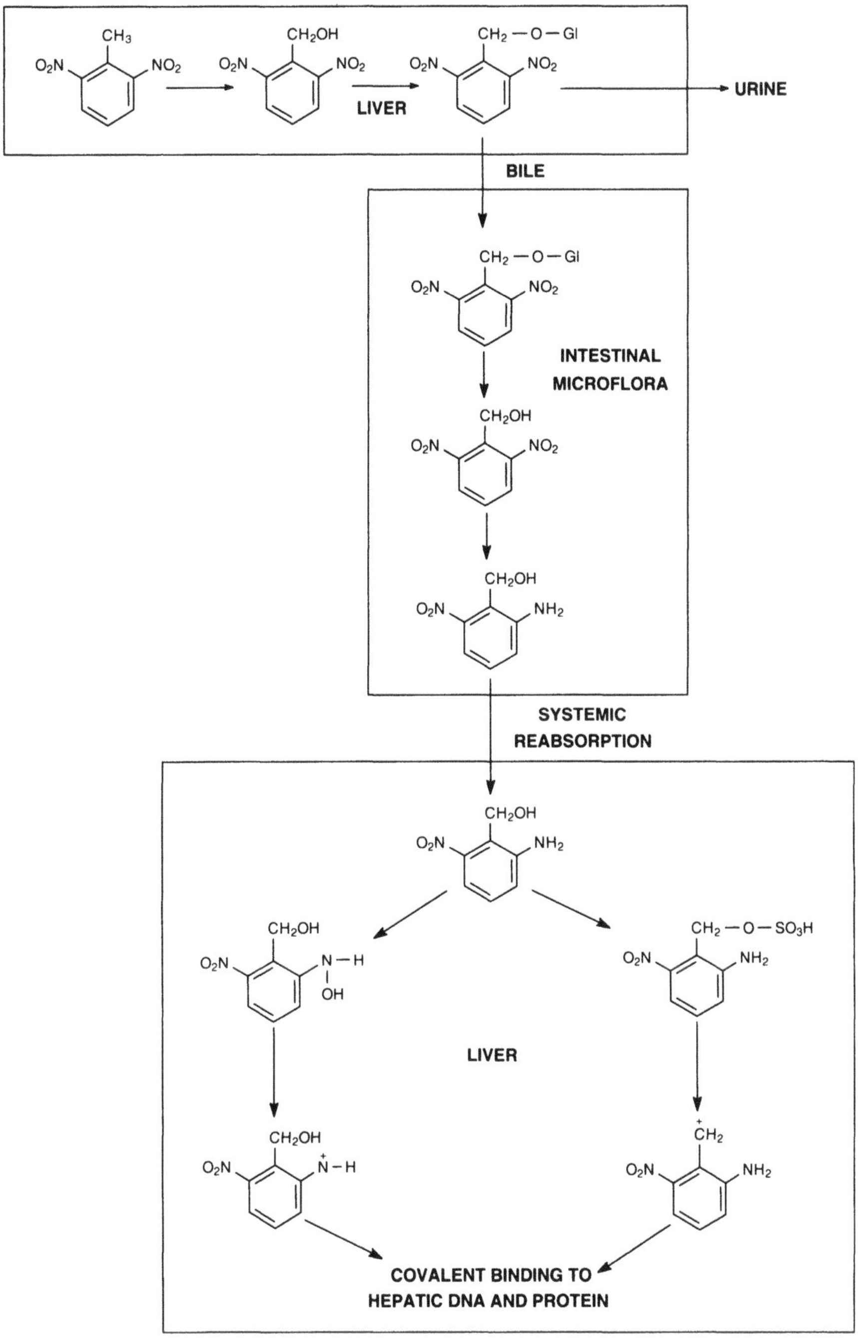

From Chism & Rickert (1985)

female rats showed dose-dependent changes in urinary excretion of 2,4-dinitrotoluene metabolites, with a smaller percentage of the dose being excreted in urine at the higher concentrations. The most significant sex-dependent difference was the greater percentage of 2,4-dinitrotoluene excreted as 2,4-dinitrobenzyl alcohol glucuronide by female rats after the 10- or 35-mg/kg dose. The finding that urine was the major route for elimination of ^{14}C is in agreement with a report by Lee et al. (1975); however, Mori et al. (1978) found that the faeces were the major route of 2,4-dinitrotoluene excretion. This difference among studies may be due to strain differences or to the use of 2,4-dinitro-^{3}H-toluene by Mori et al. (1978).

Rickert et al. (1981) explored the role of gut metabolism in the activation of 2,4-dinitrotoluene in Fischer 344 rats (80–90 days old) weighing 120–150 g (females) or 200–250 g (males). Both conventional and axenic rats received a single oral dose of 35 mg/kg bw uniformly [^{14}C]-ring-labelled 2,4-dinitrotoluene by gavage. Throughout the study, axenic rats, which are lacking in intestinal microflora, were housed in sterile isolation units. Conventional rats were housed in typical temperature- and humidity-controlled exposure rooms. Axenic males and females excreted a smaller fraction of the 2,4-dinitrotoluene dose in the urine than did conventional animals. Most notably, amounts of 4-(N-acetyl)amino-2-nitrobenzoic acid and 2-amino-4-nitrobenzoic acid excreted by axenic animals were one-tenth to one-fifth of those excreted by conventional animals. Additionally, hepatic covalent binding was decreased by one-half in axenic animals. These data suggested to the authors that intestinal microflora play a major role in the appearance of reduced urinary metabolites and of covalently bound material after 2,4-dinitrotoluene administration.

The biliary excretion and enterohepatic circulation of 2,4-dinitrotoluene was investigated in male Fischer 344 rats (200 g) given oral doses of 35, 63 or 100 mg/kg bw uniformly [^{14}C]-ring-labelled 2,4-dinitrotoluene in corn oil or female rats (160 g) given 35 mg/kg bw uniformly [^{14}C]-ring-labelled 2,4-dinitrotoluene (Medinsky & Dent, 1983). The excretion of ^{14}C in bile of male rats was related linearly to dose, with biliary elimination of ^{14}C accounting for approximately 25% of the dose. After a dose of 35 mg/kg bw 2,4-dinitrotoluene, females excreted less ^{14}C in the bile than males (18% of the dose). Over 90% of the radioactivity in the bile was identified as the glucuronide conjugate of 2,4-dinitrobenzyl alcohol. Biliary elimination half-lives for ^{14}C ranged from 3.3 to 5.3 h.

Sayama et al. (1989a) also investigated the biliary excretion of 2,4-dinitrotoluene, 2,4-dinitrobenzyl alcohol and 2,4-dinitrobenzaldehyde in male Wistar rats (180–220 g). 2-Acetylamino-4-nitrotoluene, 2,4-dinitrobenzyl alcohol, 2,4-dinitrobenzaldehyde and unchanged 2,4-dinitrotoluene were detected in the nonhydrolysed neutral basic fraction of bile from rats dosed orally with 2,4-dinitrotoluene (40 mg/kg bw in salad oil [mixed vegetable oil]). The major biliary metabolite of 2,4-dinitrotoluene in male Wistar rats was 2,4-dinitrobenzyl alcohol glucuronide (11.8% of the dose). A variety of minor metabolites, most notably 2,4-dinitrobenzaldehyde (0.27%) and 4-amino-2-nitro(2-amino-4-nitro)benzyl alcohol sulfate (1.5%), were also formed. Similar metabolites were eliminated in the bile after oral administration of 2,4-dinitrobenzyl alcohol and 2,4-dinitrobenzaldehyde.

The metabolism and excretion of the isomer 2,6-dinitrotoluene was investigated by Long and Rickert (1982) in male and female Fischer 344 rats (80–90 days old, 200 and 150 g, respectively). Uniformly [^{14}C]-ring-labelled 2,6-dinitrotoluene (> 99% radiochemically pure) was dissolved in corn oil and administered by oral gavage (10 mg/kg bw). The major route of excretion of ^{14}C after a single dose was via the urine (males, 53.6 ± 2.6%; females, 54.0 ± 4.8%). Faecal excretion accounted for 17.9% (males) and 19.8% (females) of the dose. Analysis of the urine by HPLC revealed three major metabolites that accounted for 95% of the urinary ^{14}C. These metabolites were identified as 2,6-dinitrobenzoic acid, 2,6-dinitrobenzyl alcohol glucuronide and 2-amino-6-nitrobenzoic acid. No sex-dependent difference in the total amounts of the individual metabolites was noted. These results were analogous to those found after administration of 10 mg/kg bw 2,4-dinitrotoluene to male and female Fischer 344 rats. The only major difference in the disposition of the two isomers is that no *N*-acetylaminonitrobenzoic acid was found after administration of 2,6-dinitrotoluene *in vitro*. This may reflect steric hindrance to *N*-acetylation of an amino group adjacent to a methyl group.

Mori *et al.* (1989a) also reported on the metabolism of 2,6-dinitrotoluene in male Wistar rats (180–200 g). Rats were dosed orally with a solution of 2,6-dinitrotoluene (75 mg/kg bw) in 1 ml of salad oil [type unspecified]. The bile ducts of some rats were cannulated prior to dosing. The major urinary metabolite identified by chromatography with standards was 2,6-dinitrobenzyl alcohol, which represented 1.5% of the dose excreted in 24 h. The glucuronide conjugate of 2,6-dinitrobenzyl alcohol was eliminated in the bile and accounted for 30% of the dose excreted in 24 h. Small percentages (< 0.1%) of other metabolites were also detected in bile. In these studies, no 2,6-dinitrobenzoic acid was detected in the urine. [This discrepancy between the findings of Mori *et al.* (1989a) and those of Long and Rickert (1982) may be explained by methodological differences, or may indicate strain differences in the metabolism of 2,6-dinitrotoluene.]

Chadwick *et al.* (1990) examined the gastrointestinal enzyme activity and activation of 2,6-dinitrotoluene in male CD-1 mice (38.2 g) and male Fischer 344 rats (204 g) dosed orally with 75 mg/kg bw in dimethyl sulfoxide (DMSO). Mice metabolized significantly more 2,6-dinitrotoluene to mutagenic urinary metabolites than did Fischer 344 rats. Mutagenicity was evaluated by adding β-glucuronidase-hydrolysed urine samples to cultures of *Salmonella typhimurium* strain TA98 and quantitating revertants detected in urine. The investigators noted that native intestinal nitroreductase activity was markedly higher in the CD-1 mouse than in the Fischer 344 rat and that this correlated with the increase in revertants. In contrast to these observations, a comparison of the metabolism and binding of 2,6-dinitrotoluene in Fischer 344 rats and A/J mice indicated a higher hepatic DNA binding of 2,6-dinitrotoluene in the rats (Dixit *et al.*, 1986). In addition, the metabolism of 2,6-dinitrotoluene was very slow in the small intestines of A/J mice (Schut *et al.*, 1983).

The hepatic macromolecular covalent binding and intestinal disposition of 2,6-dinitrotoluene was compared with that of 2,4-dinitrotoluene (Rickert *et al.*, 1983). Male Fischer 344 rats, 80–90 days old, received 10 or 35 mg/kg bw of each uniformly [^{14}C]-ring-labelled isomer orally by gavage in corn oil. Covalent binding of radioactivity after administration of 2,6-dinitrotoluene was always two- to five-fold higher than binding after administration of 2,4-dinitrotoluene.

Further studies comparing the hepatic macromolecular covalent binding of 2,6- and 2,4-dinitrotoluenes were conducted by Kedderis et al. (1984) in male Fischer 344 rats (180–260 g). Animals were administered intraperitoneal injections of the sulfotransferase inhibitors 2,6-dichloro-4-nitrophenol or pentachlorophenol (40 µmol/kg bw) in propane-1,2-diol or vehicle alone. Subsequently, animals were dosed orally with 2,6-dinitro[3-^{3}H]toluene or uniformly [^{14}C]-ring-labelled 2,4-dinitrotoluene dissolved in corn oil. At 12 h after dinitrotoluene administration, the livers were removed and processed for isolation of DNA, and urine was analysed by HPLC. Prior administration of the sulfotransferase inhibitors resulted in a significant decrease in the hepatic macromolecular covalent binding of both isomers with the decrease being more pronounced for 2,6-dinitrotoluene. Covalent binding to hepatic DNA was reduced by > 95% (2,6-dinitrotoluene) or > 84% (2,4-dinitrotoluene). The authors suggested that metabolites formed via a sulfotransferase-dependent pathway were responsible for the majority of the covalent binding of 2,6-dinitrotoluene to hepatic DNA.

(b) In-vitro studies

(i) *Subcellular hepatic fractions*

Postmitochondrial supernatants from the livers of CD rats, CD-1 mice, New Zealand albino rabbits, beagle dogs and rhesus monkeys were compared in their relative capacities to metabolize 1 mM 2,4-dinitro[^{14}C]toluene (Short et al., 1979). All species metabolized 2,4-dinitrotoluene. The production of dinitrobenzyl alcohol was generally greater in rabbits, dogs and monkeys than in rats and mice.

Decad et al. (1982) examined the hepatic microsomal metabolism and covalent binding of 2,4-dinitro[^{14}C]toluene in hepatic microsomes from adult male Fischer 344 rats (180–200 g). Incubations contained a microsomal protein concentration of 2 mg/mL and a substrate concentration of 200 nmol. The pattern of 2,4-dinitrotoluene metabolism was dependent on oxygen tension. Under aerobic conditions, 2,4-dinitrobenzyl alcohol was the major metabolite formed. In contrast, under anaerobic conditions no dinitrobenzyl alcohol was detected and 4-amino-2-nitrotoluene and 2-amino-4-nitrotoluene were the major metabolites. The authors suggested that oxidative metabolism of 2,4-dinitrotoluene to dinitrobenzyl alcohol was mediated by cytochrome P450-dependent mixed-function oxidases, and that the capacity to oxidize the alcohol further probably resided in the cytosol and might be catalysed by alcohol and aldehyde dehydrogenase.

Chapman et al. (1992) investigated the metabolism of 100 µM 2,6-dinitro[^{3}H]toluene by liver microsomal and cytosolic fractions obtained from male Fischer 344 rats (200–300 g) under aerobic (100% oxygen) and anaerobic (100% nitrogen) incubation conditions for 20 min. With liver microsomes, the major metabolites were 2,6-dinitrobenzyl alcohol under aerobic conditions and 2-amino-6-nitrotoluene under anaerobic conditions. According to the authors, the results using metabolic inhibitors indicated that xanthine oxidase contributes to the hypoxanthine-supported anaerobic metabolism of 2,6-dinitrotoluene in cytosol but that, in microsomes, reductive metabolism is mediated by cytochromes P450. There was a relative absence of reduced 2,6-dinitrotoluene metabolites in microsomal and cytosolic incubations under aerobic conditions, suggesting to the authors that, under fully oxygenated conditions, hepatic reduction of 2,6-dinitrotoluene probably does not contribute significantly to the activation of 2,6-dinitrotoluene *in vivo*.

Mori *et al.* (1989b) investigated the metabolism of 2,4- and 2,6-dinitrotoluenes and their dinitrobenzyl alcohols and benzaldehydes using liver microsomal and cytosolic fractions prepared from male Wistar (180–200 g) and Sprague-Dawley (180–200 g) rats. Incubation mixtures contained 1 mM of substrates and 1 or 2 mg of microsomal or cytosolic protein, respectively. The authors concluded that the dinitrobenzaldehydes were intermediates in the oxidation of dinitrobenzyl alcohols to dinitrobenzoic acids and that the oxidation of dinitrobenzyl alcohols to dinitrobenzaldehydes and the reduction of dinitrobenzaldehydes to dinitrobenzyl alcohols were reversible reactions. The investigators also determined that the oxidation of 2,6-dinitrotoluene to 2,6-dinitrobenzyl alcohol was higher than that of 2,4-dinitrotoluene to 2,4-dinitrobenzyl alcohol in both strains and that the rate of oxidation in Wistar rat microsomes was higher than that in Sprague-Dawley rat microsomes. The reduction of dinitrobenzaldehydes to dinitrobenzyl alcohols was the highest reaction for both strains, and the reduction of 2,4-dinitrobenzaldehyde to 2,4-dinitrobenzyl alcohol in Wistar rats was particularly high.

Mori *et al.* (1984b) investigated the reduction of 2,4-dinitrotoluene in male Wistar rat (190–220 g) liver microsomal and cytosol fractions. 2,4-Dinitrotoluene was reduced by nicotinamide-adenine-dinucleotide-phosphate hydrolase (NADPH)-dependent microsomal activity to 2-amino-4-nitrotoluene and 4-amino-2-nitrotoluene under anaerobic conditions. However, further reduction to 2,4-diaminotoluene could not be demonstrated, whereas 2,4-diaminotoluene could be produced in incubations using hepatic cytosol fractions. Reduction of 2,4-diaminotoluene by cytosolic enzymes was blocked by the xanthine oxidase inhibitor allopurinol. The authors suggested that diaminotoluene production by cytosol enzymes is due to both cytosolic xanthine oxidase and NADPH-dependent cytosolic enzymes.

Using hepatic microsomes obtained from 200–250-g male or 175–225-g female Fischer 344 rats, Kedderis and Rickert (1985) characterized the oxidation of 2-amino-4-nitrobenzyl alcohol and 2-amino-6-nitrobenzyl alcohol, metabolites of 2,4- and 2,6-dinitrotoluenes, respectively. The investigators found that further metabolism of 2-amino-6-nitrobenzyl alcohol resulted in the formation of two metabolites, both of which were reducing agents. One was identified as 2-hydroxylamino-6-nitrobenzyl alcohol and the other metabolite was tentatively identified as 2-amino-5-hydroxy-6-nitrobenzyl alcohol.

The in-vitro activation of 2-amino-6-nitrobenzyl alcohol was investigated using incubations of cytosol and microsomes prepared from male Fischer 344 rats (250 g, 100 days old) (Chism & Rickert, 1989). Subcellular fractions were incubated in the presence of NADPH, 3'-phosphoadenosine-5'-phosphosulfate and acetyl coenzyme A, 2 mg calf thymus DNA and 84 µM substrate. 2-Amino-6-nitrobenzyl alcohol was converted enzymatically to metabolites capable of binding to calf thymus DNA when incubated with cytosol and 3′-phosphoadenosine-5′-phosphosulfate or with microsomes and NADPH. However, when cytosol and microsomes were incubated together, activation of 2-amino-6-nitrobenzyl alcohol appeared to require only 3′-phosphoadenosine-5′-phosphosulfate, which suggested a minor role for NADPH-dependent enzymes in the activation of 2-amino-6-nitrobenzyl alcohol.

(ii) *Isolated hepatocytes and whole livers*

Bond and Rickert (1981) investigated the metabolism of 2,4-dinitrotoluene by freshly isolated primary hepatocytes from male Fischer 344 rats (200–250 g) at incubation concentrations of 100 μM. The primary metabolite formed was 2,4-dinitrobenzyl alcohol, which accounted for 75–80% of the total. Much smaller amounts of 2,4-dinitrobenzyl alcohol glucuronide or the amino or acetylamino metabolites occurred. The small amounts of these reduced dinitrotoluene metabolites formed suggested to these investigators that, under physiological conditions, the hepatic reductive metabolism of 2,4-dinitrotoluene probably plays a minor role in the overall metabolism of 2,4-dinitrotoluene.

The sex-dependent metabolism and biliary excretion of 2,4-dinitrotoluene was evaluated in isolated perfused livers from male and female Fischer 344 rats (80–90 days old, weighing approximately 200 and 160 g, respectively). Isolated livers were exposed for 90 min to 2,4-dinitro[^{14}C]toluene added to a recirculating perfusate at initial concentrations of 20 or 70 μM. Isolated perfused livers from both male and female rats displayed a capacity for oxidation, reduction, acetylation and conjugation of 2,4-dinitrotoluene or its metabolites. Oxidation of 2,4-dinitrotoluene to 2,4-dinitrobenzyl alcohol followed by glucuronidation to 2,4-dinitrobenzyl alcohol glucuronide was the major route of 2,4-dinitrotoluene metabolism (Bond *et al.*, 1981).

The metabolism of 2,6-dinitrotoluene was investigated in isolated perfused livers of Fischer 344 rats (Long & Rickert, 1982). Male and female rats (80–90 days old) weighing approximately 200 and 150 g, respectively, were anaesthetized and their livers were removed, placed in a recirculating perfusion apparatus for up to 90 min and exposed to 20 or 70 μM 2,6-dinitro[^{14}C]toluene. 2,6-Dinitrobenzyl alcohol glucuronide was the major metabolite found in both the perfusate and bile. Biliary excretion of 2,6-dinitrobenzyl alcohol glucuronide by livers from male rats was 3.3- and 8.6-fold greater than that for female rats at 20 and 70 μM, respectively. The results of studies using isolated perfused livers and the 2,6-dinitrotoluene isomer paralleled those described above for 2,4-dinitrotoluene (Bond *et al.*, 1981).

(iii) *Intestinal microflora*

The metabolism of 2,4-dinitrotoluene by caecal microflora from male Fischer 344 rats was investigated by Dent *et al.* (1981) using a substrate concentration of 2,4-dinitro-[^{14}C]-toluene of 100 μM. The authors reported that rat caecal microflora metabolized 2,4-dinitrotoluene via an ordered sequence of reductive steps. The reductive metabolic capacity of the caecal contents on a per gram weight basis exceeded that of the liver by a factor of 1000, suggesting that the caecum represents a major site for reductive metabolism of dinitrotoluene. Similarly, Sayama *et al.* (1993) investigated the intestinal biotransformation of 2,4- or 2,6-dinitrotoluene in male Wistar rats (180–200 g). The microflora were prepared from the entire intestinal (duodenum to rectum) contents of the rats. The incubation solutions contained 15 μmol 2,6- or 2,4-dinitrotoluene/4 mL incubation medium. Anaerobic incubation of 2,6-dinitrotoluene with intestinal microflora indicated that 2,6-dinitrotoluene was transformed to 2-nitroso-6-nitrotoluene, 2-hydroxylamino-6-nitrotoluene, 2-amino-6-nitrotoluene and 2,6-diaminotoluene. The presumed intermediates, the aminonitrosotoluenes and aminohydroxylaminotoluene, were not detected (Guest et al., 1982).

Mori et al. (1985) also examined the intestinal metabolism of 2,4-dinitro[^{3}H]toluene in male Wistar rats (200–220 g). These studies were carried out under anaerobic conditions with 2,4-dinitrotoluene concentrations of 25, 50 or 100 μM, using preparations of microflora from the contents of the caecum (20 mg caecal contents/mL incubation mixture). The formation of the reduced metabolites was found to proceed in an ordered fashion, with the formation of 4-hydroxylamino-2-nitrotoluene and 2-hydroxylamino-4-nitrotoluene followed by 4-amino-2-nitrotoluene and 2-amino-4-nitrotoluene. As the concentrations of these two intermediates began to disappear in the incubation mixture, the formation of the final product, 2,4-diaminotoluene, occurred.

Products detected in the anaerobic and aerobic incubation of 2,4-dinitrotoluene with *S. typhimurium* strains TA98 and TA98/1,8-DNP$_6$ were nitrosonitrotoluenes, hydroxylaminonitrotoluenes, aminonitrotoluenes and dimethyl dinitroazoxybenzene (Sayama et al., 1991).

Data on absorption, distribution, metabolism and excretion of dinitrotoluenes in whole animals suggest that, upon absorption, these compounds are metabolized extensively in the liver primarily by oxidation to dinitrobenzyl alcohol followed by conjugation with glucuronic acid. The glucuronide conjugate can be eliminated in the bile, which represents a significant route of elimination for these metabolites, or it can be eliminated in the urine. Reduction of the nitro groups occurs primarily in the intestine. The intestinal microflora are responsible for both cleavage of the glucuronic acid and reduction of one or both of the nitro groups. The resultant metabolite can be absorbed from the gut and returned to the liver where subsequent oxidation, or conjugation with sulfate, can result in the production of unstable species that can bind covalently to hepatic DNA and protein (Fig. 1).

Overall, studies with subcellular fractions and hepatocytes prepared from various species have shown that the dinitrotoluene isomers can be oxidized to dinitrobenzyl alcohol. Very little reduction of the nitro groups takes place in the hepatocytes or subcellular hepatic fractions, unless these samples are incubated under anaerobic conditions. In contrast, reductive metabolism of dinitrotoluenes occurs in incubations with intestinal microflora. The reductive capacity of intestinal microflora is approximately 1000-fold greater than that of the liver. Studies with isolated perfused livers support the hypothesis that oxidation and conjugation are the primary routes for hepatic metabolism of these isomers, and that biliary excretion of the glucuronide conjugate is an important pathway in dinitrotoluene disposition.

4.2 Toxic effects

4.2.1 *Humans*

Using medical examinations, a total of 154 workers in a military plant manufacturing powder containing technical dinitrotoluene were followed for 12 months. During that period, 112 reported complaints, and 84 had objective evidence of sickness. The most common complaints were unpleasant taste (62%), weakness (51%), headache (49%), loss of appetite (47%), dizziness (44%), nausea (37%), insomnia (37%), pains in the extremities (26%), vomiting (23%) and numbness and tingling (19%). Among the findings were pallor (36%), cyanosis (34%), anaemia (23%), leucocytosis (12%), leucopoenia (3.2%) and acute toxic hepatitis with jaundice (1.4%) (McGee *et al.*, 1942).

In the study of Levine *et al.* (1986), described in Section 2, of workers who had been employed in the 1940s and 1950s for at least one month in two munitions plants where they had had the possibility of substantial exposure to dinitrotoluene [but only marginally to nitroglycerine or ethylene glycol dinitrate], there was an increase in total mortality as compared to the general population (SMR, 1.29; $p \leq 0.001$; 164 deaths). This was mainly due to ischaemic heart disease (SMR, 1.41; $p \leq 0.01$; 64 deaths). The increase occurred more than 15 years after onset of employment (SMR, 1.54; $p \leq 0.001$). Further, such deaths occurred mainly among those with high intensity of exposure to dinitrotoluene and employment lasting longer than five months (SMRs, 2.24 and 2.05 in the two factories; both $p \leq 0.05$). There was no information on smoking habits or other established risk factors for cardiovascular disease. However, the authors stress that smoking was prohibited in the plants and that there was no increase in mortality from lung cancer or respiratory diseases. Further, at a military examination near the time of cohort entry, the workers in one of the factories did not differ from other young men in the state. Cerebrovascular mortality was not increased. There was no death from non-malignant diseases of the blood and blood-forming organs [expected number not stated], nor were there deaths attributed to diseases of the liver, besides cirrhosis (SMR, 1.03; 4 deaths). The mortality from accidents, poisonings and violence was high (SMR, 1.91; $p = 0.0007$; 28 deaths; Levine *et al.*, 1986). [The workers were subject to pre-employment medical examination and an extensive medical surveillance programme; those subjects with abnormal electrocardiogram or hypertension could be restricted from work in operations with potential exposure to dinitrotoluene (Stayner *et al.*, 1992). This may have detracted from the effect on cardiovascular disease.]

In an extended study of one of the munitions factories studied by Levine *et al.* (1986) (described in Section 2), 4989 workers were employed for more than five months in jobs with probable exposure to dinitrotoluene (> 1 day of exposure) and total deaths were as expected (SMR, 1.00; 95% CI, 0.9–1.1; 747 deaths). There was an increase in mortality from mental and personality disorders (SMR, 2.2; 95% CI, 1.2–3.9; 12 deaths), which was primarily due to alcoholism. Ischaemic heart disease was not increased (SMR, 0.98; 253 deaths). The authors stress that the results may have been affected by an extensive medical screening programme and that the apparent discrepancy versus the results obtained by Levine *at al.* (1986) in the same factory may be due to differences in the definition of exposure intensity (lower here) and the length of the follow-up (Stayner et al., 1992).

Single cases of patch test (Kanerva et al., 1991) and photopatch test (Emtestam & Forsbeck, 1985) positivity have been reported in patients with eczema of the hand and occupational exposure to dynamite.

4.2.2 Experimental systems

(a) Single-dose studies

Data on acute toxicity following a single oral dose of five dinitrotoluene isomers in rats and mice were reported by Vernot et al. (1977). LD_{50} values and 95% CIs were estimated using the moving-average technique. Values obtained for 2,4- and 2,6-dinitrotoluenes are presented below (Table 3). The animals used in these studies were male Sprague-Dawley rats (200–300 g) and CF-1 mice (22–28 g). The vehicle in which the materials was administered was not specified.

Table 3. Acute toxicity in rats and mice following single oral dose of dinitrotoluenes

Material studied	LD_{50} mg/kg (95% CI)	
	Male rats	Mice
2,4-Dinitrotoluene	270 (180–400)	1630 (1180–2240)
2,6-Dinitrotoluene	180 (130–240)	1000 (590–1700)

From Vernot et al. (1977)

Male Fischer 344 rats (180–200 g) were injected intraperitoneally with 2,6-dinitrotoluene and its metabolites 2,6-diaminotoluene and 2-amino-6-nitrotoluene dissolved in DMSO at doses of 1.2 mmol/kg bw (219 mg/kg) or 0.3 mmol/kg (55 mg/kg) (La & Froines, 1993). Intraperitoneal administration of 0.3 mmol/kg 2,6-dinitrotoluene to rats resulted in 50% lethality within two days. 2,6-Diaminotoluene and 2-amino-6-nitrotoluene did not produce the death of any animal even at 1.2 mmol/kg. Histopathological examination of livers showed that 2,6-dinitrotoluene induced extensive centrilobular haemorrhagic necrosis, whereas no evidence of necrosis was observed in rat livers at either dose level of 2,6-diaminotoluene or 2-amino-6-nitrotoluene.

Studies by La and Froines (1992a) investigated the toxicity of 2,4- and 2,6-dinitrotoluenes. All rats administered 150 mg/kg bw 2,6-dinitrotoluene either intraperitoneally or by gavage died within 24 h. None of the rats administered 375 mg/kg bw 2,4-dinitrotoluene died. Only the 2,6-isomer exhibited hepatotoxicity producing extensive centrilobular haemorrhagic necrosis. According to the authors, their findings were consistent with previous studies that compared the toxicities of 2,4- and 2,6-dinitrotoluenes. For example, Leonard et al. (1987) demonstrated increased activities of γ-glutamyltransferase and alanine aminotransferase only in rats treated with 2,6-dinitrotoluene. Increased activities of these serum enzymes are indicative of liver injury. Additionally, Mirsalis and Butterworth (1982) demonstrated differences in toxicity between 2,4- and 2,6-dinitrotoluenes *in vitro*. Hepatocytes isolated from rats pretreated with 2,6-dinitrotoluene (100 mg/kg) did not survive in culture because of its toxicity.

(b) Repeated-dose studies

Kozuka *et al.* (1979) investigated the subchronic toxicity of 2,4-dinitrotoluene incorporated into a standard commercial diet at a concentration of 0.5% and fed *ad libitum* for six months to male Wistar rats, seven weeks old at the start of the study. The average consumption of 2,4-dinitrotoluene calculated from the amount of diet consumed was approximately 66 mg per day during the first three months and 75 mg per day in the subsequent three months. [The estimated daily dose of 2,4-dinitrotoluene ranged from 330 mg/kg bw at the beginning of the study to 500 mg/kg bw at the end.] A variety of adverse clinical signs were noted in the test group, including piloerection, whitened skin colour, humpback, incoordination, decreases in spontaneous movements and general weakness. The mortality of the treated rats was about 71% after a 26-week period. There were significant reductions in body-weight gain. The relative weights of liver, spleen and kidney were increased significantly (approximately two-fold compared with controls), while testes weights were significantly decreased. The formation of puruloid matter was noted in the livers of treated rats. The percentage of methaemoglobin was significantly increased in treated rats. At the end of the study, methaemoglobin levels of 7% were noted in the treated rats compared with 1.13% in the controls. In rats fed 0.5% 2,4-dinitrotoluene for one month, triglycerides and glucose levels in serum were elevated compared with controls. Serum enzymes (GOT, lactate dehydrogenase (LDH) and alkaline and acid phosphatase) were also increased, suggesting evidence of hepatotoxicity. No histopathology was reported.

Weanling CD rats were fed 2,4-dinitrotoluene at concentrations of 0, 0.07, 0.2 or 0.7% of the diet for 13 weeks (Lee *et al.*, 1985). The low dose, estimated to be 34 mg/kg bw per day in males and 38 mg/kg bw per day in females, caused depressed weight gain. The mid dose, estimated to be 93 or 108 mg/kg bw per day day, was toxic and caused reticulocytosis, splenic haemosiderosis and decreased spermatogenesis. The high dose, 266 or 145 mg/kg bw per day, was progressively more toxic, causing death in all of the animals by the end of the 13-week period. Anaemia, reticulocytosis, excessive pigment deposits in the spleen and decreased spermatogenesis were the most marked findings.

Weanling CD-1 mice were fed 0, 0.07, 0.2 or 0.7% 2,4-dinitrotoluene in the diet for 13 weeks (Hong *et al.*, 1985). The highest dose caused mild anaemia with concurrent reticulocytosis, mild degeneration of the seminiferous tubules, mild hepatocellular dysplasia and pigmentation in the Kupffer cells of the liver in males and females. Decreased erythrocyte counts, haematocrit and haemoglobin concentration were also noted.

The chronic toxicity of 2,4-dinitrotoluene was investigated in beagle dogs (Ellis *et al.*, 1985), five to six months old at the time of study, that were given 2,4-dinitrotoluene in a hard gelatin capsule. The major adverse effect of 2,4-dinitrotoluene in dogs was a neuropathy characterized by incoordination and paralysis. Vacuolation, endothelial proliferation and gliosis of the cerebella of some of the affected dogs was observed. These effects were seen within two years in one dog given 1.5 mg/kg bw per day, within six months in all dogs given 10 mg/kg bw per day and within two months in all dogs given 25 mg/kg bw per day. Some dogs progressed to complete paralysis leading to death. Methaemoglobinaemia was also noted and the presence of Heinz bodies was common. Biliary tract hyperplasia was also noted.

In studies conducted for up to two years in which 2,4-dinitrotoluene was added to the feed of CD rats (weanlings at start of study) at doses of 0.0015, 0.01 and 0.07%, the highest dose resulted in a shortened lifespan (Lee *et al.*, 1985). Toxic anaemia, atrophy of the testes and depression of spermatogenesis were observed. After feeding for 12 months, lesions occurred in the liver. The initial lesions were small foci or areas of altered hepatocytes referred to by the authors 'hyperplastic foci'. In these animals, the liver architecture was preserved. The lesions progressed to 'hyperplastic nodules'.

Weanling CD-1 mice were fed 0.01%, 0.07% or 0.5% 2,4-dinitrotoluene in the diet for up to two years (Hong *et al.*, 1985). The authors estimated that these concentrations provided doses equal to 14, 95 or 898 mg/kg bw per day. Toxic anaemia was observed in high-dose males and females after feeding of 2,4-dinitrotoluene for 12 months. Methaemoglobin formation and the presence of numerous Heinz bodies were noted. By 21 months, all high-dose mice had died. In this group, generalized abnormal pigmentation was noted in many tissues, which was dose-dependent and increased with the length of treatment. Liver and kidney were the organs most affected. Hepatocellular dysplasia was observed. The authors use this term to characterize metabolic, degenerative and hyperplastic alterations of the cells. In males, the incidence of this lesion was increased in all three dose groups. In females, the incidence was only increased in the high-dose group.

(c) In-vitro toxicity

A reduction in the incorporation of tritiated thymidine was observed in primary cultures of aortic smooth muscle cells from dinitrotoluene-exposed animals relative to vehicle controls (Ramos *et al.*, 1991a,b). Male Sprague-Dawley rats (150–180 g) were injected intraperitoneally on five days a week for eight weeks with 2,4- or 2,6-dinitrotoluene (0.5, 5 or 10 mg/kg bw) or vehicle oil (control). Histopathological evaluation of aortae from animals exposed to either isomer showed dysplasia and rearrangement of aortic smooth muscle cells at all doses tested. Exposure of smooth muscle cells from naïve animals to dinitrotoluene *in vitro* (1, 10 or 100 μM) did not alter the extent of thymidine incorporation into DNA (Ramos *et al.*, 1991b).

The relation of various structural parameters to hepatotoxic potential was investigated by exposing isolated rat hepatocytes from young adult male Sprague-Dawley rats (320–410 g) to six dinitrotoluene isomers (Spanggord *et al.*, 1990). Dinitrotoluene-induced hepatotoxicity correlated with inhibition of protein synthesis and increases in lactate dehydrogenase release but not with lipid peroxidation. *ortho*- and *para*-Substituted isomers were more hepatotoxic at the same concentration than *meta*-substituted isomers. One group, 2,3-, 3,4- and 2,5-dinitrotoluenes, in which the nitro substituents are oriented either *ortho* or *para* to each other, was toxic at appreciably lower concentrations than the other group, which included 2,6-, 2,4- and 3,5-dinitrotoluenes in which the nitro groups are oriented *meta* to each other. It was concluded that the reducibility of the nitro groups of the parent chemicals is the principal factor for determining their hepatotoxic potential. The results indicated that dinitrotoluene isomers and/or their metabolites produced by the hepatocytes are directly cytotoxic to the cells without the need for metabolic activation by gut microflora. Comparison of the effects of 2,6-dinitrotoluene and 2,4-dinitrotoluene indicated that the inhibition of protein synthesis was similar between the two chemicals. However, at all concentrations tested, the release of LDH from the hepatocytes was higher after 2,6-dinitrotoluene administration compared to 2,4-nitrotoluene administration *in vitro*. Thus, the LDH release correlates with differences in hepatotoxicity between 2,6- and 2,4-dinitrotoluenes observed in in-vivo studies.

4.3 Reproductive and developmental effects

4.3.1 *Humans*

In the United States, an employee at a toluene diamine/dinitrotoluene facility in Kentucky voiced to his union a concern about excessive miscarriages; this led to an investigation by the United States National Institute for Occupational Safety and Health (Hamill *et al.*, 1982). The sperm counts of nine men who were exposed to the compounds (technical grade) were significantly lower than those of nine control men who had never been exposed; however, the sperm count of the control group was unaccountably high. As the authors concluded that there was a strong suggestion of a reproductive problem amongst male workers exposed to the compounds, a larger study was commissioned by the company operating the plant. This study was carried out in a different plant operated by the company in Louisiana. The operating procedures and types of exposure to the compounds were stated to be almost identical to those of the Kentucky plant and environmental monitoring showed that average levels both in area and personal samples were comparable between the two plants. Each worker was subjected to a physician's urogenital examination, an estimation of testicular volume, an assessment of serum follicle-stimulating hormone, an analysis of semen for sperm count and morphology, and an interview about reproductive history and factors related to fertility. The intensity, frequency and timing of exposure was assessed during this interview by a review of company and union work rosters and discussions with supervisors. Ninety-four men were identified by supervisors and work rosters as ever having been exposed to dinitrotoluene/toluene diamine, of whom 78 (83%) participated in the study. Of a total of 203 workers interviewed [total number eligible not specified], 200 provided blood for follicle-stimulating hormone determination and 150 of the 175 workers who had not had a vasectomy provided at least one semen specimen. Of workers classified as having the lowest level of exposure, 12% reported that they had tried to conceive for a year without success compared with 13.5% who reported 'low to high' intensity of exposure, more than six months prior to the study and 17% who reported 'low to high' exposure within six months of the study. The corresponding proportions who reported miscarriages were 19.3%, 27.0% and 13.3%. These differences were not statistically significant. There was no noteworthy difference in physical examination characteristics between the groups. Analysis of fertility rates was stated to show no decrease related to exposure to dinitrotoluene, either in the comparison of exposed workers with unexposed workers or by comparison of fertility rates of exposed workers during time periods when they held jobs that incurred exposure to dinitrotoluene and rates during other jobs held at the plant at which there was no exposure. No statistically significant difference between the different groups was observed in terms of levels of follicle-stimulating hormone, mean sperm count or mean percentage of normal morphology, although exposed workers had higher mean sperm counts and mean percentage of normal morphologies compared with colleagues who were not or slightly exposed.

4.3.2 *Experimental systems*

Groups of 13–23 pregnant Fischer 344 rats were administered 14, 35, 37.5, 75, 100 or 150 mg/kg bw per day technical-grade dinitrotoluene (76% 2,4-dinitrotoluene, 19% 2,6-dinitrotoluene, < 1% 3,5-dinitrotoluene and other isomers) by gavage on gestational days 7–20 (Price *et al.*, 1985). The mortality rates of the treated females were 4.5, 7.7, 0.0, 0.0, 4.3 and 46.2%, respectively. A group of 37 control females received the corn oil vehicle only, and 36 females received 200 mg/kg bw per day hydroxyurea and served as a positive control group for the evaluation of embryotoxic and teratogenic effects. No maternal death occurred in these groups. As the mortality rate in the group receiving the highest dose of dinitrotoluene was unexpectedly high, mated animals from the second and third of three sequential breedings were treated with 14, 37.5 or 100 mg/kg bw per day of the compound. Thus, the distribution of females was not balanced across the three breedings for this study. At sacrifice on gestational day 20, the haematological profile for dams in the 100-mg/kg bw per day group exhibited characteristic signs of dinitrotoluene toxicity [the haematological profile was not assessed for groups receiving other doses]. Maternal body-weight gain (minus the uterus) was decreased at 100 mg/kg, while there was weight loss at 150 mg/kg. Treatment-related increases in maternal liver and spleen weight as percentages of body weight during gestation were also observed in the groups treated with the higher doses of dinitrotoluene. The proportion of total implants per dam that were resorbed in the group receiving the highest dose was 46.0% (standard error, 22.3%) compared with 16.8% (standard error, 5.4%) in the control group receiving the corn oil vehicle only; this difference was not statistically significant. In litters with live fetuses, no statistically significant difference in the proportion of male fetuses per litter, the average fetal body weight per litter, the average fetal crown–rump length per litter or the average placental weight per litter was observed. No statistically significant difference was observed in the proportion of litters with one or more malformed fetuses or in the percentage of malformed fetuses per litter between the groups. Thus, dinitrotoluene was not found to produce malformations even at dose levels which produced significant maternal and embryo/fetal toxicity.

Using an abbreviated assay protocol intended to evaluate previously untested chemicals to help prioritize them for conventional testing of developmental toxicity, Hardin *et al.* (1987) administered by gavage a dose of 390 mg/kg bw per day 2,4-dinitrotoluene to 50 CD-1 mice on gestational days 6–13. This dose represents the LD_{10} predicted on the basis of dose finding. A concurrent control group of 50 mice received the corn oil vehicle only. Fifteen of the 50 treated mice died, compared with none of the 50 controls. No adverse effect on reproductive indices was observed (live births per litter, percentage survival, birth weight or weight gain). Pups were not examined systematically for malformations.

Groups of 10 adult male Sprague Dawley rats received 0, 60, 180 or 240 mg/kg bw per day 2,4-dinitrotoluene dissolved in corn oil by gavage for five days (Lane et al., 1985). A single oral dose (0.5 mg/kg bw per day) of triethylenemelamine was used as a positive control. Although a range-finding study preceded dose-selection, an unexpected excess of deaths among rats receiving the highest dose resulted in the requirement for another group of rats to receive a similar dose one week after the other groups. Of this group, 53% (8/15) died within two weeks of receiving the first dose. Each male was allowed to mate with two naïve, nulliparous females for five days each week. Mating lasted for seven weeks except for the group receiving the high dose, where the mating period was extended by six weeks to examine possible reversibility of the effects of 2,4-dinitrotoluene. No significant change in any of the indices of reproductive performance was observed in the group receiving 60 mg/kg bw per day dinitrotoluene. In the group mated with male rats receiving 180 mg/kg bw per day 2,4-dinitrotoluene, there was a significant increase in the preimplantation loss index at week 2, a significant decrease in the mating index at week 5, and a significant increase in the corpora lutea index also at week 5. In rats receiving 240 mg/kg bw per day, the mating index was low during weeks 1–6, and this made it difficult to interpret the other data for those weeks. However, by week 13, all reproductive and dominant lethal indices had recovered and were comparable to control values. Triethylenemelamine produced its classical dominant lethal effects for the first four weeks of mating. Slight cyanosis was observed in the group receiving the lowest dose, more severe cyanosis in the group receiving 180 mg/kg bw per day; in addition to the high death rates observed in the group receiving the highest dose, weight loss and cyanosis were observed. In the group receiving the highest dose, the severe reproductive effects persisted for at least eight weeks, which corresponds approximately to one spermatogenic cycle in the rat. Recovery from the toxic effects was complete by the end of two spermatogenic cycles.

Groups of nine to 10 male Sprague-Dawley rats were administered 0, 0.1 or 0.2% 2,4-dinitrotoluene in chow for three weeks (Bloch et al., 1988). An ultrastructural study of the testes was performed, serum was assayed for testosterone and gonadotropins and sperm reserve count (concentration of sperm heads in the cauda epididymides) was determined. No clinical sign of toxicity was observed in the rats. However, the final body weights were significantly reduced in animals treated with 2,4-dinitrotoluene compared with controls. No difference between testes from animals treated with 0.1% dinitrotoluene and those of controls was observed by light microscopy but electron microscopic examination showed focal alterations at this dose level. In animals receiving the higher dose, extensive disruption of spermatogenesis, irregularity of the peritubular tissue and widespread vacuolization of Sertoli cells were observed. Under electron microscopy, vesicles of varying sizes were associated with swollen mitochondria and distended endoplasmic reticula were observed. Circulating levels of follicle-stimulating hormone and luteinizing hormone were increased in the group receiving the higher dose. In this group, cauda epididymal sperm counts were reduced by 63%. There was no treatment-related effect on the levels of testosterone.

As noted above (see p. 338), Kozuka *et al.* (1979) reported testicular atrophy and Hong *et al.* (1985) reported degeneration of seminiferous tubules in rats.

4.4 Genetic and related effects

4.4.1 *Humans*

No data were available to the Working Group.

4.4.2 *Experimental systems* (see also Table 4 and Appendices 1 and 2)

(a) Macromolecular adducts

2,4-Dinitrotoluene was reported to bind covalently to hepatic DNA in Fischer 344 rats after oral administration. Upon intraperitoneal injection, 2,4-dinitrotoluene showed DNA binding in liver, lung, small intestine and large intestine of mice and rats (Dixit *et al.*, 1986). Using ^{32}P-postlabelling after intraperitoneal administration of 2,4-dinitrotoluene to Fischer 344 rats, three distinct adducts were detected in liver, kidney, lung and mammary gland. DNA binding was highest in the liver (La & Froines, 1992a,b).

2,6-Dinitrotoluene was reported to bind covalently to hepatic macromolecules, including DNA, in Fischer 344 rats after oral administration and intraperitoneal injection. Twice as much ^{14}C was found to be bound covalently to hepatic macromolecules in male than in female Fischer 344 rats after oral dosing of 10 mg/kg bw 2,6-dinitro[^{14}C]toluene. The covalent binding to rat liver macromolecules is increased in the presence of gut microflora (Long & Rickert, 1982). Hepatic macromolecular binding was increased by feeding pectin (DeBethizy *et al.*, 1983), Aroclor 1254 (Chadwick *et al.*, 1993) or coal-tar creosote (Chadwick *et al.*, 1995). After intraperitoneal injection into A/J mice, 2,6-dinitrotoluene showed DNA binding in liver but not in extrahepatic tissues such as lung, small intestine and large intestine; in rats, DNA binding was found in liver, lung and large intestine but not in small intestine (Dixit *et al.*, 1986). Using ^{32}P-postlabelling after intraperitoneal administration of 2,6-dinitrotoluene to Fischer 344 rats, four distinct adducts were detected in liver, kidney, lung and mammary gland; DNA binding was highest in the liver, and the 2,6-isomer produced a greater adduct yield than the 2,4-isomer (La & Froines, 1992a, 1993).

(b) Mutation and allied effects

2,4-Dinitrotoluene (technical grade) induced both forward and reverse mutations in *S. typhimurium* in the presence and absence of an exogenous metabolic system. 2,4-Dinitrotoluene (technical grade) did not induce gene mutation to 6-thioguanine resistance in Chinese hamster ovary cells with or without rat liver preparations for metabolic activation. Technical-grade 2,4-dinitrotoluene, both in the presence and absence of S9-mix, caused a dose-related decrease in survival but no induction of mutation in the mouse lymphoma test. Technical-grade 2,4-dinitrotoluene did not induce morphological transformation of Syrian hamster embryo (SHE) cells.

Table 4. Genetic and related effects of dinitrotoluenes

Test system	Result[a] Without exogenous metabolic system	Result[a] With exogenous metabolic system	Dose[b] (LED/HID)	Reference
2,4-Dinitrotoluene (technical grade)				
SAF, *Salmonella typhimurium* TM677, forward mutation to 8-azaguanine resistance	+	+	500	Couch *et al.* (1981)
SA0, *Salmonella typhimurium* TA100, reverse mutation	–	–	500	Couch *et al.* (1981)
SA0, *Salmonella typhimurium* TA100, reverse mutation	+	+	100	Ashby (1986)
SA5, *Salmonella typhimurium* TA1535, reverse mutation	–	–	500	Couch *et al.* (1981)
SA7, *Salmonella typhimurium* TA1537, reverse mutation	–	–	500	Couch *et al.* (1981)
SA8, *Salmonella typhimurium* TA1538, reverse mutation	+	+	500	Couch *et al.* (1981)
SA9, *Salmonella typhimurium* TA98, reverse mutation	+	+	250	Couch *et al.* (1981)
URP, Unscheduled DNA synthesis, rat primary hepatocytes	–	0	18	Bermudez *et al.* (1979)
GCO, Gene mutation, Chinese hamster ovary CHO cells *hprt* locus *in vitro*	–	–	364	Abernethy & Couch (1982)
GML, Gene mutation P388 mouse lymphoma cells *tk* locus *in vitro*	–	–	NR	Styles & Cross (1983)
TCS, Cell transformation, Syrian hamster embryo cells, clonal assay	–	0	NR	Holen *et al.* (1990)
UPR, Unscheduled DNA synthesis, rat hepatocytes *in vivo*	+		100 po × 1 with gut flora	Mirsalis *et al.* (1982a)
UPR, Unscheduled DNA synthesis, rat hepatocytes *in vivo*	–		100 po × 1 germ-free	Mirsalis *et al.* (1982a)
UPR, Unscheduled DNA synthesis, rat hepatocytes *in vivo*	+		100 po × 1	Mirsalis & Butterworth (1982)
UPR, Unscheduled DNA synthesis, AP and Fischer 344 rat hepatocytes *in vivo*	+		100 po × 1	Ashby *et al.* (1985)
UPR, Unscheduled DNA synthesis, rat hepatocytes *in vivo*	+		125 po × 1	Mirsalis *et al.* (1989)
MST, Mouse spot test, BL/6xBL/6 mice	–		100 ip × 1	Soares & Lock (1980)
MST, Mouse spot test, TxBL/6 mice	–		100 ip × 1	Soares & Lock (1980)

Table 4 (contd)

Test system	Result[a]		Dose[b] (LED/HID)	Reference
	Without exogenous metabolic system	With exogenous metabolic system		
2,4-Dinitrotoluene (technical grade) (contd)				
SVA, Sister chromatid exchange, rat hepatocytes *in vivo*	+		100 po × 1	Kligerman *et al.* (1982)
MVM, Micronucleus test, mice *in vivo*	−		400 ip × 1	Ashby *et al.* (1985)
DLM, Dominant lethal test, DBA/2J mice	−		250 po × 2	Soares & Lock (1980)
DLM, Dominant lethal test, DBA/2J mice	−		250 ip × 2	Soares & Lock (1980)
ICR, Inhibition of intercellular communication, Syrian hamster embryo cell line BPNi *in vitro*	+	0	100	Holen *et al.* (1990)
ICR, Inhibition of intercellular communication, Chinese hamster lung V79 cells *in vitro*	−	0	182	Dorman & Boreiko (1983)
2,4-Dinitrotoluene (high purity)				
PRB, Prophage induction, SOS repair test, DNA strand breaks, cross-links or related damage in *Salmonella typhimurium* TA1535/pSK1002	−	−	100	Nakamura *et al.* (1987)
PRB, Prophage induction, SOS repair test, DNA strand breaks, cross-links or related damage in *Salmonella typhimurium* TA1535/pSK1002	+	0	50	Oda *et al.* (1992)
PRB, Prophage induction, SOS repair test, DNA strand breaks, cross-links or related damage in *Salmonella typhimurium* TA1535/pSK1002	+	0	50.5	Oda *et al.* (1993)
PRB, Prophage induction, SOS repair test, DNA strand breaks, cross-links or related damage in *Salmonella typhimurium* NM1000	−	0	100	Oda *et al.* (1992)
PRB, Prophage induction, SOS repair test, DNA strand breaks, cross-links or related damage in *Salmonella typhimurium* NM1000	−	0	NR	Oda *et al.* (1993)
PRB, Prophage induction, SOS repair test, DNA strand breaks, cross-links or related damage in *Salmonella typhimurium* NM1011	+	0	25	Oda *et al.* (1992)
PRB, Prophage induction, SOS repair test, DNA strand breaks, cross-links or related damage in *Salmonella typhimurium* NM1011	+	0	2.7	Oda *et al.* (1993)

Table 4 (contd)

Test system	Result[a]		Dose[b] (LED/HID)	Reference
	Without exogenous metabolic system	With exogenous metabolic system		
2,4-Dinitrotoluene (high purity) (contd)				
PRB, Prophage induction, SOS repair test, DNA strand breaks, cross-links or related damage in *Salmonella typhimurium* NM2000	+	0	91	Oda et al. (1993)
PRB, Prophage induction, SOS repair test, DNA strand breaks, cross-links or related damage in *Salmonella typhimurium* NM2009	+	0	19	Oda et al. (1993)
PRB, Prophage induction, SOS repair test, DNA strand breaks, cross-links or related damage in *Salmonella typhimurium* NM3009	+	0	7	Oda et al. (1993)
SAF, *Salmonella typhimurium* TM677, forward mutation to 8-azaguanine resistance	+	+	500	Couch et al. (1981)
SA0, *Salmonella typhimurium* TA100, reverse mutation	–	0	910	Chiu et al. (1978)
SA0, *Salmonella typhimurium* TA100, reverse mutation	+	–	100	Couch et al. (1981)
SA0, *Salmonella typhimurium* TA100, reverse mutation	(+)	(+)	250	Tokiwa et al. (1981)
SA0, *Salmonella typhimurium* TA100, reverse mutation	(+)	0	50	Mori et al. (1982)
SA0, *Salmonella typhimurium* TA100, reverse mutation	+	–	250	Spanggord et al. (1982b)
SA0, *Salmonella typhimurium* TA100, reverse mutation	+	+	385	Haworth et al. (1983)
SA0, *Salmonella typhimurium* TA100, reverse mutation	+	+	125	Dunkel et al. (1985)
SA0, *Salmonella typhimurium* TA100, reverse mutation	+	+	100	Ashby (1986)
SA0, *Salmonella typhimurium* TA100, reverse mutation	+	+	250	Kawai et al. (1987)
SA0, *Salmonella typhimurium* TA100, reverse mutation	–	–	210	Dellarco & Prival (1989)
SA5, *Salmonella typhimurium* TA1535, reverse mutation	(+)	–	100	Couch et al. (1981)
SA5, *Salmonella typhimurium* TA1535, reverse mutation	–	–	2500	Spanggord et al. (1982b)
SA5, *Salmonella typhimurium* TA1535, reverse mutation	–	–	385	Haworth et al. (1983)
SA5, *Salmonella typhimurium* TA1535, reverse mutation	–	–	5000	Dunkel et al. (1985)

Table 4 (contd)

Test system	Result[a] Without exogenous metabolic system	With exogenous metabolic system	Dose[b] (LED/HID)	Reference
2,4-Dinitrotoluene (high purity) (contd)				
SA7, *Salmonella typhimurium* TA1537, reverse mutation	(+)	–	100	Couch *et al.* (1981)
SA7, *Salmonella typhimurium* TA1537, reverse mutation	–	–	2500	Spanggord *et al.* (1982b)
SA7, *Salmonella typhimurium* TA1537, reverse mutation	–	–	1280	Haworth *et al.* (1983)
SA7, *Salmonella typhimurium* TA1537, reverse mutation	–	–	3333	Dunkel *et al.* (1985)
SA8, *Salmonella typhimurium* TA1538, reverse mutation	+	–	100	Couch *et al.* (1981)
SA8, *Salmonella typhimurium* TA1538, reverse mutation	–	–	2500	Spanggord *et al.* (1982b)
SA8, *Salmonella typhimurium* TA1538, reverse mutation	+	–	250	Dunkel *et al.* (1985)
SA9, *Salmonella typhimurium* TA98, reverse mutation	–	0	910	Chiu *et al.* (1978)
SA9, *Salmonella typhimurium* TA98, reverse mutation	+	+	100	Couch *et al.* (1981)
SA9, *Salmonella typhimurium* TA98, reverse mutation	(+)	–	250	Tokiwa *et al.* (1981)
SA9, *Salmonella typhimurium* TA98, reverse mutation	(+)	0	50	Mori *et al.* (1982)
SA9, *Salmonella typhimurium* TA98, reverse mutation	–	–	2500	Spanggord *et al.* (1982b)
SA9, *Salmonella typhimurium* TA98, reverse mutation	–	–	1280	Haworth *et al.* (1983)
SA9, *Salmonella typhimurium* TA98, reverse mutation	(+)	(+)	125	Dunkel *et al.* (1985)
SA9, *Salmonella typhimurium* TA98, reverse mutation	–	–	1250	Kawai *et al.* (1987)
SA9, *Salmonella typhimurium* TA98, reverse mutation	–	–	210	Dellarco & Prival (1989)
SAS, *Salmonella typhimurium*, TA100NR3, reverse mutation	–	+	NR	Spanggord *et al.* (1982b)
ECW, *Escherichia coli* WP2 *uvrA*, reverse mutation	–	–	5000	Dunkel *et al.* (1985)
DMX, *Drosophila melanogaster*, sex-linked recessive lethal mutations	–		10 000 feeding	Woodruff *et al.* (1985)

Table 4 (contd)

Test system	Result[a]		Dose[b] (LED/HID)	Reference
	Without exogenous metabolic system	With exogenous metabolic system		
2,6-Dinitrotoluene (contd)				
GML, Gene mutation, P388 mouse lymphoma cells, *tk* locus *in vitro*	–	–	NR	Styles & Cross (1983)
UIA, Unscheduled DNA synthesis, rat spermatocytes *in vitro*	–	0	18	Working & Butterworth (1984)
TCS, Cell transformation, Syrian hamster embryo cells, clonal assay	–	0	NR	Holen *et al.* (1990)
UIH, Unscheduled DNA synthesis, human hepatocytes *in vitro*	–	0	182	Butterworth *et al.* (1989)
BFA, Body fluids (urine) from CD-1 mice, reverse mutation in TA98	+	0	75	George *et al.* (1991)
UPR, Unscheduled DNA synthesis, rat hepatocytes *in vivo*	+		20 po × 1	Mirsalis *et al.* (1982b)
UPR, Unscheduled DNA synthesis, rat hepatocytes *in vivo*	+		20 po × 1	Mirsalis & Butterworth (1982)
UVR, Unscheduled DNA synthesis, rat spermatocytes *in vivo*	–		20 po × 1	Working & Butterworth (1984)
BID, Binding (covalent) to DNA, A/J mouse or Fischer 344 rat hepatocytes *in vitro*	–	+	44	Dixit *et al.* (1986)
BVD, Binding (covalent) to DNA, Fischer 344 rat liver *in vivo*	+		10 po × 1	Long & Rickert (1982)
BVD, Binding (covalent) to DNA, Fischer 344 rat liver *in vivo*	+		10 po × 1	Rickert *et al.* (1983)
BVD, Binding (covalent) to DNA, BALB/c mouse liver *in vivo*	+		9.0 topical × 4	Reddy *et al.* (1984)
BVD, Binding (covalent) to DNA, Fischer 344 rat liver	+		28 po × 1	Kedderis *et al.* (1984)
BVD, Binding (covalent) to DNA, A/J mouse liver, lung, small intestine and large intestine	+		150 ip × 1	Dixit *et al.* (1986)
BVD, Binding (covalent) to DNA, Fischer 344 rat liver, lung, small intestine and large intestine	+		150 ip × 1	Dixit *et al.* (1986)

Table 4 (contd)

Test system	Result[a]		Dose[b] (LED/HID)	Reference
	Without exogenous metabolic system	With exogenous metabolic system		
2,4-Dinitrotoluene (high purity) (contd)				
BVD, Binding (covalent) to DNA, Fischer 344 rat liver, lung, small intestine and large intestine	+		150 ip × 1	Dixit et al. (1986)
BVD, Binding (covalent) to DNA, Fischer 344 rat liver, kidney, lung and mammary gland	+		150 ip × 1	La & Froines (1992a)
BVP, Binding (covalent) to RNA or protein, Fischer 344 rat liver in vivo	+		10 po × 1	Rickert et al. (1983)
BVP, Binding (covalent) to RNA or protein, animals in vivo	+		28 po × 1	Kedderis et al. (1984)
ICR, Inhibition of intercellular communication, Chinese hamster lung V79 cells, in vitro	–	0	182	Dorman & Boreiko (1983)
ICR, Inhibition of intercellular communication, Syrian hamster embryo cell line BPNi in vitro	+	0	100	Holen et al. (1990)
SPM, Sperm morphology, DBA/2J mice in vivo	–		250 po × 2	Soares & Lock (1980)
SPM, Sperm morphology, DBA/2J mice in vivo	–		250 ip × 2	Soares & Lock (1980)
2,6-Dinitrotoluene				
SAF, Salmonella typhimurium TM677, forward mutation to 8-azaguanine resistance	+	+	500	Couch et al. (1981)
SA0, Salmonella typhimurium TA100, reverse mutation	+	–	500	Couch et al. (1981)
SA0, Salmonella typhimurium TA100, reverse mutation	(+)	–	500	Tokiwa et al. (1981)
SA0, Salmonella typhimurium TA100, reverse mutation	+	+	250	Spanggord et al. (1982b)
SA0, Salmonella typhimurium TA100, reverse mutation	+	+	250	Kawai et al. (1987)
SA0, Salmonella typhimurium TA100, reverse mutation	–	–	690	Dellarco & Prival (1989)
SA0, Salmonella typhimurium TA100, reverse mutation	–	–	475	Sayama et al. (1989a)
SA0, Salmonella typhimurium TA100, reverse mutation	–	+	500	George et al. (1991)

Table 4 (contd)

Test system	Result[a]		Dose[b] (LED/HID)	Reference
	Without exogenous metabolic system	With exogenous metabolic system		

2,6-Dinitrotoluene (contd)

Test system	Without	With	Dose	Reference
SA5, *Salmonella typhimurium* TA1535, reverse mutation	+	+	500	Couch et al. (1981)
SA5, *Salmonella typhimurium* TA1535, reverse mutation	–	–	2500	Spanggord et al. (1982b)
SA7, *Salmonella typhimurium* TA1537, reverse mutation	–	+	500	Couch et al. (1981)
SA7, *Salmonella typhimurium* TA1537, reverse mutation	–	–	2500	Spanggord et al. (1982b)
SA8, *Salmonella typhimurium* TA1538, reverse mutation	+	–	500	Couch et al. (1981)
SA8, *Salmonella typhimurium* TA1538, reverse mutation	–	–	2500	Spanggord et al. (1982b)
SA9, *Salmonella typhimurium* TA98, reverse mutation	+	+	250	Couch et al. (1981)
SA9, *Salmonella typhimurium* TA98, reverse mutation	(+)	–	500	Tokiwa et al. (1981)
SA9, *Salmonella typhimurium* TA98, reverse mutation	–	–	2500	Spanggord et al. (1982b)
SA9, *Salmonella typhimurium* TA98, reverse mutation	(+)	+	500	Kawai et al. (1987)
SA9, *Salmonella typhimurium* TA98, reverse mutation	–	–	690	Dellarco & Prival (1989)
SA9, *Salmonella typhimurium* TA98, reverse mutation	–	–	475	Sayama et al. (1989a)
SA9, *Salmonella typhimurium* TA98, reverse mutation	–	+	250	George et al. (1991)
SAS, *Salmonella typhimurium*, TA100NR3, reverse mutation	–	–	2500	Spanggord et al. (1982b)
DIA, DNA strand breaks, cross-links or related damage, rat hepatocytes *in vitro*	(+)	0	546	Sina et al. (1983)
URP, Unscheduled DNA synthesis, rat primary hepatocytes *in vitro*	–	0	182	Bermudez et al. (1979)
GCO, Gene mutation, Chinese hamster ovary CHO cells, *hprt* locus *in vitro*	–	–	455	Abernethy & Couch (1982)

Table 4 (contd)

Test system	Result[a] Without exogenous metabolic system	Result[a] With exogenous metabolic system	Dose[b] (LED/HID)	Reference
2,4-Dinitrotoluene (high purity) (contd)				
DMX, *Drosophila melanogaster*, sex-linked recessive lethal mutations	+		10 000 inj.	Woodruff et al. (1985)
DMH, *Drosophila melanogaster*, heritable translocation test	−		10 000 inj.	Woodruff et al. (1985)
DIA, DNA strand breaks, cross-links or related damage, rat hepatocytes *in vitro*	(+)		546	Sina et al. (1983)
URP, Unscheduled DNA synthesis, rat primary hepatocytes *in vitro*	−	0	182	Bermudez et al. (1979)
UIA, Unscheduled DNA synthesis, rat spermatocytes *in vitro*	−	0	18	Working & Butterworth (1984)
GCO, Gene mutation, Chinese hamster ovary CHO cells, *hprt* locus *in vitro*	−	−	546	Abernethy & Couch (1982)
GML, Gene mutation, P388 mouse lymphoma cells, *tk* locus *in vitro*	+	−	160	Styles & Cross (1983)
SIC, Sister chromatid exchange, Chinese hamster CHO cells *in vitro*	−	+	1010	Loveday et al. (1989)
CIC, Chromosomal aberrations, Chinese hamster CHO cells *in vitro*	−	−	1000	Loveday et al. (1989)
TCS, Cell transformation, Syrian hamster embryo cells, clonal assay	−	0	10	Holen et al. (1990)
UIH, Unscheduled DNA synthesis, human hepatocytes *in vitro*	−	0	182	Butterworth et al. (1989)
UPR, Unscheduled DNA synthesis, rat hepatocytes *in vivo*	+		200 po × 1	Mirsalis et al. (1982b)
UPR, Unscheduled DNA synthesis, rat hepatocytes *in vivo*	(+)		100 po × 1	Mirsalis & Butterworth (1982)
DLM, Dominant lethal test, DBA/2J mice	−		250 po × 2	Soares & Lock (1980)
DLM, Dominant lethal test, DBA/2J mice	−		250 ip × 2	Soares & Lock (1980)
BVD, Binding (covalent) to DNA, Fischer 344 rat liver	+		10 po × 1	Rickert et al. (1983)
BVD, Binding (covalent) to DNA, Fischer 344 rat liver	+		28 po × 1	Kedderis et al. (1984)
BVD, Binding (covalent) to DNA, A/J mouse liver, lung, small intestine and large intestine	+		150 ip × 1	Dixit et al. (1986)

Table 4 (contd)

Test system	Result[a]		Dose[b] (LED/HID)	Reference
	Without exogenous metabolic system	With exogenous metabolic system		
2,6-Dinitrotoluene (contd)				
BVD, Binding (covalent) to DNA, Fischer 344 rat liver, kidney, lung and mammary gland	+		150 ip × 1	La & Froines (1992a)
BVD, Binding (covalent) to DNA, Fischer 344 rat liver	+		220 ip × 1	La & Froines (1993)
BVP, Binding (covalent) to RNA or protein, Fischer 344 rat liver *in vivo*	+		10 po × 1	Rickert et al. (1983)
BVP, Binding (covalent) to RNA or protein, animals *in vivo*	+		10 po × 1	DeBethizy et al. (1983)
BVP, Binding (covalent) to RNA or protein, animals *in vivo*	+		28 po × 1	Kedderis et al. (1984)
ICR, Inhibition of intercellular communication, Chinese hamster lung V79 cells *in vitro*	–	0	182	Dorman & Boreiko (1983)
ICR, Inhibition of intercellular communication, Syrian hamster embryo cell line BPNi *in vitro*	+	0	100	Holen et al. (1990)
3,5-Dinitrotoluene				
SAF, *Salmonella typhimurium* TM677, forward mutation to 8-azaguanine resistance	+	+	50	Couch et al. (1981)
SA0, *Salmonella typhimurium* TA100, reverse mutation	+	–	20	Couch et al. (1981)
SA0, *Salmonella typhimurium* TA100, reverse mutation	+	+	125	Spanggord et al. (1982b)
SA5, *Salmonella typhimurium* TA1535, reverse mutation	–	–	20	Couch et al. (1981)
SA5, *Salmonella typhimurium* TA1535, reverse mutation	–	–	550	Spanggord et al. (1982b)
SA7, *Salmonella typhimurium* TA1537, reverse mutation	+	–	20	Couch et al. (1981)
SA7, *Salmonella typhimurium* TA1537, reverse mutation	+	+	NR	Spanggord et al. (1982b)
SA8, *Salmonella typhimurium* TA1538, reverse mutation	+	+	20	Couch et al. (1981)

Table 4 (contd)

Test system	Result[a]		Dose[b] (LED/HID)	Reference
	Without exogenous metabolic system	With exogenous metabolic system		
3,5-Dinitrotoluene				
SA8, *Salmonella typhimurium* TA1538, reverse mutation	+	+	NR	Spanggord et al. (1982b)
SA9, *Salmonella typhimurium* TA98, reverse mutation	+	+	25	Couch et al. (1981)
SA9, *Salmonella typhimurium* TA98, reverse mutation	+	+	NR	Spanggord et al. (1982b)
SAS, *Salmonella typhimurium*, TA100NR3, reverse mutation	–	–	550	Spanggord et al. (1982b)
URP, Unscheduled DNA synthesis, rat primary hepatocytes *in vitro*	–	0	182	Bermudez et al. (1979)
GCO, Gene mutation, Chinese hamster ovary CHO cells, *hprt* locus *in vitro*	–	–	364	Abernethy & Couch (1982)
DLM, Dominant lethal test, DBA/2j mice	–		250 po × 2	Soares & Lock (1980)
DLM, Dominant lethal test, DBA/2j mice	–		250 ip × 2	Soares & Lock (1980)

[a] +, positive; (+), weak positive; –, negative; 0 not tested; ?, inconclusive
[b] LED, lowest effective dose; HID, highest ineffective dose. In-vitro tests, μg/mL; in-vivo tests, mg/kg bw; NR, dose not reported
[c] Positive in preincubation assay with 2 mμ flavin mononucleotide (FMN)

When administered in vivo to rats, 2,4-dinitrotoluene (technical grade) induced a dose-related increase in unscheduled DNA synthesis in hepatocytes. Treated female rats showed a much lower level than males. Unscheduled DNA synthesis was observed in rats with a normal complement of gut flora, but not in rats lacking a gut flora, which indicates that metabolism by gut flora is a necessary step in the genotoxicity of this nitro-aromatic compound. After in-vivo treatment with 2,4-dinitrotoluene (technical grade) by gavage, cultured rat lymphocytes showed a positive sister chromatid exchange response but no cell cycle inhibition or mitotic depression (Kligerman et al., 1982).

Technical-grade 2,4-dinitrotoluene inhibited intercellular communication at toxic concentrations in the Syrian hamster embryo cell line BPNi, but not in the V79 Chinese hamster lung cell assay.

Technical-grade 2,4-dinitrotoluene was negative in the mouse bone-marrow micronucleus test, the mouse dominant lethal test and the mouse spot test.

High-purity 2,4-dinitrotoluene was genotoxic in the *S. typhimurium umu* test. In the *S. typhimurium* reverse mutation assay, a number of different studies gave a heterogeneous picture. 2,4-Dinitrotoluene was negative in mutagenicity tests with the *E.* WP2 *uvrA* strain. It had little or no mutagenic activity in *S. typhimurium* TA98 using the standard plate test with or without rat liver S9, but was found to have a clear flavin mononucleotide-dependent mutagenic activity in a modified preincubation assay with hamster S9 (Dellarco & Prival, 1989).

In *Drosophila melanogaster*, high-purity 2,4-dinitrotoluene induced sex-linked recessive lethal mutations after injection, but failed to induce lethal mutations after feeding and translocations after injection.

High-purity 2,4-dinitrotoluene, at the highest dose tested, induced a small number of DNA strand breaks in rat hepatocytes *in vitro*.

High-purity 2,4-dinitrotoluene did not induce gene mutation to 6-thioguanine resistance in Chinese hamster ovary cells with or without rat liver preparations for metabolic activation. The induction of gene mutations was reported for mouse lymphoma cells in the absence of a metabolic activation system. 2,4-Dinitrotoluene produced a small but reproducible increase in sister chromatid exchange in Chinese hamster cells *in vitro* with S9, but not without S9; chromosomal aberrations were not induced with or without S9.

When high-purity 2,4-dinitrotoluene was incubated anaerobically with whole cells from rabbit intestine, culture extracts were mutagenic to *S. typhimurium* strain TA98NR (nitroreductase-deficient) (Combes & Walters, 1986).

High-purity 2,4-dinitrotoluene has been reported to be negative in a number of other mammalian genotoxicity assays: unscheduled DNA synthesis in primary rat hepatocytes and in human hepatocytes *in vitro*, the dominant lethal assay and the sperm morphology test in mice.

When administered to rats *in vivo*, high-purity 2,4-dinitrotoluene induced a weak response in unscheduled DNA synthesis in hepatocytes.

High-purity 2,4-dinitrotoluene inhibited intercellular communication at toxic concentrations in the Syrian hamster embryo cell line BPNi, but not in the V79 Chinese hamster lung cell assay. 2,4-Dinitrotoluene did not induce morphological transformation of Syrian hamster embryo (SHE) cells.

2,6-Dinitrotoluene is weakly mutagenic in the *S. typhimurium* reverse mutation test without metabolic activation and in TA1535 and TA1537 with rat liver S9. 2,6-Dinitrotoluene is metabolized by *S. typhimurium* strains TA98, TA98/1,8-DNP$_6$ and TA98NR. Results indicate that the low mutagenic activity of 2,6-dinitrotoluene is not due to low reductive metabolism by bacteria, but due to the lack of mutagenic activity of the bacterial reductive products (Sayama *et al.*, 1992). 2,6-Dinitrotoluene had little or no mutagenic activity in *S. typhimurium* TA98 using the standard plate test with or without rat liver S9, but was found to have a clear flavin mononucleotide-dependent mutagenic activity in a modified preincubation assay with hamster S9 (Dellarco & Prival, 1989).

2,6-Dinitrotoluene did not induce morphological transformation of Syrian hamster embryo (SHE) cells.

2,6-Dinitrotoluene induced DNA strand breaks in rat hepatocytes *in vitro*.

2,6-Dinitrotoluene did not induce gene mutation to 6-thioguanine resistance in Chinese hamster ovary cells with or without rat liver preparations for metabolic activation.

Oral administration of 2,6-dinitrotoluene by gavage led to excretion of mutagenic 2,6-dinitrotoluene-derived metabolites detectable in hydrolysed urines in a microsuspension bioassay (DeMarini *et al.*, 1989) using *S. typhimurium* strain TA98 without rat liver S9 activation. In a similar urinary assay, a positive response was observed with Fischer 344 rats as well as with CD-1 mice.

When administered *in vivo* to rats, 2,6-dinitrotoluene induced a strong response in unscheduled DNA synthesis in hepatocytes.

2,6-Dinitrotoluene inhibited intercellular communication at toxic concentrations in the Syrian hamster embryo cell line BPNi, but not in the V79 Chinese hamster lung cell assay.

3,5-Dinitrotoluene was mutagenic in the *S. typhimurium* reverse mutation test without metabolic activation and also in some tester strains with rat liver S9.

3,5-Dinitrotoluene did not induce gene mutation to 6-thioguanine resistance in Chinese hamster ovary cells with or without rat liver preparations for metabolic activation. It did not induce unscheduled DNA synthesis in primary rat hepatocytes *in vitro*.

5. Summary of Data Reported and Evaluation

5.1 Exposure data

2,4-, 2,6- and 3,5-Dinitrotoluenes are produced by nitration of toluene or nitrotoluenes. Dinitrotoluenes are used primarily as chemical intermediates in the production of toluene diamines and diisocyanates (mainly as the mixture of 2,4- and 2,6-isomers), while smaller amounts of the three isomers are also used to produce dyes, explosives and propellants. Human exposure to dinitrotoluenes can occur by inhalation or skin absorption during their production and use as intermediates. They have been detected in wastewater from dinitrotoluene production and use, and in surface and groundwater in the vicinity of these manufacturing facilities.

5.2 Human carcinogenicity data

A cohort study of workers from a munitions factory in the United States found an increased risk for cancer of the liver and gall-bladder among workers exposed to a mixture of 2,4- and 2,6-dinitrotoluenes, based on six cases. No such increase was detected in a previous study based on a smaller group of workers from the same and another munitions factory in the United States. These findings were not considered to be strong or consistent enough to permit a conclusion on the carcinogenicity of dinitrotoluenes in humans.

5.3 Animal carcinogenicity data

2,4-Dinitrotoluene was tested by oral administration in two adequate studies in mice and two adequate studies in rats. In one study in mice, no tumorigenic effect was reported. In the second study in mice, using higher doses, tumours of the renal tubular epithelium were observed in males. In both studies in rats, the incidence of various tumours of the integumentary system was increased in males. The incidence of hepatocellular carcinomas was increased in treated males and females in one study. The incidence of fibroadenomas of the mammary gland was increased in females in both studies.

2,6-Dinitrotoluene was tested for carcinogenicity by oral administration in two studies in male rats and increased the incidence of hepatocellular neoplastic nodules and carcinomas.

3,5-Dinitrotoluene has not been tested for carcinogenicity in experimental animals.

Technical-grade dinitrotoluene (approximately 80/20 2,4/2,6-isomers) was tested for carcinogenicity in two studies in rats by oral administration producing hepatocellular neoplastic nodules and hepatocellular carcinomas in male rats in one study and in both sexes in a second study.

5.4 Other relevant data

Dinitrotoluenes are absorbed following dermal and inhalation exposure of workers.

The most abundant metabolites of dinitrotoluenes found in urine from exposed workers were dinitrobenzoic acids. In addition, amino metabolites have been reported. The appearance of reduced metabolites suggests either that human hepatic enzymes are capable of reduction of the nitro group of dinitrotoluene or that dinitrotoluene (or its metabolites) gains access to the intestinal microflora which is capable of reduction, after which the metabolites are reabsorbed and excreted into urine. Limited data indicate a sex difference in humans as regards the urinary metabolite pattern. In humans, the elimination half-life for the urinary metabolites is 1–2.7 h.

The metabolism and excretion of dinitrotoluenes by rats seem to be qualitatively similar to those in humans. However, there are quantitative differences as regards prevalence of different metabolites. Thus, the major urinary metabolites of 2,4-dinitrotoluene are 2,4-dinitrobenzoic acid in humans and 2,4-dinitrobenzyl alcohol in rats.

Heavy human exposure to technical-grade dinitrotoluene may cause a variety of symptoms and signs, including cyanosis — presumably because of methaemoglobinaemia — anaemia and toxic hepatitis. Further, dinitrotoluenes may give rise to allergic contact dermatitis.

A variety of toxic effects are observed in animals following acute administration of various dinitrotoluene isomers. Certain dinitrotoluene isomers, most notably 2,6-dinitrotoluene, produce extensive centrilobular hepatic necrosis following administration *in vivo*.

In laboratory animals, the chronic toxic effects following exposure to dinitrotoluene include various neurotoxic effects (including paralysis), hepatotoxicity, including dysplasia, hyperplastic foci and hepatic megalocytosis, anaemia and methaemoglobinaemia.

No association was found between exposure to the compounds of male workers in a dinitrotoluene facility and the results of semen analysis, the levels of follicle-stimulating hormone or the occurrence of miscarriages or delayed conception in their partners.

In female rats, administration of technical-grade dinitrotoluene by gavage did not produce teratogenic effects even at dose levels which produce significant maternal and embryo/fetal toxicity. In studies in male rats, 2,4-dinitrotoluene induced adverse reproductive effects and anti-spermatogenic activity.

No data on the metabolism or toxicity of 3,5-dinitrotoluene were available to the Working Group.

2,4-Dinitrotoluene (technical grade) is weakly mutagenic in bacteria. It was inactive in mammalian cells *in vitro* in tests for gene mutation, unscheduled DNA synthesis and transformation, but inhibited intercellular communication at toxic concentrations. In rats *in vivo*, it induced unscheduled DNA synthesis in hepatocytes, provided the normal gut flora was present. It induced sister chromatid exchange in rat lymphocytes exposed *in vivo*. In mice, it was negative in the bone-marrow micronucleus test, the dominant lethal test and the spot test.

Purified 2,4-dinitrotoluene showed DNA binding in rats *in vivo* in several organs, the binding being highest in the liver. Three distinct adducts were identified. In bacteria, it induced DNA damage and gene mutation. In insects, it induced sex-linked recessive lethal mutations but not dominant lethal mutations or translocations. In mammalian cells *in vitro*, it induced DNA strand breaks, gene mutations in mouse lymphoma cells (without activation) but not in Chinese hamster ovary cells and a low frequency of sister chromatid exchange but not of chromosomal aberrations in Chinese hamster ovary cells. It inhibited intercellular communication but did not induce cell transformation. In mammals *in vivo*, 2,4-dinitrotoluene induced a weak response in unscheduled DNA synthesis in rat hepatocytes but was negative in the dominant lethal assay and the sperm morphology test in mice.

2,6-Dinitrotoluene is weakly mutagenic in bacteria. In mammalian cells *in vitro*, it induced DNA strand breaks but no gene mutation or cell transformation. Studies of the inhibition of intercellular communication gave equivocal results. With 2,6-dinitrotoluene, DNA adducts were found after in-vivo exposure of rats. *In vivo*, it induced unscheduled DNA synthesis in rat hepatocytes. In the urine of exposed rats, mutagenic metabolites could be detected.

Experiments indicate the following steps in the metabolic activation leading to the formation of adducts: (1) 2,6-dinitrotoluene is metabolized in the liver; (2) metabolites are excreted in the bile; (3) the biliary metabolites are hydrolysed and further metabolized in the intestine; and (4) after enterohepatic transportation of the metabolites back to the liver, the metabolites are activated further and bound to macromolecules.

3,5-Dinitrotoluene is mutagenic in bacteria but did not induce DNA damage or mutations in mammalian cells in culture.

5.5 Evaluation[1]

There is *inadequate evidence* in humans for the carcinogenicity of 2,4-, 2,6- and 3,5-dinitrotoluenes.

There is *sufficient evidence* in experimental animals for the carcinogenicity of 2,4-dinitrotoluene and 2,6-dinitrotoluene.

There is *inadequate evidence* in experimental animals for the carcinogenicity of 3,5-dinitrotoluene.

Overall evaluation

2,4- and 2,6-Dinitrotoluenes are *possibly carcinogenic to humans (Group 2B)*.

3,5-Dinitrotoluene is *not classifiable as to its carcinogenicity to humans (Group 3)*.

6. References

Abernethy, D.J. & Couch, D.B. (1982) Cytotoxicity and mutagenicity of dinitrotoluenes in Chinese hamster ovary cells. *Mutat. Res.*, **103**, 53–59

Acros Organics (1995) *Catalog of Fine Chemicals 95–96*, Pittsburgh, PA, Fisher Scientific, p. 754

Air Products and Chemicals (1995) *Material Safety Data Sheet: Dinitrotoluene*, Allentown, PA

American Conference of Governmental Industrial Hygienists (1995) *1995–1996 Threshold Limit Values for Chemical Substances and Physical Agents and Biological Exposure Indices*, Cincinnati, OH, p. 20

Arbeidsinspectie [Labour Inspection] (1994) *De Nationale MAC-Lijst 1994* [National MAC list 1994], The Hague, p. 25

---

[1] For definition of the italicized terms, see Preamble, pp. 24–27.

Ashby, J. (1986) Carcinogen/mutagen screening strategies (Letter to the Editor). *Mutagenesis*, **1**, 309–317

Ashby, J., Burlinson, B., Lefevre, P.A. & Topham, J. (1985) Non-genotoxicity of 2,4,6-trinitrotoluene (TNT) to the mouse bone marrow and the rat liver: implications for its carcinogenicity. *Arch. Toxicol.*, **58**, 14–19

Bermudez, E., Tillery, D. & Butterworth, B.E. (1979) The effect of 2,4-diaminotoluene and isomers of dinitrotoluene on unscheduled DNA synthesis in primary rat hepatocytes. *Environ. Mutag.*, **1**, 391–398

Bloch, E., Gondos, B., Gatz, M., Varma, S.K. & Thysen, B. (1988) Reproductive toxicity of 2,4-dinitrotoluene in the rat. *Toxicol. appl. Pharmacol.*, **94**, 466–472

Bond, J.A. & Rickert, D.E. (1981) Metabolism of 2,4-dinitro[^{14}C]toluene by freshly isolated Fischer-344 rat primary hepatocytes. *Drug Metab. Dispos.*, **9**, 10–14

Bond, J.A., Medinsky, M.E., Dent, J.G. & Rickert, D.E. (1981) Sex-dependent metabolism and biliary excretion of [2,4-^{14}C]dinitrotoluene in isolated perfused rat livers. *J. Pharmacol. exp. Ther.*, **219**, 598–603

Booth, G. (1991) Nitro compounds, aromatic. In: Elvers, B., Hawkins, S. & Schulz, G., eds, *Ullmann's Encyclopedia of Industrial Chemistry*, 5th rev. Ed., Vol. A17, New York, NY, VCH Publishers, pp. 411–455

Butterworth, B.E., Smith-Oliver, T., Earle, L., Loury, D.J., White, R.D., Doolittle, D.J., Working, P.K., Cattley, R.C., Jirtle, R., Michalopoulos, G. & Strom, S. (1989) Use of primary cultures of human hepatocytes in toxicology studies. *Cancer Res.*, **49**, 1075–1084

Chadwick, R.W., George, S.E., Chang, J., Kohan, M.J., Dekker, J.P., Long, J.E. & Duffy, M.C. (1990) Comparative gastrointestinal enzyme activity and activation of the promutagen 2,6-dinitrotoluene in male CD-1 mice and male Fischer 344 rats. *Cancer Lett.*, **52**, 13–19

Chadwick, R.W., George, S.E., Kohan, M.J., Williams, R.W., Allison, J.C., Hayes, Y.O. & Chang, J. (1993) Potentiation of 2,6-dinitrotoluene genotoxicity in Fischer-344 rats by pretreatment with Aroclor 1254. *Toxicology*, **80**, 153–171

Chadwick, R.W., George, S.E., Kohan, M.J., Williams, R.W., Allison, J.C., Talley, D.L., Hayes, Y.O. & Chang, J. (1995) Potentiation of 2,6-dinitrotoluene genotoxicity in Fischer 344 rats by pretreatment with coal tar creosote. *J. Toxicol. environ. Health*, **44**, 319–336

Chapman, D.E., Michener, S.R. & Powis, G. (1992) Metabolism of 2,6-dinitro[3-^{3}H]toluene by human and rat liver microsomal and cytosolic fractions. *Xenobiotica*, **22**, 1015–1028

Chemical Information Services (1994) *Directory of World Chemical Producers 1995/96 Standard Edition*, Dallas, TX, p. 306

Chism, J.P. & Rickert, D.E. (1985) Isomer- and sex-specific bioactivation of mononitrotoluenes. Role of enterohepatic circulation. *Drub Metab. Disp.*, **13**, 651–657

Chism, J.P. & Rickert, D.E. (1989) In vitro activation of 2-aminobenzyl alcohol and 2-amino-6-nitrobenzyl alcohol, metabolites of 2-nitrotoluene and 2,6-dinitrotoluene. *Chem. Res. Toxicol.*, **2**, 150–156

Chiu, C.W., Lee, L.H., Wang, C.Y. & Bryan, G.T. (1978) Mutagenicity of some commercially available nitro compounds for *Salmonella typhimurium*. *Mutat. Res.*, **58**, 11–22

Combes, R.D. & Walters, J.M. (1986) The contribution of the intestinal microflora and bacterial tester strains to metabolic activation (Abstract no. 7). *Mutagenesis*, **1**, 68

Couch, D.B., Allen, P.F. & Abernethy, D.J. (1981) The mutagenicity of dinitrotoluenes in *Salmonella typhimurium. Mutat. Res.*, **90**, 373–383

Davis, W.M., Coates, J.A., Garcia, K.L., Signorella, L.L. & Delfino, J.J. (1993) Efficient screening method for determining base/neutral and acidic semi-volatile organic priority pollutants in sediments. *J. Chromatogr.*, **643**, 341–350

DeBethizy, J.D., Sherrill, J.M., Rickert, D.E. & Hamm, T.E., Jr (1983) Effects of pectin-containing diets on the hepatic macromolecular covalent binding of 2,6-dinitro[^{3}H]toluene in Fischer-344 rats. *Toxicol. appl. Pharmacol.*, **69**, 369–376

Decad, G.M., Graichen, M.E. & Dent, J.G. (1982) Hepatic microsomal metabolism and covalent binding of 2,4-dinitrotoluene. *Toxicol. appl. Pharmacol.*, **62**, 325–334

Dellarco, V.L. & Prival, M.J. (1989) Mutagenicity of nitro compounds in *Salmonella typhimurium* in the presence of flavin mononucleotide in a preincubation assay. *Environ. Mutag.*, **13**, 116–127

DeMarini, D., Dallas, M.M. & Lewtas, J. (1989) Cytotoxicity and effect on mutagenicity of buffers in a microsuspension assay. *Teratog. Carcinog. Mutag.*, **9**, 287–295

Dent, J.G., Schnell, S.R. & Guest, D. (1981) Metabolism of 2,4-dinitrotoluene by rat hepatic microsomes and cecal flora. *Adv. exp. Med. Biol.*, **136**, 431–436

Deutsche Forschungsgemeinschaft (1995) *MAK and BAT Values 1995* (Report No. 31), Weinheim, VCH Verlagsgesellschaft, p. 48

Dixit, R., Schut, H.A.J., Klaunig, J.E. & Stoner, G.D. (1986) Metabolism and DNA binding of 2,6-dinitrotoluene in Fischer-344 rats and A/J mice. *Toxicol. appl. Pharmacol.*, **82**, 53–61

Dorman, B.H. & Boreiko, C.J. (1983) Limiting factors of the V79 cell metabolic cooperation assay for tumor promoters. *Carcinogenesis*, **4**, 873–877

Dunkel, V.C., Zeiger, E., Brusick, D., McCoy, E., McGregor, D., Mortelmans, K., Rosenkranz, H.S. & Simmon, V.F. (1985) Reproducibility of microbial mutagenicity assays: II. Testing of carcinogens and noncarcinogens in *Salmonella typhimurium* and *Escherichia coli*. *Environ. Mutag.*, **7** (Suppl. 5), 1–248

Ellis, H.V., III, Hong, C.B., Lee, C.C., Dacre, J.C. & Glennon, J.P. (1985) Subchronic and chronic toxicity studies of 2,4-dinitrotoluene. Part I. Beagle dogs. *J. Am. Coll. Toxicol.*, **4**, 233–242

Emtestam, L. & Forsbeck, M. (1985) Occupational photosensitivity to dinitrotoluene. *Photodermatology*, **2**, 120–121

Feltes, J., Levsen, K., Volmer, D. & Spiekermann, M. (1990) Gas chromatographic and mass spectrometric determination of nitroaromatics in water. *J. Chromatogr.*, **518**, 21–40

George, S.E., Chadwick, R.W., Creason, J.P., Kohan, M.J. & Dekker, J.P. (1991) Effect of pentachlorophenol on the activation of 2,6-dinitrotoluene to genotoxic urinary metabolites in CD-1 mice: a comparison of GI enzyme activities and urine mutagenicity. *Environ. mol. Mutag.*, **18**, 92–101

Girindus Corp. (1994) *Data Sheets: Girstrat® — 2,4-DNT HM (2,4-Dinitrotoluene High Melting Grade) and Girstrat® — 2,4-DNT SHM (2,4-Dinitrotoluene Super High Melting Grade)*, Palm Harbor, FL

Goldsworthy, T.L., Hamm, T.E., Jr, Rickert, D.E. & Popp, J.A. (1986) The effect of diet on 2,6-dinitrotoluene hepatocarcinogenesis. *Carcinogenesis*, **7**, 1909–1915

Guest, D., Schnell, S.R., Rickert, D.E. & Dent, J.G. (1982) Metabolism of 2,4-dinitrotoluene by intestinal microorganisms from rat, mouse, and man. *Toxicol. appl. Pharmacol.*, **64**, 160–168

Hable, M., Stern, C., Asowata, C. & Williams, K. (1991) The determination of nitroaromatics and nitramines in ground and drinking water by wide-bore capillary gas chromatography. *J. chromatogr. Sci.*, **29**, 131–135

Hall, L.W., Jr, Hall, W.S., Bushong, S.J. & Herman, R.L. (1987) In situ striped bass (*Morone saxatilis*) contaminant and water quality studies in the Potomac River. *Aquat. Toxicol.*, **10**, 73–99

Hamill, P.V.V., Steinberger, E., Levine, R.J., Rodriguez-Rigau, L.J., Lemeshow, S. & Avrunin, J.S. (1982) The epidemiologic assessment of male reproductive hazard from occupational exposure to TDA and DNT. *J. occup. Med.*, **24**, 985–993

Hansch, C., Leo, A. & Hoekman, D.H. (1995) *Exploring QSAR*, Washington DC, American Chemical Society, p. 28

Hardin, B.D., Schuler, R.L., Burg, J.R., Booth, G.M., Hazelden, K.P., MacKenzie, K.M., Piccirillo, V.J. & Smith, K.N. (1987) Evaluation of 60 chemicals in a preliminary developmental toxicity test. *Teratog. Carcinog. Mutag.*, **7**, 29–48

Hashimoto, A., Sakino, H., Yamagami, E. & Tateishi, S. (1980) Determination of dinitrotoluene isomers in sea water and industrial effluents by high-resolution electron capture gas chromatography with a glass capillary column. *Analyst*, **105**, 787–793

Hashimoto, A., Sakino, H., Kojima, T., Yamagami, E., Tateishi, S. & Akiyama, T. (1982) Sources and behaviour of dinitrotoluene isomers in sea-water. *Water Res.*, **16**, 891–897

Haworth, S., Lawlor, T., Mortelmans, K., Speck, W. & Zeiger, E. (1983) Salmonella mutagenicity test results for 250 chemicals. *Environ. Mutag.*, **Suppl. 1**, 3–142

Health and Safety Executive (1995) *Occupational Exposure Limits 1995* (Guidance Note EH 40/95), Sudbury, Suffolk, HSE Books

Holen, I., Mikalsen, S.-O. & Sanner, T. (1990) Effects of dinitrotoluenes on morphological cell transformation and intracellular communication in Syrian hamster embryo cells. *J. Toxicol. environ. Health*, **29**, 89–98

Hong, C.B., Ellis, H.V., III, Lee, C.C., Sprinz, H., Dacre, J.C. & Glennon, J.P. (1985) Subchronic and chronic toxicity studies of 2,4-dinitrotoluene. Part III. CD-1® mice. *J. Am. Coll. Toxicol.*, **4**, 257–269

Horng, J.-Y. & Huang, S.-D. (1994) Determination of the semi-volatile compounds nitrobenzene, isophorone, 2,4-dinitrotoluene and 2,6-dinitrotoluene in water using solid-phase microextraction with a polydimethylsiloxane-coated fibre. *J. Chromatogr. A*, **678**, 313–318

Howard, P.H. (1989) *Handbook of Environmental Fate and Exposure Data for Organic Chemicals*, Vol. 1, Chelsea, MI, Lewis Publishers, pp. 305–318

IARC (1978) *IARC Monographs on the Evaluation of the Carcinogenic Risk of Chemicals to Man*, Vol. 16, *Some Aromatic Amines and Related Nitro Compounds — Hair Dyes, Colouring Agents and Miscellaneous Industrial Chemicals*, Lyon, pp. 83–95

IARC (1979) *IARC Monographs on the Evaluation of the Carcinogenic Risk of Chemicals to Humans*, Vol. 19, *Some Monomers, Plastics and Synthetic Elastomers, and Acrolein*, Lyon, pp. 303–340

IARC (1986) *IARC Monographs on the Evaluation of the Carcinogenic Risk of Chemicals to Humans*, Vol. 39, *Some Chemicals Used in Plastics and Elastomers*, Lyon, pp. 287–323

IARC (1987a) *IARC Monographs on the Evaluation of Carcinogenic Risks to Humans*, Suppl. 7, *Overall Evaluations of Carcinogenicity: An Updating of* IARC Monographs *Volumes 1 to 42*, Lyon, p. 61

IARC (1987b) *IARC Monographs on the Evaluation of Carcinogenic Risks to Humans*, Suppl. 7, *Overall Evaluations of Carcinogenicity: An Updating of* IARC Monographs *Volumes 1 to 42*, Lyon, p. 72

Jenkins, T.F. & Walsh, M.E. (1992) Development of field screening methods for TNT, 2,4-DNT and RDX in soil. *Talanta*, **39**, 419–428

Jenkins, T.F., Leggett, D.C., Grant, C.L. & Bauer, C.F. (1986) Reversed-phase high-performance liquid chromatographic determination of nitroorganics in munitions wastewater. *Anal. Chem.*, **58**, 170–175

Kanerva, L., Laine, R., Jolanki, R., Tarvainen, K., Estlander, T. & Helander, I. (1991) Occupational allergic contact dermatitis caused by nitroglycerin. *Contact Derm.*, **24**, 356–362

Kawai, A., Goto, S., Matsumoto, Y. & Matsushita, H. (1987) Mutagenicity of aliphatic and aromatic nitro compounds. Industrial materials and related compounds. *Jpn. J. ind. Health*, **29**, 34–54

Kedderis, G.L. & Rickert, D.E. (1985) Characterization of the oxidation of amine metabolites of nitrotoluenes by rat hepatic microsomes. *Mol. Pharmacol.*, **28**, 207–214

Kedderis, G.L., Dyroff, M.C. & Rickert, D.E. (1984) Hepatic macromolecular covalent binding of the hepatocarcinogen 2,6-dinitrotoluene and its 2,4-isomer *in vivo*: modulation by the sulfotransferase inhibitors pentachlorophenol and 2,6-dichloro-4-nitrophenol. *Carcinogenesis*, **5**, 1199–1204

Kligerman, A.D., Wilmer, J.L. & Erexson, G.L. (1982) The use of rat and mouse lymphocytes to study cytogenetic damage after in vivo exposure to genotoxic agents. In: Bridges, B.A., Butterworth, B.E. & Weinstein, I.B., eds, *Indicators of Genotoxic Exposure* (Banbury Report 13), Cold Spring Harbor, CSH Press, pp. 277–291

Koniecki, W.B. & Linch, A.L. (1958) Determination of aromatic nitro compounds. *Anal. Chem.*, **30**, 1134–1137

Kozuka, H., Mori, M.-A. & Naruse, Y. (1979) Studies on the metabolism and toxicity of dinitrotoluenes; toxicological study of 2,4-dinitrotoluene (2,4-DNT) in rats in long term feeding. *J. toxicol. Sci.*, **4**, 221–228

La, D.K. & Froines, J.R. (1992a) Comparison of DNA adduct formation between 2,4- and 2,6-dinitrotoluene by ^{32}P-postlabelling analysis. *Arch. Toxicol.*, **66**, 633–640

La, D.K. & Froines, J.R. (1992b) ^{32}P-Postlabelling analysis of DNA adducts from Fischer-344 rats administered 2,4-diaminotoluene. *Chem.-biol. Interactions*, **83**, 121–134

La, D.K. & Froines, J.R. (1993) Comparison of DNA binding between the carcinogen 2,6-dinitrotoluene and its noncarcinogenic analog 2,6-diaminotoluene. *Mutat Res.*, **301**, 79–85

Lancaster Synthesis (1995) *Lancaster Chemicals Catalog 1995/96*, Windham, NH, p. 635

Lane, R.W., Simon, G.S., Dougherty, R.W., Egle, J.L., Jr & Borzelleca, J.F. (1985) Reproductive toxicity and lack of dominant lethal effects of 2,4-dinitrotoluene in the male rat. *Drug chem. Toxicol.*, **8**, 265–280

Lee, C.C., Dilley, J.V., Hodgson, J.R., Helton, D.D., Wiegand, W.J., Roberts, D.N., Anderson, B.S., Halfpap, L.M., Jurtz, L.D. & West, N. (1975) *Mammalian Toxicity of Munition Compounds: Phase I. Acute Oral Toxicity, Primary Skin and Eye Irritation, Dermal Sensitization and Disposition and Metabolism* (Report No. 1, MRI Project No. 3900-B), Kansas City, MI, Midwest Research Institute

Lee, C.C., Hong, C.B., Ellis, H.V., III, Dacre, J.C. & Glennon, J.P. (1985) Subchronic and chronic toxicity studies of 2,4-dinitrotoluene. Part II. CD® rats. *J. Am. Coll. Toxicol.*, **4**, 243–256

Leonard, T.B., Lyght, O. & Popp, J.A. (1983) Dinitrotoluene structure-dependent initiation of hepatocytes *in vivo*. *Carcinogenesis*, **4**, 1059–1061

Leonard, T.B., Adams, T. & Popp, J.A. (1986) Dinitrotoluene isomer-specific enhancement of the expression of diethylnitrosamine-initiated hepatocyte foci. *Carcinogenesis*, **7**, 1797–1803

Leonard, T.B., Graichen, M.E. & Popp, J.A. (1987) Dinitrotoluene isomer-specific hepatocarcinogenesis in F344 Rats. *J. natl Cancer Inst.*, **79**, 1313–1319

Levine, R.J., Turner, M.J, Crume, Y.S., Dale, M.E., Starr, T.B. & Rickert, D.E. (1985) Assessing exposure to dinitrotoluene using a biological monitor. *J. occup. Med.*, **27**, 627–638

Levine, R.J., Andjelkovich, D.A., Kersteter, S.L., Arp, E.W., Balogh, S.A., Blunden, P.B. & Stanley, J.M. (1986) Heart disease in workers exposed to dinitrotoluene. *J. occup. Med.*, **28**, 811–816

Lewalter, J. & Ellrich, D. (1991) Nitroaromatic compounds (nitrobenzene, p-nitrotoluene, p-nitrochlorobenzene, 2,6-dinitrotoluene, o-dinitrobenzene, 1-nitronaphthalene, 2-nitronaphthalene, 4-nitrobiphenyl). In: Angerer, J. & Schaller, K.H., eds, *Analyses of Hazardous Substances in Biological Materials*, Vol. 3, New York, VCH Publishers, pp. 207–229

Lewis, R.J., Sr (1993) *Hawley's Condensed Chemical Dictionary*, 12th Ed., New York, Van Nostrand Reinhold Co., p. 424

Lide, D.R., ed. (1993) *CRC Handbook of Chemistry and Physics*, 74th Ed., Boca Raton, FL, CRC Press, pp. 3-489 — 3-490

Liu, D., Thomson, K. & Anderson, A.C. (1984) Identification of nitroso compounds from biotransformation of 2,4-dinitrotoluene. *Appl. environ. Microbiol.*, **47**, 1295–1298

Lloyd, J.B.F. (1983a) High-performance liquid chromatography of organic explosives components with electrochemical detection at a pendent mercury drop electrode. *J. Chromatogr.*, **257**, 227–236

Lloyd, J.B.F. (1983b) Clean-up procedures for the examination of swabs for explosive traces by high-performance liquid chromatography with electrochemical detection at a pendent mercury drop electrode. *J. Chromatogr.*, **261**, 391–406

Long, R.M. & Rickert, D.E. (1982) Metabolism and excretion of 2,6-dinitro[^{14}C]toluene *in vivo* and in isolated perfused rat livers. *Drug Metab. Dispos.*, **10**, 455–458

Loveday, K.S., Lugo, M.H., Resnick, M.A., Anderson, B.E. & Zeiger, E. (1989) Chromosome aberration and sister chromatid exchange tests in Chinese hamster ovary cells *in vitro*: II. Results with 20 chemicals. *Environ. mol. Mutag.*, **13**, 60–94

Mabey, W.R., Smith, J.H., Podoll, R.T., Johnson, H.L., Mill, T., Chou, T.W., Gates, J., Waight-Partridge, I. & Vandenberg, D. (1982) *Aquatic Fate Process Data for Organic Priority Pollutants* (US EPA-440/4-81-014; US NTIS PB87-169090), Washington DC, Office of Water Regulations and Standards, pp. 239–243

Matson, C. (1986) Feasibility findings for in situ biodegradation of nitroaromatic compounds with specific gravities greater than that of water. In: *Proceedings of the Conference on Southwestern Ground Water Issues, Oct. 20–22, Tempe, AL*, pp. 256–268

McGee, L.M., McCausland, A., Plume, C.A. & Marlett, N.C. (1942) Metabolic disturbances in workers exposed to dinitrotoluene. *Am. J. digest. Dis.*, **9**, 329–332

Medinsky, M.A. & Dent, J.G. (1983) Biliary excretion and enterohepatic circulation of 2,4-dinitrotoluene metabolites in Fischer-344 rats. *Toxicol. appl. Pharmacol.*, **68**, 359–366

Miles (1994) *Miles Organica, The Basis for Quality*, Pittsburgh, PA, Division of Bayer AG, p. 51

Mirsalis, J.C. & Butterworth, B.E. (1982) Induction of unscheduled DNA synthesis in rat hepatocytes following in vivo treatment with dinitrotoluene. *Carcinogenesis*, **3**, 241–245

Mirsalis, J.C., Hamm, T.E., Jr & Butterworth, B.E. (1981) The role of gut flora in the induction of DNA repair in rats treated *in vivo* with dinitrotoluene (Abstract no. 307). *Proc. Am. Assoc. Cancer Res.*, **22**, 78

Mirsalis, J.C., Tyson, C.K. & Butterworth, B.E. (1982a) Detection of genotoxic carcinogens in the in vivo-in vitro hepatocyte DNA repair assay. *Environ. Mutag.*, **4**, 553–562

Mirsalis, J.C., Hamm, T.E., Jr, Sherrill, J.M. & Butterworth, B.E. (1982b) The role of gut flora in the genotoxicity of dinitrotoluene. *Nature*, **295**, 322–323

Mirsalis, J.C., Tyson, C.K., Steinmetz, K.L., Loh, E.K., Hamilton, C.M., Bakke, J.P. & Spalding, J.W. (1989) Measurement of unscheduled DNA synthesis and S-phase synthesis in rodent hepatocytes following in vivo treatment: testing of 24 compounds. *Environ. mol. Mutag.*, **14**, 155–164

Mori, M.-A., Naruse, Y. & Kozuka, H. (1977) Studies on the metabolism and toxicity of dinitrotoluenes; on the excretion and distribution of tritium-labelled 2,4-dinitrotoluene (^{3}H-2,4-DNT) in the rat. *Radioisotopes*, **26**, 780–783

Mori, M.-A., Naruse, Y. & Kozuka, H. (1978) Studies on the metabolism and toxicity of dinitrotoluenes — on the absorption and excretion of tritium-labelled 2,4-dinitrotoluene (^{3}H-2,4-DNT) in the rat. *Radioisotopes*, **27**, 715–718

Mori, M.-A., Miyahara, T., Taniguchi, K., Hasegawa, K., Kozuka, H., Miyagoshi, M. & Nagayama, T. (1982) Mutagenicity of 2,4-dinitrotoluene and its metabolites in *Salmonella typhimurium*. *Toxicol. Lett.*, **13**, 1–5

Mori, M.-A., Miyahara, T., Hasegawa, Y., Kudo, Y. & Kozuka, H. (1984a) Metabolism of dinitrotoluene isomers by *Escherichia coli* isolated from human intestine. *Chem. Pharm. Bull.*, **32**, 4070–4075

Mori, M.-A., Matsuhashi, T., Miyahara, T., Shibata, S., Izima, C. & Kozuka, H. (1984b) Reduction of 2,4-dinitrotoluene by Wistar rat liver microsomal and cytosol fractions. *Toxicol. appl. Pharmacol.*, **76**, 105–112

Mori, M.-A., Kudo, Y., Nunozawa, T., Miyahara, T. & Kozuka, H. (1985) Intestinal metabolism of 2,4-dinitrotoluene in rats. *Chem. pharm. Bull.*, **33**, 327–332

Mori, M.-A., Kawajiri, T., Sayama, M., Taniuchi, Y., Miyahara, T. & Kozuka, H. (1989a) Metabolism of 2,6-dinitrotoluene in male Wistar rat. *Xenobiotica*, **19**, 731–741

Mori, M.-A., Kawajiri, T., Sayama, M., Miyahara, T. & Kozuka, H. (1989b) Metabolism of 2,4-dinitrotoluene and 2,6-dinitrotoluene, and their dinitrobenzyl alcohols and dinitrobenzaldehydes by Wistar and Sprague-Dawley rat liver microsomal and cytosol fractions. *Chem. pharm. Bull.*, **37**, 1904–1908

Nakamura, S., Oda, Y., Shimada, T., Oki, I. & Sugimoto, K. (1987) SOS-inducing activity of chemical carcinogens and mutagens in *Salmonella typhimurium* TA1535/pSK1002: examination with 151 chemicals. *Mutat. Res.*, **192**, 239–246

Nelson, C.R. & Hites, R.A. (1980) Aromatic amines in and near the Buffalo River. *Environ. Sci. Technol.*, **14**, 1147–1150

Oda, Y., Shimada, T., Watanabe, M., Ishidate, M., Jr & Nohmi, T. (1992) A sensitive *umu* test system for the detection of mutagenic nitroarenes in *Salmonella typhimurium* NM1011 having a high nitroreductase activity. *Mutat. Res.*, **272**, 91–99

Oda, Y., Yamazaki, H., Watanabe, M., Nohmi, T. & Shimada, T. (1993) Highly sensitive *umu* test system for the detection of mutagenic nitroarenes in *Salmonella typhimurium* NM3009 having high O-acetyltransferase and nitroreductase activities. *Environ. mol. Mutag.*, **21**, 357–364

Ong, C.P., Chin, K.P., Lee, H.K. & Li, S.F.Y. (1992) Analysis of nitroaromatics in aqueous samples by capillary supercritical fluid chromatography. *Int. J. environ. Stud.*, **41**, 17–25

Pereira, W.E., Short, D.L., Manigold, D.B. & Roscio, P.K. (1979) Isolation and characterization of TNT and its metabolites in groundwater by gas chromatograph-mass spectrometer-computer techniques. *Bull. environ. Contam. Toxicol.*, **21**, 554–562

Popp, J.A. & Leonard, T.B. (1982) The use of in vivo hepatic initiation-promotion systems in understanding the hepatocarcinogenesis of technical grade dinitrotoluene. *Toxicol. Pathol.*, **10**, 190–196

Price, C.J., Tyl, R.W., Marks, T.A., Paschke, L.L., Ledoux, T.A. & Reel, J.R. (1985) Teratologic evaluation of dinitrotoluene in the Fischer 344 rat. *Fundam. appl. Toxicol.*, **5**, 948–961

Putnam, H.D., Sullivan, J.H., Jr, Pruitt, B.C., Nichols, J.C., Keirn, M.A. & Swift, D.R. (1981) Impact of trinitrotoluene wastewaters on aquatic biota in Lake Chickamauga, Tennessee. In: Bates, J.M. & Weber, C.I., eds, *Ecological Assessment of Effluent Impacts on Communities of Indigenous Aquatic Organisms* (ASTM Spec. Tech. Publ. 730), Philadelphia, PA, American Society for Testing and Materials, pp. 220–242

Ramos, K.S., McMahon, K.K., Alipui, C. & Demick, D. (1991a) Modulation of aortic smooth muscle cell proliferation by dinitrotoluene. *Adv. exp. Med. Biol.*, **283**, 805–808

Ramos, K.S., McMahon, K.K., Alipui, C. & Demick, D. (1991b) Modulation of DNA synthesis in aortic smooth muscle cells by dinitrotoluenes. *Cell Biol. Toxicol.*, **7**, 111–128

Rawlings, G.D. & Samfield, M. (1979) Toxicity of secondary effluents from textile plants. In: Burns, E.A., ed., *Symposium Proceedings: Process Measurements for Environmental Assessment* (US EPA-600/7-78-168; US NTIS PB-290331), Research Triangle Park, NC, United States Environmental Protection Agency, pp. 153–169

Reddy, M.V., Gupta, R.C., Randerath, E. & Randerath, K. (1984) ^{32}P-Postlabeling test for covalent DNA binding of chemicals *in vivo*: application to a variety of aromatic carcinogens and methylating agents. *Carcinogenesis*, **5**, 231–243

Rickert, D.E. (1987) Metabolism of nitroaromatic compounds. *Drug Metab. Rev.*, **18**, 23–53

Rickert, D.E. & Long, R.M. (1980) Tissue distribution of 2,4-dinitrotoluene and its metabolites in male and female Fischer-344 rats. *Toxicol. appl. Pharmacol.*, **56**, 286–293

Rickert, D.E. & Long, R.M. (1981) Metabolism and excretion of 2,4-dinitrotoluene in male and female Fischer 344 rats after different doses. *Drug Metab. Dispos.*, **9**, 226–232

Rickert, D.E. & Long, R.M. (1982) Metabolism and excretion of 2,6-dinitro[^{14}C]toluene *in vivo* and in isolated perfused rat livers. *Drug Metab. Dispos.*, **10**, 455–458

Rickert, D.E., Long, R.M., Krakowka, S. & Dent, J.G. (1981) Metabolism and excretion of 2,4-[^{14}C]dinitrotoluene in conventional and axenic Fischer-344 rats. *Toxicol. appl. Pharmacol.*, **59**, 574–579

Rickert, D.E., Schnell, S.R. & Long, R.M. (1983) Hepatic macromolecular covalent binding and intestinal disposition of [^{14}C]dinitrotoluenes. *J. Toxicol. environ. Health*, **11**, 555–567

Rickert, D.E., Butterworth, B.E. & Popp, J.A. (1984) Dinitrotoluene: acute toxicity, oncogenicity, genotoxicity, and metabolism. *CRC crit. Rev. Toxicol.*, **13**, 217–234

Sadtler Research Laboratories (1980) *Sadtler Standard Spectra. 1980 Cumulative Index*, Philadelphia, PA, p. 151

Sax, N.I. & Lewis, R.J., Sr (1989) *Dangerous Properties of Industrial Materials*, 7th Ed., Vol. III, New York, Van Nostrand Reinhold, pp. 1461–1462

Sayama, M., Mori, M.-A., Ishida, M., Okumura, K. & Kozuka, H. (1989a) Enterohepatic circulation of 2,4-dinitrobenzaldehyde, a mutagenic metabolite of 2,4-dinitrotoluene, in male Wistar rat. *Xenobiotica*, **19**, 83–92

Sayama, M., Mori, M.-A., Shirokawa, T., Inoue, M., Miyahara T. & Kozuka, H. (1989b) Mutagenicity of 2,6-dinitrotoluene and its metabolites, and their related compounds in *Salmonella typhimurium*. *Mutat. Res.*, **226**, 181–184

Sayama, M., Mori, M.-A., Nakada, Y., Kagamimori, S. & Kozuka, H. (1991) Metabolism of 2,4-dinitrotoluene by *Salmonella typhimurium* strains TA98, TA98NR and TA98/1,8-DNP$_6$, and mutagenicity of the metabolites of 2,4-dinitrotoluene and related compounds to strains TA98 and TA100. *Mutat. Res.*, **264**, 147–153

Sayama, M., Inoue, M., Mori, M.-A., Maruyama, Y. & Kozuka, H. (1992) Bacterial metabolism of 2,6-dinitrotoluene with *Salmonella typhimurium* and mutagenicity of the metabolites of 2,6-dinitrotoluene and related compounds. *Xenobiotica*, **22**, 633–640

Sayama, M., Mori, M.-A., Maruyama, Y., Inoue, M. & Kozuka, H. (1993) Intestinal transformation of 2,6-dinitrotoluene in male Wistar rats. *Xenobiotica*, **23**, 123–131

Schut, H.A., Loeb, T.R. & Stoner, G.D. (1982) Distribution, elimination, and test for carcinogenicity of 2,4-dinitrotoluene in strain A mice. *Toxicol. appl. Pharmacol.*, **64**, 213–220

Schut, H.A.J., Loeb, T.R., Grimes, L.A. & Stoner, C.D. (1983) Distribution, elimination and test for carcinogenicity of 2,6-dinitrotoluene after intraperitoneal and oral administration to strain A mice. *J. Toxicol. environ. Health*, **12**, 659–670

Shafer, K.H. (1982) *Determination of Nitroaromatic Compounds and Isophorone in Industrial and Municipal Wastewaters* (US EPA-600/4-82-024; US NTIS PB82-208398), Cincinnati, OH, United States Environmental Protection Agency

Shah, A.B., Combes, R.D. & Rowland, I.R. (1987) Activation to mutagenicity of dinitrotoluene and some related compounds by bacterial and mammalian enzymes (Abstract no. 12). *Mutagenesis*, **2**, 301

Short, R.D., Dacre, J.C. & Lee, C.C. (1979) A species comparison of 2,4-dinitrotoluene metabolism *in vitro*. *Experientia*, **35**, 1625–1627

Sina, J.F., Bean, C.L., Dysart, G.R., Taylor, V.I. & Bradley, M.O. (1983) Evaluation of the alkaline elution/rat hepatocyte assay as a predictor of carcinogenic/mutagenic potential. *Mutat. Res.*, **113**, 357–391

Soares, E.R. & Lock, L.F. (1980) Lack of an indication of mutagenic effects of dinitrotoluenes and diaminotoluenes in mice. *Environ. Mutag.*, **2**, 111–124

Spanggord, R.J. & Suta, B.E. (1982) Effluent analysis of wastewater generated in the manufacture of 2,4,6-trinitrotoluene. 2. Determination of a representative discharge of ether-extractable components. *Environ. Sci. Technol.*, **16**, 233–236

Spanggord, R.J., Gibson, B.W., Keck, R.G., Thomas, D.W. & Barkley, J.J., Jr (1982a) Effluent analysis of wastewater generated in the manufacture of 2,4,6-trinitrotoluene. 1. Characterization study. *Environ. Sci. Technol.*, **16**, 229–232

Spanggord, R.J., Mortelmans, K.E., Griffin, A.F. & Simmon, V.F. (1982b) Mutagenicity in *Salmonella typhimurium* and structure–activity relationships of wastewater components emanating from the manufacture of trinitrotoluene. *Environ. Mutag.*, **4**, 163–179

Spanggord, R.J., Myers, C.J., LeValley, S.E., Green, C.E. & Tyson, C.A. (1990) Structure–activity relationship for the intrinsic hepatotoxicity of dinitrotoluenes. *Chem. Res. Toxicol.*, **3**, 551–558

Stayner, L.T., Dannenberg, A.L., Thun, M., Reeve, G., Bloom, T., Boeniger, M. & Halperin, W. (1992) Cardiovascular mortality among munitions workers exposed to nitroglycerin and dinitrotoluene. *Scand. J. Work Environ. Health*, **18**, 34–43

Stayner, L.T., Dannenberg, A.L., Bloom, T. & Thun, M. (1993) Excess hepatobiliary cancer mortality among munitions workers exposed to dinitrotoluene. *J. occup. Med.*, **35**, 291–296

Steuckart, C., Berger-Preiss, E. & Levsen, K. (1994) Determination of explosives and their biodegradation products in contaminated soil and water from former ammunition plants by automated multiple development high-performance thin-layer chromatography. *Anal. Chem.*, **66**, 2570–2577

Stoner, G.D., Greisiger, E.A., Schut, H.A.J., Pereira, M.A., Loeb, T.R., Klaunig, J.E. & Branstetter, D.G. (1984) A comparison of the lung adenoma response in strain A/J mice after intraperitoneal and oral administration of carcinogens. *Toxicol. appl. Pharmacol.*, **72**, 313–323

Styles, J.A. & Cross, M.F. (1983) Activity of 2,4,6-trinitrotoluene in an in vitro mammalian gene mutation assay. *Cancer Lett.*, **20**, 103–108

Taylor, D.G. (1978) *NIOSH Manual of Analytical Methods*, 2nd Ed., Vol 4 (DHEW (NIOSH) Publ. No. 78-175), Washington DC, United States Government Printing Office, Method S215

TCI America (1994) *Organic Chemicals 94/95 Catalog*, Portland, OR, p. 587

Tokiwa, H., Nakagawa, R. & Ohnishi, Y. (1981) Mutagenic assay of aromatic nitro compounds with *Salmonella typhimurium*. *Mutat. Res.*, **91**, 321–325

Turner, M.J., Jr, Levine, R.J., Nystrom, D.D., Crume, Y.S. & Rickert, D.E. (1985) Identification and quantification of urinary metabolites of dinitrotoluenes in occupationally exposed humans. *Toxicol. appl. Pharmacol.*, **80**, 166–174

United Nations Environment Programme (1995) *International Register of Potentially Toxic Chemicals. Legal File, Dinitrotoluenes*, Geneva

United States Environmental Protection Agency (1980) *Treatability Manual* (EPA Report No. EPA-600/8-80-042), Washington DC

United States Environmental Protection Agency (1986a) Method 8090. Nitroaromatics and cyclic ketones. In: *Test Methods for Evaluating Solid Waste — Physical/Chemical Methods*, 3rd Ed., Vol. 2 (US EPA No. SW-846), Washington DC, Office of Solid Waste and Emergency Response

United States Environmental Protection Agency (1986b) Method 8250. Gas chromatography/mass spectrometry for semivolatile organics: packed column technique. In: *Test Methods for Evaluating Solid Waste — Physical/Chemical Methods*, 3rd Ed., Vol. 2 (US EPA No. SW-846), Washington DC, Office of Solid Waste and Emergency Response

United States Environmental Protection Agency (1986c) Method 8270. Gas chromatography/mass spectrometry for semivolatile organics: capillary column technique. In: *Test Methods for Evaluating Solid Waste — Physical/Chemical Methods*, 3rd Ed., Vol. 2 (US EPA No. SW-846), Washington DC, Office of Solid Waste and Emergency Response

United States Environmental Protection Agency (1994) Nitroaromatics and isophorone. *US Code fed. Regul.*, **Title 40**, Part 136, pp. 503–513 [Method 609], 574–601 [Method 625], 615–634 [Method 1625B]

United States National Cancer Institute (1978) *Bioassay of 2,4-Dinitrotoluene for Possible Carcinogenicity* (Technical Report Series No. 54; DHEW Publication No. (NIH) 78-1360), Bethesda, MD

United States National Institute for Occupational Safety and Health (1994) *Pocket Guide to Chemical Hazards* (DHHS (NIOSH) Publ. No. 94-116), Cincinnati, OH, pp. 118–119, 342

United States National Library of Medicine (1995) *Hazardous Substances Data Bank (HSDB)*, Bethesda, MD

United States Occupational Safety and Health Administration (1990) *OSHA Analytical Methods Manual*, 2nd Ed., Part 1, Vol. 2, Salt Lake City, UT, pp. 44-1–44-26

United States Occupational Safety and Health Administration (1994) Air contaminants. *US Code fed. Regul.*, **Title 29**, Part 1910.1000, p. 11

Vernot, E.H., MacEwen, J.D., Haun, C.C. & Kinkead, E.R. (1977) Acute toxicity and skin corrosion data for some organic and inorganic compounds and aqueous solutions. *Toxicol. appl. Pharmacol.*, **42**, 417–423

Woodruff, R.C., Mason, J.M., Valencia, R. & Zimmering, S. (1985) Chemical mutagenesis testing in *Drosophila*. V. Results of 53 coded compounds tested for the National Toxicology Program. *Environ. Mutag.*, **7**, 677–702

Woollen, B.H., Hall, M.G., Craig, R. & Steel, G.T. (1985) Dinitrotoluene: an assessment of occupational absorption during manufacture of blasting explosives. *Int. Arch. occup. environ. Health*, **55**, 319–330

Working, P.K. & Butterworth, B.E. (1984) An assay to detect chemically induced DNA repair in rat spermatocytes. *Environ. Mutag.*, **6**, 273–286

Zoeteman, B.C.J., Harmsen, K., Linders, J.B.H.J., Morra, C.F.H. & Slooff, W. (1980) Persistent organic pollutants in river water and ground water of the Netherlands. *Chemosphere*, **9**, 231–249

2-NITROANISOLE

1. Exposure Data

1.1 Chemical and physical data

1.1.1 *Nomenclature*

Chem. Abstr. Serv. Reg. No.: 91-23-6
Deleted CAS Reg. No.: 35973-13-8
Chem. Abstr. Name: 1-Methoxy-2-nitrobenzene
IUPAC Systematic Name: ortho-Nitroanisole
Synonyms: 2-Methoxynitrobenzene; 2-methoxy-1-nitrobenzene; *ortho*-nitroanisole; *ortho*-nitrobenzene methyl ether; 2-nitromethoxybenzene; *ortho*-nitromethoxybenzene; 1-nitro-2-methoxybenzene; *ortho*-nitrophenyl methyl ether

1.1.2 *Structural and molecular formulae and relative molecular mass*

$C_7H_7NO_3$ Relative molecular mass: 153.13

1.1.3 *Chemical and physical properties of the pure substance*

(a) *Description*: Colourless to yellowish liquid (Budavari, 1989)
(b) *Boiling-point*: 277 °C (Budavari, 1989)
(c) *Melting-point*: 10.5 °C (Lide, 1993)
(d) *Density*: 1.254 at 20 °C/4 °C (Lide, 1993)
(e) *Spectroscopy data*: Infrared (prism [5864], grating [24020]), ultraviolet [1643], nuclear magnetic resonance (proton [1887], C-13 [931]) and mass spectral data have been reported (Sadtler Research Laboratories, 1980).
(f) *Solubility*: Moderately soluble in warm water (1.69 g/L at 30 °C; BUA, 1987); soluble in ethanol and diethyl ether (Budavari, 1989)
(g) *Volatility*: Vapour pressure, 4 Pa at 30 °C (BUA, 1987)
(h) *Stability*: Explosive reaction with sodium hydroxide and zinc (Sax & Lewis, 1989)
(i) *Octanol/water partition coefficient (P)*: log P, 1.73 (Hansch *et al.*, 1995)

(j) *Conversion factor*: mg/m^3 = 6.26 × ppm[1]

1.1.4 *Technical products and impurities*

2-Nitroanisole is commercially available with a purity ranging from 98% to 99% (Aldrich Chemical Co., 1994; Fluka Chemical Corp., 1995).

1.1.5 *Analysis*

2-Nitroanisole in environmental samples can be analysed by gas chromatography with electron capture detection (Mitchell & Deveraux, 1978).

1.2 Production and use

1.2.1 *Production*

2-Nitroanisole is prepared by slowly adding methanolic sodium hydroxide to a solution of 2-chloronitrobenzene in methanol at 70 °C and then, to complete the reaction, by gradually heating the mixture under pressure to 95 °C. After dilution with water, the product is separated as an oil, at a 90% yield; methanol can be recovered from the aqueous layer (Booth, 1991; Lewis, 1993).

In western Europe in 1983, production of 2-nitroanisole was approximately 7200 tonnes per year, including approximately 4000 tonnes per year in Germany (BUA, 1987).

2-Nitroanisole is known to be produced by five companies in Japan, four in China and three in India and by one company each in Brazil, Germany and the Ukraine (Chemical Information Services, 1994).

1.2.2 *Use*

2-Nitroanisole is reduced (using a H_2-catalyst or iron-formic acid) to *ortho*-anisidine (see IARC, 1982, 1987a) or to *ortho*-dianisidine (see IARC, 1974, 1987b), both of which are important as dye intermediates. 2-Nitroanisole has also been used as an intermediate for various pharmaceuticals (Budavari, 1989; Booth, 1991; Lewis, 1993).

1.3 Occurrence

1.3.1 *Natural occurrence*

2-Nitroanisole is not known to occur as a natural product.

1.3.2 *Occupational exposure*

No information was available to the Working Group.

---

[1] Calculated from: mg/m^3 = (relative molecular mass/24.45) × ppm, assuming temperature (25 °C) and pressure (101 kPa)

1.3.3 *Environmental occurrence*

(a) Air

On 22 February 1993, approximately 10 tonnes of vapour containing 2-nitroanisole and other halogenated aromatics were accidentally released from a chemical plant in Grieshem, Germany (Anon., 1993). No concentration of 2-nitroanisole in ambient air has been reported.

(b) Water

2-Nitroanisole has been detected in water samples in Japan (0.7 µg/L) and the Netherlands (0.3–1.0 µg/L). Also, nitroanisole of unspecified isomerism has been detected in the Netherlands (0.3–1.0 µg/L) and Germany (0.1–0.9 µg/L) (BUA, 1987).

(c) Soil and sediment

2-Nitroanisole has been detected in sediment samples taken in Japan (0.01 µg/L) (BUA, 1987).

1.4 Regulations and guidelines

The former USSR has set a short-term exposure limit for 2-nitroanisole of 1 mg/m^3, with skin absorption noted as a potentially significant route of exposure (effective date, 1989) (International Labour Office, 1991; United Nations Environmental Programme, 1995).

2. Studies of Cancer in Humans

No data were available to the Working Group.

3. Studies of Cancer in Experimental Animals

3.1 Oral administration

3.1.1 *Mouse*

Groups of 60 male and 60 female B6C3F1 mice, approximately 40 days of age, were administered 0, 666, 2000 or 6000 mg/kg diet (ppm) 2-nitroanisole (purity > 99%) for 103 weeks. Groups of nine or 10 mice of each sex and from each dose group were killed at 15 months for an interim evaluation. Mean body weights of high-dose male and female mice were 33 and 43% lower than those of controls, respectively, at the end of the study; mean body weights of mid-dose male and female mice were 11 and 18% lower than those of controls, respectively, at the end of the study. Survival at 103 weeks did not differ among treated males (35/50 in controls, 43/50 at the low dose, 39/50 at the mid dose, 40/50 at the high dose) but was increased among treated females (38/50 in controls, 26/50 at the low dose, 33/50 at the mid dose, 45/50 at the high dose) when compared with controls. In treated mice, the incidence of hepatocellular adenoma was increased in both males and females (in males: 14/50 in controls, 26/50 at the low dose, 41/50 at the mid dose, 29/50 at the high dose, $p = 0.012$, logistic regression analysis for trend; in females: 14/50 in controls, 20/50 at the low dose, 36/50 at the mid dose, 18/50 at the high dose, $p < 0.001$ for the mid dose versus controls). The combined incidences of hepatocellular adenomas and carcinomas in males were 21/50 in controls, 32/50 at the low dose, 45/50 at the mid dose and 32/50 at the high dose ($p < 0.001$ for the mid dose versus controls); those in females were 17/50 in controls, 21/50 at the low dose, 37/50 at the mid dose and 20/50 at the high dose ($p < 0.001$ for the mid dose versus controls). The incidence of hepatoblastomas was also increased in male mice (0/50 in controls, 3/50 at the low dose, 17/50 at the mid dose, 9/50 at the high dose; $p < 0.001$). Other hepatic lesions of increased incidence in some or all dose groups included haemorrhage, Kupffer-cell pigmentation, eosinophilic focus, focal necrosis and cytological alteration (hepatocytic hypertrophy, nuclear enlargement and eosinophilic staining of cytoplasm) (United States National Toxicology Program, 1993).

3.1.2 Rat

Groups of 60 male and 60 female Fischer 344/N rats, approximately 40 days of age, were administered 0, 222, 666 or 2000 mg/kg diet (ppm) 2-nitroanisole (purity > 99%) for 103 weeks. Groups of nine or 10 rats of each sex and from each dose group were killed at 15 months for an interim evaluation. Body weights of treated rats were similar to those of controls. When compared with controls, survival at 103 weeks was decreased in treated males (32/50 in controls, 34/50 at the low dose, 24/50 at the mid dose, 9/50 at the high dose) but was unchanged in treated females (33/50 in controls, 41/50 at the low dose, 26/50 at the mid dose, 33/50 at the high dose). In treated rats, the incidence of mononuclear-cell leukaemia was increased in both males and females (males: 26/50 in controls, 25/50 at the low dose, 42/50 at the mid dose, 34/50 at the high dose, $p < 0.001$ life table trend test; females: 14/50 in controls, 11/50 at the low dose, 14/50 at the mid dose, 26/50 at the high dose, $p = 0.001$, life table trend test) (United States National Toxicology Program, 1993).

Groups of 60 male and 60 female Fischer 344/N rats, approximately 40 days of age, were administered 0, 6000 or 18 000 mg/kg diet (ppm) 2-nitroanisole (purity > 99%) for 27 weeks (see Table 1). Groups of 10 animals of each sex and from each dose group were killed at three, six, nine or 15 months for interim evaluation. The remaining animals were killed 77 weeks after cessation of treatment (after 104 weeks of study). The body weights of the treated rats were markedly lower than those of controls during the 27-week treatment period and did not recover after cessation of treatment. Survival at 104 weeks was decreased in treated males and females, with the respective Kaplan Meier probabilities of survival at the end of the study being 63% and 68% for control males and females, 4% and 23% for low-dose males and females, and 0% for both high-dose males and high-dose females. Increased incidences of tumours of the urinary bladder, the large intestine and the kidney occurred in treated males and females. The incidences of selected tumours at the interim kills are given in Table 1. The overall incidences of

Table 1. Incidence of selected tumours in rats fed 2-nitroanisole in the diet for 27 weeks

	Males			Females		
	0 ppm	6000 ppm	18 000 ppm	0 ppm	6000 ppm	18 000 ppm
Urinary bladder						
Transitional-cell carcinoma						
3-month interim	0/9	0/9	1/10	0/10	0/10	0/10
6-month interim	0/10	0/10	10/10*	0/10	0/10	10/10*
9-month interim	0/10	3/10	6/6*	0/10	1/9	6/6*
15-month interim	0/9	1/3	–	0/8	9/10*	–
Large intestine						
Carcinoma						
3-month interim	0/10	0/10	0/10	0/10	0/10	0/10
6-month interim	0/10	0/10	0/10	0/10	0/10	0/10
9-month interim	0/10	0/10	1/6	0/10	0/10	0/6
15-month interim	0/9	0/3	–	0/8	0/10	–
Kidney						
Transitional-cell carcinoma						
3-month interim	0/10	0/10	0/10	0/10	0/10	0/10
6-month interim	0/10	0/10	0/10	0/10	0/10	0/10
9-month interim	0/10	0/10	2/6	0/10	0/10	0/6
15-month interim	0/9	0/3	–	0/8	0/10	–

From United States National Toxicology Program (1993)
*, $p < 0.01$, Fisher's exact test

urinary bladder neoplasms were as follows: transitional-cell papilloma — males: 0/59 in controls, 9/59 at the low dose ($p < 0.01$, Fisher's exact test) and 1/60 at the high dose; females: 0/58 in controls, 2/59 at the low dose and 1/60 at the high dose; transitional-cell carcinoma — males: 0/59 in controls, 27/59 at the low dose ($p < 0.01$) and 50/60 at the high dose ($p < 0.01$); females: 0/58 in controls, 28/59 at the low dose ($p < 0.01$) and 48/60 at the high dose ($p < 0.01$). A few squamous-cell papillomas and carcinomas of the urinary bladder also occurred in high-dose males and females, and urinary bladder sarcomas occurred with overall incidences of 0/59, 2/59 and 9/59 ($p < 0.01$) in control, low-dose and high-dose males and 0/58, 2/59 and 14/60 ($p < 0.01$) in control, low-dose and high-dose females, respectively. Squamous metaplasias and connective tissue proliferation also occurred with increased frequency in the urinary bladders of treated males and females. The overall incidences of adenomatous polyps of the large intestine were 0/60, 26/60 ($p < 0.01$) and 30/60 ($p < 0.01$) in control, low-dose and high-dose males and 0/60, 8/60 ($p < 0.01$) and 18/60 ($p < 0.01$) in control, low-dose and high-dose females, respectively. Carcinomas of the large intestine occurred in 5/60 ($p < 0.05$) high-dose males and 2/60 high-dose females. The overall incidence of transitional-cell tumours of the kidney was also increased in treated males and females compared to controls. No tumour of this type was found in male or female controls; transitional-cell carcinomas were found in 1/60 low-dose males and 8/60 ($p < 0.01$) high-dose males; 4/60 high-dose males had a transitional-cell papilloma. Of high-dose female rats, 1/60 had a transitional-cell papilloma and 1/60 had a transitional-cell carcinoma. The incidence of hyperplasia of the transitional epithelium was significantly increased in all treated males and females. The severity of nephropathy was increased in treated males at three and six months (United States National Toxicology Program, 1993).

3.2 Carcinogenicity of metabolites

One of the metabolites of 2-nitroanisole, *ortho*-anisidine, when tested as the hydrochloride by oral administration in the diet, was found to produce transitional-cell tumours of the urinary bladder in mice and rats and transitional-cell carcinomas of the renal pelvis in rats (IARC, 1982).

4. Other Data Relevant for an Evaluation of Carcinogenicity and its Mechanisms

4.1 Absorption, distribution, metabolism and excretion

4.1.1 *Humans*

No data were available to the Working Group.

4.1.2 *Experimental systems*

The pharmacokinetics and metabolism of 2-nitroanisole were studied in male Fischer 344 rats by Miller *et al.* (1985). Three dose levels of [^{14}C]2-nitroanisole (5, 50 or 500 mg/kg bw) were administered orally to rats, and daily excreta were analysed for ^{14}C. 2-Nitroanisole was readily absorbed from the stomach at the 50-mg/kg bw dose, as less than 10% of the initial dose remained in the stomach at 6 h. At the dose of 500 mg/kg bw, 36% of the initial dose remained in the stomach at 6 h. Peak blood levels reflect the dose-dependence of absorption, with parent 2-nitroanisole reaching maximal concentrations at 3 h (0.9% of dose) after a 50-mg/kg bw dose and at 6 h (0.9% of dose) after a 500-mg/kg bw dose. Within seven days, 7% of the dose had been excreted in the faeces for all dose levels and about 70% of the dose had been eliminated in the urine. For doses of 50 and 500 mg/kg bw, the urinary metabolite profiles at 8 h were not substantially different. The predominant route of elimination was through metabolism to 2-nitrophenol (5–8% of urinary radioactivity), subsequent sulfation to 2-nitrophenyl sulfate (64–68% of urinary radioactivity) and glucuronidation to 2-nitrophenyl glucuronide (13–15% of urinary radioactivity). Seven days after oral administration of 2-nitroanisole, less than 0.5% of the administered dose remained in the carcass.

Since the two highest oral doses saturated the urinary excretion rate of 2-nitroanisole, an intravenous dose of 25 mg/kg bw was used for pharmacokinetic studies. Following a 25 mg/kg bw intravenous injection of [^{14}C]2-nitroanisole, blood, tissues and excreta were collected at times ranging from 15 min to seven days. The distribution of 2-nitroanisole-derived ^{14}C to tissues (muscle, 20%; skin, 10%; fat, 6.8%; blood, 6.5%; liver, 4.8%; plasma, 3.1%; kidney, 2.8%; and small intestine, 1.9%) occurred rapidly following administration. Peak tissue concentrations were reached in all tissues within 15 min. Urinary and faecal elimination were similar to that found after oral administration (urine, 86% by seven days; faeces, 9% by seven days). The subsequent elimination of ^{14}C was rapid and biphasic. The initial elimination phase in all tissues had a half-life of 1–2 h, and the terminal phase half-lives for all tissues ranged from 2.5 to 6.2 days. Elimination of parent 2-nitroanisole from the blood was biphasic with initial and terminal half-lives of 30 min and 2.2 h, respectively. Monophasic elimination of 2-nitroanisole from the liver, kidneys and small intestine occurred with half-lives of 0.35, 0.55, and 0.68 h, respectively. Biliary excretion was similar to faecal elimination, indicating a lack of enterohepatic recirculation. Urine collected for 24 h after intraperitoneal administration of 25 mg/kg bw 2-nitroanisole had a profile similar to that observed after oral administration (63% 2-nitrophenyl sulfate, 11% 2-nitrophenyl glucuronide, 1.5% 2-nitrophenol and 0.6% *ortho*-anisidine) (Miller *et al.*, 1985).

In a study conducted by Yuan *et al.* (1991), male Fischer 344 rats were fed freshly prepared NIH-07 feed or NIOH-07 feed stored for 30 days containing 0.25 mg/g feed 2-nitroanisole. The animals were dosed daily for 3 h for seven consecutive days; on day 7, they were placed in metabolism cages and their urine was collected for 18 h. After the urine collection, the rats were fed control diet for three days. The treated groups then exchanged regimens and were treated for an additional seven days following the same dosing schedule. Again, an 18-h urine sample was collected from each rat and analysed for total 2-nitrophenol. Extraction and analysis indicated that stored feed bound 2-nitroanisole more tightly than freshly prepared feed. Eighteen-hour urine samples collected on day 7 indicated that this binding did not affect systemic bioavailability. Approximately 2.5 mg 2-nitrophenol were excreted in the urine during the 18-h collection period. Total 2-nitrophenol (2-nitrophenol and its conjugates) found in urine decreased in the second week of 2-nitroanisole feeding despite increased feed consumption. Thus, the metabolism of 2-nitroanisole may be affected by continuous exposure.

4.2 Toxic effects

4.2.1 *Humans*

No data were available to the Working Group.

4.2.2 *Experimental systems*

(*a*) *Single-dose studies*

The oral LD_{50} of 2-nitroanisole is 740 mg/kg bw in rats and 1300 mg/kg bw in mice (United States National Institute for Occupational Safety and Health, 1994).

(b) Repeated-dose studies

The United States National Toxicology Program (1993) reported toxic effects after dietary administration of 2-nitroanisole to male and female Fischer 344 rats (583, 1166, 2332, 4665 or 9330 ppm) and B6C3F1 mice (250, 500, 1000, 2000 or 4000 ppm) for 14 days. Mean body-weight gains were depressed in male rats fed 4665 or 9330 ppm, in male mice fed 250 ppm or more and in female mice fed 4000 ppm. Absolute liver weight was increased in male rats fed 1166 ppm and in female rats fed 583 ppm or more. Erythrocyte counts, haematocrit values and haemoglobin concentrations in all exposed male rats were significantly lower than those in controls. Methaemoglobin concentrations were significantly increased in male rats fed 1166 ppm or more. With the exception of depressed body-weight gain, no treatment-related effect was observed in mice.

In 13-week studies, male and female B6C3F1 mice and Fischer 344 rats were fed with diets including 60 (mice only), 200, 600, 2000, 6000 or 18 000 (rats only) ppm 2-nitroanisole (United States National Toxicology Program, 1993). Male and female rats receiving diets containing 6000 or 18 000 ppm and mice receiving diets containing 6000 ppm 2-nitroanisole exhibited lower mean body weights than the controls. Lower haemoglobin and haematocrit values were observed in male and female rats receiving 2000, 6000 or 18 000 ppm and male and female mice receiving 2000 or 6000 ppm 2-nitroanisole. Methaemoglobin increases were observed in male and female rats (6000 and 18 000 ppm) and male mice (6000 ppm). In rats, the principal lesions observed were in the urinary bladder (hyperplasia, 6000 and 18 000 ppm), spleen (congestion, 6000 and 18 000 ppm), kidney (renal tubule necrosis, 600–6000 ppm) and liver (hepatocytic hypertrophy, 18 000 ppm). In male mice, hepatocytic hypertrophy was observed at 200 ppm and above.

4.3 Reproductive and developmental effects

No data were available to the Working Group.

4.4 Genetic and related effects

4.4.1 Humans

No data were available to the Working Group.

4.4.2 Experimental systems (see also Table 2 and Appendices 1 and 2)

2-Nitroanisole was positive in the *rec* assay in *Bacillus subtilis* strains H17 and M45.

2-Nitroanisole was tested in several laboratories for the induction of gene mutations in *Salmonella typhimurium*. Positive responses were obtained consistently with the strain TA100. Variable responses were obtained with some other strains.

In single studies with cultured mammalian cells, 2-nitroanisole induced sister chromatid exchange and chromosomal aberrations in Chinese hamster ovary (CHO) cells and mutation at the *tk* locus of mouse lymphoma L5178Y cells. The clastogenic activity was weak and observed only in the presence of S9, whereas sister chromatid exchange and *tk* mutations were induced in the absence of S9.

Table 2. Genetic and related effects of 2-nitroanisole

Test system	Result[a] Without exogenous metabolic system	Result[a] With exogenous metabolic system	Dose[b] (LED/HID)	Reference
BSD, *Bacillus subtilis rec* H97 and M45 strains, differential toxicity	+	0	625	Shimizu & Yano (1986)
SA0, *Salmonella typhimurium* TA100, reverse mutation	(+)	0	1530	Chiu et al. (1978)
SA0, *Salmonella typhimurium* TA100, reverse mutation	+	+	256	Haworth et al. (1983)
SA0, *Salmonella typhimurium* TA100, reverse mutation	+	0	480	Shimizu & Yano (1986)
SA0, *Salmonella typhimurium* TA100, reverse mutation	0	+	580	Dellarco & Prival (1989)
SA0, *Salmonella typhimurium* TA100, reverse mutation	+	+	128	US National Toxicology Program (1993)
SA5, *Salmonella typhimurium* TA1535, reverse mutation	–	–	385	Haworth et al. (1983)
SA5, *Salmonella typhimurium* TA1535, reverse mutation	–	0	2400	Shimizu & Yano (1986)
SA5, *Salmonella typhimurium* TA1535, reverse mutation	(+)	–	1280	US National Toxicology Program (1993)
SA7, *Salmonella typhimurium* TA1537, reverse mutation	–	–	385	Haworth et al. (1983)
SA7, *Salmonella typhimurium* TA1537, reverse mutation	–	0	2400	Shimizu & Yano (1986)
SA8, *Salmonella typhimurium* TA1538, reverse mutation	+	0	480	Shimizu & Yano (1986)
SA9, *Salmonella typhimurium* TA98, reverse mutation	(+)	0	765	Chiu et al. (1978)
SA9, *Salmonella typhimurium* TA98, reverse mutation	–	–	385	Haworth et al. (1983)
SA9, *Salmonella typhimurium* TA98, reverse mutation	+	0	480	Shimizu & Yano (1986)
SA9, *Salmonella typhimurium* TA98, reverse mutation	–	–	385	US National Toxicology Program (1993)
SAS, *Salmonella typhimurium* TA97, reverse mutation	–	–	1280	US National Toxicology Program (1993)
G5T, Gene mutation, mouse lymphoma L5178Y cells *in vitro*	+	0	250	US National Toxicology Program (1993)
SIC, Sister chromatid exchange, Chinese hamster cells *in vitro*	+	+	123	Galloway et al. (1987)
CIC, Chromosomal aberrations, Chinese hamster cells *in vitro*	–	(+)	1060	Galloway et al. (1987)

[a] +, positive; (+), weak positive; –, negative; 0, not tested; ?, inconclusive
[b] LED, lowest effective dose; HID, highest ineffective dose. In-vitro tests, μg/mL; in-vivo tests, mg/kg bw

5. Summary of Data Reported and Evaluation

5.1 Exposure data

2-Nitroanisole is produced by the reaction of methanolic sodium hydroxide with 2-chloronitrobenzene. It is mainly used in the production of the dye intermediates *ortho*-anisidine and *ortho*-dianisidine. Human exposure may occur during its production and use.

5.2 Human carcinogenicity data

No data on the carcinogenicity of 2-nitrosanisole in humans were available to the Working Group.

5.3 Animal carcinogenicity studies

2-Nitroanisole was tested for carcinogenicity by oral administration in one study in mice and in two studies in rats. In mice, the incidence of hepatocellular adenomas was increased in males and females, and that of hepatoblastomas was increased in males. In one study in rats, the incidence of mononuclear-cell leukaemia was increased in males and females. In the second study, which used a shorter duration of treatment but higher doses, increases were seen in the incidences of tumours of the urinary bladder, the large intestine and the kidney.

5.4 Other relevant data

No human data were available on the metabolism of 2-nitroanisole.

In rats, 2-nitroanisole is absorbed after oral administration, and the major route of its rapid elimination is the urine. The predominant metabolic pathway involves the formation of 2-nitrophenol, with its subsequent conjugation with sulfate and glucuronic acid. 2-Nitroanisole causes methaemoglobinaemia following dietary administration of high doses to rats and mice. Pathological lesions observed in rats occurred in the urinary bladder, spleen, kidney and liver. In mice, 2-nitroanisole causes hypertrophy in the liver.

2-Nitroanisole is mutagenic in bacteria. In single studies, it induced mutations, sister chromatid exchange and a low frequency of chromosomal aberrations in cultured mammalian cells.

5.5 Evaluation[1]

There is *inadequate evidence* in humans for the carcinogenicity of 2-nitroanisole.

There is *sufficient evidence* in experimental animals for the carcinogenicity of 2-nitroanisole.

---

[1]For definition of the italicized terms, see Preamble, pp. 24–27.

Overall evaluation

2-Nitroanisole is *possibly carcinogenic to humans (Group 2B)*.

6. References

Aldrich Chemical Co. (1994) *Aldrich Catalog/Handbook of Fine Chemicals 1994–1995*, Milwaukee, WI, p. 1037

Anon. (1993) Hoechst head responds in aftermath of release. *Chem. Mark. Rep.*, **243**, 5

Booth, G. (1991) Nitro compounds, aromatic. In: Elvers, B., Hawkins, S. & Schulz, G., eds, *Ullmann's Encyclopedia of Industrial Chemistry*, 5th rev. Ed., Vol. A17, New York, NY, VCH Publishers, pp. 411–455

BUA (1987) *GDCh-Advisory Committee on Existing Chemicals of Environmental Relevance (BUA): o-Nitroanisole (1-Methoxy-2-nitrobenzene) (BUA Report 9) (June 1987)*, Weinheim, VCH Verlagsgesellschaft mbH

Budavari, S., ed. (1989) *The Merck Index*, 11th Ed., Rahway, NJ, Merck & Co., p. 1042

Chemical Information Services (1994) *Directory of World Chemical Producers 1995/96 Edition*, Dallas, TX, p. 524

Chiu, C.W., Lee, L.H., Wang, C.Y. & Bryan, G.T. (1978) Mutagenicity of some commercially available nitro compounds for *Salmonella typhimurium*. *Mutat. Res.*, **58**, 11–22

Dellarco, V.L. & Prival, M.J. (1989) Mutagenicity of nitro compounds in *Salmonella typhimurium* in the presence of flavin mononucleotide in a preincubation assay. *Environ. mol. Mutag.*, **13**, 116–127

Fluka Chemical Corp. (1995) *Fluka Chemika–BioChemika Analytika 1995/96*, Buchs, Fluka Chemie AG, p. 1065

Galloway, S.M., Armstrong, M.J., Reuben, C., Colman, S., Brown, B., Cannon, C., Bloom, A.D., Nakamura, F., Ahmed. M., Duk, S., Rimpo, J., Margolin, B.H., Resnick, M.A., Anderson, B. & Zeiger, E. (1987) Chromosome aberrations and sister chromatid exchanges in Chinese hamster ovary cells: evaluation of 108 chemicals. *Environ. mol. Mutag.*, **10** (Suppl. 10), 1–175

Hansch, C., Leo, A. & Hoekman, D.H. (1995) *Exploring QSAR*, Washington DC, American Chemical Society, p. 30

Haworth, S., Lawlor, T., Mortelmans, K., Speck, W. & Zeiger, E. (1983) *Salmonella* mutagenicity test results for 250 chemicals. *Environ. Mutag.*, **Suppl. 1**, 3–142

IARC (1974) *IARC Monographs on the Evaluation of Carcinogenic Risk of Chemicals to Man, Vol. 4, Some Aromatic Amines, Hydrazine and Related Substances, N-Nitroso Compounds and Miscellaneous Alkylating Agents*, Lyon, pp. 41–47

IARC (1982) *IARC Monographs on the Evaluation of the Carcinogenic Risk of Chemicals to Humans, Vol. 27, Some Aromatic Amines, Anthraquinones and Nitroso Compounds*, pp. 63–80

IARC (1987a) *IARC Monographs on the Evaluation of Carcinogenic Risks to Humans, Suppl. 7, Overall Evaluations of Carcinogenicity: An Updating of* IARC Monographs *Volumes 1 to 42*, Lyon, p. 57

IARC (1987b) *IARC Monographs on the Evaluation of Carcinogenic Risks to Humans*, Suppl. 7, *Overall Evaluations of Carcinogenicity: An Updating of IARC Monographs Volumes 1 to 42*, Lyon, pp. 198–199

International Labour Office (1991) *Occupational Exposure Limits for Airborne Toxic Substances: Values of Selected Countries* (Occupational Safety and Health Series No. 37), 3rd Ed., Geneva, pp. 292–293

Lewis, R.J., Sr (1993) *Hawley's Condensed Chemical Dictionary*, 12th Ed., New York, Van Nostrand Reinhold Co., pp. 824–825

Lide, D.R., ed. (1993) *CRC Handbook of Chemistry and Physics*, 74th Ed., Boca Raton, FL, CRC Press, p. 3–46

Miller, M.J., Sipes, I.G., Perry, D.F. & Carter, D.E. (1985) Pharmacokinetics of o-nitroanisole in Fischer 344 rats. *Drug Metab. Dispos.*, **13**, 527–531

Mitchell, J., Jr & Deveraux, H.D. (1978) Determination of traces of organic compounds in the atmosphere: role of detectors in gas chromatography. *Anal. chim. Acta*, **100**, 45–52

Sadtler Research Laboratories (1980) *Sadtler Standard Spectra. 1980 Cumulative Index*, Philadelphia, PA, p. 156

Sax, N.I. & Lewis, R.J., Sr (1989) *Dangerous Properties of Industrial Materials*, 7th ed., Vol. 3, New York, Van Nostrand Reinhold Co., p. 2502

Shimizu, M. & Yano, E. (1986) Mutagenicity of mono-nitrobenzene derivatives in the Ames test and rec assay. *Mutat. Res.*, **170**, 11–22

United Nations Environment Programme (1995) *International Register of Potentially Toxic Chemicals, Legal File, 2-Nitroanisole*, Geneva

United States National Toxicology Program (1993) *Toxicology and Carcinogenesis Studies of o-Nitroanisole (CAS No. 91-23-6) in F344 rats and B6C3F$_1$ Mice (Feed Studies)* (Technical Report Series 416; NIH Publication 93-3147), Research Triangle Park, NC

United States National Institute for Occupational Safety and Health (1994) *RTECs Chem.-Bank*, Cincinnati, OH

Yuan, J., Jameson, C.W., Goehl, T.J., Collins, B.J., Corniffe, G., Kuhn, G. & Castro, C. (1991) Effects of physical binding of o-nitroanisole with feed upon its systemic availability in male F344 rats. *Bull. environ. Contam. Toxicol.*, **47**, 152–159

NITROBENZENE

1. Exposure Data

1.1 Chemical and physical data

1.1.1 Nomenclature

Chem. Abstr. Serv. Reg. No.: 98-95-3

Chem. Abstr. Name: Nitrobenzene

IUPAC Systematic Name: Nitrobenzene

Synonyms: Essence of mirbane; essence of myrbane; mirbane oil; nitrobenzol; oil of mirbane; oil of myrbane

1.1.2 Structural and molecular formulae and relative molecular mass

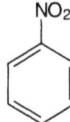

$C_6H_5NO_2$ Relative molecular mass: 123.11

1.1.3 Chemical and physical properties of the pure substance

(a) *Description*: Greenish yellow crystals or yellow oily liquid with an odour of bitter almonds (Booth, 1991; Lewis, 1993)

(b) *Boiling-point*: 210.8 °C (Lide, 1993)

(c) *Melting-point*: 5.8 °C (Dunlap, 1981)

(d) *Density*: 1.2037 at 20 °C/4 °C (Lide, 1993)

(e) *Spectroscopy data*: Infrared (prism [12], grating [10]), ultraviolet [8], nuclear magnetic resonance (proton [4], C-13 [1401]) and mass spectral data have been reported (Sadtler Research Laboratories, 1980).

(f) *Solubility*: Moderately soluble in water (1.9 g/L at 20 °C); soluble in acetone, benzene, diethyl ether and ethanol (Booth, 1991; Lide, 1993)

(g) *Volatility*: Vapour pressure, 0.15 mm Hg [20 Pa] at 20 °C; relative vapour density (air = 1), 4.1 (Verschueren, 1983; Booth, 1991; Lide, 1993)

(h) *Stability*: Moderate explosion hazard when exposed to heat or flame; explosive reaction with solid or concentrated alkali (e.g. sodium hydroxide or potassium hydroxide) and heat, with aluminium chloride and phenol (at 120 °C), with aniline, glycerol and sulfuric acid, and with nitric and sulfuric acids and heat;

forms explosive mixtures with aluminium chloride, oxidants, phosphorous pentachloride, potassium and sulfuric acid (Sax & Lewis, 1989)

(i) *Octanol/water partition coefficient (P)*: log P, 1.85 (Hansch et al., 1995)

(j) *Conversion factor*: mg/m^3 = 5.04 × ppm[1]

1.1.4 *Technical products and impurities*

Nitrobenzene is available commercially at a purity of 99.9% (First Chemical Corp., 1993).

1.1.5 *Analysis*

Selected methods for the analysis of nitrobenzene in various media are identified in Table 1.

The physicochemical properties (volatility, water solubility and partition coefficient) of nitrobenzene determine the manner in which it is analysed in biological samples. Typically, the routine determination of nitrobenzene in the urine, at concentrations in the range of 5–50 mg/L, is based on colorimetric analysis. This is achieved through acidification of the urine and zinc reduction of the nitro group of nitrobenzene. Subsequent diazotization and coupling to 1-amino-8-naphthol-2,4-disulfonic acid (Chicago acid) allows spectrophotometric determination of the primary aromatic amine as a red azo dye. An alternative method, in which reduction of the aromatic nitro compounds is carried out under alkaline conditions through the use of formamidine sulfinic acid (thiourea dioxide), has been described (Koniecki & Linch, 1958).

The difficulty of analysis of nitrobenzene and its metabolite aniline in animals has been discussed (Albrecht & Neumann, 1985). Excretion of the parent compound and metabolites in urine has been determined, but, for practical reasons, this type of biological monitoring has not yet produced satisfactory results. Nitrobenzene metabolites are bound to blood proteins, both in haemoglobin and in plasma. Acute poisoning by nitrobenzene is usually monitored by measuring levels of methaemoglobin, which is produced by the metabolic products of nitrobenzene, but this is a relatively non-specific method since many toxicants produce methaemoglobin. Determination of total 4-nitrophenol in urine specimens collected at the end of the work week has also been recommended for monitoring nitrobenzene exposure (Albrecht & Neumann, 1985; Agency for Toxic Substances and Disease Registry, 1990; American Conference of Governmental Industrial Hygienists, 1995).

Pendergrass (1994) reported an approach for estimating workplace exposure to nitrobenzene. A qualitative estimate of potential dermal exposure to nitrobenzene in the workplace was obtained using gauze surface wipes, a dermal badge sampler was developed to estimate potential worker dermal exposure to nitrobenzene via splashes, spills and aerosol vapours, and an air sampling train, consisting of an acid-treated glass fibre filter in series with a large silica gel tube, allowed airborne workplace exposures to nitro-

---

[1]Calculated from: mg/m^3 = (relative molecular mass/24.45) × ppm, assuming temperature (25 °C) and pressure (101 kPa)

benzene to be quantified. All samples were desorbed with ethanol followed by analysis using capillary gas chromatography with flame ionization detection.

Table 1. Methods for the analysis of nitrobenzene

Sample matrix	Sample preparation	Assay procedure	Limit of detection	Reference
Air	Draw air through solid sorbent tube; desorb with methanol	GC/FID	0.02 mg	Eller (1994) [Method 2005]
	Trap in ethanol or isopropanol; reduce to aniline with zinc/hydrochloric acid; couple with 1,2-naphthoquinone-4-sulfonic acid, disodium salt; extract with carbon tetrachloride; read at 450 nm	Colorimetry	10 µg	Dangwal & Jethani (1980)
Water	Extract sample with dichloromethane or adsorb on Amberlite XAD resin and elute with dichloromethane	GC/ECD	NR	Feltes et al. (1990)
	Purge sample with helium; trap on solid sorbent; desorb thermally (capillary column method)	GC/MS	1.2 µg/L	Munch & Eichelberger (1992)
	Solid-phase microextraction with a polydimethylsiloxane-coated fibre; desorb thermally	GC/FID	9 µg/L	Horng & Huang (1994)
Municipal and industrial discharges	Add sodium oxalate/EDTA/perchloric acid solutions; filter; adjust to pH 3.0 with perchloric acid	LC/UV	NR	Nielen et al. (1985)
	Extract with dichloromethane; dry; exchange to hexane	GC/ECD GC/FID	13.7 µg/L 3.6 µg/L	US Environmental Protection Agency (1986a, 1994) [Methods 8090 & 609]
	Extract with dichloromethane at pH > 11 and at pH < 2; dry (packed column method)	GC/MS	1.9 µg/L	US Environmental Protection Agency (1986b, 1994) [Methods 8250 & 625]
	Add isotopically labelled analogue to sample; extract with methylene chloride at pH 12–13 and at pH < 2; dry	GC/MS	10 µg/L	US Environmental Protection Agency (1994) [Method 1625B]
Water, soil, sediment, waste	Extract with dichloromethane (capillary column method)	GC/MS	PQL[a]	US Environmental Protection Agency (1986c) [Method 8270]

Table 1 (contd)

Sample matrix	Sample preparation	Assay procedure	Limit of detection	Reference
Soil	Extract with methanol; clean-up with solid-phase extraction	HPLC/UV	NR	Grob & Cao (1990)
Urine	Reduce to aniline with zinc/hydrochloric acid; couple with 1,2-naphthoquinone-4-sulfonic acid, disodium salt; extract with carbon tetrachloride; read at 450 nm	Colorimetry	0.8 mg/L	Dangwal & Jethani (1980)
Blood	Extract from separated plasma and concentrate with 2,2,4-trimethylpentane	GC/ECD	10 µg/L	Lewalter & Ellrich (1991)

GC, gas chromatography; FID, flame ionization detection; ECD, electron capture detection; NR, not reported; MS, mass spectrometry; EDTA, ethylenediaminotetraacetic acid; LC, liquid chromatography; UV, ultraviolet detection; HPLC, high-performance liquid chromatography

"PQL, practical quantitation limit: groundwater, 10 µg/L; low soil/sediment, 660 µg/kg; medium level soil and sludges by sonicator, 4.95 mg/kg; non-water-miscible waste, 49.5 mg/kg

1.2 Production and use

1.2.1 Production

Nitrobenzene was first synthesized in 1834 by treating benzene with fuming nitric acid, and was first produced commercially in England in 1856 (Dunlap, 1981).

Nitrobenzene is manufactured commercially by the direct nitration of benzene using what is known as 'mixed acid' or 'nitrating acid' (27–32% HNO_3, 56–60% H_2SO_4, 8–17% H_2O). Historically, it was produced by a batch process. With a typical batch process, the reactor was charged with benzene at a temperature of 50–55 °C. The mixed acid was then added slowly below the surface of the benzene and the temperature raised to 80–90 °C. The reaction mixture was fed into a separator where the spent acid settled to the bottom and was drawn off to be refortified. The crude nitrobenzene was drawn from the top of the separator and was washed in several steps with dilute sodium carbonate and then water. Depending on the desired purity of the nitrobenzene, the product was then distilled. Today, nitrobenzene is made by a continuous process, but the sequence of operations is basically the same as for batch processing; however, for a given rate of production, the size of the reactors is much smaller in a continuous process. A 120-L continuous reactor has been reported to give the same output of nitrobenzene as a 6000-L batch reactor (Dunlap, 1981; Booth 1991).

The annual world production capacity for nitrobenzene in 1985 was approximately 1.7 million tonnes, with about one-third of this production located in western Europe and one-third in the United States of America (Booth, 1991). The increasing production/demand for nitrobenzene in the United States since 1960 is presented in Table 2.

Table 2. Production/demand levels for nitrobenzene in the United States

Year	Production/demand (thousand tonnes)
1960	73
1965	127
1970	249
1975	259
1980	277
1984	431
1986	435
1987	422[a]
1989	533[a]
1990	533[a]
1992	612[a]
1993	671[a]

From Mannsville Chemical Products Corp. (1984); Anon. (1987, 1990); American Conference of Governmental Industrial Hygienists (1991); Anon. (1993)
[a] Demand

Nitrobenzene is known to be produced by seven companies in China, six companies in the United States, five companies each in Brazil and Japan, four companies each in Germany and India, two companies each in Italy and Russia, and one company each in Argentina, Armenia, Belgium, Czech Republic, France, Hungary, Mexico, Portugal, Republic of Korea, Romania, Spain and the United Kingdom (Chemical Information Services, 1994).

1.2.2 Use

Nitrobenzene has a wide variety of uses. Most significantly, and accounting for 95% or more of nitrobenzene use, is the manufacture of aniline (see IARC, 1982, 1987) through the reduction of the nitro group of nitrobenzene. Aniline is a major chemical intermediate in the production of dyestuffs and other products.

Lower-volume industrial uses of nitrobenzene include electrolytic reduction to 4-aminophenol, nitration to give 1,3-dinitrobenzene, chlorination to give 3-chloronitrobenzene (see this volume), sulfonation to give 3-nitrobenzenesulfonic acid and chlorosulfonation to give 3-nitrobenzenesulfonyl chloride. The last three products are consumed mainly as their reduction products, 3-chloroaniline, metanilic acid and 3-aminobenzenesulfonamide, respectively. Nitrobenzene is also used as a solvent for cellulose ethers, in modifying the esterification of cellulose acetate, in the preparation of nitrocellulose (pyroxylin) derivatives and in refining lubricating oils (Parmeggiani, 1983;

Budavari, 1989; Booth, 1991). It is also used in the production of various pharmaceutical products, in rubber industry applications and as an industrial solvent (Mannsville Chemical Products Corp., 1984).

Nitrobenzene is also used as a constituent of soap and polishes, as a solvent for some paints and as a preservative in spray paints. Owing to its musk-like odour, nitrobenzene can be used to mask unpleasant smells (Parmeggiani, 1983; Budavari, 1989; Booth, 1991; Lewis, 1993). In addition, it is reportedly used as a substitute for almond essence in the perfume industry. It is registered as an insecticide for use on cadavers (United States Environmental Protection Agency, 1988).

Nigrosin (CI Solvent Black 5), where it is still produced and used (e.g. in certain inks), is the crude mixture obtained by reacting nitrobenzene with aniline and aniline hydrochloride at 200 °C in the presence of iron or copper (Booth, 1991).

1.3 Occurrence

1.3.1 *Natural occurrence*

Nitrobenzene is not known to occur as a natural product.

1.3.2 *Occupational exposure*

No data on occupational exposures to nitrobenzene were available to the Working Group.

The National Occupational Exposure Survey conducted between 1981 and 1983 indicated that 5080 employees in the United States were potentially exposed to nitrobenzene. The estimate is based on a survey of United States companies and did not involve measurements of actual exposure (United States National Institute for Occupational Safety and Health, 1995).

1.3.3 *Environmental occurrence*

(a) Air

In 1981, sites in three industrialized cities in New Jersey, United States, were monitored continuously for six weeks for a number of airborne toxic substances. The results for nitrobenzene were as follows: Newark, 81% of 37 samples positive; geometric mean concentration, 0.07 ppb [0.35 µg/m^3]; Elizabeth, 86% of 36 samples positive; geometric mean concentration, 0.10 ppb [0.5 µg/m^3]; and Camden, 86% of 37 samples positive; geometric mean concentration, 0.07 ppb [0.35 µg/m^3] (Harkov *et al.*, 1983). Trace levels of nitrobenzene were found in two of 13 air samples from the Lipari and BFI landfills in New Jersey (Howard, 1989). Mean air concentrations of nitrobenzene at five abandoned hazardous landfill sites in New Jersey ranged from 0.01 to 1.32 ppb [0.05–6.65 µg/m^3], with a maximum concentration of 3.46 ppb [17.4 µg/m^3] (Harkov *et al.*, 1985).

The United States Environmental Protection Agency assessed volatile organic compounds in the atmosphere at 595 urban/suburban sites using available data. Nitrobenzene was found to have had a maximal concentration of 2.8 ppb [14 µg/m^3] and a mean concentration of 0.17 ppb [0.86 µg/m^3]; 75% of the samples contained less than 0.09 ppb [0.45 µg/m^3] (Brodzinsky & Singh, 1982).

(b) Water

Among United States water supplies, nitrobenzene was detected but not quantified in finished water from the Carrollton Water Plant in Louisiana, and in drinking-water in Cincinnati, OH. Also, in a survey of 14 treated drinking-water supplies of varied sources in the United Kingdom, nitrobenzene was detected in one supply which came from an upland reservoir (Howard, 1989).

Ambient surface water and industrial effluents have been monitored for nitrobenzene at 836 and 1245 stations, respectively, in the United States for the United States Environmental Protection Agency STORET database. Of these, 0.4% and 1.8% reported detectable levels (< 10 µg/L) of nitrobenzene, respectively (Staples *et al.*, 1985). In the Netherlands, average and maximal levels of nitrobenzene were 1.7 and 13.8 µg/L in the River Wall and < 0.1 and 0.3 µg/L in the River Maas (Meijers & van der Leer, 1976); in another study in the Netherlands, water in the River Rhine contained 0.5 µg/L nitrobenzene (Zoeteman *et al.*, 1980). In water samples collected in 1986 from the Scheldt Estuary, located in the South-west Netherlands and North-west Belgium, the dissolved concentration of nitrobenzene was 0.13 µg/L (van Zoest & van Eck, 1991). A two-week composite water sample taken in 1984 from the River Rhine near Dusseldorf, Germany, contained a mean nitrobenzene concentration of 0.42 g/L (Sontheimer *et al.*, 1985). In the late 1980s, the concentrations of nitrobenzene in the River Elbe, Germany, were 0.1 µg/L in a sample collected near Lauenberg, 0.03 µg/L in a sample collected near Brokdorf and 0.02 µg/L in the sample collected near Brunsbüttel (Feltes *et al.*, 1990). Samples of river water and seawater from various locations in Japan contained 0.16–0.99 ppb [µg/L] nitrobenzene (Sugiyama *et al.*, 1978).

In groundwater samples collected from January to March 1987 in Degrémont, France, nitrobenzene was identified as a pollutant at concentrations ranging from 3 to 12 µg/L (Duguet *et al.*, 1988).

A comprehensive survey of wastewater from 4000 industrial and publicly owned treatment works, carried out by the United States Environmental Protection Agency, identified nitrobenzene in discharges from the following industrial categories (frequency of occurrence, median concentration in µg/L): organic chemicals (36, 43.7); organics and plastics (13, 3876.7); explosives (8, 51.7); inorganic chemicals (3, 1995.3); leather tanning (1, 3.7); petroleum refining (1, 7.7); nonferrous metals (1, 47.7); pulp and paper (1, 124.3); auto and other laundries (1, 40.4); and pesticides manufacture (1, 16.3). The highest effluent concentration for a single sample was 100 mg/L in the organics and plastics industry (Howard, 1989). Nitrobenzene was also detected in the final effluent of three wastewater treatment works and an oil refinery in Illinois (Ellis *et al.*, 1982), and two samplings of the final effluent of the Los Angeles County Municipal Wastewater Treatment Plant collected in 1978 and 1980 contained mean concentrations of 20 and < 10 µg/L nitrobenzene, respectively (Young *et al.*, 1983).

Wastewaters discharged from a nitrobenzene manufacturing plant in India were found to contain 55–138 mg/L (mean, 107 mg/L) nitrobenzene in an acidic stream and 52–93 mg/L (mean, 67 mg/L) nitrobenzene in an alkaline stream. Wastewaters discharged from a chloronitrobenzene manufacturing plant in India were found to contain 4–17 mg/L (mean, 9 mg/L) nitrobenzene (Swaminathan et al., 1987).

(c) *Soil and sediments*

Nitrobenzene was detected at a concentration of 8 mg/kg (ppm) in one out of two soil samples along the Buffalo River in Buffalo, NY, United States, but was not detected in three samples of bottom sediment from the river (Nelson & Hites, 1980). None of the 349 stations monitoring for nitrobenzene in sediment in the United States Environmental Protection Agency STORET database reported detectable levels of nitrobenzene (Staples et al., 1985).

1.3.4 *Food*

None of the 122 monitoring stations analysing for nitrobenzene in fish in the United States Environmental Protection Agency STORET database reported detectable levels in any sample (Staples et al., 1985)

1.4 Regulations and guidelines

Occupational exposure limits and guidelines for nitrobenzene in several countries are presented in Table 3.

Table 3. Occupational exposure limits and guidelines for nitrobenzene

Country	Year	Concentration (mg/m³)	Interpretation
Argentina	1991	5 (Sk)	TWA
Australia	1993	5 (Sk)	TWA
Belgium	1993	5 (Sk)	TWA
Bulgaria[a]	1995	5 (Sk)	TWA
Canada	1991	5 (Sk)	TWA
Colombia[a]	1995	5 (Sk)	TWA
Czech Republic	1993	5 (Sk)	TWA
		25	STEL
Denmark	1993	5 (Sk)	TWA
Egypt	1993	5 (Sk)	TWA
Finland	1993	5 (Sk)	TWA
		15	STEL (15 min)
France	1993	5 (Sk)	TWA
Germany	1995	5 (Sk)[b]	MAK
Hungary	1993	3 (Sk)	TWA
		6	STEL
Japan	1993	5 (Sk)	TWA

Table 3 (contd)

Country	Year	Concentration (mg/m³)	Interpretation
Jordan[a]	1995	5 (Sk)	TWA
Mexico	1991	5 (Sk)	TWA
		10	STEL
Netherlands	1994	5 (Sk)	TWA
New Zealand[a]	1995	5 (Sk)	TWA
Republic of Korea[a]	1995	5 (Sk)	TWA
Poland	1993	3	TWA
Russia	1993	3 (Sk)	STEL
Singapore[a]	1995	5 (Sk)	TWA
Sweden	1993	5 (Sk)	TWA
		10	STEL
Switzerland	1993	5 (Sk)	TWA
		10	STEL
Turkey	1993	5 (Sk)	TWA
United Kingdom	1995	5 (Sk)	TWA
		10	STEL (15 min)
USA			
ACGIH (TLV)	1995	5 (Sk)[b]	TWA
OSHA (PEL)	1994	5 (Sk)	TWA
NIOSH (REL)	1994	5 (Sk)	TWA
Viet Nam[a]	1993	5 (Sk)	TWA

From Arbeidsinspectie (1994); United States National Institute for Occupational Safety and Health (NIOSH) (1994a,b); United States Occupational Safety and Health Administration (OSHA) (1994); Health and Safety Executive (1995); American Conference of Governmental Industrial Hygienists (ACGIH) (1995); Deutsche Forschungsgemeinschaft (1995); United Nations Environment Program (1995)

TWA, time-weighted average; Sk, absorption through the skin may be a significant source of exposure; STEL, short-term exposure limit; MAK, maximal workplace concentration; TLV, threshold limit value; PEL, permissible exposure limit; REL, recommended exposure limit

[a] Follows ACGIH TLVs
[b] Substance identified in the BEI (Biological Exposure Indices) documentation for inducers of methaemoglobin

Two methods for biological monitoring of nitrobenzene exposures have been adopted. Total 4-nitrophenol in urine is measured as a biological marker for exposure to nitrobenzene with a Biological Exposure Index of 5 mg/g creatinine (Lauwerys, 1991;

American Conference of Governmental Industrial Hygienists, 1995). A less specific biological marker for exposure to nitrobenzene is methaemoglobin level in the blood, with a maximal permissible value of 1.5% of haemoglobin (American Conference of Governmental Industrial Hygienists, 1995).

In Germany, a BAT (Biological Tolerance Value) at the workplace of 100 µg aniline/L blood has been established (aniline released from aniline-haemoglobin conjugate) (Deutsche Forschungsgemeinschaft, 1995).

2. Studies of Cancer in Humans

No data were available to the Working Group.

3. Studies of Cancer in Experimental Animals

3.1 Inhalation exposure

3.1.1 *Mouse*

Groups of 70 male and 70 female B6C3F1 mice, 63 days of age, were exposed by inhalation to air containing target concentrations of 0, 5, 25 or 50 ppm [0, 25, 125 or 250 mg/m^3] nitrobenzene (> 99.8% pure) for 6 h per day on five days per week for 24 months. Body weights of high-dose male mice were approximately 5–8% lower than those of controls throughout the study. Probability of survival at 24 months was 60% for males and 45% for females and was not affected by exposure to nitrobenzene, except that mid-dose females had better survival than controls (70%). The incidence of alveolar–bronchiolar neoplasms was increased in treated males (alveolar–bronchiolar adenomas and carcinomas: 9/68 in controls, 21/67 at the low dose, 21/65 at the mid dose and 23/66 at the high dose; $p < 0.05$, Cochran-Armitage trend test). The incidence of alveolar–bronchiolar hyperplasia was also increased in mid- and high-dose males and in mid-dose females. The incidence of thyroid follicular-cell adenomas was increased in treated males (0/65 in controls, 4/65 at the low dose, 1/65 at the mid dose, 7/64 at the high dose; $p < 0.05$ trend test) and that of thyroid follicular-cell hyperplasia was increased in mid- and high-dose males. The incidence of hepatocellular adenomas was increased in treated females (6/51 in controls, 5/61 at the low dose, 5/64 at the mid dose, 13/62 at the high dose; $p < 0.05$ trend test), although the incidence of hepatocellular adenomas and carcinomas combined was not increased (7/51, 7/61, 7/64, 14/62, respectively). Mammary gland adenocarcinomas were found in 5/60 ($p < 0.05$) high-dose females compared to 0/48 controls (Cattley *et al.*, 1994).

3.1.2 *Rat*

Groups of 70 male and 70 female Fischer 344 rats, 62 days of age, were exposed by inhalation to air containing target concentrations of 0, 1, 5 or 25 ppm [0, 5, 25 or

125 mg/m^3] nitrobenzene (> 99.8% pure) for 6 h per day on five days per week for 24 months. Groups of 10 rats per sex and per group were killed for an interim evaluation at 15 months. Body weights of high-dose males were slightly lower than those of controls during the study. Probability of survival at 24 months was 75% for males and 80% for females and was not affected by exposure to nitrobenzene. Increased incidences were noted for hepatic eosinophilic foci in mid- and high-dose males and in high-dose females, and for hepatocellular neoplasms in both treated males (adenomas and carcinomas: 1/69 in controls, 4/69 at the low dose, 5/70 at the mid dose, 16/70 at the high dose; $p < 0.05$, Cochran-Armitage trend test) and treated females (0/70 in controls, 2/66 at the low dose, 0/66 at the mid dose, 4/70 at the high dose; $p < 0.05$ trend test). Thyroid follicular-cell hyperplasia occurred with a positive exposure-related trend in males and the incidences of thyroid follicular-cell adenomas and adenocarcinomas were increased in exposed males (2/69 in controls, 1/69 at the low dose, 5/70 at the mid dose, 8/70 at the high dose; $p < 0.05$ trend test). The incidence of endometrial stromal polyps was increased in exposed females (11/69 in controls, 17/65 at the low dose, 15/65 at the mid dose, 25/69 at the high dose; $p < 0.05$); that of renal tubular-cell adenomas was increased in exposed males (0/69 in controls, 0/68 at the low dose, 0/70 at the mid dose, 5/70 at the high dose; $p < 0.05$, Fisher exact test) and one renal tubular-cell carcinoma occurred in another high-dose male. There was an increased severity of nephropathy in exposed males and females (Cattley *et al.*, 1994).

Groups of 70 male Charles River CD rats, 62 days of age, were exposed by inhalation to air containing target concentrations of 0, 1, 5 or 25 ppm [0, 5, 25 or 125 mg/m^3] nitrobenzene (> 99.8% pure) for 6 h per day on five days per week for 24 months. Groups of 10 rats per sex and per group were killed for an interim evaluation at 15 months. Body weights and survival were not affected by exposure to nitrobenzene during the study. The incidence of hepatocellular neoplasms was increased in treated groups (adenomas and carcinomas: 2/63 in controls, 1/67 at the low dose, 4/70 at the mid dose, 9/65 at the high dose; $p < 0.05$, Cochran-Armatage trend test). The incidence of spongiosis hepatis was increased in high-dose rats, and that of centrilobular hepatocytomegaly was increased in mid- and high-dose groups. The incidence of Kupffer-cell pigmentation was increased in all treated groups (Cattley *et al.*, 1994, 1995).

4. Other Data Relevant for an Evaluation of Carcinogenicity and its Mechanisms

4.1 Absorption, distribution, metabolism and excretion

4.1.1 *Humans*

The metabolism of nitrobenzene has been reviewed (Beauchamp *et al.*, 1982; Rickert, 1987).

Salmowa *et al.* (1963) exposed seven volunteers to nitrobenzene vapours (5–30 µg/L) [5–30 mg/m^3, 1–6 ppm] for 6 h. Nitrobenzene was readily absorbed, initially at an average of approximately 87% in the first hour to 73% in the sixth hour, probably because of saturation. 4-Nitrophenol rapidly appeared in the urine, and its maximal excretion, at a concentration of 5 mg/L urine, occurred approximately 2 h after the end of exposure. Of the absorbed dose, 6–37% were recovered as 4-nitrophenol, and the level of this metabolite could be used as an index of exposure. 4-Aminophenol was not found in the urine. [However, the method used to analyse 4-aminophenol was fairly insensitive and low concentrations may have occurred.] After the end of exposure, there was a biphasic decline in 4-nitrophenol concentration in the urine. The initial half-life was about 5 h and the terminal half-life was about 60 h, which indicates that, potentially, accumulation could occur during a working week.

In vitro, nitrobenzene is rapidly absorbed through excised human skin in a diffusion cell (Bronaugh & Maibach, 1985). Feldmann and Maibach (1970) applied [^{14}C]-nitrobenzene on the forearm of volunteers, who did not wash the area of skin for 24 h. Over five days, excretion in urine was only 1.5% of the applied dose. Piotrowski (1967) used a technique to expose the whole skin surface to nitrobenzene vapour without inhalation of the compound. During the first day, at an air level of 1 ppm [5 mg/m^3], about 7 mg nitrobenzene were absorbed through the skin, of which about 20% were excreted into the urine.

Feldmann and Maibach (1970) injected [^{14}C]nitrobenzene intravenously into volunteers. Excretion in the urine was 60.5% of the dose over five days. The elimination half-life was 20 h.

In a case of nitrobenzene poisoning in a woman using a paint containing nitrobenzene as solvent (99.7% nitrobenzene, 0.27% benzene in distillate), the urinary level of 4-nitrophenol was 1056 nmol/mL [142 mg/L] one day after the end of exposure. Simultaneously, a concentration of 400 nmol/mL [39.6 mg/L] 4-aminophenol was detected. The levels decreased with an estimated half-life of a few days (Ikeda & Kita, 1964).

4.1.2 *Experimental systems*

Bronaugh and Maibach (1985) studied the percutaneous absorption of 4 µg/cm^2 nitrobenzene in an acetone vehicle in monkeys. *In vitro*, 6.2 ± 1.0% of the applied dose was absorbed percutaneously, and *in vivo*, 4.2 ± 0.5% of the applied dose was excreted in urine after five days. Loss of nitrobenzene due to volatilization could have affected the amount of nitrobenzene absorbed.

Schmieder and Henry (1988) studied the equilibrium binding of nitrobenzene to plasma proteins *in vitro* in pooled plasma samples from Sprague-Dawley rats [sex unspecified]. Aliquots of pooled plasma samples spiked with 0.32–933 mg/L nitrobenzene were allowed to equilibrate for 30–90 min at 25 °C. Of the nitrobenzene, 72.0 ± 4.5% were bound to the rat plasma proteins.

Parke (1956) administered 250 mg/kg bw [^{14}C]nitrobenzene by stomach tube to rabbits and measured metabolites in expired air, urine and faeces. Nearly 70% of the administered radioactivity was excreted within five days. Major metabolic products were 3- and 4-nitrophenols and 4-aminophenol. Minor metabolites included aniline, 2-aminophenol, 3-aminophenol, 4-nitrocatechol and 4-nitrophenyl mercapturic acid.

Rickert et al. (1983) administered [^{14}C]nitrobenzene to male Fischer 344 rats (22.5 or 225 mg/kg bw, 20 µCi, in corn oil orally or 225 mg/kg bw intraperitoneally), male CD rats (22.5 or 225 mg/kg bw orally), male B6C3F1 mice (225 mg/kg bw orally) and germ-free male Fischer 344 rats (225 mg/kg bw orally). No significant effect of route of administration or strain was observed for the excretion of radioactivity in urine, faeces or expired air following administration of 225 mg/kg bw nitrobenzene. Following oral administration of 225 mg/kg bw to Fischer 344 rats, excretion of radioactivity was distributed as follows: urine, 63.2%; faeces, 14.2%; expired air, 1.6%. At this same dose, but following intraperitoneal administration, the distribution of excretion of radioactivity was very similar: urine, 56.8%; faeces, 13.7%; expired air, 1.4%. A smaller dose of 22.5 mg/kg bw administered orally to Fischer 344 rats, resulted in a significantly higher proportion of radioactivity excreted in faeces (21.4%). Following a similar treatment pattern, B6C3F1 mice excreted a smaller percentage of the dose in urine (34.7%) than did rats, but similar percentages in faeces (18.8%) and expired air (0.8%). Four major metabolites were found in the urine of Fischer 344 rats: 4-hydroxyacetanilide sulfate; 4-nitrophenol sulfate; 3-nitrophenol sulfate; and an unidentified metabolite. 4-Hydroxyacetanilide sulfate and 4-nitrophenol sulfate were excreted in approximately equal proportions (20% of dose). 3-Nitrophenol sulfate and the unidentified metabolite each made up 10% of the dose. 4-Hydroxyacetanilide, 4-nitrophenol and 3-nitrophenol were found in the urine of B6C3F1 mice and CD rats but not of Fischer 344 rats. B6C3F1 mice and CD rats also excreted each of the above metabolites as glucuronides (except 3-nitrophenol in mice) and sulfates. Mice excreted nearly 10% of the dose as 4-aminophenol sulfate, whereas rats did not excrete this metabolite.

Bile collected from Fischer 344 and CD rats over the first 12 h after oral administration of 225 mg/kg bw nitrobenzene contained 1.8% and 3.8% of the dose, respectively. Of six peaks detected, three co-eluted with 4-hydroxy-3-methylthioacetanilide, 2-acetamido-3-(5'-acetanido-2'-hydroxyphenylthio)propanoic acid and S-(5'-acetamido-2'-hydroxyphenyl)glutathione. Another co-eluted with glutathione sulfinanilide. None of the metabolites recovered in bile of conventional Fischer 344 rats was found in bile of germ-free Fischer 344 rats (Rickert et al., 1983).

In-vivo experiments determined the role of microflora in nitrobenzene metabolism in control and animals treated with antibiotics (Levin & Dent, 1982). Antibiotic treatment totally inhibited in-vitro metabolism of nitrobenzene by caecal contents and decreased the expected level of methaemoglobin formation after a single oral dose of 300 mg/kg bw nitrobenzene. The excretion of ^{14}C was not altered by antibiotic treatment; however, the pattern of urinary metabolites was changed. Antibiotic treatment decreased the urinary excretion of the reduced metabolite, 4-hydroxyacetanilide, to 6% of control values and that of an unidentified metabolite to 14% of control values; excretion of 3-nitrophenol was increased over control values.

Nitrobenzene is reduced to aniline in in-vitro hepatic microsome systems via the intermediate products nitrosobenzene and phenylhydroxylamine (Harada & Omura, 1980). Blaauboer and Van Holsteijn (1983) investigated the formation and disposition of N-hydroxylated metabolites of nitrobenzene (phenylhydroxylamine and nitrosobenzene) by isolated rat hepatocytes. Apparent kinetic parameters for nitrobenzene reduction by hepatocytes, as measured by secretion of N-oxygenated products into the incubation medium, were V_{max} 1.44 ± 0.21 nmol/min/mL and K_m 4.2 ± 1.4 mM. Phenobarbital pretreatment stimulated the secretion of hydroxylated metabolites 2.8-fold.

Levin and Dent (1982) studied the metabolism of nitrobenzene using hepatic microsomes and caecal microflora from male Fischer 344 rats *in vitro*. Oxidative metabolism of 100 μM [^{14}C]nitrobenzene occurred at a rate of 0.008 ± 0.003 nmol/mg protein/min. The major product was unidentified and accounted for nearly 40% of the metabolites formed. Metabolism of nitrobenzene was also studied under anaerobic conditions, in which microsomal reduction occurred much more rapidly than did oxidation (0.33 versus 0.022 nmol/mg protein/min). The rate of reduction by caecal contents was 150-fold that in microsomes.

Protein binding

Albrecht and Neumann (1985) measured tissue dosimetry and haemoglobin binding in Wistar rats following a 0.20 mmol/kg bw [24.6 mg/kg bw] oral dose of [^{14}C]nitrobenzene. Radioactivity in tissues (pmol/mg/dose [mmol/kg]) after one day was as follows: blood, 229 ± 48; liver, 129 ± 9.5; kidney, 204 ± 27; and lung, 62 ± 14. The binding index (mmol/mol haemoglobin/dose [mmol/kg]) was 72.8 ± 10. Specific binding (pmol/mg/dose) was 1030 ± 137 for haemoglobin and 136 ± 34 for plasma proteins.

Goldstein and Rickert (1984) determined species differences in the covalent binding of [^{14}C]nitrobenzene to erythrocytes and spleen of male Fischer 344 and male B6C3F1 mice following an oral dose of 75, 150, 200 or 300 mg/kg bw nitrobenzene in corn oil. Total radioactivity in erythrocytes, as a percentage of dose, averaged 0.57 ± 0.11% and 0.08 ± 0.01% in rats and mice, respectively, following treatment with 200 mg/kg nitrobenzene. In both species, total and bound concentrations of ^{14}C were four to six times greater in erythrocytes than in spleen. All of the covalently bound nitrobenzene-related material in haemoblogin was recovered in the protein fraction, suggesting that nitrobenzene or its metabolites bind specifically to the globin moiety.

Reddy *et al.* (1976) administered nitrobenzene to normal and germ-free rats to determine the role of gut flora and tissues with regard to nitrobenzene reduction and formation of methaemoglobin. When nitrobenzene (200 mg/kg bw) was administered intraperitoneally to normal male Sprague-Dawley rats, 30–40% of the blood haemoglobin was converted to methaemoglobin. No measurable formation was observed 7 h after administration to germ-free rats. Suzuki *et al.* (1989) administered orally 0.5 mmol/kg bw nitrobenzene in a corn oil solution to male Sprague-Dawley rats; 48 h after administration, haemoglobin binding was 657.0 ± 36.7 nmol/g haemoglobin. Pretreatment with antibiotics decreased haemoglobin binding to 88.2 ± 10.5 nmol/g haemoglobin. The effect of dietary pectin on methaemogolobin formation from nitrobenzene was studied in male CDF rats (Goldstein *et al.*, 1984). Rats were held on one of three dietary regimens (0%, 5% or 8.4% pectin) for 28 days, after which they received 600 mg/kg bw nitrobenzne orally in corn oil. Animals fed the 8.4% pectin diet had the highest methaemoglobin content (64 ± 1%) and those fed the pectin-free diet had the lowest (20 ± 5%). The total number of caecal anaerobes was elevated (2–2.5 fold) and the metabolism of nitrobenzene by the caecal contents was also greater in animals fed the diets containing pectin.

4.2 Toxic effects

4.2.1 *Humans*

The toxic effects of nitrobenzene have been reviewed (Agency for Toxic Substances and Disease Registry, 1990).

Cases of severe poisoning were reported as early as 1886 in infants exposed to dye-stamped diapers and persons wearing freshly dyed shoes. The condition was often referred to as 'nitrobenzene poisoning', although exposure to nitrobenzene had not necessarily occurred; the conditions may have been caused by aniline (Agency for Toxic Substances and Disease Registry, 1990).

Methaemoglobinaemia, with cyanosis, headache, dyspnoea, weakness and ultimately coma and death, is the main characteristic of acute nitrobenzene poisoning. Nitrobenzene may also induce haemolysis, which is, however, usually mild (Hunter, 1943).

Methaemoglobinaemia was reported in three-week-old twins (Stevens, 1928) and in a 12-month-old girl (Stevenson & Forbes, 1942) exposed to nitrobenzene from insect-exterminator sprays for several hours. Moreover, a woman who worked under bad hygienic conditions in a cable insulation factory for three months developed serious poisoning. Her methaemoglobin level in the blood was 29.5% (37 g/L) up to 36 h after the end of exposure. [Lethal at about 80%; 'normally' about 1% or 1 g/L; 'normal' half-life 15–20 h.] She also developed haemolysis, as well as slight toxic hepatitis and peripheral neuropathy. It was discovered that she had a hereditary deficiency of NADH-methaemoglobin reductase, which may have made her particularly sensitive, and which also probably explained the high methaemoglobin level a long time after exposure (Kokal *et al.*, 1984). Development of toxic hepatitis after acute episodes of methaemoglobinaemia has been reported repeatedly (Ajmani *et al.*, 1986).

There is little quantitative information on the relationship between toxic effects and exposure. A man who had ingested about 7 g nitrobenzene developed methaemoglobinaemia (78% of methaemoglobin; Schimelman *et al.*, 1978). Pacséri *et al.* (1958) found air concentrations of nitrobenzene averaging 6 ppm [30 mg/m^3] in a plant producing nitroaromatic compounds. There was no obvious case of poisoning, although 'one or two' cases of headache and vertigo were mentioned. Examination of the blood revealed low concentrations of methaemoglobin. Previously, concentrations of nearly 40 ppm [200 mg/m^3] were stated to have caused poisonings.

Ikeda and Kita (1964) reported methaemoglobinaemia (about 30 g/L) two days after the end of exposure in a woman who for 17 months had used a paint containing nitrobenzene as solvent. The urinary level of 4-nitrophenol was 1056 µmol/mL [142 mg/mL] and that of *para*-aminophenol was about 400 µmol/mL [39.6 mg/mL] one day after the end of exposure. The patient also had clinical and laboratory signs of haemolytic anaemia and toxic hepatitis. Salmowa *et al.* (1963) found no increase of methaemoglobin concentration in blood in seven volunteers exposed for 6 h to air levels of up to 30 µg/L [6 ppm] and excreting up to 5 mg/L 4-nitrophenol in urine.

4.2.2 *Experimental systems*

(a) *Single dose studies*

The single oral LD_{50} for nitrobenzene in rats was 600 mg/kg bw (Agency for Toxic Substances and Disease Registry, 1990). Single acute exposures of male Fischer 344 rats to ≥ 200 mg/kg bw nitrobenzene resulted in significantly elevated (> 20%) methaemoglobin (Goldstein *et al.*, 1984), while higher single oral exposures (550 mg/kg bw) resulted in encephalomalacia and haemorrhage of the brainstem and cerebellum in male Fischer 344 rats (Morgan *et al.*, 1985). Necrosis of seminiferous tubules and hepatocellular nucleolar enlargement in male Fischer 344 rats following single oral exposure have also been reported (Bond *et al.*, 1981). The latter liver lesions were observed at doses as low as 110 mg/kg bw whereas the testicular lesions occurred at doses ≥ 300 mg/kg bw. Acute exposure by injection of nitrobenzene has been reported to cause methaemoglobinaemia, neurotoxicity and death in a variety of animal species (reviewed in Beauchamp *et al.*, 1982).

(b) *Repeated-dose studies*

Male and female Fischer 344 rats, Sprague-Dawley (CD) rats and B6C3F1 mice (9–10 weeks old) were exposed by inhalation to 10, 35 or 125 ppm [50, 175 or 625 mg/m^3] nitrobenzene vapours for 6 h per day on five days per week for up to two weeks (Medinsky & Irons, 1985). Animals were sacrificed at three or 14 days following the last exposure. Early morbidity among male and female mice exposed to 125 ppm necessitated euthanasia between two and four days of exposure. Some male and female Sprague-Dawley rats were found dead after the fourth day of exposure, but the remaining animals in the group exhibited rapid shallow breathing, wheezing and an orange discoloration around the urogenital orifice. In contrast, Fischer 344 rats exposed to 125 ppm exhibited no adverse clinical signs over the entire two-week period. The presumptive cause of death of the Sprague-Dawley rats exposed to 125 ppm nitrobenzene was perivascular haemorrhage in the cerebellar peduncle. Species and sex-related differences in liver pathology were also observed in animals exposed to 125 ppm nitrobenzene. Male mice exhibited centrilobular necrosis, superimposed on severe central lobular hydropic degeneration. In contrast, no necrosis was observed in livers from female mice at the same concentration. Liver pathology observed in Sprague-Dawley rats was similar but not as severe as that described for the mice. Livers from Sprague-Dawley rats that died early exhibited centrilobular hydropic degeneration and basophilic hepatocytic degeneration in periportal areas. No significant histological findings was observed in the livers from male and female Fischer 344 rats.

Moderate bronchiolar hyperplasia was observed in male and female mice exposed to 125 ppm nitrobenzene; mild hyperplasia was present in animals examined three days after the last exposure to 35 ppm. Perivascular oedema and vascular congestion were found in lungs taken from dead or moribund Sprague-Dawley rats after three to five days of exposure to 125 ppm nitrobenzene. No histopathology was found in the lungs from Fischer 344 rats exposed to 125 ppm nitrobenzene. Sprague-Dawley rats also exhibited moderate-to-severe hydropic degeneration of cortical tubular cells. Minimal degenerative changes were noted in the kidneys of some mice. The only renal lesion in Fischer 344 rats was a moderate to severe hyaline nephrosis in males that regressed in animals allowed to recover for 14 days. Splenic lesions were evident in all rats and mice in all groups exposed to nitrobenzene. Lesions consisted of increased extramedullary haematopoiesis and acute congestion. Thus for nitrobenzene, the most sensitive organ after 14-day inhalation exposure was the spleen (Medinsky & Irons, 1985).

The effects of chronic (two-year) inhalation exposure to nitrobenzene in B6C3F1 mice and Fischer 344 and Charles River (CD) rats have been described (Cattley *et al.*, 1994). Methaemoglobinaemia and anaemia were observed in both species at $\geq$ 25 ppm [100 mg/m^3] exposure concentrations. Other effects included lesions of the nose, liver, testis and lung. In mice, degeneration and loss of olfactory epithelium were observed at $\geq$ 5 ppm; the incidence of pigment deposition in olfactory epithelium was increased in mice and rats. Cytomegaly of centrilobular hepatocytes was induced in mice and rats, particularly males, at $\geq$ 5 ppm; in male mice multinucleation of hepatocytes was also induced. An increased incidence of testicular atrophy and epididymal hypospermia was observed in male CD (but not Fischer 344) rats at 25 ppm. In mice, an unusual pulmonary lesion, alveolar bronchialization, was frequently induced by exposure to $\geq$ 5 ppm nitrobenzene.

4.3 Reproductive and developmental effects

4.3.1 *Humans*

No data were available to the Working Group.

4.3.2 *Experimental systems*

Groups of 26 pregnant Sprague-Dawley rats were exposed by inhalation to 0, 1, 10 and 40 ppm [5, 50 and 200 mg/m^3] nitrobenzene vapour for 6 h per day on gestational days 6–15. Maternal weight gain was reduced during exposure to 40 ppm, with full recovery by gestational day 21. Absolute and relative spleen weights were increased at 10 and 40 ppm. There was no effect of treatment on resorptions or dead fetuses, on the sex ratio of live fetuses or on fetal body weights per litter. No treatment-related effect on the incidence of fetal malformations or variations was observed (Tyl *et al.*, 1987).

In an accompanying paper, a two-generation reproduction study was described, again involving exposure of Sprague-Dawley rats to nitrobenzene vapour (Dodd *et al.*, 1987). Groups of 30 male and 30 female rats were exposed to concentrations of 0, 1, 10 or 40 ppm nitrobenzene vapour for 6 h per day on five days per week for 10 weeks. F_1 rats were produced from the F_0 rats and at least one male and one female were picked randomly from each litter to form a group size of 30 per sex. F_1 rats remained in the same exposure group as their F_0 parents. Additional female rats were used for a second mating with the recovery group high-dose and control F_1 males. No effect on reproduction was observed at doses of 1 or 10 ppm nitrobenzene. At 40 ppm, a decrease in the fertility index of the F_0 and F_1 generations occurred, and this was associated with reduced testicular and epididymal weight, atrophy of the seminiferous tubules, spermatocytic degeneration and the presence of giant syncytial spermatocytes. The only significant observation in the litter derived from rats exposed to 40 ppm was an approximate 12% decrease in the mean body weights of F_1 rats on postnatal day 21. Survival indices were unaltered. In the F_1 rats, males of the high-dose and control groups were allowed a nine-week nonexposure recovery period. At the end of this period, the F_1 males were mated with virgin females, which had never been exposed to nitrobenzene. An almost five-fold increase in the fertility index was observed, indicating at least partial functional reversibility upon removal from nitrobenzene exposure. In addition, the numbers of giant syncytial spermatocytes and degenerated spermatocytes were greatly reduced; testicular seminiferous tubule atrophy persisted.

In a study reported as an abstract, groups of 22 pregnant rabbits were exposed by inhalation to 0, 9.9, 41 and 101 ppm [50, 207 and 509 mg/m^3] nitrobenzene for 6 h per day on gestational days 7–19. The dams were sacrificed on gestational day 30 and the fetuses were evaluated for external, visceral and skeletal malformations. No adverse effect was associated with the lowest dose. At the two higher doses (41 and 101 ppm), liver weights were slightly higher and methaemoglobin levels were significantly increased compared with controls. At the highest dose, a slight increase in fetal resorption was observed. No teratogenic effect was apparent at any of the exposure levels investigated (Schroeder *et al.*, 1986).

Necrosis of seminiferous tubules has been described in Fischer 344 rats after exposure to nitrobenzene (see Section 4.2.2).

4.4 Genetic and related effects

4.4.1 *Humans*

No data were available to the Working Group.

4.4.2 *Experimental systems* (see also Table 4 and Appendices 1 and 2)

No standard reverse mutation test with *Salmonella typhimurium* showed mutagenic activity of nitrobenzene. Only a few *Salmonella* tests in the presence of S9 and norharman were positive.

In cultures of primary human hepatocytes *in vitro*, no unscheduled DNA synthesis was observed.

In Fischer 344 rats, no significant increase in sister chromatid exchange frequency or chromosomal aberrations was found in peripheral blood lymphocytes. No significant increase in sister chromatid exchange was observed in the isolated splenic lymphocytes after in-vivo exposure to up to 50 ppm nitrobenzene for 6 h per day for 21 days during a 29-day period; the toxicity of the dosing regimen was demonstrated by cell cycle inhibition and mitotic depression in the lymphocytes.

Table 4. Genetic and related effects of nitrobenzene

Test system	Result[a] Without exogenous metabolic system	Result[a] With exogenous metabolic system	Dose[b] (LED/HID)	Reference
SA0, *Salmonella typhimurium* TA100, reverse mutation	0	–	1250	Anderson & Styles (1978)
SA0, *Salmonella typhimurium* TA100, reverse mutation	–	0	615	Chiu et al. (1978)
SA0, *Salmonella typhimurium* TA100, reverse mutation	–	–	385	Haworth et al. (1983)
SA0, *Salmonella typhimurium* TA100, reverse mutation	–	–	2355	Shimizu et al. (1983)
SA0, *Salmonella typhimurium* TA100, reverse mutation	–	0	50	Suzuki et al. (1983)
SA0, *Salmonella typhimurium* TA100, reverse mutation	–	–	NR	Nohmi et al. (1984)
SA0, *Salmonella typhimurium* TA100, reverse mutation	–	0	500	Vance & Levin (1984)
SA0, *Salmonella typhimurium* TA100, reverse mutation	–	–	NR	Kawai et al. (1987)
SA0, *Salmonella typhimurium* TA100, reverse mutation	–	–	465	Dellarco & Prival (1989)
SA5, *Salmonella typhimurium* TA1535, reverse mutation	0	–	1250	Anderson & Styles (1978)
SA5, *Salmonella typhimurium* TA1535, reverse mutation	–	–	128	Haworth et al. (1983)
SA5, *Salmonella typhimurium* TA1535, reverse mutation	–	–	2355	Shimizu et al. (1983)
SA5, *Salmonella typhimurium* TA1535, reverse mutation	–	0	500	Vance & Levin (1984)
SA7, *Salmonella typhimurium* TA1537, reverse mutation	–	–	128	Haworth et al. (1983)
SA7, *Salmonella typhimurium* TA1537, reverse mutation	–	–	2355	Shimizu et al. (1983)
SA7, *Salmonella typhimurium* TA1537, reverse mutation	–	0	500	Vance & Levin (1984)
SA8, *Salmonella typhimurium* TA1538, reverse mutation	0	–	1250	Anderson & Styles (1978)
SA8, *Salmonella typhimurium* TA1538, reverse mutation	–	–	2355	Shimizu et al. (1983)
SA8, *Salmonella typhimurium* TA1538, reverse mutation	–	0	500	Vance & Levin (1984)
SA9, *Salmonella typhimurium* TA98, reverse mutation	0	–	1250	Anderson & Styles (1978)
SA9, *Salmonella typhimurium* TA98, reverse mutation	–	0	615	Chiu et al. (1978)
SA9, *Salmonella typhimurium* TA98, reverse mutation	–	–	385	Haworth et al. (1983)
SA9, *Salmonella typhimurium* TA98, reverse mutation	–	–	2355	Shimizu et al. (1983)
SA9, *Salmonella typhimurium* TA98, reverse mutation	–	–[d]	50	Suzuki et al. (1983)
SA9, *Salmonella typhimurium* TA98, reverse mutation	–	–	NR	Nohmi et al. (1984)

Table 4 (contd)

Test system	Result[a]		Dose[b] (LED/HID)	Reference
	Without exogenous metabolic system	With exogenous metabolic system		
SA9, *Salmonella typhimurium* TA98, reverse mutation	–	0	500	Vance & Levin (1984)
SA9, *Salmonella typhimurium* TA98, reverse mutation	–	–	NR	Kawai et al. (1987)
SA9, *Salmonella typhimurium* TA98, reverse mutation	0	–[d]	100	Suzuki et al. (1987)
SA9, *Salmonella typhimurium* TA98, reverse mutation	–	–[c]	465	Dellarco & Prival (1989)
SAS, *Salmonella typhimurium* TA100NR, reverse mutation	–	0	500	Vance & Levin (1984)
SAS, *Salmonella typhimurium* TA1537NR, reverse mutation	–	0	500	Vance & Levin (1984)
SAS, *Salmonella typhimurium* TA98NR, reverse mutation	–	0	500	Vance & Levin (1984)
SAS, *Salmonella typhimurium* TA98a, reverse mutation	–	0	500	Vance & Levin (1984)
SAS, *Salmonella typhimurium* TA98NR reverse mutation	0	–[d]	500	Suzuki et al. (1987)
SAS, *Salmonella typhimurium* TA98/1,8-DNP₆, reverse mutation	0	–[d]	100	Suzuki et al. (1987)
UIH, Unscheduled DNA synthesis, human hepatocytes *in vitro*	–	0	123	Butterworth et al. (1989)
UPR, Unscheduled DNA synthesis, rat hepatocytes *in vitro*	–		500 po × 1	Mirsalis et al. (1982)
SVA, Sister chromatid exchange, peripheral blood lymphocytes, male F344 rats *in vivo*	–		53 inh 6h/d × 21	Kligerman et al. (1983)
SVA, Sister chromatid exchange, splenic lymphocytes, male F344 rats *in vivo*	–		53 inh 6h/d × 21	Kligerman et al. (1983)
CVA, Chromosomal aberrations, peripheral blood lymphocytes, male F344 rats *in vivo*	–		53 inh 6h/d × 21	Kligerman et al. (1983)

[a] +, positive; (+), weak positive; –, negative; 0, not tested; ?, inconclusive
[b] LED, lowest effective dose; HID, highest ineffective dose. In-vitro tests, µg/mL; in-vivo tests, mg/kg bw; NR, dose not reported
[c] Negative with or without 2mM flavin mononucleotide (FMN) in preincubation mix
[d] Positive in the presence of 200 µg/plate norharman

Oral administration of 500 mg/kg bw nitrobenzene to rats did not induce unscheduled DNA synthesis in hepatocytes cultured from the exposed animals.

5. Summary of Data Reported and Evaluation

5.1 Exposure data

Nitrobenzene has been produced commercially since the early nineteenth century by nitration of benzene. It is a major chemical intermediate used mainly in the production of aniline, itself a major chemical intermediate in the production of dyes. Human exposure may occur both by inhalation and by skin absorption during its production and use. Nitrobenzene has been detected in surface and groundwater.

5.2 Human carcinogenicity data

No data were available to the Working Group.

5.3 Animal carcinogenicity data

Nitrobenzene was tested by inhalation exposure in one study in mice and in two studies in rats. In mice, the incidences of alveolar–bronchiolar neoplasms and thyroid follicular-cell adenomas were increased in males. In one study in rats, the incidences of hepatocellular neoplasms, thyroid follicular-cell adenomas and adenocarcinomas and renal tubular-cell adenomas were increased in treated males. In treated females, the incidences of hepatocellular neoplasms and endometrial stromal polyps were increased. In a study using male rats only, the incidence of hepatocellular neoplasms was increased.

5.4 Other relevant data

In humans, nitrobenzene is readily absorbed by inhalation. Penetration through the skin also occurs. A major part of the absorbed dose is excreted into the urine: 10–20% of the dose is excreted as 4-nitrophenol, the concentration of which may be used for biological monitoring. A smaller fraction is excreted as 4-aminophenol. The elimination kinetics contains at least two compartments, the first with a half-life of hours and the second with a half-life of days.

In rodents and rabbits, 4-nitrophenol and 4-aminophenol are major urinary metabolites.

There is limited information on the toxic effects of exposure to nitrobenzene in humans. However, it is clear that both accidental ingestion and occupational exposure may cause methaemoglobinaemia, haemolytic anaemia and toxic hepatitis.

Following inhalation of nitrobenzene, liver, lung and splenic toxicity is observed in both rats and mice, although mice appear to be more sensitive than rats to the toxic effects of this chemical. Methaemoglobinaemia and anaemia are also observed in both rats and mice.

In female rats, no teratogenic or reproductive effect of exposure to nitrobenzene was observed. Testicular atrophy has been observed in rats. In a two-generation reproduction study in rats, a decrease in the fertility index of the F_0 and F_1 generations occurred. No teratogenic effect has been observed in rabbits.

Nitrobenzene was non-genotoxic in bacteria and mammalian cells *in vitro*. In mammals *in vivo*, it was inactive.

5.5 Evaluation[1]

There is *inadequate evidence* in humans for the carcinogenicity of nitrobenzene.

There is *sufficient evidence* in experimental animals for the carcinogenicity of nitrobenzene.

Overall evaluation

Nitrobenzene is *possibly carcinogenic to humans (Group 2B)*.

6. References

Agency for Toxic Substances and Disease Registry (1990) *Toxicological Profile for Nitrobenzene* (TP-90-19), Atlanta, GA, United States Public Health Service

Ajmani, A., Prakash, S.K., Jain, S.K. & Shah, P. (1986) Aquired methemoglobinemia following nitrobenzene poisoning. *J. Assoc. Phys. India*, **34**, 891–892

Albrecht, W. & Neumann, H.-G. (1985) Biomonitoring of aniline and nitrobenzene: hemoglobin binding in rats and analysis of adducts. *Arch. Toxicol.*, **57**, 1–5

American Conference of Governmental Industrial Hygienists (1991) *Documentation of the Threshold Limit Values and Biological Exposure Indices*, 6th Ed., Vol. 2, Cincinnati, OH, pp. 1096–1099

American Conference of Governmental Industrial Hygienists (1995) *1995–1996 Threshold Limit Values for Chemical Substances and Physical Agents and Biological Exposure Indices*, Cincinnati, OH, pp. 28, 65

Anderson, D. & Styles, J.A. (1978) The bacterial mutation test. *Br. J. Cancer*, **37**, 924–930

Anon. (1987) Chemical profile: nitrobenzene. *Chem. Mark. Rep.*, **232**, 50

Anon. (1990) Chemical profile: nitrobenzene. *Chem. Mark. Rep.*, **238**, 50

Anon. (1993) Chemical profile: nitrobenzene. *Chem. Mark. Rep.*, **244**, 57

Arbeidsinspectie [Labour Inspection] (1994) *De Nationale MAC-Lijst 1994* [National MAC list 1994], The Hague, p. 35

Beauchamp, R.O., Jr, Irons, R.D., Rickert, D.E., Couch, D.B. & Hamm, T.E., Jr (1982) A critical review of the literature on nitrobenzene toxicity. *CRC crit. Rev. Toxicol.*, **11**, 33–84

---

[1]For definition of the italicized terms, see Preamble, pp. 24–27.

Blaauboer, B.J. & Van Holsteijn, C.W. M. (1983) Formation and disposition of N-hydroxylated metabolites of aniline and nitrobenzene by isolated rat hepatocytes. *Xenobiotica*, **13**, 295–302

Bond, J.A., Chism, J.P., Rickert, D.E. & Popp, J.A. (1981) Induction of hepatic and testicular lesions in Fischer-344 rats by single oral doses of nitrobenzene. *Fundam. appl. Toxicol.*, **1**, 389–394

Booth, G. (1991) Nitro compounds, aromatic. In: Elvers, B., Hawkins, S. & Schulz, G., eds, *Ullmann's Encyclopedia of Industrial Chemistry*, 5th rev. Ed., Vol. A17, New York, NY, VCH Publishers, pp. 411–455

Brodzinsky, R. & Singh, H.B. (1982) *Volatile Organic Chemicals in the Atmosphere: An Assessment of Available Data* (US EPA Report No. EPA-600/3-83027a; US NTIS PB83-195503), Research Triangle Park, NC, United States Environmental Protection Agency

Bronaugh, R.L. & Maibach, H.I. (1985) Percutaneous absorption of nitroaromatic compounds: in vivo and in vitro studies in the human and monkey. *J. invest. Dermatol.*, **84**, 180–183

Budavari, S., ed. (1989) *The Merck Index*, 11th Ed., Rahway, NJ, Merck & Co., p. 1042

Butterworth, B.E., Smith-Oliver, T., Earle, L., Loury, D.J., White, R.D., Doolittle, D.J., Working, P.K., Cattley, R.C., Jirtle, R., Michalopoulos, G. & Strom, S. (1989) Use of primary cultures of human hepatocytes in toxicology studies. *Cancer Res.*, **49**, 1079–1084

Cattley, R.C., Everitt, J.I., Gross, E.A., Moss, O.R., Hamm, T.E., Jr & Popp, J.A. (1994) Carcinogenicity and toxicity of inhaled nitrobenzene in B6C3F1 mice and F344 and CD rats. *Fundam. appl. Toxicol.*, **22**, 328–340

Cattley, R.C., Everitt, J.I., Gross, E.A., Moss, O.R., Hamm, T.E., Jr & Popp, J.A. (1995) Erratum. *Fundam. appl. Toxicol.*, **25**, 159

Chemical Information Services (1994) *Directory of World Chemical Producers 1995/96 Standard Edition*, Dallas, TX, pp. 524–525

Chiu, C.W., Lee, L.H., Wang, C.Y. & Bryan, G.T. (1978) Mutagenicity of some commercially available nitro compounds for *Salmonella typhimurium*. *Mutat. Res.*, **58**, 11–22

Dangwal, S.K. & Jethani, B.M. (1980) A simple method of determination of nitrobenzene and nitrochlorobenzene in air and urine. *Am. ind. Hyg. Assoc. J.*, **41**, 847–850

Dellarco, V.L. & Prival, M.J. (1989) Mutagenicity of nitro compounds in *Salmonella typhimurium* in the presence of flavin mononucleotide in a preincubatiuon assay. *Environ. mol. Mutag.*, **13**, 116–127

Deutsche Forschungsgemeinschaft (1995) *MAK and BAT Values 1995* (Report No. 31), Weinheim, VCH Verlagsgesellschaft, pp. 49, 145

Dodd, D.E., Fowler, E.H., Snellings, W.M., Pritts, I.M., Tyl, R.W., Lyon, J.P., O'Neal, F.O. & Kimmerle, G. (1987) Reproduction and fertility evaluations in CD rats following nitrobenzene inhalation. *Fundam. appl. Toxicol.*, **8**, 493–505

Duguet, J.P., Anselme, C., Mazounie, P. & Mallevialle, J. (1988) Application of the ozone–hydrogen peroxide combination for the removal of toxic compounds from a groundwater. In: Angeletti, G. & Bjorseth, A., eds, *Organic Micropollutants in the Aquatic Environment, Proceedings of the Fifth European Symposium*, Dordrecht, Kluwer Academic Publishers, pp. 299–309

Dunlap, K.L. (1981) Nitrobenzene and nitrotoluenes. In: Mark, H.F., Othmer, D.F., Overberger, C.G., Seaborg, G.T. & Grayson, N., eds, *Kirk-Othmer Encyclopedia of Chemical Technology*, 3rd Ed., Vol. 15, New York, John Wiley & Sons, pp. 916–932

Eller, P.M., ed. (1994) Nitrobenzenes — Method 2005. In: *NIOSH Manual of Analytical Methods*, 4th Ed., Vol 2 (E-N) (DHHS (NIOSH) Publ. No. 94-113), Washington DC, United States Government Printing Office

Ellis, D.D., Jone, C.M., Larson, R.A. & Schaeffer, D.J. (1982) Organic constitutents of mutagenic secondary effluents from wastewater treatment plants. *Arch. environ. Contam. Toxicol.*, **11**, 373–382

Feldmann, R.J. & Maibach, H.I. (1970) Absorption of some organic compounds through the skin in man. *J. invest. Dermatol.*, **54**, 399–404

Feltes, J., Levsen, K., Volmer, D. & Spiekermann, M. (1990) Gas chromatographic and mass spectrometric determination of nitroaromatics in water. *J. Chromatogr.*, **518**, 21–40

First Chemical Corp. (1993) *Material Safety Data Sheet: Nitrobenzene*, Pascagoula, MS

Goldstein, R.S. & Rickert, D.E. (1984) Macromolecular covalent binding of [^{14}C]nitrobenzene in the erythrocyte and spleen of rats and mice. *Chem.-biol. Interactions*, **50**, 27–37

Goldstein, R.S., Chism, J.P., Sherrill, J.M. & Hamm, T.E., Jr (1984) Influence of dietary pectin on intestinal microfloral metabolism and toxicity of nitrobenzene. *Toxicol. appl. Pharmacol.*, **75**, 547–553

Grob, R.L. & Cao, K.B. (1990) High performance liquid chromatographic study of the recovery of aromatic amine and nitro compounds from soil. *J. environ. Sci. Health*, **A25**, 117–136

Hansch, C., Leo, A. & Hoekman, D.H. (1995) *Exploring QSAR*, Washington DC, American Chemical Society, p. 18

Harada, N. & Omura, T. (1980) Participation of cytochrome P450 in the reduction of nitro compounds by rat liver microsomes. *J. Biochem. Tokyo*, **87**, 1539–1544

Harkov, R., Kebbekus, B., Bozzelli, J.W. & Lioy, P. (1983) Measurement of selected volatile compounds at three locations in New Jersey during the summer season. *J. Air Pollut. Control Assoc.*, **33**, 1177–1183

Harkov, R., Gianti, S.J., Jr, Bozzelli, J.W. & LaRegina, J.E. (1985) Monitoring volatile organic compounds at hazardous and sanitary landfills in New Jersey. *J. environ. Sci. Health*, **A20**, 491–501

Haworth, S., Lawlor, T., Mortelsmans, K., Speck, W. & Zeiger, E. (1983) *Salmonella* mutagenicity test results for 250 chemicals. *Environ. Mutag.*, **Suppl. 1**, 3–142

Health and Safety Executive (1995) *Occupational Exposure Limits 1995* (Guidance Note EH 40/95), Sudbury, Suffolk, HSE Books, p. 35

Horng, J.-Y. & Huang, S.-D. (1994) Determination of the semi-volatile compounds nitrobenzene, isophorone, 2,4-dinitrotoluene and 2,6-dinitrotoluene in water using solid-phase microextraction with a polydimethylsiloxane-coated fibre. *J. Chromatogr. A*, **678**, 313–318

Howard, P.H. (1989) *Handbook of Environmental Fate and Exposure Data for Organic Chemicals*, Vol. 1, Chelsea, Michigan, Lewis Publishers, pp. 421–430

Hunter, D. (1943) Industrial toxicology. *Q. J. Med.*, **12**, 185–258

IARC (1982) *IARC Monographs on the Evaluation of the Carcinogenic Risk of Chemicals to Humans, Vol. 27, Some Aromatic Amines, Anthraquinones and Nitroso Compounds, and Inorganic Fluorides Used in Drinking Water and Dental Preparations*, Lyon, pp. 39–61

IARC (1987) *IARC Monographs on the Evaluation of Carcinogenic Risks to Humans, Suppl. 7, Overall Evaluations of Carcinogenicity: An Updating of* IARC Monographs *Volumes 1 to 42*, Lyon, pp. 99–100

Ikeda, M. & Kita, A. (1964) Excretion of *p*-nitrophenol and *p*-aminophenol in the urine of a patient exposed to nitrobenzene. *Br. J. ind. Med.*, **21**, 210–213

Kawai, A., Goto, S., Matsumoto, Y. & Matsushita, H. (1987) Mutagenicity of aliphatic and aromatic nitro compounds. Industrial materials and related compounds. *Jpn. J. ind. Health*, **29**, 34–54

Kligerman, A.D., Erexson, G.L., Wilmer, J.L. & Phelps, M.C. (1983) Analysis of cytogenetic damage in rat lymphocytes following in vivo exposure to nitrobenzene. *Toxicol. Lett.*, **18**, 219–226

Kokal, K.C., Khanna, S.S., Retnam, V.J. & Dastur, F.D. (1984) Methemoglobinemia: an unusual presentaion. *J. Assoc. Phys. India*, **32**, 833–834

Koniecki, W.B. & Linch, A.L. (1958) Determination of aromatic nitro compounds. *Anal. Chem.*, **30**, 1134–1137

Lauwerys, R.R. (1991) Occupational toxicology. In: Amdur, M.O., Doull, J. & Klaassen, C.D., eds, *Casarett and Doull's Toxicology — The Basic Science of Poisons*, 4th Ed., New York, Pergamon Press, pp. 947–969

Levin, A.A. & Dent, J.G. (1982) Comparison of the metabolism of nitrobenzene by hepatic microsomes and cecal microflora from Fischer-344 rats *in vitro* and the relative importance of each *in vivo*. *Drug Metab. Dispos.*, **10**, 450–454

Lewalter, J. & Ellrich, D. (1991) Nitroaromatic compounds (nitrobenzene, p-nitrotoluene, p-nitrochlorobenzene, 2,6-dinitrotoluene, o-dinitrobenzene, 1-nitronaphthalene, 2-nitronaphthalene, 4-nitrobiphenyl). In: Angerer, J. & Schaller, K.H., eds, *Analyses of Hazardous Substances in Biological Materials*, Vol. 3, New York, VCH Publishers, pp. 207–229

Lewis, R.J., Sr (1993) *Hawley's Condensed Chemical Dictionary*, 12th Ed., New York, Van Nostrand Reinhold Co., p. 825

Lide, D.R., ed. (1993) *CRC Handbook of Chemistry and Physics*, 74th Ed., Boca Raton, FL, CRC Press, pp. 3–85, 6–76

Mannsville Chemical Products Corp. (1984) *Chemical Products Synopsis: Nitrobenzene*, Cortland, NY

Medinsky, M.A. & Irons, R.D. (1985) Sex, strain, and species differences in the response of rodents to nitrobenzene vapors. In: Rickert, D.E., ed., *The Toxicity of Nitroaromatic Compounds*, New York, Hemisphere Publishing, pp. 35–51

Meijers, A.P. & van der Leer, R.C. (1976) The occurrence of organic micropollutants in the River Rhine and the River Maas in 1974. *Water Res.*, **10**, 597–604

Mirsalis, J.C., Tyson, C.K. & Butterworth, B.E. (1982) Detection of genotoxic carcinogens in the *in vivo–in vitro* hepatocyte DNA repair assay. *Environ. Mutag.*, **4**, 553–562

Morgan, K.T., Gross, E.A., Lyght, O. & Bond, J.A. (1985) Morphologic and biochemical studies of a nitrobenzene-induced encephalopathy in rats. *Neurotoxicology*, **6**, 105–116

Munch, J.W. & Eichelberger, J.W. (1992) Evaluation of 48 compounds for possible inclusion in the U.S. EPA Method 524.2, Revision 3.0: expansion of the method analyte list to a total of 83 compounds. *J. chromatogr. Sci.*, **30**, 471–477

Nelson, C.R. & Hites, R.A. (1980) Aromatic amines in and near the Buffalo River. *Environ. Sci. Technol.*, **14**, 1147–1150

Nielen, M.W.F., Brinkman, U.A.T. & Frei, R.W. (1985) Industrial wastewater analysis by liquid chromatography with precolumn technology and diode-array detection. *Anal. Chem.*, **57**, 806–810

Nohmi, T., Yoshikawa, K., Nakadate, M., Miyata, R. & Ishidate M., Jr (1984) Mutations in *Salmonella typhimurium* and inactivation of *Bacillus subtilis* transforming DNA induced by phenylhydroxylamine derivatives. *Mutat. Res.*, **136**, 159–168

Pacséri, L., Magos, L. & Batskor, A. (1958) Threshold and toxic limits of some amino and nitro compounds. *Arch. ind. Health*, **18**, 1–18

Parke, D.V. (1956) Detoxication. LXVIII. The metabolism of C^{14}-nitrobenzene in the rabbit and guinea pig. *Biochem. J.*, **62**, 339–346

Parmeggiani, L., ed. (1983) *Encyclopedia of Occupational Health and Safety*, 3rd. Ed., Vol. 2, L-Z, Geneva, International Labour Office, p. 1448

Pendergrass, S.M. (1994) An approach for estimating workplace exposure to o-toluidine, aniline and nitrobenzene. *Am. ind. Hyg. Assoc. J.*, **55**, 733–737

Piotrowski, J. (1967) Further investigations on the evaluation of exposure to nitrobenzene. *Br. J. ind. Med.*, **24**, 60–65

Reddy, B.G., Pohl, T.R. & Krishna, G. (1976) The requirement of gut flora in nitrobenzene-induced methemoglobinemia in rats. *Biochem. Pharmacol.*, **25**, 1119–1122

Rickert, D.E. (1987) Metabolism of nitroaromatic compounds. *Drug Metab. Rev.*, **18**, 23–53

Rickert, D.E., Bond, J.A., Long, R.M. & Chism, J.P. (1983) Metabolism and excretion of nitrobenzene by rats and mice. *Toxicol. appl. Pharmacol.*, **67**, 206–214

Sadtler Research Laboratories (1980) *Sadtler Standard Spectra. 1980 Cumulative Index*, Philadelphia, PA, p. 92

Salmowa, J., Piotrowski, J. & Neuhorn, U. (1963) Evaluation of exposure to nitrobenzene. Absorption of nitrobenzene vapour through lungs and excretion of *p*-nitrophenol in urine. *Br. J. ind. Med.*, **20**, 41–46

Sax, N.I. & Lewis, R.J., Sr (1989) *Dangerous Properties of Industrial Materials*, 7th Ed., Vol. 3, New York, Van Nostrand Reinhold Co., pp. 2504–2505

Schimelman, M.A., Soler, J.M. & Muller, H.A. (1978) Methemoglobinemia: nitrobenzene ingestion. *JACEP*, **7**, 406–408

Schmieder, P.K. & Henry, T.R. (1988) Plasma binding of 1-butanol, phenol, nitrobenzene, and pentachlorophenol in the rainbow trout and rat: a comparative study. *Comp. Biochem. Physiol.*, **91C**, 413–418

Schroeder, R.E., Terrill, J.B., Lyon, J.P., Kaplan, A.M. & Kimmerle, G. (1986) An inhalation teratology study in the rabbit with nitrobenzene. *Toxicologist*, **6**, 93

Shackelford, W.M., Cline, D.M., Faas, L. & Kurth, G. (1983) An evaluation of automated spectrum matching for survey identification of wastewater components by gas chromatography-mass spectrometry. *Anal. chim. Acta*, **146**, 15–27

Shimizu, M., Yasui, Y. & Matsumoto, N. (1983) Structural specificity of aromatic compounds with special reference to mutagenic activity in *Salmonella typhimurium* — a series of chloro- and fluoro-nitrobenzene derivatives. *Mutat. Res.*, **116**, 217–238

Sontheimer, H., Brauch, H.-J. & Kühn, W. (1985) Impact of different types of organic micropollutants present on sources of drinking water on the quality of drinking water. *Sci. total Environ.*, **47**, 27–44

Staples, C.A., Werner, A.F. & Hoogheem, T.J. (1985) Assessment of priority pollutant concentration in the United States using STORET database. *Environ. Toxicol. Chem.*, **4**, 131–142

Stevens, A.M. (1928) Cyanosis in infants from nitrobenzene. *J. Am. med. Assoc.*, **90**, 116

Stevenson, A. & Forbes, R.P. (1942) Nitrobenzene poisoning. *J. Pediat.*, **21**, 224

Sugiyama, H., Tanaka, K., Fukaya, K., Nishiyama, N. & Wada, Y. (1978) Studies on environmental pollution by chemical substances. II. Determination of aromatic nitro compounds in river and sea water. *Eisei Kagaku*, **24**, 11–18

Suzuki, J., Koyama, T. & Suzuki, S. (1983) Mutagenicities of mono-nitrobenzene derivatives in the presence of norharman. *Mutat. Res.*, **120**, 105–110

Suzuki, J., Takahashi, N., Kobayashi, Y., Miyamae, R., Ohsawa, M. & Suzuki, S. (1987) Dependence on *Salmonella typhimurium* enzymes of mutagenicities of nitrobenzene and its derivatives in the presence of rat-liver S9 and norharman. *Mutat. Res.*, **178**, 187–193

Suzuki, J., Meguro, S.-I., Morita, O., Hirayama, S. & Suzuki, S. (1989) Comparison of *in vivo* binding of aromatic nitro and amino compounds to rat hemoglobin. *Biochem. Pharmacol.*, **38**, 3511–3519

Swaminathan, K., Kondawar, V.K., Chakrabarti, T. & Subrahmanyam, P.V.R. (1987) Identification and quantification of organics in nitro aromatic manufacturing wastewaters. *Indian J. environ. Health*, **29**, 32–38

Tyl, R.W., France, K.A., Fisher, L.C., Dodd, D.E., Pritis, I.M., Lyon, J.P., O'Neal, F.O. & Kimmerle, G. (1987) Developmental toxicity evaluation of inhaled nitrobenzene in CD rats. *Fundam. appl. Toxicol.*, **8**, 482–492

United Nations Environment Programme (1995) *International Register of Potentially Toxic Chemicals, Legal File, Nitrobenzene*, Geneva

United States Environmental Protection Agency (1986a) Method 8090. Nitroaromatics and cyclic ketones. In: *Test Methods for Evaluating Solid Wast — Physical/Chemical Methods*, 3rd Ed., Vol. 2 (US EPA No. SW-846), Washington DC, Office of Solid Waste and Emergency Response

United States Environmental Protection Agency (1986b) Method 8250. Gas chromatography/-mass spectrometry for semivolatile organics: packed column technique. In: *Test Methods for Evaluating Solid Waste — Physical/Chemical Methods*, 3rd Ed., Vol. 2 (US EPA No. SW-846), Washington DC, Office of Solid Waste and Emergency Response

United States Environmental Protection Agency (1986c) Method 8270. Gas chromatography/-mass spectrometry for semivolatile organics: capillary column technique. In: *Test Methods for Evaluating Solid Waste — Physical/Chemical Methods*, 3rd Ed., Vol. 2 (US EPA No. SW-846), Washington DC, Office of Solid Waste and Emergency Response

United States Environmental Protection Agency (1988) *Extremely Hazardous Substances*, Vol. 2, Park Ridge, NJ, Noyes Data Corp., pp. 1145–1151

United States Environmental Protection Agency (1994) Appendix A to Part 136 — Methods for organic chemical analysis of municipal and industrial wastewater. *US Code fed. Regul.*, **Title 40**, Part. 136, pp. 503–513 [Method 609], 574–601 [Method 625], 615–634 [Method 1625B]

United States National Institute for Occupational Safety and Health (1994a) *Pocket Guide to Chemical Hazards* (DHHS (NIOSH) Publ. No. 94-116), Cincinnati, OH, pp. 226–227

United States National Institute for Occupational Safety and Health (1994b) *RTECs Chem.-Bank*, Cincinnati, OH

United States National Institute for Occupational Safety and Health (1995) *National Occupational Exposure Survey (1981–1983)*, Cincinnati, OH

United States National Library of Medicine (1995) *Registry of Toxic Effects of Chemical Substances (RTECS)*, Bethesda, MD

United States Occupational Safety and Health Administration (1994) Air contaminants. *US Code fed. Regul.*, **Title 29**, Part 1910.1000, p. 14

Vance, W.A. & Levin, D.E. (1984) Structural features of nitroaromatics that determine mutagenic activity in *Salmonella typhimurium*. *Environ. Mutag.*, **6**, 797–811

Verschueren, K. (1983) *Handbook of Environmental Data on Organic Chemicals*, 2nd Ed., New York, Van Nostrand Reinhold Co., pp. 910–912

Zoeteman, B.C.J., Harmsen, K., Linders, J.B.H.J., Morra, C.F.H. & Slooff, W. (1980) Persistent organic pollutants in river water and ground water of the Netherlands. *Chemosphere*, **9**, 231–249

van Zoest, R. & van Eck, G.T.M. (1991) Occurrence and behaviour of several groups of organic micropollutants in the Scheldt estuary. *Sci. total Environ.*, **103**, 57–71

Young, D.R., Gossett, R.W., Baird, R.B., Brown, D.A., Taylor, P.A. & Mille, M.J. (1983) Wastewater inputs and marine bioaccumulation of priority pollutant organics off southern California. In: Jolley, R.L., Brungs, W.A., Cotruvo, J.A., Cumming, R.B., Mattice, J.S. & Jacobs, V.A., eds, *Water Chlorination, Environmental Impact and Health Effects*, Vol. 4, Book 2, *Environment, Health, and Risk*, Ann Arbor, Ann Arbor Sciences, pp. 871–884

2-NITROTOLUENE, 3-NITROTOLUENE AND 4-NITROTOLUENE

1. Exposure Data

1.1 Chemical and physical data

1.1.1 Nomenclature

2-Nitrotoluene

Chem. Abstr. Serv. Reg. No.: 88-72-2

Deleted CAS Reg. No.: 57158-05-1

Chem. Abstr. Name: 1-Methyl-2-nitrobenzene

IUPAC Systematic Name: *ortho*-Nitrotoluene

Synonyms: 2-Methylnitrobenzene; *ortho*-methylnitrobenzene; 2-methyl-1-nitrobenzene; *ortho*-mononitrotoluene; 2-nitrotoluol; *ortho*-nitrotoluol

3-Nitrotoluene

Chem. Abstr. Serv. Reg. No.: 99-08-1

Chem. Abstr. Name: 1-Methyl-3-nitrobenzene

IUPAC Systematic Name: *meta*-Nitrotoluene

Synonyms: 3-Methylnitrobenzene; *meta*-methylnitrobenzene; 3-methyl-1-nitrobenzene; *meta*-mononitrotoluene; 3-nitrotoluol; *meta*-nitrotoluol

4-Nitrotoluene

Chem. Abstr. Serv. Reg. No.: 99-99-0

Chem. Abstr. Name: 1-Methyl-4-nitrobenzene

IUPAC Systematic Name: *para*-Nitrotoluene

Synonyms: 4-Methylnitrobenzene; *para*-methylnitrobenzene; 4-methyl-1-nitrobenzene; *para*-nitrotoluene; 4-nitrotoluol; *para*-nitrotoluol

1.1.2 *Structural and molecular formulae and relative molecular mass*

2-Nitrotoluene   3-Nitrotoluene   4-Nitrotoluene

$C_7H_7NO_2$ Relative molecular mass: 137.15

1.1.3 *Chemical and physical properties of the pure substance*

2-Nitrotoluene

(a) *Description*: Pale yellowish liquid which crystallizes at lower temperatures to solid α- and β-forms (Budavari, 1989; Lide, 1993)

(b) *Boiling-point*: 221.7 °C (Lide, 1993)

(c) *Melting-point*: −9.5 °C (needles; α-form), −2.9 °C (crystals, β-form) (Lide, 1993)

(d) *Density*: 1.1629 at 20 °C/4 °C (Lide, 1993)

(e) *Spectroscopy data*: Infrared (prism [4692], grating [437]), ultraviolet (UV) [1292], nuclear magnetic resonance (proton [676], C-13 [857]) and mass spectral data have been reported (Sadtler Research Laboratories, 1980).

(f) *Solubility*: Slightly soluble in water (0.54 g/L at 20 °C); soluble in benzene, diethyl ether, ethanol and petroleum ether (Budavari, 1989; Hoechst Chemicals, 1989; Lide, 1993)

(g) *Volatility*: Vapour pressure, 0.1 mm Hg [13 Pa] at 20 °C; relative vapour density (air = 1), 4.72 (Verschueren, 1983; Booth, 1991)

(h) *Stability*: Combustible when exposed to heat or open flame; potentially explosive reaction with alkali (Sax & Lewis, 1989)

(i) *Octanol/water partition coefficient (P)*: log P, 2.30 (Hansch *et al.*, 1995)

(j) *Conversion factor*: $mg/m^3 = 5.6 \times ppm^1$

3-Nitrotoluene

(a) *Description*: Pale yellow liquid (Budavari, 1989; Lide, 1993)

(b) *Boiling-point*: 232.6 °C (Lide, 1993)

(c) *Melting-point*: 16 °C (Lide, 1993)

(d) *Density*: 1.1571 at 20 °C/4 °C (Lide, 1993)

---

[1] Calculated from: mg/m^3 = (relative molecular mass/24.45) × ppm, assuming temperature (25 °C) and pressure (101 kPa)

(e) *Spectroscopy data*: Infrared (prism [183], grating [61]), UV [73], nuclear magnetic resonance (proton [22], C-13 [2002]) and mass spectral data have been reported (Sadtler Research Laboratories, 1980).

(f) *Solubility*: Slightly soluble in water (0.5 g/L at 30 °C); soluble in benzene, diethyl ether and ethanol (Budavari, 1989; Booth, 1991; Lewis, 1993)

(g) *Volatility*: Vapour pressure, 0.1 mm Hg [13 Pa] at 20 °C; relative vapour density (air = 1), 4.72 (Verschueren, 1983; Booth, 1991)

(h) *Stability*: Combustible when exposed to heat, flame or oxidizers (Sax & Lewis, 1989)

(i) *Octanol/water partition coefficient (P)*: log P, 2.42 (Hansch *et al*., 1995)

(j) *Conversion factor*: $mg/m^3 = 5.61 \times ppm^1$

4-Nitrotoluene

(a) *Description*: Yellowish orthorhombic crystals (from diethyl ether or ethanol) (Budavari, 1989; Lide, 1993)

(b) *Boiling-point*: 238.3 °C (Lide 1993)

(c) *Melting-point*: 54.5 °C (Lide, 1993)

(d) *Spectroscopy data*: Infrared (prism [4693], grating [438]), UV [1293], nuclear magnetic resonance (proton [677], C-13 [632]) and mass spectral data have been reported (Sadtler Research Laboratories, 1980).

(e) *Solubility*: Slightly soluble in water (0.26 g/L at 20 °C); soluble in acetone, benzene, chloroform, diethyl ether and ethanol (Budavari, 1989; Booth, 1991; Lide, 1993)

(f) *Volatility*: Vapour pressure, 0.1 mm Hg [13 Pa] at 20 °C; relative vapour density (air = 1), 4.72 (Verschueren, 1983; Booth, 1991)

(g) *Stability*: Combustible when exposed to heat or flame; mixtures with tetranitromethane are sensitive high explosives (Sax & Lewis, 1989)

(h) *Octanol/water partition coefficient (P)*: log P, 2.37 (Hansch *et al*., 1995)

(i) *Conversion factor*: $mg/m^3 = 5.61 \times ppm^1$

1.1.4 *Technical products and impurities*

2-Nitrotoluene is available commercially at a purity of 99.2%–99.5% and containing the following typical impurities: 3- and 4-nitrotoluenes, 0.8%; water, 0.2%; toluene, 0.1% (Hoechst Chemicals, 1989; DuPont Chemicals, 1994; First Chemical Corp., 1995a).

3-Nitrotoluene is available commercially at a purity of 99.0% with the following typical impurities: 2- and 4-nitrotoluenes, 1.0%; and water, 0.1% (First Chemical Corp., 1995b).

---

[1] Calculated from: mg/m^3 = (relative molecular mass/24.45) × ppm, assuming temperature (25 °C) and pressure (101 kPa)

4-Nitrotoluene is available commercially at a purity of 99.5% with the following typical impurities (max.): 2- and 3-nitrotoluenes, 0.5%; dinitrotoluene, 0.1%; and water, 0.1% (First Chemical Corp., 1995c).

1.1.5 Analysis

The routine determination of urinary nitrotoluenes (2- and 4-isomers) at concentrations of 5–50 mg/L is achieved by colorimetric analysis. In acidified urine, the nitro group of the nitrotoluenes is zinc reduced and, after diazotization and coupling to 1-amino-8-naphthol-2,4-disulfonic acid (Chicago acid), the primary aromatic amine is determined as a red azo dye. An alternative method, which uses formamidine sulfinic acid (thiourea dioxide) to reduce the aromatic nitro compounds under alkaline conditions, has been described (Koniecki & Linch, 1958).

Selected methods for the analysis of 2-, 3- and 4-nitrotoluenes in various media are identified in Table 1.

Table 1. Methods for the analysis of nitrotoluenes[a,b,c]

Sample matrix	Sample preparation	Assay procedure	Limit of detection	Reference
Air	Draw air through solid sorbent tube; desorb with methanol	GC/FID	0.8 µg/sample[a,b,c]	Eller (1994) [Method 2005]
Water	Extract with dichloromethane or adsorb on Amberlite XAD resin and elute with dichloromethane	GC/ECD	NR[b]	Feltes et al. (1990)
	Extract with diethyl ether; dry over anhydrous magnesium sulfate; filter	GC/FID	NR[a,c]	Spanggord et al. (1982a)
	Liquid–liquid extraction with dichloromethane; dry with anhydrous sodium sulfate; evaporate to dryness; redissolve in methanol	SFC/FID	20 ppm[c] (mg/L)	Ong et al. (1992)
Blood	Extract from separated plasma and concentrate using 2,2,4-trimethylpentane	GC/ECD	15 µg/L[c]	Lewalter & Ellrich (1991)

GC, gas chromatography; FID, flame ionization detection; ECD, electron capture detection; NR, not reported; SFC, capillary supercritical fluid chromatography

[a] 2-Nitrotoluene

[b] 3-Nitrotoluene

[c] 4-Nitrotoluene

1.2 Production and use

1.2.1 Production

The nitrotoluenes are produced commercially by the nitration of toluene with mixed acids: nitric acid (HNO_3)/sulfuric acid, HNO_3/aromatic sulfonic acid or HNO_3/phosphoric acid. The resultant isomer ratio depends on the catalyst and conditions but generally is in the range of 45–62% 2-nitrotoluene, 2–5% 3-nitrotoluene and 33–50% 4-nitrotoluene (Booth, 1991).

Large-scale nitration is carried out under typical mixed acid conditions at 25–40 °C and with a nitric acid:toluene molar ratio close to one. A 2-nitrotoluene : 4-nitrotoluene ratio of 1.6 and an isolated yield of 96% total nitrotoluenes are typical in a continuous reactor, similar to that described for nitrobenzene (see the monograph on p. 384), but operating at as low a temperature as possible to avoid the more-readily formed by-products. 2-, 3- and 4-Nitrotoluenes are separated and purified by fractional distillation and crystallization (Booth, 1991).

It has been estimated that 13 000 tonnes of 2-nitrotoluene, very small quantities of 3-nitrotoluene and 6800 tonnes of 4-nitrotoluene are used annually in the United States of America (American Conference of Governmental Industrial Hygienists, 1991). In 1984, the yearly production capacity for mononitrotoluene (all isomers) in the western world was approximately 200 000 tonnes (Booth, 1991).

2-Nitrotoluene and 4-nitrotoluene are produced by six companies in China, three companies in Japan, two companies each in Germany, India, the Republic of Korea and the United States, and one company each in the Czech Republic, Italy, Romania, Sweden and the United Kingdom. 3-Nitrotoluene is produced by five companies in China, and one company each in Germany, India, Italy, Japan, the Republic of Korea, Sweden, the United Kingdom and the United States (Chemical Information Services, 1994).

1.2.2 Use

2-Nitrotoluene is used for the production of derivatives that are principally colourant intermediates. For example, the following derivatives are all intermediates in the preparation of various azo dyes: *ortho*-toluidine (see IARC, 1982, 1987), 2-amino-4-chlorotoluene (Fast Scarlet TR Base, by reduction of 4-chloro-2-nitrotoluene), 2-amino-6-chlorotoluene (Fast Red KB Base, by reduction of 2-chloro-6-nitrotoluene) and *ortho*-toluidine-4-sulfonic acid (by reduction of 2-nitrotoluene-4-sulfonic acid). A more recent use for *ortho*-toluidine is its conversion to 2-ethyl-6-methylaniline, an intermediate in the manufacture of agricultural chemicals. This is an important outlet for the typical surplus of 2-nitrotoluene. 2-Nitrotoluene is also used in the manufacture of rubber chemicals and in various azo and sulfur dyes for cotton, wool, silk, leather and paper (American Conference of Governmental Industrial Hygienists, 1991; Booth, 1991).

3-Nitrotoluene is used in the manufacture of *meta*-toluidine and nitrobenzoic acids, and in the manufacture of agricultural, pharmaceutical, photographic and rubber chemicals (Budavari, 1989; American Conference of Governmental Industrial Hygienists, 1991; Booth, 1991; First Chemical Corp., 1995b).

4-Nitrotoluene is used for the production of derivatives that are used principally as colourant intermediates. Important derivatives include *para*-toluidine, 4-nitrobenzoic acid (by oxidation of 4-nitrotoluene), 4-amino-2-chlorotoluene (by reduction of 2-chloro-4-nitrotoluene) and 4-nitrotoluene-2-sulfonic acid, the last of which is of importance in forming stilbene intermediates for fluorescent whitening agents. 4-Nitrotoluene is also used in the manufacture of agricultural, pharmaceutical and rubber chemicals, and in various azo and sulfur dyes for cotton, wool, silk, leather and paper (American Conference of Governmental Industrial Hygienists, 1991; Booth, 1991).

1.3 Occurrence

1.3.1 *Natural occurrence*

Nitrotoluenes are not known to occur as natural products.

1.3.2 *Occupational exposure*

The most probable routes of human exposure to nitrotoluenes are inhalation and dermal contact of workers involved in the production and use of nitrotoluenes, dinitrotoluenes and trinitrotoluene. Very few data on occupational exposures to nitrotoluene exist. Ahlborg *et al.* (1985) reported a maximal air concentration of 2.0 mg/m^3 2-nitrotoluene in the nitrotoluene production area of a chemical plant producing pharmaceuticals and explosives.

1.3.3 *Environmental occurrence*

(a) Air

2-Nitrotoluene was detected in the ambient air at a plant in Deepwater, NJ, United States, at a concentration of 47 ng/m^3; 4-nitrotoluene was detected at concentrations ranging from 59 to 89 ng/m^3 (Pellizzari, 1978).

(b) Water

2-Nitrotoluene has been identified qualitatively in German drinking-water (Kool *et al.*, 1982).

In the Netherlands, during 1974, 2- and 4-nitrotoluenes were detected in the River Waal at an average concentration of 4.5 µg/L (max., 18.1 mg/L) and in the River Maas at a maximal concentration of 0.3 µg/L (Meijers & van der Leer, 1976). Also in the Netherlands, 2- and 4-nitrotoluenes were detected in the River Rhine at a concentration of 10 µg/L (Piet & Morra, 1983).

2-Nitrotoluene concentrations of 0.4 and 7.4 µg/L were detected in surface-water samples collected from two brooks near Hischagen/Waldhof, Germany, in the vicinity of a former munitions manufacturing plant closed after the Second World War; the brooks fed into the River Losse, which had a concentration of 1.2 µg/L. Two ponds in the Clausthal-Zellerfeld region of Germany, again near previous munitions manufacturing plants, had levels of 0.4 and 22.0 µg/L; these ponds fed into the River Oder, which had a level of < 0.01 µg/L. Concentrations at three locations (Brunsbüttel, Brokdorf, Lauenburg) along the River Elbe ranged from 0.05 to 0.4 µg/L. Corresponding levels of

3-nitrotoluene were 0.1 and 0.9 µg/L in the brooks, 0.1 µg/L in the River Losse, 0.1 and 1.5 µg/L in the ponds and < 0.01 µg/L in the River Oder. Corresponding levels of 4-nitrotoluene were 0.3 and 4.5 µg/L in the brooks, 0.4 µg/L in the River Losse, 0.2 and 4.8 µg/L in the ponds and < 0.01 µg/L in the River Oder (Feltes *et al.*, 1990).

3-Nitrotoluene was measured at a concentration of 1 µg/L in water from the River Rhine and detected qualitatively in the River Rhine on several other occasions (United States National Library of Medicine, 1995).

2- and 4-Nitrotoluenes were detected in the effluent from a plant manufacturing trinitrotoluene in Radford, VA, United States, at concentrations ranging from 0.32 to 16 mg/L and from 0.12 to 9.2 mg/L, respectively (Howard, 1989). These compounds were also detected in the wastewater resulting from the production and purification of 2,4,6-trinitrotoluene at concentrations ranging from 0.02 to 0.14 mg/L (in 21/54 samples) and from 0.01 to 0.17 mg/L (in 23/54 samples), respectively (Spanggord *et al.*, 1982a), and in the raw effluent from a plant manufacturing dinitrotoluene at 7.8 mg/L and 8.8 mg/L, respectively. 2-Nitrotoluene was detected in a waste-treatment lagoon of a paper mill and 4-nitrotoluene in a waste-treatment lagoon of a chemical company at 0.04 mg/L (Howard, 1989). An acidic stream of wastewater discharged from a nitrotoluene-manufacturing plant in India was found to contain 87–102 mg/L, 0.4–3.1 mg/L and 8–67 mg/L 2-, 3- and 4-nitrotoluenes, respectively; corresponding levels in an alkaline stream were 53–80 mg/L, 6–11 mg/L and 84–145 mg/L (Swaminathan *et al.*, 1987).

Nitrotoluenes were identified as pollutants in groundwater samples collected from January to March 1987 in Degrémont, France. Concentrations ranged from 90 to 165 µg/L for 2-nitrotoluene, 9 to 19 µg/L for 3-nitrotoluene and from 3 to 20 µg/L for 4-nitrotoluene (Duguet *et al.*, 1988).

1.4 Regulations and guidelines

Occupational exposure limits and guidelines for nitrotoluenes in several countries are presented in Table 2.

2. Studies of Cancer in Humans

No data were available to the Working Group.

3. Studies of Cancer in Experimental Animals[1]

There have been no reports of long-term studies of the carcinogenicity of 2-, 3- or 4-nitrotoluene in experimental animals.

---

[1] The Working Group noted that additional studies, including short-term and carcinogenicity studies, are currently being conducted on 2-nitrotoluene (IARC, 1994).

Table 2. Occupational exposure limits and guidelines for nitrotoluenes (all isomers)

Country	Year	Concentration (mg/m³)	Interpretation
Australia	1993	11 (Sk)	TWA
Belgium	1993	11 (Sk)	TWA
Bulgaria[a]	1995	11 (Sk)	TWA
Colombia[a]	1995	11 (Sk)	TWA
Czech Republic[b]	1993	5	TWA
		20	STEL (15 min)
Denmark	1993	12 (Sk)	TWA
Finland	1993	30 (Sk)	TWA
		60	STEL (15 min)
France[b]	1991	11 (Sk)	TWA
Germany	1995	30 (Sk) (3- and 4-isomers)	MAK
		None (Sk), IIIA2 (2-isomer)	
Jordan[a]	1995	11 (Sk)	TWA
Netherlands[b]	1994	11 (Sk)	TWA
New Zealand[a]	1995	11 (Sk)	TWA
Poland	1991	3	TWA
Republic of Korea[a]	1995	11 (Sk)	TWA
Russia	1991	3 (Sk)	STEL
Singapore[a]	1995	11 (Sk)	TWA
Switzerland	1991	11 (Sk)	TWA
		22	STEL (15 min)
United Kingdom	1995	30 (Sk)	TWA
		60	STEL (15 min)
USA			
ACGIH (TLV)	1995	11 (Sk)[c]	TWA
OSHA (PEL)	1994	30 (Sk)	TWA
NIOSH (REL)	1994	11 (Sk)	TWA
Viet Nam[a]	1995	11 (Sk)	TWA

From International Labour Office (1991); Työministeriö (1993); Arbeitsinspectie (1994); United States National Institute for Occupational Safety and Health (NIOSH) (1994a,b); United States Occupational Safety and Health Administration (OSHA) (1994); American Conference of Governmental Industrial Hygienists (ACGIH) (1995); Deutsche Forschungsgemeinschaft (1995); Health and Safety Executive (1995)

Sk, absorption through the skin may be a significant source of exposure; TWA, time-weighted average; STEL, short-term exposure limit; IIIA2, substances shown to be clearly carcinogenic only in animal studies but under conditions indicative of carcinogenic potential at the workplace; TLV, threshold limit value; PEL, permissible exposure limit; REL, recommended exposure limit

[a] Follows ACGIH TLVs
[b] 3-Nitrotoluene only
[c] Substances identified in the BEI (Biological Exposure Indices) documentation as inducers of methaemoglobin

Oral administration

Rat: Groups of 10 male Fischer 344/N rats, six to eight weeks of age, received diets containing 0, 625, 1250, 2500, 5000 or 10 000 mg/kg diet (ppm) 2-nitrotoluene (> 96% pure) for 13 weeks. Animals consumed an estimated average of 0, 45, 89, 179, 353 or 694 mg/kg bw 2-nitrotoluene daily. All animals survived to the end of the 13-week study, at which time they were killed for complete histopathological evaluation. Body-weight gains were reduced by 12, 28 and 44% relative to controls in the 2500-, 5000- and 10 000-ppm dose groups, respectively. Two rats receiving 10 000 ppm had mesothelial-cell hyperplasia of the tunica vaginalis on the surface of the epididymis and three receiving 5000 ppm had mesotheliomas at the same location. Mesotheliomas had not been observed previously in treated or control male rats in approximately 435 other 13-week studies (United States National Toxicology Program, 1992; Dunnick *et al.*, 1994). The historical incidence of mesotheliomas of all organs in control male rats in two-year studies of the United States National Toxicology Program is 2.7% (Haseman *et al.*, 1990).

4. Other Data Relevant for an Evaluation of Carcinogenicity and its Mechanisms

4.1 Absorption, distribution, metabolism and excretion

4.1.1 *Humans*

In the chemical-processing department of a plant producing pharmaceuticals and explosives, Ahlborg *et al.* (1988) analysed diazo-positive metabolites in the hydrolysed urine from workers exposed to aromatic nitroamino compounds, including nitrotoluenes, 2,4,6-trinitrotoluene (see monograph, p. 460) and nitro- and aminobenzoic acids. In the post-shift samples of 45 exposed workers, levels were higher than in 25 unexposed workers. After a holiday, the levels in the exposed workers were lower than in their own post-shift samples. In the unexposed group, no such difference was noted. [The workers were also exposed to compounds other than nitrotoluenes that may have given rise to diazo-positive urinary excretion.]

4.1.2 *Experimental systems*

(a) *Metabolites*

Following the administration of a single oral dose of 200 mg/kg bw nitro[^{14}C]toluene isomers to male Fischer 344 rats, a total of 80–90% was eliminated within 72 h in urine, faeces and expired air. A higher percentage of 2-nitrotoluene (85.8%) than 3-nitrotoluene (67.8%) or 4-nitrotoluene (76.7%) was excreted in urine. Nine metabolites were found in the urine of rats given 2-nitrotoluene. For three rats 72 h after treatment, the mean percentage of the dose administered plus or minus the standard error of the mean for each metabolite was: nitrobenzoic acid (28.6 ± 0.3); 2-nitrobenzyl glucuronide (14.1 ± 0.7);

S-(2-nitrobenzyl)-N-acetylcysteine (11.6 ± 0.2); S-(2-nitrobenzyl)glutathione (3.9 ± 0.4); 2-aminobenzoic acid (1.8 ± 0.2); 2-nitrobenzyl sulfate (0.5 ± 0.2); 2-nitrobenzyl alcohol (0.4 ± 0.3); and two unidentified metabolites (15.9 ± 1.5 and 6.0 ± 0.6). Eight metabolites were found in the urine of rats given 3-nitrotoluene. For three rats 72 h after treatment, the mean percentage of the dose administered plus or minus the standard error of the mean for each metabolite was: 3-nitrohippuric acid (23.6 ± 2.0); 3-nitrobenzoic acid (21.1 ± 1.1); 3-acetamidobenzoic acid (11.6 ± 0.4); 3-nitrobenzyl glucuronide (2.0 ± 0.1); S-(3-nitrobenzyl)glutathione (1.3 ± 0.1); 3-aminobenzoic acid (1.2 ± 0.3); and two unidentified metabolites (2.6 ± 0.3 and 2.2 ± 0.3). Eight metabolites were found in urine following administration of 4-nitrotoluene. Again, for three rats 72 h after treatment, the mean percentage of the dose administered plus or minus the standard error the mean for each metabolite was: 4-nitrobenzoic acid (28.0 ± 2.6); 4-acetamidobenzoic acid (27.1 ± 3.0); 4-nitrohippuric acid (13.0 ± 0.7); S-(4-nitrobenzyl)-N-acetylcysteine (3.7 ± 0.1); 4-nitrobenzyl glucuronide (1.4 ± 0.1); 4-aminobenzoic acid (0.8 ± 0.1); S-methyl-2-nitrophenyl glucuronide (0.3 ± 0.0); and 5-methyl-2-nitrophenyl sulfate (0.2 ± 0.0) (Chism et al., 1984).

In bile duct-cannulated male Fischer 344 rats administered 200 mg/kg bw nitro[^{14}C]toluene isomers, 28.6% of the 2-nitrotoluene dose was excreted in bile in males, compared to 9.6% in females. The major metabolite in bile following 2-nitrotoluene administration was 2-nitrobenzyl glucuronide. Males excreted 22% of the dose as this metabolite and females excreted 8.3%. Following 3-nitrotoluene administration, 10.8% of the dose was excreted in the bile of male rats, compared to 4.3% in female rats. 3-Nitrobenzoic acid was the most abundant biliary metabolite of 3-nitrotoluene (3.4 and 1.7% of the dose in males and females, respectively), followed by 3-nitrobenzyl glucuronide (2.8 and 0.7% of the dose in males and females, respectively). In male rats, 9.8% of the 4-nitrotoluene dose was excreted in bile, compared to 1.3% in females. The most abundant biliary metabolite following 4-nitrotoluene administration was 4-nitrobenzoic acid (2.8 and 0.8% of the dose in males and females, respectively) (Chism & Rickert, 1985). Compared to controls, sham operation decreased the urinary excretion of 2-nitro-[^{14}C]toluene (200 mg/kg) by 20–30% and bile duct cannulation decreased urinary excretion of radioactivity by 52–59%.

(b) Macromolecular binding

Following an oral dose (200 mg/kg bw) to male and female Fischer 344 rats, more 2-nitrotoluene than 3-nitrotoluene or 4-nitrotoluene was bound covalently to hepatic macromolecules in both males (2-, 3- and 4-nitrotoluenes: 36.6, 6.9 and 10.1 nmol nitrotoluene equivalents/g protein, respectively) and females (2-, 3- and 4-nitrotoluenes: 11.2, 7.9 and 8.5 nmol nitrotoluene equivalents/g protein, respectively). Three times more 2-nitrotoluene was bound to hepatic macromolecules in males than in females. Interruption of enterohepatic circulation by bile duct cannulation decreased hepatic macromolecular covalent binding of 2-nitrotoluene in male rats by 93% when compared to sham-operated controls and by 98% when compared to intact controls. In female rats, after bile duct cannulation, binding of 2-nitrotoluene was decreased by 85% compared to intact controls and by 78% compared to sham-operated controls (Chism & Rickert,

1985). In-vitro studies suggest that the metabolite of 2-nitrotoluene responsible for covalent binding to DNA is 2-aminobenzyl sulfate (Chism & Rickert, 1989).

4.2 Toxic effects

4.1.1 Humans

No data were available to the Working Group.

4.2.2 Experimental systems

(a) *Single-dose studies*

The single oral LD_{50}s determined in male Sprague-Dawley rats, male and female Wistar rats and male CF-1 mice were: 2-nitrotoluene — 891, 2100, 2100 and 2463 mg/kg bw, respectively; 3-nitrotoluene — 1072, 2200, 2000 and 330 mg/kg bw, respectively; and 4-nitrotoluene — 2144, 4700, 3200 and 1231 mg/kg bw, respectively (Ciss *et al.*, 1980a; Registry of Toxic Effects of Chemical Substances, 1991).

(b) *Repeated-dose studies*

In 14-day studies, 2-, 3- and 4-nitrotoluenes were administered to male and female Fischer 344/N rats at concentrations ranging from 625 to 20 000 mg/kg diet (ppm). Estimated consumption based on food intake was 56–696 mg/kg bw 2-nitrotoluene, 61–881 mg/kg bw 3-nitrotoluene and 106–869 mg/kg bw 4-nitrotoluene for males, and 55–779 mg/kg bw 2-nitrotoluene, 58–754 mg/kg bw 3-nitrotoluene and 105–611 mg/kg bw 4-nitrotoluene for females. Body-weight gains in male rats were decreased in groups receiving the isomers at concentrations of 5000 ppm and above; male rats receiving 3-nitrotoluene at 2500 ppm (259 mg/kg bw) gained less weight than controls and male rats receiving 4-nitrotoluene at 20 000 ppm (849 mg/kg bw) lost weight. Body weights of female rats were affected at higher concentrations than those of males. No gross lesion related to 2-nitrotoluene treatment was observed, but in the livers of 4/5 male rats in the 10 000-ppm (696 mg/kg bw) group minimal oval-cell hyperplasias, which consisted of proliferations of small cells with pale-staining oval-shaped nuclei, were found. These cells were dispersed between hepatocytes in the portal areas. No lesion was observed in the livers of female rats. Following administration of 4-nitrotoluene, no treatment-related gross lesion was observed either. However, increased congestion and extramedullary haematopoiesis were seen in the spleen of one male rat at 5000 ppm (446 mg/kg bw) and in most males and females at 10 000 (431 and 420 mg/kg bw) and 20 000 ppm (869 and 611 mg/kg bw). Lymphoid depletion occurred in the thymus and spleen of a few rats in the 10 000- and 20 000-ppm groups; this was attributed to the marked reduction in body-weight gain or body-weight loss during the study (United States National Toxicology Program, 1992).

In 14-day studies, the three isomers were given in the diet to male and female B6C3F1 mice at doses of 388–10 000 mg/kg (ppm). Estimated consumption was 63–854 mg/kg bw 2-nitrotoluene, 66–779 mg/kg bw 3-nitrotoluene and 202–1548 mg/kg bw 4-nitrotoluene for males, and 134–1224 mg/kg bw 2-nitrotoluene, 92–901 mg/kg bw

3-nitrotoluene and 388–2010 mg/kg bw 4-nitrotoluene for females. In male mice, liver weights were increased in the group administered the highest dose of 2-nitrotoluene. Relative liver weights were also increased in female mice that received all but the lowest dose of 3-nitrotoluene and in males that received diets containing 2500 ppm (409 mg/kg bw) and 5000 ppm (779 mg/kg bw) 3-nitrotoluene. Following 4-nitrotoluene administration, relative liver weights were increased in a dose-related manner in all males and all but the low-dose females (United States National Toxicology Program, 1992).

In a 13-week feeding study in male and female Fischer 344 rats (625–10 000 ppm in the diet; about 40–700 mg/kg bw per day), all three nitrotoluene isomers caused an increase in the incidence of hyaline-droplet nephropathy in male rats, with 2-nitrotoluene being the most toxic. Hyaline droplets were associated with accumulation of $\alpha_{2\mu}$-globulin in the kidney and were characterized by an accumulation of eosinophilic crystalline-like inclusions and globular droplets within the cytoplasm and lumen of the renal tubules. Other kidney lesions included minimal to mild enlargement (karomegaly) of the proximal tubule epithelium in male and female rats receiving 4-nitrotoluene, and yellow- to-brown pigment (possibly lipofuscin) in the cytoplasm of renal tubule epithelial cells of male and female rats receiving 2- or 4-nitrotoluene. A treatment-related increase in extramedullary haematopoiesis, haemosiderin pigment and/or congestion occurred in the splenic pulp of male and female rats treated with either isomer. At higher doses, there was minimal to mild thickening of the splenic capsule accompanied by mesothelial-cell hypertrophy. Some decreases in haematocrit, erythrocyte counts and haemoglobin concentrations and small increases in platelet and lymphocyte counts were observed at the highest-dose level (United States National Toxicology Program, 1992; Dunnick et al., 1994).

In the livers of male rats given 2500 mg/kg diet (179 mg/kg bw) 2-nitrotoluene or more for 14 weeks, various non-neoplastic lesions were observed, including cytoplasmic vacuolization, oval-cell hyperplasia and inflammation. Cytoplasmic vacuolization was characterized by multiple round clear spaces of varying size in hepatocytes throughout the liver lobule, most prominantly in the portal area. Oval-cell hyperplasia consisted of increased numbers of small cells with pale staining cytoplasm and round-to-oval nuclei. These cells were generally interspersed between hepatocytes in single or double rows, but sometimes formed small nodules or ductular structures in the portal area of the liver lobule (United States National Toxicology Program, 1992; Dunnick et al., 1994).

In a 13-week feeding study of the three isomers in male and female B6C3F1 mice (625–10 000 ppm in the diet; about 100–1700 mg/kg bw per day), the only evidence of toxicity was degeneration and metaplasia of the olfactory epithelium following administration of 2-nitrotoluene. No hepatic toxicity was noted in mice, although all three isomers caused increased liver weights (United States National Toxicology Program, 1992; Dunnick et al., 1994).

An increase in liver foci (placental glutathione-S-transferase-positive) was observed in male Fischer 344 rats fed 5000 mg/kg diet 2-nitrotoluene for 13 or 26 weeks (Ton et al., 1995).

4.3 Reproductive and developmental effects

4.3.1 *Humans*

No data were available to the Working Group.

4.3.2 *Experimental systems*

Groups of 10 male and 10 female Wistar rats were administered daily by gavage 0.2 g/kg bw 2-nitrotoluene in olive oil, 0.3 g/kg bw 3-nitrotoluene in olive oil, 0.4 g/kg bw 4-nitrotoluene suspended in 1% methylcellulose, olive oil alone or methylcellulose alone on five days per week for three months. At the end of this period, the rats were paired in four groups: (i) five exposed males with five exposed females; (ii) five exposed males with five unexposed females; (iii) five unexposed males with five unexposed females; (iv) five unexposed males and five exposed females. The treatment was continued for three more months. Two deaths were observed in the exposed groups; these were due to errors in performing the intubation procedure. Splenic enlargement was observed in males exposed to each of the three isomers. With the 2-isomer, renal tubules were dilated and contained hyaline droplets. These were more frequent in females (7/9) than in males (3/8). With the 4-isomer, all nine exposed males had testicular atrophy and, in five of these, this was associated with necrosis of the seminiferous tubules. No adverse effect on reproduction or on the offspring was observed for any of the isomers (Ciss *et al.*, 1980b). [The Working Group noted the absence of descriptions of animal randomization and husbandry.]

Male and female Fischer 344 rats were administered 3-nitrotoluene at doses of 625–10 000 mg/kg diet in a 14-day study (see Section 4.2). At necropsy, chemically related gross lesions were described as a reduction in size of the testis and uterus in rats from the 10 000-ppm (881 and 754 mg/kg bw for males and females, respectively) groups. The testis, epididymis, uterus and liver from all rats were examined microscopically. All males in the highest-dose group (881 mg/kg bw) had mild to moderate degeneration of the testis characterized by a loss of germinal epithelium and the presence of abnormal (syncytia) spermatids in the lumen of the seminiferous tubules and ducts of the epididymis. One of the males in the 5000-ppm group (431 mg/kg bw) had moderate testicular degeneration, but the lesion was unilateral and the relationship to treatment was uncertain. Compared to controls and lower-dose groups, the uteri of female rats in the highest-dose group (754 mg/kg bw) had thinner muscular walls and less developed endometrium (United States National Toxicology Program, 1992).

Groups of 10 male and 10 female Fischer 344/N and B6C3F1 mice were administered doses of 0, 2500, 5000 or 10 000 ppm 2-, 3,- or 4-nitrotoluene in their feed for 13 weeks (United States National Toxicology Program, 1992). [The concentration of 10 000 ppm corresponds to an estimated dose, based on measures of food consumption, of about 700 mg/kg bw per day for rats and about 1500 mg/kg bw per day for mice.] Treatment had no effect on survival, and clinical signs of toxicity were limited to decreases in food consumption. Decreased body-weight gain was observed in both species for each of the isomers at the higher-dose levels and was most pronounced in rats receiving

4-nitrotoluene. All three isomers impaired testicular function in rats, as shown microscopically and by measurement of sperm density, motility and number. The three isomers also increased the length of the oestrus cycle in rats. The 2-isomer appeared generally to be more toxic than the other isomers. Degeneration of the testes occurred in all male rats receiving 5000 or 10 000 ppm of the 2-isomer. Virtually no sperm was present in the epididymides of rats receiving 2-nitrotoluene at 10 000 ppm. Sperm counts were also diminished in the group of rats receiving 5000 ppm of this isomer. Only 4/10 female rats receiving the highest dose had a measurable oestrus cycle. With the 3- and 4-isomers, degeneration of the testes occurred only in rats receiving doses of 10 000 ppm. The severity of this lesion was less than that observed with the 2-isomer at this level of dose. Among females receiving 3-nitrotoluene, there was a dose-related increase in the length of the oestrus cycle; concurrently, the number of cycling animals diminished. With the 4-isomer, 9/10 females in the group receiving 10 000 ppm had no discernible oestrus cycle. No growth or histopathological change in the uterus or ovaries was associated with treatment with any of the three isomers. Except for a significant decrease in sperm motility in mice receiving 10 000 ppm 2-nitrotoluene, no change was noted in the reproductive system evaluations in male or female mice for any of the three isomers (United States National Toxicology Program, 1992).

Administration of single intraperitoneal doses of 30 and 100 mg/kg bw 4-nitrotoluene to female Sprague-Dawley rats increased uterine weights without producing overt toxicity. Doses of 1000 mg/kg were toxic (Smith & Quinn, 1992).

4.4 Genetic and related effects

4.4.1 *Humans*

No data were available to the Working Group.

4.4.2 *Experimental systems* (see also Table 3 and Appendices 1 and 2)

2-Nitrotoluene did not induce unscheduled DNA synthesis in primary cultures of human hepatocytes treated *in vitro* or in rat spermatogonic cells exposed *in vitro*. It did not induce chromosomal aberrations, but induced sister chromatid exchange in the presence of S9 in Chinese hamster ovary cells *in vitro*.

In vivo in rats, 2-nitrotoluene bound covalently to hepatic macromolecules, including DNA.

In hepatocytes of male Fischer 344 rats, 2-nitrotoluene induced unscheduled DNA-synthesis after dosing by gavage *in vivo*, but not after treatment of the hepatocytes *in vitro*. That 2-nitrotoluene did not induce unscheduled DNA synthesis in germ-free animals suggests an obligatory role of intestinal bacteria in the metabolic activation (Butterworth *et al.*, 1982; Doolittle *et al.*, 1983). The in-vivo activity is dependent upon the sex — in male and female Fischer 344 rats having similar populations of intestinal bacteria, 2-nitrotoluene induced unscheduled DNA synthesis only in males. This difference might be explained (Doolittle *et al.*, 1983) by sex differences in biliary excretion (Chism & Rickert, 1985).

Table 3. Genetic and related effects of nitrotoluenes

Test system	Result[a]		Dose[b] (LED/HID)	Reference
	Without exogenous metabolic system	With exogenous metabolic system		
2-Nitrotoluene				
BSD, *Bacillus subtilis rec* strains, differential toxicity	–	0	NR	Shimizu & Yano (1986)
SA0, *Salmonella typhimurium* TA100, reverse mutation	–	0	685	Chiu et al. (1978)
SA0, *Salmonella typhimurium* TA100, reverse mutation	–	–	500	Miyata et al. (1981)
SA0, *Salmonella typhimurium* TA100, reverse mutation	–	–	500	Tokiwa et al. (1981)
SA0, *Salmonella typhimurium* TA100, reverse mutation	–	–	2500	Spanggord et al. (1982b)
SA0, *Salmonella typhimurium* TA100, reverse mutation	–	–	256	Haworth et al. (1983)
SA0, *Salmonella typhimurium* TA100, reverse mutation	–	0	50	Suzuki et al. (1983)
SA0, *Salmonella typhimurium* TA100, reverse mutation	–	–	450	Shimizu & Yano (1986)
SA5, *Salmonella typhimurium* TA1535, reverse mutation	–	–	500	Miyata et al. (1981)
SA5, *Salmonella typhimurium* TA1535, reverse mutation	–	–	2500	Spanggord et al. (1982b)
SA5, *Salmonella typhimurium* TA1535, reverse mutation	–	–	256	Haworth et al. (1983)
SA5, *Salmonella typhimurium* TA1535, reverse mutation	–	–	450	Shimizu & Yano (1986)
SA7, *Salmonella typhimurium* TA1537, reverse mutation	–	–	150	Miyata et al. (1981)
SA7, *Salmonella typhimurium* TA1537, reverse mutation	–	–	2500	Spanggord et al. (1982b)
SA7, *Salmonella typhimurium* TA1537, reverse mutation	–	–	256	Haworth et al. (1983)
SA7, *Salmonella typhimurium* TA1537, reverse mutation	–	–	450	Shimizu & Yano (1986)
SA8, *Salmonella typhimurium* TA1538, reverse mutation	–	–	2500	Spanggord et al. (1982b)
SA8, *Salmonella typhimurium* TA1538, reverse mutation	–	–	450	Shimizu & Yano (1986)
SA9, *Salmonella typhimurium* TA98, reverse mutation	–	0	685	Chiu et al. (1978)
SA9, *Salmonella typhimurium* TA98, reverse mutation	–	–	150	Miyata et al. (1981)
SA9, *Salmonella typhimurium* TA98, reverse mutation	–	–	500	Tokiwa et al. (1981)
SA9, *Salmonella typhimurium* TA98, reverse mutation	–	–	2500	Spanggord et al. (1982a)
SA9, *Salmonella typhimurium* TA98, reverse mutation	–	–	256	Haworth et al. (1983)
SA9, *Salmonella typhimurium* TA98, reverse mutation	–	–	50	Suzuki et al. (1983)
SA9, *Salmonella typhimurium* TA98, reverse mutation	–	–	450	Shimizu & Yano (1986)

Table 3 (contd)

Test system	Result[a] Without exogenous metabolic system	Result[a] With exogenous metabolic system	Dose[b] (LED/HID)	Reference
2-Nitrotoluene (contd)				
SAS, *Salmonella typhimurium* TA92, reverse mutation	–	–	150	Miyata *et al.* (1981)
SAS, *Salmonella typhimurium* TA94, reverse mutation	–	–	500	Miyata *et al.* (1981)
SIC, Sister chromatid exchange, Chinese hamster CHO cells *in vitro*	?	+	355	Galloway *et al.* (1987)
CIC, Chromosomal aberrations, Chinese hamster CHO cells *in vitro*	–	–	420	Galloway *et al.* (1987)
CIC, Chromosomal aberrations, Chinese hamster CHL cells *in vitro*	–	0	250	Ishidate *et al.* (1988)
URP, Unscheduled DNA synthesis, rat primary hepatocytes *in vitro*	–	0	13.7	Doolittle *et al.* (1983)
UIA, Unscheduled DNA synthesis, rat pachytene spermatocytes and round spermatids *in vitro*	–	0	13.7	Working & Butterworth (1984)
UIH, Unscheduled DNA synthesis, human hepatocytes *in vitro*	–	0	137	Butterworth *et al.* (1989)
UPR, Unscheduled DNA synthesis, male rat hepatocytes *in vivo*	+		200 po × 1	Doolittle *et al.* (1983)
UPR, Unscheduled DNA synthesis, male germ-free rat hepatocytes *in vivo*	–		500 po × 1	Doolittle *et al.* (1983)
UPR, Unscheduled DNA synthesis, female rat hepatocytes *in vivo*	–		200 po × 1	Doolittle *et al.* (1983)
UPR, Unscheduled DNA synthesis, male rat hepatocytes *in vivo*	+		200 po × 1	US National Toxicology Program (1992)
UPR, Unscheduled DNA synthesis, female rat hepatocytes *in vivo*	+		750 po × 1	US National Toxicology Program (1992)
UVM, Unscheduled DNA synthesis, male mouse hepatocytes *in vivo*	–		750 po × 1	US National Toxicology Program (1992)
UVM, Unscheduled DNA synthesis, female mouse hepatocytes *in vivo*	+		750 po × 1	US National Toxicology Program (1992)
BVD, Binding (covalent) to DNA, male rat liver *in vivo*	+		200 po × 1	Rickert *et al.* (1984)
BVP, Binding (covalent) to RNA or protein, male rat liver *in vivo*	+		200 po × 1	Rickert *et al.* (1984, 1986)

Table 3 (contd)

Test system	Result[a]		Dose[b] (LED/HID)	Reference
	Without exogenous metabolic system	With exogenous metabolic system		
3-Nitrotoluene				
BSD, *Bacillus subtilis rec* strains, differential toxicity	–	0	NR	Shimizu & Yano (1986)
SA0, *Salmonella typhimurium* TA100, reverse mutation	–	–	150	Miyata et al. (1981)
SA0, *Salmonella typhimurium* TA100, reverse mutation	–	–	50	Tokiwa et al. (1981)
SA0, *Salmonella typhimurium* TA100, reverse mutation	–	–	2500	Spanggord et al. (1982b)
SA0, *Salmonella typhimurium* TA100, reverse mutation	–	0	50	Suzuki et al. (1983)
SA0, *Salmonella typhimurium* TA100, reverse mutation	–	–	38	Haworth et al. (1983)
SA0, *Salmonella typhimurium* TA100, reverse mutation	–	–	445	Shimizu & Yano (1986)
SA5, *Salmonella typhimurium* TA1535, reverse mutation	–	–	150	Miyata et al. (1981)
SA5, *Salmonella typhimurium* TA1535, reverse mutation	–	–	2500	Spanggord et al. (1982b)
SA5, *Salmonella typhimurium* TA1535, reverse mutation	–	–	128	Haworth et al. (1983)
SA5, *Salmonella typhimurium* TA1535, reverse mutation	–	–	225	Shimizu & Yano (1986)
SA7, *Salmonella typhimurium* TA1537, reverse mutation	–	–	150	Miyata et al. (1981)
SA7, *Salmonella typhimurium* TA1537, reverse mutation	–	–	2500	Spanggord et al. (1982b)
SA7, *Salmonella typhimurium* TA1537, reverse mutation	–	–	128	Haworth et al. (1983)
SA7, *Salmonella typhimurium* TA1537, reverse mutation	–	–	225	Shimizu & Yano (1986)
SA8, *Salmonella typhimurium* TA1538, reverse mutation	–	–	2500	Spanggord et al. (1982b)
SA8, *Salmonella typhimurium* TA1538, reverse mutation	–	–	445	Shimizu & Yano (1986)
SA9, *Salmonella typhimurium* TA98, reverse mutation	–	–	2500	Spanggord et al. (1982b)
SA9, *Salmonella typhimurium* TA98, reverse mutation	–	–	128	Haworth et al. (1983)
SA9, *Salmonella typhimurium* TA98, reverse mutation	–	–	150	Miyata et al. (1981)
SA9, *Salmonella typhimurium* TA98, reverse mutation	–	–	50	Tokiwa et al. (1981)
SA9, *Salmonella typhimurium* TA98, reverse mutation	–	[d]	50	Suzuki et al. (1983)
SA9, *Salmonella typhimurium* TA98, reverse mutation	–	–	445	Shimizu & Yano (1986)
SAS, *Salmonella typhimurium* TA92, reverse mutation	–	–	500	Miyata et al. (1981)
SAS, *Salmonella typhimurium* TA94, reverse mutation	–	–	500	Miyata et al. (1981)

Table 3 (contd)

Test system	Result[a]		Dose[b] (LED/HID)	Reference
	Without exogenous metabolic system	With exogenous metabolic system		
3-Nitrotoluene (contd)				
URP, Unscheduled DNA synthesis, rat primary hepatocytes in vitro	–	0	13.7	Doolittle et al. (1983)
UIA, Unscheduled DNA synthesis, rat pachytene spermatocytes and round spermatids in vitro	–	0	13.7	Working & Butterworth (1984)
SIC, Sister chromatid exchange, Chinese hamster CHO cells in vitro	+	–	150	Galloway et al. (1987)
CIC, Chromosomal aberrations, Chinese hamster CHO cells in vitro	–	–	483	Galloway et al. (1987)
CIC, Chromosomal aberrations, Chinese hamster CHL cells in vitro	–	0	250	Ishidate et al. (1988)
UPR, Unscheduled DNA synthesis, male rat hepatocytes in vivo	–		500 po × 1	Doolittle et al. (1983)
UPR, Unscheduled DNA synthesis, male rat hepatocytes in vivo	–		500 po × 1	US National Toxicology Program (1983)
BVD, Binding (covalent) to DNA, male rat liver in vivo	–		200 po × 1	Rickert et al. (1984)
BVP, Binding (covalent) to RNA or protein, male rat liver in vivo	+		200 po × 1	Rickert et al. (1984, 1986)
4-Nitrotoluene				
BSD, Bacillus subtilis rec strains, differential toxicity	+	0	NR	Shimizu & Yano (1986)
SA0, Salmonella typhimurium TA100, reverse mutation	–	0	685	Chiu et al. (1978)
SA0, Salmonella typhimurium TA100, reverse mutation	–	–	500	Miyata et al. (1981)
SA0, Salmonella typhimurium TA100, reverse mutation	–	–	250	Tokiwa et al. (1981)
SA0, Salmonella typhimurium TA100, reverse mutation	+	+	NR	Spanggord et al. (1982b)
SA0, Salmonella typhimurium TA100, reverse mutation	–	–	128	Haworth et al. (1983)
SA0, Salmonella typhimurium TA100, reverse mutation	–	0	50	Suzuki et al. (1983)
SA0, Salmonella typhimurium TA100, reverse mutation	+	0	385	Shimizu & Yano (1986)
SA5, Salmonella typhimurium TA1535, reverse mutation	–	–	500	Miyata et al. (1981)
SA5, Salmonella typhimurium TA1535, reverse mutation	–	–	2500	Spanggord et al. (1982b)
SA5, Salmonella typhimurium TA1535, reverse mutation	–	–	128	Haworth et al. (1983)

Table 3 (contd)

Test system	Result[a] Without exogenous metabolic system	With exogenous metabolic system	Dose[b] (LED/HID)	Reference
4-Nitrotoluene (contd)				
SA5, *Salmonella typhimurium* TA1535, reverse mutation	–	0	1925	Shimizu & Yano (1986)
SA7, *Salmonella typhimurium* TA1537, reverse mutation	–	–	150	Miyata et al. (1981)
SA7, *Salmonella typhimurium* TA1537, reverse mutation	–	–	2500	Spanggord et al. (1982b)
SA7, *Salmonella typhimurium* TA1537, reverse mutation	–	–	128	Haworth et al. (1983)
SA7, *Salmonella typhimurium* TA1537, reverse mutation	–	0	1925	Shimizu & Yano (1986)
SA8, *Salmonella typhimurium* TA1538, reverse mutation	–	–	2500	Spanggord et al. (1982b)
SA8, *Salmonella typhimurium* TA1538, reverse mutation	–	0	1925	Shimizu & Yano (1986)
SA9, *Salmonella typhimurium* TA98, reverse mutation	–	0	685	Chiu et al. (1978)
SA9, *Salmonella typhimurium* TA98, reverse mutation	–	–	150	Miyata et al. (1981)
SA9, *Salmonella typhimurium* TA98, reverse mutation	–	–	500	Tokiwa et al. (1981)
SA9, *Salmonella typhimurium* TA98, reverse mutation	–	–	2500	Spanggord et al. (1982b)
SA9, *Salmonella typhimurium* TA98, reverse mutation	–	–	128	Haworth et al. (1983)
SA9, *Salmonella typhimurium* TA98, reverse mutation	–	–[d]	50	Suzuki et al. (1983)
SA9, *Salmonella typhimurium* TA98, reverse mutation	–	0	1925	Shimizu & Yano (1986)
SAS, *Salmonella typhimurium* TA92, reverse mutation	–	–	1500	Miyata et al. (1981)
SAS, *Salmonella typhimurium* TA92, reverse mutation	–	–	500	Miyata et al. (1981)
SCG, *Saccharomyces cerevisiae*, gene conversion	–	0	100	Marquardt et al. (1970)
G5T, Gene mutation, mouse L5178Y cells, *tk* locus *in vitro*	–	+	75	US National Toxicology Program (1992)
SIC, Sister chromatid exchange, Chinese hamster ovary CHO cells *in vitro*	+	+	200	Galloway et al. (1987)
CIC, Chromosomal aberrations, Chinese hamster ovary CHO cells *in vitro*	–	(+)	550	Galloway et al. (1987)
CIC, Chromosomal aberrations, Chinese hamster lung CHL cells *in vitro*	–	0	250	Ishidate et al. (1988)
URP, Unscheduled DNA synthesis, rat primary hepatocytes *in vitro*	–	0	13.7	Doolittle et al. (1983)
UIA, Unscheduled DNA synthesis, rat pachytene spermatocytes and round spermatids *in vitro*	–		13.7	Working & Butterworth (1984)

428   IARC MONOGRAPHS VOLUME 65

Table 3 (contd)

Test system	Result[a]		Dose[b] (LED/HID)	Reference
	Without exogenous metabolic system	With exogenous metabolic system		
4-Nitrotoluene (contd)				
UPR, Unscheduled DNA synthesis, male rat hepatocytes *in vivo*	–		500 po × 1	Doolittle *et al.* (1983)
UPR, Unscheduled DNA synthesis, rat hepatocytes *in vivo*	–		1000 po × 1	Mirsalis *et al.* (1989)
UPR, Unscheduled DNA synthesis, male rat hepatocytes *in vivo*	–		500 po × 1	US National Toxicology Program (1992)
MVM, Micronucleus test, mice *in vivo*	–		NR	Ohuchida *et al.* (1989)
BVD, Binding (covalent) to DNA, male rat liver *in vivo*	–		200 po × 1	Rickert *et al.* (1984)
BVP, Binding (covalent) to RNA or protein, male rat liver *in vivo*	+		200 po × 1	Rickert *et al.* (1984, 1986)

[a] +, positive; (+), weak positive; –, negative; 0, not tested; ?, inconclusive
[b] LED, lowest effective dose; HID, highest ineffective dose. In-vitro tests, μg/mL; in-vivo tests, mg/kg bw; NR, dose not reported
[c] Positive in the presence of 200 μg/plate norharman
[d] Negative also in the presence of 200 μg/plate norharman

In vivo in rats, 3-nitrotoluene bound covalently to hepatic macromolecules, but not to DNA. 3-Nitrotoluene was shown to be a weak inducer of sister chromatid exchange in Chinese hamster ovary cells *in vitro* but did not induce chromosomal aberrations. 3-Nitrotoluene did not induce unscheduled DNA synthesis in Fischer 344 rat hepatocytes, either after in-vitro treatment or after in-vivo treatment. It did not induce unscheduled DNA synthesis in rat spermatogonia exposed *in vitro*.

In the yeast *Saccharomyces cerevisiae*, 4-nitrotoluene did not induce mitotic gene conversion.

In vivo in rats, 4-nitrotoluene bound covalently to hepatic macromolecules, but not to DNA. 4-Nitrotoluene induced sister chromatid exchange and chromosomal aberrations in Chinese hamster ovary cells *in vitro* in one study. 4-Nitrotoluene did not induce unscheduled DNA synthesis in Fischer 344 rat primary hepatocyte cultures after in-vitro or in-vivo treatment. It did not induce unscheduled DNA synthesis in rat spermatogonia exposed *in vitro*.

4-Nitrotoluene did not induce micronuclei in polychromatic erythrocytes after intraperitoneal injection into male BDF1 mice.

5. Summary of Data Reported and Evaluation

5.1 Exposure data

2-, 3- and 4-Nitrotoluenes are produced commercially, as a mixture, by nitration of toluene. 2- and 4-Nitrotoluenes are used mainly to produce intermediates in the production of colourants. All of these isomers are also used in much smaller quantities in the production of agricultural, pharmaceutical and rubber chemicals. Human exposure to nitrotoluenes can occur during their production and use, although few data are available. Nitrotoluenes have been detected in effluents from the manufacture or use nitrotoluenes and in surface and groundwater.

5.2 Human carcinogenicity data

No data were available to the Working Group.

5.3 Animal carcinogenicity data

No long-term study of the carcinogenicity of 2-, 3- or 4-nitrotoluene was available to the Working Group.

Rare mesotheliomas of the tunica vaginalis were reported in male rats receiving 2-nitrotoluene in the diet for 13 weeks.

5.4 Other relevant data

No relevant data on absorption, distribution, metabolism or excretion in humans were available to the Working Group.

Urinary elimination is the major route of excretion in rats exposed to nitrotoluene isomers. Male rats excrete more of an administered dose of nitrotoluene in the bile compared with female rats. All three nitrotoluene isomers cause an increase in the incidence of hyaline droplet nephropathy in male rats: the hyaline droplets were associated with $\alpha_{2\mu}$-globulin. Liver toxicity was observed in rats exposed to 2-nitrotoluene. In mice, the only evidence of toxicity was degeneration and metaplasia of the olfactory epithelium.

In rats, no adverse effect on reproduction or on the offspring was observed following administration of 2-, 3- or 4-nitrotoluene by gavage. All three isomers impaired testicular function and increased the length of the oestrus cycle. The 2-isomer decreased sperm motility in mice.

2-Nitrotoluene was not genotoxic in bacteria, but induced sister chromatid exchange in cultured mammalian cells. *In vivo* in rats, 2-nitrotoluene bound to macromolecules and, in males, induced unscheduled DNA synthesis in liver cells. In-vivo activity depends on the presence of intestinal bacteria.

3-Nitrotoluene produced a weak induction of sister chromatid exchange, but not of chromosomal aberrations or unscheduled DNA synthesis in mammalian cells *in vitro*. *In vivo*, it bound to macromolecules but not to DNA and did not induce unscheduled DNA synthesis.

4-Nitrotoluene was not genotoxic in yeast. In mammalian cells *in vitro*, it induced sister chromatid exchange and chromosomal aberrations. It did not induce unscheduled DNA synthesis in rat cells exposed either *in vitro* or *in vivo*. *In vivo* in rats, it bound to macromolecules, but not to DNA. It did not induce micronuclei in mouse bone marrow *in vivo*.

5.5 Evaluation[1]

There is *inadequate evidence* in humans for the carcinogenicity of nitrotoluenes.

There is *limited evidence* in experimental animals for the carcinogenicity of 2-nitrotoluene.

There is *inadequate evidence* in experimental animals for the carcinogenicity of 3- and 4-nitrotoluenes.

Overall evaluation

Nitrotoluenes are *not classifiable as to their carcinogenicity to humans (Group 3)*.

---

[1] For definition of the italicized terms, see Preamble, pp. 24–27.

6. References

Ahlborg, G., Bergström, B., Hogstedt, C., Einistö, P. & Sorsa, M. (1985) Urinary screening for potentially genotoxic exposures in a chemical industry. *Br. J. ind. Med.*, **42**, 691–699

Ahlborg, G., Jr, Ulander, A., Bergström, B. & Oliv, Å. (1988) Diazo-positive metabolites in urine from workers exposed to aromatic nitro-amino compounds. *Int. Arch. occup. environ. Health*, **60**, 51–54

American Conference of Governmental Industrial Hygienists (ACGIH) (1991) *Documentation of the Threshold Limit Values and Biological Exposure Indices*, 6th Ed., Cincinnati, OH, pp. 1131–1133

American Conference of Governmental Industrial Hygienists (ACGIH) (1995) *1995–1996 Threshold Limit Values for Chemical Substances and Physical Agents and Biological Exposure Indices*, Cincinnati, OH, p. 28

Arbeidsinspectie [Labour Inspection] (1994) *De Nationale MAC-Lijst 1994* [National MAC list 1994], The Hague, p. 35

Booth, G. (1991) Nitro compounds, aromatic. In: Elvers, B., Hawkins, S. & Schulz, G., eds, *Ullmann's Encyclopedia of Industrial Chemistry*, 5th rev. Ed., Vol. A17, New York, VCH Publishers, pp. 411–455

Budavari, S., ed. (1989) *The Merck Index*, 11th Ed., Rahway, NJ, Merck & Co., p. 1051

Butterworth, B.E., Doolittle, D.J., Working, P.K., Strom, S.C., Jirtle, R.L. & Michalopoulos, G. (1982) Chemically-induced DNA repair in rodent and human cells. In: Bridges, B.A., Butterworth, B.E. & Weinstein, I.B., eds, *Indicators of Genotoxic Exposure* (Banbury Report 13), Cold Spring Harbor, NY, CSH Press, pp. 101–114

Butterworth, B.E., Smith-Oliver, T., Earle, L., Loury, D.J., White, R.D., Doolittle, D.J., Working, P.K., Cattley, R.C., Jirtle, R., Michalopoulos, G. & Strom, S. (1989) Use of primary cultures of human hepatocytes in toxicology studies. *Cancer Res.*, **49**, 1075–1084

Chemical Information Services (1994) *Directory of World Chemical Producers 1995/96*, Oceanside, NY, p. 530

Chism, J.P. & Rickert, D.E. (1985) Isomer- and sex-specific bioactivation of mononitrotoluenes: role of enterohepatic circulation. *Drug Metab. Dispos.*, **13**, 651–657

Chism, J.P. & Rickert, D.E. (1989) In vitro activation of 2-aminobenzyl alcohol and 2-amino-6-nitrobenzyl alcohol, metabolites of 2-nitrotoluene and 2,6-dinitrotoluene. *Chem. Res. Toxicol.*, **2**, 150–156

Chism, J.P., Turner, M.J., Jr & Rickert, D.E. (1984) The metabolism and excretion of mononitrotoluenes by Fischer 344 rats. *Drug Metab. Dispos.*, **12**, 596–602

Chiu, C.W., Lee, L.H., Wang, C.Y. & Bryan, G.T. (1978) Mutagenicity of some commercially available nitro compounds for *Salmonella typhimurium*. *Mutat. Res.*, **58**, 11–22

Ciss, M., Dutertre, H., Huyen, N., Phu-Lich, N. & Truhaut, R. (1980a) Toxicological study of nitrotoluenes: acute and subacute toxicity. *Dakar méd.*, **25**, 303–311 (in French)

Ciss, M., Huyen, N., Dutertre, H., Phu-Lich, N. & Truhaut, R. (1980b) Toxicological study of nitrotoluenes: long-term toxicity. *Dakar méd.*, **25**, 293–302 (in French)

Deutsche Forschungsgemeinschaft (1995) *MAK and BAT Values 1995* (Report No. 31), Weinheim, VCH Verlagsgesellschaft, p. 71

Doolittle, D.J., Sherrill, J.M. & Butterworth, B.E. (1983) Influence of intestinal bacteria, sex of the animal, and position of the nitro group on the hepatic genotoxicity of nitrotoluene isomers *in vivo*. *Cancer Res.*, **43**, 2836–2842

Duguet, J.P., Anselme, C., Mazounie, P. & Mallevialle, J. (1988) Application of the ozone–hydrogen peroxide combination for the removal of toxic compounds from a groundwater. In: Angeletti, G. & Bjorseth, A., eds, *Organic Micropollutants in the Aquatic Environment, Proceedings of the Fifth European Symposium*, Dordrecht, Kluwer Academic Publishers, pp. 299–309

Dunnick, J.K., Elwell, M.R. & Bucher, J.R. (1994) Comparative toxicities of *o*-, *m*-, and *p*-nitrotoluene in 13-week feed studies in F344 rats and B6C3F1 mice. *Fundam. appl. Toxicol.*, **22**, 411–421

DuPont Chemicals (1994) *Material Safety Data Sheet: o-Nitrochlorobenzene*, Wilmington, DE

Eller, P.M., ed. (1994) Nitrobenzenes — Method 2005. In: *NIOSH Manual of Analytical Methods*, 4th Ed., Vol 2 (E-N) (DHHS (NIOSH) Publ. No. 94-113), Washington DC, United States Government Printing Office

Feltes, J., Levsen, K., Volmer, D. & Spiekermann, M. (1990) Gas chromatographic and mass spectrometric determination of nitroaromatics in water. *J. Chromatogr.*, **518**, 21–40

First Chemical Corp. (1995a) *Product Data Sheet: ortho-Nitrotoluene*, Pascagoula, MS

First Chemical Corp. (1995b) *Product Data Sheet: meta-Nitrotoluene*, Pascagoula, MS

First Chemical Corp. (1995c) *Product Data Sheet: para-Nitrotoluene*, Pascagoula, MS

Galloway, S.M., Armstrong, M.J., Reuben, C., Colman, S., Brown, B., Cannon, C., Bloom, A.D., Nakamura, F., Ahmed. M., Duk, S., Rimpo, J., Margolin, B.H., Resnick, M.A., Anderson, B. & Zeiger, E. (1987) Chromosome aberrations and sister chromatid exchanges in Chinese hamster ovary cells: evaluation of 108 chemicals. *Environ. mol. Mutag.*, **10** (Suppl. 10), 1–175

Hansch, C., Leo, A. & Hoekman, D.H. (1995) *Exploring QSAR*, Washington DC, American Chemical Society, p. 30

Haseman, J.K., Arnold, J. & Eustis, S.L. (1990) Tumor incidence in Fischer 344 rats. In: Boorman, G.A., Eustis, S.L., Elwell, M.R., Montgomery, C.A., Jr & MacKenzie, W.F., eds, *Pathology of the Fischer Rat*, San Diego, CA, Academic Press, pp. 557–564

Haworth, S., Lawlor, T., Mortelmans, K., Speck, W. & Zeiger, E. (1983) *Salmonella* mutagenicity test results for 250 chemicals. *Environ. Mutag.*, **Suppl. 1**, 3–142

Health and Safety Executive (1995) *Occupational Exposure Limits 1995* (Guidance Note EH 40/95), Sudbury, Suffolk, HSE Books, p. 35

Hoechst Chemicals (1989) *Product Information: o-Nitrotoluene*, Charlotte, NC

Howard, P.H. (1989) *Handbook of Environmental Fate and Exposure Data for Organic Chemicals*, Vol. 1, Chelsea, MI, Lewis Publishers, pp. 454–467

IARC (1982) *IARC Monographs on the Evaluation of the Carcinogenic Risk of Chemicals to Humans*, Vol. 27, *Some Aromatic Amines, Anthraquinones and Nitroso Compounds, and Inorganic Fluorides Used in Drinking Water and Dental Preparations*, Lyon, pp. 155–175

IARC (1987) *IARC Monographs on the Evaluation of Carcinogenic Risks to Humans*, Suppl. 7, *Overall Evaluations of Carcinogenicity: An Updating of* IARC Monographs *Volumes 1 to 42*, Lyon, pp. 362–363

IARC (1994) *Directory of Agents Being Tested for Carcinogenicity*, Vol. 16, Lyon, p. 162

International Labour Office (1991) *Occupational Exposure Limits for Airborne Toxic Substances: Values of Selected Countries* (Occupational Safety and Health Series No. 37), 3rd Ed., Geneva, pp. 300–301

Ishidate, M., Jr, Harnois, M.C. & Sofuni, T. (1988) A comparative analysis of data on the clastogenicity of 951 chemical substances tested in mammalian cell cultures. *Mutat. Res.*, **195**, 151–213

Koniecki, W.B. & Linch, A.L. (1958) Determination of aromatic nitro compounds. *Anal. Chem.*, **30**, 1134–1137

Kool, H.J., van Kreijl, C.F. & Zoeteman, B.C.J. (1982) Toxicology assessment of organic compounds in drinking water. *Criteria Rev. environ. Control*, **12**, 307–357

Lewalter, J. & Ellrich, D. (1991) Nitroaromatic compounds (nitrobenzene, p-nitrotoluene, p-nitrochlorobenzene, 2,6-dinitrotoluene, o-dinitrobenzene, 1-nitronaphthalene, 2-nitronaphthalene, 4-nitrobiphenyl). In: Angerer, J. & Schaller, K.H., eds, *Analyses of Hazardous Substances in Biological Materials*, Vol. 3, New York, VCH Publishers, pp. 207–229

Lewis, R.J., Sr (1993) *Hawley's Condensed Chemical Dictionary*, 12th Ed., New York, Van Nostrand Reinhold Co., pp. 832–833

Lide, D.R., ed. (1993) *CRC Handbook of Chemistry and Physics*, 74th Ed., Boca Raton, FL, CRC Press, p. 3–490

Marquardt, H., Zimmermann, F.K., Dannenberg, H., Neumann, H.-G., Bodenberger, A. & Metzler, M. (1970) Genetic activity of aromatic amines and their derivatives: induction of mitotic conversion in the yeast *Saccharomyces cerevisiae*. *Z. Krebsforsch.*, **74**, 412–433 (in German)

Meijers, A.P. & van der Leer, C.R. (1976) The occurrence of organic micropollutants in the River Rhine and the River Maas in 1974. *Water Res.*, **10**, 597–604

Mirsalis, J.C., Tyson, C.K., Steinmetz, K.L., Loh, E.K.., Hamilton, C.M., Bakke, J.P. & Spalding, J.W. (1989) Measurement of unscheduled DNA synthesis and S-phase synthesis in rodent hepatocytes following *in vivo* treatment: testing of 24 compounds. *Environ. mol. Mutag.*, **14**, 155–164

Miyata, R., Nohmi, T., Yoshikawa, K. & Ishidate, M., Jr (1981) Metabolic activation of p-nitrotoluene and trichloroethylene by rat-liver S9 or mouse-liver S9 fractions in *Salmonella typhimurium* strains. *Eisei Shikendo Hokoku*, **99**, 60–65

Ohuchida, A., Furukawa, A. & Yoshida, R. (1989) Micronucleus test of polyploidy inducers (Abstract no. 36). *Mutat. Res.*, **216**, 371–372

Ong, C.P., Chin, K.P., Lee, H.K. & Li, S.F.Y. (1992) Analysis of nitroaromatics in aqueous samples by capillary supercritical fluid chromatography. *Int. J. environ. Stud.*, **41**, 17–25

Pellizzari, E.D. (1978) *Quanitification of Chlorinated Hydrocarbons in Previously Collected Air Samples* (US EPA Report No. EPA-450/3-78-112), Washington DC, United States Environmental Protection Agency

Piet, G.J. & Morra, C.F. (1983) *Artificial Groundwater Recharge* (Water Resources Eng. Series), Pitman Publ., pp. 31–42

Registry of Toxic Effects of Chemical Substances (1991) *LD50 Ratings for o-, m-, and p-Nitrotoluene*, Research Triangle Park, NC, United States National Institute of Environmental Health Sciences

Rickert, D.E., Long, R.M., Dyroff, M.C. & Kedderis, G.L. (1984) Hepatic macromolecular covalent binding of mononitrotoluenes in Fischer-344 rats. *Chem.-biol. Interactions*, **52**, 131–139

Rickert, D.E., Chism, J.P. & Kedderis, G.L. (1986) Metabolism and carcinogenicity of nitrotoluenes. *Adv. exp. Med. Biol.*, **197**, 536–571

Sadtler Research Laboratories (1980) *Sadtler Standard Spectra. 1980 Cumulative Index*, Philadelphia, PA

Sax, N.I. & Lewis, R.J., Jr (1989) *Dangerous Properties of Industrial Materials*, 7th Ed., New York, Van Nostrand Reinhold Co., pp. 2344–2345

Shimizu, M. & Yano, E. (1986) Mutagenicity of mono-nitrobenzene derivatives in the Ames test and rec assay. *Mutat. Res.*, **170**, 11–22

Smith, E.R. & Quinn, M.M. (1992) Uterotrophic action in rats of amsonic acid and three of its synthetic precursors. *J. Toxicol. environ. Health*, **36**, 13–25

Spanggord, R.J., Gibson, B.W., Keck, R.G., Thomas, D.W. & Barkley, J.J., Jr (1982a) Effluent analysis of wastewater generated in the manufacture of 2,4,6-trinitrotoluene. 1. Characterization study. *Environ. Sci. Technol.*, **16**, 229–232

Spanggord, R.J., Mortelmans, K.E., Griffin, A.F. & Simmon, V.F. (1982b) Mutagenicity in *Salmonella typhimurium* and structure–activity relationships of wastewater components emanating from the manufacture of trinitrotoluene. *Environ. Mutag.*, **4**, 163–179

Suzuki, J., Koyama, T. & Suzuki, S. (1983) Mutagenicities of mono-nitrobenzene derivatives in the presence of norharman. *Mutat. Res.*, **120**, 105–110

Swaminathan, K., Kondawar, V.K., Chakrabarti, T. & Subrahmanyam, P.V.R. (1987) Identification and quantification of organics in nitro aromatic manufacturing wastewaters. *Indian J. environ. Health*, **29**, 32–38

Tokiwa, H., Nakagawa, R. & Ohnishi, Y. (1981) Mutagenic assay of aromatic nitro compounds with *Salmonella typhimurium*. *Mutat. Res.*, **91**, 321–325

Ton, T.T., Elwell, M.R., Morris, R.W. & Maronpot, R.R. (1995) Development and persistence of placental glutathione-*S*-transferase-positive foci in livers of male F344 rats exposed to *o*-nitrotoluene. *Cancer Lett.*, **95**, 167–173

Työministeriö [Ministry of Labour] (1993) *Limit Values 1993*, Helsinki, p. 16 (in Finnish)

United States National Institute for Occupational Safety and Health (1994a) *Pocket Guide to Chemical Hazards* (DHHS (NIOSH) Publ. No. 94-116), Cincinnati, OH, pp. 232–233

United States National Institute for Occupational Safety and Health (1994b) *RTECs Chem.-Bank*, Cincinnati, OH

United States National Library of Medicine (1995) *Hazardous Substances Data Bank (HSDB)*, Bethesda, MD

United States National Toxicology Program (1992) *NTP Technical Report on Toxicity Studies of o-, m-, and p-Nitrotoluenes Administered in Dosed Feed to F344/N Rats and B6C3F$_1$ Mice* (Toxicity Report Series No. 23), Research Triangle Park, NC

United States Occupational Safety and Health Administration (1994) Air contaminants.*US Code fed. Regul.*, **Title 29**, Part 1910.1000, p. 14

Verschueren, K. (1983) *Handbook of Environmental Data on Organic Chemicals*, 2nd Ed., New York, Van Nostrand Reinhold Co., pp. 865–866

Working, P.K. & Butterworth, B.E. (1984) An assay to detect chemically induced DNA repair in rat spermatocytes. *Environ. Mutag.*, **6**, 273–286

TETRANITROMETHANE

1. Exposure Data

1.1 Chemical and physical data

1.1.1 *Nomenclature*

Chem. Abstr. Serv. Reg. No.: 509-14-8
Chem. Abstr. Name: Tetranitromethane
IUPAC Systematic Name: Tetranitromethane
Synonyms: Tetan; TNM

1.1.2 *Structural and molecular formulae and relative molecular mass*

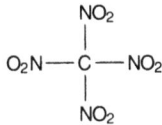

CN_4O_8 Relative molecular mass: 196.04

1.1.3 *Chemical and physical properties of the pure substance*

(a) *Description*: Pale yellow liquid with a pungent odour (Budavari, 1989; Lewis, 1993)

(b) *Boiling-point*: 126 °C (Lide, 1993)

(c) *Melting-point*: 14.2 °C (Lide, 1993)

(d) *Density*: 1.6380 at 20 °C/4 °C (Lide, 1993)

(e) *Spectroscopy data*: Infrared (prism [2492], grating [29557]) and mass spectral data have been reported (Sadtler Research Laboratories, 1980)

(f) *Solubility*: Insoluble in water; soluble in alcoholic potassium hydroxide, diethyl ether and ethanol (Budavari, 1989; Davis, 1993; Lide, 1993)

(g) *Volatility*: Vapour pressure, 8.4 mm Hg [1.1 kPa] at 20 °C; relative vapour density (air = 1), 0.8 (Davis, 1993)

(h) *Stability*: Highly explosive when exposed to heat or shock; mixtures with amines (e.g. aniline) ignite spontaneously and may explode. Mixtures with cotton or toluene may explode when ignited. Forms sensitive and powerful explosive mixtures with nitrobenzene, 1-nitrotoluene, 4-nitrotoluene, 1,3-dinitrobenzene,

1-nitronaphthalene, other oxygen-deficient explosives and hydrocarbons (Sax & Lewis, 1989)

(i) *Octanol/water partition coefficient (P)*: log P, –0.791 (United States National Library of Medicine, 1995)

(j) *Conversion factor*: mg/m^3 = 8.02 × ppm[1]

1.1.4 *Technical products and impurities*

Tetranitromethane is available commercially at an unknown purity (Aldrich Chemical Co., 1994).

1.1.5 *Analysis*

Early methods (1940–59) of determining nitroalkanes used colorimetric procedures. Since 1970, these have been replaced by instrumental methods, primarily gas chromatography, mass spectrometry and infrared spectroscopy (Davis, 1993).

Tetranitromethane can be determined in air using gas chromatography with alkali-flame ionization detection (Taylor, 1977).

1.2 Production and use

1.2.1 *Production*

Tetranitromethane can be prepared by nitration of acetic anhydride with anhydrous nitric acid (Budavari, 1989) or by reacting fuming nitric acid with benzene or acetylene (Lewis, 1993). It has also been prepared by nitration of acetylene with excess nitric acid to form a mixture of trinitromethane and nitric acid, which can be converted to tetranitromethane by sulfuric acid at elevated temperatures (Stockinger, 1982).

Tetranitromethane is produced by one company in the United States of America and one company in Russia (Chemical Information Services, 1994).

1.2.2 *Use*

Tetranitromethane has been used as an oxidizer in rocket propellants, as an explosive in admixture with toluene (see IARC, 1989a) and to increase the cetane number of diesel fuels (see IARC, 1989b), it has also been used as a reagent for detecting the presence of double bonds in organic compounds and for nitration of tyrosine in proteins and peptides (Budavari, 1989; American Conference of Governmental Industrial Hygienists, 1991; Lewis, 1993).

---

[1]Calculated from: mg/m^3 = (relative molecular mass/24.45) × ppm, assuming temperature (25 °C) and pressure (101 kPa)

1.3 Occurrence

1.3.1 *Natural occurrence*

Tetranitromethane is not known to occur as a natural product.

1.3.2 *Occupational exposure*

The National Occupational Exposure Survey conducted between 1981 and 1983 indicated that 4350 employees in the United States were potentially exposed to tetranitromethane. The estimate is based on a survey of companies and did not involve measurements of actual exposure (United States National Institute for Occupational Safety and Health, 1995).

1.3.3 *Environmental occurrence*

No information was available to the Working Group.

1.4 Regulations and guidelines

Occupational exposure limits and guidelines for tetranitromethane in several countries are presented in Table 1.

Table 1. Occupational exposure limits and guidelines for tetranitromethane

Country	Year	Concentration (mg/m^3)	Interpretation
Argentina	1991	8	TWA
Australia	1993	8	TWA
Belgium	1993	8	TWA
Bulgaria[a]	1995	0.04	TWA
Canada	1991	8	TWA
Colombia[a]	1995	0.04	TWA
Denmark	1993	8	TWA
Finland	1993	8	TWA
		24	STEL
France	1993	8	TWA
Germany	1995	None, IIIA2[b]	MAK
Jordan[a]	1995	0.04	TWA
Mexico	1991	8	TWA
Netherlands	1994	8	TWA
New Zealand[a]	1995	0.04	TWA
Philippines	1993	8	TWA
Republic of Korea[a]	1995	0.04	TWA
Russia	1993	0.3	STEL
Singapore[a]	1995	0.04	TWA
Switzerland	1993	8	TWA
Turkey	1993	8	TWA

Table 1 (contd)

Country	Year	Concentration (mg/m³)	Interpretation
USA			
ACGIH (TLV)	1995	0.04, A2[b]	TWA
OSHA (PEL)	1994	8	TWA
NIOSH (REL)	1994	8	TWA
Viet Nam[a]	1995	0.04	TWA

From Työministeriö (1993); Arbeitsinspectie (1994); US National Institute for Occupational Safety and Health (NIOSH) (1994a,b); US Occupational Safety and Health Administration (OSHA) (1994); American Conference of Governmental Industrial Hygienists (ACGIH) (1995); Deutsche Forschungsgemeinschaft (1995); United Nations Environment Programme (1995)

TWA, time-weighted average; STEL, short-term exposure limit; MAK, maximum workplace concentration; TLV, threshold limit value; PEL, permissible exposure limit; REL, recommended exposure limit

[a] Follows ACGIH values
[b] IIIA2, substances shown to be clearly carcinogenic only in animal studies but under conditions indicative of carcinogenic potential at the workplace; A2, suspected human carcinogen

2. Studies of Cancer in Humans

No data were available to the Working Group.

3. Studies of Cancer in Experimental Animals

3.1 Inhalation exposure

3.1.1 *Mouse*

Groups of 50 male and 50 female B6C3F1 mice, eight to 10 weeks of age, were exposed to air containing target concentrations of 0, 0.5 or 2 ppm [0, 4 or 16 mg/m³] tetranitromethane (approximately 100% pure) for 6 h per day on five days per week for 103 weeks. Additional groups of six male mice were exposed to 0 or 2 ppm and 10 male mice were exposed to 0.5 ppm for 52 weeks. All animals received complete histopathological evaluation. Monitoring indicated that the mean daily chamber concentrations were within ± 10% of the target concentrations on more than 96% of the study days. In the 52-week study, multiple alveolar–bronchiolar adenomas were found in the lung of one mouse in the 2-ppm group. Hepatocellular adenomas were found in the livers of four mice in the 0.5-ppm group. In the 103-week study, body weights of exposed mice were generally 5–10% lower than those of controls. Survival at 103 weeks was decreased in exposed males (37/50 in controls, 26/50 at the low dose, 15/50 at the high dose) but was unchanged in females (31/50 in controls, 28/50 at the low dose, 24/50 at the high dose). The incidences of alveolar–bronchiolar neoplasms were increased in exposed males and females; those of both adenomas and carcinomas were markedly increased over controls (alveolar–bronchiolar adenomas and carcinomas: males — 12/50 in controls, 27/50 at the low dose, 47/50 at the high dose; females — 4/49 in controls, 24/50 at the low dose, 49/50 at the high dose; $p < 0.001$, logistic regression trend analysis). The incidences of hyperplasia of the alveolar epithelium and of the bronchioles were significantly increased in all treated males and females. No increase in the incidence of hepatocellular adenomas was seen in the 103-week study (United States National Toxicology Program, 1990; Bucher *et al.*, 1991).

3.1.2 *Rat*

Groups of 50 male and 50 female Fischer 344/N rats, six to seven weeks of age, were exposed to air containing target concentrations of 0, 2 or 5 ppm [0, 16 or 40 mg/m³] tetranitromethane (approximately 100% pure) for 6 h per day on five days per week for 103 weeks. Monitoring indicated that the mean daily chamber concentrations were within ± 10% of the target concentrations on more than 99% of the study days. All animals received complete histopathological examination. Body weights of exposed rats were generally 5–10% lower than those of controls. Survival at 103 weeks was decreased in high-dose males (18/50 in controls, 17/50 at the low dose, 4/50 at the high dose) but was unchanged in females (25/50 in controls, 34/50 at the low dose, 15/50 at the high dose). The incidences of alveolar–bronchiolar neoplasms were increased in treated males and females; those of both adenomas and carcinomas were markedly increased over controls (alveolar–bronchiolar adenomas and carcinomas: males — 1/50 in controls, 33/50 at the low dose, 46/50 at the high dose; females — 0/50 in controls, 22/50 at the low dose, 50/50 at the high dose; $p < 0.001$, logistic regression analysis). The incidence of squamous-cell carcinomas of the lung was also significantly increased in high-dose males and females (males — 0/50 in controls, 1/50 at the low dose, 19/50 at the high dose; females — 0/50 in controls, 1/50 at the low dose, 12/50 at the high dose; $p < 0.001$, life-table analysis). The incidences of hyperplasia of the alveolar epithelium and of the bronchioles were significantly increased in all treated males and females (United States National Toxicology Program, 1990; Bucher *et al.*, 1991).

4. Other Data Relevant for an Evaluation of Carcinogenicity and its Mechanisms

4.1 Absorption, distribution, metabolism and excretion

4.1.1 *Humans*

No data were available to the Working Group.

4.1.2 *Experimental systems*

No studies have been published on the pharmacokinetics and metabolism of tetranitromethane in laboratory animals.

4.2 Toxic effects

4.2.1 *Humans*

No data were available to the Working Group.

4.2.2 *Experimental systems*

(a) *Single-dose studies*

In experiments of inhalation exposure of rats [strain unspecified] at different concentrations of tetranitromethane, the times to 50% lethality were reported to be 36 min at 1230 ppm [9840 mg/m^3], 60 min at 300 ppm [2400 mg/m^3] and 5.8 h at 33 ppm [264 mg/m^3] (Horn, 1954).

(b) *Repeated-dose studies*

Male and female Fischer 344/N rats were exposed by inhalation to 2, 5, 10 or 25 ppm [16, 40, 80 or 200 mg/m^3] tetranitromethane for 6 h per day on five days per week for 14 days (United States National Toxicology Program, 1990). All exposed rats were lethargic. All rats exposed to 25 ppm tetranitromethane died on the first day and two had pulmonary oedema. At 10 ppm, body-weight depression was observed in males and females.

Male and female B6C3F1 mice were exposed by inhalation to 2, 5, 10, 25 or 50 ppm [16, 40, 80, 200 or 400 mg/m^3] tetranitromethane for 6 h per day on five days per week for 14 days (United States National Toxicology Program, 1990). All mice exposed to 50 ppm died on day two. Three of five males and all females exposed to 25 ppm died before the end of the study. Body-weight depression was observed in males exposed to 5 ppm or above and in females exposed to 10 ppm. Mice that survived exposures of 10 or 25 ppm were observed to have inflamed, reddened lungs.

Male and female B6C3F1 mice and Fischer 344/N rats were exposed by inhalation to 0.2, 0.7, 2, 5 and 10 ppm [0.8, 5.6, 16, 40 and 80 mg/m^3] tetranitromethane for 13 weeks. Mean body-weight depression, lethargy and serous exudate in the nasal passage were observed in male and female rats and mice exposed to 10 ppm. In addition, chronic lung inflammation was present in rats. No compound-related death occurred. Focal squamous metaplasia of the respiratory epithelium of the nasal mucosa was observed in 40% of the female rats exposed to 10 ppm, but not in those exposed to 5 ppm. Focal squamous metaplasia (mild) of the respiratory epithelium of the nasal mucosa was also observed in mice. Dyspnoea (at 10 ppm) and inflammation of the nasal mucosa (at 5 and 10 ppm) were observed in mice. Increases in relative liver weight were seen at all dose levels in male and female rats and male mice (United States National Toxicology Program, 1990).

Male and female F344/N rats and B6C3F1 mice were exposed by inhalation to 2 or 5 ppm and 0.5 or 2 ppm tetranitromethane, respectively, for two years (United States

National Toxicology Program, 1990). The incidence of inflammation of the nasal mucosa in rats was increased, relative to controls, at 5 ppm. In mice, significant increases in nasal lumen exudate were associated with exposure of males and females to 2 ppm and 0.5 ppm, respectively.

(c) In-vitro systems

The ability of tetranitromethane to nitrate tyrosine residues in proteins selectively has been used to investigate enzyme and receptor–active sites. Incubation of horse erythrocyte glutathione transferase resulted in a complete loss of enzyme activity (Del Boccio *et al.*, 1991). Approximately 25 µM cytochrome P450 LM4 (CYP1A2; prepared from liver microsomes of phenobarbital-pretreated rabbits) was incubated with 50 µM tetranitromethane at 25 °C, pH 7.5 (Jänig *et al.*, 1987). The modified enzyme possessed reduced 4-nitrophenetole *O*-deethylase activity (20% of control), increased binding affinity for 4-nitrophenetole, decreased binding affinity for α-naphthoflavone and metyrapone and a reduced rate of electron transfer from NADPH-cytochrome P450 reductase.

Brain and heart muscarinic receptors have been modified by incubation of tetranitromethane (50 µM) with homogenates (0.5 mg protein/mL) from male CD rats for 20 min at 25 °C and pH 8.1. An increased affinity for the agonist acetylcholine was observed in membranes from the striatum, hippocampus, cerebral cortex and heart atrium, but not the brain stem (Gurwitz & Sokolovsky, 1985a,b). Similar results were obtained by incubating cortical slices with 300 µM tetranitromethane (Gurwitz & Sokolovsky, 1985b). Incubation of bovine adrenal medullary microsomes with tetranitromethane (10–100 µM for 20 min at 25 °C and pH 8) caused a concentration-dependent, irreversible decrease in the number of muscarinic binding sites and binding affinity (Yamanaka *et al.*, 1988).

4.3 Reproductive and developmental effects

No data were available to the Working Group.

4.4 Genetic and related effects

4.4.1 *Humans*

No data were available to the Working Group.

4.4.2 *Experimental systems* (see also Table 2 and Appendices 1 and 2)

(a) Mutation and allied effects

Tetranitromethane with and without metabolic activation by S9-mix was mutagenic in *Salmonella typhimurium* TA98, TA100 and TA1535 but negative in TA1537. Tetranitromethane did not induce deletion mutation; it induced chromosomal aberrations and sister chromatid exchange in Chinese hamster ovary cells.

Table 2. Genetic and related effects of tetranitromethane

Test system	Result[a]		Dose[b] (LED/HID)	Reference
	Without exogenous metabolic system	With exogenous metabolic system		
SA0, *Salmonella typhimurium* TA100, reverse mutation	+	(+)	1.5	Kawai et al. (1987)
SA0, *Salmonella typhimurium* TA100, reverse mutation	(+)	+	1.25	Zeiger et al. (1987)
SA0, *Salmonella typhimurium* TA100, reverse mutation	+	+	4.0	Würgler et al. (1990)
SA2, *Salmonella typhimurium* TA102, reverse mutation	–	+	33	Würgler et al. (1990)
SA5, *Salmonella typhimurium* TA1535, reverse mutation	+	+	0.75	Zeiger et al. (1987)
SA7, *Salmonella typhimurium* TA1537, reverse mutation	–	–	38	Zeiger et al. (1987)
SA9, *Salmonella typhimurium* TA98, reverse mutation	–	–	500	Kawai et al. (1987)
SA9, *Salmonella typhimurium* TA98, reverse mutation	+	(+)	4.0	Würgler et al. (1990)
SA9, *Salmonella typhimurium* TA98, reverse mutation	(+)	(+)	1.0	Zeiger et al. (1992)
SAS, *Salmonella typhimurium* TA97, reverse mutation	+	(+)	8.0	Würgler et al. (1990)
SAS, *Salmonella typhimurium* LT2, deletion mutation	–	0	4000	Alper & Ames (1975)
ECW, *Escherichia coli* WP2 *uvrA*, reverse mutation	+	+	1.0	Kawai et al. (1987)
SIC, Sister chromatid exchange, Chinese hamster ovary CHO cells *in vitro*	+	–	1.7	US National Toxicology Program (1990)
CIC, Chromosomal aberrations, Chinese hamster ovary CHO cells *in vitro*	–	(+)	20	US National Toxicology Program (1990)

[a] +, positive; (+), weak positive; –, negative; 0, not tested; ?, inconclusive
[b] LED, lowest effective dose; HID, highest ineffective dose. In-vitro tests, μg/mL; in-vivo tests, mg/kg bw

(b) *Mutational spectra*

Dominant transforming genes were studied in lung tumours from Fischer 344 rats and C57Bl/6 × C3H F_1 mice exposed chronically by inhalation to tetranitromethane. All of the 10 rat tumours tested and all of the nine mouse tumours tested had a GC → AT transition in the second base of codon 12 of the K-*ras* oncogene (Stowers *et al.*, 1987; You *et al.*, 1991).

5. Summary of Data Reported and Evaluation

5.1 Exposure data

Tetranitromethane is produced commercially by nitration of acetic anhydride or acetylene. Tetranitromethane has been used as an oxidizer in rocket propellants and explosives. Few data are available on human exposure to tetranitromethane.

5.2 Human carcinogenicity data

No data were available to the Working Group.

5.3 Animal carcinogenicity data

Tetranitromethane was tested for carcinogenicity in one study in mice and in one study in rats by inhalation exposure. The incidence of alveolar–bronchiolar adenomas and carcinomas was markedly increased in treated mice and rats, and the incidence of squamous-cell carcinomas of the lung was increased in treated rats.

5.4 Other relevant data

No human or experimental animal data were available to the Working Group on the absorption, distribution, metabolism or excretion of tetranitromethane.

No human data were available on the toxic effects of tetranitromethane.

Tetranitromethane induced lethargy, reduced body-weight gain and inflammatory changes in the upper and lower respiratory tract of rats and mice.

Tetranitromethane is genotoxic in bacteria and cultured mammalian cells. Tumours from tetranitromethane-treated rats and mice had a GC → AT transition in the second base of codon 12 of the K-*ras* oncogene.

5.5 Evaluation[1]

There is *inadequate evidence* in humans for the carcinogenicity of tetranitromethane.

---

[1]For definition of the italicized terms, see Preamble, pp. 24–27.

There is *sufficient evidence* in experimental animals for the carcinogenicity of tetranitromethane.

Overall evaluation

Tetranitromethane is *possibly carcinogenic to humans (Group 2B)*.

6. References

Aldrich Chemical Co. (1994) *Aldrich Catalog/Handbook of Fine Chemicals 1994–1995*, Milwaukee, WI, p. 1334

Alper, M.D. & Ames, B.N. (1975) Positive selection of mutants with deletions of the *gal-chl* region of the *Salmonella* chromosome as a screening procedure for mutagens that cause deletions. *J. Bacteriol.*, **121**, 259–266

American Conference of Governmental Industrial Hygienists (1991) *Documentation of the Threshold Limit Values and Biological Exposure Indices*, 6th Ed., Vol. 3, Cincinnati, OH, pp. 1526–1528

American Conference of Governmental Industrial Hygienists (1995) *1995–1996 Threshold Limit Values for Chemical Substances and Physical Agents and Biological Exposure Indices*, Cincinnati, OH, p. 33

Arbeidsinspectie [Labour Inspection] (1994) *De Nationale MAC-Lijst 1994* [National MAC list 1994], The Hague, p. 39

Bucher, J.R., Huff, J.E., Jokinen, M., Haseman, J.K., Steadhan, M. & Cholakis, J.M. (1991) Inhalation of tetranitromethane causes nasal passage irritation and pulmonary carcinogenesis in rodents. *Cancer Lett.*, **57**, 95–101

Budavari, S., ed. (1989) *The Merck Index*, 11th Ed., Rahway, NJ, Merck & Co., p. 9171

Chemical Information Services (1994) *Directory of World Chemical Producers 1995/96 Standard Edition*, Dallas, TX, p. 661

Davis, R.A. (1993) Aliphatic nitro, nitrate, and nitrite compounds. In: Clayton, G.D. & Clayton, F.E., eds, *Patty's Industrial Hygiene and Toxicology*, Vol. II, Part A, Toxicology, New York, John Wiley & Sons, pp. 599–605, 614–615

Del Boccio, G., Pennelli, A., Whitehead, E.P., Lo Bello, M., Petruzzelli, R., Federici, G. & Ricci, G. (1991) Interaction of glutathione transferase from horse erythrocytes with 7-chloro-4-nitrobenzo-2-oxa-1,3-diazole. *J. biol. Chem.*, **266**, 13777–13782

Deutsche Forschungsgemeinschaft (1995) *MAK and BAT Values 1995* (Report No. 31), Weinheim, VCH Verlagsgesellschaft, p. 83

Gurwitz, D. & Sokolovsky, M. (1985a) Rat brain and heart muscarinic receptors: modification with tetranitromethane. *Biochem. biophys. Res. Comm.*, **131**, 1124–1131

Gurwitz, D. & Sokolovsky, M. (1985b) Increased agonist affinity is induced in tetranitromethane-modified muscarinic receptors. *Biochemistry*, **24**, 8086–8093

Horn, H.J. (1954) Inhalation toxicology of tetranitromethane. *Arch. ind. Hyg. occup. Med.*, **10**, 213–222

IARC (1989a) *IARC Monographs on the Evaluation of Carcinogenic Risks to Humans*, Vol. 47, *Some Organic Solvents, Resin Monomers and Related Compounds, Pigments and Occupational Exposures in Paint Manufacture and Painting*, Lyon, pp. 79–123

IARC (1989b) *IARC Monographs on the Evaluation of Carcinogenic Risks to Humans*, Vol. 45, *Occupational Exposures in Petroleum Refining; Crude Oil and Major Petroleum Fuels*, Lyon, pp. 219–237

Jänig, G.-R., Kraft, R., Blanck, J., Ristau, O., Rabe, H. & Ruckpaul, K. (1987) Chemical modification of cytochrome P-450 LM4. Identification of functionally linked tyrosine residues. *Biochim. biophys. Acta*, **916**, 512–523

Kawal, A., Goto, S., Matsumoto, Y. & Matsushita, H. (1987) Mutagenicity of aliphatic and aromatic nitro compounds. *Jpn J. ind. Health*, **29**, 34–54

Lewis, R.J., Sr (1993) *Hawley's Condensed Chemical Dictionary*, 12th Ed., New York, Van Nostrand Reinhold Co., p. 1135

Lide, D.R., ed. (1993) *CRC Handbook of Chemistry and Physics*, 74th Ed., Boca Raton, FL, CRC Press, pp. 3–320

Sadtler Research Laboratories (1980) *Sadtler Standard Spectra. 1980 Cumulative Index*, Philadelphia, PA, p. 1

Sax, N.I. & Lewis, R.J., Sr (1989) *Dangerous Properties of Industrial Materials*, 7th Ed., Vol. 3, New York, Van Nostrand Reinhold Co., pp. 3230

Stockinger, H.E. (1982) Aliphatic nitro compounds, nitrates, and nitrites. In: Clayton, G.D. & Clayton, F.E., eds, *Pattys's Industrial Hygiene and Toxicology*, Vol. II, Part C, Toxicology, New York, John Wiley & Sons, pp. 4155–4157

Stowers, S.J., Glover, P.L., Reynolds, S.H., Boone, L.R., Maronpot, R.R. & Anderson, M.W. (1987) Activation of the K-*ras* protooncogene in lung tumors from rats and mice chronically exposed to tetranitromethane. *Cancer Res.*, **47**, 3212–3219

Taylor, D.G. (1977) *NIOSH Manual of Analytical Methods*, 2nd Ed., Vol. 3 (DHEW (NIOSH) Publ. No. 77-157-C), Washington DC, United States Government Printing Office, pp. S224-1–S224-7

Työministeriö [Ministry of Labour] (1993) *Limit Values 1993*, Helsinki, p. 18 (in Finnish)

United Nations Environment Programme (1995) *International Register of Potentially Toxic Chemicals, Legal File, Tetranitromethane*, Geneva

United States National Institute for Occupational Safety and Health (1994a) *Pocket Guide to Chemical Hazards* (DHHS (NIOSH) Publ. No. 94-116). Cincinnati, OH, pp. 304–305

United States National Institute for Occupational Safety and Health (1994b) *RTECs Chem.-Bank*, Cincinnati, OH

United States National Institute for Occupational Safety and Health (1995) *National Occupational Exposure Survey (1981–1983)*, Cincinnati, OH

United States National Library of Medicine (1995) *Hazardous Substances Data Bank (HSDB)*, Bethesda, MD

United States National Toxicology Program (1990) *Toxicology and Carcinogenesis Studies of Tetranitromethane (CAS No. 509-14-8) in F344/N rats and B6C3F$_1$ Mice (Inhalation Studies)* (NTP TR 386; NIH Publication 90-2841), Research Triangle Park, NC

United States Occupational Safety and Health Administration (1994) Air contaminants. *US Code fed. Regul.*, **Title 29**, Part 1910.1000, p. 16

Würgler, F.E., Friederich, U., Fürer, E. & Ganss, M. (1990) *Salmonella*/mammalian microsome assay with tetranitromethane and 3-nitro-L-tyrosine. *Mutat. Res.*, **244**, 7–14

Yamanaka, K., Muramatsu, I. & Kigoshi, S. (1988) Tetranitromethane modification of the muscarinic receptors in bovine adrenal medulla. *Jpn. J. Pharmacol.*, **48**, 67–76

You, M., Wang, Y., Lineen, A., Stoner, G.D., You, L., Maronpot, R.R. & Anderson, M.W. (1991) Activation of protooncogenes in mouse lung tumors. *Exp. Lung Res.*, **17**, 389–400

Zeiger, E., Anderson, B., Haworth, S., Lawlor, T., Mortelmans, K. & Speck, W. (1987) *Salmonella* mutagenicity tests. III. Results from the testing of 255 chemicals. *Environ. Mutag.*, **9** (Suppl. 9), 1–110

2,4,6-TRINITROTOLUENE

1. Exposure Data

1.1 Chemical and physical data

1.1.1 *Nomenclature*

Chem. Abstr. Serv. Reg. No.: 118-96-7
Chem. Abstr. Name: 2-Methyl-1,3,5-trinitrobenzene
IUPAC Systematic Name: 2,4,6-Trinitrotoluene
Synonyms: Methyltrinitrobenzene; 1-methyl-2,4,6-trinitrobenzene; TNT; α-TNT; trinitrotoluene; α-trinitrotoluol; s-trinitrotoluene; s-trinitrotoluol; *sym*-trinitrotoluene; *sym*-trinitrotoluol

1.1.2 *Structural and molecular formulae and relative molecular mass*

$C_7H_5N_3O_6$

Relative molecular mass: 227.13

1.1.3 *Chemical and physical properties of the pure substance*

(a) *Description*: Yellow monoclinic needles or orthorhombic crystals from alcohol (Lewis, 1993; Lide, 1993)

(b) *Boiling-point*: 240 °C (explodes) (Lide, 1993)

(c) *Melting-point*: 82 °C (Lide, 1993)

(e) *Spectroscopy data*: Infrared (prism [21886], grating [32803]), nuclear magnetic resonance (C-13 [18215, V486]) and mass spectral data have been reported (Sadtler Research Laboratories, 1980)

(e) *Solubility*: Slightly soluble in water (0.01% (0.10 g/L) at 25 °C); soluble in acetone, benzene, oils and greases, and diethyl ether (McConnell & Flinn, 1946; Budavari, 1989; Lide, 1993)

(f) *Volatility*: Vapour pressure, 0.0002 mm Hg [0.027 Pa] at 20 °C; relative vapour density (air = 1), 7.85 (Verschueren, 1983; Boublík *et al.*, 1984)

(g) *Stability*: Moderate explosion risk; the pure chemical will detonate only if vigorously shocked or heated to > 200 °C (Lewis, 1993). Reacts with nitric acid and metals (e.g. lead or iron) to form explosive products more sensitive to shock or friction. Bases (e.g. sodium hydroxide, potassium iodide, tetramethyl ammonium octahydrotriborate) induce deflagration in molten trinitrotoluene. Can react vigorously with reducing materials (Sax & Lewis, 1989).

(h) *Octanol/water partition coefficient (P)*: log P, 1.60 (Hansch et al., 1995)

(i) *Conversion factor*: $mg/m^3 = 9.29 \times ppm^1$

1.1.4 Technical products and impurities

2,4,6-Trinitrotoluene is available commercially in the following forms: dry or wetted with < 10%, < 30% or > 30% water by weight. Military-grade flaked trinitrotoluene is available with the following specifications: setting-point, 80.2 °C; water, 0.10% max.; acidity (as H_2SO_4), 0.02% max.; alkalinity, none; materials insoluble in benzene, 0.05% max.; and sodium, 0.001% max. Commercial-grade trinitrotoluene (Nitropel) is available with the following specifications: setting-point, 80.1 °C; and water, 1.2% max. (United States National Library of Medicine, 1995; ICI Explosives Canada, undated).

Trade names for 2,4,6-trinitrotoluene include Entsufon, Gradetol, Nitropel, Tolit, Tolite, Trilit, Tritol, Trotyl and Trotyl oil.

1.1.5 Analysis

Selected methods for the analysis of 2,4,6-trinitrotoluene in various media are presented in Table 1.

Table 1. Methods for the analysis of 2,4,6-trinitrotoluene

Sample matrix	Sample preparation	Assay procedure	Limit of detection	Reference
Air	Draw air through modified Tenax-GC tube; desorb with acetone	GC/TEA	21 µg/m³	US Occupational Safety and Health Administration (1990) [Method 44]
	Draw air through glass wool–charcoal; desorb with benzene	GC/ECD	< 0.05 ppb [< 5 µg/m³]	Pella (1976)
	Direct incorporation of sample into glow discharge chamber	GDMS	~1.4 ppt [13 ng/m³]	McLuckey et al. (1988)
	Direct incorporation of sample into reaction chamber	IMS	0.01 ppb	Spangler et al. (1983)

---

[1] Calculated from: mg/m^3 = (relative molecular mass/24.45) × ppm, assuming temperature (25 °C) and pressure (101 kPa)

Table 1 (contd)

Sample matrix	Sample preparation	Assay procedure	Limit of detection	Reference
Incinerator emission	Draw air through Amberlite XAD-2; desorb with toluene	GC/ECD	0.025 µg/mL	Van Slyke et al. (1985)
Water	Collect sample on SEP-PAK cartridges and elute with methanol; concentrate; elute from reverse-phase column with methanol/water	HPLC/UV	0.5–1.0 µg/L	Kaplan & Kaplan (1981)
	Extract sample with dichloromethane or adsorb on Amberlite XAD resin and elute with dichloromethane	GC/ECD	NR	Feltes et al. (1990)
	Extract sample with toluene	GC/ECD	0.06 µg/L	Hable et al. (1991)
	Solid-phase extraction	HPLC	0.1 µg/L	Roberts (1986)
	Collect sample on Amberlite XAD-2/4/8; dry; desorb with dichloromethane; dry over anhydrous Na_2SO_4; exchange solvent to methanol; concentrate; elute from reverse-phase column with methanol/water	HPLC/UV	50 ng/L	Feltes & Levsen (1989)
	Collect sample on Hayesep-R; elute with acetone; concentrate; add internal standard; exchange solvent to methanol/water	HPLC/UV/ PC	1 µg/L	US Army (1989)
Wastewater, groundwater	Dilute sample with methanol/acetonitrile; filter; elute from reverse-phase column with methanol/acetonitrile/water	HPLC/UV	14 µg/L	Jenkins et al. (1984)
Soil	Air dry, grind, homogenize sample; extract with acetonitrile in ultrasonic bath; dilute with aqueous $CaCl_2$; filter; elute from reverse-phase column with methanol/water	HPLC/UV	0.08 µg/g	Jenkins et al. (1989); Bauer et al. (1990)
	Extract with methanol; filter extract; add $CaCl_2$ and refilter; pump through indicator tube	Indicator tube	0.5 µg/g	Jenkins & Schumacher (1990)
	Extract with acetone; react supernatant with potassium hydroxide/sodium sulfite; read absorbance at 540 nm	Colorimetry	1 µg/g	Jenkins & Walsh (1992)
Urine	Acidify sample to hydrolyse; neutralize and extract with toluene; add Na_2SO_4 and filter; evaporate and redissolve in acetone or acetonitrile	HPLC/MS	0.1 µg/L	Yinon & Hwang (1985, 1986a)

Table 1 (contd)

Sample matrix	Sample preparation	Assay procedure	Limit of detection	Reference
Urine (contd)	Acidify and heat sample; neutralize and extract with diethyl ether; evaporate and redissolve in acetone; develop silica gel plate with benzene/diethyl ether/methanol	TLC/ if possible densitometry	100 ng/spot	Liu et al. (1991)
Blood	Centrifuge sample; dilute serum with water; extract with dichloromethane; centrifuge and add Na_2SO_4; filter and evaporate; redissolve in dichloromethane; evaporate and redissolve in acetonitrile	HPLC/MS	NR	Yinon & Hwang (1986b)
Plasma, kidney	Add NaCl/acetic acid solution to sample; extract with toluene; add water and evaporate organic phase; add acetonitrile containing internal standard; filter	HPLC/UV	248 µg/L (plasma) 211 ng/g (kidney)	Lakings & Gan (1981)
Muscle, fat	Homogenize sample; extract with acetonitrile; concentrate; add internal standard and water; filter	HPLC/UV	66 ng/g	Lakings & Gan (1981)
Liver	Homogenize sample; add NaCl/acetic acid solution; extract with toluene; evaporate and redissolve in acetonitrile containing internal standard; filter	HPLC/UV	50 ng/g	Lakings & Gan (1981)
Handswabs	Wipe hand with swab soaked in acetone; squeeze out acetone and concentrate	HPLC/TEA	10 pg	Fine et al. (1984)
	Wipe hand with swab soaked in MTBE and extract with MTBE in pentane; centrifuge to remove debris; clean-up on Amberlite XAD-7 and elute with ethyl acetate	GC/ECD	< 2 ng/swab	Douse (1985, 1987); Douse & Smith (1986)
	Wipe hand with swab soaked in ethanol; extract in water/buffer solution with vortexing; add aliquots to antibody-coated microtitre plates	ELISA	15 ng/swab	Fetteroff et al. (1991)

GC, gas chromatography; TEA, thermal energy analysers; ECD, electron capture detection; GDMS, glow-discharge mass spectrometry; IMS, ion-mobilization spectrometry; HPLC, high-performance liquid chromatography; UV, ultraviolet detection; NR, not reported; PC, photoconductivity detection; MS, mass spectrometry; TLC, thin-layer chromatography; MTBE, methyl tert-butyl ether; ELISA, enzyme-linked immunosorbent assay

1.2 Production and use

1.2.1 *Production*

2,4,6-Trinitrotoluene has been produced by nitration of toluene with 'mixed acid' (HNO_3 and H_2SO_4) either in three steps or by continuous flow according to the Schmid-Meissner and Biazi processes. Small amounts of the 2,3,4- and 2,4,5-isomers are produced, which can be removed by washing with aqueous sodium sulfite solution. The Bofors-Norell process includes both continuous nitration of toluene or mononitrotoluene to trinitrotoluene, and continuous crystallization from dilute nitric acid (Ryon, 1987; Budavari, 1989; Lewis, 1993).

Typically, the production process begins with continuous nitration of toluene and the subsequent crystallization of 2,4,6-trinitrotoluene. The product is then washed and neutralized and then dried in a tank at up to 100 °C. Dry 2,4,6-trinitrotoluene is crushed and flaked and packed in cloth or containers. A small fraction of total trinitrotoluene production is further distilled, washed, dried and packed (ICI Explosives Canada, undated).

2,4,6-Trinitrotoluene is produced by two companies in Japan, and one company each in Argentina, Brazil, Canada, China, Egypt, Finland, Portugal, Taiwan, Turkey and the United Kingdom (Chemical Information Services, 1994).

1.2.2 *Use*

2,4,6-Trinitrotoluene is used as a high explosive in military and industrial applications. It has been widely used for filling shells, grenades and airborne demolition bombs, as it is sufficiently insensitive to the shock of ejection from a gun barrel but can be exploded on impact by a detonator mechanism. It has been used either as the pure explosive or in binary mixtures. The most common binary mixtures of 2,4,6-trinitrotoluene are cyclotols (mixtures with RDX (cyclotrimethylenetrinitramine or 1,3,5-trinitrohexahydro-1,3,5-triazine)), octols (mixtures with HMX (cyclotetramethylenetetranitramine or 1,3,5,7-tetranitro-1,3,5,7-tetraazocyclooctane)), amatols (mixtures with ammonium nitrate) and tritonals (mixtures with aluminium). In addition to military use, small amounts of 2,4,6-trinitrotoluene have been used for industrial explosive applications, such as deep-well and underwater blasting. It has also been used as a chemical intermediate in dyestuffs and photographic chemicals (Gibbs & Popolato, 1980; Budavari, 1989; Kline, 1990; Lewis, 1993).

1.3 Occurrence

1.3.1 *Natural occurrence*

2,4,6-Trinitrotoluene is not known to occur as a natural product.

1.3.2 *Occupational exposure*

Exposures to 2,4,6-trinitrotoluene may occur during its primary production, in munitions manufacture and loading, and during blasting operations. 2,4,6-Trinitrotoluene is readily absorbed through the skin, so measurements of personal airborne concentrations will underestimate exposures when the opportunity for dermal uptake is present. Exposures to airborne 2,4,6-trinitrotoluene may occur when 2,4,6-trinitrotoluene as a dust is mixed with other ingredients or as 2,4,6-trinitrotoluene vapour.

(a) Manufacture of trinitrotoluene

El Ghawabi et al. (1974) described exposures to 2,4,6-trinitrotoluene during its manufacture. Mean summer concentrations of trinitrotoluene in mg/m^3 (range) for the following operations are: nitration, 0.62 (0.15–1.2); crystallization, 0.5 (0.25–0.7); filtration, 0.75 (0.4–0.9); washing, 0.5 (0.25–0.75); crushing, 7.5 (6–10); and distillation, 0.5 (0.4–0.6). Mean winter concentrations of 2,4,6-trinitrotoluene were slightly lower. The highest exposures occurred during crushing operations, which were intermittent.

In a Finnish 2,4,6-trinitrotoluene production plant, the mean 2,4,6-trinitrotoluene air concentrations were 0.35 mg/m^3 (range, 0.31–0.39 mg/m^3) in the synthesis process room and 0.1 mg/m^3 (range, 0.02–0.19 mg/m^3) in the packing room (Savolainen et al., 1985).

(b) Munitions production

Munitions production begins with the mixing of 2,4,6-trinitrotoluene with other ingredients, where 2,4,6-trinitrotoluene may be 30% of the bulk weight (Woollen et al., 1986). After milling or mixing, the mixture is transported to a filling area or shed where it is transferred into metal containers or cardboard tubes that, when filled, are crimped closed. The contents of the containers are then fused or 'melted' by heating or dipping into hot wax. The final step is packaging.

Early reports of 2,4,6-trinitrotoluene exposures during munitions production were usually associated with reports of adverse health effects, and described air concentrations of 0.5–3.5 mg/m^3 2,4,6-trinitrotoluene (Cone, 1944; Eddy, 1944; Stewart et al., 1945; Ermakov et al, 1969). Stewart et al. (1945) reported 2,4,6-trinitrotoluene exposures of 0.3–0.6 mg/m^3 for filling area workers in a munitions loading plant and 0.3–1.3 mg/m^3 for workers who worked in the melt houses; these workers wore overalls and caps but no gloves or masks.

Air sampling for 2,4,6-trinitrotoluene was conducted in 1952 at an ammunition plant in the United States during a variety of operations (Goodwin, 1972). Air concentrations (in mg/m^3) for the following operations (personal samples) were: mixing 2,4,6-trinitrotoluene, 0.8–1.4; melting 2,4,6-trinitrotoluene, 0.9–2.9; screening 2,4,6-trinitrotoluene, 0.2–1.3; assembling grenades, 0.5–9.5; pouring 2,4,6-trinitrotoluene, 0.5–3.1; and pellet insertion, 1.6–4.7. Workers wore respirators in 'dusty areas' and wore overalls, disposable head coverings, socks and gloves. They were required to bathe with potassium sulfite soap, which turned red when in contact with 2,4,6-trinitrotoluene, until the red colour disappeared.

Exposures during intermittent 2,4,6-trinitrotoluene bagging operations ranged from 0.62 to 4.00 mg/m^3 in a study at a United States military munitions washout plant (Friedlander et al., 1974). After engineering controls were introduced at this plant in

1974, personal 8-h time-weighted average (TWA) exposures ranged from 0.08 to 0.59 mg/m^3 (Hathaway, 1977).

Exposures at a shell-loading plant in the United States ranged from 0.3 to 0.8 mg/m^3 (8-h TWA) and increased as production rate increased (Morton *et al.*, 1976). Engineering controls were instituted after air concentrations higher than 1.5 mg/m^3 were found at some operations.

In a study of 533 2,4,6-trinitrotoluene-exposed munitions workers, 8-h TWA personal exposures ranged from not detected to 1.84 mg/m^3, with 12% of workers exposed to more than 0.5 mg/m^3 (Buck & Wilson, 1975).

In a series of Czech studies of a plant manufacturing ammunition, mean workroom air concentrations of 2,4,6-trinitrotoluene were 0.22–9.6 mg/m^3 in the 1950s, 0.03–4.2 mg/m^3 in the 1960s and 0.04–0.76 mg/m^3 in the 1970s. The highest air concentrations were found in pressing and filling operations. In another plant producing powdered explosives for mines and quarries, mean workroom air concentrations of 2,4,6-trinitrotoluene ranged from 0.05 to 6.3 mg/m^3 in the 1960s and 1970s. The highest air concentrations were found during cartridge- and sack-filling operations (Hassman, 1979).

A study of explosives production in the United Kingdom compared air concentrations of 2,4,6-trinitrotoluene with post-shift urinary dinitroaminotoluene metabolites (Woollen *et al.*, 1986). Personal exposure (mg/m^3) for the following operations were: milling, 0.2 (range, < 0.01–0.71); filling, 0.04 (range, < 0.01–0.22); crimping, 0.39; and packing, 0.05. This study showed substantial interindividual variability in post-shift concentrations of dinitroaminotoluene from day to day. An important finding of this study was that personal inhalation exposures of 2,4,6-trinitrotoluene did not account for observed excretion of dinitroaminotoluene, thus, dermal uptake must have been an important exposure route.

In a Finnish study of both trinitrotoluene production and munitions assembly, the highest concentrations of 2,4,6-trinitrotoluene vapour were in casting and cooling and those of 2,4,6-trinitrotoluene dust in the sieve house (Ahlborg *et al.*, 1988a). Workers wore respirators and protective clothing during operations with higher exposure potential. Personal air concentrations of 2,4,6-trinitrotoluene for the following departments in mg/m^3 in 1983 were: trotyl (2,4,6-trinitrotoluene) foundry, 0.2–0.5; sieve house, 0.5; test foundry, 0.2–0.3; octol-hexotol (mixture of explosives including 2,4,6-trinitrotoluene) foundry, 0.1–0.2; and grenade assembly, 0.1. Based on urine concentrations of 2,4,6-trinitrotoluene metabolites, the authors concluded that dermal absorption contributed significantly to 2,4,6-trinitrotoluene uptake. Dermal uptake most probably occurs among the workers exposed to high dust concentrations such as, for example, those in the sieve house.

In a munitions plant in the United States, 2,4,6-trinitrotoluene and RDX were monitored in several areas (Bishop *et al.*, 1988). In the kettle area, where 2,4,6-trinitrotoluene was transferred from boxes to kettles, air concentrations averaged 0.02 mg/m^3. In the incorporation area where 2,4,6-trinitrotoluene was melted and transferred to kettles for combination with RDX, 2,4,6-trinitrotoluene air concentrations averaged 0.207 mg/m^3. In a bagging area where the final product was packaged, air concentrations averaged 0.006 mg/m^3.

(c) Blasting operations

During explosive blasting operations in Ukrainian pit mines, exposures as high as 12.5 ± 3.31 mg/m^3 2,4,6-trinitrotoluene were measured in the breathing zone of workers filling dry drill holes with explosives containing 21% trinitrotoluene. Contamination of the hands and uncovered parts of the body was greater in the pit mine blasters than in the warehouse loaders. It occurred through contact with the contaminated surface of the bags, shaking them out and collecting explosive material which had spilled (Melnichenko, 1976).

1.3.3 Environmental occurrence

(a) Water

Although for many years waste munitions were discarded at sea, 2,4,6-trinitrotoluene has not been detected (< 2 ng/L) in ocean waters or sediment near several dump sites off the coasts of South Carolina, Florida, California or Washington, United States (Hoffsommer & Rosen, 1972; Hoffsommer *et al.*, 1972).

2,4,6-Trinitrotoluene has been detected in surface-water and groundwater samples collected in several monitoring studies in the vicinity of munitions facilities. It was detected in contaminated groundwater both beneath and originating from the disposal beds of a demilitarization facility in Nevada, United States, at a maximum concentration of 620 µg/L in 1976; in 1977, it was detected in groundwater samples collected at the same place at concentrations of 320 µg/L 200 ft [61 m] away from the facility and 1 µg/L at 1070 ft [326 m] away (Goerlitz & Franks, 1989). 2,4,6-Trinitrotoluene concentrations of 12.0 and 19.0 µg/L were detected in surface-water samples collected from two brooks near Hirschagen/Waldhof, Germany, in the vicinity of what was a munitions manufacturing plant during the Second World War; the river into which the brooks fed (River Losse) had a concentration of 0.7 µg/L. Two ponds in the Clausthal-Zellerfeld region of Germany, again near a former munitions manufacturing plant, had levels of 0.5 µg/L; the ponds feed into the River Oder, which had a level of < 0.01 µg/L (Feltes *et al.*, 1990). Concentrations of 2,4,6-trinitrotoluene ranged from 690 to 1370 µg/L in groundwater samples collected near a former explosives factory in Elsnig, Germany (Steuckart *et al.*, 1994).

In samples of wastewaters generated in the manufacture of 2,4,6-trinitrotoluene over a 12-month period in the United States, 2,4,6-trinitrotoluene was detected in 20% (11/54 samples) at a concentration range of 0.1–3.4 mg/L (Spanggord *et al.*, 1982a). It was also detected in the effluent water from a 2,4,6-trinitrotoluene manufacturing plant in Virginia, United States, at concentrations ranging from 101 to 143 ppm (mg/L) (Nay *et al.*, 1972). 2,4,6-Trinitrotoluene has also been found in 'pink-water effluents' (wastewater from one of the purification steps in the manufacture of 2,4,6-trinitrotoluene) at concentrations of 774–998 µg/L in lagoon water and 2900–6400 µg/L in groundwater

(Triegel et al., 1983) and at 1–178 mg/L in effluents from loading, assembling and packaging plants in the United States (Patterson et al., 1977).

(b) *Soil and sediments*

At a waste-disposal site in Missouri, United States, where 2,4,6-trinitrotoluene explosives were burned in the 1940s, 2,4,6-trinitrotoluene has been detected in surface-soil samples at an average concentration of 13 g/kg (Haroun et al., 1990). In West Virginia, United States, at burning sites at a munitions plant, 2,4,6-trinitrotoluene and other nitroaromatics were detected in surface soils at concentrations of up to 4% (40 g/kg). Nitroaromatics, principally 2,4,6-trinitrotoluene, were detected at up to 20 g/kg within 5–10 m of the foundations of processing and refining facilities (Kraus et al., 1985). Concentrations of 2,4,6-trinitrotoluene ranged from 0.1 to 38.6 g/kg in soil samples collected from an army depot in Oregon, United States (Jenkins & Walsh, 1992). A soil sample collected near a former ammunition plant in Brandenburg, Germany, had a 2,4,6-trinitrotoluene level of 234 µg/kg (Steuckart et al., 1994). At a munitions plant located in Texarkana, TX, United States, 2,4,6-trinitrotoluene has been detected at a concentration of about 15% in samples of sludge taken from ponds used as solids-settling areas for pink-water effluent; 2,4,6-trinitrotoluene concentrations were highest in surface-soil samples (e.g. 18.8 mg/kg at a depth of 0.2–0.6 m), and decreased with depth (e.g. < 3 mg/kg below 4.5 m) (Phung & Bulot, 1981). 2,4,6-Trinitrotoluene concentrations of 200–56 700 ppm (mg/kg) were found in sludge samples from pink-water lagoons and at 18.9–158 ppm [mg/kg] in surface-soil samples collected from directly beneath the lagoon in United States (Triegel et al., 1983).

1.4 Regulations and guidelines

Occupational exposure limits and guidelines in several countries are given in Table 2.

Table 2. Occupational exposure limits and guidelines for 2,4,6-trinitrotoluene

Country	Year	Concentration (mg/m^3)	Interpretation
Argentina	1991	0.5 (Sk)	TWA
Australia	1993	0.5 (Sk)	TWA
Belgium	1993	0.5 (Sk)	TWA
Bulgaria[a]	1995	0.5 (Sk)	TWA
Canada	1991	0.5 (Sk)	TWA
Colombia[a]	1995	0.5 (Sk)	TWA
Czech Republic	1993	0.5 (Sk)	TWA
		2.5	STEL
Denmark	1993	0.5 (Sk)	STEL
Egypt	1993	0.5	TWA
Finland	1993	0.5 (Sk)	TWA
		3	STEL
France	1993	0.5 (Sk)	TWA

Table 2 (contd)

Country	Year	Concentration (mg/m³)	Interpretation
Germany	1995	0.1 (Sk, III, IIIB)	MAK
Hungary	1993	0.3 (Sk)	TWA
		0.5	STEL
Jordan[a]	1995	0.5 (Sk)	TWA
Mexico	1991	0.5	TWA
		3	STEL (15 min)
Netherlands	1994	0.1 (Sk)	TWA
New Zealand[a]	1995	0.5 (Sk)	TWA
Philippines	1993	1.5 (Sk)	TWA
Republic of Korea[a]	1995	0.5 (Sk)	TWA
Russia	1993	0.1 (Sk)	TWA
		0.5	STEL
Singapore[a]	1995	0.5 (Sk)	TWA
Switzerland	1993	0.1 (Sk)	TWA
		0.2	STEL
Turkey	1993	1.5 (Sk)	TWA
United Kingdom	1995	0.5 (Sk)	TWA
USA			
ACGIH (TLV)	1995	0.5 (Sk)[b]	TWA
OSHA (PEL)	1994	1.5 (Sk)	TWA
NIOSH (REL)	1994	0.5 (Sk)	TWA
Viet Nam[a]	1995	0.5 (Sk)	TWA

From Arbeidsinspectie (1994); US National Institute for Occupational Safety and Health (NIOSH) (1994a,b); US Occupational Safety and Health Administration (OSHA) (1994); American Conference of Governmental Industrial Hygienists (ACGIH) (1995); Deutsche Forschungsgemeinschaft (1995); Health and Safety Executive (1995); United Nations Environment Programme (1995)

Sk, absorption through the skin may be a significant source of exposure; TWA, time-weighted average; STEL, short-term exposure limit; III, substances with systemic effects (half-life < 2h); IIIB, suspected of having carcinogenic potential; TLV, threshold limit values; PEL, permissible exposure limit; REL, recommended exposure limit

[a] Follows ACGIH TLVs

[b] Substance identified by other sources as a suspected or confirmed human carcinogen

2. Studies of Cancer in Humans

2.1 Case report

Garfinkel *et al.* (1988) reported a case of liver cancer in a 61-year-old engineer who had been exposed daily to 2,4,6-trinitrotoluene for 35 years. He had no past history of infectious hepatitis or alcohol abuse, which are known risk factors for liver cancer.

2.2 Descriptive study

Kolb *et al.* (1993) conducted an ecological study in which comparisons were made between the incidence of acute and chronic myelogenous leukaemias in two counties in Central Hesse, Germany. Contamination of the soil with 2,4,6-trinitrotoluene had been documented in one of the counties (Marburg-Biedenkopf) resulting from an underground plant in the city of Stadtallendorf that had produced 2,4,6-trinitrotoluene during the Second World War: this study was initiated following the observation of what was believed to be an unusually high number of leukaemias in the city of Stadtallendorf. The incidence of leukaemia in Marburg-Biedenkopf during 1983–89 was compared with that of the neighbouring county of Giessen which was not contaminated with 2,4,6-trinitrotoluene. Cases of leukaemia among individuals over the age of 18 were identified in the medical centres covering the two counties, and the population at risk was identified through residential registries. A statistically significant excess of acute and chronic myelogenous leukaemia was observed among male and female residents of Marburg-Biedenkopf. Stratification of the analysis by age revealed that excess risk was predominantly among residents over 65 years of age.

3. Studies of Cancer in Experimental Animals

No adequate data were available to the Working Group.

4. Other Data Relevant for an Evaluation of Carcinogenicity and its Mechanisms

4.1 Absorption, distribution, metabolism and excretion

4.1.1 *Humans*

The toxicology of 2,4,6-trinitrotoluene has been reviewed (Zakhari *et al.*, 1978; Rickert, 1987).

2,4,6-Trinitrotoluene is absorbed through the skin in exposed workers (Voegtlin *et al.*, 1921; Neal *et al.*, 1944).

In humans exposed orally to 2,4,6-trinitrotoluene, dinitroaminotoluenes (2,4-dinitro-6-aminotoluene and 2,6-dinitro-4-aminotoluene) were found in the urine (Horecker & Snyder, 1944; Lemberg & Callaghan, 1945).

Hassman (1971a,b) carried out two surveys of 2,4,6-trinitrotoluene-exposed workers. Air concentrations of 2,4,6-trinitrotoluene was found to range from 0.6 to 4.0 mg/m^3. The urinary excretion of 2,6-dinitro-4-aminotoluene was 2.5 and 6.5 mg/L, respectively.

Absorption and excretion of 2,4,6-trinitrotoluene was assessed in groups of workers in two explosives factories. 2,6-Dinitroaminotoluene (and 2,4-dinitro-6-aminotoluene) were found in most post-shift urine samples (mean level, 9.7 mg/L; range, 0.1–44 mg/L). (In unhydrolysed samples, much lower levels were detected, indicating that only a small fraction is present as free amine, and that the remainder is bound, probably as a N-glucuronide conjugate). There was a decrease in concentrations over night and an increase over shift, and the highest levels were usually found a few hours after the end of exposure; thus, relatively fast absorption and excretion occur. However, dinitroaminotoluenes were still present after 36 h and even after 17 days, indicating a fraction with slow metabolism. No association was found between levels of urinary dinitroaminotoluenes and air levels of 2,4,6-trinitrotoluene (< 0.01–0.29 mg/m^3 and 0.05–0.71 mg/m^3 by personal sampling in the two factories), and the urinary levels varied considerably between days. Calculations showed that inhalation did not account for the urinary dinitroaminotoluenes and that dermal uptake must have been considerable (Woollen *et al.*, 1986).

In urinary samples from a group of 2,4,6-trinitrotoluene workers, 2,6-dinitro-4-aminotoluene was the main metabolite (at 0.24–9.65 mg/L), followed by 4,6-dinitro-2-aminotoluene. Unchanged 2,4,6-trinitrotoluene, 2,6-diamino-4-nitrotoluene and 2,4-diamino-6-nitrotoluene were present at lower concentrations (Yinon & Hwang, 1986a).

Ahlborg *et al.* (1988b) analysed diazo-positive metabolites in hydrolysed urine from 2,4,6-trinitrotoluene workers. In workers with low, medium and high exposure, the levels after a holiday were lower (means, 0.36–0.49 µmol/mol creatinine in the three groups) than post-shift levels (0.56, 0.84 and 1.06 µmol/mol creatinine, respectively; the corresponding air levels were undetectable, < 0.3 and up to 0.5–0.6 mg/m^3, respectively). After the holiday, there was no difference in urinary excretion between workers with varying intensity of exposure.

4.1.2 *Experimental systems*

Rabbits administered oral doses of about 100–300 mg/kg bw 2,4,6-trinitrotoluene excreted the following in the urine: 2,6-dinitro-4-hydroxylaminotoluene, 2,6-dinitro-4-aminotoluene and 2,4-dinitro-6-aminotoluene (Channon *et al.*, 1944). The blood of male rabbits given 100 mg 2,4,6-trinitrotoluene by gavage contained 2,6-dinitro-4-aminotoluene and 2,4-dinitro-6-aminotoluene, as well as the parent compound (Yinon & Hwang, 1987). The urine of male SPD rats given 20 mg 2,4,6-trinitrotoluene by gavage contained the parent compound, 2,4-diamino-6-nitrotoluene, 2,6-dinitro-4-aminotoluene and 2,4-dinitro-6-aminotoluene (Yinon & Hwang, 1985). In male SPD rats receiving a skin application of 20 mg 2,4,6-trinitrotoluene for 2 h, 2,4-diamino-6-aminotoluene and 2,6-dinitro-4-aminotoluene were seen in urine (Yinon & Hwang, 1987).

2,6-Dinitro-4-aminotoluene and 2,6-dinitro-4-hydroxylaminotoluene were found in the urine of dogs given 2,4,6-trinitrotoluene orally (Snyder, 1946).

In the above studies, no data exist to assess the percentage of the dose that was eliminated as each metabolite. 2,4,6-Trinitrotoluene undergoes both oxidative and reductive metabolism in animals. The nitro groups are reduced through intermediate hydroxylamines to amines. The methyl group can be oxidized to an alcohol and an acid, both of which can be conjugated with glucuronic acid and excreted in the urine.

Zwirner-Baier *et al.* (1994) treated female Wistar rats by oral gavage with 0.5 mmol [114 mg]/kg bw 2,4,6-trinitrotoluene and found that, 24 h after dosing, 2,4,6-trinitrotoluene was bound to haemoglobin.

4.2 Toxic effects

4.2.1 *Humans*

The toxicity of 2,4,6-trinitrotoluene to humans has been reviewed (Hathaway, 1977; Zakhari *et al.*, 1978; Ryon & Ross, 1990).

2,4,6-Trinitrotoluene has several types of effect on the haematological system. Exposure to 2,4,6-trinitrotoluene may cause methaemoglobinaemia, with cyanosis. In the bone marrow, hypercellularity and hypocellularity have been reported, the latter resulting in a reduction of circulating red and white blood cells and platelets and, in severe cases, aplastic anaemia (24 cases in the United Kingdom, 14 in the United States), which has symptoms such as pallor, fatigue, bleeding and infection (Sievers *et al.*, 1946).

Djerassi and Vitany (1975) reported three cases of acute haemolytic disease in glucose-6-phosphate dehydrogenase-deficient workers filling shells with a 2,4,6-trinitrotoluene mixture. The onset of the disease was within two to four days after the start of exposure. The air concentration was not known, but had earlier been 1.8–2.95 mg/m^3.

Nine workers in a factory producing 2,6,6-trinitrotoluene had somewhat lower activities than 25 unexposed controls of the enzymes δ-aminolevulinic acid synthase (EC 2.3.1.37) and haeme synthase (EC 4.99.1.1) in red blood cells (presumably in reticulocytes). The 2,4,6-trinitrotoluene concentration in the process room air was 0.35 (range, 0.31–0.39) mg/m^3 and that in the packing department was 0.10 (0.02–0.19) mg/m^3 (Savolainen *et al.*, 1985).

Liver damage is the second main toxic effect of exposure to 2,4,6-trinitrotoluene. Initial symptoms in acute poisoning include jaundice, excretion of bile pigments in urine, epigastric pain, nausea and, in some cases, eventual coma and death. In 10 acute fatal cases, pathological examination revealed reduced liver weight, destruction of parenchymal cells, haemorrhagic areas, perivascular infiltration of lymphocytes and polymorphonuclear lymphocytes, and fat-infiltration of cells (McConnel & Flinn, 1946). A single case of macronodular liver cirrhosis (and hepatocellular carcinoma) was diagnosed in an engineer who had been exposed to 2,4,6-trinitrotoluene daily for 35 years and had had no history of viral hepatitis or alcohol abuse (Garfinkel *et al.*, 1988).

Stewart *et al.* (1945) studied students who worked in a munitions loading plant in which both inhalation and skin exposure occurred. For 52 students who worked in filling areas during the four-to-11-week study period, the air levels of 2,4,6-trinitrotoluene ranged from 0.3 to 0.6 mg/m^3 for an average of 33 days. These students also had significant skin exposure and 27 of them developed skin rashes. Ten students working in both filling areas (average, 18 days) and melt houses (average, 15 days) had exposures ranging from 0.3 to 1.3 mg/m^3 and probably had only minimal skin exposure. Compared to pre-employment values, over 80% of the students had decreases in their blood haemoglobin levels. Only minimal changes in reticulocyte counts occurred during exposure, but there was an increase 48 h after termination of 2,4,6-trinitrotoluene exposure. In addition, students in both filling areas and melt houses had significant increases in blood bilirubin levels.

El Ghawabi *et al.* (1974) examined 38 (three were excluded because of previous diseases) workers in a 2,4,6-trinitrotoluene production and shell-loading plant and 20 unexposed control workers. The 2,4,6-trinitrotoluene exposure ranged from 0.1 to 1.2 mg/m^3, with peaks of up to 10 mg/m^3. The 2,4,6-trinitrotoluene workers had higher prevalences of respiratory (sneezing, sore throat and cough) and gastrointestinal (stomach ache, anorexia, constipation, flatulence, nausea and vomiting) complaints than the controls. Further, they had lower average blood haemoglobin levels: 85% (100% is 15 g/100 mL) versus 96% in controls. However, there were no significant differences in liver tests and no case of cataract was recorded.

In a shell-loading plant, 43 workers were examined before employment and then followed monthly for five months. During this period, the time-weighted exposure increased from 0.3 to 0.8 mg/m^3 2,4,6-trinitrotoluene as a result of increased production. At the end of the five months, there were statistically significant increases in serum lactate dehydrogenase and SGOT (serum glutamic oxalacetic transaminase). Further, blood haemoglobin levels decreased, although not statistically significantly (Morton *et al.*, 1976).

Cataracts have been reported in 2,4,6-trinitrotoluene workers, even at low exposure levels. Opacities are bilateral, symmetric and typical. They are initially noted only in the peripheral parts of the lens, where they do not interfere with the visual fields. Six cases of cataract were recorded in 12 munitions factory workers (mean age, 39.5 years) exposed to air levels of 2,4,6-trinitrotoluene ranging from 0.14 to 0.58 mg/m^3 (Härkönen *et al.*, 1983). Of 413 workers (mean age, 38 years) exposed to 2,4,6-trinitrotoluene (for three months to 29 years; air levels stated to have been below 1 mg/m^3), cataracts were found in 34.6%; among workers exposed to more than 20 years, the figure rose to 88.7%. There was said to be no association between cataracts and liver tests (Anshou, 1990). The changes in the lens may be virtually irreversible.

Allergic contact dermatitis has been reported in skin areas exposed to 2,4,6-trinitrotoluene (Goh & Rajan, 1983; Goh, 1984).

4.2.2 Experimental systems

(a) Single-dose studies

The acute toxicity of 2,4,6-trinitrotoluene has been reported in laboratory animals including rats, mice and dogs (Dilley et al., 1982). The reported oral LD_{50} values for 2,4,6-trinitrotoluene in male and female Sprague-Dawley rats were 1320 mg/kg bw and 795 mg/kg bw, respectively. The oral LD_{50} in both male and female Swiss-Webster mice was 660 mg/kg bw.

(b) Repeated-dose studies

Levine et al. (1984) investigated the subchronic toxicity of 2,4,6-trinitrotoluene in male and female Fischer 344 rats treated with 1, 5, 25, 125 or 300 mg/kg bw per day for 13 weeks by administration in the diet. Anaemia was observed in all treated rats, the severity of which was dose-dependent. Splenomegaly, hepatomegaly/hepatocytomegaly and testicular atrophy were seen at doses of 125 and 300 mg/kg per day.

Dilley et al. (1982) administered gelatin capsules containing 0, 0.2, 2.0 and 20 mg/kg bw 2,4,6-trinitrotoluene to beagle dogs daily for up to 13 weeks. In the same study, Sprague-Dawley rats received 0, 0.002, 0.01, 0.05 or 0.25% and Swiss-Webster mice 0, 0.001, 0.005, 0.025 or 0.125% 2,4,6-trinitrotoluene in their diets over the same period. At the highest doses, all species exhibited anaemia. Additionally, the authors reported enlarged spleens and livers and depressed body weight and/or depressed body-weight gain (temporary in dogs and mice) for all species. 2,4,6-Trinitrotoluene induced elevated cholesterol and depressed serum glutamic pyruvic transaminase (SGPT) activity in dogs and rats. SGPT depression in rats appeared after 13 weeks. Reduced testes size was observed in rats at the highest dose. Most of the toxic effects were reversible; however, in rats, testicular atrophy did not reverse within a four-week recovery period after treatment.

Levine et al. (1990) evaluated the oral toxicity of 2,4,6-trinitrotoluene in male and female beagle dogs administered 0, 0.5, 2, 8 or 32 mg/kg bw daily for 26 weeks. Dose-dependent anaemia was observed in treated dogs. Dose-dependent decreases in SGPT were seen in both males and females. In dogs receiving 8 or 32 mg/kg per day, hepatomegaly was observed and hepatocytomegaly increased in severity as a function of dose. Spleen enlargement was also observed. The major toxic effects, therefore, included haemolytic anaemia, methaemoglobinaemia, liver injury, splenomegaly and death (only at 32 mg/kg bw per day).

(c) Biochemical alterations

Tenhunen et al. (1984) injected male Wistar rats intraperitoneally with 100 mg/kg bw 2,4,6-trinitrotoluene in olive oil and the animals were killed 48 h after injection. δ-Aminolevulinic acid synthase activity in reticulocytes was approximately 70% that of control values, and red blood cell coproporphyrin was marginally below that of control values. Liver haeme synthase activity was approximately 60% that of control values and no effect was noted on δ-aminolevulinic acid synthase or biliverdin reductase activity in livers.

Jiang et al. (1991) treated male Wistar rats by oral gavage with 200 mg/kg bw 2,4,6-trinitrotoluene per day on six days per week for six weeks. Blood, testes and liver were obtained from animals killed after two, four and six weeks of treatment. At six weeks of treatment, copper concentrations in testes were decreased by 30% compared to controls and remained so for two weeks after the end of the six-week treatment regimen. Zinc concentrations in testes were significantly depressed throughout, and two weeks beyond the six-week treatment period.

Lingyuan et al. (1989) treated Chinese rhesus monkeys with 60 and 120 mg/kg bw 2,4,6-trinitrotoluene orally once a day, four times a week for three months. Forty-eight hours after the last treatment, the monkeys were killed and mitochondria and microsomes were prepared for measurement of superoxide anion and hydrogen peroxide. Hydrogen peroxide was quantified by measuring the conversion of methanol to formaldehyde. The results revealed that 2,4,6-trinitrotoluene increased the formation of hydrogen peroxide.

In studies by Short and Lee (1980) to assess whether 2,4,6-trinitrotoluene could modify the biotransformation of model xenobiotics, male CD rats were administered orally 0.4 mmol [91 mg]/kg bw 2,4,6-trinitrotoluene twice a day for three days and once on the fourth day. The results indicated that 2,4,6-trinitrotoluene was not effective in inducing in-vivo biotransformation of xenobiotics (as measured by zoxazolamine paralysis or hexobarbital sleeping time).

4.3 Reproductive and developmental effects

No data were available to the Working Group.

4.4 Genetic and related effects

4.4.1 *Humans*

Urine concentrates from workers exposed to 2,4,6-trinitrotoluene showed, compared to urine concentrates from unexposed controls, an increased mutagenic activity in *Salmonella typhimurium* strain TA98 in the absence of rat-liver S9. The same strain responded only weakly when the S9 mix was used, while, with *Escherichia coli* WP2 *uvrA* in the presence of S9, no effect of worker exposure was observed (Ahlborg et al., 1985). Since the response with *S. typhimurium* strains TA98 and TA98NR (deficient in nitroreductase activity) was about the same, bacterial nitroreductase activity is not significantly responsible for the mutagenicity of the urine samples (Ahlborg et al., 1988a).

4.4.2 *Experimental systems* (see also Table 3 and Appendices 1 and 2)

2,4,6-Trinitrotoluene was mutagenic in *S. typhimurium* strains TA1538 and TA98 (with and without rat S9) as well as in TA98 and TA100, with and without human placenta S9 and with and without rat-liver S9. In addition, 2,4,6-trinitrotoluene was reported to be negative in strains TA1535 and TA100NR3 (nitroreductase-deficient) and positive in TA1537 (with and without rat S9). In the P388 mouse lymphoma gene mutation assay, 2,4,6-trinitrotoluene induced mutations in the absence, but not in the presence of S9-mix.

Table 3. Genetic and related effects of 2,4,6-trinitrotoluene

Test system	Result[a] Without exogenous metabolic system	Result[a] With exogenous metabolic system	Dose[b] (LED/HID)	Reference
SA0, Salmonella typhimurium TA100, reverse mutation	+	+	50	Spanggord et al. (1982b)
SA0, Salmonella typhimurium TA100, reverse mutation	+	0	25	Whong & Edwards (1984)
SA0, Salmonella typhimurium TA100, reverse mutation	+	+	10	Tan et al. (1992)
SA0, Salmonella typhimurium TA100, reverse mutation	+, −	+	0.5	Karamova et al. (1994)
SA5, Salmonella typhimurium TA1535, reverse mutation	−	−	2500	Spanggord et al. (1982b)
SA5, Salmonella typhimurium TA1535, reverse mutation	−	0	100	Whong & Edwards (1984)
SA7, Salmonella typhimurium TA1537, reverse mutation	+	+	NR	Spanggord et al. (1982b)
SA7, Salmonella typhimurium TA1537, reverse mutation	+	0	25	Whong & Edwards (1984)
SA8, Salmonella typhimurium TA1538, reverse mutation	+	+	25	Kaplan & Kaplan (1982)
SA8, Salmonella typhimurium TA1538, reverse mutation	+	+	NR	Spanggord et al. (1982b)
SA8, Salmonella typhimurium TA1538, reverse mutation	+	0	12.5	Whong & Edwards (1984)
SA9, Salmonella typhimurium TA98, reverse mutation	+	0	2.5	Won et al. (1976)
SA9, Salmonella typhimurium TA98, reverse mutation	+	0	250	Kaplan & Kaplan (1982)
SA9, Salmonella typhimurium TA98, reverse mutation	+	+	NR	Spanggord et al. (1982b)

Table 3 (contd)

Test system	Result[a]		Dose[b] (LED/HID)	Reference
	Without exogenous metabolic system	With exogenous metabolic system		
SA9, *Salmonella typhimurium* TA98, reverse mutation	+	0	12.5	Whong & Edwards (1984)
SA9, *Salmonella typhimurium* TA98, reverse mutation	+	(+)	10	Tan et al. (1992)
SA9, *Salmonella typhimurium* TA98, reverse mutation	+	+	0.5	Karamova et al. (1994)
SAS, *Salmonella typhimurium* TA100NR3, reverse mutation	–	–	250	Spanggord et al. (1982b)
SAS, *Salmonella typhimurium* TA100NR (nitroreductase deficient), reverse mutation	–	–	50	Karamova et al. (1994)
SAS, *Salmonella typhimurium* TA100/1,8-DNP (o-acetyltransferase deficient), reverse mutation	–	–	50	Karamova et al. (1994)
GML, Gene mutation, mouse lymphoma cells P388, *tk* locus *in vitro*	+	–	40	Styles & Cross (1983)
UPR, Unscheduled DNA synthesis, male rat hepatocytes *in vivo*	–		1000 po × 1	Ashby et al. (1985)
MVM, Micronucleus test, male mouse bone marrow cells *in vivo*	–		80 ip × 1	Ashby et al. (1985)

[a] +, positive; (+), weak positive; –, negative; 0, not tested; ?, inconclusive (variable response within several experiments within an adequate study)

[b] LED, lowest effective dose; HID, highest ineffective dose. In-vitro tests, μg/mL; in-vivo tests, mg/kg bw; NR, dose not reported

2,4,6-Trinitrotoluene was negative in the mouse bone-marrow micronucleus assay and in an in-vivo/in-vitro rat liver assay for unscheduled DNA synthesis.

The mutagenicity of urine of rats exposed to 2,4,6-trinitrotoluene by intraperitoneal injection was studied in the *S. typhimurium* assay. In the absence of rat-liver S9, only a weakly positive response was observed in strain TA98, but a strong response was seen with strains YG1021 (nitroreductase-overproducing) and YG1024 (*O*-acetyltransferase-overproducing). The strains TA98NR and TA98/1,8-DNP$_6$ (*O*-acetyltransferase-deficient) showed no effect. In the presence of S9, strains YG1021 and YG1024 gave a weak effect. Thus, high levels of both nitroreductase and *O*-acetyltransferase significantly increase the sensitivity of the indicator strain to the mutagenicity of urine caused by 2,4,6-trinitrotoluene exposure (Einistö, 1991).

5. Summary of Data Reported and Evaluation

5.1 Exposure data

2,4,6-Trinitrotoluene is produced commercially by the nitration of toluene. It is used mainly as a high explosive in military and industrial applications. Exposures to 2,4,6-trinitrotoluene both through inhalation and skin absorption can occur during its production, during munitions manufacturing and loading, and during blasting operations. 2,4,6-Trinitrotoluene has been detected in wastewater, surface and groundwater, and in soils and sediments near plants manufacturing 2,4,6-trinitrotoluene and explosives.

5.2 Human carcinogenicity data

One ecological study was available that noted an association between leukaemia and residence in an area contaminated with 2,4,6-trinitrotoluene.

5.3 Animal carcinogenicity data

No adequate study on the carcinogenicity of 2,4,6-trinitrotoluene in experimental animals was available to the Working Group.

5.4 Other relevant data

In humans, absorption of 2,4,6-trinitrotoluene both through the skin and the gastrointestinal route had been demonstrated. 2,4,6-Trinitrotoluene is also probably absorbed in the respiratory tract. However, the dermal route is the commonest in occupational settings.

In humans exposed to 2,4,6-trinitrotoluene, mainly dinitroaminotoluenes and also diaminonitrotoluenes, probably mainly as conjugates, as well as unchanged 2,4,6-trinitrotoluene were found in the urine.

In humans, exposure to 2,4,6-trinitrotoluene has been found to cause haematological disorders, including aplastic anaemia, haemolytic anaemia and methaemoglobinaemia. 2,4,6-Trinitrotoluene may cause toxic hepatitis. Moreover, allergic contact dermatitis and cataracts may occur, as well as gastritis and respiratory mucous membrane and conjunctival irritation.

2,4,6-Trinitrotoluene undergoes both oxidative and reductive metabolism in animals. It causes anaemia and hepatotoxicity in rats and dogs. Testicular atrophy occurs in rats following exposure to 2,4,6-trinitrotoluene.

In workers exposed to 2,4,6-trinitrotoluene, increased bacterial mutagenic activity was found in the urine.

2,4,6-Trinitrotoluene is mutagenic in bacteria with and without a metabolic activation system. In cultured mammalian cells, it is mutagenic only in the absence of a metabolic activation system. Although 2,4,6-trinitrotoluene was negative in mammals *in vivo* for unscheduled DNA synthesis in the liver and micronuclei induction in bone marrow, the urine of rats is mutagenic after intraperitoneal injection of 2,4,6-trinitrotoluene.

5.5 Evaluation[1]

There is *inadequate evidence* in humans for the carcinogenicity of 2,4,6-trinitrotoluene.

There is *inadequate evidence* in experimental animals for the carcinogenicity of 2,4,6-trinitrotoluene.

Overall evaluation

2,4,6-Trinitrotoluene is *not classifiable as to its carcinogenicity to humans (Group 3)*.

6. References

Ahlborg, G., Jr, Bergström, B., Hogstedt, C., Einistö, P. & Sorsa, M. (1985) Urinary screening for potentially genotoxic exposures in a chemical industry. *Br. J. ind. Med.*, **42**, 691–699

Ahlborg, G., Jr, Einistö, P. & Sorsa, M. (1988a) Mutagenic activity and metabolites in the urine of workers exposed to trinitrotoluene (TNT). *Br. J. ind. Med.*, **45**, 353–358

Ahlborg, G., Jr, Ulander, A., Bergström, B. & Oliv, Å. (1988b) Diazo-positive metabolites in urine from workers exposed to aromatic nitro-amino compounds. *Int. Arch. occup. environ. Health*, **60**, 51–54

American Conference of Governmental Industrial Hygienists (1995) *1995–1996 Threshold Limit Values for Chemical Substances and Physical Agents and Biological Exposure Indices*, Cincinnati, OH, p. 35

---

[1]For definition of the italicized terms, see Preamble, pp. 24–27.

Anshou, Z. (1990) A clinical study of trinitrotoluene cataract. *Pol. J. occup. Med.*, **3**, pp. 171–176

Arbeidsinspectie [Labour Inspection] (1994) *De Nationale MAC-Lijst 1994* [National MAC list 1994], The Hague, p. 41

Ashby, J., Burlinson, B., Lefevre, P.A. & Topham, J. (1985) Non-genotoxicity of 2,4,6-trinitrotoluene (TNT) to the mouse bone marrow and the rat liver: implications for its carcinogenicity. *Arch. Toxicol.*, **58**, 14–19

Bauer, C.F., Koza, S.M. & Jenkins, T.F. (1990) Liquid chromatographic method for determination of explosives residues in soil: collaborative study. *J. Assoc. off. anal. Chem.*, **73**, 541–552

Bishop, R.W., Kennedy, J., Podolak, G. & Ryea, J., Jr (1988) A field evaluation of air sampling methods for TNT and RDX. *Am. ind. Hyg. Assoc. J.*, **49**, 635–638

Boublík, T., Fried, V. & Hála, E., eds (1984) *The Vapour Pressures of Pure Substances. Selected Values of the Temperature Dependence of the Vapour Pressures of Some Pure Substances in the Normal and Low Pressure Region*, Vol. 17, Amsterdam, Elsevier

Buck, C.R. & Wilson, S.E. (1975) *Adverse Health Effects of Selected Explosives (TNT, RDX)* (Occupational Health Special Study No. 32-049-75 76), Washington DC, United States National Technical Information Services

Budavari, S., ed. (1989) *The Merck Index*, 11th Ed., Rahway, NJ, Merck & Co., pp. 1530–1531

Channon, H.J., Mills, G.T. & Williams, R.T. (1944) The metabolism of 2,4,6-trinitrotoluene. *Biochem. J.*, **38**, 70–85

Chemical Information Services (1994) *Directory of World Chemical Producers 1995/96 Standard Edition*, Dallas, TX, p. 693

Cone, T.E., Jr (1944) A review of the effect of trinitrotoluene (TNT) on the formed elements of the blood. *J. ind. Hyg. Toxicol.*, **26**, 260–263

Deutsche Forschungsgemeinschaft (1995) *MAK and BAT Values 1995* (Report No. 31), Weinheim, VCH Verlagsgesellschaft, p. 87

Dilley, J.V., Tyson, C.A., Spanggord, R.J., Sasmore, D.P., Newell, G.W. & Dacre, J.C. (1982) Short-term oral toxicity of 2,4,6-trinitrotoluene in mice, rats and dogs. *J. Toxicol. environ. Health*, **9**, 565–585

Djerassi, L. & Vitany, L. (1975) Haemolytic episode in G6 PD deficient workers exposed to TNT. *Br. J. ind. Med.*, **32**, 54–58

Douse, J.M.F. (1985) Trace analysis of explosives at the low nanogram level in handswab extracts using columns of Amberlite XAD-7 porous polymer beads and silica capillary column gas chromatography with thermal energy analysis and electron-capture detection. *J. Chromatogr.*, **328**, 155–165

Douse, J.M.F. (1987) Improved method for the trace analysis of explosives by silica capillary gas chromatography with thermal energy analysis detection. *J. Chromatogr.*, **410**, 181–189

Douse, J.M.F. & Smith, R.N. (1986) Trace analysis of explosives and firearm discharge residues in the metropolitan police forensic science laboratory. *J. energ. Mat.*, **4**, 169–186

Eddy, J.H., Jr (1944) Aplastic anemia following TNT experience. A report of 3 cases. *J. Am. med. Assoc.*, **125**, 1169–1172

Einistö, P. (1991) Role of bacterial nitroreductase and O-acetyltransferase in urine mutagenicity assay of rats exposed to 2,4,6-trinitrotoluene (TNT). *Mutat. Res.*, **262**, 167–169

El Ghawabi, S.H., Ibrahim, G.A., Gaber, M.F., El Rahman Harooni, A., Mansoor, M.B., El Owny, R. & Soudi, M.M. (1974) Trinitrotoluene exposure. *Ain Shams med. J.*, **25**, 545–549

Ermakov, E.V., Ajzenschadt, V.S. & Ventçenostcev, B.B. (1969) Chronic trinitrotoluene poisoning (clinical picture and pathogenesis of disorders of the central and autonomic nervous systems). *Sovet. Med.*, **32**, 119–122 (in Russian)

Feltes, J. & Levsen, K. (1989) Reversed phase high performance liquid chromatographic determination with photodiode-array detection of nitroaromatics from former ammunition plants in surface waters. *J. high Resolut. Chromatogr.*, **12**, 613–619

Feltes, J., Levsen, K., Volmer, D. & Spiekermann, M. (1990) Gas chromatographic and mass spectrometric determination of nitroaromatics in water. *J. Chromatogr.*, **518**, 21–40

Fetteroff, D.D., Mudd, J.L. & Teten, K. (1991) An enzyme-linked immunosorbent assay (ELISA) for trinitrotoluene (TNT) residue on hands. *J. forensic Sci.*, **36**, 343–349

Fine, D.H., Yu, W.C., Goff, E.U., Bender, E.C. & Reutrer, D.J. (1984) Picogram analyses of explosive residues using the thermal energy analyzer (TEA). *J. forensic Sci.*, **29**, 732–746

Friedlander, B.R., Vorphal, K.W., Glenn, R.E. & Jordan, P.T. (1974) *Occupational Health Special Study No. 99-020-74, APE 1300 Wash-out Plant*, Washington DC, United States National Technical Information Service

Garfinkel, D., Sidi, Y., Steier, M., Rothem, A., Marilus, R., Atsmon, A. & Pinkhas, J. (1988) Liver cirrhosis and hepatocellular carcinoma after prolonged exposure to TNT: causal relationship or mere coincidence? *Rev. roum. Med. med. Int.*, **26**, 287–290

Gibbs, T.R. & Popolato, A., eds (1980) *LASL Explosive Property Data*, Berkeley, CA, University of California Press, pp. 172–187

Goerlitz, D.F. & Franks, B.J. (1989) Use of on-site high performance liquid chromatography to evaluate the magnitude and extent of organic contaminants in aquifers. *Ground Water Monit. Rev.*, **9**, 122–129

Goh, C.L. (1984) Allergic contact dermatitis from tetryl and trinitrotoluene. *Contact Derm.*, **10**, 108

Goh, C. & Rajan, V. (1983) Contact sensitivity to trinitrotoluene. *Contact Derm.*, **9**, 433–434

Goodwin, J.W. (1972) Twenty years handling TNT in a shell loading plant. *Am. ind. Hyg. Assoc. J.*, **33**, 41–44

Hable, M., Stern, C., Asowata, C. & Williams, K. (1991) The determination of nitroaromatic and nitramines in ground and drinking water by wide-bore capillary gas chromatography. *J. chromatogr. Sci.*, **29**, 131–135

Hansch, C., Leo, A. & Hoekman, D.H. (1995) *Exploring QSAR*, Washington DC, American Chemical Society, p. 27

Härkönen, H., Karki, M., Lahti, A. & Savolainen, H. (1983) Early equatorial cataracts in workers exposed to trinitrotoluene. *Am. J. Ophthal.*, **95**, 807–810

Haroun, L.A., MacDonell, M.M., Peterson, J.M. & Fingleton, D.J. (1990) Multimedia assessment of health risks for the Weldon Spring site remedial action project. In: *Proceedings of the 83rd Annual Meeting and Exhibition, Pittsburgh, PN, June 24–29 1990*, Air and Waste Management Association, pp. 1–19

Hassman, P. (1971a) Trinitrotoluene. *Prac. Lék*, **23**, 285–294 (in Czech)

Hassman, P. (1971b) Correlation between Webster's reaction and urine 2,6-dinitro-4-aminotoluene in trinitrotoluene exposed workers. *Prac. Lék.*, **23**, 312–314 (in Czech)

Hassman, P. (1979) Health status of workers after long-lasting contact with trinitrotoluene. In: *Collection of Scientific Works of the Charles University Faculty of Medicine in Hradec Kralove*, Vol. 22, No. 1

Hassman, P. & Juran, J. (1971) Trinitrotoluene cataract. *Sb. Ved. Prac.*, **14**, 261–274

Hathaway. J. (1977) Trinitrotoluene: a review of reported dose related effects providing documentation for a workplace standard. *J. occup. Med.*, **19**, 341–345

Health and Safety Executive (1995) *Occupational Exposure Limits 1995* (Guidance Note EH 40/95), Sudbury, Suffolk, HSE Books, p. 38

Hoffsommer, J.C. & Rosen, J.M. (1972) Analysis of explosives in sea water. *Bull. environ. Contam. Toxicol.*, **7**, 177–181

Hoffsommer, J.C., Glover, D.J. & Rosen, J.M. (1972) *Analysis of Explosives in Sea Water and in Ocean Floor Sediment and Fauna* (Report No. NOLTR-72-215; US NTIS AD-757778), Silver Spring, MD, Naval Ordonance Laboratory

Horecker, B.L. & Snyder, R.K. (1944) IX. Effect of ingestion of small quantities of TNT to humans. *Public Health Bull.*, **285**, 50–52

ICI Explosives Canada (undated) *Product Data Sheet: TNT — Trinitrotoluene*, North York, Ontario

Jenkins, T.F. & Schumacher, P.W. (1990) *Evaluation of a Field Kit for Detection of TNT in Water and Soils* (Report No. CETHA-TE-CR-90056), Aberdeen, MD, United States Army Toxic and Hazardous Materials Agency

Jenkins, T.F. & Walsh, M.E. (1992) Development of field screening methods for TNT, 2,4-DNT and RDX in soil. *Talanta*, **39**, 419–428

Jenkins, T.F., Bauer, C.F., Leggett, D.C. & Grant, C.L. (1984) *Reverse Phase HPLC Method for Analysis of TNT, RDX, HMX, and 2,4-DNT in Munitions Wastewater* (Report No. DRXTH-TE-TR-8430), Aberdeen, MD, United States Army Toxic and Hazardous Materials Agency

Jenkins, T.F., Walsh, M.E., Schumacher, P.W., Miyares, P.H., Bauer, C.F. & Grant, C.L. (1989) Liquid chromatographic method for determination of extractable nitroaromatic and nitramine residues in soil. *J. Assoc. off. anal. Chem.*, **72**, 890–899

Jiang, Q.-G., Sun, J.-G. & Qin, X.-F. (1991) The effects of trinitrotoluene toxicity on zinc and copper metabolism. *Toxicol. Lett.*, **55**, 343–349

Kaplan, D.L. & Kaplan, A.M. (1981) *Analytical Method for Concentration of Trace Organics from Water* (Report No. NATICK/TR-81-014), Aberdeen, MD, United States Army Toxic and Hazardous Materials Agency

Kaplan, D.L. & Kaplan, A.M. (1982) Mutagenicity of 2,4,6-trinitrotoluene-surfactant complexes. *Bull. environ. Contam. Toxicol.*, **28**, 33–38

Karamova, N.S., Il'inskaya, O.N. & Ivanchenko, O.B. (19..) Mutagenic activity of 2,4,6-trinitrotoluene: the role of metabolizing enzymes. *Genetica*, **30**, 898–902 (in Russian)

Kline, C. (1990) *Kline Guide to the U.S. Chemical Industry*, 5th Ed., Fairfield, NJ, Kline and Co., pp. 106–109

Kolb, G., Becker, N., Scheller, S., Zugmaier, G., Pralle, H., Wahrendorf, J. & Havemann, K. (1993) Increased risk of acute myelogenous leukemia (AML) and chronic myelogenous leukemia (CML) in a county of Hesse, Germany. *Soz. Präventivmed.*, **38**, 190–195

Kraus, D.L., Hendry, C.D. & Keirn, M.A. (1985) US Department of Defense superfund implementation at a former TNT manufacturing facility. In: *Sixth National Conference on Management of Uncontrolled Hazardous Waste Sites*, Silver Spring, MD, Hazardous Materials Control Research Institute, pp. 314–318

Lakings, D.B. & Gan, O. (1981) *Identification or Development of Chemical Analysis Methods for Plants and Animal Tissues* (Report No. DRXTH-TE-CR-80086), Aberdeen, MD, United States Army Toxic and Hazardous Materials Agency

Lemberg, R. & Callaghan, J.P. (1945) Metabolism of aromatic nitro compounds. 1. Estimation of diazotisable amines in rats and human urine after inatake of 2,4,6-trinitrotoluene. *Aust. J. exp. Biol. med. Sci.*, **23**, 1–5

Levine, B.S., Furedi, E.M., Gordon, D.E., Lish, P.M. & Barkley, J.J. (1984) Subchronic toxicity of trinitrotoluene in Fischer 344 rats. *Toxicology*, **32**, 253–265

Levine, B.S., Rust, J.H., Barkley, J.J., Furedi, E.M. & Lish, P.M. (1990) Six month oral toxicity study of trinitrotoluene in beagle dogs. *Toxicology*, **63**, 233–244

Lewis, R.J., Sr (1993) *Hawley's Condensed Chemical Dictionary*, 12th Ed., New York, Van Nostrand Reinhold Co., p. 1185

Lide, D.R., ed. (1993) *CRC Handbook of Chemistry and Physics*, 74th Ed., Boca Raton, FL, CRC Press, p. 3–492

Lingyuan, K., Quanguan, J. & Qingshan, Q. (1989) Formation of superoxide radical and hydrogen peroxide enhanced by trinitrotoluene in rat liver, brain, kidney, and testicle *in vitro* and monkey liver *in vivo*. *Biomed. environ. Sciences*, **2**, 72–77

Liu, Y., Wei, W., Wang, M., Pu, Y., Lin, L. & Zhang, J. (1991) Simultaneous determination of the residues of TNT and its metabolites in human urine by thin-layer chromatography. *J. planar Chromatogr.*, **4**, 146–149

McConnell, W.J. & Flinn, R.H. (1946) Summary of twenty-two trinitrotoluene fatalities in World War II. *J. ind. Hyg. Toxicol.*, **28**, 76–86

McLuckey, S.A., Glish, G.L. & Grant, B.C. (1988) Atmospheric sampling glow discharge ionization source for the determination of trace organic compounds in ambient air. *Anal. Chem.*, **60**, 2220–2227

Melnichenko, R. (1976) Working conditions of miners handling explosives containing trinitrotoluene and prevention of their poisonings. *Gig. Tr. prof. Zabol.*, **20**, 10–13 (in Russian)

Morton, A.R., Ranadive, M.V. & Hathaway, J.A. (1976) Biological effects of trinitrotoluene from exposure below the threshold limit value. *Am. ind. Hyg. Assoc. J.*, **37**, 56–60

Nay, M.W., Jr, Randall, C.W. & King, P.H. (1972) Factors affecting color development during treatment of TNT wastes. *Ind. Wastes*, **18**, 20–29

Neal, P.A., von Oettingen, W.F. & Snyder, R.K. (1944) XI. Absorption of TNT through the intact skin of human subjects. *Public Health Bull.*, **285**, 55

Patterson, J.W., Shapira, N.I. & Brown, J. (1977) Pollution abatement in the military explosives industry. In: *Proceedings of the 31st Industrial Waste Conference, May 4–6, 1976, Purdue University, Lafayet, IN*, Ann Arbor, Ann Arbor Science, pp. 385–394

Pella, P.A. (1976) Generator for producing trace vapor concentrations of 2,4,6-trinitrotoluene, 2,4-dinitrotoluuene, and ethylene glycol dinitrate for calibrating explosives vapor detectors. *Anal. Chem.*, **48**, 1632–1637

Phung, H.T. & Bulot, M.W. (1981) Subsurface investigation of metal sludge and explosive disposal pond areas. In: Conway, R.A. & Malloy, D.C., eds, *Hazardous Solid Waste Testing, First Conference* (ASTM STP 760), Philadelphia, PA, American Society for Testing and Materials, pp. 305–320

Rickert, D.E. (1987) Metabolism of nitroaromatic compounds. *Drug Metab. Rev.*, **18**, 23–53

Roberts, W.C. (1986) *Data Summary for Trinitrotoluene* (Document No. AD-A199118), Frederick, MD, United States Medical Research and Development Laboratory

Ryon, M.G. (1987) *Water Quality Criteria for 2,4,6-Trinitrotoluene (TNT). Final Report.* (Report No. AD-ORNL-6304; US NTIS AD-A188-951), Frederick, MD, United States Medical Research and Development Laboratory

Ryon, M.G. & Ross, R.H. (1990) Water quality criteria for 2,4,6-trinitrotoluene. *Regul. Toxicol. Pharmacol.*, **11**, 104–113

Sadtler Research Laboratories (1980) *Sadtler Standard Spectra. 1980 Cumulative Index*, Philadelphia, PA, p. 146

Savolainen, H., Tenhunen, R. & Härkönen, H. (1985) Reticulocyte haem synthesis in occupational exposure to trinitrotoluene. *Br. J. ind. Med.*, **42**, 354–355

Sax, N.I. & Lewis, R.J., Sr (1989) *Dangerous Properties of Industrial Materials*, 7th Ed., Vol. 3, New York, Van Nostrand Reinhold Co., pp. 3405–3406

Short, R.D. & Lee, C.C. (1980) Effect of some nitrotoluenes on the biotransformation of xenobiotics in rats. *Experienta*, **36**, 100–101

Sievers, R. (1947) *An Evaluation of Neurologic Symptoms and Findings Occurring Among TNT Workers* (Supplement No. 196 to the Public Health Reports), Washington DC, Division of Federal Security Agency, United States Public Health Service

Sievers, R.E., Stump, R.L. & Monaco, A.R. (1946) Aplastic anemia following exposure to trinitrotoluene. *Occup. Med.*, **1**, 351–362

Snyder, R.K. (1946) Metabolites of 2,4,6-trinitrotoluene (TNT) excreted in the urine of dogs. *J. ind. Hyg. Toxicol.*, **28**, 59–75

Spanggord, R.J., Gibson, B.W., Keck, R.G., Thomas, D.W. & Barkley, J.J., Jr (1982a) Effluent analysis of wastewater generated in the manufacture of 2,4,6-trinitrotoluene. 1. Characterization study. *Environ. Sci. Technol.*, **16**, 229–232

Spanggord, R.J., Mortelmans, K.E., Griffin, A.F. & Simmon, V.F. (1982b) Mutagenicity in *Salmonella typhimurium* and structure–activity relationships of wastewater components emanating from the manufacture of trinitrotoluene. *Environ. Mutag.*, **4**, 163–179

Spangler, G.E., Carrico, J.P. & Kim, S.H. (1983) Analysis of explosives and explosive residues with ion mobility spectrometry (IMS). In: *Proceedings of the International Symposium on Analytical Detection of Explosives*, Quantico, VA, United States Federal Bureau of Investigation, pp. 267–282

Steuckart, C., Berger-Preiss, E. & Levsen, K. (1994) Determination of explosives and their biodegradation products in contaminated soil and water from former ammunition plants by automated multiple development high-performance thin-layer chromatography. *Anal. Chem.*, **66**, 2570–2577

Stewart, A., Witts, L.J., Higgins, G. & O'Brien, J.R.P. (1945) Some early effects of exposure to trinitrotoluene. *Br. J. ind. Med.*, **2**, 74–82

Styles, J.A. & Cross, M.F. (1983) Activity of 2,4,6-trinitrotoluene in an in vitro mammalian gene mutation assay. *Cancer Lett.*, **20**, 103–108

Tan, E.L., Ho, C.H., Griest, W.H. & Tyndall, R.L. (1992) Mutagenicity of trinitrotoluene and its metabolites formed during composting. *J. Toxicol. environ. Health*, **36**, 165–175

Tenhunen, R., Zitting, A., Nickels, J. & Savolainen, H. (1984) Trinitrotoluene-induced effects on rat heme metabolism. *Exp. mol. Pathol.*, **40**, 362–366

Triegel, E.K., Kolmer, J.R. & Ounanian, D.W. (1983) Solidification and thermal degradation of TNT waste sludges using asphalt encapsulation. In: *National Conference on Management of Uncontrolled Hazard Waste Sites*, Silver Spring, MD, Hazardous Materials Controls Research Institute, pp. 270–274

United Nations Environment Programme (1995) *International Register of Potentially Toxic Chemicals, Legal File, 2,4,6-Trinitrotoluene*, Geneva

United States Army (1989) *Validation of Sorbent Tube/High Performance Liquid Chromatographic Procedure for the Determination of Eight Explosives in Water* (Document No. AD-A210777), Aberdeen, MD, United States Army Environmental Hygiene Agency/ United States Army Toxic and Hazardous Materials Agency

United States National Institute for Occupational Safety and Health (1994a) *RTECs Chem.-Bank*, Cincinnati, OH

United States National Institute for Occupational Safety and Health (1994b) *Pocket Guide to Chemical Hazards* (DHHS (NIOSH) Publ. No. 94-116), Cincinnati, OH, pp. 322–323

United States National Institute for Occupational Safety and Health (1995) *National Occupational Exposure Survey (1981–1983)*, Cincinnati, OH

United States National Library of Medicine (1995) *Hazardous Substances Data Bank (HSDB)*, Bethesda, MD

United States Occupational Safety and Health Administration (1990) *Method 44. OSHA Analytical Methods Manual*, 2nd Ed., Part 1, Vol. 2, Salt Lake City, UT

United States Occupational Safety and Health Administration (1994) Air contaminants. *US Code fed. Regul.*, **Title 29**, Part 1910.1000, p. 17

Van Slyke, S.M., Scheibler, S.T. & Williams, K.E. (1985) Sampling and analytical techniques for air pollution source tests of incinerators of explosive materials. In: *Proceedings of the APCA 78th Annual Meeting* (Vol. 6), Pittsburgh, PA, Air Pollution Control Association, pp. 85–83.3

Verschueren, K. (1983) *Handbook of Environmental Data on Organic Chemicals*, 2nd Ed., New York, Van Nostrand Reinhold Co., pp. 1169–1170

Voegtlin, C., Hooper, K.W. & Johnson, J.M. (1921) Trinitrotoluene poisoning — Its nature, diagnosis and prevention. *J. ind. Hyg.*, **3**, 239–254

Whong, W.-Z. & Edwards G.S. (1984) Genotoxic activity of nitroaromatic explosives and related compounds in *Salmonella typhimurium*. *Mutat. Res.*, **136**, 209–215

Won, W.D., DiSalvo, L.H. & Ng, J. (1976) Toxicity and mutagenicity of 2,4,6-trinitrotoluene and its microbial metabolites. *Appl. environ. Microbiol.*, **31**, 576–580

Woollen, B., Hall, M., Craig, R. & Steel, G. (1986) Trinitrotoluene: assessment of occupational absorption during manufacture of explosives. *Br. J. ind. Med.*, **43**, 465–473

Yinon, J. & Hwang, D.-G. (1985) Identification of urinary metabolites of 2,4,6-trinitrotoluene in rats by liquid chromatography-mass spectrometry. *Toxicol. Lett.*, **26**, 205–209

Yinon, J. & Hwang, D.-G. (1986a) Metabolic studies of explosives: Part 5. Detection and analysis of 2,4,6-trinitrotoluene and its metabolites in urine of munition workers by micro liquid chromatography/mass spectrometry. *Biomed. Chromatogr.*, **1**, 123–125

Yinon, J. & Hwang, D.-G. (1986b) Metabolic studies of explosives. Part 4. Determination of 2,4,6-trinitrotoluene and its metabolites in blood of rabbits by high-performance liquid chromatography–mass spectrometry. *J. Chromatogr.*, **375**, 154–158

Yinon, J. & Hwang, D.-G. (1987) Applications of liquid chromatography–mass spectrometry in metabolic studies of explosives. *J. Chromatogr.*, **394**, 253–257

Zakhari, S., Villaume, J.E. & Craig, P.N. (1978) *A Literature Review — Problem Definition Studies on Selected Toxic Chemicals, Vol. 3, Occupational Health and Safety Aspects of 2,4,6-Trinitrotoluene (TNT)* (Report No. DAMD17-77-C-7020), Philadelphia, PN, Science Information Services Department, The Franklin Institute Research Laboratories

Zwirner-Baier, I., Kordowich, F.-J. & Neumann, H.-G. (1994) Hydrolysable hemoglobin adducts of polyfunctional monocyclic *N*-substituted arenes as dosimeters of exposure and markers of metabolism. *Environ. Health Perspectives*, **102** (Suppl. 6), 43–45

MUSK AMBRETTE AND MUSK XYLENE

1. Exposure Data

1.1 Chemical and physical data

1.1.1 Nomenclature

Musk ambrette

Chem. Abstr. Serv. Reg. No.: 83-66-9
Chem. Abstr. Name: 1-(1,1-Dimethylethyl)-2-methoxy-4-methyl-3,5-dinitrobenzene
IUPAC Systematic Name: 6-*tert*-Butyl-3-methyl-2,4-dinitroanisole
Synonyms: Amber musk; artificial musk ambrette; 5-*tert*-butyl-1,3-dinitro-4-methoxy-2-methylbenzene; 4-*tert*-butyl-3-methoxy-2,6-dinitrotoluene; 2,6-dinitro-3-methoxy-4-*tert*-butyltoluene; synthetic musk ambrette

Musk xylene

Chem. Abstr. Serv. Reg. No.: 81-15-2
Chem. Abstr. Name: 1-(1,1-Dimethylethyl)-3,5-dimethyl-2,4,6-trinitrobenzene
IUPAC Systematic Name: 5-*tert*-Butyl-2,4,6-trinitro-*meta*-xylene
Synonyms: 1-*tert*-Butyl-3,5-dimethyl-2,4,6-trinitrobenzene; musk xylol; 2,4,6-trinitro-1,3-dimethyl-5-*tert*-butylbenzene; 2,4,6-trinitro-3,5-dimethyl-*tert*-butylbenzene; xylene musk

1.1.2 Structural and molecular formulae and relative molecular mass

Musk ambrette

$C_{12}H_{16}N_2O_5$　　　　　　　　　　　Relative molecular mass: 268.30

Musk xylene

$C_{12}H_{15}N_3O_6$ Relative molecular mass: 297.30

1.1.3 *Chemical and physical properties of the pure substance*

Musk ambrette

(a) *Description*: Pale yellowish, whitish yellow or yellow granular crystals, leaves (from alcohol) or pale yellow powder with a sweet, heavy floral-musky odour (Lide, 1993; Flavor and Extract Manufacturers' Association, 1995a; Penta Manufacturing Co., 1995a)

(b) *Boiling-point*: > 200 °C (Flavor and Extract Manufacturers' Association, 1995a)

(c) *Melting-point*: 84–86 °C (Lide, 1993)

(d) *Spectroscopy data*: Infrared (prism [18719], grating [8877]), ultraviolet (UV) [6029], nuclear magnetic resonance (proton [2085]) and mass spectral data have been reported (Sadtler Research Laboratories, 1980)

(e) *Solubility*: Virtually insoluble in water; soluble in 95% ethanol (3.3 g/100 g), methyl carbitol (16.4 g/100 g), benzyl benzoate (50.0 g/100 g), diethyl phthalate (36.7 g/100 g) and diethyl ether (Lide, 1993; Research Institute for Fragrance Materials, 1994a; Penta Manufacturing Co., 1995a)

Musk xylene

(a) *Description*: Yellow crystals, plates or needles (from alcohol) with a musky odour (Lide, 1993; Penta Manufacturing Co., 1995b)

(b) *Boiling-point*: > 200 °C (Penta Manufacturing Co., 1995b)

(c) *Melting-point*: 110 °C (Lide, 1993)

(d) *Spectroscopy data*: Infrared (prism [1479], grating [250]), UV [422], nuclear magnetic resonance (proton [6497], C-13 [4212]) and mass spectral data have been reported (Sadtler Research Laboratories, 1980).

(e) *Solubility*: Virtually insoluble in water; soluble in diethyl ether and ethanol (Lide, 1993; Research Institute for Fragrance Materials, 1994b)

(f) *Volatility*: Vapour pressure, < 0.001 mm Hg [0.13 Pa] at 20 °C (Flavor and Extract Manufacturers' Association, 1995b)

(g) *Octanol/water partition coefficient (P)*: log P, 5.20 (Helbling *et al.*, 1994)

1.1.4 Technical products and impurities

Musk ambrette and musk xylene are available commercially (Penta Manufacturing Co., 1995a,b).

1.1.5 Analysis

Several methods for the determination of nitro musks in fragrance products have been developed based on gas chromatography (GC) (electron capture detection (ECD)/GC, Betts *et al.*, 1982; capillary GC, Spanedda *et al.*, 1986; ECD/GC, Porcu & Spanedda, 1988), liquid chromatography (Bruze *et al.*, 1985) and thin-layer chromatography (Bruze *et al.*, 1985; Goh & Kwok, 1986). Wisneski *et al.* (1994) described a method for the determination of musk ambrette in fragrance products by ECD/GC.

Capillary GC with atomic emission detection using programmed temperature vaporization has been used to detect nitro musks in human fat. The limits of detection for the nitro musks using this method were 1.0–1.6 ng (Linkerhägner *et al.*, 1994).

Nitro musks have been analysed in human adipose and fish tissues and human milk samples by capillary GC/ECD with confirmation by mass spectrometry (MS). The detection limit was 10 µg/kg fat (Rimkus & Wolf, 1993a,b; Rimkus *et al.*, 1994; Rimkus & Wolf, 1995).

Liebl and Ehrenstorfer (1993) used a similar GC/ECD method for the analysis of nitro musks in human milk samples.

Helbling *et al.* (1994) described a capillary GC/MS method for the determination of musk xylene in blood. The detection limit for musk xylene was 5 pg/g plasma or 1 ng/g lipids.

Similar methods have been developed for the determination and quantitation of nitro musks in cosmetics and detergents (Sommer, 1993).

1.2 Production and use

1.2.1 Production

Probably the earliest report on the synthesis of compounds having musk-like odour appeared in 1759 in the *Actes de l'Académie de Berlin* which contained Morggraf's statement that, when oil of amber is treated with fuming nitric acid, a resinous material is obtained that possesses a musk odour (Bedoukian, 1986).

Although Kelbe was probably the first to prepare a synthetic nitro musk and characterize it, Baur is credited with the discovery and commercialization of nitrated compounds having strong musk odours. In his 1889 German patent, Baur described a process whereby toluene was butylated with butyl halide in the presence of aluminium chloride and the product, boiling at 170–200 °C, was nitrated to give a crystalline substance possessing a strong musk odour (Bedoukian, 1986).

In 1892, Baur obtained another patent in which he identified his original musk as being trinitro-butyltoluene. At the same time he described a new product obtained by nitrating butylated *meta*-cresol methyl ether. This compound later became known as musk ambrette. In 1894, Baur patented another very important musk compound known today as musk ketone. This compound was prepared by nitrating acetylated *tert*-butyl-*meta*-xylene (Bedoukian, 1986).

Musk ambrette is now prepared commercially by the following multistep synthesis. The potassium salt of *meta*-cresol is methylated with dimethyl sulfate to give the methyl ether, which is then butylated using isobutyl chloride in the presence of aluminium chloride. The resulting *tert*-butylcresyl methyl ether is obtained in yields of 55–60% and is purified with fractionation. Nitration at temperatures below 0 °C with fuming nitric acid leads to the formation of the dinitro derivatives in yields of 45–60%. The pure product is obtained by crystallization from 95% ethanol. The by-products in this case consist of the mononitro derivative, 4,6-dinitro-*meta*-cresol methyl ether and smaller amounts of trinitro derivatives (Bedoukian, 1986).

Musk ambrette can also be produced by the methylation of 5-methyl-2-*tert*-butyl-phenol to the corresponding anisole (ambrogen), which on nitration gives musk ambrette (Reed, 1978).

Musk xylene is prepared by the nitration of *tert*-butyl-*meta*-xylene. The *tert*-butyl-*meta*-xylene is prepared by the Friedel-Crafts alkylation of *meta*-xylene with *tert*-butyl or isobutyl chloride in the presence of anhydrous aluminium chloride. The yield is typically around 70–80%. The product is nitrated with fuming nitric acid or with a mixture of sulfuric and nitric acids (70 : 30), and crude musk xylene crystallizes from the heated reaction mixture upon cooling. The crystals are filtered and washed with water and dilute sodium carbonate, and the dried product is purified by recrystallizing from 95% ethanol, with a yield (based on *tert*-butyl-*meta*-xylene) of about 88% (Bedoukian, 1986).

The aromatic class of musks consists of macrocyclics, polycyclics and nitro musks. In 1987, nitro musks constituted about 35% of the worldwide production volume of about 7000 tonnes per year of aromatic musk chemicals. Most musk compounds were produced in western Europe, where capacity exceeded demand by about 25%. The United Kingdom ranked number one in aromatic musk production, with 28% of the total worldwide. In 1987, demand for musk in the United States of America exceeded domestic production by 100%; almost 60% of the volume consumed was imported, with about 40% of nitro musk imports coming from China. Until the mid-1980s, China and India produced only nitro musks (Anon., 1988; Barbetta *et al.*, 1988).

By the early 1990s, annual worldwide production of nitro musks had declined to approximately 1000 tonnes, of which 67% was musk xylene, 21% musk ketone and 12% musk ambrette (Qinghua, 1993; Ippen, 1994). Musk ambrette was produced mainly in China and India for internal markets (Topfer, 1992).

1.2.2 *Use*

Since the mid-1980s, nitro musks have begun to be replaced in many uses by other aromatic musks, notably the polycyclics. This is due to the superior fragrance qualities of the newer materials and concerns about potential toxicity of the nitro musks (Anon., 1989; Topfer, 1990, 1992).

Musk ambrette was a fragrance ingredient used for a wide variety of applications. However, by 1992, it was reportedly no longer used in the United States and its use was very limited in Europe (Topfer, 1992). Musk ambrette has been used as a fragrance in products at the following typical concentrations (%): soap, 0.03 (max., 0.2); detergent, 0.003 (max., 0.02); creams/lotions, 0.01 (max., 0.07); and perfume, 0.2 (max., 2.0) (Opdyke, 1975). Musk ambrette also has been used in certain beverages and foods at the following concentrations (ppm) (mg/kg): alcoholic beverages, 0.10; non-alcoholic beverages, 0.18 (max., 0.42); gelatin pudding, 0.45 (max., 1.32); chewing gum, 36.0; and hard candy, 423.0 (Flavor and Extract Manufacturers' Association, 1995a).

Musk xylene is a fragrance ingredient used in fragrance compounds for a wide variety of applications. It has been in use since the early 1900s and its use in the European Union is in the region of 200 tonnes per annum. In a survey of major fragrance companies, the Research Institute for Fragrance Materials found the estimated upper 90th percentile concentrations of musk xylene in cosmetic products to be (%): toilet soap, 0.04; shampoo, 0.01; skin cream, 0.0075; deodorant, 0.0075; aftershave, 0.03; cologne/toilet water, 0.075; and fine fragrance, 0.05–0.1 (Research Institute for Fragrance Materials, 1994b).

1.3 Occurrence

1.3.1 *Natural occurrence*

None of the nitro musks (musk ambrette and musk xylene) are known to occur as natural products.

1.3.2 *Occupational exposure*

The National Occupational Exposure Survey conducted between 1981 and 1983 indicated that 22 735 employees in the United States were potentially exposed to musk ambrette and 134 410 were potentially exposed to musk xylene. The estimate is based on a survey of companies and did not involve measurements of actual exposure (United States National Institute for Occupational Safety and Health, 1995).

1.3.3 *Environmental occurrence*

(a) *Water*

In recent studies, nitroaromatic compounds, including nitro musks, were detected in unfiltered water samples of the North Sea (German Bight) and the Rivers Elbe and Stör. The highest concentrations of nitro musks were found in the effluents of a wastewater treatment plant of the city of Hamburg. These values were about one order of magnitude higher than those in the River Elbe near Hamburg. The concentrations in water from various sampling points in the Rivers Elbe and Stör were about 1 ng/L musk xylene (Rimkus & Wolf, 1995). The lowest contamination levels were found in water from various stations in the North Sea, with musk xylene ranging from < 0.03 to 0.17 ng/L (Gatermann *et al.*, 1995).

Musk xylene was detected in the River Tama (river water and dam water) in Japan in 1981 at a mean concentration of 4.1 ng/L (18 samples), in the flowing water in the tributaries which discharged into the River Tama at a mean concentration of 15 ng/L (13 samples) and in the wastewater from the sewage of three treatment plants at a mean concentration of 32 ng/L (3 samples) (Yamagishi et al., 1983).

(b) Other

Musk xylene was detected (by GC/flame ionization) as one of the components of Japanese incense sticks and was attributed to the synthetic perfumes used in the sticks (Takiura et al., 1973).

Nitro musks are used as fragrance ingredients in products commonly used both at home and in the laboratory. In the early 1980s, when musk xylene had not yet been reported as a contaminant in foods or the environment, it was detected in one of three fish samples caught in a particular lake. However, the sample was suspected to have been contaminated outside the aquatic environment; qualitative ECD/GC or GC/MS analyses of a sample of soap and three samples of hand lotions used in the laboratory showed the presence of musk xylene in each product. Musk ambrette was also found in two of the three hand lotions examined (Yurawecz & Puma, 1983).

Goh and Kwok (1986) analysed 32 men's colognes for the presence of nitro musks using thin-layer chromatography. The concentrations of musk ambrette varied from 0.02% to 0.39% w/v in 14 colognes and that of musk xylene from 0.02% to 0.78% w/v in 11 colognes.

Using GC coupled with thermal energy analysis, Nair et al. (1986) reported that, during analysis of extracts of betel quid with tobacco and of saliva of chewers of betel quid with tobacco for N-nitrosamines, two unknown compounds were detected. These were subsequently identified as musk ambrette and musk xylene by GC/MS and Fourier transform nuclear magnetic resonance spectroscopy. In samples of betel quid and tobacco, musk ambrette concentrations ranged from 0.82 to 1.44 mg/g wet weight and musk xylene concentrations from 0.45 to 0.79 mg/g wet weight; in samples of perfumed chewing tobacco, the concentrations ranged from 11.22 to 23.51 mg/g wet weight and 'not detected' to 0.60 mg/g wet weight, respectively.

Nitro musks have been identified and quantified in cosmetics and detergents. In a study in Germany, a total of 60 cosmetic products and 41 detergents were analysed; 53% of them contained nitro musks, with musk xylene being present mainly in detergents and musk ambrette detected in only two samples. Results from the various categories of products (number of samples with detectable levels of the nitro musk/number of samples analysed, maximum concentration found) were as follows: perfume — musk xylene (4/23, 13 mg/kg), musk ambrette (none detected); shampoo — musk xylene (2/13, 300 mg/kg), musk ambrette (1/13, 18 mg/kg); lotion and creme samples — musk xylene (1/24, 16 mg/kg), musk ambrette (none detected); liquid and powder detergents — musk xylene (14/30, 100 mg/kg), musk ambrette (1/30, 5.3 mg/kg); fabric softener — musk xylene (3/11, 7.2 mg/kg), musk ambrette (none detected) (Sommer, 1993).

The United States Food and Drug Administration screened 125 finished fragrances in 1985 and 1986 and found that, in both years, over 40% of the products contained musk ambrette. Perfume and cosmetic products in the market-place were also surveyed for the presence of musk ambrette in 1989, 1990 and 1992. In this study, musk ambrette was detected in 41% (29/41) of the products assayed in 1989, 8% (3/36) of the products assayed in 1990 and 11% (2/18) of the products assayed in 1992. Musk ambrette levels in the products ranged between 0.045% and 0.35% (Anon., 1987; Jackson, 1993).

1.3.4 Food

Musk xylene was detected in 40 samples collected from several sampling stations along the River Tama at a dam and in Tokyo Bay, Japan, during July and October 1980 and 1981 (three species of freshwater fish and four species of marine shellfish). The average concentrations of musk xylene were 53.9 µg/kg (ppb) in the viscera of freshwater fish, 16.0 µg/kg (ppb) in the fish muscle and 2.7 µg/kg (ppb) in marine shellfish (Yamagishi et al., 1983).

In 1991–92, residues of musk xylene were identified in tissues of farmed fish (mainly trout) and in tissues of fish from the River Lauchert, Germany. The residue levels ranged from 5 to 82 µg/kg fresh weight in 40/44 samples (when the concentration of the river water was measured at the same time as that in fish). The author noted that the river water contamination was caused by musk xylene that was added to washing powders as a perfuming agent (Hahn, 1993).

During the German Food Contamination Monitoring Programme in 1990–92 and in an extension of that study, 142 samples of fish, mussels and shrimp were analysed for nitro musks. Low levels of musk xylene (0.01–0.04 mg/g fat) were detected in the mussel samples. In trout samples from aquaculture ponds in Schleswig-Holstein, low musk xylene levels (max., 0.1 mg/kg fat) were determined, although some samples of imported trout contained high concentrations of musk xylene (max., 1.06 mg/kg fat or 0.048 mg/kg fish). Very low levels of musk ambrette were also tentatively identified, but not confirmed, in a small number of samples. Fish samples from waters in northern Germany were found to contain concentrations of musk xylene that varied with the pollution level of the water source (max., 0.35 mg/kg fat) (Rimkus & Wolf, 1993b, 1995).

1.3.5 Biological monitoring

Musk xylene has been found in human adipose tissue and breast milk. The quantity found in human fat (32 samples, 13 in women, 19 in men) varied between 0.02 and 0.09 mg/kg fat in men and 0.02 and 0.22 mg/kg fat in women. The quantity present did not vary with age as did the quantities of other substances investigated. The quantity in human breast milk (23 samples) varied between 0.02 and 0.19 mg/kg fat. The average fat content of the milk (where given) was 2.2% (range, 0.1–5.1%) (Rimkus & Wolf, 1993a; Rimkus et al., 1994).

Using GC/ECD, Liebl and Ehrenstorfer (1993) analysed 391 milk samples (48 in 1991, 343 in 1992) of nursing mothers living in southern Bavaria, Germany, for nitro musks. Musk ambrette was detected at concentrations ranging from < 0.01 to 0.29 mg/kg fat (mean concentration, 0.04 mg/kg fat). Concentrations of musk xylene were about two to three times higher, ranging from 0.01 to 1.22 mg/kg fat, with a mean content of 0.1 mg/kg fat.

Helbling *et al.* (1994) reported on the levels of musk xylene in 11 blood samples from three individuals. Musk xylene concentrations ranged from 66 to 270 pg/g plasma or 12 to 49 ng/g blood lipids. Potential laboratory sources of contamination during analysis, including paper tissues, latex gloves, the surface of a worker's hands and laboratory solvents, contributed about 50 pg/g to plasma levels and 10 ng/g to blood lipid levels.

1.4 Regulations and guidelines

Spurred by reports in 1979 and 1980 that musk ambrette was a photosensitizer, the International Fragrance Association (IFRA) in Geneva, Switzerland, issued non-enforceable guidelines in 1981 limiting the use of musk ambrette to 4% in new fragrance compounds (Anon., 1983). Since 1983, IFRA has recommended that musk ambrette should not be used in fragrance products for cosmetics, toiletries or other products which under normal conditions of use will come into contact with the skin. This includes rinse-off products. For other applications, musk ambrette should not be used as a fragrance ingredient at a level over 4% in fragrance compounds. This restriction should not be exceeded irrespective of the end-use concentration. The low-level use (< 1%) of fragrance compounds in those products in which use is allowed would result in final product concentrations of less than 0.04% (Research Institute for Fragrance Materials, 1994a).

Musk ambrette was removed from the Generally Recognized As Safe (GRAS) list in the United States in 1984 and does not have any other food use status. Musk xylene does not have food use status (Oser *et al.*, 1984; Research Institute for Fragrance Materials, 1994a). For this reason, they should not be used in lip products or flavours for oral hygiene products (Research Institute for Fragrance Materials, 1994a,b).

Due to its toxicity profile, musk xylene has not been used in Japanese products on the basis of a voluntary restriction since 1982 (Minegishi *et al.*, 1991).

2. Studies of Cancer in Humans

No data were available to the Working Group.

3. Studies of Cancer in Experimental Animals

Musk ambrette

No data were available to the Working Group.

Musk xylene

3.1 Oral administration

Mouse: Groups of 50 male and 50 female B6C3F1 mice, six weeks of age, were administered 0, 0.075 or 0.15% musk xylene (purity, > 96%) in the diet for 80 weeks, after which they were maintained on basal diet until week 90 when all survivors were killed. Dietary intakes were 0.091 (range, 0.07–0.125) and 0.170 (0.141–0.228) g/kg bw per day for males and 0.101 (0.080–0.143) and 0.192 (0.166–0.259) g/kg bw per day for females in the low- and high-dose groups, respectively. Musk xylene intake had a significant inhibitory effect on growth in high-dose males, and this was apparent from week 4 to week 80. By the end of the study, there was no longer any difference between the groups. In females, no significant difference in growth occurred throughout the experiment. There was no significant difference in cumulative mortality between controls and treated males or females. Complete histopathological examination was carried out on all animals. The overall tumour incidences (number of mice with tumours) in treated males and females at both dose levels were significantly higher than those in controls (males — 22/49 in controls; 37/50 at the low dose and 40/47 at the high dose; females — 9/46 in controls; 30/50 at the low dose; and 30/49 at the high dose. Increased tumour incidences were observed in the liver and Harderian gland (see Table 1) (Maekawa *et al.*, 1990).

Table 1. Summary of main neoplastic lesions in B6C3F1 mice given musk xylene in the diet for 80 weeks

Tumour site and type	Number of male mice with tumours			Number of female mice with tumours		
Dose	0%	0.075%	0.15%	0%	0.075%	0.15%
Effective number of mice	49	50	47	46	50	49
Liver						
Adenoma	9	19*	20**	1	14***	13***
Carcinoma	2	8*	13**	0	1	2
Adenoma/carcinoma	11	27**	33***	1	15***	15***
Harderian gland						
Adenoma	2	9*	10*	3	3	5
Carcinoma	1	1	0	0	0	0
Adenoma/carcinoma	3	10*	10*	3	3	5

From Maekawa *et al.* (1990)
*$p < 0.05$, **$p < 0.01$; ***$p < 0.001$; χ^2-test

4. Other Data Relevant to an Evaluation of Carcinogenicity and its Mechanisms

4.1 Absorption, distribution, metabolism and excretion

4.1.1 Humans

Musk xylene (^{15}N-labelled) was given to three volunteers and the elimination from blood plasma was followed for up to 162 days. The elimination half-life ranged from 63 to 107 days (Kokot-Helbling et al., 1995).

Musk xylene has been found in human fatty tissue (Rimkus & Wolf, 1993a; Rimkus et al., 1994) and breast milk (Liebl & Ehrenstorfer, 1993; Rimkus & Wolf, 1993a; Rimkus et al., 1994).

4.1.2 Experimental systems

When musk xylene (^{3}H-labelled in the 5-*tert*-butyl group) was administered intragastrically to male Wistar rats, approximately 50% of the dose was excreted into urine and faeces by 24 h and almost 87% by seven days. The proportion of the dose excreted into urine and faeces was 10.3% and 75.5%, respectively. The main metabolites observed were derived from the reduction of the 2-nitro group (2-amino-5-*tert*-butyl-4,6-dinitroxylene; 2-amino-5-*tert*-butyl-1-methyl-3-hydroxymethyl-4,6-dinitrobenzene; 2-amino-5-*tert*-hydroxybutyl-4,6-dinitroxylene) while reduction at the 4-nitro position proceeded less effectively (4-amino-5-*tert*-butyl-2,6-dinitroxylene; 4-amino-5-*tert*-butyl-1-methyl-3-hydroxymethyl-4,6-dinitrobenzene) (Minegishi et al., 1991).

4.2 Toxic effects

4.2.1 Humans

Musk ambrette can cause photoallergic contact dermatitis (Raugi et al., 1979). Most cases were in men, although contact dermatitis in a woman whose husband used a cologne containing musk ambrette has also been reported (Fisher, 1995). Patients presented with patches of eczema on the cheeks, chin and neck — the light-exposed areas on which their aftershaves had been applied. A few individuals have a more widespread eczematous reaction. There are several probable reasons why men are particularly affected: the concentration of musk ambrette in aftershaves was previously very high (as high as 15%), and the aftershave was applied in relatively large volumes to the thin skin of the face, which was often freshly abraded by shaving and is usually a maximally light-exposed area (Wojnarowska & Calnan, 1986).

Patients exhibited a positive photopatch test, which is provoked by UVB and sometimes also by UVA radiations (Ramsay, 1984). In addition, some of the patients displayed patch-test positivity (without light) to musk ambrette (Wojnarowska & Calnan, 1986) and, in one case, to its photodecomposition products as well (Bruze & Gruvberger, 1985).

Some patients develop a persistent light reaction/chronic actinic dermatitis (pruritic dermatitis with lichenification on the light-exposed areas), which can persist for years, in spite of the patient having removed the exposure (Cronin, 1984). The mechanism behind this reaction is not known.

Over a six-year period (1985–90) in New York City, United States, photopatch tests were carried out on 187 patients (76 males and 111 females) with a history of photosensitivity. Ten of the relevant responses were due to musk ambrette (DeLeo et al., 1992). Other musks are less sensitizing than musk ambrette, although positive photopatch tests in patients have also been obtained with musk xylene and moskene, but not with musk ketone or musk tibetine (Cronin, 1984).

4.2.2 *Experimental systems*

Musk ambrette

An oral LD_{50} of 339 mg/kg bw was reported for musk ambrette in rats (Jenner et al., 1964); a value of 4.8 g/kg has also been reported. The acute dermal LD_{50} of musk ambrette exceeded 2 g/kg in rabbits (Opdyke, 1975).

Musk ambrette induced photosensitivity in guinea-pigs after application to abraded skin or by using occlusion (Kochever et al., 1979; Jordan, 1982; Bueler et al., 1985); it was also positive in the mouse ear-swelling model (Gerberick & Ryan, 1990a). UVB irradiation did not enhance the photoallergic reaction of mouse ear to musk ambrette caused by UVA (as it did for the model photoallergen, 6-methylcoumarin) (Gerberick & Ryan, 1990b); musk ambrette did not elicit a positive photoallergic response in the local lymph node assay (as did the strong photoallergens tetrachlorosalicylanilide and fentichlor) (Scholes et al., 1991).

No effect was observed in rats after a 12-week feeding of 0.76 mg/kg musk ambrette in the diet (Bär & Griepentrog, 1967). After feeding 0.5–4 mg/g of diet musk ambrette to rats, growth retardation, testicular atrophy (at 2.5 mg/g) and progressive paralysis of hind limbs (at 1.5 mg/g) were observed after 12–15 weeks. At the higher doses, complete hind limb paralysis was observed within 16–40 weeks. In female rats, depressed erythrocyte counts and haemoglobin values were observed at $\geq$ 1.5 mg/g musk ambrette and icterus, indicating haemolysis, at all dose levels. Histopathological investigation revealed muscular and testicular atrophy in males and enlarged adrenal glands in females (Davis et al., 1967 (abstract); Spencer et al., 1984). Neuropathological changes included primary demyelination and distal axonal degeneration (Ford et al., 1990).

In rats, musk ambrette induced CYP1A2 but much less CYP1A1. At a daily intraperitoneal dose of 0.1 mmol (28 mg)/kg bw for five days, it caused a 50% increase in the hepatic activity of UDP-glucuronosyl transferase but did not affect the activities of DT-diaphorase or glutathione S-transferase (Iwata et al., 1993a).

Musk xylene

The acute oral LD_{50} of musk xylene was reported to exceed 10 g/kg, and the acute dermal LD_{50} in rabbits to exceed 15 g/kg (Opdyke, 1975).

Musk xylene was a weak inducer of contact hypersensitivity (occluded patch test) in guinea-pigs; the reaction was not affected by exposure to UV irradiation (Parker *et al.*, 1986).

Of five male and five female B6C3F1 mice given a single oral dose of 4000 mg/kg bw musk xylene, one female mouse died within 14 days. When musk xylene was added to the diet of B6C3F1 mice (8 males and 8 females) at concentration levels of 0.3, 0.6, 1.25, 2.5 or 5% for 14 days, all mice given 0.6% musk xylene, except one female, died within two to four days. All animals given 0.3% musk xylene survived to the end of the experiment. Histological examination revealed haemorrhagic erosions in the glandular stomach only. In a 17-week study with dietary levels of 0.0375, 0.075, 0.15, 0.3 or 0.6% musk xylene, all mice at the highest-dose level and all females and 8/10 males given 0.3% musk xylene died; no death occurred in the other groups. At dose levels of 0.15% and less, no difference in body-weight development or organ weights between treated and control animals was observed. Enlargement and irregularity of liver cells were observed in animals fed 0.15% musk xylene (Maekawa *et al.*, 1990).

No excess mortality, decrease in body-weight gain, clinical chemistry abnormality or gross or microscopic change were observed in Sprague-Dawley rats after daily dermal applications of 240 mg/kg bw musk xylene for 90 days. The only abnormal finding was an increase in the relative liver weight (Ford *et al.*, 1990).

Musk xylene increased the liver weight and the total cytochrome P450 content, and induced CYP1A2 and, to a far lesser extent, CYP1A1 in male Wistar rats (Iwata *et al.*, 1992, 1993a,b). At a daily intraperitoneal dose of 0.1 mmol [30 mg]/kg bw for five days, musk xylene did not affect the hepatic activities of DT-diaphorase, glutathione *S*-transferase or UPD-glucuronosyl transferase (Iwata *et al.*, 1993b). At higher doses, induction of these enzyme activities was observed (Iwata *et al.*, 1993a).

In male B6C3F1 mice, musk xylene given at 200 mg/kg bw per day by gavage for seven days increased liver weight by 40%, caused hepatocellular hypertrophy and increased cytochrome P450 content with a concomitant induction of CYP1A1, 1A2 and 2B proteins. However, while the activities of CYP1A1 and 1A2 were elevated, that of 2B was not. This was explained by the fact that it inhibited murine CYP2B enzymes *in vitro* (IC_{50} approx 1 µmol [300 µg]/L) (Lehman-McKeeman *et al.*, 1995). As reported in an abstract, a single oral dose (200 mg/kg bw) of musk xylene given to B6C3F1 mice pretreated with phenobarbital decreased the measurable CYP2B activity by 90%; this inhibition was not seen in mice also treated with a combination of neomycin, tetracycline and bacitracin. The authors interpreted this to mean that the inhibition of CYP2B was not due to musk xylene itself but to a metabolite formed by intestinal microflora (Caudill *et al.*, 1995; Lehman-McKeeman *et al.*, 1995).

4.3 Reproductive and developmental effects

No data were available to the Working Group.

4.4 Genetic and related effects

4.4.1 *Humans*

No data were available to the Working Group.

4.4.2 *Experimental systems* (see also Table 2 and Appendices 1 and 2)

Musk ambrette was mutagenic in *Salmonella typhimurium* TA100 requiring metabolic activation by rat-liver S9.

In *Drosophila*, musk ambrette induced sex-linked recessive lethal mutations in mature sperm. After intraperitoneal injection or after oral dosing musk ambrette did not induce micronuclei in the bone marrow of male or female NMRI mice.

Musk xylene gave uniformly negative results in a series of short-term genotoxicity tests that included the *S. typhimurium* mutation test, the mouse lymphoma assay, an in-vitro cytogenetics assay in Chinese hamster ovary (CHO) cells, the in-vitro unscheduled DNA synthesis assay in primary rat hepatocytes and an in-vivo unscheduled DNA synthesis assay.

5. Summary of Data Reported and Evaluation

5.1 Exposure data

Musk ambrette and musk xylene are nitro musks, which are prepared by nitration of *tert*-butylcresol methyl ether and *tert*-butyl-*meta*-xylene, respectively. Musk xylene and, in lower amounts, musk ambrette have been used since the early 1900s as fragrance ingredients in perfumes, soaps, detergents and cosmetics. Musk ambrette has also been used at low levels in foods such as candy, chewing gum and beverages. Nitro musks have been detected in surface waters and in fish and shellfish.

5.2 Human carcinogenicity data

No data were available to the Working Group.

5.3 Animal carcinogenicity data

No data were available on the carcinogenicity of musk ambrette.

Musk xylene was tested for carcinogenicity in mice by oral administration in the diet in one experiment and induced increased incidences of hepatocellular adenomas and carcinomas and Harderian gland tumours in males and hepatocellular adenomas in females.

Table 2. Genetic and related effects of nitro musks

Test system	Result[a] Without exogenous metabolic system	Result[a] With exogenous metabolic system	Dose[b] (LED/HID)	Reference
Musk ambrette				
SA0, *Salmonella typhimurium* TA100, reverse mutation	–	+	200	Wild et al. (1983)
SA0, *Salmonella typhimurium* TA100, reverse mutation	–	+	100	Nair et al. (1986)
SA0, *Salmonella typhimurium* TA100, reverse mutation	–	–	128	Zeiger et al. (1987)
SA5, *Salmonella typhimurium* TA1535, reverse mutation	–	–	128	Zeiger et al. (1987)
SA7, *Salmonella typhimurium* TA1537, reverse mutation	–	–	128	Zeiger et al. (1987)
SA9, *Salmonella typhimurium* TA98, reverse mutation	–	–	200	Nair et al. (1986)
SA9, *Salmonella typhimurium* TA98, reverse mutation	–	–	128	Zeiger et al. (1987)
DMX, *Drosophila melanogaster*, sex-linked recessive lethal mutations	(+)		2680 adult feeding	Wild et al. (1983)
MVM, Micronucleus test, mouse bone-marrow cells *in vivo*	–		1072 ip × 2	Wild et al. (1983)
MVM, Micronucleus test, mouse bone-marrow cells *in vivo*	–		2948 po × 1	Wild et al. (1983)
Musk xylene				
SA0, *Salmonella typhimurium* TA100, reverse mutation	–	–	200	Nair et al. (1986)
SA0, *Salmonella typhimurium* TA100, reverse mutation	–	–	100	Api et al. (1995)
SA5, *Salmonella typhimurium* TA1535, reverse mutation	–	–	100	Api et al. (1995)
SA7, *Salmonella typhimurium* TA1537, reverse mutation	–	–	100	Api et al. (1995)
SA8, *Salmonella typhimurium* TA1538, reverse mutation	–	–	100	Api et al. (1995)
SA9, *Salmonella typhimurium* TA98, reverse mutation	–	–	200	Nair et al. (1986)
SA9, *Salmonella typhimurium* TA98, reverse mutation	–	–	100	Api et al. (1995)
URP, Unscheduled DNA synthesis, rat primary hepatocytes *in vitro*	–	0	30	Api et al. (1995)
G5T, Gene mutation, mouse lymphoma L5178Y cells, tk locus	–	–	400	Api et al. (1995)
CIC, Chromosomal aberrations, Chinese hamster CHO cells *in vitro*	–	–	30	Api et al. (1995)
UPR, Unscheduled DNA synthesis, rat hepatocytes *in vivo*	–		5000	Api et al. (1995)

[a] +, positive; (+), weak positive; –, negative; 0, not tested; ?, inconclusive (variable response within several experiments within an adequate study)

[b] LED, lowest effective dose; HID, highest ineffective dose. In-vitro tests, μg/mL; in-vivo tests, mg/kg bw

5.4 Other relevant data

Application of musk ambrette on the skin may cause photocontact dermatitis and chronic actinic dermatitis.

Musk ambrette was mutagenic in *Salmonella* and *Drosophila*. It did not induce micronuclei in the bone marrow of mice *in vivo*.

In humans, musk xylene is absorbed from the gastrointestinal tract. It is distributed to the adipose tissue and its half-time in blood plasma is two to three months. It is excreted in human milk.

Musk xylene is metabolized in the rat by nitroreduction. Musk xylene is a phenobarbital-type inducer of cytochromes P450 in rats and mice.

Musk xylene did not induce genetic damage in bacteria, cultured mammalian cells or, in one study, in mammals *in vivo*.

5.5 Evaluation[1]

There is *inadequate evidence* in humans for the carcinogenicity of musk ambrette and musk xylene.

There is *inadequate evidence* in experimental animals for the carcinogenicity of musk ambrette.

There is *limited evidence* in experimental animals for the carcinogenicity of musk xylene.

Overall evaluation

Musk ambrette and musk xylene are *not classifiable as to their carcinogenicity to humans (Group 3)*.

6. References

Anon. (1983) Musk ambrette photoallergy safety question still unresolved. *Chem. Mark. Rep.*, **223**, 34

Anon. (1987) Musk ambrette's persistent presence in fragrances. *FDC Reports Rose Sheet*, **(May 25)**, 4–5

Anon. (1988) Study predicts geographical shift in musk aroma chemical production. *Soap Cosmet. chem. Special.*, **64**, 88

Anon. (1989) Polycyclic musks gain with decline in use of nitromusks. *Chem. Mark. Rep.*, **235**, 21

Api, A.M., Ford, R.A. & San, R.H.C. (1995) An evaluation of musk xylene in a battery of genotoxicity tests. *Food chem. Toxicol.*, **33**, 1039–1045

---

[1]For definition of the italicized terms, see Preamble, pp. 24–27.

Bär, F. & Griepentrog, F. (1967) State of safety evaluation of aromatics for food. *Med. Ernähr.*, **8**, 244–251 (in German)

Barbetta, L., Trowbridge, T. & Eldib, I.A. (1988) Musk aroma chemical industry. *Perfum. Flavor.*, **13**, 60–61

Bedoukian, P.Z. (1986) *Perfumery and Flavoring Synthetics*, 3rd Ed., Wheaton, IL, Allured Publishing, pp. 322–333

Betts, T.J., Tai, G.M. & Turner, R.A. (1982) Evaluation of toiletries for possible allergenic concentrations of nitromusks using electron-capture gas chromatography. *J. Chromatogr.*, **244**, 381–384

Bruze, M. & Gruvberger, B. (1985) Contact allergy to photodecomposition products of musk ambrette. *Photodermatology,* **2**, 310–314

Bruze, M., Edman, B., Niklasson, B. & Möller, H. (1985) Thin layer chromatography and high pressure liquid chromatography of musk ambrette and other nitromusk compounds including photopatch studies. *Photodermatology*, **2**, 295–302

Buehler, E.V., Newmann, E.A. & Parker, R.D. (1985) Use of the occlusive patch to evaluate the photosensitive properties of chemicals in guinea pigs. *Food chem. Toxicol.*, **7**, 689–694

Caudill, D., Johnson, D.R. & Lehman-McKeeman, L.D. (1995) Musk xylol (MX) induces and inhibits mouse cytochrome P-450 2B enzymes. *Toxicologist*, **15**, 117

Cronin, E. (1984) Photosensitivity to musk ambrette. *Contact Derm.*, **11**, 88–92

Davis, D.A., Taylor, J.M., Jones, W.I. & Brouwer, J.B. (1967) Toxicity to musk ambrette (Abstract no. 71). *Toxicol. appl. Pharmacol.*, **10**, 405

DeLeo, V.A., Suarez, S.M. & Maso, M.J. (1992) Photoallergic contact dermatitis. Results of photopatch testing in New York, 1985 to 1990. *Arch. Dermatol.*, **128**, 1513–1518

Fisher, A.A. (1995) Consort contact dermatitis due to musk ambrette. *Curr. Contact News*, **55**, 199–200

Flavor and Extract Manufacturers' Association (1995a) *Monograph: Musk Ambrette*, Washington DC

Flavor and Extract Manufacturers' Association (1995b) *Monograph: Musk Xylene*, Washington DC

Ford, R.A., Api, A.M. & Newberne, P.M. (1990) 90-Day dermal toxicity study and neurotoxicity evaluation of nitromusks in the albino rat. *Food chem. Toxicol.*, **28**, 55–61

Gatermann, R., Hühnerfuss, H., Rimkus, G., Wolf, M. & Franke, S. (1995) The distribution of nitrobenzene and other nitroaromatic compounds in the North Sea. *Mar. Pollut. Bull.*, **30**, 221–227

Gerberick, G.F. & Ryan, C.A. (1990a) A predictive mouse ear-swelling model for investigating topical photoallergy. *Food chem. Toxicol.*, **28**, 361–368

Gerberick, G.F. & Ryan, C.A. (1990b) Use of UVB and UVA to induce and elicit contact photoallergy in the mouse. *Photodermatol. Photoimmunol. Photomed.*, **7**, 13–19

Goh, C.L. & Kwok, S.F. (1986) A simple method of qualitative analysis of musk ambrette, musk ketone and musk xylene in cologne. *Contact Derm.*, **14**, 53–56

Hahn, J. (1993) Occurrence of musk xylene in fish. *Dtsch. Lebensm. Rundsch.*, **89**, 175–177 (in German)

Helbling, K.S., Schmid, P. & Schlatter, C. (1994) The trace analysis of musk xylene in biological samples: problems associated with its ubiquitous occurrence. *Chemosphere*, **29**, 477–484

Ippen, H. (1994) Nitro musk. *Int. Arch. occup. environ. Health*, **66**, 283–285

Iwata, N., Minegishi, K.-I., Suzuki, K., Ohno, Y., Kawanishi, T. & Takahashi, A. (1992) Musk xylene is a novel specific inducer of cytochrome P-450IA2. *Biochem. biophys. Res. Commun.*, **184**, 149–153

Iwata, N., Minegishi, K.-I., Suzuki, K., Ohno, Y., Igarashi, T., Satoh, T. & Takahashi, A. (1993a) An unusual profile of musk xylene-induced drug-metabolizing enzymes in rat liver. *Biochem. Pharmacol.*, **45**, 1659–1665

Iwata, N., Suzuki, K., Minegishi, K.-I., Kawanishi, T., Hara, S., Endo, T. & Takahashi, A. (1993b) Induction of cytochrome P450 1A2 by musk analogues and other inducing agents in rat liver. *Eur. J. Pharmacol.*, **248**, 243–250

Jackson, E.M. (1993) Substantiating the safety of fragrances and fragranced products. *Cosmet. Toiletries*, **108**, 43–46

Jenner, P.M., Hagan, E.C., Taylor, J.M., Cook, E.L. & Fitzhugh, O.G. (1964) Food flavourings and compounds of related structure. *Food Cosmet. Toxicol.*, **2**, 327–343

Jordan, W.P., Jr (1982) The guinea pig as a model for predicting photoallergic contact dermatitis. *Contact Derm.*, **8**, 109–116

Kim, Y.E., Fornwald, L.W., Anderson, L.M. & Beebe, L.E. (1995) Delayed induction of pulmonary CYP2B in mice after coadministration of nitrosodimethylamine (NDMA) and ethanol (abstract). *Toxicologist*, **15**, 117

Kochever, I.E., Zalar, G.L., Einbinder, J. & Harber, L.C. (1979) Assay of contact photosensitivity to musk ambrette in guinea pigs. *J. invest. Dermatol.*, **73**, 144–146

Kokot-Helbling, K., Schmid, P. & Schlatter, C. (1995) Human exposure to musk xylene — absorption, pharmacokinetics and toxicology. *Mitt. geb. Lebensmitt. Hyg.*, **86**, 1–13 (in German)

Lehman-McKeeman, L.D., Caudill, D., Young, J.A. & Dierckman, T.A. (1995) Musk xylene induces and inhibits mouse hepatic cytochrome P-450 2B enzymes. *Biochem. biophys. Res. Commun.*, **206**, 975–980

Lide, D.R., ed. (1993) *CRC Handbook of Chemistry and Physics*, 74th Ed., Boca Raton, FL, CRC Press, p. 3-79

Liebl, B. & Ehrenstorfer, S. (1993) Nitro musks in human milk. *Chemosphere*, **27**, 2253–2260

Linkerhägner, M., Stan, H.-J. & Rimkus, G. (1994) Detection of nitro musks in human fat by capillary gas chromatography with atomic emission detection (AED) using programmed temperature vaporization (PTV). *J. high Resolut. Chromatogr.*, **17**, 821–826

Maekawa, A., Matsushima, Y., Onodera, H., Shibutani, M., Ogasawara, H., Kodama, Y., Kurokawa, Y. & Hayashi, Y. (1990) Long-term toxicity/carcinogenicity of musk xylol in B6C3F mice. *Food chem. Toxicol.*, **28**, 581–586

Minegishi, K.-I., Nambaru, S., Fukuoka, M., Tanaka, A. & Nishimaki-Mogami, T. (1991) Distribution, metabolism, and excretion of musk xylene in rats. *Arch. Toxicol.*, **65**, 273–282

Nair, J., Ohshima, H., Malaveille, C., Friesen, M., O'Neill, I.K., Hautefeuille, A. & Bartsch, H. (1986) Identification, occurrence and mutagenicity in *Salmonella typhimurium* of two synthetic nitroarenes, musk ambrette and musk xylene, in Indian chewing tobacco and betel quid. *Food chem. Toxicol.*, **24**, 27–31

Opdyke, D.L.J. (1975) Fragrance raw material monographs: musk ambrette; musk ketone; musk xylol. *Food chem. Toxicol.*, **13** (Suppl.), 875–878, 881

Oser, B.L., Ford, R.A. & Bernard, B.K. (1984) Recent progress in the consideration of flavoring ingredients under the food additives amendment. 13. GRAS substances. *Food Technol.*, **38**, 66-89

Parker, R.D., Buehler, E.V. & Newmann, E.A. (1986) Phototoxicity, photoallergy, and contact sensitization of nitro musk perfume raw materials. *Contact Derm.*, **14**, 103–109

Penta Manufacturing Co. (1995a) *Technical Data Sheet: Musk Ambrette*, Livingston, NJ

Penta Manufacturing Co. (1995b) *Technical Data Sheet: Musk Xylene*, Livingston, NJ

Porcu, M. & Spanedda, L. (1988) Gas chromatographic determination of synthetic musks in alcohol-containing perfumes. *Riv. Merceol.*, **27**, 175–185 (in Italian)

Qinghua, Z. (1993) China's perfumery industry picks up. *Perfum. Flavor.*, **18**, 47–48

Ramsay, C.A. (1984) Transient and persistent photosensitivity due to musk ambrette. Clinical and photobiological studies. *Br. J. Dermatol.*, **111**, 423–429

Raugi, G.J., Storrs, F.J. & Larsen, W.G. (1979) Photoallergic contact dermatitis to men's perfumes. *Contact Dermatol.*, **5**, 251–260

Reed, H.W.B. (1978) Alkylphenols. In: Mark, H.F., Othmer, D.F., Overberger, C.G., Seaborg, G.T. & Grayson, N., eds, *Kirk-Othmer Encyclopedia of Chemical Technology*, 3rd Ed., Vol. 2, New York, John Wiley & Sons, p. 75

Research Institute for Fragrance Materials (1994a) *Safety Evaluation of Musk Ambrette — Summary*, Hackensack, NJ

Research Institute for Fragrance Materials (1994b) *Safety Evaluation of Musk Xylene — Summary*, Hackensack, NJ

Rimkus, G. & Wolf, M. (1993a) Occurrence of nitro musks in human milk and human fat. *Dtsch. Lebensm. Rundsch.*, **89**, 103–107 (in German)

Rimkus, G. & Wolf, M. (1993b) Residues and contaminants in fish in aqua culture. *Dtsch. Lebensm. Rundsch.*, **89**, 171–175 (in German)

Rimkus, G.G. & Wolf, M. (1995) Nitro musk fragrances in biota from freshwater and marine environment. *Chemosphere*, **30**, 641–651

Rimkus, G., Rimkus, B. & Wolf, M. (1994) Nitro musks in human adipose tissue and breast milk. *Chemosphere*, **28**, 421–432

Sadtler Research Laboratories (1980) *Sadtler Standard Spectra. 1980 Cumulative Index*, Philadelphia, PA, pp. 595, 600

Scholes, E.W., Basketter, D.A., Lovell, W.W., Sarll, A.E. & Pendlington, R.U. (1991) The identification of photoallergic potential in the local lymph node assay. *Photodermatol. Photoimmunol. Photomed.*, **8**, 249–254

Sommer, C. (1993) Gas chromatography determination of nitro musks in cosmetics and detergents. *Dtsch. Lebensm. Rundsch.*, **89**, 108–111 (in German)

Spanedda, L., Melis, M. & Roni, C. (1986) Separation of nitro musks by capillary gas chromatography. *J. Chromatogr.*, **362**, 278–280

Spencer, P.S., Bischoff-Fenton, M.C., Moreno, O., Opdyke, D.L. & Ford, R.A. (1984) Neurotoxic properties of musk ambrette. *Toxicol. appl. Pharmacol.*, **75**, 571–575

Takiura, K., Yamaji, A., Iwasaki, K. & Yuki, H. (1973) Analysis of Japanese incense sticks by gas chromatography. *Bunseki Kagaku*, **22**, 916–918 (in Japanese)

Topfer, K. (1990) New aromatic musks may displace most nitrocyclics. *Chem. Mark. Rep.*, **238**, 35–36

Topfer, K. (1992) Chinese musk prices drop as polycyclic use grows. *Chem. Mark. Rep.*, **242**, 26

United States National Institute for Occupational Safety and Health (1995) *National Occupational Exposure Survey (1981–1983)*, Cincinnati, OH

Wild, D., King, M.-T., Gocke, E. & Eckhardt, K. (1983) Study of artificial flavouring substances for mutagenicity in the *Salmonella*/microsome, Basc and micronucleus tests. *Food chem. Toxicol.*, **21**, 707–719

Wisneski, H.H., Yates, R.L. & Havery, D.C. (1994) Determination of musk ambrette in fragrance products by capillary gas chromatography with electron capture detection: interlaboratory study. *J. Assoc. off. Anal. Chem. int.*, **77**, 1467–1471

Wojnarowska, F. & Calnan, C.D. (1986) Contact and photoallergy to musk ambrette. *Br. J. Dermatol.*, **114**, 667–675

Yamagishi, T., Miyazaki, T., Horii, S. & Akiyama, K. (1983) Synthetic musk residues in biota and water from Tama River and Tokyo Bay (Japan). *Arch. environ. Contam. Toxicol.*, **12**, 83–89

Yurawecz, M.P. & Puma, B.J. (1983) Nitro musk fragrances as potential contaminants in pesticide residue analysis. *J. Assoc. off. anal. Chem.*, **66**, 241–247

Zeiger, E., Anderson, B., Haworth, S., Lawlor, T., Mortelmans, K. & Speck, W. (1987) *Salmonella* mutagenicity tests: III. Results from the testing of 255 chemicals. *Environ. Mutag.*, **9** (Suppl. 9), 1–110

SUMMARY OF FINAL EVALUATIONS

Agent (occupation)	Degree of evidence of carcinogenicity		Overall evaluation of carcinogenicity to humans
	Human	Animal	
Carbon black	I	S	2B
Carbon black extracts		S	
Chloronitrobenzenes	I	I	3
3,7-Dinitrofluoranthene	I	S	2B
3,9-Dinitrofluoranthene	I	S	2B
2,4-Dinitrotoluene	I	S	2B
2,6-Dinitrotoluene	I	S	2B
3,5-Dinitrotoluene	I	I	3
Musk ambrette	I	I	3
Musk xylene	I	L	3
2-Nitroanisole	I	S	2B
Nitrobenzene	I	S	2B
Nitrotoluenes	I		3
2-Nitrotoluene		L	
3-Nitrotoluene		I	
4-Nitrotoluene		I	
Printing inks	I	I	3
Printing processes (occupational exposures in)	L		2B
Tetranitromethane	I	S	2B
2,4,6-Trinitrotoluene	I	I	3

S, sufficient evidence; L, limited evidence; I, inadequate evidence; for definitions of criteria for degrees of evidence and groups, see preamble, pp. 24–27

APPENDIX 1

SUMMARY TABLES OF
GENETIC AND RELATED EFFECTS

APPENDIX 1

Summary table of genetic and related effects of carbon black

Non-mammalian systems				Mammalian systems			
Proka-ryotes	Lower eukaryotes	Plants	Insects	In vitro			In vivo
				Animal cells	Human cells	Animals	Humans
D G	D R G A	D G C	R G C A	D G S M C A T I	D G S M C A T I	D G S M C DL A	D S M C A
–			– –	– – – –		+!	

A, aneuploidy; C, chromosomal aberrations; D, DNA damage; DL, dominant lethal mutation; G, gene mutation; I, inhibition of intercellular communication; M, micronuclei; R, mitotic recombination and gene conversion; S, sister chromatid exchange; T, cell transformation

In completing the table, the following symbols indicate the consensus of the Working Group with regard to the results for each end-point:

+ considered to be positive for the specific end-point and level of biological complexity
+! considered to be positive, but only one valid study was available to the Working Group
– considered to be negative
–! considered to be negative, but only one valid study was available to the Working Group
? considered to be equivocal or inconclusive (e.g. there were contradictory results from different laboratories; there were confounding exposures; the results were equivocal)

Summary table of genetic and related effects of 2-chloronitrobenzene

Non-mammalian systems				Mammalian systems			
Proka-ryotes	Lower eukaryotes	Plants	Insects	In vitro			In vivo
				Animal cells	Human cells	Animals	Humans
D G	D R G A	D G C	R G C A	D G S M C A T I	D G S M C A T I	D G S M C DL A	D S M C A
–¹ +			–	+¹ +¹		+	

A, aneuploidy; C, chromosomal aberrations; D, DNA damage; DL, dominant lethal mutation; G, gene mutation; I, inhibition of intercellular communication; M, micronuclei; R, mitotic recombination and gene conversion; S, sister chromatid exchange; T, cell transformation

In completing the table, the following symbols indicate the consensus of the Working Group with regard to the results for each end-point:

+   considered to be positive for the specific end-point and level of biological complexity
+¹ considered to be positive, but only one valid study was available to the Working Group
–   considered to be negative
–¹ considered to be negative, but only one valid study was available to the Working Group
?   considered to be equivocal or inconclusive (e.g. there were contradictory results from different laboratories; there were confounding exposures; the results were equivocal)

Summary table of genetic and related effects of 3-chloronitrobenzene

Non-mammalian systems				Mammalian systems			
Proka-ryotes	Lower eukaryotes	Plants	Insects	In vitro			In vivo
				Animal cells	Human cells	Animals	Humans
D G	D R G A	D G C	R G C A	D G S M C A T I	D G S M C A T I	D G S M C DL A	D S M C A
– –				– –			

A, aneuploidy; C, chromosomal aberrations; D, DNA damage; DL, dominant lethal mutation; G, gene mutation; I, inhibition of intercellular communication; M, micronuclei; R, mitotic recombination and gene conversion; S, sister chromatid exchange; T, cell transformation

In completing the table, the following symbols indicate the consensus of the Working Group with regard to the results for each end-point:

+ considered to be positive for the specific end-point and level of biological complexity
+¹ considered to be positive, but only one valid study was available to the Working Group
– considered to be negative
–¹ considered to be negative, but only one valid study was available to the Working Group
? considered to be equivocal or inconclusive (e.g. there were contradictory results from different laboratories; there were confounding exposures; the results were equivocal)

Summary table of genetic and related effects of 4-chloronitrobenzene

Non-mammalian systems				Mammalian systems			
Proka-ryotes	Lower eukaryotes	Plants	Insects	In vitro		In vivo	
				Animal cells	Human cells	Animals	Humans
D G	D R G A	D G C	R G C A	D G S M C A T I	D G S M C A T I	D G S M C DL A	D S M C A
−¹ +			−	+ +¹ +		+¹	

A, aneuploidy; C, chromosomal aberrations; D, DNA damage; DL, dominant lethal mutation; G, gene mutation; I, inhibition of intercellular communication; M, micronuclei; R, mitotic recombination and gene conversion; S, sister chromatid exchange; T, cell transformation

In completing the table, the following symbols indicate the consensus of the Working Group with regard to the results for each end-point:

+   considered to be positive for the specific end-point and level of biological complexity
+¹   considered to be positive, but only one valid study was available to the Working Group
−   considered to be negative
−¹   considered to be negative, but only one valid study was available to the Working Group
?   considered to be equivocal or inconclusive (e.g. there were contradictory results from different laboratories; there were confounding exposures; the results were equivocal)

APPENDIX 1

Summary table of genetic and related effects of 3,7-dinitrofluoranthene

Non-mammalian systems				Mammalian systems			
Proka-ryotes	Lower eukaryotes	Plants	Insects	In vitro			In vivo		Humans
				Animal cells	Human cells	Animal cells	Human cells	Animals	Humans
D G	D R G A	D G C	R G C A	D G S M C A T I	D G S M C A T I	D G S M C DL A	D S M C A
+ +				– ^{1} + ^{1}		+ ^{1}	

A, aneuploidy; C, chromosomal aberrations; D, DNA damage; DL, dominant lethal mutation; G, gene mutation; I, inhibition of intercellular communication; M, micronuclei; R, mitotic recombination and gene conversion; S, sister chromatid exchange; T, cell transformation

In completing the table, the following symbols indicate the consensus of the Working Group with regard to the results for each end-point:

+   considered to be positive for the specific end-point and level of biological complexity
+ ^{1} considered to be positive, but only one valid study was available to the Working Group
–   considered to be negative
– ^{1} considered to be negative, but only one valid study was available to the Working Group
?   considered to be equivocal or inconclusive (e.g. there were contradictory results from different laboratories; there were confounding exposures; the results were equivocal)

Summary table of genetic and related effects of 3,9-dinitrofluoranthene

Non-mammalian systems				Mammalian systems			
Prokaryotes	Lower eukaryotes	Plants	Insects	In vitro		In vivo	
				Animal cells	Human cells	Animals	Humans
D G	D R G A	D G C	R G C A	D G S M C A T I	D G S M C A T I	D G S M C DL A	D S M C A
+ +				–¹ +¹		+¹	

A, aneuploidy; C, chromosomal aberrations; D, DNA damage; DL, dominant lethal mutation; G, gene mutation; I, inhibition of intercellular communication; M, micronuclei; R, mitotic recombination and gene conversion; S, sister chromatid exchange; T, cell transformation

In completing the table, the following symbols indicate the consensus of the Working Group with regard to the results for each end-point:

+   considered to be positive for the specific end-point and level of biological complexity
+¹   considered to be positive, but only one valid study was available to the Working Group
–   considered to be negative
–¹   considered to be negative, but only one valid study was available to the Working Group
?   considered to be equivocal or inconclusive (e.g. there were contradictory results from different laboratories; there were confounding exposures; the results were equivocal)

APPENDIX 1

Summary table of genetic and related effects of 2,4-dinitrotoluene (technical grade)

Non-mammalian systems				Mammalian systems			
Proka-ryotes	Lower eukaryotes	Plants	Insects	In vitro		In vivo	
				Animal cells	Human cells	Animals	Humans
D G	D R G A	D G C	R G C A	D G S M C A T I	D G S M C A T I	D G S M C DL A	D S M C A
+				$–^?$ $–$ $+^1$		+ $+^1$ $–^1$ $–$	$–$

A, aneuploidy; C, chromosomal aberrations; D, DNA damage; DL, dominant lethal mutation; G, gene mutation; I, inhibition of intercellular communication; M, micronuclei; R, mitotic recombination and gene conversion; S, sister chromatid exchange; T, cell transformation

In completing the table, the following symbols indicate the consensus of the Working Group with regard to the results for each end-point:

+ considered to be positive for the specific end-point and level of biological complexity
$+^1$ considered to be positive, but only one valid study was available to the Working Group
– considered to be negative
$–^1$ considered to be negative, but only one valid study was available to the Working Group
? considered to be equivocal or inconclusive (e.g. there were contradictory results from different laboratories; there were confounding exposures; the results were equivocal)

Summary table of genetic and related effects of 2,4-dinitrotoluene (high purity)

Non-mammalian systems				Mammalian systems			
Prokaryotes	Lower eukaryotes	Plants	Insects	In vitro			In vivo
				Animal cells	Human cells		Animals	Humans
D G	D R G A	D G C	R G C A	D G S M C A T I	D G S M C A T I	D G S M C DL A	D S M C A
+ +			? –	– ? +¹ – – ?	–¹	+ –	

A, aneuploidy; C, chromosomal aberrations; D, DNA damage; DL, dominant lethal mutation; G, gene mutation; I, inhibition of intercellular communication; M, micronuclei; R, mitotic recombination and gene conversion; S, sister chromatid exchange; T, cell transformation

In completing the table, the following symbols indicate the consensus of the Working Group with regard to the results for each end-point:
+   considered to be positive for the specific end-point and level of biological complexity
+¹ considered to be positive, but only one valid study was available to the Working Group
–   considered to be negative
–¹ considered to be negative, but only one valid study was available to the Working Group
?   considered to be equivocal or inconclusive (e.g. there were contradictory results from different laboratories; there were confounding exposures; the results were equivocal)

Summary table of genetic and related effects of 2,6-dinitrotoluene

Non-mammalian systems				Mammalian systems			
Proka-ryotes	Lower eukaryotes	Plants	Insects	In vitro		In vivo	
				Animal cells	Human cells	Animals	Humans
D G	D R G A	D G C	R G C A	D G S M C A T I	D G S M C A T I	D G S M C DL A	D S M C A
+				+¹ –	–¹ ?	+	

A, aneuploidy; C, chromosomal aberrations; D, DNA damage; DL, dominant lethal mutation; G, gene mutation; I, inhibition of intercellular communication; M, micronuclei; R, mitotic recombination and gene conversion; S, sister chromatid exchange; T, cell transformation

In completing the table, the following symbols indicate the consensus of the Working Group with regard to the results for each end-point:

+   considered to be positive for the specific end-point and level of biological complexity
+¹   considered to be positive, but only one valid study was available to the Working Group
–   considered to be negative
–¹   considered to be negative, but only one valid study was available to the Working Group
?   considered to be equivocal or inconclusive (e.g. there were contradictory results from different laboratories; there were confounding exposures; the results were equivocal)

Summary table of genetic and related effects of 3,5-dinitrotoluene

Non-mammalian systems				Mammalian systems			
Prokaryotes	Lower eukaryotes	Plants	Insects	In vitro			In vivo
				Animal cells	Human cells	Animals	Humans
D G	D R G A	D G C	R G C A	D G S M C A T I	D G S M C A T I	D G S M C DL A	D S M C A
+				– –		–	

A, aneuploidy; C, chromosomal aberrations; D, DNA damage; DL, dominant lethal mutation; G, gene mutation; I, inhibition of intercellular communication; M, micronuclei; R, mitotic recombination and gene conversion; S, sister chromatid exchange; T, cell transformation

In completing the table, the following symbols indicate the consensus of the Working Group with regard to the results for each end-point:

+   considered to be positive for the specific end-point and level of biological complexity
+⁻¹ considered to be positive, but only one valid study was available to the Working Group
–   considered to be negative
–⁻¹ considered to be negative, but only one valid study was available to the Working Group
?   considered to be equivocal or inconclusive (e.g. there were contradictory results from different laboratories; there were confounding exposures; the results were equivocal)

APPENDIX 1

Summary table of genetic and related effects of 2-nitroanisole

Non-mammalian systems				Mammalian systems			
Proka-ryotes	Lower eukaryotes	Plants	Insects	In vitro		In vivo	
				Animal cells	Human cells	Animals	Humans
D G	D R G A	D G C	R G C A	D G S M C A T I	D G S M C A T I	D G S M C DL A	D S M C A
$+^1$ +				$+^1$ $+^1$ $+^1$			

A, aneuploidy; C, chromosomal aberrations; D, DNA damage; DL, dominant lethal mutation; G, gene mutation; I, inhibition of intercellular communication; M, micronuclei; R, mitotic recombination and gene conversion; S, sister chromatid exchange; T, cell transformation

In completing the table, the following symbols indicate the consensus of the Working Group with regard to the results for each end-point:

+ considered to be positive for the specific end-point and level of biological complexity
$+^1$ considered to be positive, but only one valid study was available to the Working Group
– considered to be negative
$–^1$ considered to be negative, but only one valid study was available to the Working Group
? considered to be equivocal or inconclusive (e.g. there were contradictory results from different laboratories; there were confounding exposures; the results were equivocal)

Summary table of genetic and related effects of nitrobenzene

Non-mammalian systems				Mammalian systems			
Prokaryotes	Lower eukaryotes	Plants	Insects	In vitro		In vivo	
				Animal cells	Human cells	Animals	Humans
D G	D R G A	D G C	R G C A	D G S M C A T I	D G S M C A T I	D G S M C DL A	D S M C A
–					–	– – –	

A, aneuploidy; C, chromosomal aberrations; D, DNA damage; DL, dominant lethal mutation; G, gene mutation; I, inhibition of intercellular communication; M, micronuclei; R, mitotic recombination and gene conversion; S, sister chromatid exchange; T, cell transformation

In completing the table, the following symbols indicate the consensus of the Working Group with regard to the results for each end-point:

+ considered to be positive for the specific end-point and level of biological complexity
+¹ considered to be positive, but only one valid study was available to the Working Group
– considered to be negative
–¹ considered to be negative, but only one valid study was available to the Working Group
? considered to be equivocal or inconclusive (e.g. there were contradictory results from different laboratories; there were confounding exposures; the results were equivocal)

APPENDIX 1

Summary table of genetic and related effects of 2-nitrotoluene

Non-mammalian systems				Mammalian systems			
				In vitro		In vivo	
Proka- ryotes	Lower eukaryotes	Plants	Insects	Animal cells	Human cells	Animals	Humans
D G	D R G A	D G C	R G C A	D G S M C A T I	D G S M C A T I	D G S M C DL A	D S M C A
– –				– +¹ – –	–¹	+	

A, aneuploidy; C, chromosomal aberrations; D, DNA damage; DL, dominant lethal mutation; G, gene mutation; I, inhibition of intercellular communication; M, micronuclei; R, mitotic recombination and gene conversion; S, sister chromatid exchange; T, cell transformation

In completing the table, the following symbols indicate the consensus of the Working Group with regard to the results for each end-point:

+   considered to be positive for the specific end-point and level of biological complexity
+   considered to be positive, but only one valid study was available to the Working Group
–   considered to be negative
–   considered to be negative, but only one valid study was available to the Working Group
?   considered to be equivocal or inconclusive (e.g. there were contradictory results from different laboratories; there were confounding exposures; the results were equivocal)

Summary table of genetic and related effects of 3-nitrotoluene

Non-mammalian systems				Mammalian systems			
Proka-ryotes	Lower eukaryotes	Plants	Insects	In vitro		In vivo	
				Animal cells	Human cells	Animals	Humans
D G	D R G A	D G C	R G C A	D G S M C A T I	D G S M C A T I	D G S M C DL A	D S M C A
–¹ –				– – +¹ –		–	

A, aneuploidy; C, chromosomal aberrations; D, DNA damage; DL, dominant lethal mutation; G, gene mutation; I, inhibition of intercellular communication; M, micronuclei; R, mitotic recombination and gene conversion; S, sister chromatid exchange; T, cell transformation

In completing the table, the following symbols indicate the consensus of the Working Group with regard to the results for each end-point:

+ considered to be positive for the specific end-point and level of biological complexity
+¹ considered to be positive, but only one valid study was available to the Working Group
– considered to be negative
–¹ considered to be negative, but only one valid study was available to the Working Group
? considered to be equivocal or inconclusive (e.g. there were contradictory results from different laboratories; there were confounding exposures; the results were equivocal)

Summary table of genetic and related effects of 4-nitrotoluene

Non-mammalian systems				Mammalian systems			
Proka-ryotes	Lower eukaryotes	Plants	Insects	In vitro			In vivo
				Animal cells	Human cells	Animals	Humans
D G	D R G A	D G C	R G C A	D G S M C A T I	D G S M C A T I	D G S M C DL A	D S M C A
$+^1$ +	$-^1$			$-$ $+^1$ $+^1$ $+^1$		$-^1$ $-^1$	

A, aneuploidy; C, chromosomal aberrations; D, DNA damage; DL, dominant lethal mutation; G, gene mutation; I, inhibition of intercellular communication; M, micronuclei; R, mitotic recombination and gene conversion; S, sister chromatid exchange; T, cell transformation

In completing the table, the following symbols indicate the consensus of the Working Group with regard to the results for each end-point:

+ considered to be positive for the specific end-point and level of biological complexity
$+^1$ considered to be positive, but only one valid study was available to the Working Group
− considered to be negative
$-^1$ considered to be negative, but only one valid study was available to the Working Group
? considered to be equivocal or inconclusive (e.g. there were contradictory results from different laboratories; there were confounding exposures; the results were equivocal)

Summary table of genetic and related effects of tetranitromethane

Non-mammalian systems				Mammalian systems			
Proka-ryotes	Lower eukaryotes	Plants	Insects	In vitro			In vivo
				Animal cells	Human cells	Animals	Humans
D G	D R G A	D G C	R G C A	D G S M C A T I	D G S M C A T I	D G S M C DL A	D S M C A
+				+¹ +¹			

A, aneuploidy; C, chromosomal aberrations; D, DNA damage; DL, dominant lethal mutation; G, gene mutation; I, inhibition of intercellular communication; M, micronuclei; R, mitotic recombination and gene conversion; S, sister chromatid exchange; T, cell transformation

In completing the table, the following symbols indicate the consensus of the Working Group with regard to the results for each end-point:

+ considered to be positive for the specific end-point and level of biological complexity
+¹ considered to be positive, but only one valid study was available to the Working Group
– considered to be negative
–¹ considered to be negative, but only one valid study was available to the Working Group
? considered to be equivocal or inconclusive (e.g. there were contradictory results from different laboratories; there were confounding exposures; the results were equivocal)

APPENDIX 1

Summary table of genetic and related effects of 2,4,6-trinitrotoluene

Non-mammalian systems				Mammalian systems			
Proka-ryotes	Lower eukaryotes	Plants	Insects	In vitro			In vivo
				Animal cells	Human cells	Animals	Humans
D G	D R G A	D G C	R G C A	D G S M C A T I	D G S M C A T I	D G S M C DL A	D S M C A
+				+¹		−¹ −¹	

A, aneuploidy; C, chromosomal aberrations; D, DNA damage; DL, dominant lethal mutation; G, gene mutation; I, inhibition of intercellular communication; M, micronuclei; R, mitotic recombination and gene conversion; S, sister chromatid exchange; T, cell transformation

In completing the table, the following symbols indicate the consensus of the Working Group with regard to the results for each end-point:

+ considered to be positive for the specific end-point and level of biological complexity
+¹ considered to be positive, but only one valid study was available to the Working Group
− considered to be negative
−¹ considered to be negative, but only one valid study was available to the Working Group
? considered to be equivocal or inconclusive (e.g. there were contradictory results from different laboratories; there were confounding exposures; the results were equivocal)

Summary table of genetic and related effects of musk ambrette

Non-mammalian systems				Mammalian systems			
Proka-ryotes	Lower eukaryotes	Plants	Insects	*In vitro*		*In vivo*	
				Animal cells	Human cells	Animals	Humans
D G	D R G A	D G C	R G C A	D G S M C A T I	D G S M C A T I	D G S M C DL A	D S M C A
+			+[1]			–	

A, aneuploidy; C, chromosomal aberrations; D, DNA damage; DL, dominant lethal mutation; G, gene mutation; I, inhibition of intercellular communication; M, micronuclei; R, mitotic recombination and gene conversion; S, sister chromatid exchange; T, cell transformation

In completing the table, the following symbols indicate the consensus of the Working Group with regard to the results for each end-point:

+   considered to be positive for the specific end-point and level of biological complexity
+[1] considered to be positive, but only one valid study was available to the Working Group
–   considered to be negative
–[1] considered to be negative, but only one valid study was available to the Working Group
?   considered to be equivocal or inconclusive (e.g. there were contradictory results from different laboratories; there were confounding exposures; the results were equivocal)

APPENDIX 1

Summary table of genetic and related effects of musk xylene

Non-mammalian systems				Mammalian systems			
Proka-ryotes	Lower eukaryotes	Plants	Insects	In vitro			In vivo
				Animal cells	Human cells	Animals	Humans
D G	D R G A	D G C	R G C A	D G S M C A T I	D G S M C A T I	D G S M C DL A	D S M C A
–				–¹ –¹ –¹		–¹	

A, aneuploidy; C, chromosomal aberrations; D, DNA damage; DL, dominant lethal mutation; G, gene mutation; I, inhibition of intercellular communication; M, micronuclei; R, mitotic recombination and gene conversion; S, sister chromatid exchange; T, cell transformation

In completing the table, the following symbols indicate the consensus of the Working Group with regard to the results for each end-point:

+ considered to be positive for the specific end-point and level of biological complexity
+¹ considered to be positive, but only one valid study was available to the Working Group
– considered to be negative
–¹ considered to be negative, but only one valid study was available to the Working Group
? considered to be equivocal or inconclusive (e.g. there were contradictory results from different laboratories; there were confounding exposures; the results were equivocal)

APPENDIX 2

ACTIVITY PROFILES FOR
GENETIC AND RELATED EFFECTS

APPENDIX 2

ACTIVITY PROFILES FOR GENETIC AND RELATED EFFECTS

Methods

The x-axis of the activity profile (Waters *et al.*, 1987, 1988) represents the bioassays in phylogenetic sequence by end-point, and the values on the y-axis represent the logarithmically transformed lowest effective doses (LED) and highest ineffective doses (HID) tested. The term 'dose', as used in this report, does not take into consideration length of treatment or exposure and may therefore be considered synonymous with concentration. In practice, the concentrations used in all the in-vitro tests were converted to µg/ml, and those for in-vivo tests were expressed as mg/kg bw. Because dose units are plotted on a log scale, differences in the relative molecular masses of compounds do not, in most cases, greatly influence comparisons of their activity profiles. Conventions for dose conversions are given below.

Profile-line height (the magnitude of each bar) is a function of the LED or HID, which is associated with the characteristics of each individual test system — such as population size, cell-cycle kinetics and metabolic competence. Thus, the detection limit of each test system is different, and, across a given activity profile, responses will vary substantially. No attempt is made to adjust or relate responses in one test system to those of another.

Line heights are derived as follows: for negative test results, the highest dose tested without appreciable toxicity is defined as the HID. If there was evidence of extreme toxicity, the next highest dose is used. A single dose tested with a negative result is considered to be equivalent to the HID. Similarly, for positive results, the LED is recorded. If the original data were analysed statistically by the author, the dose recorded is that at which the response was significant ($p < 0.05$). If the available data were not analysed statistically, the dose required to produce an effect is estimated as follows: when a dose-related positive response is observed with two or more doses, the lower of the doses is taken as the LED; a single dose resulting in a positive response is considered to be equivalent to the LED.

In order to accommodate both the wide range of doses encountered and positive and negative responses on a continuous scale, doses are transformed logarithmically, so that effective (LED) and ineffective (HID) doses are represented by positive and negative numbers, respectively. The response, or logarithmic dose unit (LDUij), for a given test system i and chemical j is represented by the expressions

$$LDU_{ij} = -\log_{10}(\text{dose}), \text{ for HID values; LDU} \leq 0$$
and (1)
$$LDU_{ij} = -\log_{10}(\text{dose} \times 10^{-5}), \text{ for LED values; LDU} \geq 0.$$

These simple relationships define a dose range of 0 to –5 logarithmic units for ineffective doses (1–100 000 µg/ml or mg/kg bw) and 0 to +8 logarithmic units for effective doses (100 000–0.001 µg/ml or mg/kg bw). A scale illustrating the LDU values is shown in Figure 1. Negative responses at doses less than 1 µg/ml (mg/kg bw) are set equal to 1. Effectively, an LED value ≥ 100 000 or an HID value ≤ 1 produces an LDU = 0; no quantitative information is gained from such extreme values. The dotted lines at the levels of log dose units 1 and –1 define a 'zone of uncertainty' in which positive results are reported at such high doses (between 10 000 and 100 000 mg/ml or mg/kg bw) or negative results are reported at such low doses (1 to 10 mg/ml or mg/kg bw) as to call into question the adequacy of the test.

Fig. 1. Scale of log dose units used on the y-axis of activity profiles

Positive (µg/ml or mg/kg bw)		Log dose units	
0.001		8	----
0.01		7	--
0.1		6	--
1.0		5	--
10		4	--
100		3	--
1000		2	--
10 000		1	--
100 000	1	0	----
	10	–1	--
	100	–2	--
	1000	–3	--
	10 000	–4	--
	100 000	–5	----

Negative (µg/ml or mg/kg bw)

In practice, an activity profile is computer generated. A data entry programme is used to store abstracted data from published reports. A sequential file (in ASCII) is created for each compound, and a record within that file consists of the name and Chemical Abstracts Service number of the compound, a three-letter code for the test system (see below), the qualitative test result (with and without an exogenous metabolic system), dose (LED or HID), citation number and additional source information. An abbreviated citation for each publication is stored in a segment of a record accessing both the test data file and the citation file. During processing of the data file, an average of the logarithmic values of the data subset is calculated, and the length of the profile line represents this average value. All dose values are plotted for each profile line, regardless of whether results are positive or negative. Results obtained in the absence of an exogenous metabolic system are indicated by a bar (−), and results obtained in the presence of an exogenous metabolic system are indicated by an upward-directed arrow (↑). When all results for a given assay are either positive or negative, the mean of the LDU values is plotted as a solid line; when conflicting data are reported for the same assay (i.e. both positive and negative results), the majority data are shown by a solid line and the minority data by a dashed line (drawn to the extreme conflicting response). In the few cases in which the numbers of positive and negative results are equal, the solid line is drawn in the positive direction and the maximal negative response is indicated with a dashed line. Profile lines are identified by three-letter code words representing the commonly used tests. Code words for most of the test systems in current use in genetic toxicology were defined for the US Environmental Protection Agency's GENE-TOX Program (Waters, 1979; Waters & Auletta, 1981). For *IARC Monographs* Supplement 6, Volume 44 and subsequent volumes, including this publication, codes were redefined in a manner that should facilitate inclusion of additional tests. Naming conventions are described below.

Data listings are presented in the text and include end-point and test codes, a short test code definition, results, either with (M) or without (NM) an exogenous activation system, the associated LED or HID value and a short citation. Test codes are organized phylogenetically and by end-point from left to right across each activity profile and from top to bottom of the corresponding data listing. End-points are defined as follows: A, aneuploidy; C, chromosomal aberrations; D, DNA damage; F, assays of body fluids; G, gene mutation; H, host-mediated assays; I, inhibition of intercellular communication; M, micronuclei; P, sperm morphology; R, mitotic recombination or gene conversion; S, sister chromatid exchange; and T, cell transformation.

Dose conversions for activity profiles

Doses are converted to μg/ml for in-vitro tests and to mg/kg bw per day for in-vivo experiments.

1. In-vitro test systems
    (*a*) Weight/volume converts directly to μg/ml.
    (*b*) Molar (M) concentration × molecular weight = mg/ml = 10^3 mg/ml; mM concentration × molecular weight = μg/ml.
    (*c*) Soluble solids expressed as % concentration are assumed to be in units of mass per volume (i.e. 1% = 0.01 g/ml = 10 000 μg/ml; also, 1 ppm = 1 μg/ml).
    (*d*) Liquids and gases expressed as % concentration are assumed to be given in units of volume per volume. Liquids are converted to weight per volume using the density (D) of the solution (D = g/ml). Gases are converted from volume to mass using the ideal gas law, PV = nRT. For exposure at 20–37 °C at standard

atmospheric pressure, 1% (v/v) = 0.4 µg/ml × molecular weight of the gas. Also, 1 ppm (v/v) = 4 × 10^5 µg/ml × molecular weight.

(e) In microbial plate tests, it is usual for the doses to be reported as weight/plate, whereas concentrations are required to enter data on the activity profile chart. While remaining cognisant of the errors involved in the process, it is assumed that a 2-ml volume of top agar is delivered to each plate and that the test substance remains in solution within it; concentrations are derived from the reported weight/plate values by dividing by this arbitrary volume. For spot tests, a 1-ml volume is used in the calculation.

(f) Conversion of particulate concentrations given in µg/cm^2 is based on the area (A) of the dish and the volume of medium per dish; i.e. for a 100-mm dish: $A = \pi R^2 = \pi \times (5 \text{ cm})^2 = 78.5 \text{ cm}^2$. If the volume of medium is 10 ml, then 78.5 cm^2 = 10 ml and 1 cm^2 = 0.13 ml.

2. In-vitro systems using in-vivo activation

For the body fluid-urine (BF-) test, the concentration used is the dose (in mg/kg bw) of the compound administered to test animals or patients.

3. In-vivo test systems

(a) Doses are converted to mg/kg bw per day of exposure, assuming 100% absorption. Standard values are used for each sex and species of rodent, including body weight and average intake per day, as reported by Gold *et al.* (1984). For example, in a test using male mice fed 50 ppm of the agent in the diet, the standard food intake per day is 12% of body weight, and the conversion is dose = 50 ppm × 12% = 6 mg/kg bw per day.

Standard values used for humans are: weight—males, 70 kg; females, 55 kg; surface area, 1.7 m^2; inhalation rate, 20 L/min for light work, 30 L/min for mild exercise.

(b) When reported, the dose at the target site is used. For example, doses given in studies of lymphocytes of humans exposed *in vivo* are the measured blood concentrations in µg/ml.

Codes for test systems

For specific nonmammalian test systems, the first two letters of the three-symbol code word define the test organism (e.g. SA- for *Salmonella typhimurium*, EC- for *Escherichia coli*). If the species is not known, the convention used is -S-. The third symbol may be used to define the tester strain (e.g. SA8 for *S. typhimurium* TA1538, ECW for *E. coli* WP2*uvr*A). When strain designation is not indicated, the third letter is used to define the specific genetic end-point under investigation (e.g. --D for differential toxicity, --F for forward mutation, --G for gene conversion or genetic crossing-over, --N for aneuploidy, --R for reverse mutation, --U for unscheduled DNA synthesis). The third letter may also be used to define the general end-point under investigation when a more complete definition is not possible or relevant (e.g. --M for mutation, --C for chromosomal aberration). For mammalian test systems, the first letter of the three-letter code word defines the genetic end-point under investigation: A-- for aneuploidy, B-- for binding, C-- for chromosomal aberration, D-- for DNA strand breaks, G-- for gene mutation, I-- for inhibition of intercellular communication, M-- for micronucleus formation, R-- for DNA repair, S-- for sister chromatid exchange, T-- for cell transformation and U-- for unscheduled DNA synthesis.

For animal (i.e. non-human) test systems *in vitro*, when the cell type is not specified, the code letters -IA are used. For such assays *in vivo*, when the animal species is not specified, the code letters -VA are used. Commonly used animal species are identified by the third letter (e.g. --C for Chinese hamster, --M for mouse, --R for rat, --S for Syrian hamster).

For test systems using human cells *in vitro*, when the cell type is not specified, the code letters -IH are used. For assays on humans *in vivo*, when the cell type is not specified, the code letters -VH are used. Otherwise, the second letter specifies the cell type under investigation (e.g. -BH for bone marrow, -LH for lymphocytes).

Some other specific coding conventions used for mammalian systems are as follows: BF- for body fluids, HM- for host-mediated, --L for leukocytes or lymphocytes *in vitro* (-AL, animals; -HL, humans), -L- for leukocytes *in vivo* (-LA, animals; -LH, humans), --T for transformed cells.

Note that these are examples of major conventions used to define the assay code words. The alphabetized listing of codes must be examined to confirm a specific code word. As might be expected from the limitation to three symbols, some codes do not fit the naming conventions precisely. In a few cases, test systems are defined by first-letter code words, for example: MST, mouse spot test; SLP, mouse specific locus mutation, postspermatogonia; SLO, mouse specific locus mutation, other stages; DLM, dominant lethal mutation in mice; DLR, dominant lethal mutation in rats; MHT, mouse heritable translocation.

The genetic activity profiles and listings were prepared in collaboration with Environmental Health Research and Testing Inc. (EHRT) under contract to the United States Environmental Protection Agency; EHRT also determined the doses used. The references cited in each genetic activity profile listing can be found in the list of references in the appropriate monograph.

References

Garrett, N.E., Stack, H.F., Gross, M.R. & Waters, M.D. (1984) An analysis of the spectra of genetic activity produced by known or suspected human carcinogens. *Mutat. Res.*, **134**, 89–111

Gold, L.S., Sawyer, C.B., Magaw, R., Backman, G.M., de Veciana, M., Levinson, R., Hooper, N.K., Havender, W.R., Bernstein, L., Peto, R., Pike, M.C. & Ames, B.N. (1984) A carcinogenic potency database of the standardized results of animal bioassays. *Environ. Health Perspect.*, **58**, 9–319

Waters, M.D. (1979) *The GENE-TOX program*. In: Hsie, A.W., O'Neill, J.P. & McElheny, V.K., eds, *Mammalian Cell Mutagenesis: The Maturation of Test Systems* (Banbury Report 2), Cold Spring Harbor, NY, CSH Press, pp. 449–467

Waters, M.D. & Auletta, A. (1981) The GENE-TOX program: genetic activity evaluation. *J. chem. Inf. comput. Sci.*, **21**, 35–38

Waters, M.D., Stack, H.F., Brady, A.L., Lohman, P.H.M., Haroun, L. & Vainio, H. (1987) Appendix 1: Activity profiles for genetic and related tests. In: *IARC Monographs on the Evaluation of the Carcinogenic Risk of Chemicals to Humans*, Suppl. 6, *Genetic and Related Effects: An Updating of Selected* IARC Monographs *from Volumes 1 to 42*, Lyon, IARC, pp. 687–696

Waters, M.D., Stack, H.F., Brady, A.L., Lohman, P.H.M., Haroun, L. & Vainio, H. (1988) Use of computerized data listings and activity profiles of genetic and related effects in the review of 195 compounds. *Mutat. Res.*, **205**, 295–312

APPENDIX 2

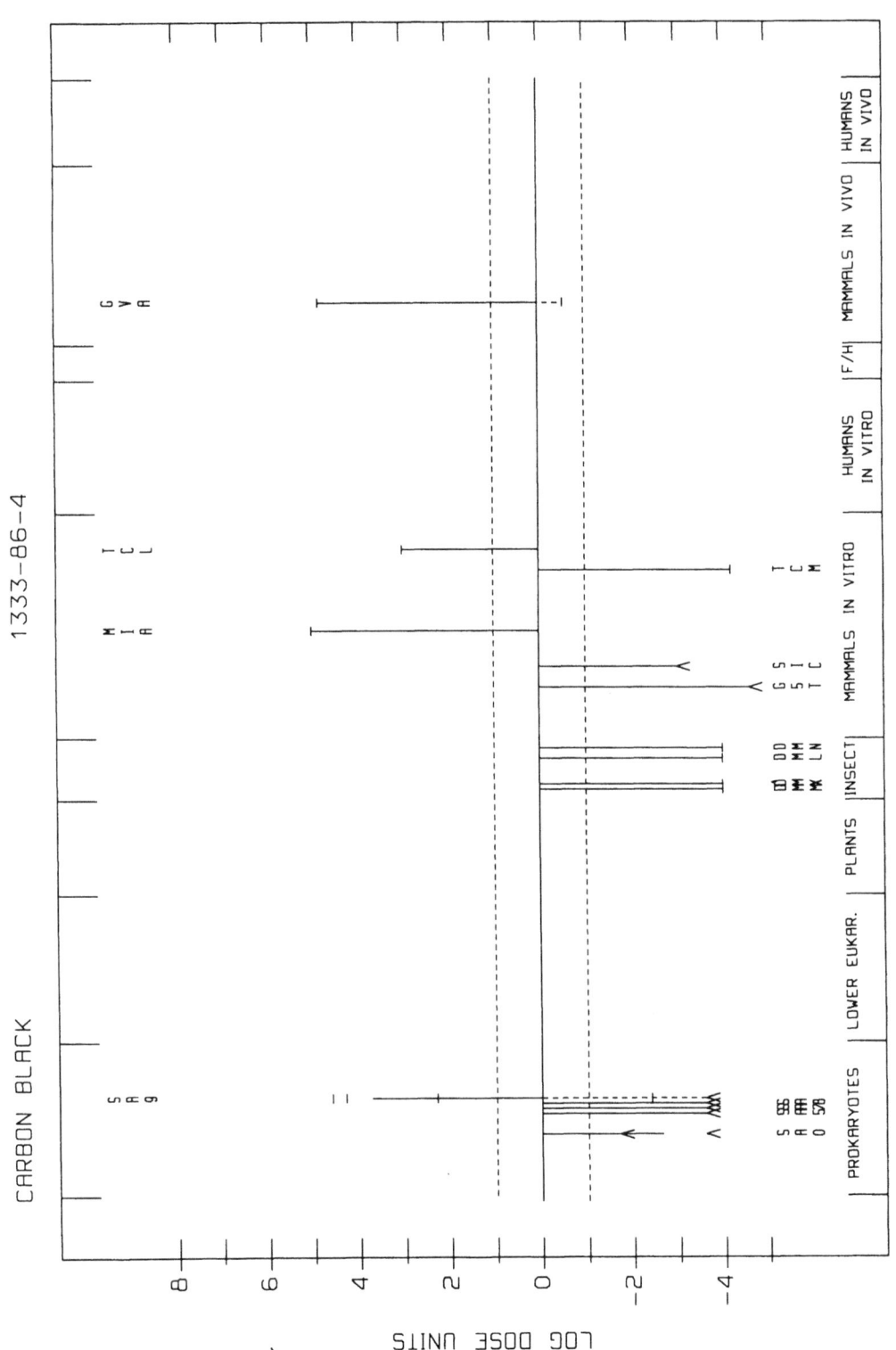

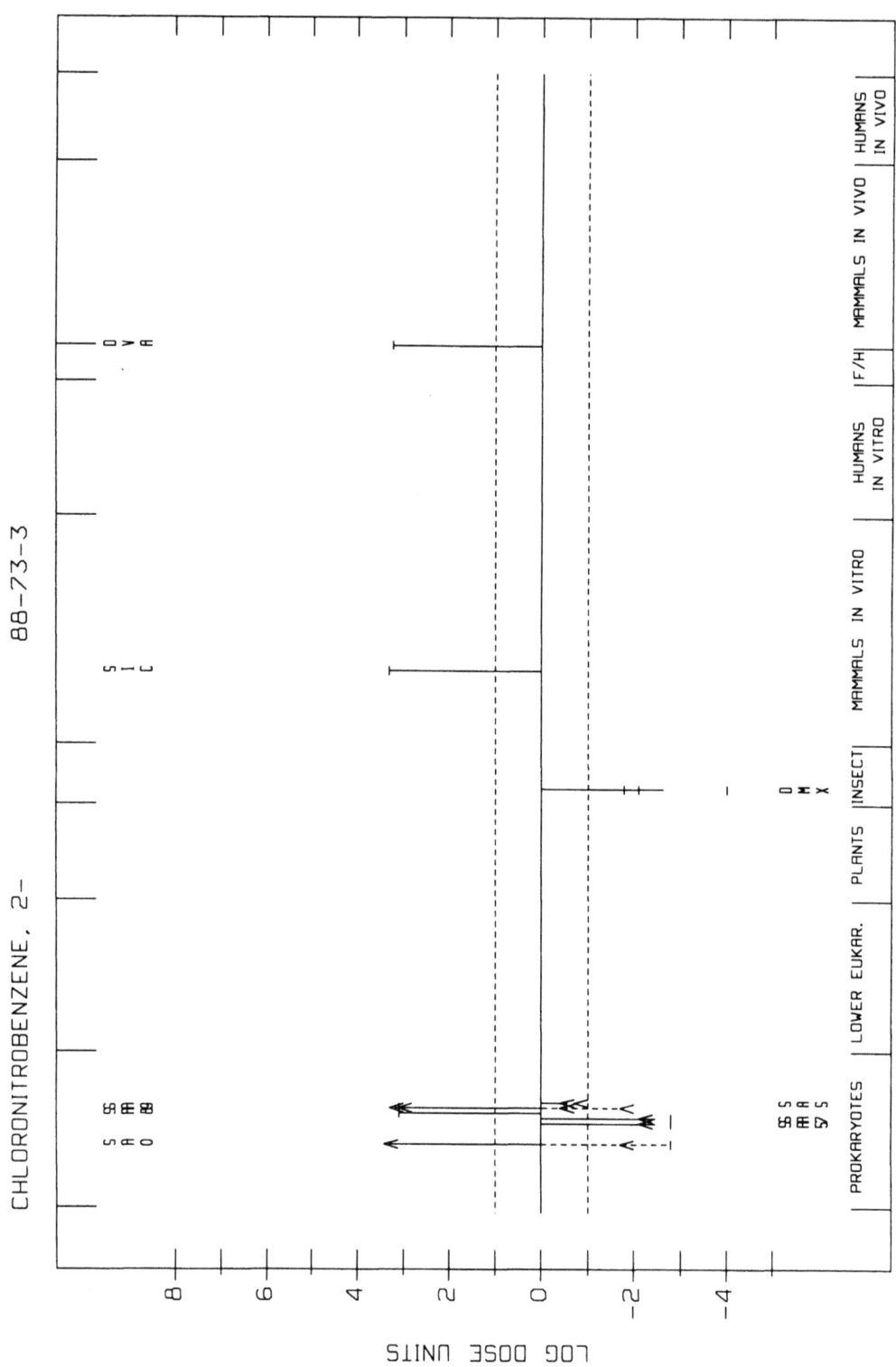

APPENDIX 2

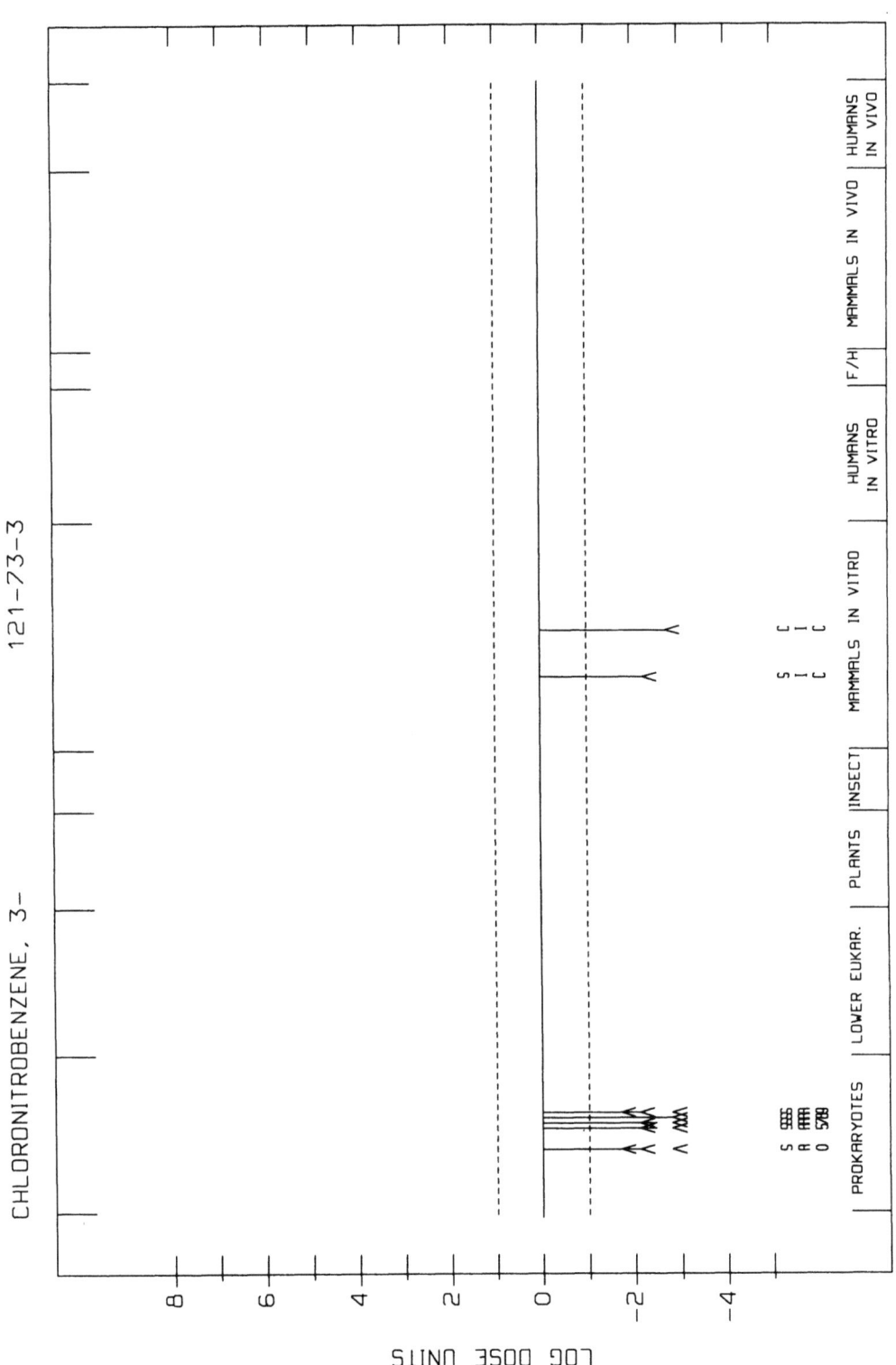

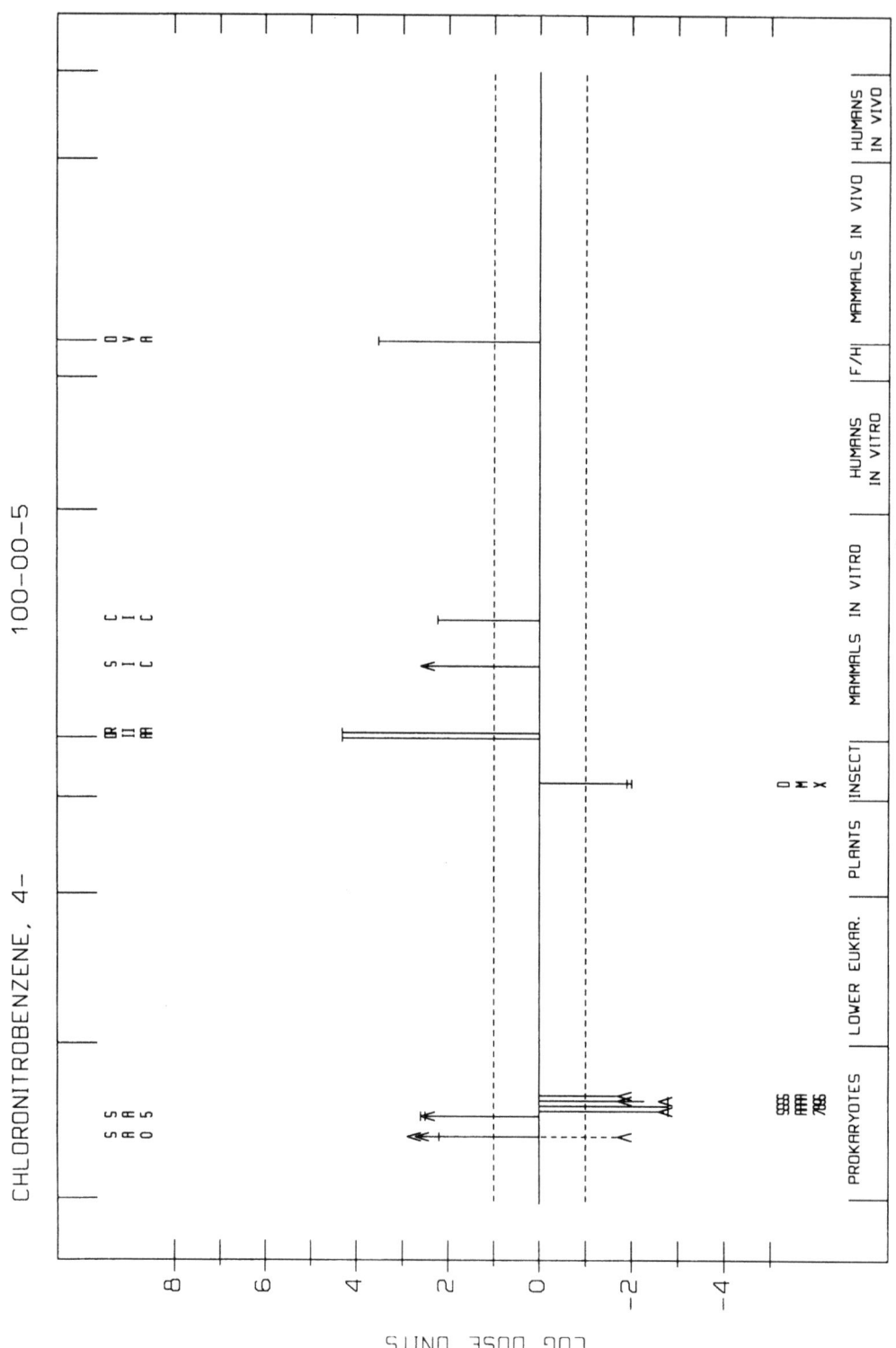

APPENDIX 2

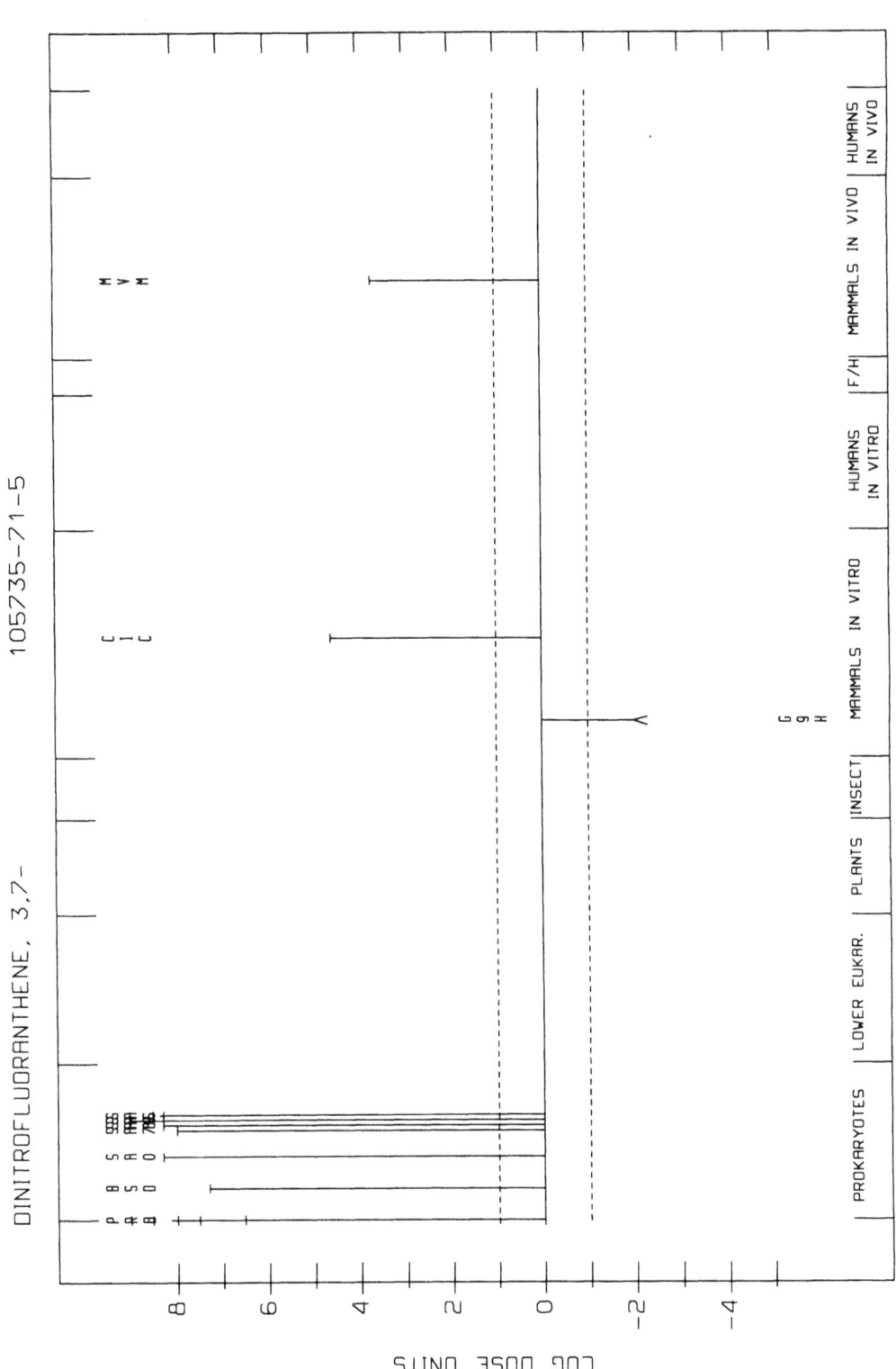

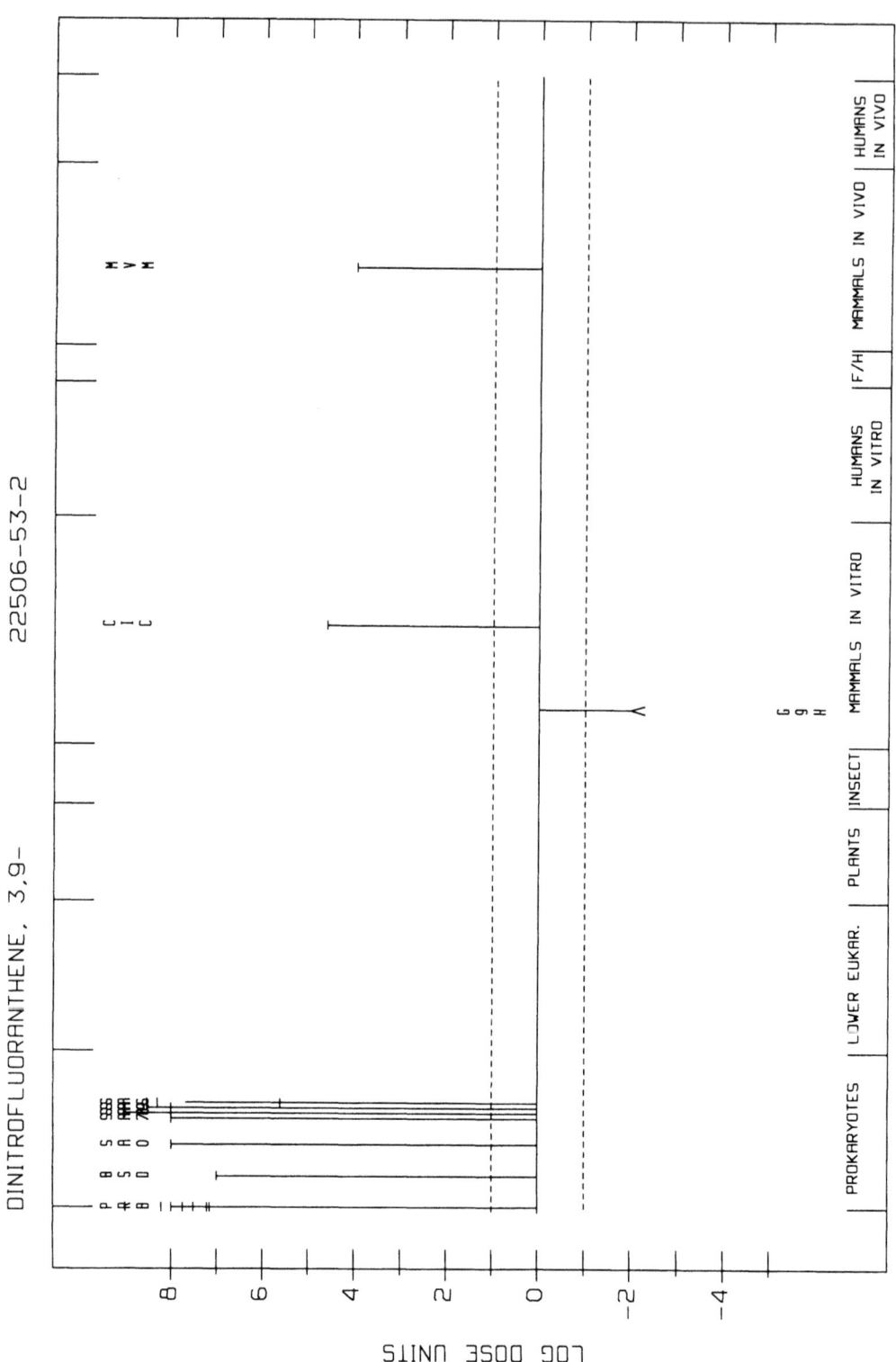

APPENDIX 2

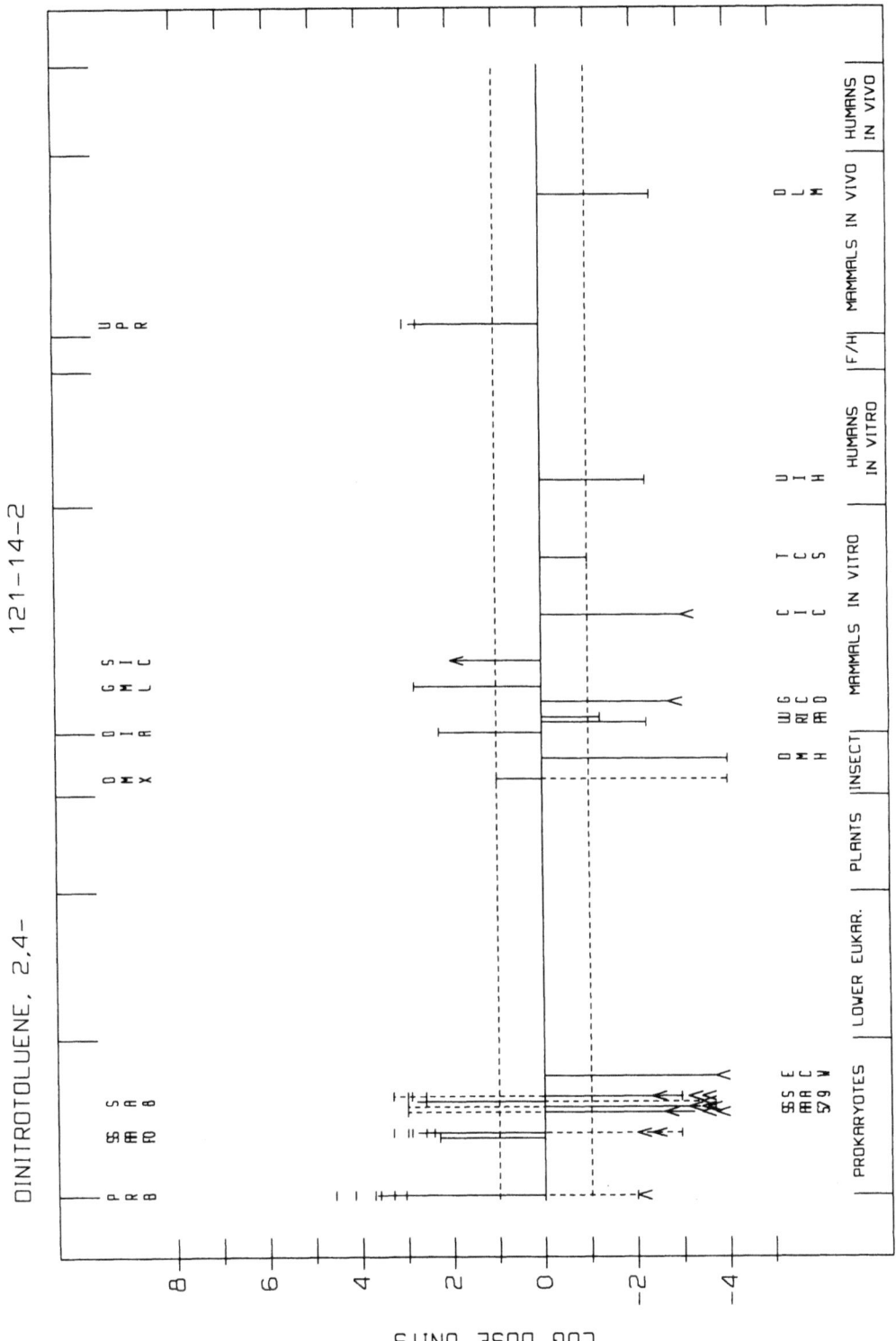

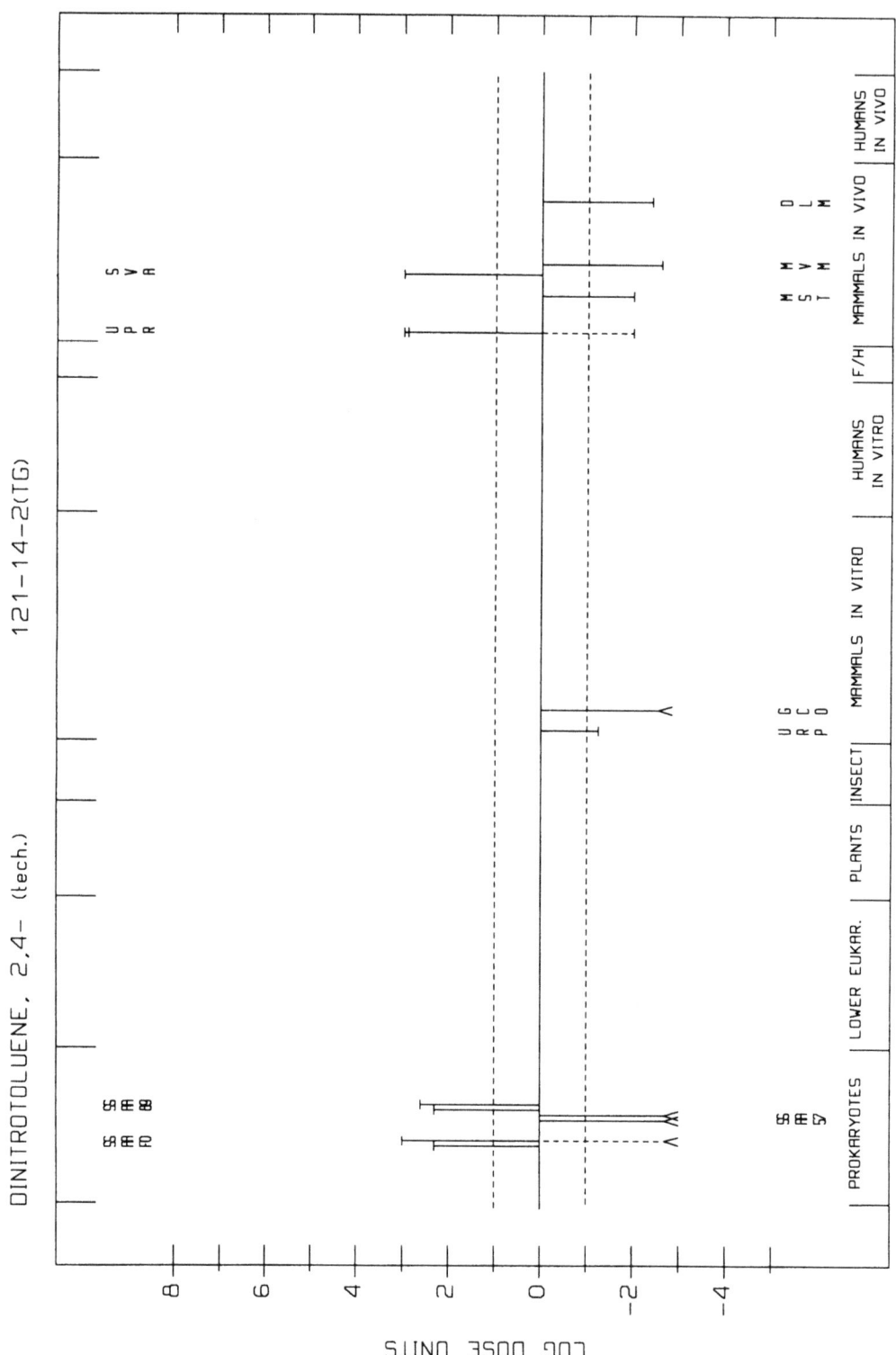

APPENDIX 2

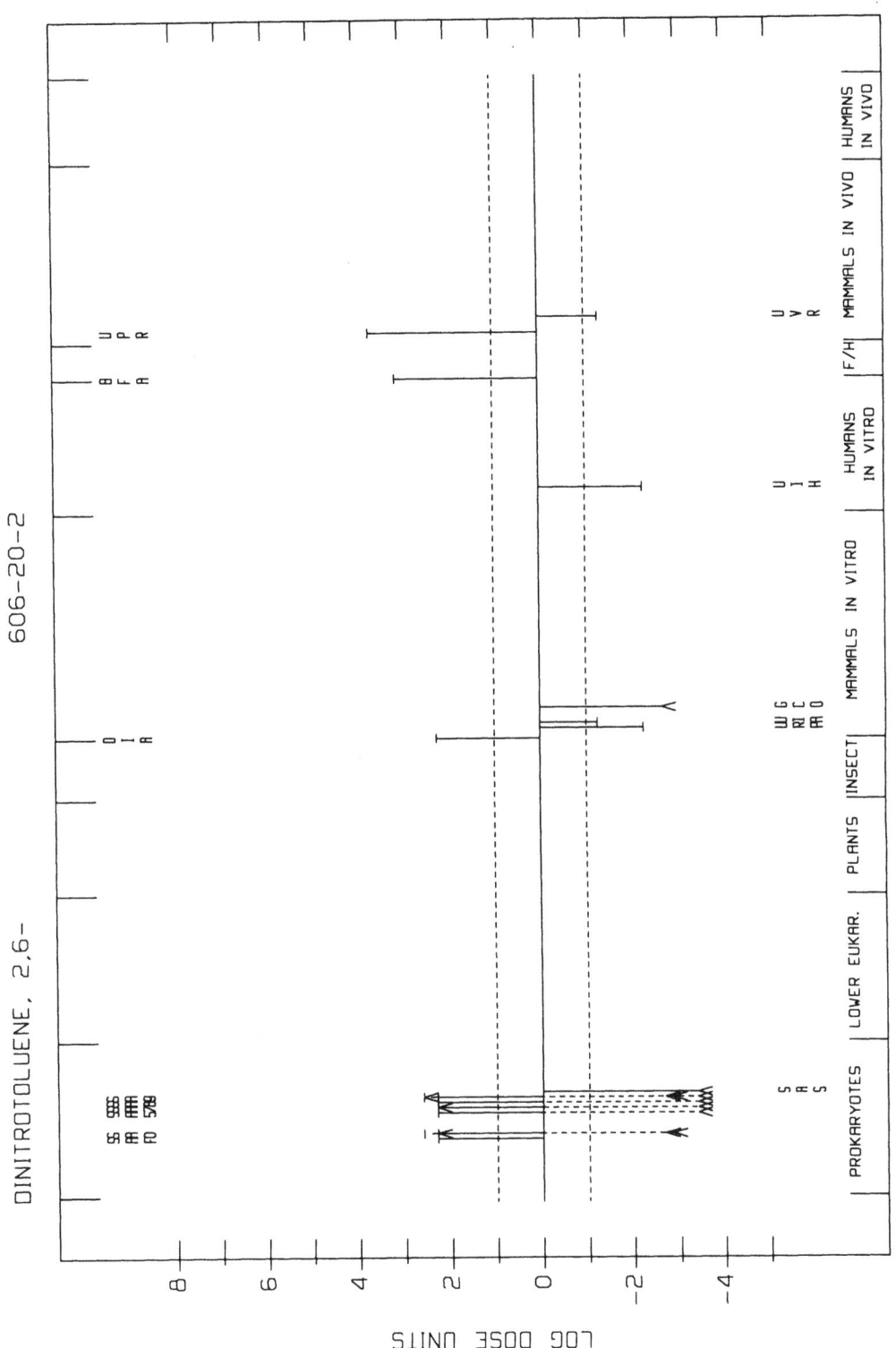

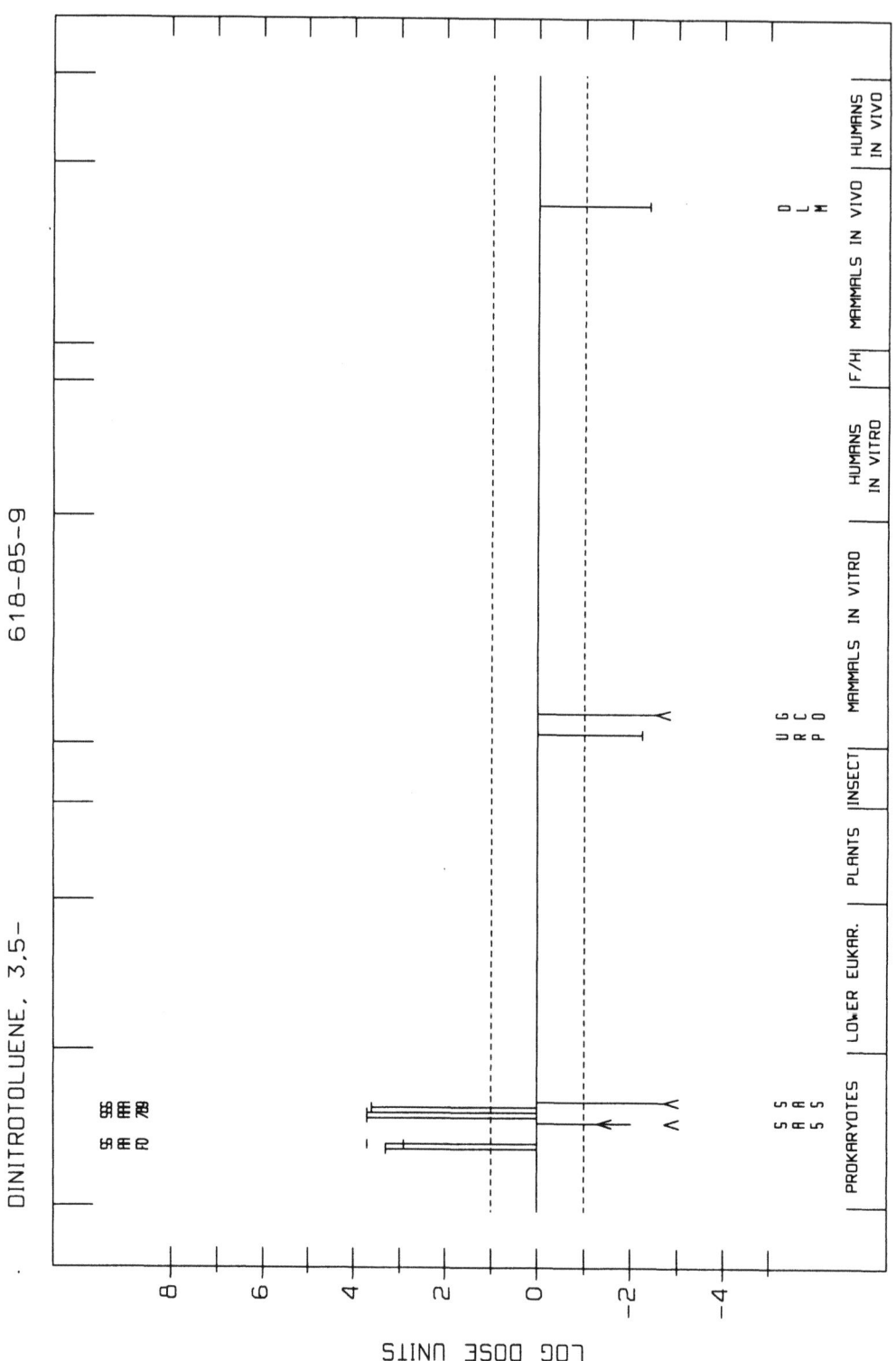

APPENDIX 2

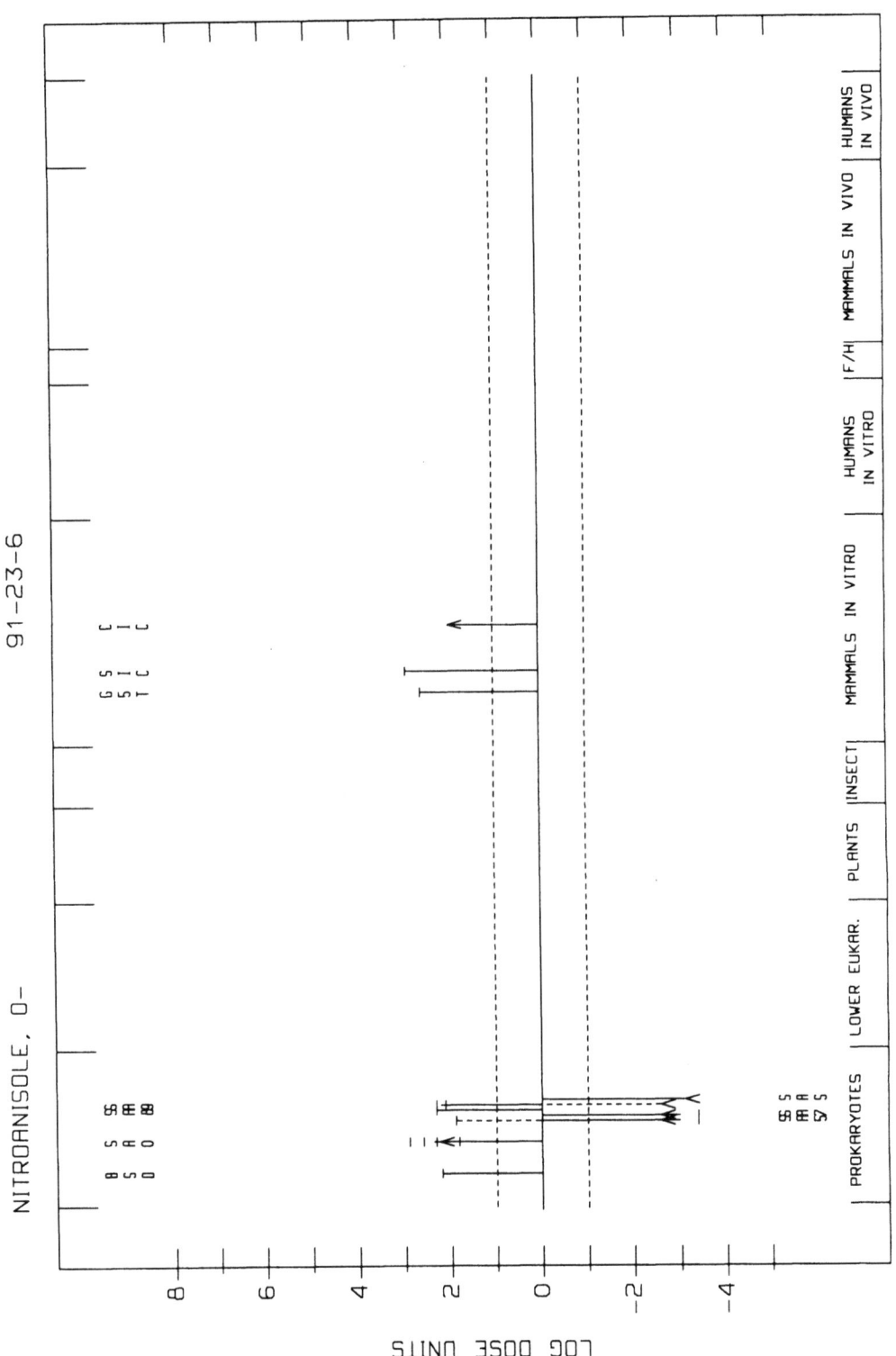

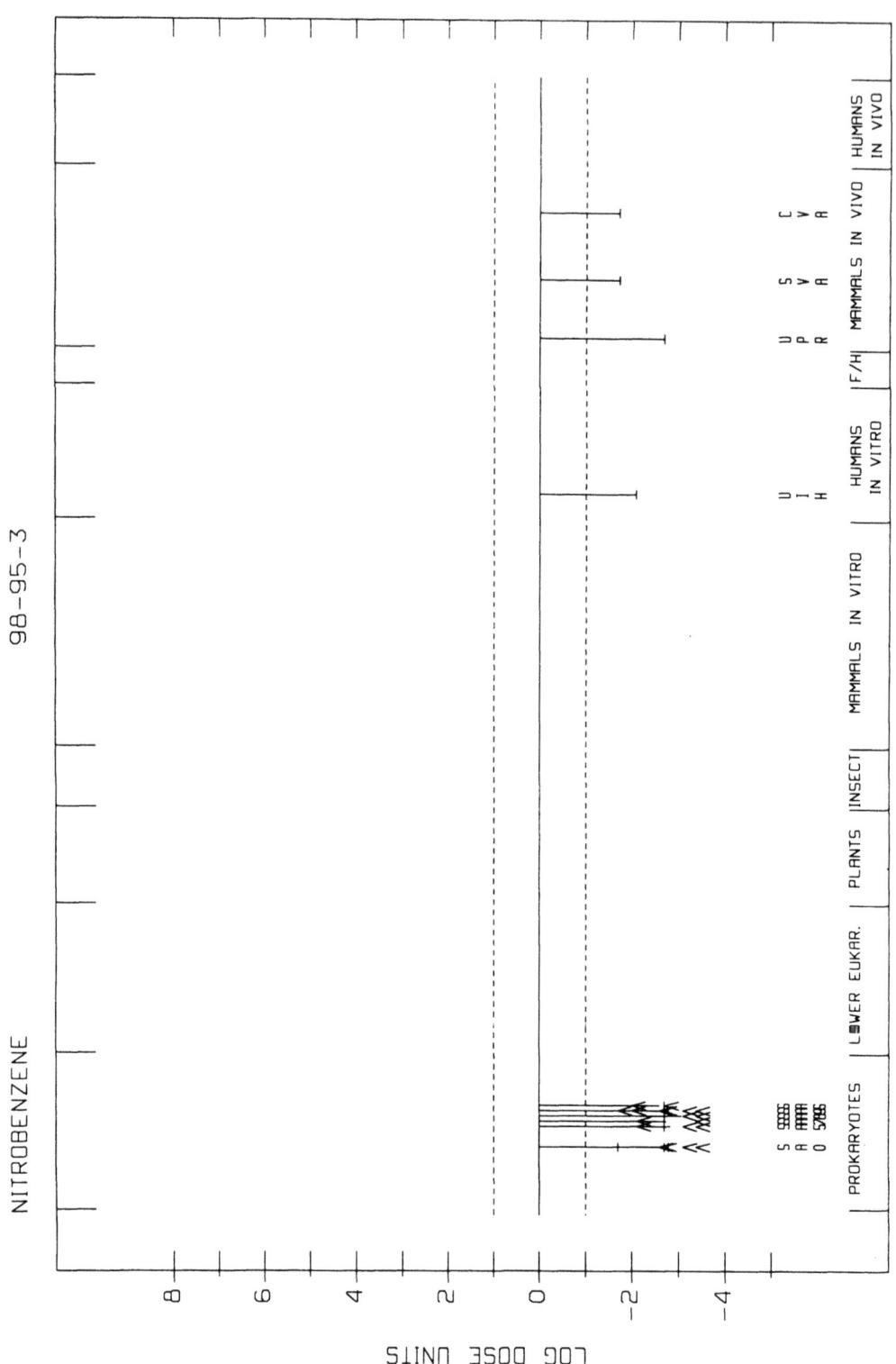

APPENDIX 2

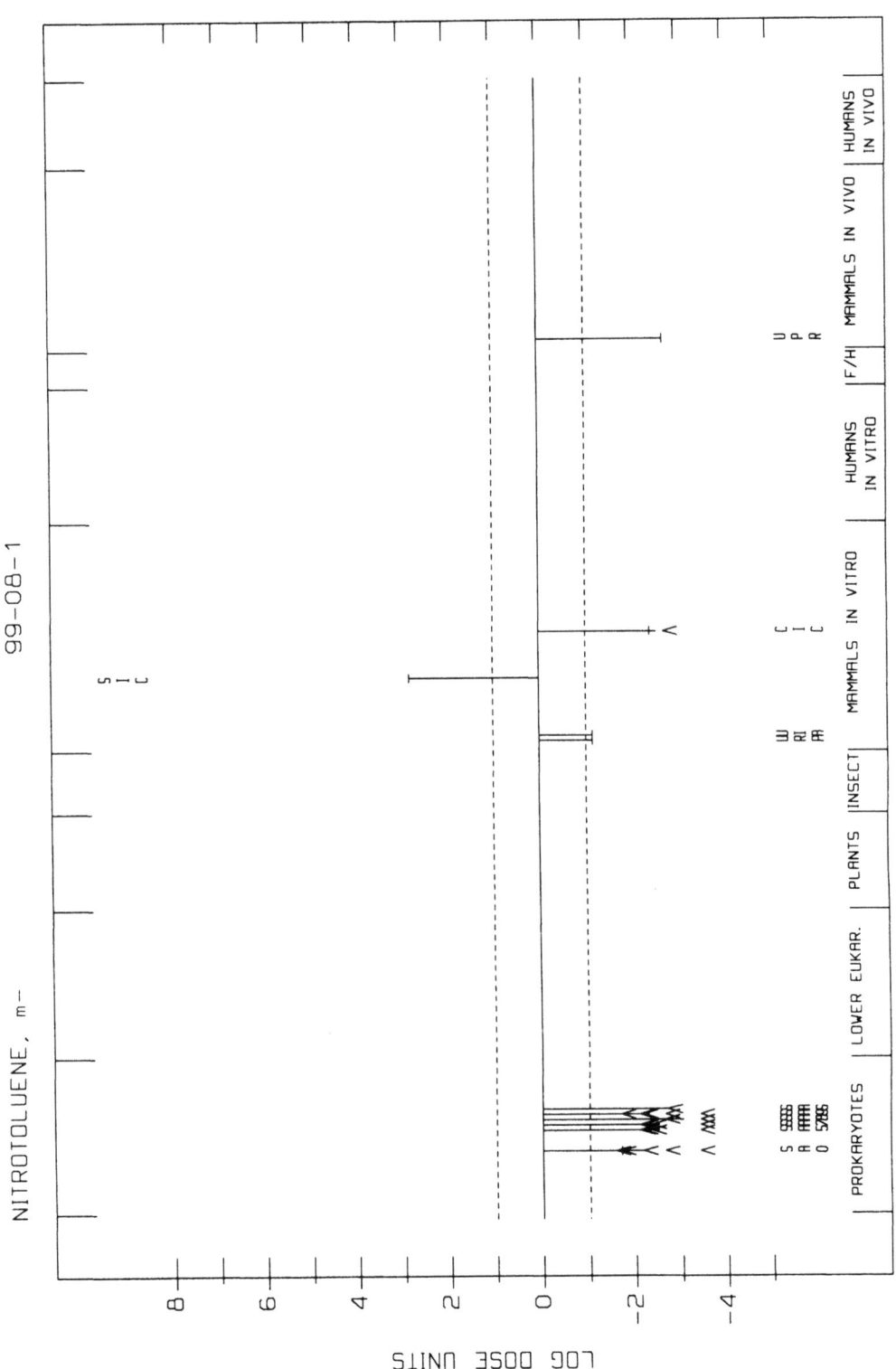

542   IARC MONOGRAPHS VOLUME 65

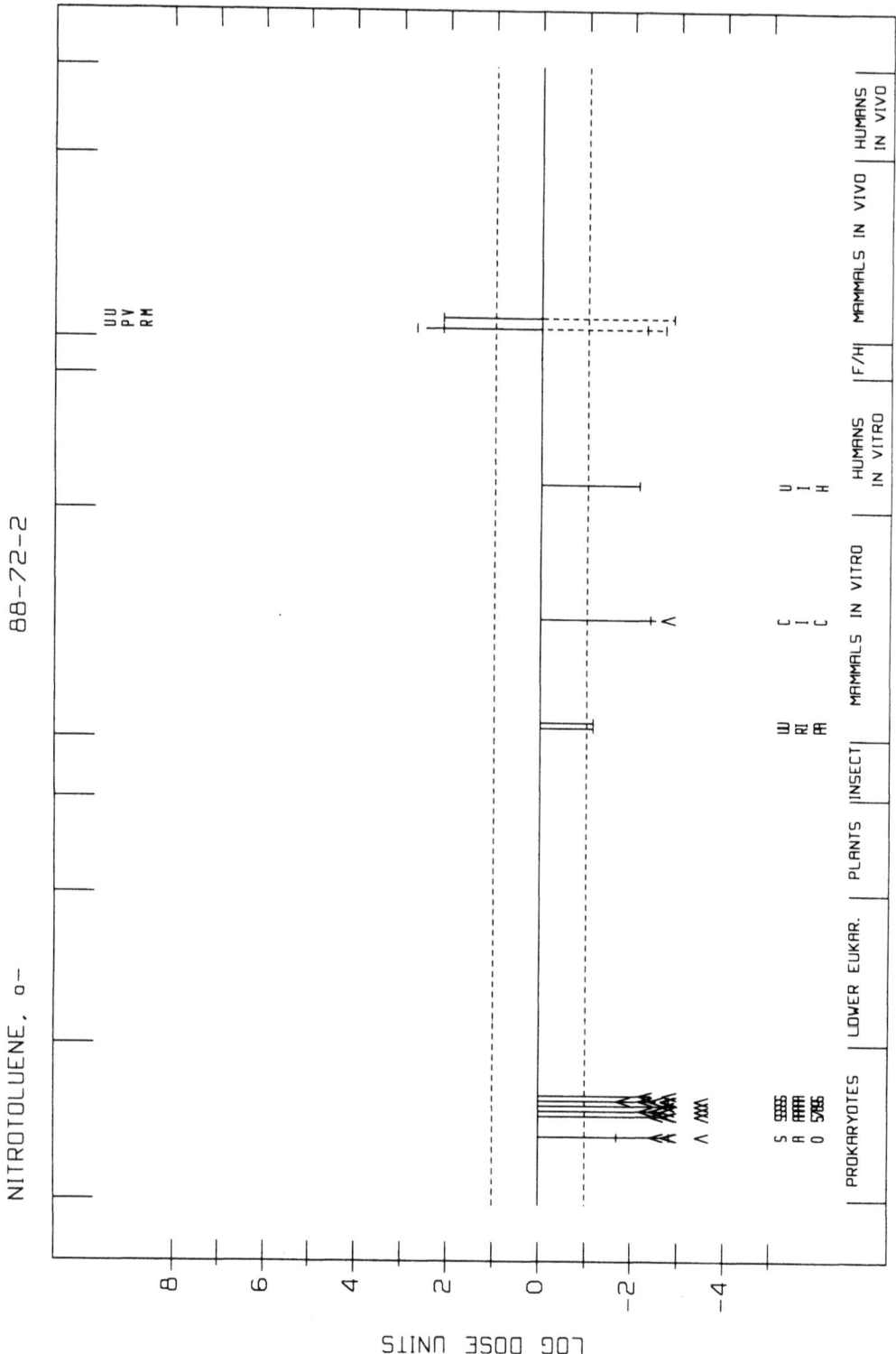

APPENDIX 2

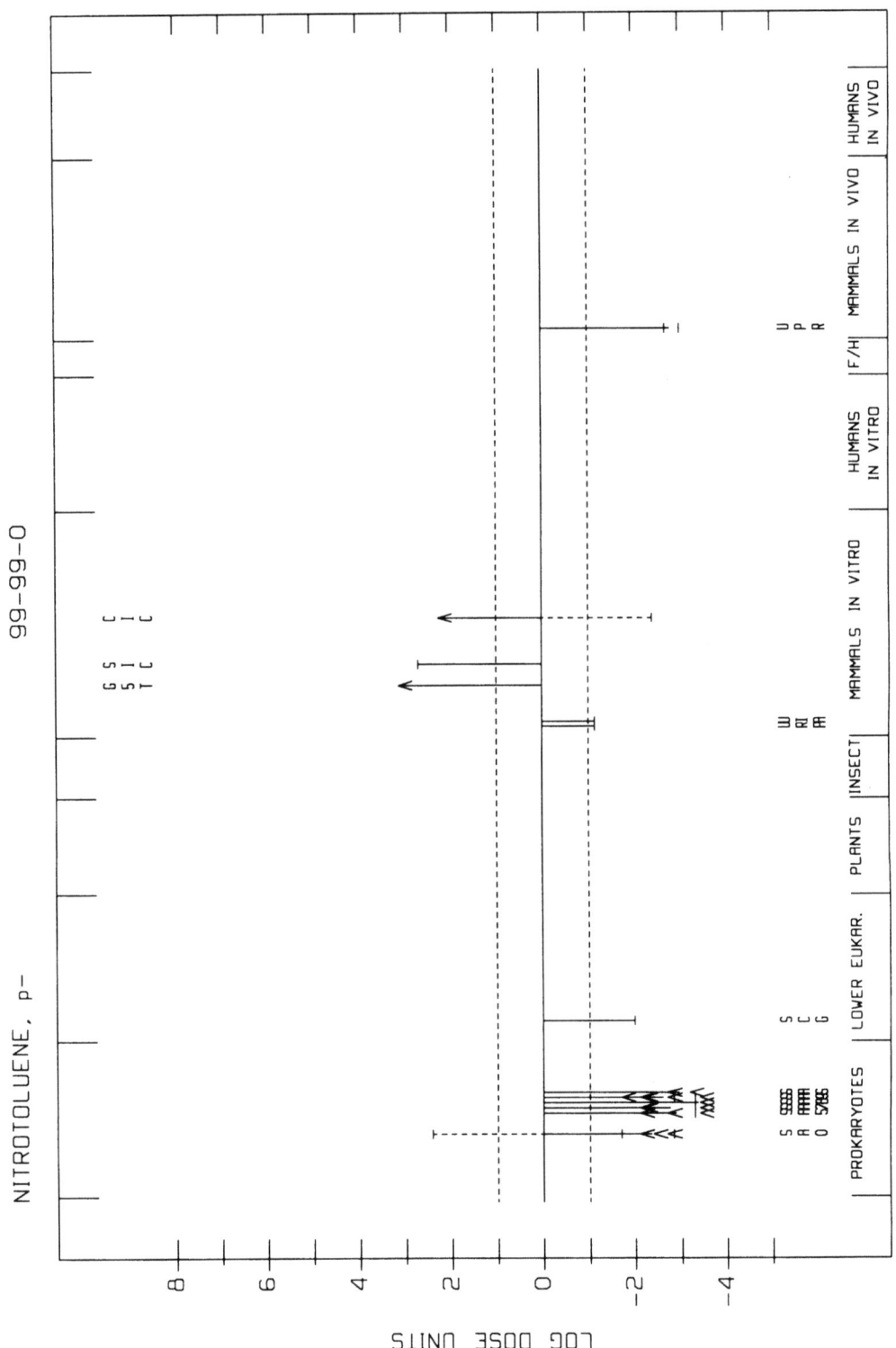

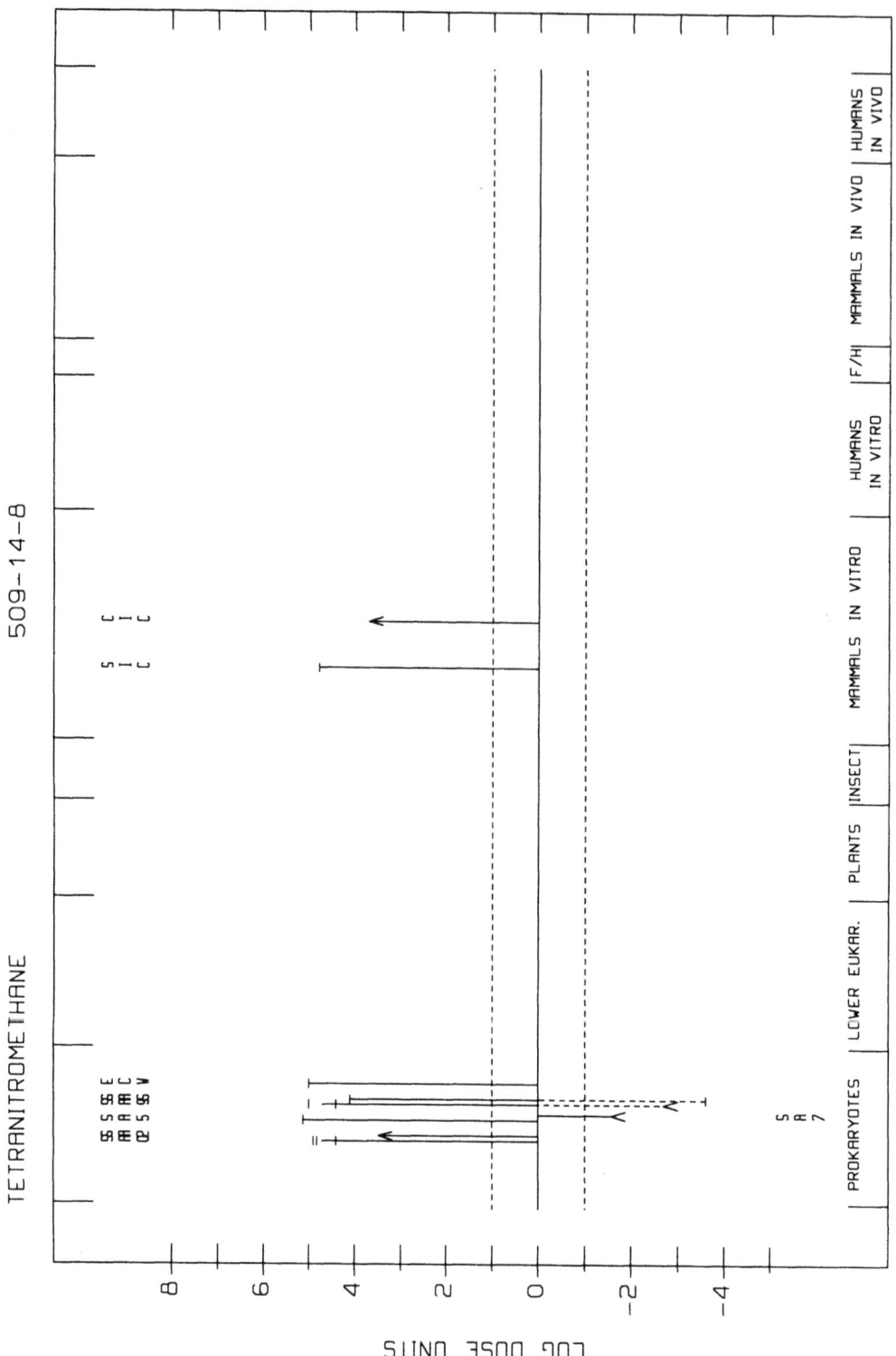

APPENDIX 2

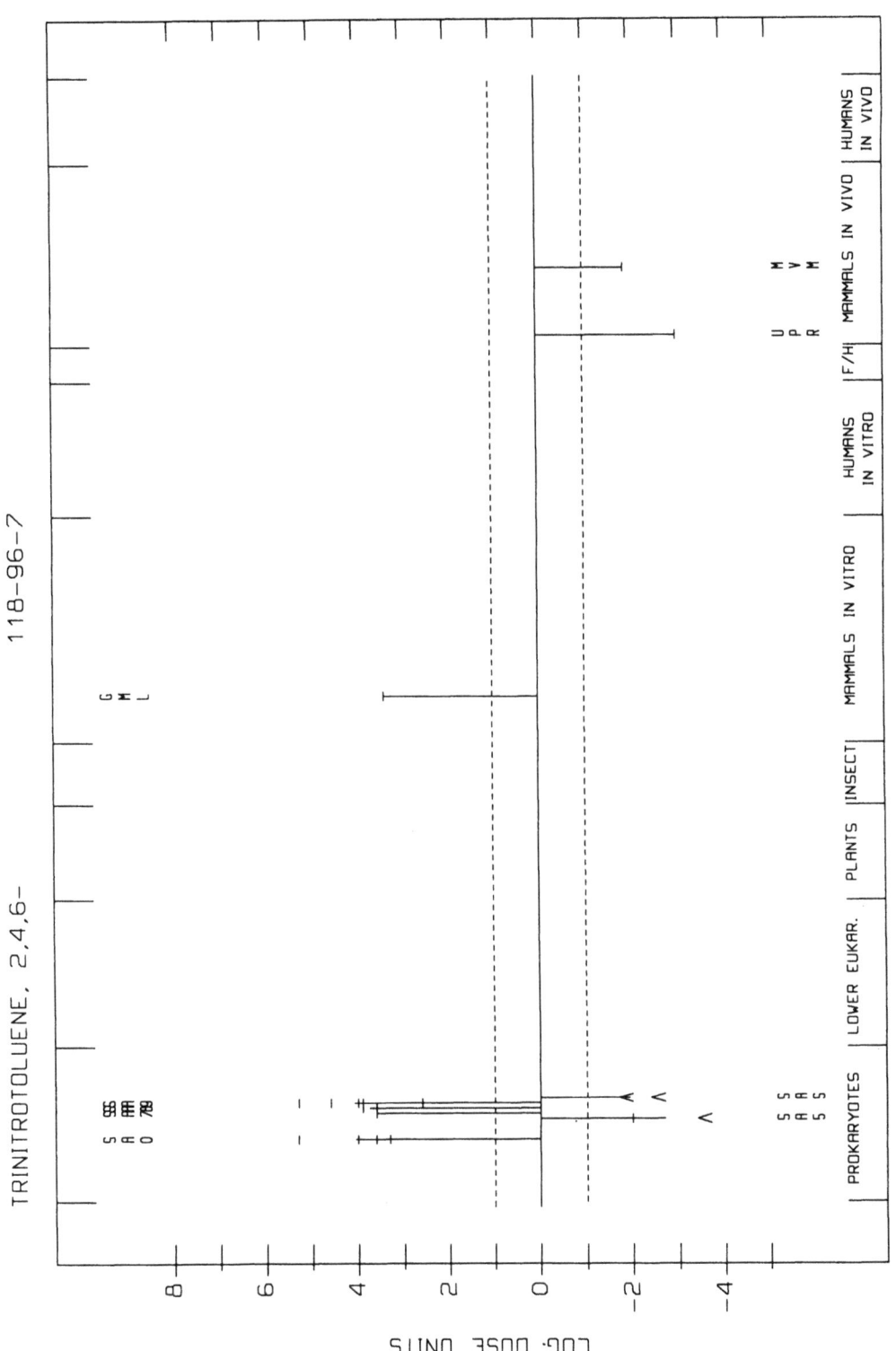

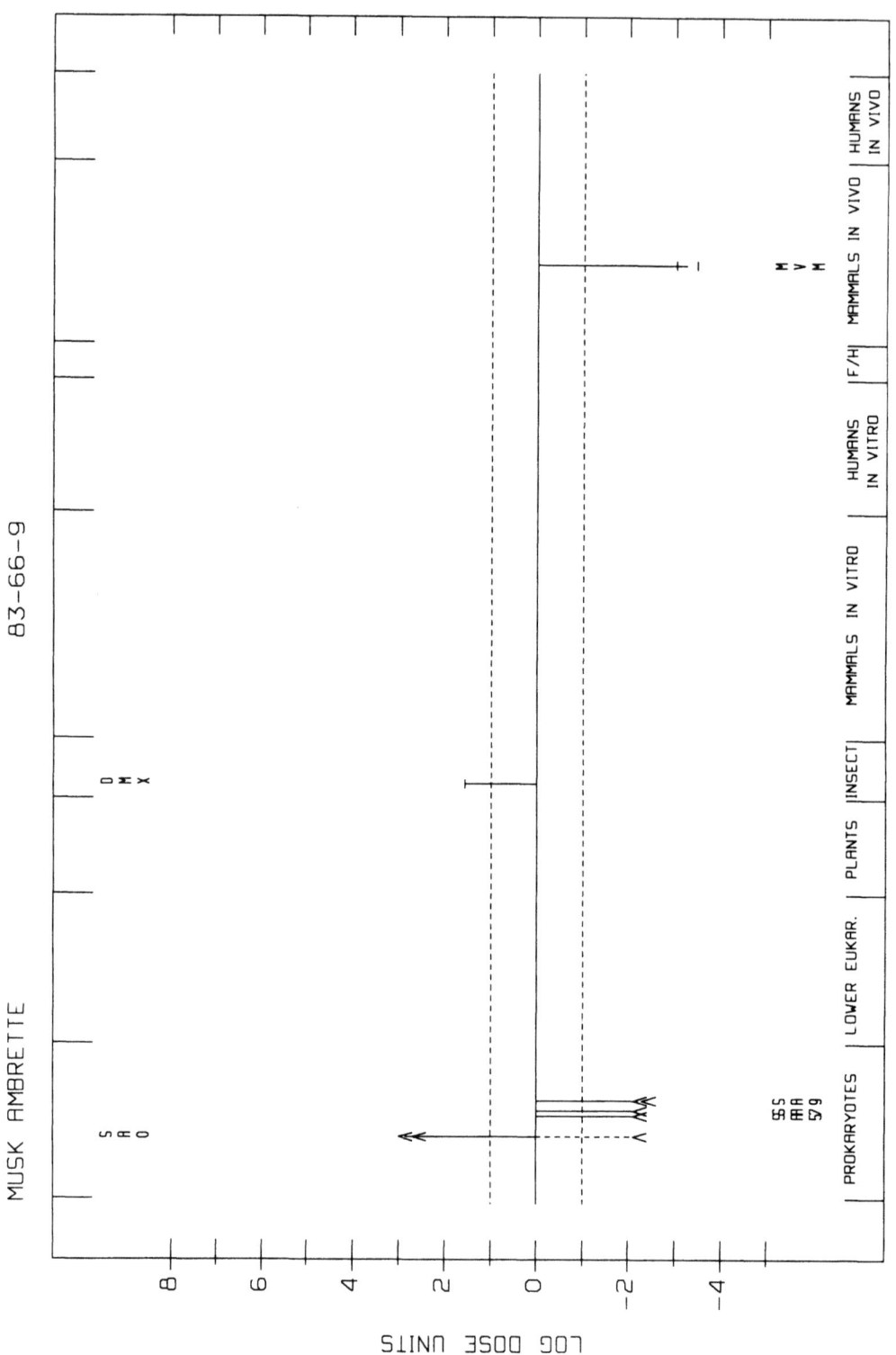

APPENDIX 2

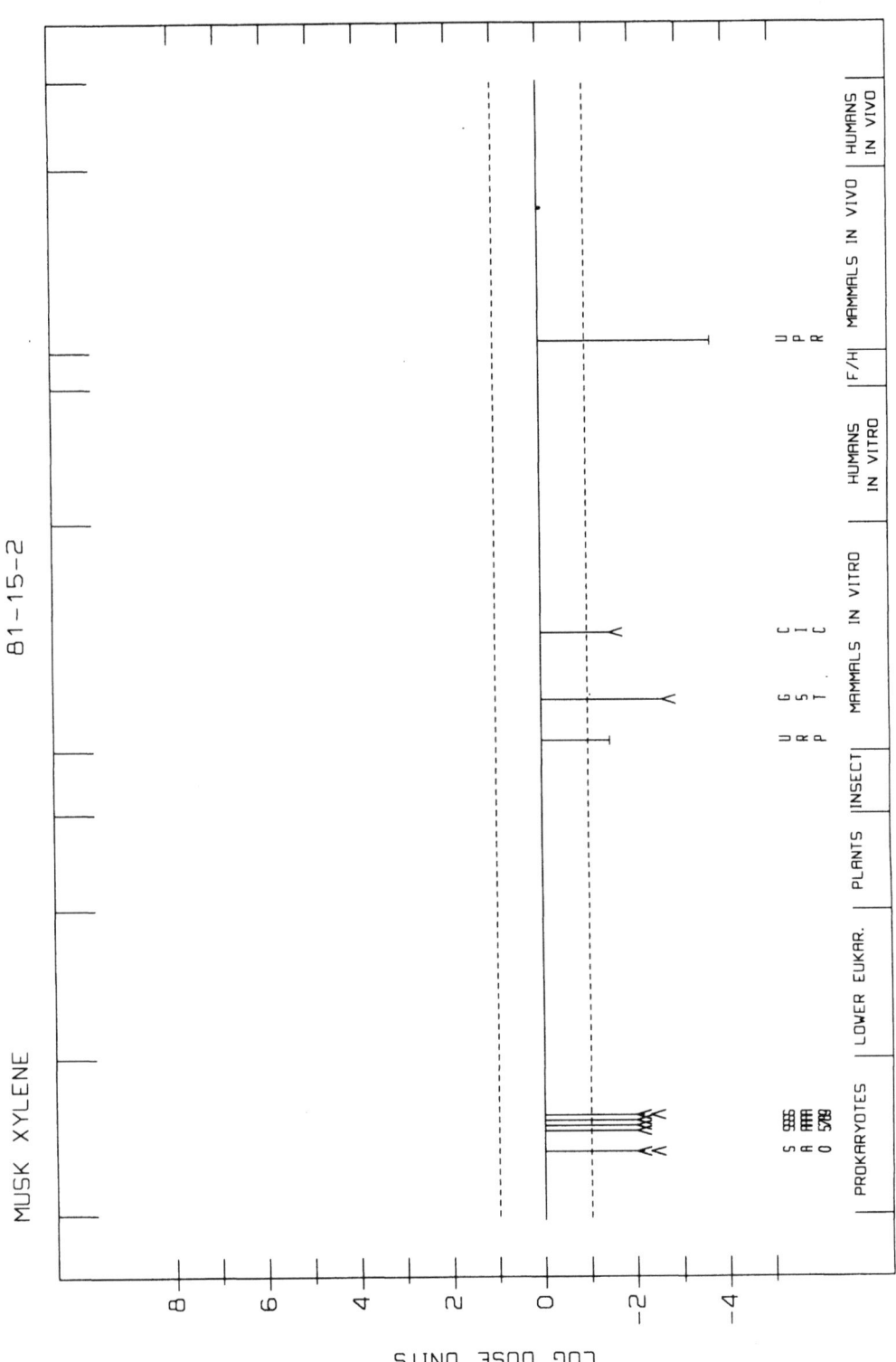

SUPPLEMENTARY CORRIGENDA TO VOLUMES 1–64

Volume 60

p. 276, section 4.2.1, 3 rd paragraph, 2nd line
replace above 100 mg/m^3 [433 ppm] *by* above 100 ppm [433 mg/m^3]

Volume 62

end of p. 325 - Add the following entries to table:

SCF, *Saccharomyces cerevisiae*, forward mutation	+/0	1800	Chanet & von Borstel (1979)
SCR, *Saccharomyces cerevisiae*, reverse mutation	+/0	6	Chanet & von Borstel (1979)
SGR, *Saccharomyces griseoflavus*, reverse mutation	–/0	216	Mashima & Ikeda (1958)

Volume 63

p. 126 - Add:

MVM, Micronucleus induction, mouse bone marrow, *in vivo*	–/0	9800	Kligerman *et al.* (1994)

p. 128 - For Shahin & Von Borstel (1977), replace *SCG* with *SCR*

p. 199 - For Bronzetti *et al.* (1983), results are   –/0

p. 304 - Replace *ECR* with *EC2*

p. 358 - For Grafström *et al.* (1986), DIH, DNA strand breaks, results should be:
+/0

pp. 382–383 - For Haworth *et al.* (1983), SA5, SA7, SA9, dose should be 128.

-549-

CUMULATIVE CROSS INDEX TO *IARC MONOGRAPHS ON THE EVALUATION OF CARCINOGENIC RISKS TO HUMANS*

The volume, page and year of publication are given. References to corrigenda are given in parentheses.

A

A-α-C	40, 245 (1986); *Suppl. 7*, 56 (1987)
Acetaldehyde	36, 101 (1985) (*corr. 42*, 263); *Suppl. 7*, 77 (1987)
Acetaldehyde formylmethylhydrazone (*see* Gyromitrin)	
Acetamide	7, 197 (1974); *Suppl. 7*, 389 (1987)
Acetaminophen (*see* Paracetamol)	
Acridine orange	16, 145 (1978); *Suppl. 7*, 56 (1987)
Acriflavinium chloride	13, 31 (1977); *Suppl. 7*, 56 (1987)
Acrolein	19, 479 (1979); 36, 133 (1985); *Suppl. 7*, 78 (1987); 63, 337 (1995) (*corr. 65*, 549)
Acrylamide	39, 41 (1986); *Suppl. 7*, 56 (1987); 60, 389 (1994)
Acrylic acid	19, 47 (1979); *Suppl. 7*, 56 (1987)
Acrylic fibres	19, 86 (1979); *Suppl. 7*, 56 (1987)
Acrylonitrile	19, 73 (1979); *Suppl. 7*, 79 (1987)
Acrylonitrile-butadiene-styrene copolymers	19, 91 (1979); *Suppl. 7*, 56 (1987)
Actinolite (*see* Asbestos)	
Actinomycins	10, 29 (1976) (*corr. 42*, 255); *Suppl. 7*, 80 (1987)
Adriamycin	10, 43 (1976); *Suppl. 7*, 82 (1987)
AF-2	31, 47 (1983); *Suppl. 7*, 56 (1987)
Aflatoxins	1, 145 (1972) (*corr. 42*, 251); 10, 51 (1976); *Suppl. 7*, 83 (1987); 56, 245 (1993)
Aflatoxin B$_1$ (*see* Aflatoxins)	
Aflatoxin B$_2$ (*see* Aflatoxins)	
Aflatoxin G$_1$ (*see* Aflatoxins)	
Aflatoxin G$_2$ (*see* Aflatoxins)	
Aflatoxin M$_1$ (*see* Aflatoxins)	
Agaritine	31, 63 (1983); *Suppl. 7*, 56 (1987)
Alcohol drinking	44 (1988)
Aldicarb	53, 93 (1991)
Aldrin	5, 25 (1974); *Suppl. 7*, 88 (1987)
Allyl chloride	36, 39 (1985); *Suppl. 7*, 56 (1987)
Allyl isothiocyanate	36, 55 (1985); *Suppl. 7*, 56 (1987)
Allyl isovalerate	36, 69 (1985); *Suppl. 7*, 56 (1987)
Aluminium production	34, 37 (1984); *Suppl. 7*, 89 (1987)

Amaranth	8, 41 (1975); *Suppl. 7*, 56 (1987)
5-Aminoacenaphthene	*16*, 243 (1978); *Suppl. 7*, 56 (1987)
2-Aminoanthraquinone	27, 191 (1982); *Suppl. 7*, 56 (1987)
para-Aminoazobenzene	8, 53 (1975); *Suppl. 7*, 390 (1987)
ortho-Aminoazotoluene	8, 61 (1975) (*corr. 42*, 254); *Suppl. 7*, 56 (1987)
para-Aminobenzoic acid	*16*, 249 (1978); *Suppl. 7*, 56 (1987)
4-Aminobiphenyl	*1*, 74 (1972) (*corr. 42*, 251); *Suppl. 7*, 91 (1987)
2-Amino-3,4-dimethylimidazo[4,5-*f*]quinoline (*see* MeIQ)	
2-Amino-3,8-dimethylimidazo[4,5-*f*]quinoxaline (*see* MeIQx)	
3-Amino-1,4-dimethyl-5*H*-pyrido[4,3-*b*]indole (*see* Trp-P-1)	
2-Aminodipyrido[1,2-*a*:3′,2′-*d*]imidazole (*see* Glu-P-2)	
1-Amino-2-methylanthraquinone	27, 199 (1982); *Suppl. 7*, 57 (1987)
2-Amino-3-methylimidazo[4,5-*f*]quinoline (*see* IQ)	
2-Amino-6-methyldipyrido[1,2-*a*:3′,2′-*d*]imidazole (*see* Glu-P-1)	
2-Amino-1-methyl-6-phenylimidazo[4,5-*b*]pyridine (*see* PhIP)	
2-Amino-3-methyl-9*H*-pyrido[2,3-*b*]indole (*see* MeA-α-C)	
3-Amino-1-methyl-5*H*-pyrido[4,3-*b*]indole (*see* Trp-P-2)	
2-Amino-5-(5-nitro-2-furyl)-1,3,4-thiadiazole	7, 143 (1974); *Suppl. 7*, 57 (1987)
2-Amino-4-nitrophenol	57, 167 (1993)
2-Amino-5-nitrophenol	57, 177 (1993)
4-Amino-2-nitrophenol	*16*, 43 (1978); *Suppl. 7*, 57 (1987)
2-Amino-5-nitrothiazole	*31*, 71 (1983); *Suppl. 7*, 57 (1987)
2-Amino-9*H*-pyrido[2,3-*b*]indole (*see* A-α-C)	
11-Aminoundecanoic acid	39, 239 (1986); *Suppl. 7*, 57 (1987)
Amitrole	7, 31 (1974); *41*, 293 (1986) (*corr. 52*, 513; *Suppl. 7*, 92 (1987)
Ammonium potassium selenide (*see* Selenium and selenium compounds)	
Amorphous silica (*see also* Silica)	42, 39 (1987); *Suppl. 7*, 341 (1987)
Amosite (*see* Asbestos)	
Ampicillin	50, 153 (1990)
Anabolic steroids (*see* Androgenic (anabolic) steroids)	
Anaesthetics, volatile	*11*, 285 (1976); *Suppl. 7*, 93 (1987)
Analgesic mixtures containing phenacetin (*see also* Phenacetin)	*Suppl. 7*, 310 (1987)
Androgenic (anabolic) steroids	*Suppl. 7*, 96 (1987)
Angelicin and some synthetic derivatives (*see also* Angelicins)	40, 291 (1986)
Angelicin plus ultraviolet radiation (*see also* Angelicin and some synthetic derivatives)	*Suppl. 7*, 57 (1987)
Angelicins	*Suppl. 7*, 57 (1987)
Aniline	4, 27 (1974) (*corr. 42*, 252); 27, 39 (1982); *Suppl. 7*, 99 (1987)
ortho-Anisidine	27, 63 (1982); *Suppl. 7*, 57 (1987)
para-Anisidine	27, 65 (1982); *Suppl. 7*, 57 (1987)
Anthanthrene	32, 95 (1983); *Suppl. 7*, 57 (1987)
Anthophyllite (*see* Asbestos)	
Anthracene	32, 105 (1983); *Suppl. 7*, 57 (1987)
Anthranilic acid	*16*, 265 (1978); *Suppl. 7*, 57 (1987)
Antimony trioxide	47, 291 (1989)
Antimony trisulfide	47, 291 (1989)
ANTU (*see* 1-Naphthylthiourea)	
Apholate	9, 31 (1975); *Suppl. 7*, 57 (1987)
Aramite*	5, 39 (1974); *Suppl. 7*, 57 (1987)
Areca nut (*see* Betel quid)	
Arsanilic acid (*see* Arsenic and arsenic compounds)	

Arsenic and arsenic compounds	*1*, 41 (1972); *2*, 48 (1973); *23*, 39 (1980); *Suppl. 7*, 100 (1987)
Arsenic pentoxide (*see* Arsenic and arsenic compounds)	
Arsenic sulfide (*see* Arsenic and arsenic compounds)	
Arsenic trioxide (*see* Arsenic and arsenic compounds)	
Arsine (*see* Arsenic and arsenic compounds)	
Asbestos	*2*, 17 (1973) (*corr. 42*, 252); *14* (1977) (*corr. 42*, 256); *Suppl. 7*, 106 (1987) (*corr. 45*, 283)
Atrazine	*53*, 441 (1991)
Attapulgite	*42*, 159 (1987); *Suppl. 7*, 117 (1987)
Auramine (technical-grade)	*1*, 69 (1972) (*corr. 42*, 251); *Suppl. 7*, 118 (1987)
Auramine, manufacture of (*see also* Auramine, technical-grade)	*Suppl. 7*, 118 (1987)
Aurothioglucose	*13*, 39 (1977); *Suppl. 7*, 57 (1987)
Azacitidine	*26*, 37 (1981); *Suppl. 7*, 57 (1987); *50*, 47 (1990)
5-Azacytidine (*see* Azacitidine)	
Azaserine	*10*, 73 (1976) (*corr. 42*, 255); *Suppl. 7*, 57 (1987)
Azathioprine	*26*, 47 (1981); *Suppl. 7*, 119 (1987)
Aziridine	*9*, 37 (1975); *Suppl. 7*, 58 (1987)
2-(1-Aziridinyl)ethanol	*9*, 47 (1975); *Suppl. 7*, 58 (1987)
Aziridyl benzoquinone	*9*, 51 (1975); *Suppl. 7*, 58 (1987)
Azobenzene	*8*, 75 (1975); *Suppl. 7*, 58 (1987)

B

Barium chromate (*see* Chromium and chromium compounds)	
Basic chromic sulfate (*see* Chromium and chromium compounds)	
BCNU (*see* Bischloroethyl nitrosourea)	
Benz[*a*]acridine	*32*, 123 (1983); *Suppl. 7*, 58 (1987)
Benz[*c*]acridine	*3*, 241 (1973); *32*, 129 (1983); *Suppl. 7*, 58 (1987)
Benzal chloride (*see also* α-Chlorinated toluenes)	*29*, 65 (1982); *Suppl. 7*, 148 (1987)
Benz[*a*]anthracene	*3*, 45 (1973); *32*, 135 (1983); *Suppl. 7*, 58 (1987)
Benzene	*7*, 203 (1974) (*corr. 42*, 254); *29*, 93, 391 (1982); *Suppl. 7*, 120 (1987)
Benzidine	*1*, 80 (1972); *29*, 149, 391 (1982); *Suppl. 7*, 123 (1987)
Benzidine-based dyes	*Suppl. 7*, 125 (1987)
Benzo[*b*]fluoranthene	*3*, 69 (1973); *32*, 147 (1983); *Suppl. 7*, 58 (1987)
Benzo[*j*]fluoranthene	*3*, 82 (1973); *32*, 155 (1983); *Suppl. 7*, 58 (1987)
Benzo[*k*]fluoranthene	*32*, 163 (1983); *Suppl. 7*, 58 (1987)
Benzo[*ghi*]fluoranthene	*32*, 171 (1983); *Suppl. 7*, 58 (1987)
Benzo[*a*]fluorene	*32*, 177 (1983); *Suppl. 7*, 58 (1987)
Benzo[*b*]fluorene	*32*, 183 (1983); *Suppl. 7*, 58 (1987)
Benzo[*c*]fluorene	*32*, 189 (1983); *Suppl. 7*, 58 (1987)
Benzofuran	*63*, 431 (1995)
Benzo[*ghi*]perylene	*32*, 195 (1983); *Suppl. 7*, 58 (1987)
Benzo[*c*]phenanthrene	*32*, 205 (1983); *Suppl. 7*, 58 (1987)

Benzo[a]pyrene	3, 91 (1973); 32, 211 (1983); Suppl. 7, 58 (1987)
Benzo[e]pyrene	3, 137 (1973); 32, 225 (1983); Suppl. 7, 58 (1987)
para-Benzoquinone dioxime	29, 185 (1982); Suppl. 7, 58 (1987)
Benzotrichloride (see also α-Chlorinated toluenes)	29, 73 (1982); Suppl. 7, 148 (1987)
Benzoyl chloride	29, 83 (1982) (corr. 42, 261); Suppl. 7, 126 (1987)
Benzoyl peroxide	36, 267 (1985); Suppl. 7, 58 (1987)
Benzyl acetate	40, 109 (1986); Suppl. 7, 58 (1987)
Benzyl chloride (see also α-Chlorinated toluenes)	11, 217 (1976) (corr. 42, 256); 29, 49 (1982); Suppl. 7, 148 (1987)
Benzyl violet 4B	16, 153 (1978); Suppl. 7, 58 (1987)
Bertrandite (see Beryllium and beryllium compounds)	
Beryllium and beryllium compounds	1, 17 (1972); 23, 143 (1980) (corr. 42, 260); Suppl. 7, 127 (1987); 58, 41 (1993)
Beryllium acetate (see Beryllium and beryllium compounds)	
Beryllium acetate, basic (see Beryllium and beryllium compounds)	
Beryllium-aluminium alloy (see Beryllium and beryllium compounds)	
Beryllium carbonate (see Beryllium and beryllium compounds)	
Beryllium chloride (see Beryllium and beryllium compounds)	
Beryllium-copper alloy (see Beryllium and beryllium compounds)	
Beryllium-copper-cobalt alloy (see Beryllium and beryllium compounds)	
Beryllium fluoride (see Beryllium and beryllium compounds)	
Beryllium hydroxide (see Beryllium and beryllium compounds)	
Beryllium-nickel alloy (see Beryllium and beryllium compounds)	
Beryllium oxide (see Beryllium and beryllium compounds)	
Beryllium phosphate (see Beryllium and beryllium compounds)	
Beryllium silicate (see Beryllium and beryllium compounds)	
Beryllium sulfate (see Beryllium and beryllium compounds)	
Beryl ore (see Beryllium and beryllium compounds)	
Betel quid	37, 141 (1985); Suppl. 7, 128 (1987)
Betel-quid chewing (see Betel quid)	
BHA (see Butylated hydroxyanisole)	
BHT (see Butylated hydroxytoluene)	
Bis(1-aziridinyl)morpholinophosphine sulfide	9, 55 (1975); Suppl. 7, 58 (1987)
Bis(2-chloroethyl)ether	9, 117 (1975); Suppl. 7, 58 (1987)
N,N-Bis(2-chloroethyl)-2-naphthylamine	4, 119 (1974) (corr. 42, 253); Suppl. 7, 130 (1987)
Bischloroethyl nitrosourea (see also Chloroethyl nitrosoureas)	26, 79 (1981); Suppl. 7, 150 (1987)
1,2-Bis(chloromethoxy)ethane	15, 31 (1977); Suppl. 7, 58 (1987)
1,4-Bis(chloromethoxymethyl)benzene	15, 37 (1977); Suppl. 7, 58 (1987)
Bis(chloromethyl)ether	4, 231 (1974) (corr. 42, 253); Suppl. 7, 131 (1987)
Bis(2-chloro-1-methylethyl)ether	41, 149 (1986); Suppl. 7, 59 (1987)
Bis(2,3-epoxycyclopentyl)ether	47, 231 (1989)
Bisphenol A diglycidyl ether (see Glycidyl ethers)	
Bisulfites (see Sulfur dioxide and some sulfites, bisulfites and metabisulfites)	
Bitumens	35, 39 (1985); Suppl. 7, 133 (1987)
Bleomycins	26, 97 (1981); Suppl. 7, 134 (1987)
Blue VRS	16, 163 (1978); Suppl. 7, 59 (1987)
Boot and shoe manufacture and repair	25, 249 (1981); Suppl. 7, 232 (1987)
Bracken fern	40, 47 (1986); Suppl. 7, 135 (1987)

Brilliant Blue FCF, disodium salt	*16*, 171 (1978) (*corr. 42*, 257); *Suppl. 7*, 59 (1987)
Bromochloroacetonitrile (*see* Halogenated acetonitriles)	
Bromodichloromethane	*52*, 179 (1991)
Bromoethane	*52*, 299 (1991)
Bromoform	*52*, 213 (1991)
1,3-Butadiene	*39*, 155 (1986) (*corr. 42*, 264 *Suppl. 7*, 136 (1987); *54*, 237 (1992)
1,4-Butanediol dimethanesulfonate	*4*, 247 (1974); *Suppl. 7*, 137 (1987)
n-Butyl acrylate	*39*, 67 (1986); *Suppl. 7*, 59 (1987)
Butylated hydroxyanisole	*40*, 123 (1986); *Suppl. 7*, 59 (1987)
Butylated hydroxytoluene	*40*, 161 (1986); *Suppl. 7*, 59 (1987)
Butyl benzyl phthalate	*29*, 193 (1982) (*corr. 42*, 261); *Suppl. 7*, 59 (1987)
β-Butyrolactone	*11*, 225 (1976); *Suppl. 7*, 59 (1987)
γ-Butyrolactone	*11*, 231 (1976); *Suppl. 7*, 59 (1987)

C

Cabinet-making (*see* Furniture and cabinet-making)	
Cadmium acetate (*see* Cadmium and cadmium compounds)	
Cadmium and cadmium compounds	*2*, 74 (1973); *11*, 39 (1976) (*corr. 42*, 255); *Suppl. 7*, 139 (1987); *58*, 119 (1993)
Cadmium chloride (*see* Cadmium and cadmium compounds)	
Cadmium oxide (*see* Cadmium and cadmium compounds)	
Cadmium sulfate (*see* Cadmium and cadmium compounds)	
Cadmium sulfide (*see* Cadmium and cadmium compounds)	
Caffeic acid	*56*, 115 (1993)
Caffeine	*51*, 291 (1991)
Calcium arsenate (*see* Arsenic and arsenic compounds)	
Calcium chromate (see Chromium and chromium compounds)	
Calcium cyclamate (*see* Cyclamates)	
Calcium saccharin (*see* Saccharin)	
Cantharidin	*10*, 79 (1976); *Suppl. 7*, 59 (1987)
Caprolactam	*19*, 115 (1979) (*corr. 42*, 258); *39*, 247 (1986) (*corr. 42*, 264); *Suppl. 7*, 390 (1987)
Captafol	*53*, 353 (1991)
Captan	*30*, 295 (1983); *Suppl. 7*, 59 (1987)
Carbaryl	*12*, 37 (1976); *Suppl. 7*, 59 (1987)
Carbazole	*32*, 239 (1983); *Suppl. 7*, 59 (1987)
3-Carbethoxypsoralen	*40*, 317 (1986); *Suppl. 7*, 59 (1987)
Carbon black	*3*, 22 (1973); *33*, 35 (1984); *Suppl. 7*, 142 (1987); *65*, 149 (1996)
Carbon tetrachloride	*1*, 53 (1972); *20*, 371 (1979); *Suppl. 7*, 143 (1987)
Carmoisine	*8*, 83 (1975); *Suppl. 7*, 59 (1987)
Carpentry and joinery	*25*, 139 (1981); *Suppl. 7*, 378 (1987)
Carrageenan	*10*, 181 (1976) (*corr. 42*, 255); *31*, 79 (1983); *Suppl. 7*, 59 (1987)
Catechol	*15*, 155 (1977); *Suppl. 7*, 59 (1987)
CCNU (*see* 1-(2-Chloroethyl)-3-cyclohexyl-1-nitrosourea)	
Ceramic fibres (see Man-made mineral fibres)	

Chemotherapy, combined, including alkylating agents (*see* MOPP and
    other combined chemotherapy including alkylating agents)
Chloral   *63*, 245 (1995)
Chloral hydrate   *63*, 245 (1995)
Chlorambucil   *9*, 125 (1975); *26*, 115 (1981);
                                                                            Suppl. 7, 144 (1987)
Chloramphenicol   *10*, 85 (1976); *Suppl. 7*, 145 (1987);
                                                                            50, 169 (1990)
Chlordane (*see also* Chlordane/Heptachlor)   *20*, 45 (1979) (*corr. 42*, 258)
Chlordane/Heptachlor   *Suppl. 7*, 146 (1987); *53*, 115 (1991)
Chlordecone   *20*, 67 (1979); *Suppl. 7*, 59 (1987)
Chlordimeform   *30*, 61 (1983); *Suppl. 7*, 59 (1987)
Chlorendic acid   *48*, 45 (1990)
Chlorinated dibenzodioxins (other than TCDD)   *15*, 41 (1977); *Suppl. 7*, 59 (1987)
Chlorinated drinking-water   *52*, 45 (1991)
Chlorinated paraffins   *48*, 55 (1990)
α-Chlorinated toluenes   *Suppl. 7*, 148 (1987)
Chlormadinone acetate (*see also* Progestins; Combined oral   *6*, 149 (1974); *21*, 365 (1979)
    contraceptives)
Chlornaphazine (*see N,N*-Bis(2-chloroethyl)-2-naphthylamine)
Chloroacetonitrile (*see* Halogenated acetonitriles)
para-Chloroaniline   *57*, 305 (1993)
Chlorobenzilate   *5*, 75 (1974); *30*, 73 (1983);
                                                                            Suppl. 7, 60 (1987)
Chlorodibromomethane   *52*, 243 (1991)
Chlorodifluoromethane   *41*, 237 (1986) (*corr. 51*, 483);
                                                                            Suppl. 7, 149 (1987)
Chloroethane   *52*, 315 (1991)
1-(2-Chloroethyl)-3-cyclohexyl-1-nitrosourea (*see also* Chloroethyl   *26*, 137 (1981) (*corr. 42*, 260);
    nitrosoureas)   *Suppl. 7*, 150 (1987)
1-(2-Chloroethyl)-3-(4-methylcyclohexyl)-1-nitrosourea (*see also*   *Suppl. 7*, 150 (1987)
    Chloroethyl nitrosoureas)
Chloroethyl nitrosoureas   *Suppl. 7*, 150 (1987)
Chlorofluoromethane   *41*, 229 (1986); *Suppl. 7*, 60 (1987)
Chloroform   *1*, 61 (1972); *20*, 401 (1979)
                                                                            Suppl. 7, 152 (1987)
Chloromethyl methyl ether (technical-grade) (*see also*   *4*, 239 (1974); *Suppl. 7*, 131 (1987)
    Bis(chloromethyl)ether)
(4-Chloro-2-methylphenoxy)acetic acid (*see* MCPA)
1-Chloro-2-methylpropene   *63*, 315 (1995)
3-Chloro-2-methylpropene   *63*, 325 (1995)
2-Chloronitrobenzene   *65*, 263 (1996)
3-Chloronitrobenzene   *65*, 263 (1996)
4-Chloronitrobenzene   *65*, 263 (1996)
Chlorophenols   *Suppl. 7*, 154 (1987)
Chlorophenols (occupational exposures to)   *41*, 319 (1986)
Chlorophenoxy herbicides   *Suppl. 7*, 156 (1987)
Chlorophenoxy herbicides (occupational exposures to)   *41*, 357 (1986)
4-Chloro-*ortho*-phenylenediamine   *27*, 81 (1982); *Suppl. 7*, 60 (1987)
4-Chloro-*meta*-phenylenediamine   *27*, 82 (1982); *Suppl. 7*, 60 (1987)
Chloroprene   *19*, 131 (1979); *Suppl. 7*, 160 (1987)
Chloropropham   *12*, 55 (1976); *Suppl. 7*, 60 (1987)
Chloroquine   *13*, 47 (1977); *Suppl. 7*, 60 (1987)
Chlorothalonil   *30*, 319 (1983); *Suppl. 7*, 60 (1987)

para-Chloro-ortho-toluidine and its strong acid salts (see also Chlordimeform)	16, 277 (1978); 30, 65 (1983); Suppl. 7, 60 (1987); 48, 123 (1990)
Chlorotrianisene (see also Nonsteroidal oestrogens)	21, 139 (1979)
2-Chloro-1,1,1-trifluoroethane	41, 253 (1986); Suppl. 7, 60 (1987)
Chlorozotocin	50, 65 (1990)
Cholesterol	10, 99 (1976); 31, 95 (1983); Suppl. 7, 161 (1987)
Chromic acetate (see Chromium and chromium compounds)	
Chromic chloride (see Chromium and chromium compounds)	
Chromic oxide (see Chromium and chromium compounds)	
Chromic phosphate (see Chromium and chromium compounds)	
Chromite ore (see Chromium and chromium compounds)	
Chromium and chromium compounds	2, 100 (1973); 23, 205 (1980); Suppl. 7, 165 (1987); 49, 49 (1990) (corr. 51, 483)
Chromium carbonyl (see Chromium and chromium compounds)	
Chromium potassium sulfate (see Chromium and chromium compounds)	
Chromium sulfate (see Chromium and chromium compounds)	
Chromium trioxide (see Chromium and chromium compounds)	
Chrysazin (see Dantron)	
Chrysene	3, 159 (1973); 32, 247 (1983); Suppl. 7, 60 (1987)
Chrysoidine	8, 91 (1975); Suppl. 7, 169 (1987)
Chrysotile (see Asbestos)	
CI Acid Orange 3	57, 121 (1993)
CI Acid Red 114	57, 247 (1993)
CI Basic Red 9	57, 215 (1993)
Ciclosporin	50, 77 (1990)
CI Direct Blue 15	57, 235 (1993)
CI Disperse Yellow 3 (see Disperse Yellow 3)	
Cimetidine	50, 235 (1990)
Cinnamyl anthranilate	16, 287 (1978); 31, 133 (1983); Suppl. 7, 60 (1987)
CI Pigment Red 3	57, 259 (1993)
CI Pigment Red 53:1 (see D&C Red No. 9)	
Cisplatin	26, 151 (1981); Suppl. 7, 170 (1987)
Citrinin	40, 67 (1986); Suppl. 7, 60 (1987)
Citrus Red No. 2	8, 101 (1975) (corr. 42, 254) Suppl. 7, 60 (1987)
Clofibrate	24, 39 (1980); Suppl. 7, 171 (1987)
Clomiphene citrate	21, 551 (1979); Suppl. 7, 172 (1987)
Clonorchis sinensis (infection with)	61, 121 (1994)
Coal gasification	34, 65 (1984); Suppl. 7, 173 (1987)
Coal-tar pitches (see also Coal-tars)	35, 83 (1985); Suppl. 7, 174 (1987)
Coal-tars	35, 83 (1985); Suppl. 7, 175 (1987)
Cobalt[III] acetate (see Cobalt and cobalt compounds)	
Cobalt-aluminium-chromium spinel (see Cobalt and cobalt compounds)	
Cobalt and cobalt compounds	52, 363 (1991)
Cobalt[II] chloride (see Cobalt and cobalt compounds)	
Cobalt-chromium alloy (see Chromium and chromium compounds)	
Cobalt-chromium-molybdenum alloys (see Cobalt and cobalt compounds)	
Cobalt metal powder (see Cobalt and cobalt compounds)	
Cobalt naphthenate (see Cobalt and cobalt compounds)	
Cobalt[II] oxide (see Cobalt and cobalt compounds)	
Cobalt[II,III] oxide (see Cobalt and cobalt compounds)	

Cobalt[II] sulfide (see Cobalt and cobalt compounds)
Coffee   51, 41 (1991) (corr. 52, 513)
Coke production   34, 101 (1984); Suppl. 7, 176 (1987)
Combined oral contraceptives (see also Oestrogens, progestins   Suppl. 7, 297 (1987)
    and combinations)
Conjugated oestrogens (see also Steroidal oestrogens)   21, 147 (1979)
Contraceptives, oral (see Combined oral contraceptives;
    Sequential oral contraceptives)
Copper 8-hydroxyquinoline   15, 103 (1977); Suppl. 7, 61 (1987)
Coronene   32, 263 (1983); Suppl. 7, 61 (1987)
Coumarin   10, 113 (1976); Suppl. 7, 61 (1987)
Creosotes (see also Coal-tars)   35, 83 (1985); Suppl. 7, 177 (1987)
meta-Cresidine   27, 91 (1982); Suppl. 7, 61 (1987)
para-Cresidine   27, 92 (1982); Suppl. 7, 61 (1987)
Crocidolite (see Asbestos)
Crotonaldehyde   63, 373 (1995) (corr. 65, 549)
Crude oil   45, 119 (1989)
Crystalline silica (see also Silica)   42, 39 (1987); Suppl. 7, 341 (1987)
Cycasin   1, 157 (1972) (corr. 42, 251); 10,
                                                                    121 (1976); Suppl. 7, 61 (1987)
Cyclamates   22, 55 (1980); Suppl. 7, 178 (1987)
Cyclamic acid (see Cyclamates)
Cyclochlorotine   10, 139 (1976); Suppl. 7, 61 (1987)
Cyclohexanone   47, 157 (1989)
Cyclohexylamine (see Cyclamates)
Cyclopenta[cd]pyrene   32, 269 (1983); Suppl. 7, 61 (1987)
Cyclopropane (see Anaesthetics, volatile)
Cyclophosphamide   9, 135 (1975); 26, 165 (1981);
                                                                    Suppl. 7, 182 (1987)

D

2,4-D (see also Chlorophenoxy herbicides; Chlorophenoxy   15, 111 (1977)
    herbicides, occupational exposures to)
Dacarbazine   26, 203 (1981); Suppl. 7, 184 (1987)
Dantron   50, 265 (1990) (corr. 59, 257)
D&C Red No. 9   8, 107 (1975); Suppl. 7, 61 (1987);
                                                                    57, 203 (1993)
Dapsone   24, 59 (1980); Suppl. 7, 185 (1987)
Daunomycin   10, 145 (1976); Suppl. 7, 61 (1987)
DDD (see DDT)
DDE (see DDT)
DDT   5, 83 (1974) (corr. 42, 253);
                                                                    Suppl. 7, 186 (1987); 53, 179 (1991)
Decabromodiphenyl oxide   48, 73 (1990)
Deltamethrin   53, 251 (1991)
Deoxynivalenol (see Toxins derived from Fusarium graminearum,
    F. culmorum and F. crookwellense)
Diacetylaminoazotoluene   8, 113 (1975); Suppl. 7, 61 (1987)
N,N'-Diacetylbenzidine   16, 293 (1978); Suppl. 7, 61 (1987)
Diallate   12, 69 (1976); 30, 235 (1983);
                                                                    Suppl. 7, 61 (1987)
2,4-Diaminoanisole   16, 51 (1978); 27, 103 (1982);
                                                                    Suppl. 7, 61 (1987)

CUMULATIVE INDEX

4,4'-Diaminodiphenyl ether	*16*, 301 (1978); *29*, 203 (1982); *Suppl. 7*, 61 (1987)
1,2-Diamino-4-nitrobenzene	*16*, 63 (1978); *Suppl. 7*, 61 (1987)
1,4-Diamino-2-nitrobenzene	*16*, 73 (1978); *Suppl. 7*, 61 (1987); *57*, 185 (1993)
2,6-Diamino-3-(phenylazo)pyridine (*see* Phenazopyridine hydrochloride)	
2,4-Diaminotoluene (*see also* Toluene diisocyanates)	*16*, 83 (1978); *Suppl. 7*, 61 (1987)
2,5-Diaminotoluene (*see also* Toluene diisocyanates)	*16*, 97 (1978); *Suppl. 7*, 61 (1987)
ortho-Dianisidine (*see* 3,3'-Dimethoxybenzidine)	
Diazepam	*13*, 57 (1977); *Suppl. 7*, 189 (1987)
Diazomethane	*7*, 223 (1974); *Suppl. 7*, 61 (1987)
Dibenz[*a,h*]acridine	*3*, 247 (1973); *32*, 277 (1983); *Suppl. 7*, 61 (1987)
Dibenz[*a,j*]acridine	*3*, 254 (1973); *32*, 283 (1983); *Suppl. 7*, 61 (1987)
Dibenz[*a,c*]anthracene	*32*, 289 (1983) (*corr. 42*, 262); *Suppl. 7*, 61 (1987)
Dibenz[*a,h*]anthracene	*3*, 178 (1973) (*corr. 43*, 261); *32*, 299 (1983); *Suppl. 7*, 61 (1987)
Dibenz[*a,j*]anthracene	*32*, 309 (1983); *Suppl. 7*, 61 (1987)
7*H*-Dibenzo[*c,g*]carbazole	*3*, 260 (1973); *32*, 315 (1983); *Suppl. 7*, 61 (1987)
Dibenzodioxins, chlorinated (other than TCDD) [*see* Chlorinated dibenzodioxins (other than TCDD)]	
Dibenzo[*a,e*]fluoranthene	*32*, 321 (1983); *Suppl. 7*, 61 (1987)
Dibenzo[*h,rst*]pentaphene	*3*, 197 (1973); *Suppl. 7*, 62 (1987)
Dibenzo[*a,e*]pyrene	*3*, 201 (1973); *32*, 327 (1983); *Suppl. 7*, 62 (1987)
Dibenzo[*a,h*]pyrene	*3*, 207 (1973); *32*, 331 (1983); *Suppl. 7*, 62 (1987)
Dibenzo[*a,i*]pyrene	*3*, 215 (1973); *32*, 337 (1983); *Suppl. 7*, 62 (1987)
Dibenzo[*a,l*]pyrene	*3*, 224 (1973); *32*, 343 (1983); *Suppl. 7*, 62 (1987)
Dibromoacetonitrile (*see* Halogenated acetonitriles)	
1,2-Dibromo-3-chloropropane	*15*, 139 (1977); *20*, 83 (1979); *Suppl. 7*, 191 (1987)
Dichloroacetic acid	*63*, 271 (1995)
Dichloroacetonitrile (*see* Halogenated acetonitriles)	
Dichloroacetylene	*39*, 369 (1986); *Suppl. 7*, 62 (1987)
ortho-Dichlorobenzene	*7*, 231 (1974); *29*, 213 (1982); *Suppl. 7*, 192 (1987)
para-Dichlorobenzene	*7*, 231 (1974); *29*, 215 (1982); *Suppl. 7*, 192 (1987)
3,3'-Dichlorobenzidine	*4*, 49 (1974); *29*, 239 (1982); *Suppl. 7*, 193 (1987)
trans-1,4-Dichlorobutene	*15*, 149 (1977); *Suppl. 7*, 62 (1987)
3,3'-Dichloro-4,4'-diaminodiphenyl ether	*16*, 309 (1978); *Suppl. 7*, 62 (1987)
1,2-Dichloroethane	*20*, 429 (1979); *Suppl. 7*, 62 (1987)
Dichloromethane	*20*, 449 (1979); *41*, 43 (1986); *Suppl. 7*, 194 (1987)
2,4-Dichlorophenol (*see* Chlorophenols; Chlorophenols, occupational exposures to)	
(2,4-Dichlorophenoxy)acetic acid (*see* 2,4-D)	
2,6-Dichloro-*para*-phenylenediamine	*39*, 325 (1986); *Suppl. 7*, 62 (1987)

1,2-Dichloropropane	*41*, 131 (1986); *Suppl. 7*, 62 (1987)
1,3-Dichloropropene (technical-grade)	*41*, 113 (1986); *Suppl. 7*, 195 (1987)
Dichlorvos	*20*, 97 (1979); *Suppl. 7*, 62 (1987); *53*, 267 (1991)
Dicofol	*30*, 87 (1983); *Suppl. 7*, 62 (1987)
Dicyclohexylamine (*see* Cyclamates)	
Dieldrin	*5*, 125 (1974); *Suppl. 7*, 196 (1987)
Dienoestrol (*see also* Nonsteroidal oestrogens)	*21*, 161 (1979)
Diepoxybutane	*11*, 115 (1976) (*corr. 42*, 255); *Suppl. 7*, 62 (1987)
Diesel and gasoline engine exhausts	*46*, 41 (1989)
Diesel fuels	*45*, 219 (1989) (*corr. 47*, 505)
Diethyl ether (*see* Anaesthetics, volatile)	
Di(2-ethylhexyl)adipate	*29*, 257 (1982); *Suppl. 7*, 62 (1987)
Di(2-ethylhexyl)phthalate	*29*, 269 (1982) (*corr. 42*, 261); *Suppl. 7*, 62 (1987)
1,2-Diethylhydrazine	*4*, 153 (1974); *Suppl. 7*, 62 (1987)
Diethylstilboestrol	*6*, 55 (1974); *21*, 173 (1979) (*corr. 42*, 259); *Suppl. 7*, 273 (1987)
Diethylstilboestrol dipropionate (*see* Diethylstilboestrol)	
Diethyl sulfate	*4*, 277 (1974); *Suppl. 7*, 198 (1987); *54*, 213 (1992)
Diglycidyl resorcinol ether	*11*, 125 (1976); *36*, 181 (1985); *Suppl. 7*, 62 (1987)
Dihydrosafrole	*1*, 170 (1972); *10*, 233 (1976) *Suppl. 7*, 62 (1987)
1,8-Dihydroxyanthraquinone (*see* Dantron)	
Dihydroxybenzenes (*see* Catechol; Hydroquinone; Resorcinol)	
Dihydroxymethylfuratrizine	*24*, 77 (1980); *Suppl. 7*, 62 (1987)
Diisopropyl sulfate	*54*, 229 (1992)
Dimethisterone (*see also* Progestins; Sequential oral contraceptives	*6*, 167 (1974); *21*, 377 (1979))
Dimethoxane	*15*, 177 (1977); *Suppl. 7*, 62 (1987)
3,3′-Dimethoxybenzidine	*4*, 41 (1974); *Suppl. 7*, 198 (1987)
3,3′-Dimethoxybenzidine-4,4′-diisocyanate	*39*, 279 (1986); *Suppl. 7*, 62 (1987)
para-Dimethylaminoazobenzene	*8*, 125 (1975); *Suppl. 7*, 62 (1987)
para-Dimethylaminoazobenzenediazo sodium sulfonate	*8*, 147 (1975); *Suppl. 7*, 62 (1987)
trans-2-[(Dimethylamino)methylimino]-5-[2-(5-nitro-2-furyl)-vinyl]-1,3,4-oxadiazole	*7*, 147 (1974) (*corr. 42*, 253); *Suppl. 7*, 62 (1987)
4,4′-Dimethylangelicin plus ultraviolet radiation (*see also* Angelicin and some synthetic derivatives)	*Suppl. 7*, 57 (1987)
4,5′-Dimethylangelicin plus ultraviolet radiation (*see also* Angelicin and some synthetic derivatives)	*Suppl. 7*, 57 (1987)
2,6-Dimethylaniline	*57*, 323 (1993)
N,N-Dimethylaniline	*57*, 337 (1993)
Dimethylarsinic acid (*see* Arsenic and arsenic compounds)	
3,3′-Dimethylbenzidine	*1*, 87 (1972); *Suppl. 7*, 62 (1987)
Dimethylcarbamoyl chloride	*12*, 77 (1976); *Suppl. 7*, 199 (1987)
Dimethylformamide	*47*, 171 (1989)
1,1-Dimethylhydrazine	*4*, 137 (1974); *Suppl. 7*, 62 (1987)
1,2-Dimethylhydrazine	*4*, 145 (1974) (*corr. 42*, 253); *Suppl. 7*, 62 (1987)
Dimethyl hydrogen phosphite	*48*, 85 (1990)
1,4-Dimethylphenanthrene	*32*, 349 (1983); *Suppl. 7*, 62 (1987)
Dimethyl sulfate	*4*, 271 (1974); *Suppl. 7*, 200 (1987)
3,7-Dinitrofluoranthene	*46*, 189 (1989); *65*, 297 (1996)

3,9-Dinitrofluoranthene	46, 195 (1989); 65, 297 (1996)
1,3-Dinitropyrene	46, 201 (1989)
1,6-Dinitropyrene	46, 215 (1989)
1,8-Dinitropyrene	33, 171 (1984); Suppl. 7, 63 (1987); 46, 231 (1989)
Dinitrosopentamethylenetetramine	11, 241 (1976); Suppl. 7, 63 (1987)
2,4-Dinitrotoluene	65, 309 (1996)
2,6-Dinitrotoluene	65, 309 (1996)
3,5-Dinitrotoluene	65, 309 (1996)
1,4-Dioxane	11, 247 (1976); Suppl. 7, 201 (1987)
2,4'-Diphenyldiamine	16, 313 (1978); Suppl. 7, 63 (1987)
Direct Black 38 (see also Benzidine-based dyes)	29, 295 (1982) (corr. 42, 261)
Direct Blue 6 (see also Benzidine-based dyes)	29, 311 (1982)
Direct Brown 95 (see also Benzidine-based dyes)	29, 321 (1982)
Disperse Blue 1	48, 139 (1990)
Disperse Yellow 3	8, 97 (1975); Suppl. 7, 60 (1987); 48, 149 (1990)
Disulfiram	12, 85 (1976); Suppl. 7, 63 (1987)
Dithranol	13, 75 (1977); Suppl. 7, 63 (1987)
Divinyl ether (see Anaesthetics, volatile)	
Dry cleaning	63, 33 (1995)
Dulcin	12, 97 (1976); Suppl. 7, 63 (1987)

E

Endrin	5, 157 (1974); Suppl. 7, 63 (1987)
Enflurane (see Anaesthetics, volatile)	
Eosin	15, 183 (1977); Suppl. 7, 63 (1987)
Epichlorohydrin	11, 131 (1976) (corr. 42, 256); Suppl. 7, 202 (1987)
1,2-Epoxybutane	47, 217 (1989)
1-Epoxyethyl-3,4-epoxycyclohexane (see 4-Vinylcyclohexene diepoxide)	
3,4-Epoxy-6-methylcyclohexylmethyl-3,4-epoxy-6-methyl-cyclohexane carboxylate	11, 147 (1976); Suppl. 7, 63 (1987)
cis-9,10-Epoxystearic acid	11, 153 (1976); Suppl. 7, 63 (1987)
Erionite	42, 225 (1987); Suppl. 7, 203 (1987)
Ethinyloestradiol (see also Steroidal oestrogens)	6, 77 (1974); 21, 233 (1979)
Ethionamide	13, 83 (1977); Suppl. 7, 63 (1987)
Ethyl acrylate	19, 57 (1979); 39, 81 (1986); Suppl. 7, 63 (1987)
Ethylene	19, 157 (1979); Suppl. 7, 63 (1987); 60, 45 (1994)
Ethylene dibromide	15, 195 (1977); Suppl. 7, 204 (1987)
Ethylene oxide	11, 157 (1976); 36, 189 (1985) (corr. 42, 263); Suppl. 7, 205 (1987); 60, 73 (1994)
Ethylene sulfide	11, 257 (1976); Suppl. 7, 63 (1987)
Ethylene thiourea	7, 45 (1974); Suppl. 7, 207 (1987)
2-Ethylhexyl acrylate	60, 475 (1994)
Ethyl methanesulfonate	7, 245 (1974); Suppl. 7, 63 (1987)
N-Ethyl-N-nitrosourea	1, 135 (1972); 17, 191 (1978); Suppl. 7, 63 (1987)
Ethyl selenac (see also Selenium and selenium compounds)	12, 107 (1976); Suppl. 7, 63 (1987)
Ethyl tellurac	12, 115 (1976); Suppl. 7, 63 (1987)

Ethynodiol diacetate (*see also* Progestins; Combined oral contraceptives)   6, 173 (1974); 21, 387 (1979)
Eugenol   36, 75 (1985); *Suppl. 7*, 63 (1987)
Evans blue   8, 151 (1975); *Suppl. 7*, 63 (1987)

F

Fast Green FCF   16, 187 (1978); *Suppl. 7*, 63 (1987)
Fenvalerate   53, 309 (1991)
Ferbam   12, 121 (1976) (*corr. 42*, 256); *Suppl. 7*, 63 (1987)
Ferric oxide   1, 29 (1972); *Suppl. 7*, 216 (1987)
Ferrochromium (*see* Chromium and chromium compounds)
Fluometuron   30, 245 (1983); *Suppl. 7*, 63 (1987)
Fluoranthene   32, 355 (1983); *Suppl. 7*, 63 (1987)
Fluorene   32, 365 (1983); *Suppl. 7*, 63 (1987)
Fluorescent lighting (exposure to) (*see* Ultraviolet radiation)
Fluorides (inorganic, used in drinking-water)   27, 237 (1982); *Suppl. 7*, 208 (1987)
5-Fluorouracil   26, 217 (1981); *Suppl. 7*, 210 (1987)
Fluorspar (*see* Fluorides)
Fluosilicic acid (*see* Fluorides)
Fluroxene (*see* Anaesthetics, volatile)
Formaldehyde   29, 345 (1982); *Suppl. 7*, 211 (1987); 62, 217 (1995) (*corr. 65*, 549)
2-(2-Formylhydrazino)-4-(5-nitro-2-furyl)thiazole   7, 151 (1974) (*corr. 42*, 253); *Suppl. 7*, 63 (1987)
Frusemide (*see* Furosemide)
Fuel oils (heating oils)   45, 239 (1989) (*corr. 47*, 505)
Fumonisin B_1 (*see* Toxins derived from Fusarium moniliforme)
Fumonisin B_2 (*see* Toxins derived from Fusarium moniliforme)
Furan   63, 393 (1995)
Furazolidone   31, 141 (1983); *Suppl. 7*, 63 (1987)
Furfural   63, 409 (1995)
Furniture and cabinet-making   25, 99 (1981); *Suppl. 7*, 380 (1987)
Furosemide   50, 277 (1990)
2-(2-Furyl)-3-(5-nitro-2-furyl)acrylamide (*see* AF-2)
Fusarenon-X (*see* Toxins derived from *Fusarium graminearum, F. culmorum* and *F. crookwellense*)
Fusarenone-X (*see* Toxins derived from *Fusarium graminearum, F. culmorum* and *F. crookwellense*)
Fusarin C (*see* Toxins derived from *Fusarium moniliforme*)

G

Gasoline   45, 159 (1989) (corr. 47, 505)
Gasoline engine exhaust (see Diesel and gasoline engine exhausts)
Glass fibres (see Man-made mineral fibres)
Glass manufacturing industry, occupational exposures in   58, 347 (1993)
Glasswool (*see* Man-made mineral fibres)
Glass filaments (*see* Man-made mineral fibres)
Glu-P-1   40, 223 (1986); *Suppl. 7*, 64 (1987)
Glu-P-2   40, 235 (1986); *Suppl. 7*, 64 (1987)

L-Glutamic acid, 5-[2-(4-hydroxymethyl)phenylhydrazide]
  (see Agaritine)
Glycidaldehyde   *11*, 175 (1976); *Suppl. 7*, 64 (1987)
Glycidyl ethers   *47*, 237 (1989)
Glycidyl oleate   *11*, 183 (1976); *Suppl. 7*, 64 (1987)
Glycidyl stearate   *11*, 187 (1976); *Suppl. 7*, 64 (1987)
Griseofulvin   *10*, 153 (1976); *Suppl. 7*, 391 (1987)
Guinea Green B   *16*, 199 (1978); *Suppl. 7*, 64 (1987)
Gyromitrin   *31*, 163 (1983); *Suppl. 7*, 391 (1987)

H

Haematite   *1*, 29 (1972); *Suppl. 7*, 216 (1987)
Haematite and ferric oxide   *Suppl. 7*, 216 (1987)
Haematite mining, underground, with exposure to radon   *1*, 29 (1972); *Suppl. 7*, 216 (1987)
Hairdressers and barbers (occupational exposure as)   *57*, 43 (1993)
Hair dyes, epidemiology of   *16*, 29 (1978); *27*, 307 (1982);
Halogenated acetonitriles   *52*, 269 (1991)
Halothane (see Anaesthetics, volatile)
HC Blue No. 1   *57*, 129 (1993)
HC Blue No. 2   *57*, 143 (1993)
α-HCH (see Hexachlorocyclohexanes)
β-HCH (see Hexachlorocyclohexanes)
γ-HCH (see Hexachlorocyclohexanes)
HC Red No. 3   *57*, 153 (1993)
HC Yellow No. 4   *57*, 159 (1993)
Heating oils (see Fuel oils)
Helicobacter pylori (infection with)   *61*, 177 (1994)
Hepatitis B virus   *59*, 45 (1994)
Hepatitis C virus   *59*, 165 (1994)
Hepatitis D virus   *59*, 223 (1994)
Heptachlor (see also Chlordane/Heptachlor)   *5*, 173 (1974); *20*, 129 (1979)
Hexachlorobenzene   *20*, 155 (1979); *Suppl. 7*, 219 (1987)
Hexachlorobutadiene   *20*, 179 (1979); *Suppl. 7*, 64 (1987)
Hexachlorocyclohexanes   *5*, 47 (1974); *20*, 195 (1979)
                                                          (*corr. 42*, 258); *Suppl. 7*, 220 (1987)
Hexachlorocyclohexane, technical-grade (see Hexachlorocyclohexanes)
Hexachloroethane   *20*, 467 (1979); *Suppl. 7*, 64 (1987)
Hexachlorophene   *20*, 241 (1979); *Suppl. 7*, 64 (1987)
Hexamethylphosphoramide   *15*, 211 (1977); *Suppl. 7*, 64 (1987)
Hexoestrol (see Nonsteroidal oestrogens)
Human papillomaviruses   *64* (1995)
Hycanthone mesylate   *13*, 91 (1977); *Suppl. 7*, 64 (1987)
Hydralazine   *24*, 85 (1980); *Suppl. 7*, 222 (1987)
Hydrazine   *4*, 127 (1974); *Suppl. 7*, 223 (1987)
Hydrochloric acid   *54*, 189 (1992)
Hydrochlorothiazide   *50*, 293 (1990)
Hydrogen peroxide   *36*, 285 (1985); *Suppl. 7*, 64 (1987)
Hydroquinone   *15*, 155 (1977); *Suppl. 7*, 64 (1987)
4-Hydroxyazobenzene   *8*, 157 (1975); *Suppl. 7*, 64 (1987)
17α-Hydroxyprogesterone caproate (see also Progestins)   *21*, 399 (1979) (*corr. 42*, 259)
8-Hydroxyquinoline   *13*, 101 (1977); *Suppl. 7*, 64 (1987)
8-Hydroxysenkirkine   *10*, 265 (1976); *Suppl. 7*, 64 (1987)
Hypochlorite salts   *52*, 159 (1991)

I

Indeno[1,2,3-*cd*]pyrene	*3*, 229 (1973); *32*, 373 (1983); *Suppl. 7*, 64 (1987)
Inorganic acids (see Sulfuric acid and other strong inorganic acids, occupational exposures to mists and vapours from)	
Insecticides, occupational exposures in spraying and application of	*53*, 45 (1991)
IQ	*40*, 261 (1986); *Suppl. 7*, 64 (1987); *56*, 165 (1993)
Iron and steel founding	*34*, 133 (1984); *Suppl. 7*, 224 (1987)
Iron-dextran complex	*2*, 161 (1973); *Suppl. 7*, 226 (1987)
Iron-dextrin complex	*2*, 161 (1973) (*corr. 42*, 252); *Suppl. 7*, 64 (1987)
Iron oxide (*see* Ferric oxide)	
Iron oxide, saccharated (*see* Saccharated iron oxide)	
Iron sorbitol-citric acid complex	*2*, 161 (1973); *Suppl. 7*, 64 (1987)
Isatidine	*10*, 269 (1976); *Suppl. 7*, 65 (1987)
Isoflurane (*see* Anaesthetics, volatile)	
Isoniazid (*see* Isonicotinic acid hydrazide)	
Isonicotinic acid hydrazide	*4*, 159 (1974); *Suppl. 7*, 227 (1987)
Isophosphamide	*26*, 237 (1981); *Suppl. 7*, 65 (1987)
Isoprene	*60*, 215 (1994)
Isopropanol	*15*, 223 (1977); *Suppl. 7*, 229 (1987)
Isopropanol manufacture (strong-acid process) (*see* also Isopropanol; Sulfuric acid and other strong inorganic acids, occupational exposures to mists and vapours from)	*Suppl. 7*, 229 (1987)
Isopropyl oils	*15*, 223 (1977); *Suppl. 7*, 229 (1987)
Isosafrole	*1*, 169 (1972); *10*, 232 (1976); *Suppl. 7*, 65 (1987)

J

Jacobine	*10*, 275 (1976); *Suppl. 7*, 65 (1987)
Jet fuel	*45*, 203 (1989)
Joinery (*see* Carpentry and joinery)	

K

Kaempferol	*31*, 171 (1983); *Suppl. 7*, 65 (1987)
Kepone (*see* Chlordecone)	

L

Lasiocarpine	*10*, 281 (1976); *Suppl. 7*, 65 (1987)
Lauroyl peroxide	*36*, 315 (1985); *Suppl. 7*, 65 (1987)
Lead acetate (*see* Lead and lead compounds)	
Lead and lead compounds	*1*, 40 (1972) (*corr. 42*, 251); *2*, 52, 150 (1973); *12*, 131 (1976); *23*, 40, 208, 209, 325 (1980); *Suppl. 7*, 230 (1987)
Lead arsenate (*see* Arsenic and arsenic compounds)	

Lead carbonate (see Lead and lead compounds)
Lead chloride (see Lead and lead compounds)
Lead chromate (see Chromium and chromium compounds)
Lead chromate oxide (see Chromium and chromium compounds)
Lead naphthenate (see Lead and lead compounds)
Lead nitrate (see Lead and lead compounds)
Lead oxide (see Lead and lead compounds)
Lead phosphate (see Lead and lead compounds)
Lead subacetate (see Lead and lead compounds)
Lead tetroxide (see Lead and lead compounds)

Leather goods manufacture	25, 279 (1981); Suppl. 7, 235 (1987)
Leather industries	25, 199 (1981); Suppl. 7, 232 (1987)
Leather tanning and processing	25, 201 (1981); Suppl. 7, 236 (1987)
Ledate (see also Lead and lead compounds)	12, 131 (1976)
Light Green SF	16, 209 (1978); Suppl. 7, 65 (1987)
d-Limonene	56, 135 (1993)

Lindane (see Hexachlorocyclohexanes)
Liver flukes (see Clonorchis sinensis, Opisthorchis felineus and Opisthorchis viverrini)

The lumber and sawmill industries (including logging)	25, 49 (1981); Suppl. 7, 383 (1987)
Luteoskyrin	10, 163 (1976); Suppl. 7, 65 (1987)
Lynoestrenol (see also Progestins; Combined oral contraceptives)	21, 407 (1979)

M

Magenta	4, 57 (1974) (corr. 42, 252); Suppl. 7, 238 (1987); 57, 215 (1993)
Magenta, manufacture of (see also Magenta)	Suppl. 7, 238 (1987); 57, 215 (1993)
Malathion	30, 103 (1983); Suppl. 7, 65 (1987)
Maleic hydrazide	4, 173 (1974) (corr. 42, 253); Suppl. 7, 65 (1987)
Malonaldehyde	36, 163 (1985); Suppl. 7, 65 (1987)
Maneb	12, 137 (1976); Suppl. 7, 65 (1987)
Man-made mineral fibres	43, 39 (1988)
Mannomustine	9, 157 (1975); Suppl. 7, 65 (1987)
Mate	51, 273 (1991)
MCPA (see also Chlorophenoxy herbicides; Chlorophenoxy herbicides, occupational exposures to)	30, 255 (1983)
MeA-α-C	40, 253 (1986); Suppl. 7, 65 (1987)
Medphalan	9, 168 (1975); Suppl. 7, 65 (1987)
Medroxyprogesterone acetate	6, 157 (1974); 21, 417 (1979) (corr. 42, 259); Suppl. 7, 289 (1987)

Megestrol acetate (see also Progestins; Combined oral contraceptives)

MeIQ	40, 275 (1986); Suppl. 7, 65 (1987); 56, 197 (1993)
MeIQx	40, 283 (1986); Suppl. 7, 65 (1987); 56, 211 (1993)
Melamine	39, 333 (1986); Suppl. 7, 65 (1987)
Melphalan	9, 167 (1975); Suppl. 7, 239 (1987)
6-Mercaptopurine	26, 249 (1981); Suppl. 7, 240 (1987)

Mercuric chloride (see Mercury and mercury compounds)

Mercury and mercury compounds	58, 239 (1993)
Merphalan	9, 169 (1975); Suppl. 7, 65 (1987)

Mestranol (*see also* Steroidal oestrogens) 6, 87 (1974); *21*, 257 (1979)
 (*corr. 42*, 259)

Metabisulfites (*see* Sulfur dioxide and some sulfites, bisulfites and metabisulfites)
Metallic mercury (*see* Mercury and mercury compounds)
Methanearsonic acid, disodium salt (*see* Arsenic and arsenic compounds)
Methanearsonic acid, monosodium salt (*see* Arsenic and arsenic compounds
Methotrexate 26, 267 (1981); *Suppl. 7*, 241 (1987)
Methoxsalen (*see* 8-Methoxypsoralen)
Methoxychlor 5, 193 (1974); *20*, 259 (1979); *Suppl. 7*, 66 (1987)

Methoxyflurane (*see* Anaesthetics, volatile)
5-Methoxypsoralen *40*, 327 (1986); *Suppl. 7*, 242 (1987)
8-Methoxypsoralen (*see also* 8-Methoxypsoralen plus ultraviolet radiation) 24, 101 (1980)
8-Methoxypsoralen plus ultraviolet radiation *Suppl. 7*, 243 (1987)
Methyl acrylate *19*, 52 (1979); *39*, 99 (1986); *Suppl. 7*, 66 (1987)

5-Methylangelicin plus ultraviolet radiation (*see also* Angelicin and some synthetic derivatives) *Suppl. 7*, 57 (1987)
2-Methylaziridine 9, 61 (1975); *Suppl. 7*, 66 (1987)
Methylazoxymethanol acetate *1*, 164 (1972); *10*, 131 (1976); *Suppl. 7*, 66 (1987)
Methyl bromide *41*, 187 (1986) (*corr. 45*, 283); *Suppl. 7*, 245 (1987)
Methyl carbamate *12*, 151 (1976); *Suppl. 7*, 66 (1987)
Methyl-CCNU [*see* 1-(2-Chloroethyl)-3-(4-methylcyclohexyl)-1-nitrosourea]
Methyl chloride *41*, 161 (1986); *Suppl. 7*, 246 (1987)
1-, 2-, 3-, 4-, 5- and 6-Methylchrysenes *32*, 379 (1983); *Suppl. 7*, 66 (1987)
N-Methyl-*N*,4-dinitrosoaniline 1, 141 (1972); *Suppl. 7*, 66 (1987)
4,4'-Methylene bis(2-chloroaniline) 4, 65 (1974) (*corr. 42*, 252); *Suppl. 7*, 246 (1987); *57*, 271 (1993)
4,4'-Methylene bis(*N*,*N*-dimethyl)benzenamine 27, 119 (1982); *Suppl. 7*, 66 (1987)
4,4'-Methylene bis(2-methylaniline) 4, 73 (1974); *Suppl. 7*, 248 (1987)
4,4'-Methylenedianiline 4, 79 (1974) (*corr. 42*, 252); *39*, 347 (1986); *Suppl. 7*, 66 (1987)
4,4'-Methylenediphenyl diisocyanate *19*, 314 (1979); *Suppl. 7*, 66 (1987)
2-Methylfluoranthene *32*, 399 (1983); *Suppl. 7*, 66 (1987)
3-Methylfluoranthene *32*, 399 (1983); *Suppl. 7*, 66 (1987)
Methylglyoxal *51*, 443 (1991)
Methyl iodide *15*, 245 (1977); *41*, 213 (1986); *Suppl. 7*, 66 (1987)

Methylmercury chloride (*see* Mercury and mercury compounds)
Methylmercury compounds (*see* Mercury and mercury compounds)
Methyl methacrylate *19*, 187 (1979); *Suppl. 7*, 66 (1987); *60*, 445 (1994)

Methyl methanesulfonate 7, 253 (1974); *Suppl. 7*, 66 (1987)
2-Methyl-1-nitroanthraquinone 27, 205 (1982); *Suppl. 7*, 66 (1987)
N-Methyl-*N*'-nitro-*N*-nitrosoguanidine 4, 183 (1974); *Suppl. 7*, 248 (1987)
3-Methylnitrosaminopropionaldehyde [*see* 3-(*N*-Nitrosomethylamino)-propionaldehyde]
3-Methylnitrosaminopropionitrile [*see* 3-(*N*-Nitrosomethylamino)-propionitrile]

4-(Methylnitrosamino)-4-(3-pyridyl)-1-butanal-[see 4-(N-Nitrosomethyl-
    amino)-4-(3-pyridyl)-1-butanal]
4-(Methylnitrosamino)-1-(3-pyridyl)-1-butanone [see 4-(-Nitrosomethyl-
    amino)-1-(3-pyridyl)-1-butanone]
N-Methyl-N-nitrosourea   *1*, 125 (1972); *17*, 227 (1978);
                                                Suppl. 7, 66 (1987)
N-Methyl-N-nitrosourethane   *4*, 211 (1974); Suppl. 7, 66 (1987)
N-Methylolacrylamide   *60*, 435 (1994)
Methyl parathion   *30*, 131 (1983); Suppl. 7, 392 (1987)
1-Methylphenanthrene   *32*, 405 (1983); Suppl. 7, 66 (1987)
7-Methylpyrido[3,4-c]psoralen   *40*, 349 (1986); Suppl. 7, 71 (1987)
Methyl red   *8*, 161 (1975); Suppl. 7, 66 (1987)
Methyl selenac (see also Selenium and selenium compounds)   *12*, 161 (1976); Suppl. 7, 66 (1987)
Methylthiouracil   *7*, 53 (1974); Suppl. 7, 66 (1987)
Metronidazole   *13*, 113 (1977); Suppl. 7, 250 (1987)
Mineral oils   *3*, 30 (1973); *33*, 87 (1984)
                                                (corr. 42, 262); Suppl. 7, 252 (1987)
Mirex   *5*, 203 (1974); *20*, 283 (1979)
                                                (corr. 42, 258); Suppl. 7, 66 (1987)
Mitomycin C   *10*, 171 (1976); Suppl. 7, 67 (1987)
MNNG [see N-Methyl-N'-nitro-N-nitrosoguanidine]
MOCA [see 4,4'-Methylene bis(2-chloroaniline)]
Modacrylic fibres   *19*, 86 (1979); Suppl. 7, 67 (1987)
Monocrotaline   *10*, 291 (1976); Suppl. 7, 67 (1987)
Monuron   *12*, 167 (1976); Suppl. 7, 67 (1987);
                                                53, 467 (1991)
MOPP and other combined chemotherapy including   Suppl. 7, 254 (1987)
    alkylating agents
Morpholine   *47*, 199 (1989)
5-(Morpholinomethyl)-3-[(5-nitrofurfurylidene)amino]-2-   *7*, 161 (1974); Suppl. 7, 67 (1987)
    oxazolidinone
Musk ambrette   *65*, 477 (1996)
Musk xylene   *65*, 477 (1996)
Mustard gas   *9*, 181 (1975) (corr. 42, 254);
                                                Suppl. 7, 259 (1987)
Myleran (see 1,4-Butanediol dimethanesulfonate)

N

Nafenopin   *24*, 125 (1980); Suppl. 7, 67 (1987)
1,5-Naphthalenediamine   *27*, 127 (1982); Suppl. 7, 67 (1987)
1,5-Naphthalene diisocyanate   *19*, 311 (1979); Suppl. 7, 67 (1987)
1-Naphthylamine   *4*, 87 (1974) (corr. 42, 253);
                                                Suppl. 7, 260 (1987)
2-Naphthylamine   *4*, 97 (1974); Suppl. 7, 261 (1987)
1-Naphthylthiourea   *30*, 347 (1983); Suppl. 7, 263 (1987)
Nickel acetate (see Nickel and nickel compounds)
Nickel ammonium sulfate (see Nickel and nickel compounds)
Nickel and nickel compounds   *2*, 126 (1973) (corr. 42, 252); *11*, 75
                                                (1976); Suppl. 7, 264 (1987)
                                                (corr. 45, 283); *49*, 257 (1990)
Nickel carbonate (see Nickel and nickel compounds)
Nickel carbonyl (see Nickel and nickel compounds)
Nickel chloride (see Nickel and nickel compounds)

Nickel-gallium alloy (see Nickel and nickel compounds)
Nickel hydroxide (see Nickel and nickel compounds)
Nickelocene (see Nickel and nickel compounds)
Nickel oxide (see Nickel and nickel compounds)
Nickel subsulfide (see Nickel and nickel compounds)
Nickel sulfate (see Nickel and nickel compounds)

Niridazole	13, 123 (1977); Suppl. 7, 67 (1987)
Nithiazide	31, 179 (1983); Suppl. 7, 67 (1987)
Nitrilotriacetic acid and its salts	48, 181 (1990)
5-Nitroacenaphthene	16, 319 (1978); Suppl. 7, 67 (1987)
5-Nitro-*ortho*-anisidine	27, 133 (1982); Suppl. 7, 67 (1987)
2-Nitroanisole	65, 369 (1996)
9-Nitroanthracene	33, 179 (1984); Suppl. 7, 67 (1987)
7-Nitrobenz[*a*]anthracene	46, 247 (1989)
Nitrobenzene	65, 381 (1996)
6-Nitrobenzo[*a*]pyrene	33, 187 (1984); Suppl. 7, 67 (1987); 46, 255 (1989)
4-Nitrobiphenyl	4, 113 (1974); Suppl. 7, 67 (1987)
6-Nitrochrysene	33, 195 (1984); Suppl. 7, 67 (1987); 46, 267 (1989)
Nitrofen (technical-grade)	30, 271 (1983); Suppl. 7, 67 (1987)
3-Nitrofluoranthene	33, 201 (1984); Suppl. 7, 67 (1987)
2-Nitrofluorene	46, 277 (1989)
Nitrofural	7, 171 (1974); Suppl. 7, 67 (1987); 50, 195 (1990)
5-Nitro-2-furaldehyde semicarbazone (see Nitrofural)	
Nitrofurantoin	50, 211 (1990)
Nitrofurazone (see Nitrofural)	
1-[(5-Nitrofurfurylidene)amino]-2-imidazolidinone	7, 181 (1974); Suppl. 7, 67 (1987)
N-[4-(5-Nitro-2-furyl)-2-thiazolyl]acetamide	1, 181 (1972); 7, 185 (1974); Suppl. 7, 67 (1987)
Nitrogen mustard	9, 193 (1975); Suppl. 7, 269 (1987)
Nitrogen mustard N-oxide	9, 209 (1975); Suppl. 7, 67 (1987)
1-Nitronaphthalene	46, 291 (1989)
2-Nitronaphthalene	46, 303 (1989)
3-Nitroperylene	46, 313 (1989)
2-Nitro-*para*-phenylenediamine (see 1,4-Diamino-2-nitrobenzene)	
2-Nitropropane	29, 331 (1982); Suppl. 7, 67 (1987)
1-Nitropyrene	33, 209 (1984); Suppl. 7, 67 (1987); 46, 321 (1989)
2-Nitropyrene	46, 359 (1989)
4-Nitropyrene	46, 367 (1989)
N-Nitrosatable drugs	24, 297 (1980) (corr. 42, 260)
N-Nitrosatable pesticides	30, 359 (1983)
N'-Nitrosoanabasine	37, 225 (1985); Suppl. 7, 67 (1987)
N'-Nitrosoanatabine	37, 233 (1985); Suppl. 7, 67 (1987)
N-Nitrosodi-*n*-butylamine	4, 197 (1974); 17, 51 (1978); Suppl. 7, 67 (1987)
N-Nitrosodiethanolamine	17, 77 (1978); Suppl. 7, 67 (1987)
N-Nitrosodiethylamine	1, 107 (1972) (corr. 42, 251); 17, 83 (1978) (corr. 42, 257); Suppl. 7, 67 (1987)
N-Nitrosodimethylamine	1, 95 (1972); 17, 125 (1978) (corr. 42, 257); Suppl. 7, 67 (1987)
N-Nitrosodiphenylamine	27, 213 (1982); Suppl. 7, 67 (1987)

para-Nitrosodiphenylamine	*27*, 227 (1982) (*corr. 42*, 261); *Suppl. 7*, 68 (1987)
N-Nitrosodi-*n*-propylamine	*17*, 177 (1978); *Suppl. 7*, 68 (1987)
N-Nitroso-*N*-ethylurea (*see N*-Ethyl-*N*-nitrosourea)	
N-Nitrosofolic acid	*17*, 217 (1978); *Suppl. 7*, 68 (1987)
N-Nitrosoguvacine	*37*, 263 (1985); *Suppl. 7*, 68 (1987)
N-Nitrosoguvacoline	*37*, 263 (1985); *Suppl. 7*, 68 (1987)
N-Nitrosohydroxyproline	*17*, 304 (1978); *Suppl. 7*, 68 (1987)
3-(*N*-Nitrosomethylamino)propionaldehyde	*37*, 263 (1985); *Suppl. 7*, 68 (1987)
3-(*N*-Nitrosomethylamino)propionitrile	*37*, 263 (1985); *Suppl. 7*, 68 (1987)
4-(*N*-Nitrosomethylamino)-4-(3-pyridyl)-1-butanal	*37*, 205 (1985); *Suppl. 7*, 68 (1987)
4-(*N*-Nitrosomethylamino)-1-(3-pyridyl)-1-butanone	*37*, 209 (1985); *Suppl. 7*, 68 (1987)
N-Nitrosomethylethylamine	*17*, 221 (1978); *Suppl. 7*, 68 (1987)
N-Nitroso-*N*-methylurea (*see N*-Methyl-*N*-nitrosourea)	
N-Nitroso-*N*-methylurethane (*see N*-Methyl-*N*-nitrosourethane)	
N-Nitrosomethylvinylamine	*17*, 257 (1978); *Suppl. 7*, 68 (1987)
N-Nitrosomorpholine	*17*, 263 (1978); *Suppl. 7*, 68 (1987)
N'-Nitrosonornicotine	*17*, 281 (1978); *37*, 241 (1985); *Suppl. 7*, 68 (1987)
N-Nitrosopiperidine	*17*, 287 (1978); *Suppl. 7*, 68 (1987)
N-Nitrosoproline	*17*, 303 (1978); *Suppl. 7*, 68 (1987)
N-Nitrosopyrrolidine	*17*, 313 (1978); *Suppl. 7*, 68 (1987)
N-Nitrososarcosine	*17*, 327 (1978); *Suppl. 7*, 68 (1987)
Nitrosoureas, chloroethyl (*see* Chloroethyl nitrosoureas)	
5-Nitro-*ortho*-toluidine	*48*, 169 (1990)
2-Nitrotoluene	*65*, 409 (1996)
3-Nitrotoluene	*65*, 409 (1996)
4-Nitrotoluene	*65*, 409 (1996)
Nitrous oxide (*see* Anaesthetics, volatile)	
Nitrovin	*31*, 185 (1983); *Suppl. 7*, 68 (1987)
Nivalenol (*see* Toxins derived from *Fusarium graminearum, F. culmorum* and *F. crookwellense*)	
NNA [*see* 4-(*N*-Nitrosomethylamino)-4-(3-pyridyl)-1-butanal]	
NNK [*see* 4-(*N*-Nitrosomethylamino)-1-(3-pyridyl)-1-butanone]	
Nonsteroidal oestrogens (*see also* Oestrogens, progestins and combinations)	*Suppl. 7*, 272 (1987)
Norethisterone (*see also* Progestins; Combined oral contraceptives)	*6*, 179 (1974); *21*, 461 (1979)
Norethynodrel (*see also* Progestins; Combined oral contraceptives	*6*, 191 (1974); *21*, 461 (1979) (*corr. 42*, 259)
Norgestrel (*see also* Progestins, Combined oral contraceptives)	*6*, 201 (1974); *21*, 479 (1979)
Nylon 6	*19*, 120 (1979); *Suppl. 7*, 68 (1987)

O

Ochratoxin A	*10*, 191 (1976); *31*, 191 (1983) (*corr. 42*, 262); *Suppl. 7*, 271 (1987); *56*, 489 (1993)
Oestradiol-17β (*see also* Steroidal oestrogens)	*6*, 99 (1974); *21*, 279 (1979)
Oestradiol 3-benzoate (*see* Oestradiol-17β)	
Oestradiol dipropionate (*see* Oestradiol-17β)	
Oestradiol mustard	*9*, 217 (1975); *Suppl. 7*, 68 (1987)
Oestradiol-17β-valerate (*see* Oestradiol-17β)	
Oestriol (*see also* Steroidal oestrogens)	*6*, 117 (1974); *21*, 327 (1979); *Suppl. 7*, 285 (1987)

Oestrogen-progestin combinations (*see* Oestrogens, progestins and combinations)
Oestrogen-progestin replacement therapy (*see also* Oestrogens, progestins and combinations)   *Suppl. 7*, 308 (1987)
Oestrogen replacement therapy (*see also* Oestrogens, progestins and combinations)   *Suppl. 7*, 280 (1987)
Oestrogens (*see* Oestrogens, progestins and combinations)
Oestrogens, conjugated (*see* Conjugated oestrogens)
Oestrogens, nonsteroidal (*see* Nonsteroidal oestrogens)
Oestrogens, progestins and combinations   6 (1974); *21* (1979); *Suppl. 7*, 272 (1987)
Oestrogens, steroidal (*see* Steroidal oestrogens)
Oestrone (*see* also Steroidal oestrogens)   6, 123 (1974); *21*, 343 (1979) (*corr. 42*, 259)
Oestrone benzoate (*see* Oestrone)
Oil Orange SS   *8*, 165 (1975); *Suppl. 7*, 69 (1987)
Opisthorchis felineus (infection with)   *61*, 121 (1994)
Opisthorchis viverrini (infection with)   *61*, 121 (1994)
Oral contraceptives, combined (*see* Combined oral contraceptives)
Oral contraceptives, investigational (*see* Combined oral contraceptives)
Oral contraceptives, sequential (*see* Sequential oral contraceptives)
Orange I   *8*, 173 (1975); *Suppl. 7*, 69 (1987)
Orange G   *8*, 181 (1975); *Suppl. 7*, 69 (1987)
Organolead compounds (*see also* Lead and lead compounds)   *Suppl. 7*, 230 (1987)
Oxazepam   *13*, 58 (1977); *Suppl. 7*, 69 (1987)
Oxymetholone [*see also* Androgenic (anabolic) steroids]   *13*, 131 (1977)
Oxyphenbutazone   *13*, 185 (1977); *Suppl. 7*, 69 (1987)

P

Paint manufacture and painting (occupational exposures in)   *47*, 329 (1989)
Panfuran S (*see also* Dihydroxymethylfuratrizine)   *24*, 77 (1980); *Suppl. 7*, 69 (1987)
Paper manufacture (*see* Pulp and paper manufacture)
Paracetamol   *50*, 307 (1990)
Parasorbic acid   *10*, 199 (1976) (*corr. 42*, 255); *Suppl. 7*, 69 (1987)
Parathion   *30*, 153 (1983); *Suppl. 7*, 69 (1987)
Patulin   10, 205 (1976); 40, 83 (1986); *Suppl. 7*, 69 (1987)
Penicillic acid   *10*, 211 (1976); *Suppl. 7*, 69 (1987)
Pentachloroethane   *41*, 99 (1986); *Suppl. 7*, 69 (1987)
Pentachloronitrobenzene (see Quintozene)
Pentachlorophenol (*see also* Chlorophenols; Chlorophenols, occupational exposures to)   20, 303 (1979); *53*, 371 (1991)
Permethrin   53, 329 (1991)
Perylene   *32*, 411 (1983); *Suppl. 7*, 69 (1987)
Petasitenine   *31*, 207 (1983); *Suppl. 7*, 69 (1987)
Petasites japonicus (*see* Pyrrolizidine alkaloids)
Petroleum refining (occupational exposures in)   *45*, 39 (1989)
Some petroleum solvents   *47*, 43 (1989)
Phenacetin   *13*, 141 (1977); *24*, 135 (1980); *Suppl. 7*, 310 (1987)
Phenanthrene   *32*, 419 (1983); *Suppl. 7*, 69 (1987)

CUMULATIVE INDEX

Phenazopyridine hydrochloride	8, 117 (1975); 24, 163 (1980)
	(*corr. 42*, 260); *Suppl. 7*, 312 (1987)
Phenelzine sulfate	24, 175 (1980); *Suppl. 7*, 312 (1987)
Phenicarbazide	12, 177 (1976); *Suppl. 7*, 70 (1987)
Phenobarbital	13, 157 (1977); *Suppl. 7*, 313 (1987)
Phenol	47, 263 (1989) (*corr. 50*, 385)
Phenoxyacetic acid herbicides (*see* Chlorophenoxy herbicides)	
Phenoxybenzamine hydrochloride	9, 223 (1975); 24, 185 (1980);
	Suppl. 7, 70 (1987)
Phenylbutazone	13, 183 (1977); *Suppl. 7*, 316 (1987)
meta-Phenylenediamine	16, 111 (1978); *Suppl. 7*, 70 (1987)
para-Phenylenediamine	16, 125 (1978); *Suppl. 7*, 70 (1987)
Phenyl glycidyl ether (*see* Glycidyl ethers)	
N-Phenyl-2-naphthylamine	16, 325 (1978) (*corr. 42*, 257);
	Suppl. 7, 318 (1987)
ortho-Phenylphenol	30, 329 (1983); *Suppl. 7*, 70 (1987)
Phenytoin	13, 201 (1977); *Suppl. 7*, 319 (1987)
PhIP	56, 229 (1993)
Pickled vegetables	56, 83 (1993)
Picloram	53, 481 (1991)
Piperazine oestrone sulfate (*see* Conjugated oestrogens)	
Piperonyl butoxide	30, 183 (1983); *Suppl. 7*, 70 (1987)
Pitches, coal-tar (*see* Coal-tar pitches)	
Polyacrylic acid	19, 62 (1979); *Suppl. 7*, 70 (1987)
Polybrominated biphenyls	18, 107 (1978); 41, 261 (1986);
	Suppl. 7, 321 (1987)
Polychlorinated biphenyls	7, 261 (1974); 18, 43 (1978)
	(*corr. 42*, 258); *Suppl. 7*, 322 (1987)
Polychlorinated camphenes (*see* Toxaphene)	
Polychloroprene	19, 141 (1979); *Suppl. 7*, 70 (1987)
Polyethylene	19, 164 (1979); *Suppl. 7*, 70 (1987)
Polymethylene polyphenyl isocyanate	19, 314 (1979); *Suppl. 7*, 70 (1987)
Polymethyl methacrylate	19, 195 (1979); *Suppl. 7*, 70 (1987)
Polyoestradiol phosphate (*see* Oestradiol-17β)	
Polypropylene	19, 218 (1979); *Suppl. 7*, 70 (1987)
Polystyrene	19, 245 (1979); *Suppl. 7*, 70 (1987)
Polytetrafluoroethylene	19, 288 (1979); *Suppl. 7*, 70 (1987)
Polyurethane foams	19, 320 (1979); *Suppl. 7*, 70 (1987)
Polyvinyl acetate	19, 346 (1979); *Suppl. 7*, 70 (1987)
Polyvinyl alcohol	19, 351 (1979); *Suppl. 7*, 70 (1987)
Polyvinyl chloride	7, 306 (1974); 19, 402 (1979);
	Suppl. 7, 70 (1987)
Polyvinyl pyrrolidone	19, 463 (1979); *Suppl. 7*, 70 (1987)
Ponceau MX	8, 189 (1975); *Suppl. 7*, 70 (1987)
Ponceau 3R	8, 199 (1975); *Suppl. 7*, 70 (1987)
Ponceau SX	8, 207 (1975); *Suppl. 7*, 70 (1987)
Potassium arsenate (*see* Arsenic and arsenic compounds)	
Potassium arsenite (*see* Arsenic and arsenic compounds)	
Potassium bis(2-hydroxyethyl)dithiocarbamate	12, 183 (1976); *Suppl. 7*, 70 (1987)
Potassium bromate	40, 207 (1986); *Suppl. 7*, 70 (1987)
Potassium chromate (*see* Chromium and chromium compounds)	
Potassium dichromate (*see* Chromium and chromium compounds)	
Prednimustine	50, 115 (1990)
Prednisone	26, 293 (1981); *Suppl. 7*, 326 (1987)
Printing processes and printing inks	65, 33 (1996)

Procarbazine hydrochloride	26, 311 (1981); *Suppl. 7*, 327 (1987)
Proflavine salts	24, 195 (1980); *Suppl. 7*, 70 (1987)
Progesterone (*see also* Progestins; Combined oral contraceptives)	6, 135 (1974); 21, 491 (1979) (*corr. 42*, 259)
Progestins (*see also* Oestrogens, progestins and combinations)	*Suppl. 7*, 289 (1987)
Pronetalol hydrochloride	13, 227 (1977) (*corr. 42*, 256); *Suppl. 7*, 70 (1987)
1,3-Propane sultone	4, 253 (1974) (*corr. 42*, 253); *Suppl. 7*, 70 (1987)
Propham	12, 189 (1976); *Suppl. 7*, 70 (1987)
β-Propiolactone	4, 259 (1974) (*corr. 42*, 253); *Suppl. 7*, 70 (1987)
n-Propyl carbamate	12, 201 (1976); *Suppl. 7*, 70 (1987)
Propylene	19, 213 (1979); *Suppl. 7*, 71 (1987); 60, 161 (1994)
Propylene oxide	11, 191 (1976); 36, 227 (1985) (*corr. 42*, 263); *Suppl. 7*, 328 (1987); 60, 181 (1994)
Propylthiouracil	7, 67 (1974); *Suppl. 7*, 329 (1987)
Ptaquiloside (*see also* Bracken fern)	40, 55 (1986); *Suppl. 7*, 71 (1987)
Pulp and paper manufacture	25, 157 (1981); *Suppl. 7*, 385 (1987)
Pyrene	32, 431 (1983); *Suppl. 7*, 71 (1987)
Pyrido[3,4-*c*]psoralen	40, 349 (1986); *Suppl. 7*, 71 (1987)
Pyrimethamine	13, 233 (1977); *Suppl. 7*, 71 (1987)
Pyrrolizidine alkaloids (*see* Hydroxysenkirkine; Isatidine; Jacobine; Lasiocarpine; Monocrotaline; Retrorsine; Riddelliine; Seneciphylline; Senkirkine)	

Q

Quercetin (*see also* Bracken fern)	31, 213 (1983); *Suppl. 7*, 71 (1987)
para-Quinone	15, 255 (1977); *Suppl. 7*, 71 (1987)
Quintozene	5, 211 (1974); *Suppl. 7*, 71 (1987)

R

Radon	43, 173 (1988) (*corr. 45*, 283)
Reserpine	10, 217 (1976); 24, 211 (1980) (*corr. 42*, 260); *Suppl. 7*, 330 (1987)
Resorcinol	15, 155 (1977); *Suppl. 7*, 71 (1987)
Retrorsine	10, 303 (1976); *Suppl. 7*, 71 (1987)
Rhodamine B	16, 221 (1978); *Suppl. 7*, 71 (1987)
Rhodamine 6G	16, 233 (1978); *Suppl. 7*, 71 (1987)
Riddelliine	10, 313 (1976); *Suppl. 7*, 71 (1987)
Rifampicin	24, 243 (1980); *Suppl. 7*, 71 (1987)
Rockwool (*see* Man-made mineral fibres)	
The rubber industry	28 (1982) (*corr. 42*, 261); *Suppl. 7*, 332 (1987)
Rugulosin	40, 99 (1986); *Suppl. 7*, 71 (1987)

S

Saccharated iron oxide	2, 161 (1973); *Suppl. 7*, 71 (1987)
Saccharin	22, 111 (1980) (*corr. 42*, 259); *Suppl. 7*, 334 (1987)
Safrole	*1*, 169 (1972); *10*, 231 (1976); *Suppl. 7*, 71 (1987)
Salted fish	56, 41 (1993)
The sawmill industry (including logging) [*see* The lumber and sawmill industry (including logging)]	
Scarlet Red	8, 217 (1975); *Suppl. 7*, 71 (1987)
Schistosoma haematobium (infection with)	*61*, 45 (1994)
Schistosoma japonicum (infection with)	*61*, 45 (1994)
Schistosoma mansoni (infection with)	*61*, 45 (1994)
Selenium and selenium compounds	9, 245 (1975) (*corr. 42*, 255); *Suppl. 7*, 71 (1987)
Selenium dioxide (*see* Selenium and selenium compounds)	
Selenium oxide (*see* Selenium and selenium compounds)	
Semicarbazide hydrochloride	*12*, 209 (1976) (*corr. 42*, 256); *Suppl. 7*, 71 (1987)
Senecio jacobaea L. (*see* Pyrrolizidine alkaloids)	
Senecio longilobus (*see* Pyrrolizidine alkaloids)	
Seneciphylline	*10*, 319, 335 (1976); *Suppl. 7*, 71 (1987)
Senkirkine	*10*, 327 (1976); *31*, 231 (1983); *Suppl. 7*, 71 (1987)
Sepiolite	42, 175 (1987); *Suppl. 7*, 71 (1987)
Sequential oral contraceptives (*see also* Oestrogens, progestins and combinations)	*Suppl. 7*, 296 (1987)
Shale-oils	35, 161 (1985); *Suppl. 7*, 339 (1987)
Shikimic acid (*see also* Bracken fern)	40, 55 (1986); *Suppl. 7*, 71 (1987)
Shoe manufacture and repair (*see* Boot and shoe manufacture and repair)	
Silica (*see also* Amorphous silica; Crystalline silica)	42, 39 (1987)
Simazine	53, 495 (1991)
Slagwool (*see* Man-made mineral fibres)	
Sodium arsenate (*see* Arsenic and arsenic compounds)	
Sodium arsenite (*see* Arsenic and arsenic compounds)	
Sodium cacodylate (*see* Arsenic and arsenic compounds)	
Sodium chlorite	52, 145 (1991)
Sodium chromate (*see* Chromium and chromium compounds)	
Sodium cyclamate (*see* Cyclamates)	
Sodium dichromate (*see* Chromium and chromium compounds)	
Sodium diethyldithiocarbamate	*12*, 217 (1976); *Suppl. 7*, 71 (1987)
Sodium equilin sulfate (*see* Conjugated oestrogens)	
Sodium fluoride (*see* Fluorides)	
Sodium monofluorophosphate (*see* Fluorides)	
Sodium oestrone sulfate (*see* Conjugated oestrogens)	
Sodium *ortho*-phenylphenate (*see also* ortho-Phenylphenol)	30, 329 (1983); *Suppl. 7*, 392 (1987)
Sodium saccharin (*see* Saccharin)	
Sodium selenate (*see* Selenium and selenium compounds)	
Sodium selenite (*see* Selenium and selenium compounds)	
Sodium silicofluoride (*see* Fluorides)	
Solar radiation	55 (1992)

Soots	3, 22 (1973); 35, 219 (1985); Suppl. 7, 343 (1987)
Spironolactone	24, 259 (1980); Suppl. 7, 344 (1987)
Stannous fluoride (*see* Fluorides)	
Steel founding (*see* Iron and steel founding)	
Sterigmatocystin	1, 175 (1972); 10, 245 (1976); Suppl. 7, 72 (1987)
Steroidal oestrogens (*see also* Oestrogens, progestins and combinations)	Suppl. 7, 280 (1987)
Streptozotocin	4, 221 (1974); 17, 337 (1978); Suppl. 7, 72 (1987)
Strobaner (*see* Terpene polychlorinates)	
Strontium chromate (*see* Chromium and chromium compounds)	
Styrene	19, 231 (1979) (*corr.* 42, 258); Suppl. 7, 345 (1987); 60, 233 (1994) (*corr.* 65, 549)
Styrene-acrylonitrile-copolymers	19, 97 (1979); Suppl. 7, 72 (1987)
Styrene-butadiene copolymers	19, 252 (1979); Suppl. 7, 72 (1987)
Styrene-7,8-oxide	11, 201 (1976); 19, 275 (1979); 36, 245 (1985); Suppl. 7, 72 (1987); 60, 321 (1994)
Succinic anhydride	15, 265 (1977); Suppl. 7, 72 (1987)
Sudan I	8, 225 (1975); Suppl. 7, 72 (1987)
Sudan II	8, 233 (1975); Suppl. 7, 72 (1987)
Sudan III	8, 241 (1975); Suppl. 7, 72 (1987)
Sudan Brown RR	8, 249 (1975); Suppl. 7, 72 (1987)
Sudan Red 7B	8, 253 (1975); Suppl. 7, 72 (1987)
Sulfafurazole	24, 275 (1980); Suppl. 7, 347 (1987)
Sulfallate	30, 283 (1983); Suppl. 7, 72 (1987)
Sulfamethoxazole	24, 285 (1980); Suppl. 7, 348 (1987)
Sulfites (*see* Sulfur dioxide and some sulfites, bisulfites and metabisulfites)	
Sulfur dioxide and some sulfites, bisulfites and metabisulfites	54, 131 (1992)
Sulfur mustard (*see* Mustard gas)	
Sulfuric acid and other strong inorganic acids, occupational exposures to mists and vapours from	54, 41 (1992)
Sulfur trioxide	54, 121 (1992)
Sulphisoxazole (*see* Sulfafurazole)	
Sunset Yellow FCF	8, 257 (1975); Suppl. 7, 72 (1987)
Symphytine	31, 239 (1983); Suppl. 7, 72 (1987)

T

2,4,5-T (*see also* Chlorophenoxy herbicides; Chlorophenoxy herbicides, occupational exposures to)	15, 273 (1977)
Talc	42, 185 (1987); Suppl. 7, 349 (1987)
Tannic acid	10, 253 (1976) (*corr.* 42, 255); Suppl. 7, 72 (1987)
Tannins (*see also* Tannic acid)	10, 254 (1976); Suppl. 7, 72 (1987)
TCDD (*see* 2,3,7,8-Tetrachlorodibenzo-*para*-dioxin)	
TDE (*see* DDT)	
Tea	51, 207 (1991)
Terpene polychlorinates	5, 219 (1974); Suppl. 7, 72 (1987)
Testosterone (*see also* Androgenic (anabolic) steroids)	6, 209 (1974); 21, 519 (1979)
Testosterone oenanthate (*see* Testosterone)	

Testosterone propionate (see Testosterone)	
2,2′,5,5′-Tetrachlorobenzidine	27, 141 (1982); Suppl. 7, 72 (1987)
2,3,7,8-Tetrachlorodibenzo-*para*-dioxin	15, 41 (1977); Suppl. 7, 350 (1987)
1,1,1,2-Tetrachloroethane	41, 87 (1986); Suppl. 7, 72 (1987)
1,1,2,2-Tetrachloroethane	20, 477 (1979); Suppl. 7, 354 (1987)
Tetrachloroethylene	20, 491 (1979); Suppl. 7, 355 (1987); 63, 159 (1995) (*corr.* 65, 549)
2,3,4,6-Tetrachlorophenol (see Chlorophenols; Chlorophenols, occupational exposures to)	
Tetrachlorvinphos	30, 197 (1983); Suppl. 7, 72 (1987)
Tetraethyllead (see Lead and lead compounds)	
Tetrafluoroethylene	19, 285 (1979); Suppl. 7, 72 (1987)
Tetrakis(hydroxymethyl) phosphonium salts	48, 95 (1990)
Tetramethyllead (see Lead and lead compounds)	
Tetranitromethane	65, 437 (1996)
Textile manufacturing industry, exposures in	48, 215 (1990) (*corr.* 51, 483)
Theobromine	51, 421 (1991)
Theophylline	51, 391 (1991)
Thioacetamide	7, 77 (1974); Suppl. 7, 72 (1987)
4,4′-Thiodianiline	16, 343 (1978); 27, 147 (1982); Suppl. 7, 72 (1987)
Thiotepa	9, 85 (1975); Suppl. 7, 368 (1987); 50, 123 (1990)
Thiouracil	7, 85 (1974); Suppl. 7, 72 (1987)
Thiourea	7, 95 (1974); Suppl. 7, 72 (1987)
Thiram	12, 225 (1976); Suppl. 7, 72 (1987); 53, 403 (1991)
Titanium dioxide	47, 307 (1989)
Tobacco habits other than smoking (see Tobacco products, smokeless)	
Tobacco products, smokeless	37 (1985) (*corr.* 42, 263; 52, 513); Suppl. 7, 357 (1987)
Tobacco smoke	38 (1986) (*corr.* 42, 263); Suppl. 7, 357 (1987)
Tobacco smoking (see Tobacco smoke)	
ortho-Tolidine (see 3,3′-Dimethylbenzidine)	
2,4-Toluene diisocyanate (see also Toluene diisocyanates)	19, 303 (1979); 39, 287 (1986)
2,6-Toluene diisocyanate (see also Toluene diisocyanates)	19, 303 (1979); 39, 289 (1986)
Toluene	47, 79 (1989)
Toluene diisocyanates	39, 287 (1986) (*corr.* 42, 264); Suppl. 7, 72 (1987)
Toluenes, α-chlorinated (see α-Chlorinated toluenes)	
ortho-Toluenesulfonamide (see Saccharin)	
ortho-Toluidine	16, 349 (1978); 27, 155 (1982); Suppl. 7, 362 (1987)
Toxaphene	20, 327 (1979); Suppl. 7, 72 (1987)
T-2 Toxin (see Toxins derived from *Fusarium sporotrichioides*)	
Toxins derived from *Fusarium graminearum*, *F. culmorum* and *F. crookwellense*	11, 169 (1976); 31, 153, 279 (1983); Suppl. 7, 64, 74 (1987); 56, 397 (1993)
Toxins derived from *Fusarium moniliforme*	56, 445 (1993)
Toxins derived from *Fusarium sporotrichioides*	31, 265 (1983); Suppl. 7, 73 (1987); 56, 467 (1993)
Tremolite (see Asbestos)	
Treosulfan	26, 341 (1981); Suppl. 7, 363 (1987)
Triaziquone [see Tris(aziridinyl)-*para*-benzoquinone]	

Trichlorfon	30, 207 (1983); Suppl. 7, 73 (1987)
Trichlormethine	9, 229 (1975); Suppl. 7, 73 (1987); 50, 143 (1990)
Trichloroacetic acid	63, 291 (1995) (corr. 65, 549)
Trichloroacetonitrile (see Halogenated acetonitriles)	
1,1,1-Trichloroethane	20, 515 (1979); Suppl. 7, 73 (1987)
1,1,2-Trichloroethane	20, 533 (1979); Suppl. 7, 73 (1987); 52, 337 (1991)
Trichloroethylene	11, 263 (1976); 20, 545 (1979); Suppl. 7, 364 (1987); 63, 75 (1995) (corr. 65, 549)
2,4,5-Trichlorophenol (see also Chlorophenols; Chlorophenols occupational exposures to)	20, 349 (1979)
2,4,6-Trichlorophenol (see also Chlorophenols; Chlorophenols, occupational exposures to)	20, 349 (1979)
(2,4,5-Trichlorophenoxy)acetic acid (see 2,4,5-T)	
1,2,3-Trichloropropane	63, 223 (1995)
Trichlorotriethylamine-hydrochloride (see Trichlormethine)	
T$_2$-Trichothecene (see Toxins derived from Fusarium sporotrichioides)	
Triethylene glycol diglycidyl ether	11, 209 (1976); Suppl. 7, 73 (1987)
Trifluralin	53, 515 (1991)
4,4',6-Trimethylangelicin plus ultraviolet radiation (see also Angelicin and some synthetic derivatives)	Suppl. 7, 57 (1987)
2,4,5-Trimethylaniline	27, 177 (1982); Suppl. 7, 73 (1987)
2,4,6-Trimethylaniline	27, 178 (1982); Suppl. 7, 73 (1987)
4,5',8-Trimethylpsoralen	40, 357 (1986); Suppl. 7, 366 (1987)
Trimustine hydrochloride (see Trichlormethine)	
2,4,6-Trinitrotoluene	65, 449 (1996)
Triphenylene	32, 447 (1983); Suppl. 7, 73 (1987)
Tris(aziridinyl)-para-benzoquinone	9, 67 (1975); Suppl. 7, 367 (1987)
Tris(1-aziridinyl)phosphine-oxide	9, 75 (1975); Suppl. 7, 73 (1987)
Tris(1-aziridinyl)phosphine-sulphide (see Thiotepa)	
2,4,6-Tris(1-aziridinyl)-s-triazine	9, 95 (1975); Suppl. 7, 73 (1987)
Tris(2-chloroethyl) phosphate	48, 109 (1990)
1,2,3-Tris(chloromethoxy)propane	15, 301 (1977); Suppl. 7, 73 (1987)
Tris(2,3-dibromopropyl)phosphate	20, 575 (1979); Suppl. 7, 369 (1987)
Tris(2-methyl-1-aziridinyl)phosphine-oxide	9, 107 (1975); Suppl. 7, 73 (1987)
Trp-P-1	31, 247 (1983); Suppl. 7, 73 (1987)
Trp-P-2	31, 255 (1983); Suppl. 7, 73 (1987)
Trypan blue	8, 267 (1975); Suppl. 7, 73 (1987)
Tussilago farfara L. (see Pyrrolizidine alkaloids)	

U

Ultraviolet radiation	40, 379 (1986); 55 (1992)
Underground haematite mining with exposure to radon	1, 29 (1972); Suppl. 7, 216 (1987)
Uracil mustard	9, 235 (1975); Suppl. 7, 370 (1987)
Urethane	7, 111 (1974); Suppl. 7, 73 (1987)

V

Vat Yellow 4	48, 161 (1990)
Vinblastine sulfate	26, 349 (1981) (corr. 42, 261);

	Suppl. 7, 371 (1987)
Vincristine sulfate	26, 365 (1981); *Suppl. 7*, 372 (1987)
Vinyl acetate	*19*, 341 (1979); *39*, 113 (1986);
	Suppl. 7, 73 (1987); *63*, 443 (1995)
Vinyl bromide	*19*, 367 (1979); *39*, 133 (1986);
	Suppl. 7, 73 (1987)
Vinyl chloride	*7*, 291 (1974); *19*, 377 (1979)
	(*corr. 42*, 258); *Suppl. 7*, 373 (1987)
Vinyl chloride-vinyl acetate copolymers	*7*, 311 (1976); *19*, 412 (1979)
	(*corr. 42*, 258); *Suppl. 7*, 73 (1987)
4-Vinylcyclohexene	*11*, 277 (1976); *39*, 181 (1986)
	Suppl. 7, 73 (1987); *60*, 347 (1994)
4-Vinylcyclohexene diepoxide	*11*, 141 (1976); *Suppl. 7*, 63 (1987);
	60, 361 (1994)
Vinyl fluoride	*39*, 147 (1986); *Suppl. 7*, 73 (1987);
	63, 467 (1995)
Vinylidene chloride	*19*, 439 (1979); *39*, 195 (1986);
	Suppl. 7, 376 (1987)
Vinylidene chloride-vinyl chloride copolymers	*19*, 448 (1979) (*corr. 42*, 258);
	Suppl. 7, 73 (1987)
Vinylidene fluoride	*39*, 227 (1986); *Suppl. 7*, 73 (1987)
N-Vinyl-2-pyrrolidone	*19*, 461 (1979); *Suppl. 7*, 73 (1987)
Vinyl toluene	*60*, 373 (1994)

W

Welding	*49*, 447 (1990) (*corr. 52*, 513)
Wollastonite	*42*, 145 (1987); *Suppl. 7*, 377 (1987)
Wood dust	*62*, 35 (1995)
Wood industries	25 (1981); *Suppl. 7*, 378 (1987)

X

Xylene	*47*, 125 (1989)
2,4-Xylidine	*16*, 367 (1978); *Suppl. 7*, 74 (1987)
2,5-Xylidine	*16*, 377 (1978); *Suppl. 7*, 74 (1987)
2,6-Xylidine (*see* 2,6-Dimethylaniline)	

Y

Yellow AB	*8*, 279 (1975); *Suppl. 7*, 74 (1987)
Yellow OB	*8*, 287 (1975); *Suppl. 7*, 74 (1987)

Z

Zearalenone (*see* Toxins derived from *Fusarium graminearum*,	
F. culmorum and *F. crookwellense*)	
Zectran	*12*, 237 (1976); *Suppl. 7*, 74 (1987)
Zinc beryllium silicate (*see* Beryllium and beryllium compounds)	
Zinc chromate (*see* Chromium and chromium compounds)	
Zinc chromate hydroxide (*see* Chromium and chromium compounds)	

Zinc potassium chromate (see Chromium and chromium compounds)
Zinc yellow (see Chromium and chromium compounds)
Zineb — *12*, 245 (1976); *Suppl. 7*, 74 (1987)
Ziram — *12*, 259 (1976); *Suppl. 7*, 74 (1987); *53*, 423 (1991)

PUBLICATIONS OF THE INTERNATIONAL AGENCY FOR RESEARCH ON CANCER
Scientific Publications Series

No. 1 Liver Cancer
1971; 176 pages (*out of print*)

No. 2 Oncogenesis and Herpesviruses
Edited by P.M. Biggs, G. de-Thé and L.N. Payne
1972; 515 pages (*out of print*)

No. 3 N-Nitroso Compounds: Analysis and Formation
Edited by P. Bogovski, R. Preussman and E.A. Walker
1972; 140 pages (*out of print*)

No. 4 Transplacental Carcinogenesis
Edited by L. Tomatis and U. Mohr
1973; 181 pages (*out of print*)

No. 5/6 Pathology of Tumours in Laboratory Animals, Volume 1, Tumours of the Rat
Edited by V.S. Turusov
1973/1976; 533 pages (*out of print*)

No. 7 Host Environment Interactions in the Etiology of Cancer in Man
Edited by R. Doll and I. Vodopija
1973; 464 pages (*out of print*)

No. 8 Biological Effects of Asbestos
Edited by P. Bogovski, J.C. Gilson, V. Timbrell and J.C. Wagner
1973; 346 pages (*out of print*)

No. 9 N-Nitroso Compounds in the Environment
Edited by P. Bogovski and E.A. Walker
1974; 243 pages (*out of print*)

No. 10 Chemical Carcinogenesis Essays
Edited by R. Montesano and L. Tomatis
1974; 230 pages (*out of print*)

No. 11 Oncogenesis and Herpesviruses II
Edited by G. de-Thé, M.A. Epstein and H. zur Hausen
1975; Part I: 511 pages
Part II: 403 pages (*out of print*)

No. 12 Screening Tests in Chemical Carcinogenesis
Edited by R. Montesano, H. Bartsch and L. Tomatis
1976; 666 pages (*out of print*)

No. 13 Environmental Pollution and Carcinogenic Risks
Edited by C. Rosenfeld and W. Davis
1975; 441 pages (*out of print*)

No. 14 Environmental N-Nitroso Compounds. Analysis and Formation
Edited by E.A. Walker, P. Bogovski and L. Griciute
1976; 512 pages (*out of print*)

No. 15 Cancer Incidence in Five Continents, Volume III
Edited by J.A.H. Waterhouse, C. Muir, P. Correa and J. Powell
1976; 584 pages (*out of print*)

No. 16 Air Pollution and Cancer in Man
Edited by U. Mohr, D. Schmähl and L. Tomatis
1977; 328 pages (*out of print*)

No. 17 Directory of On-going Research in Cancer Epidemiology 1977
Edited by C.S. Muir and G. Wagner
1977; 599 pages (*out of print*)

No. 18 Environmental Carcinogens. Selected Methods of Analysis. Volume 1: Analysis of Volatile Nitrosamines in Food
Editor-in-Chief: H. Egan
1978; 212 pages (*out of print*)

No. 19 Environmental Aspects of N-Nitroso Compounds
Edited by E.A. Walker, M. Castegnaro, L. Griciute and R.E. Lyle
1978; 561 pages (*out of print*)

No. 20 Nasopharyngeal Carcinoma: Etiology and Control
Edited by G. de-Thé and Y. Ito
1978; 606 pages (*out of print*)

No. 21 Cancer Registration and its Techniques
Edited by R. MacLennan, C. Muir, R. Steinitz and A. Winkler
1978; 235 pages (*out of print*)

No. 22 Environmental Carcinogens. Selected Methods of Analysis. Volume 2: Methods for the Measurement of Vinyl Chloride in Poly(vinyl chloride), Air, Water and Foodstuffs
Editor-in-Chief: H. Egan
1978; 142 pages (*out of print*)

No. 23 Pathology of Tumours in Laboratory Animals. Volume II: Tumours of the Mouse
Editor-in-Chief: V.S. Turusov
1979; 669 pages (*out of print*)

No. 24 Oncogenesis and Herpesviruses III
Edited by G. de-Thé, W. Henle and F. Rapp
1978; Part I: 580 pages, Part II: 512 pages (*out of print*)

Prices are subject to change without notice. Limited supplies of certain books marked '*out of print*' are available directly from IARC.

List of IARC Publications

No. 25 Carcinogenic Risk. Strategies for Intervention
Edited by W. Davis and C. Rosenfeld
1979; 280 pages (*out of print*)

No. 26 Directory of On-going Research in Cancer Epidemiology 1978
Edited by C.S. Muir and G. Wagner
1978; 550 pages (*out of print*)

No. 27 Molecular and Cellular Aspects of Carcinogen Screening Tests
Edited by R. Montesano, H. Bartsch and L. Tomatis
1980; 372 pages £30.00

No. 28 Directory of On-going Research in Cancer Epidemiology 1979
Edited by C.S. Muir and G. Wagner
1979; 672 pages (*out of print*)

No. 29 Environmental Carcinogens. Selected Methods of Analysis. Volume 3: Analysis of Polycyclic Aromatic Hydrocarbons in Environmental Samples
Editor-in-Chief: H. Egan
1979; 240 pages (*out of print*)

No. 30 Biological Effects of Mineral Fibres
Editor-in-Chief: J.C. Wagner
1980; Volume 1: 494 pages Volume 2: 513 pages (*out of print*)

No. 31 N-Nitroso Compounds: Analysis, Formation and Occurrence
Edited by E.A. Walker, L. Griciute, M. Castegnaro and M. Börzsönyi
1980; 835 pages (*out of print*)

No. 32 Statistical Methods in Cancer Research. Volume 1. The Analysis of Case-control Studies
By N.E. Breslow and N.E. Day
1980; 338 pages £18.00

No. 33 Handling Chemical Carcinogens in the Laboratory
Edited by R. Montesano et al.
1979; 32 pages (*out of print*)

No. 34 Pathology of Tumours in Laboratory Animals. Volume III. Tumours of the Hamster
Editor-in-Chief: V.S. Turusov
1982; 461 pages (*out of print*)

No. 35 Directory of On-going Research in Cancer Epidemiology 1980
Edited by C.S. Muir and G. Wagner
1980; 660 pages (*out of print*)

No. 36 Cancer Mortality by Occupation and Social Class 1851-1971
Edited by W.P.D. Logan
1982; 253 pages (*out of print*)

No. 37 Laboratory Decontamination and Destruction of Aflatoxins B_1, B_2, G_1, G_2 in Laboratory Wastes
Edited by M. Castegnaro et al.
1980; 56 pages (*out of print*)

No. 38 Directory of On-going Research in Cancer Epidemiology 1981
Edited by C.S. Muir and G. Wagner
1981; 696 pages (*out of print*)

No. 39 Host Factors in Human Carcinogenesis
Edited by H. Bartsch and B. Armstrong
1982; 583 pages (*out of print*)

No. 40 Environmental Carcinogens. Selected Methods of Analysis. Volume 4: Some Aromatic Amines and Azo Dyes in the General and Industrial Environment
Edited by L. Fishbein, M. Castegnaro, I.K. O'Neill and H. Bartsch
1981; 347 pages (*out of print*)

No. 41 N-Nitroso Compounds: Occurrence and Biological Effects
Edited by H. Bartsch, I.K. O'Neill, M. Castegnaro and M. Okada
1982; 755 pages (*out of print*)

No. 42 Cancer Incidence in Five Continents, Volume IV
Edited by J. Waterhouse, C. Muir, K. Shanmugaratnam and J. Powell
1982; 811 pages (*out of print*)

No. 43 Laboratory Decontamination and Destruction of Carcinogens in Laboratory Wastes: Some N-Nitrosamines
Edited by M. Castegnaro et al.
1982; 73 pages £7.50

No. 44 Environmental Carcinogens. Selected Methods of Analysis. Volume 5: Some Mycotoxins
Edited by L. Stoloff, M. Castegnaro, P. Scott, I.K. O'Neill and H. Bartsch
1983; 455 pages (*out of print*)

No. 45 Environmental Carcinogens. Selected Methods of Analysis. Volume 6: N-Nitroso Compounds
Edited by R. Preussmann, I.K. O'Neill, G. Eisenbrand, B. Spiegelhalder and H. Bartsch
1983; 508 pages (*out of print*)

No. 46 Directory of On-going Research in Cancer Epidemiology 1982
Edited by C.S. Muir and G. Wagner
1982; 722 pages (*out of print*)

No. 47 Cancer Incidence in Singapore 1968–1977
Edited by K. Shanmugaratnam, H.P. Lee and N.E. Day
1983; 171 pages (*out of print*)

No. 48 Cancer Incidence in the USSR (2nd Revised Edition)
Edited by N.P. Napalkov, G.F. Tserkovny, V.M. Merabishvili, D.M. Parkin, M. Smans and C.S. Muir
1983; 75 pages (*out of print*)

No. 49 Laboratory Decontamination and Destruction of Carcinogens in Laboratory Wastes: Some Polycyclic Aromatic Hydrocarbons
Edited by M. Castegnaro et al.
1983; 87 pages (*out of print*)

No. 50 Directory of On-going Research in Cancer Epidemiology 1983
Edited by C.S. Muir and G. Wagner
1983; 731 pages (*out of print*)

No. 51 Modulators of Experimental Carcinogenesis
Edited by V. Turusov and R. Montesano
1983; 307 pages (*out of print*)

List of IARC Publications

No. 52 Second Cancers in Relation to Radiation Treatment for Cervical Cancer: Results of a Cancer Registry Collaboration
Edited by N.E. Day and J.C. Boice, Jr
1984; 207 pages (*out of print*)

No. 53 Nickel in the Human Environment
Editor-in-Chief: F.W. Sunderman, Jr
1984; 529 pages (*out of print*)

No. 54 Laboratory Decontamination and Destruction of Carcinogens in Laboratory Wastes: Some Hydrazines
Edited by M. Castegnaro et al.
1983; 87 pages (*out of print*)

No. 55 Laboratory Decontamination and Destruction of Carcinogens in Laboratory Wastes: Some N-Nitrosamides
Edited by M. Castegnaro et al.
1984; 66 pages (*out of print*)

No. 56 Models, Mechanisms and Etiology of Tumour Promotion
Edited by M. Börzsönyi, N.E. Day, K. Lapis and H. Yamasaki
1984; 532 pages (*out of print*)

No. 57 N-Nitroso Compounds: Occurrence, Biological Effects and Relevance to Human Cancer
Edited by I.K. O'Neill, R.C. von Borstel, C.T. Miller, J. Long and H. Bartsch
1984; 1013 pages (*out of print*)

No. 58 Age-related Factors in Carcinogenesis
Edited by A. Likhachev, V. Anisimov and R. Montesano
1985; 288 pages (*out of print*)

No. 59 Monitoring Human Exposure to Carcinogenic and Mutagenic Agents
Edited by A. Berlin, M. Draper, K. Hemminki and H. Vainio
1984; 457 pages (*out of print*)

No. 60 Burkitt's Lymphoma: A Human Cancer Model
Edited by G. Lenoir, G. O'Conor and C.L.M. Olweny
1985; 484 pages (*out of print*)

No. 61 Laboratory Decontamination and Destruction of Carcinogens in Laboratory Wastes: Some Haloethers
Edited by M. Castegnaro et al.
1985; 55 pages (*out of print*)

No. 62 Directory of On-going Research in Cancer Epidemiology 1984
Edited by C.S. Muir and G. Wagner
1984; 717 pages (*out of print*)

No. 63 Virus-associated Cancers in Africa
Edited by A.O. Williams, G.T. O'Conor, G.B. de-Thé and C.A. Johnson
1984; 773 pages (*out of print*)

No. 64 Laboratory Decontamination and Destruction of Carcinogens in Laboratory Wastes: Some Aromatic Amines and 4-Nitrobiphenyl
Edited by M. Castegnaro et al.
1985; 84 pages (*out of print*)

No. 65 Interpretation of Negative Epidemiological Evidence for Carcinogenicity
Edited by N.J. Wald and R. Doll
1985; 232 pages (*out of print*)

No. 66 The Role of the Registry in Cancer Control
Edited by D.M. Parkin, G. Wagner and C.S. Muir
1985; 152 pages £10.00

No. 67 Transformation Assay of Established Cell Lines: Mechanisms and Application
Edited by T. Kakunaga and H. Yamasaki
1985; 225 pages (*out of print*)

No. 68 Environmental Carcinogens. Selected Methods of Analysis. Volume 7. Some Volatile Halogenated Hydrocarbons
Edited by L. Fishbein and I.K. O'Neill
1985; 479 pages (*out of print*)

No. 69 Directory of On-going Research in Cancer Epidemiology 1985
Edited by C.S. Muir and G. Wagner
1985; 745 pages (*out of print*)

No. 70 The Role of Cyclic Nucleic Acid Adducts in Carcinogenesis and Mutagenesis
Edited by B. Singer and H. Bartsch
1986; 467 pages (*out of print*)

No. 71 Environmental Carcinogens. Selected Methods of Analysis. Volume 8: Some Metals: As, Be, Cd, Cr, Ni, Pb, Se, Zn
Edited by I.K. O'Neill, P. Schuller and L. Fishbein
1986; 485 pages (*out of print*)

No. 72 Atlas of Cancer in Scotland, 1975–1980. Incidence and Epidemiological Perspective
Edited by I. Kemp, P. Boyle, M. Smans and C.S. Muir
1985; 285 pages (*out of print*)

No. 73 Laboratory Decontamination and Destruction of Carcinogens in Laboratory Wastes: Some Antineoplastic Agents
Edited by M. Castegnaro et al.
1985; 163 pages £13.50

No. 74 Tobacco: A Major International Health Hazard
Edited by D. Zaridze and R. Peto
1986; 324 pages £24.00

No. 75 Cancer Occurrence in Developing Countries
Edited by D.M. Parkin
1986; 339 pages £24.00

No. 76 Screening for Cancer of the Uterine Cervix
Edited by M. Hakama, A.B. Miller and N.E. Day
1986; 315 pages £31.50

No. 77 Hexachlorobenzene: Proceedings of an International Symposium
Edited by C.R. Morris and J.R.P. Cabral
1986; 668 pages (*out of print*)

No. 78 Carcinogenicity of Alkylating Cytostatic Drugs
Edited by D. Schmähl and J.M. Kaldor
1986; 337 pages (*out of print*)

No. 79 Statistical Methods in Cancer Research. Volume III: The Design and Analysis of Long-term Animal Experiments
By J.J. Gart, D. Krewski, P.N. Lee, R.E. Tarone and J. Wahrendorf
1986; 213 pages £23.50

List of IARC Publications

No. 80 Directory of On-going Research in Cancer Epidemiology 1986
Edited by C.S. Muir and G. Wagner
1986; 805 pages (*out of print*)

No. 81 Environmental Carcinogens: Methods of Analysis and Exposure Measurement. Volume 9: Passive Smoking
Edited by I.K. O'Neill, K.D. Brunnemann, B. Dodet and D. Hoffmann
1987; 383 pages £37.00

No. 82 Statistical Methods in Cancer Research. Volume II: The Design and Analysis of Cohort Studies
By N.E. Breslow and N.E. Day
1987; 404 pages £25.00

No. 83 Long-term and Short-term Assays for Carcinogens: A Critical Appraisal
Edited by R. Montesano, H. Bartsch, H. Vainio, J. Wilbourn and H. Yamasaki
1986; 575 pages £37.00

No. 84 The Relevance of N-Nitroso Compounds to Human Cancer: Exposure and Mechanisms
Edited by H. Bartsch, I.K. O'Neill and R. Schulte-Hermann
1987; 671 pages (*out of print*)

No. 85 Environmental Carcinogens: Methods of Analysis and Exposure Measurement. Volume 10: Benzene and Alkylated Benzenes
Edited by L. Fishbein and I.K. O'Neill
1988; 327 pages £42.00

No. 86 Directory of On-going Research in Cancer Epidemiology 1987
Edited by D.M. Parkin and J. Wahrendorf
1987; 676 pages (*out of print*)

No. 87 International Incidence of Childhood Cancer
Edited by D.M. Parkin, C.A. Stiller, C.A. Bieber, G.J. Draper, B. Terracini and J.L. Young
1988; 401 pages £35.00

No. 88 Cancer Incidence in Five Continents Volume V
Edited by C. Muir, J. Waterhouse, T. Mack, J. Powell and S. Whelan
1987; 1004 pages £58.00

No. 89 Method for Detecting DNA Damaging Agents in Humans: Applications in Cancer Epidemiology and Prevention
Edited by H. Bartsch, K. Hemminki and I.K. O'Neill
1988; 518 pages £50.00

No. 90 Non-occupational Exposure to Mineral Fibres
Edited by J. Bignon, J. Peto and R. Saracci
1989; 500 pages £52.50

No. 91 Trends in Cancer Incidence in Singapore 1968–1982
Edited by H.P. Lee, N.E. Day and K. Shanmugaratnam
1988; 160 pages (*out of print*)

No. 92 Cell Differentiation, Genes and Cancer
Edited by T. Kakunaga, T. Sugimura, L. Tomatis and H. Yamasaki
1988; 204 pages £29.00

No. 93 Directory of On-going Research in Cancer Epidemiology 1988
Edited by M. Coleman and J. Wahrendorf
1988; 662 pages (*out of print*)

No. 94 Human Papillomavirus and Cervical Cancer
Edited by N. Muñoz, F.X. Bosch and O.M. Jensen
1989; 154 pages £22.50

No. 95 Cancer Registration: Principles and Methods
Edited by O.M. Jensen, D.M. Parkin, R. MacLennan, C.S. Muir and R. Skeet
1991; 288 pages £28.00

No. 96 Perinatal and Multigeneration Carcinogenesis
Edited by N.P. Napalkov, J.M. Rice, L. Tomatis and H. Yamasaki
1989; 436 pages £52.50

No. 97 Occupational Exposure to Silica and Cancer Risk
Edited by L. Simonato, A.C. Fletcher, R. Saracci and T. Thomas
1990; 124 pages £24.00

No. 98 Cancer Incidence in Jewish Migrants to Israel, 1961–1981
Edited by R. Steinitz, D.M. Parkin, J.L. Young, C.A. Bieber and L. Katz
1989; 320 pages £37.00

No. 99 Pathology of Tumours in Laboratory Animals, Second Edition, Volume 1, Tumours of the Rat
Edited by V.S. Turusov and U. Mohr
740 pages £90.00

No. 100 Cancer: Causes, Occurrence and Control
Editor-in-Chief L. Tomatis
1990; 352 pages £25.50

No. 101 Directory of On-going Research in Cancer Epidemiology 1989/90
Edited by M. Coleman and J. Wahrendorf
1989; 818 pages £42.00

No. 102 Patterns of Cancer in Five Continents
Edited by S.L. Whelan, D.M. Parkin & E. Masuyer
1990; 162 pages £26.50

No. 103 Evaluating Effectiveness of Primary Prevention of Cancer
Edited by M. Hakama, V. Beral, J.W. Cullen and D.M. Parkin
1990; 250 pages £34.00

No. 104 Complex Mixtures and Cancer Risk
Edited by H. Vainio, M. Sorsa and A.J. McMichael
1990; 442 pages £40.00

No. 105 Relevance to Human Cancer of N-Nitroso Compounds, Tobacco Smoke and Mycotoxins
Edited by I.K. O'Neill, J. Chen and H. Bartsch
1991; 614 pages £74.00

No. 106 Atlas of Cancer Incidence in the Former German Democratic Republic
Edited by W.H. Mehnert, M. Smans, C.S. Muir, M. Möhner & D. Schön
1992; 384 pages £52.50

List of IARC Publications

No. 107 Atlas of Cancer Mortality in the European Economic Community
Edited by M. Smans, C.S. Muir and P. Boyle
1992; 280 pages £35.00

No. 108 Environmental Carcinogens: Methods of Analysis and Exposure Measurement. Volume 11: Polychlorinated Dioxins and Dibenzofurans
Edited by C. Rappe, H.R. Buser, B. Dodet and I.K. O'Neill
1991; 426 pages £47.50

No. 109 Environmental Carcinogens: Methods of Analysis and Exposure Measurement. Volume 12: Indoor Air Contaminants
Edited by B. Seifert, H. van de Wiel, B. Dodet and I.K. O'Neill
1993; 384 pages £45.00

No. 110 Directory of On-going Research in Cancer Epidemiology 1991
Edited by M. Coleman and J. Wahrendorf
1991; 753 pages £40.00

No. 111 Pathology of Tumours in Laboratory Animals, Second Edition, Volume 2, Tumours of the Mouse
Edited by V.S. Turusov and U. Mohr
1993; 776 pages; £90.00

No. 112 Autopsy in Epidemiology and Medical Research
Edited by E. Riboli and M. Delendi
1991; 288 pages £26.50

No. 113 Laboratory Decontamination and Destruction of Carcinogens in Laboratory Wastes: Some Mycotoxins
Edited by M. Castegnaro, J. Barek, J.-M. Frémy, M. Lafontaine, M. Miraglia, E.B. Sansone and G.M. Telling
1991; 64 pages £12.00

No. 114 Laboratory Decontamination and Destruction of Carcinogens in Laboratory Wastes: Some Polycyclic Heterocyclic Hydrocarbons
Edited by M. Castegnaro, J. Barek J. Jacob, U. Kirso, M. Lafontaine, E.B. Sansone, G.M. Telling and T. Vu Duc
1991; 50 pages £8.00

No. 115 Mycotoxins, Endemic Nephropathy and Urinary Tract Tumours
Edited by M. Castegnaro, R. Plestina, G. Dirheimer, I.N. Chernozemsky and H Bartsch
1991; 340 pages £47.50

No. 116 Mechanisms of Carcinogenesis in Risk Identification
Edited by H. Vainio, P.N. Magee, D.B. McGregor & A.J. McMichael
1992; 616 pages £69.00

No. 117 Directory of On-going Research in Cancer Epidemiology 1992
Edited by M. Coleman, J. Wahrendorf & E. Démaret
1992; 773 pages £44.50

No. 118 Cadmium in the Human Environment: Toxicity and Carcinogenicity
Edited by G.F. Nordberg, R.F.M. Herber & L. Alessio
1992; 470 pages £60.00

No. 119 The Epidemiology of Cervical Cancer and Human Papillomavirus
Edited by N. Muñoz, F.X. Bosch, K.V. Shah & A. Meheus
1992; 288 pages £29.50

No. 120 Cancer Incidence in Five Continents, Volume VI
Edited by D.M. Parkin, C.S. Muir, S.L. Whelan, Y.T. Gao, J. Ferlay & J. Powell
1992; 1080 pages £120.00

No. 121 Trends in Cancer Incidence and Mortality
M.P. Coleman, J. Estève, P. Damiecki, A. Arslan and H. Renard
1993; 806 pages, £120.00

No. 122 International Classification of Rodent Tumours. Part 1. The Rat
Editor-in-Chief: U. Mohr
1992/95; 10 fascicles of 60–100 pages, £120.00

No. 123 Cancer in Italian Migrant Populations
Edited by M. Geddes, D.M. Parkin, M. Khlat, D. Balzi and E. Buiatti
1993; 292 pages, £40.00

No. 124 Postlabelling Methods for Detection of DNA Adducts
Edited by D.H. Phillips, M. Castegnaro and H. Bartsch
1993; 392 pages; £46.00

No. 125 DNA Adducts: Identification and Biological Significance
Edited by K. Hemminki, A. Dipple, D. Shuker, F.F. Kadlubar, D. Segerbäck and H. Bartsch
1994; 480 pages; £52.00

No. 127 Butadiene and Styrene: Assessment of Health Hazards
Edited by M. Sorsa, K. Peltonen, H. Vainio and K. Hemminki
1993; 412 pages; £54.00

No. 128 Statistical Methods in Cancer Research. Volume IV. Descriptive Epidemiology
By J. Estève, E. Benhamou & L. Raymond
1994; 302 pages; £25.00

No. 129 Occupational Cancer in Developing Countries
Edited by N. Pearce, E. Matos, H. Vainio, P. Boffetta & M. Kogevinas
1994; 192 pages £20.00

No. 130 Directory of On-going Research in Cancer Epidemiology 1994
Edited by R. Sankaranarayanan, J. Wahrendorf and E. Démaret
1994; 792 pages, £46.00

No. 132 Survival of Cancer Patients in Europe. The EUROCARE Study
Edited by F. Berrino, M. Sant, A. Verdecchia, R. Capocaccia, T. Hakulinen and J. Estève
1994; 463 pages; £45.00

List of IARC Publications

IARC MONOGRAPHS ON THE EVALUATION OF CARCINOGENIC RISKS TO HUMANS

Volume 1 Some Inorganic Substances, Chlorinated Hydrocarbons, Aromatic Amines, *N*-Nitroso Compounds, and Natural Products
1972; 184 pages (*out of print*)

Volume 2 Some Inorganic and Organometallic Compounds
1973; 181 pages (*out of print*)

Volume 3 Certain Polycyclic Aromatic Hydrocarbons and Heterocyclic Compounds
1973; 271 pages (*out of print*)

Volume 4 Some Aromatic Amines, Hydrazine and Related Substances, *N*-Nitroso Compounds and Miscellaneous Alkylating Agents
1974; 286 pages Sw. fr. 18.−

Volume 5 Some Organochlorine Pesticides
1974; 241 pages (*out of print*)

Volume 6 Sex Hormones
1974; 243 pages (*out of print*)

Volume 7 Some Anti-Thyroid and Related Substances, Nitrofurans and Industrial Chemicals
1974; 326 pages (*out of print*)

Volume 8 Some Aromatic Azo Compounds
1975; 357 pages Sw. fr. 44.−

Volume 9 Some Aziridines, *N*-, *S*- and *O*-Mustards and Selenium
1975; 268 pages Sw.fr. 33.−

Volume 10 Some Naturally Occurring Substances
1976; 353 pages (*out of print*)

Volume 11 Cadmium, Nickel, Some Epoxides, Miscellaneous Industrial Chemicals and General Considerations on Volatile Anaesthetics
1976; 306 pages (*out of print*)

Volume 12 Some Carbamates, Thiocarbamates and Carbazides
1976; 282 pages Sw. fr. 41.−

Volume 13 Some Miscellaneous Pharmaceutical Substances
1977; 255 pages Sw. fr. 36.−

Volume 14 Asbestos
1977; 106 pages (*out of print*)

Volume 15 Some Fumigants, The Herbicides 2,4-D and 2,4,5-T, Chlorinated Dibenzodioxins and Miscellaneous Industrial Chemicals
1977; 354 pages (*out of print*)

Volume 16 Some Aromatic Amines and Related Nitro Compounds − Hair Dyes, Colouring Agents and Miscellaneous Industrial Chemicals
1978; 400 pages Sw. fr. 60.−

Volume 17 Some *N*-Nitroso Compounds
1978; 365 pages Sw. fr. 60.−

Volume 18 Polychlorinated Biphenyls and Polybrominated Biphenyls
1978; 140 pages Sw. fr. 24.−

Volume 19 Some Monomers, Plastics and Synthetic Elastomers, and Acrolein
1979; 513 pages (*out of print*)

Volume 20 Some Halogenated Hydrocarbons
1979; 609 pages (*out of print*)

Volume 21 Sex Hormones (II)
1979; 583 pages Sw. fr. 72.−

Volume 22 Some Non-Nutritive Sweetening Agents
1980; 208 pages Sw. fr. 30.−

Volume 23 Some Metals and Metallic Compounds
1980; 438 pages (*out of print*)

Volume 24 Some Pharmaceutical Drugs
1980; 337 pages Sw. fr. 48.−

Volume 25 Wood, Leather and Some Associated Industries
1981; 412 pages Sw. fr. 72.−

Volume 26 Some Antineoplastic and Immunosuppressive Agents
1981; 411 pages Sw. fr. 75.−

Volume 27 Some Aromatic Amines, Anthraquinones and Nitroso Compounds, and Inorganic Fluorides Used in Drinking Water and Dental Preparations
1982; 341 pages Sw. fr. 48.−

Volume 28 The Rubber Industry
1982; 486 pages Sw. fr. 84.−

Volume 29 Some Industrial Chemicals and Dyestuffs
1982; 416 pages Sw. fr. 72.−

Volume 30 Miscellaneous Pesticides
1983; 424 pages Sw. fr. 72.−

Volume 31 Some Food Additives, Feed Additives and Naturally Occurring Substances
1983; 314 pages Sw. fr. 66.−

Volume 32 Polynuclear Aromatic Compounds, Part 1: Chemical, Environmental and Experimental Data
1983; 477 pages Sw. fr. 88.−

Volume 33 Polynuclear Aromatic Compounds, Part 2: Carbon Blacks, Mineral Oils and Some Nitroarenes
1984; 245 pages (*out of print*)

Volume 34 Polynuclear Aromatic Compounds, Part 3: Industrial Exposures in Aluminium Production, Coal Gasification, Coke Production, and Iron and Steel Founding
1984; 219 pages Sw. fr. 53.−

Volume 35 Polynuclear Aromatic Compounds, Part 4: Bitumens, Coal-tars and Derived Products, Shale-oils and Soots
1985; 271 pages Sw. fr. 77.−

List of IARC Publications

Volume 36 **Allyl Compounds, Aldehydes, Epoxides and Peroxides**
1985; 369 pages Sw. fr. 77.–

Volume 37 **Tobacco Habits Other than Smoking: Betel-quid and Areca-nut Chewing; and some Related Nitrosamines**
1985; 291 pages Sw. fr. 77.–

Volume 38 **Tobacco Smoking**
1986; 421 pages Sw. fr. 83.–

Volume 39 **Some Chemicals Used in Plastics and Elastomers**
1986; 403 pages Sw. fr. 83.–

Volume 40 **Some Naturally Occurring and Synthetic Food Components, Furocoumarins and Ultraviolet Radiation**
1986; 444 pages Sw. fr. 83.–

Volume 41 **Some Halogenated Hydrocarbons and Pesticide Exposures**
1986; 434 pages Sw. fr. 83.–

Volume 42 **Silica and Some Silicates**
1987; 289 pages Sw. fr. 72.

Volume 43 **Man-Made Mineral Fibres and Radon**
1988; 300 pages Sw. fr. 72.–

Volume 44 **Alcohol Drinking**
1988; 416 pages Sw. fr. 83.

Volume 45 **Occupational Exposures in Petroleum Refining; Crude Oil and Major Petroleum Fuels**
1989; 322 pages Sw. fr. 72.–

Volume 46 **Diesel and Gasoline Engine Exhausts and Some Nitroarenes**
1989; 458 pages Sw. fr. 83.–

Volume 47 **Some Organic Solvents, Resin Monomers and Related Compounds, Pigments and Occupational Exposures in Paint Manufacture and Painting**
1989; 535 pages Sw. fr. 94.–

Volume 48 **Some Flame Retardants and Textile Chemicals, and Exposures in the Textile Manufacturing Industry**
1990; 345 pages Sw. fr. 72.–

Volume 49 **Chromium, Nickel and Welding**
1990; 677 pages Sw. fr. 105.-

Volume 50 **Pharmaceutical Drugs**
1990; 415 pages Sw. fr. 93.-

Volume 51 **Coffee, Tea, Mate, Methylxanthines and Methylglyoxal**
1991; 513 pages Sw. fr. 88.-

Volume 52 **Chlorinated Drinking-water; Chlorination By-products; Some Other Halogenated Compounds; Cobalt and Cobalt Compounds**
1991; 544 pages Sw. fr. 88.-

Volume 53 **Occupational Exposures in Insecticide Application and some Pesticides**
1991; 612 pages Sw. fr. 105.-

Volume 54 **Occupational Exposures to Mists and Vapours from Strong Inorganic Acids; and Other Industrial Chemicals**
1992; 336 pages Sw. fr. 72.-

Volume 55 **Solar and Ultraviolet Radiation**
1992; 316 pages Sw. fr. 65.-

Volume 56 **Some Naturally Occurring Substances: Food Items and Constituents, Heterocyclic Aromatic Amines and Mycotoxins**
1993; 600 pages Sw. fr. 95.-

Volume 57 **Occupational Exposures of Hairdressers and Barbers and Personal Use of Hair Colourants; Some Hair Dyes, Cosmetic Colourants, Industrial Dyestuffs and Aromatic Amines**
1993; 428 pages Sw. fr. 75.-

Volume 58 **Beryllium, Cadmium, Mercury and Exposures in the Glass Manufacturing Industry**
1993; 426 pages Sw. fr. 75.-

Volume 59 **Hepatitis Viruses**
1994; 286 pages Sw. fr. 65.-

Volume 60 **Some Industrial Chemicals**
1994; 560 pages Sw. fr. 90.-

Volume 61 **Schistosomes, Liver Flukes and Helicobacter pylori**
1994; 270 pages Sw. fr. 70.-

Volume 62 **Wood Dust and Formaldehyde**
1995; 406 pages Sw. fr. 80.–

Volume 63 **Dry Cleaning, some Chlorinated Solvents and other Industrial Chemicals**
1995; 558 pages Sw. fr. 90.–

Volume 64 **Human Papilloma Viruses**
1995; 409 pages Sw. fr. 80.–

Supplement No. 1
Chemicals and Industrial Processes Associated with Cancer in Humans (IARC Monographs, Volumes 1 to 20)
1979; 71 pages (*out of print*)

Supplement No. 2
Long-term and Short-term Screening Assays for Carcinogens: A Critical Appraisal
1980; 426 pages Sw. fr. 40.-

Supplement No. 3
Cross Index of Synonyms and Trade Names in Volumes 1 to 26
1982; 199 pages (*out of print*)

Supplement No. 4
Chemicals, Industrial Processes and Industries Associated with Cancer in Humans (IARC Monographs, Volumes 1 to 29)
1982; 292 pages (*out of print*)

Supplement No. 5
Cross Index of Synonyms and Trade Names in Volumes 1 to 36
1985; 259 pages (*out of print*)

Supplement No. 6
Genetic and Related Effects: An Updating of Selected IARC Monographs from Volumes 1 to 42
1987; 729 pages Sw. fr. 80.-

Supplement No. 7
Overall Evaluations of Carcinogenicity: An Updating of IARC Monographs Volumes 1-42
1987; 440 pages Sw. fr. 65.-

Supplement No. 8
Cross Index of Synonyms and Trade Names in Volumes 1 to 46
1990; 346 pages Sw. fr. 60.-

List of IARC Publications

IARC TECHNICAL REPORTS

No. 1 **Cancer in Costa Rica**
Edited by R. Sierra, R. Barrantes,
G. Muñoz Leiva, D.M. Parkin, C.A.
Bieber and N. Muñoz Calero
1988; 124 pages Sw. fr. 30.-

No. 2 **SEARCH: A Computer Package to Assist the Statistical Analysis of Case-control Studies**
Edited by G.J. Macfarlane,
P. Boyle and P. Maisonneuve
1991; 80 pages (*out of print*)

No. 3 **Cancer Registration in the European Economic Community**
Edited by M.P. Coleman and
E. Démaret
1988; 188 pages Sw. fr. 30.-

No. 4 **Diet, Hormones and Cancer: Methodological Issues for Prospective Studies**
Edited by E. Riboli and R. Saracci
1988; 156 pages Sw. fr. 30.-

No. 5 **Cancer in the Philippines**
Edited by A.V. Laudico,
D. Esteban and D.M. Parkin
1989; 186 pages Sw. fr. 30.-

No. 6 **La genèse du Centre International de Recherche sur le Cancer**
Par R. Sohier et A.G.B. Sutherland
1990; 104 pages Sw. fr. 30.-

No. 7 **Epidémiologie du cancer dans les pays de langue latine**
1990; 310 pages Sw. fr. 30.-

No. 8 **Comparative Study of Anti-smoking Legislation in Countries of the European Economic Community**
Edited by A. Sasco, P. Dalla Vorgia and P. Van der Elst
1992; 82 pages Sw. fr. 30.-

No. 9 **Epidémiologie du cancer dans les pays de langue latine**
1991 346 pages Sw. fr. 30.-

No. 10 **Manual for Cancer Registry Personnel**
Edited by S. Whelan et al.
1995; 370 pages Sw. fr. 45.-

No. 11 **Nitroso Compounds: Biological Mechanisms, Exposures and Cancer Etiology**
Edited by I.K. O'Neill & H. Bartsch
1992; 149 pages Sw. fr. 30.-

No. 12 **Epidémiologie du cancer dans les pays de langue latine**
1992; 375 pages Sw. fr. 30.-

No. 13 **Health, Solar UV Radiation and Environmental Change**
By A. Kricker, B.K. Armstrong, M.E. Jones and R.C. Burton
1993; 216 pages Sw.fr. 30.-

No. 14 **Epidémiologie du cancer dans les pays de langue latine**
1993; 385 pages Sw. fr. 30.-

No. 15 **Cancer in the African Population of Bulawayo, Zimbabwe, 1963–1977: Incidence, Time Trends and Risk Factors**
By M.E.G. Skinner, D.M. Parkin, A.P. Vizcaino and A. Ndhlovu
1993; 123 pages Sw. fr. 30.-

No. 16 **Cancer in Thailand, 1988–1991**
By V. Vatanasapt, N. Martin, H. Sriplung, K. Vindavijak, S. Sontipong, S. Sriamporn, D.M. Parkin and J. Ferlay
1993; 164 pages Sw. fr. 30.-

No. 18 **Intervention Trials for Cancer Prevention**
By E. Buiatti
1994; 52 pages Sw. fr. 30.-

No. 19 **Comparability and Quality Control in Cancer Registration**
By D.M. Parkin, V.W. Chen, J. Ferlay, J. Galceran, H.H. Storm and S.L. Whelan
1994; 110 pages plus diskette
Sw. fr. 40.-

No. 20 **Epidémiologie du cancer dans les pays de langue latine**
1994; 346 pages Sw. fr. 30.-

No. 21 **ICD Conversion Programs for Cancer**
By J. Ferlay
1994; 24 pages plus diskette
Sw. fr. 30.-

No. 22 **Cancer Incidence by Occupation and Industry in Tianjin, China, 1981–1987**
By Q.S. Wang, P. Boffetta, M. Kogevinas and D.M. Parkin
1994; 96 pages Sw. fr. 30.–

No. 23 **An Evaluation Programme for Cancer Preventive Agents**
By B.W. Stewart
1995; 40 pages Sw. fr. 20.-

No. 24 **Peroxisome Proliferation and its role in Carcinogenesis**
Views and expert opinions of an IARC Working Group 7–11 December 1994
1995; 90 pages Sw. fr. 30.-

No. 25 **Combined Analysis of Cancer Mortality in Nuclear Industry Workers in Canada, the United Kingdom and the United States**
E. Cardis et al.
1995; 160 pages Sw. fr. 30.-

DIRECTORY OF AGENTS BEING TESTED FOR CARCINOGENICITY

(Until Vol. 13 Information Bulletin on the Survey of Chemicals Being Tested for Carcinogenicity)

No. 8 Edited by M.-J. Ghess, H. Bartsch and L. Tomatis
1979; 604 pages Sw. fr. 40.-

No. 9 Edited by M.-J. Ghess, J.D. Wilbourn, H. Bartsch and L. Tomatis
1981; 294 pages Sw. fr. 41.-

No. 10 Edited by J. Ghess, J.D. Wilbourn and H. Bartsch
1982; 362 pages Sw. fr. 42.-

No. 11 Edited by M.-J. Ghess, J.D. Wilbourn, H. Vainio and H. Bartsch
1984; 362 pages Sw. fr. 50.-

No. 12 Edited by M.-J. Ghess, J.D. Wilbourn, A. Tossavainen and H. Vainio
1986; 385 pages Sw. fr. 50.-

No. 13 Edited by M.-J. Ghess, J.D. Wilbourn and A. Aitio 1988; 404 pages Sw. fr. 43.-

No. 14 Edited by M.-J. Ghess, J.D. Wilbourn and H. Vainio
1990; 370 pages Sw. fr. 45.-

No. 15 Edited by M.-J. Ghess, J.D. Wilbourn and H. Vainio
1992; 318 pages Sw. fr. 45.-

No. 16 Edited by M.-J. Ghess, J.D. Wilbourn and H. Vainio
1994; 294 pages Sw. fr. 50.-

List of IARC Publications

NON-SERIAL PUBLICATIONS

Alcool et Cancer
By A. Tuyns (in French only)
1978; 42 pages Fr. fr. 35.-

Cancer Morbidity and Causes of Death Among Danish Brewery Workers
By O.M. Jensen
1980; 143 pages Fr. fr. 75.-

Directory of Computer Systems Used in Cancer Registries
By H.R. Menck and D.M. Parkin
1986; 236 pages Fr. fr. 50.-

Facts and Figures of Cancer in the European Community
Edited by J. Estève, A. Kricker, J. Ferlay and D.M. Parkin
1993; 52 pages Sw. fr. 10.-

www.ingramcontent.com/pod-product-compliance
Ingram Content Group UK Ltd.
Pitfield, Milton Keynes, MK11 3LW, UK
UKHW051257180426
11947UKWH00020B/1766